CHEMICAL DYNAMICS

ADVANCES IN CHEMICAL PHYSICS

VOLUME 21

ADVANCES IN CHEMICAL PHYSICS

EDITORS

I. Prigogine

Faculté des Sciences, Université Libre de Bruxelles, Brussels, Belgium

S. Rice

*James Franck Institute, University of Chicago,
Chicago, Illinois*

EDITORIAL BOARD

CHEMICAL DYNAMICS

Papers in Honor of Henry Eyring

Edited by **JOSEPH O. HIRSCHFELDER**
The University of Wisconsin
Madison, Wisconsin

DOUGLAS HENDERSON
International Business
Machines Corporation
San Jose, California

WILEY-INTERSCIENCE

a Division of John Wiley & Sons, Inc.

New York · London · Sydney · Toronto

LIBRARY OF CONGRESS CATALOG CARD NUMBER: 77-136715
ISBN   0-471-40066-1

PRINTED IN THE UNITED STATES OF AMERICA
10   9   8   7   6   5   4   3   2   1

AUTHORS

P. M. ALLEN, Faculté des Sciences, Université Libre de Bruxelles, Brussels, Belgium

R. J. BAXTER, Department of Mathematics, Massachusetts Institute of Technology, Cambridge, Massachusetts

G. BLYHOLDER, Department of Chemistry, University of Arkansas, Fayetteville

DONALD D. BODE, JR., Lawrence Radiation Laboratory, University of California, Livermore, California

LOUIS L. BURTON, Department of Chemistry, University of Alabama, Tuscaloosa, Alabama

CHARLES M. CARLSON, Boeing Scientific Research Laboratories, Seattle, Washington

FELIX CERNUSCHI, Departamento de Fisica, Facultad de Ingenieria, Buenos Aires, Argentina

P. CHAKRABORTI, Chemistry Department, Purdue University, Lafayette, Indiana

SEIHUN CHANG, Department of Chemistry, Seoul National University, Seoul, Korea

TERESA REE CHAY, Department of Chemistry, University of California, San Diego, La Jolla, California

C. J. CHRISTENSEN, Department of Chemistry, University of Utah, Salt Lake City, Utah

H. F. COFFEY, The Lilly Research Laboratories, Eli Lilly & Co., Indianapolis, Indiana

MELVIN A. COOK, IRECO Chemicals, West Jordan, Utah

C. A. COULSON, Mathematical Institute, Oxford University, Oxford, England

I. B. CUTLER, College of Engineering, University of Utah, Salt Lake City, Utah

GEORGE H. DUFFEY, Department of Physics, South Dakota State University, Brookings

JOHN V. DUGAN, JR., NASA Lewis Research Center, Cleveland, Ohio

MICHAEL P. EASTMAN, University of California, Los Alamos Scientific Laboratory, Los Alamos, New Mexico

EDWARD M. EYRING, Department of Chemistry, University of Utah, Salt Lake City

LEROY EYRING, Department of Chemistry, Arizona State University, Tempe, Arizona

ARTHUR A. FROST, Department of Chemistry, Northwestern University, Evanston, Illinois

MARIO GIAMBIAGI, Departamento de Fisica, Facultad de Ingenieria, Buenos Aires, Argentina

RICHARD C. GRAHAM, University of Utah, Salt Lake City, Utah

H. TRACY HALL, Brigham Young University, Provo, Utah

GEORGE HALSEY, Department of Chemistry, University of Washington, Seattle Washington

HARRY HECHT, University of California, Los Alamos Scientific Laboratory, Los Alamos, New Mexico

IZUMI HIGUCHI, Department of Synthetic Chemistry, Faculty of Engineering, Shizouka University, Japan

JOEL A. HOUTMAN, Department of Physics, South Dakota State University, Brookings

MU SHIK JHON, Department of Chemistry, University of Virginia, Charlottesville, Virginia

WALTER KAUZMANN, Department of Chemistry, Princeton University, Princeton, New Jersey

MITSUHIKO KODAIRA, Department of Applied Physics, University of Tokyo, Bunkyo-ku, Tokyo, Japan

DENNIS E. KONASEWICH, School of Chemistry, University of Minnesota, Minneapolis, Minnesota

MAURICE M. KREEVOY, School of Chemistry, University of Minnesota, Minneapolis, Minnesota

SOOK-IL KWUN, Department of Physics, Seoul National University, Seoul, Korea

KEITH J. LAIDLER, Department of Chemistry, University of Ottawa, Canada

P. W. LANDIS, The Lilly Research Laboratories, Eli Lilly & Co., Indianapolis, Indiana

W. BURTON LEWIS, University of California, Los Alamos Scientific Laboratory, Los Alamos, New Mexico

S. H. LIN, Department of Chemistry, Arizona State University, Tempe

C. Y. LIN MA, Department of Chemistry, Arizona State University, Tempe

P. V. MCALLISTER, College of Engineering, University of Utah, Salt Lake City, Utah

LAYTON L. MCCOY, Department of Chemistry, University of Missouri, Kansas City, Missouri

P. MCGUIRE, Chemistry Department, Purdue University, Lafayette, Indiana

DAN MCLACHLAN, JR., Department of Mineralogy, Ohio State University, Columbus, Ohio

R. G. A. R. MACLAGAN, Oxford University, Mathematical Institute, Oxford, England

JOHN L. MAGEE, Chemistry Department and Radiation Laboratory, University of Notre Dame, Notre Dame, Indiana

F. A. MATSEN, The Molecular Physics Group, University of Texas, Austin, Texas

MARY MONCRIEFF-YEATES, Textile Research Institute, Princeton, New Jersey

LARRY G. MORGAN, Pacific Northwest Laboratories, Battelle Memorial Institute, Richland, Washington

JOHN R. MORREY, Pacific Northwest Laboratories, Battelle Memorial Institute, Richland, Washington

EARL M. MORTENSEN, Department of Chemistry, Cleveland State University, Cleveland

C. R. MUELLER, Chemistry Department, Purdue University, Lafayette, Indiana

G. NICOLIS, Faculté des Sciences, Université Libre de Bruxelles, Brussels, Belgium

HYUNGSUK PAK, Department of Chemistry, Seoul National University, Seoul, Korea

J. PENTA, Chemistry Department, Purdue University, Lafayette, Indiana

I. PRIGOGINE, Faculté des Sciences, Université Libre de Bruxelles, Brussells, Belgium

FRANCIS H. REE, Lawrence Radiation Laboratory, University of California, Livermore, California

TAIKYUE REE, Department of Chemistry, University of Utah, Salt Lake City

STUART A. RICE, Department of Chemistry and The James Franck Institute, University of Chicago, Chicago, Illinois

LARRY D. RICH, University of Utah, Salt Lake City, Utah

E. T. SAMULSKI, Textile Research Institute, Princeton, New Jersey

MYRIAM SEGRE, Departmento de Fisica, Facultad de Ciencias Exactas, Buenos Aires, Argentina

B. SMITH, Chemistry Department, Purdue University, Lafayette, Indiana

MASAAKI TAMAYAMA, Tokyo, Japan

SIR HUGH TAYLOR, Woodrow Wilson National Fellowship Foundation, Princeton, New Jersey

NEWELL TAYLOR, University of Utah, Salt Lake City, Utah

L. G. TENSMEYER, The Lilly Research Laboratories, Eli Lilly & Co., Indianapolis, Indiana

T. R. THOMSON, Department of Chemistry, University of Utah, Salt Lake City, Utah

A. V. TOBOLSKY, Department of Chemistry, Princeton University, Princeton, New Jersey

ALAN TWEEDALE, Department of Chemistry, University of Ottawa, Canada

HIROSH UTSUGI, Department of Applied Science, Faculty of Engineering, Tohoku University, Sendai, Japan

E. R. VAN ARTSDALEN, Department of Chemistry, University of Alabama, Tuscaloosa, Alabama

ANDREW VANHOOK, Department of Chemistry, College of the Holy Cross, Worcester, Massachusetts

TSUTOMU WATANABE, Department of Applied Physics, University of Tokyo, Bunkyo-ku, Tokyo, Japan

R. O. WATTS, Diffusion Research Unit, Research School of Physical Sciences, Australian National University, Canberra, ACT, Australia

GARY R. WEBER, Department of Chemistry, Arizona State University, Tempe, Arizona

HOWARD J. WHITE, JR., Textile Research Institute, Princeton, New Jersey

FERD WILLIAMS, Physics Department, University of Delaware, Newark, Delaware

W. WILLIAMS, Chemistry Department, Purdue University, Lafayette, Indiana

HANS F. WINTERKORN, Department of Civil and Geological Engineering, Princeton University, Princeton, New Jersey

KANG YANG, Central Research Division, Continental Oil Company, Ponca City, Oklahoma

SHANG J. YAO, Thermodynamics Research Center, Department of Chemistry, Texas A&M University, College Station

M. E. ZANDLER, Department of Chemistry, University of Utah, Salt Lake City, Utah

BRUNO J. ZWOLINSKI, Thermodynamics Research Center, Department of Chemistry, Texas A&M University, College Station

PREFACE

This book is a tribute to Henry Eyring from his students and associates on the occasion of his seventieth birthday. For more than four decades Henry Eyring has played an important role in the development of theoretical chemistry and its application to many fields. His genius and enthusiasm have inspired everyone who has ever worked with him (or even heard him lecture).

Because of Eyring's broad interests and research in many fields the technical portion of this volume is arranged into six sections: Quantum Mechanics, Reaction Rates, Properties of Molecules, Liquids, Biological Applications and Engineering Applications. Hugh Taylor, Edward Eyring, LeRoy Eyring, and Henry himself furnished the biographical data. All of Eyring's students and co-authors (whose whereabouts could be ascertained) were invited to contribute an article and/or reminiscence.

Thanks are due to the Section Editors who carefully studied the manuscripts and suggested revision: A. A. Frost, K. J. Laidler, W. Kauzman, D. Henderson, F. H. Johnson, and C. J. Christensen. Special thanks are due to Edward Eyring who provided valuable assistance and advice. We are also grateful to Henry Eyring's secretary, Mary Lou Harris, who prepared the list of Eyring's students and associates together with those addresses that she knew. Thanks are also due to Mrs. Gloria Lawton who furnished additional addresses by going through American Men of Science and the address lists of members of various technical societies.

JOSEPH O. HIRSCHFELDER
DOUGLAS HENDERSON

CONTENTS

PART III. PROPERTIES OF MOLECULES
        EDITED BY WALTER KAUZMANN

PART IV. THEORY OF LIQUIDS
        EDITED BY DOUGLAS HENDERSON

PART V. BIOLOGICAL APPLICATIONS
EDITED BY FRANK H. JOHNSON

PART VI. ENGINEERING APPLICATIONS
EDITED BY C. J. CHRISTENSEN

HENRY EYRING

American chemist, born Feb. 20, 1901,
Colonia Juarez, Chihuahua, Mexico

Eyring pioneered in the application of quantum and statistical mechanics to chemistry and developed the Theory of Absolute Reaction Rates and the Significant Structure Theory of Liquids. Absolute reaction rate theory provides a basis for treating all chemical reactions. The theory of liquids provides the basis of a quantitative formulation of the thermodynamic properties of liquids and, together with rate theory, provides a similar basis for transport properties. Eyring also developed theories of optical activity, mass spectrography, the addition of dipoles and bond lengths in flexible high polymers, and bioluminescence. For his contributions to theoretical chemistry, Eyring was awarded the American Chemical Society's Gilbert Newton Lewis Medal in 1963.

Rate Theory. Quantum mechanics, by providing a theory for calculating the interactions between atoms and molecules, makes it possible to calculate the energy of any configuration of atoms. If this energy is plotted "vertically," and if appropriate interatomic distances are chosen in such a way as to specify the atomic configuration, the resulting multidimensional surface is called the potential surface in configuration space. The low regions of this surface correspond to compounds, and a reaction may be described as the passage of a point from one minimum through the saddle point into a neighboring valley. Fritz London, in 1928, called attention to the possibility of constructing such surfaces and gave approximate formulas for simple three- and four-atom systems.

In 1929 and 1930 Eyring and Michael Polanyi calculated the potential surface for three hydrogen atoms. Subsequently, Eyring with his collaborators extended the quantum-mechanical calculations to many atoms and constructed surfaces for a wide variety of molecular systems. In the meantime E. P. Wigner and Pelzer, using the Eyring-Polanyi surface, calculated the rate

"Modern Men of Science," Vol. I, McGraw-Hill, New York, 1964.

of the reaction

$$H + H_2 \text{ (para)} \rightarrow H_2 \text{ (ortho)} + H \tag{1}$$

In 1935 Eyring formulated the general rate expression. A system at the saddle point, the activated complex, is like any other molecule in all of its degrees of freedom except for the reaction coordinate, that is, the coordinate normal to the potential barrier. The activated complex is next assumed to be in equilibrium with the reactants. Using the theory of small vibrations to calculate the normal modes, it is possible to arrive at the explicit expression for the specific rate of reaction

$$k' = \kappa \frac{kT}{h} e^{-\Delta G^{\ddagger}/RT} \tag{2}$$

Here $\Delta G^{\ddagger}$ is the work required to assemble the activated complex from the reactants. The transmission coefficient κ is ordinarily near unity and is the factor that corrects for quantum mechanical effects and any departure from equilibrium. The symbols k, R, h, and T are the Boltzmann constant, gas constant, Planck's constant, and absolute temperature, respectively.

Theory of Liquids.   To remove a molecule from a condensed phase to the vapor state involves breaking all the bonds holding the molecule, whereas to vaporize a molecule only involves breaking half the bonds. It follows that the creation of a vacancy costs the same energy as the vaporization of a molecule. If an isolated vacancy is formed in a solid it is locked in and provides only a positional degeneracy. At the melting point enough vacancies go into a liquid, acting cooperatively, to mobilize the vacancies, giving them gaslike properties.

The result is that the liquid contains a mixture of gaslike and solidlike degrees of freedom. This model leads to an expression for the Helmholtz free energy of the liquid in agreement with experiment. For simple liquids all parameters are calculable from the models. For complicated liquids the appropriate model may be selected by comparing calculated and observed thermodynamic properties much as structure can be determined in x-ray analysis.

Absolute rate theory has been applied successfully to all types of problems. Current developments are concerned with explicit calculations of the transmission coefficient κ. Liquid theory, although newer, is also widely applicable and has already had many successes in calculating the properties of ordinary liquids, molten metals, molten salts, and water.

Born in northern Mexico, Eyring rode the range with his father until the Mexican revolution forced the American colonists to leave Mexico in 1912. After a year in El Paso the family moved to Arizona. Eyring attended the University of Arizona, where he obtained a B.S. in mining engineering (1923) and an M.S. in metallurgy (1924). He obtained his Ph.D. in chemistry at the University of California, Berkeley, in 1927; taught two years at the University of Wisconsin, where he collaborated with Farrington Daniels; spent a year at Fritz Haber's laboratory in Berlin working with Polanyi; returned to Berkeley for a year as lecturer; and for fifteen years (1931–46) was on the faculty of Princeton University. In 1946 he became dean of the Graduate School and professor of chemistry and metallurgy at the University of Utah. He was elected to the National Academy of Sciences in 1945.

Eyring has published more than 450 scientific papers and coauthored six books: *The Theory of Rate Processes* (1941), *Quantum Chemistry* (1944), *The Kinetic Basis of Molecular Biology* (1954), *Modern Chemical Kinetics* (1963), *Statistical Mechanics and Dynamics* (1964), and *Significant Liquid Structures* (1969).

Biographical Data of Henry Eyring

Present Position   Professor of Chemistry and Metallurgy, University of Utah, Salt Lake City, Utah 84112

Born   February 20, 1907, Colonial Juarez, Chihuahua, Mexico

Parents   Edward Christian and Caroline Romney Eyring

Education   B.S. (Mining Engineering), University of Arizona, 1923
M.S. (Metallurgical Engineering), University of Arizona, 1924
Ph.D. (Chemistry), University of California, 1927
Honorary degrees from University of Utah, Northwestern University, Princeton University, Seoul National University, Indiana Central College, Brigham Young University, University of California (Davis), Western Reserve University, Denison University, Marquette University, and Notre Dame University

Married   Mildred Bennion, August 25, 1928

Children   Edward Marcus Eyring
Henry Bennion Eyring
Harden Romney Eyring

Experience   Instructor in Chemistry, University of Arizona, Tucson, Arizona, 1924–1925
Instructor in Chemistry, University of Wisconsin, 1927–1928
Research Associate, Chemistry, University of Wisconsin, 1928–1929
National Research Fellow, Kaiser-Wilhelm Institute, Berlin, 1929–1930
Lecturer, Chemistry, University of California, 1930–1931
Assistant and Associate Professor of Chemistry, Princeton University 1931–1938
Professor of Chemistry, Princeton University, 1938–1946

Dean of Graduate School and Professor of Chemistry, University of Utah, 1946–1966
Distinguished Professor of Chemistry and Metallurgy, 1966–

Organizational Membership

Member, National Science Board (1962–1968); Am. Assoc. Adv. of Sci. (Pres. 1965, Board of Dir. 1961–1966) (Pres. Pacific Div. 1958, V. Pres. 1946); Amer. Chem. Soc. (Pres. 1963); Amer. Acad. of Arts & Sci.; Nat'l Acad of Sci.; Amer. Phil. Soc.; Utah Acad. of Sci.; Sigma Xi; Phi Kappa Phi; Soc. Rheol. (Vice Pres. 1946); Board of Directors of Annual Reviews, 1954–1970; Editor of Annual Reviews of Physical Chemistry, 1956–; Scientific Advisory Committee of Welch Foundation, 1954–; Member of General Sunday School Board, Church of Jesus Christ of Latter Day Saints; Hon. Member of Phi Lambda Upsilon; Hon. Member of The Chemists Club; Hon. Member of The Chemical Society; Gamma Alpha; Chairman, Intern'l Science Exhibition—Molecule Class, 1958; International Academy of Quantum Molecular Science (1969–)

Publications

The Theory of Rate Processes, with Samuel Glasstone and Keith J. Laidler, McGraw-Hill, 1941; *Quantum Chemistry*, with John Walter and George Kimball, John Wiley and Sons, 1944; *The Kinetic Basis of Molecular Biology*, with Frank H. Johnson and Milton J. Polissar, John Wiley and Sons, 1954; *Modern Chemical Kinetics*, with Edward M. Eyring, Reinhold Publishing Co., 1963; *Statistical Mechanics & Dynamics*, with Henderson, Stover and E. M. Eyring, John Wiley and Sons, 1964; *Significant Liquid Structures*, with Muh Shik Jhon. Has published more than 435 papers in National Journals

Special Interests

Radioactivity, application of quantum mechanics to chemistry; theory of reaction rates; theory of liquids, rheology; molecular biology, theory of flames, optical rotation

Awards

Ninth Annual Prize of A.A.A.S. (Newcomb Medal), 1932
University of Arizona Alumni Achievement Award, 1947
Research Corporation Award, 1949
Second Bingham Medal from Society of Rheology, 1949

William H. Nichols Medal from N.Y. Section of ACS, 1951
Local ACS Award (Salt Lake City), 1959
James E. Talmage Award, Brigham Young University, 1959
Award of Merit, University of Arizona, 1960
Distinguished Alumni Award, University of Utah, 1961
G. N. Lewis Award, California Section, 1963
Peter Debye Award, 1964
National Medal of Science, 1966
Irving Langmuir Award, 1967
Willard Gibbs Medal, 1968
Madison Marshall Award, Alabama Sect. of A.C.S., 1968
Elliott Cresson Medal, Franklin Institute, 1969
Linus Pauling Medal, Puget Sound ACS Section, 1969

The Scientific Researches of

HENRY EYRING

during five decades, 1925 *to* 1970
at Berkeley, Madison, Berlin, Princeton, and Utah

by

Hugh Taylor
with the assistance of **Edward M. Eyring**

In September 1930 Gilbert N. Lewis of the University of California invited me to meet Henry Eyring at the laboratories of the University of California, Berkeley. Eyring had just returned from a year in Berlin, 1929–1930, as a National Research Fellow, where he had been working with Michael Polanyi and had been associated with Eugene Wigner in his then current studies of reaction kinetics. At that time Eyring was about 30 years old. He had taken his undergraduate degree (B.S.) at the University of Arizona in 1923 and his M.S. in 1924. He took his Ph.D. at the University of California, Berkeley, in 1927. His thesis was in the field of radioactivity, his professor, G. E. Gibson, a Scot born in Edinburgh, educated at Edinburgh University and at Breslau in Germany, one of Lewis' early importations to the rising school of physical chemistry on the Berkeley campus. Eyring's thesis dealt with The Ionization and Stopping Power of Various Gases for Alpha Particles from Polonium.

On completion of his graduate training he became an instructor in the Chemistry Department of the University of Wisconsin, 1927–1928, and research associate in the following year. There, in association with Professor Farrington Daniels, he made his first major contribution to reaction kinetics—and that an experimental one—the demonstration that in liquid solvents, as in the gaseous phase, the decomposition of nitrogen pentoxide was a unimolecular reaction. How puzzling unimolecular processes were at that time can be fully appreciated only by those who then learned of their existence. In the closing years of the second decade of the twentieth century it led to a radiation theory of chemical reaction, espoused by J. Perrin in France and W. C. McC. Lewis in England, which became the topic of a famous Faraday Society Discussion in 1922, during which Lindemann, in a paragraph of discussion, provided the essential clue to the existence of unimolecular homogeneous collisional processes. He emphasized that if the decomposition of the activated, energy-rich molecule of reactant were slow compared with the rate of deactivation (essentially the rate of activation) the kinetic expression would approximate to first order above a certain limiting reaction pressure. Christiansen in the same year reached a similar conclusion. Daniels and Hunt had shown that in the case of gaseous nitrogen pentoxide decomposition the first order constant was essentially constant between 278 and 9.6 mm and only fell by 20 percent at pressures of 0.0185 mm. Eyring's work extended this result to decomposition in several solvents. That collisions and not the ambient radiation were important was demonstrated in 1925 by G. N. Lewis and D. F. Smith, who showed that reaction at a given temperature was suppressed almost entirely in beams of reactants without collisions, whereas ready decomposition occurred at the same temperature when collisions were permitted.

In Berlin, Eyring had been concerned with the rate of interaction of hydrogen atoms with hydrogen molecules, since this had been established as the probable mechanism of the *para–ortho* hydrogen conversion. A system consisting of three hydrogen atoms seemed to promise possible theoretical treatment. Expectations were realized, and it was shown that a linear approach of a hydrogen atom to a hydrogen molecule required a potential energy barrier of $\sim$8 kcal if exchange of partners in the three atom system were to be achieved. The observed experimental activation energy was 7.25 kcal.

In Berkeley, in 1930–1931, Eyring turned his attention to calculations of the activation energies of interactions between hydrogen molecules and the halogens. The experimental data on the kinetics with chlorine, bromine, and iodine were already well known. The results of the theoretical calculations confirmed the known experimental findings: hydrogen and iodine reacted bimolecularly. Bromine and chlorine reacted by chain mechanisms involving reaction of atomic halogen with hydrogen molecules. (That iodine might also react thus was a much later experimental conclusion.) The outstanding result of the theoretical calculations came in the examination of the reaction of hydrogen and fluorine. The computation showed that reaction between these species should be *slow* with an activation energy of 50 kcal due to the smallness of the molecules and the nearness of approach requisite to bring about reaction. Eyring reported these startling results at the Spring Meeting of the American Chemical Society in Indianapolis in 1931. I was aware at that time of a recent finding of von Wartenburg and J. Taylor that pure hydrogen and fluorine were not very reactive, even when irradiated with ultraviolet light, in sharp contrast to statements in the textbooks that these species were so reactive that solid fluorine and liquid hydrogen combined explosively at liquid hydrogen temperatures. This observation dated back to a celebrated collaboration between Dewar, the English chemist, and Moissan, the French chemist. They showed that fluorine was frozen solid in liquid hydrogen and that when the two condensed species were brought together an explosion resulted. It is on record that Eyring, in order to test his own calculations, made a mixture of hydrogen and fluorine and noted little reaction over a period of time. Anxious to end the experiment, he turned a stream of nitrogen through the apparatus. An immediate explosion occurred! It may have been due to the catalytic effect of dust from the rubber tubing connecting the nitrogen tank to the hydrogen-fluorine container. This was the background of an invitation to Eyring to become a research associate at Princeton University. It was the beginning of a phase of research extending from 1931 to 1945 in which he progressed from instructor to professor at Princeton.

Among the problems to which Eyring directed early attention was the theoretical examination of the phenomenon of activated adsorption, or chemisorption, then under active experimental study at Princeton. With A. Sherman he studied the chemisorption of hydrogen on carbon surfaces. The calculations revealed that at large and small carbon–carbon distances on the surface the activation energy of adsorption should be high; but there was a most-favored C–C distance, 2.54 Å, where the activation energy was a minimum. This confirmed the findings of Beech and others concerning the significance of lattice distance in surface atoms of catalysts.

When heavy hydrogen became available, Eyring was ready with the significance of zero-point energy in both the separation and the reactions of isotopic species. The extension of these studies to a particularization of the activated complex at the height of the potential energy barrier led to the absolute rate theory of reaction velocity. Such a complex was regarded as an ordinary molecule, with all the usual thermodynamic properties, save in one respect. In the complex, one degree of freedom, which in a normal molecule would be a vibrational degree of freedom, consists in this case of a motion in the direction of the reaction coordinate and leads to a decomposition of the complex at a definite rate to yield the products. The rate of reaction is governed by the concentration $C^{\pm}$ of activated complexes at the top of the potential energy barrier multiplied by the frequency of crossing the barrier. Application of the normal statistical treatment to this problem, with the restriction just indicated as to one degree of freedom, resulted in the revolutionary equation

$$\text{rate of reaction} = C \cdot \frac{\pm}{} \frac{kT}{h}.$$

The effective velocity of crossing the energy barrier is the universal frequency kT/h, dependent on temperature and independent of the nature of the reactants and type of reaction. For a reaction $A + B$ giving products the equation became

$$\text{rate of reaction} = k' C_A C_B,$$

where

$$k' = \frac{kT}{h} \cdot \frac{C^{\pm}_{\doteq}}{C_A C_B} = \frac{kT}{h} K^{\pm}_{\doteq},$$

where $K^{\pm}_{\doteq}$ is the equilibrium in the process

$$A + B \rightleftharpoons M,$$

and where M is the activated complex assumed in equilibrium with the reactants.

Off the main stream of this research were calculations of the binding energies in the growth of crystal nuclei from metallic atoms (1933) and the theoretical treatment of chemical reactions produced by ionization processes (1936), notably the *ortho–para* hydrogen conversion by α-particles and the radiochemical synthesis and decomposition of hydrogen bromide. In this it was shown that the concept of ion clusters in such reactions was unnecessary and that quantitative calculations of reaction rates could be secured without such an assumption.

The major breakthrough in the theory of absolute reaction rates came in 1936 when the theory was extended to all processes occurring in time. Viscosity, plasticity, diffusion, and electrical conduction, all fell magically under the spell of kT/h. The observation of the essential constancy over a range of temperature of the mean density of a liquid and its saturated vapor observed by Cailletet and Mathias in 1885 received an immediate explanation when Eyring pointed out that if, when a molecule of liquid escaped to the vapor phase, it left behind in the liquid a hole, or vacancies equivalent in volume to the escaping molecule, the observed behavior would result. A saturated vapor consisted then of a given number of molecules in a given volume of vapor, and the liquid contained an equivalent volume of holes. Holes in liquids became as real as the molecules in the vapor, were responsible for many of the properties of the liquids, and were capable of statistical treatment. Such treatment revealed the reciprocal properties of viscosity and diffusion, accurately accounting for all the best available experimental data. Articles dealing with the theory of the liquid state in which J. Hirschfelder, R. H. Ewell, D. Stevenson, and J. F. Kincaid were collaborators emerged in 1937–1939. With W. J. Moore the theory of viscosity was extended to the viscosity of unimolecular films.

The organic chemist got his due share of help with studies of optical rotation (W. Kauzmann, 1941), directed substitution, and the statistics of long-chain molecules (with R. E. Powell). With A. E. Stearn we note the application of absolute rate theory to proteins (1939). In the same year a collaboration with S. Glasstone and K. J. Laidler extended the theory to the treatment of overvoltage. The major issue of this collaboration was "The Theory of Rate Processes" published by McGraw-Hill in 1941, a saddle-point in the whole history of all types of change, chemical and physical, with time.

In 1942 the application of the rate theory to biochemical systems was initiated by an examination of bacterial luminescence (with J. L. Magee) and the nature of enzyme inhibitions in bacterial luminescence (with F. H. Johnson and R. W. Williams). The collaboration with Johnson yielded a successful interpretation of the influence of temperature and pressure on bioluminescent systems and ultimately the major contribution to biological science, "Kinetic Basis of Molecular Biology" by Eyring with F. H. Johnson and M. J. Polissar.

Just before American entrance into World War II a group of papers was written on liquid sulphur and other polymeric materials with R. E. Powell, H. M. Hulburt, H. A. Herman, and A. Tobolsky as collaborators. In the war years the further progress of rate theory was retarded by reason of special services to the war effort of the United States, principally in the theory of explosions. During these war years, 1944–1946, he also accepted the responsibility of directing the research effort of the Textile Research Institute at Princeton and initiated a vigorous program, applying his theory of rate processes to various properties of fibers. Shortly after the end of the war the decision was reached by Eyring to accept the offer of the University of Utah to become Professor of Chemistry and Dean of the Graduate School.

Recognizing my inability to interpret the Utah phase of Eyring's career, I invited his son, Professor Edward M. Eyring, who shared in this whole activity in Utah, to contribute his estimate of the scientific outcome of the years from 1945 to the present. It is with considerable pleasure that I append his report in the following brief summary:

In the years since his arrival at the University of Utah in 1946 Eyring has continuously maintained a research group of 20 or more graduate and postdoctoral students working on an exotically diverse array of problems. The tensile properties of metals, ceramics, and fibers, diffusion through biological membranes, aqueous chromic salts, potential change in illuminated chloroplasts, mass loss in meteors, dipole moments, mass spectroscopy, synthesis of diamonds, photosynthesis, enzyme catalysis, inorganic redox reactions in solutions, chromatography, inflammation of biological tissues, thixotropy, strain electrometry, and corrosion of metals are just a few of the subjects Eyring and his collaborators explored in his first decade at Utah.

The range of Eyring's interests in the years since has been no less dramatic: biogenesis, dislocations in crystals, scientific creativity, graphite oxidation, nerve action, fused salts, electron spin correlation, the electrochemical theory of smelting, sticking coefficients for nitrogen on solids, sapphire, ions in flames, radon emanation from building materials, absolute configuration in optical rotation, liquid hydrogen, protein synthesis, differential thermal analysis, electrode passivation, the structure of liquid water, porphyrins under high pressure, optical rotatory dispersion, and the Faraday effect are among the topics Eyring has tackled since about 1956. Of these many subjects the one that has probably engendered his greatest and most sustained enthusiasm has been his "significant structure theory of liquids."

In assessing Eyring's impact on international science since 1946 it would be accurate to say that he has beautifully elaborated his work at Princeton on absolute rate theory, optical rotation, and liquid theory in such a way as to inspire imitation as well as occasional controversy. His impact on science of all kinds in Utah has been equally dramatic. In addition to about 70 Ph.D. theses in chemistry that he has personally directed at Utah, in his capacity as Dean of the Graduate School for 20 years, he also nurtured to maturity doctoral programs in more than 50 different academic departments. Finally, and by no means least, he has taught 27 hours of graduate courses in quantum mechanics, statistical mechanics, and rate processes each calendar year since coming to Utah. The 50 or more students enrolled each year from quite a variety of science and engineering departments besides chemistry always profit enormously from the opportunity of observing Eyring's bold inquisitiveness regarding every new problem. Nothing is too hard for Eyring, and in a very real way this fortifies the courage of the students and faculty who have been exposed to his audacious imagination. In fact, it is probably true that the most significant impact that Eyring has had on science in Utah has been in providing an accessible standard of what a first-rate chemist is like.

MY BROTHER HENRY

by

LeRoy Eyring
Department of Chemistry
Arizona State University, Tempe

This is one view of Henry Eyring from within his family. I did not grow up with Henry but came along 13 children later, and in my earliest recollections he is a shadowy idol off somewhere in the great world succeeding. He was away in college when I was born, but on his occasional return visits to the ranch he augmented my education in algebra when I was 10 and chemistry when I was 12. We did not really get acquainted, however, until I was grown.

Henry was the third child and first son of a very large family. As it turned out, the life of his family divided itself into two periods. A prosperous one in the Mormon colonies in Mexico and one in Arizona in which the family struggled to re-establish itself but everyone slipped away before it could be accomplished.

An early family story relates that one night Father came home and after unsaddling he led the horse down to the river to drink with the 3-year-old Henry riding bareback. While the horse was drinking he shook himself and Henry tumbled off into the river. His mother reported that when he was snatched from the stream his first words were, "Put me back on the horse," which was done.

By the time he was five he was riding the range with his father. For whatever cause there grew into this relationship a special understanding. Henry's early and growing sense of responsibility and his achievement, together with his lifelong appreciation of the values of his church and family, earned his father's unqualified regard.

Henry was 11 at the time of the family's exodus from Mexico to the United States. He found a job as a cash boy in an El Paso store until the decision was made not to return to Mexico. The family finally settled on a piece of land just south of the Gila River near the town of Pima, Arizona.

There were a few acres of land under irrigation, but in the first dozen or more years much more had to be cleared and brought into cultivation.

Mesquite trees were first cut and their stumps grubbed. The chaparral was uprooted from the desert silt by a steel rail drawn between teams of horses hooked to each end. The debris was stacked in great piles and burned. Ditches were dug and water brought to level and finish fields suitable for cotton or alfalfa. Nevertheless the time came when Henry knew he must leave this for a more intellectually stimulating environment.

Going to college meant finding the closest place and enrolling. He took the $500 scholarship he had won and went to the University of Arizona at Tucson. Not only did he make his own way at school but managed to send money home. Indeed, when Henry was established in his first job he raised a critical sum needed to save the family farm.

In the summer vacations Henry worked in the copper mines at Globe and Miami about 90 miles from Pima. This no doubt directed him toward a mining engineering degree. Later, however, when he was shaken by blasting in a deep mine, he was persuaded that his newly won engineering degree was a mistake. He returned to Tucson to take an M.S. degree in metallurgical engineering and then went on to Berkeley for his Ph.D. in chemistry.

"A prophet is not without honour save in his own country and in his own house," does not apply to Henry. When he retired as Dean at the University of Utah, all his 15 brothers and sisters, their spouses, and many of their children and their children's children, making a group of more than 90, gathered in Salt Lake City to honor him. We had not all been together as a single family group in more than 40 years. It was an unselfconscious acknowledgement of his family's affection.

Whenever Henry came to any of the Mormon communities in southern Arizona on Sunday School business he attracted a huge audience which contained a cross section of the congregation. These admirers regarded him as the resident scientist of God's Kingdom here on earth if not beyond. They listened with rapt attention as each heard a confirmation of his own highest hopes in language he understood. On one occasion my sister innocently observed that Henry talked like he didn't know anything.

Henry has a private religion just as he has a private science and he regards each as corroborating absolute truth. He carries his religion like a tent, and where he pitches it is the sanctuary that shelters him and all others he can entreat to enter.

Henry tried to win souls for chemistry everywhere and I was an easy target. Because of his persuasiveness I don't remember seriously considering any other profession. In recent years when I decided to return to Arizona he supported me by coming to Arizona State University each year for several in a row as a guest speaker who could not be scheduled for too many talks.

He did this at a time when he was deeply committed with national as well as local obligations.

At Chemistry meetings I was always warmly drawn into any group surrounding him. Afterward we found an opportunity for walking and talking together while he gallantly and good naturedly sought to bring me under his tent of moral and religious conviction.

At a recent ACS meeting in New York he suggested we walk, carrying our suitcases, the 50 blocks to the airport terminal. I agreed but was reluctant, since, unlike Henry's, my suitcase is always heavy. Much to my relief he called early in the morning to say he had forgotten a scheduled talk he was to give at our arranged departure time which made it necessary for us to take a cab. On the way to the airport Henry discovered that the taxi driver had been a leader in a noisy demonstration against the hotel the day before. This led to a spirited discussion lasting nearly an hour and a half while we fought traffic to the airport just as the rush hour was beginning. The taxi driver was extraordinarily able in his own defense; nevertheless he received a sustained lecture on the virtues of the free enterprize system and its importance in preserving human initiative and also of the evils of smoking and drinking. They parted with broad mutual respect.

One early morning several years ago we ran into each other on a back street in Atlantic City, neither suspecting that the other was in town, and had breakfast together. It pleased him to think that a Divine benevolence had caused our paths to cross. We differed in our belief concerning this, but I am grateful for the crossing of our lives and am content to leave this out of the realm of chance.

LeRoy Eyring

Tempe, Arizona
December 16, 1970

THE DAILY DOUBLE

There are so many aspects of Henry Eyring's attitudes and philosophy of life that invite reminiscences and analysis that it is difficult to pick the outstanding one. His kindness and lack of recrimination toward people he completely disagrees with and who disagree with him is one of which the best examples had better not be made public.

One that I have long admired, however, and occasionally been exasperated with, is his desire for *everybody* to understand science in general and Eyring's current science in particular. It was so characteristic, when working with him, to have a guest arrive to be greeted by the words "What George and I are working on is . . . ." It made little difference whether the visitor was of the level of Debye or of a church visitor who, up to that moment at least, was a stranger to science. A sincere effort was made to bring the visitor into the problem, no matter how far back one had to start.

A particularly amusing example developed when one of the janitors at the Frick laboratory turned the conversation on to the odds in the daily double. Eyring was eager to prove, for mainly moral reasons, that it was almost a waste of money to bet. The janitor in turn was very interested and gathered his fellow janitors for an impromptu lecture on the probability of gambling. Unfortunately, the numbers refused to behave, and Eyring was left in a frantic effort to repair the damage done to the janitors, who were convinced that the great professor had come up with better than even-money odds.

GEORGE HALSEY

Department of Chemistry
University of Washington, Seattle

Helium Molecular States[*]

F. A. MATSEN

The Molecular Physics Group, The University of Texas at Austin

Henry Eyring, through his lectures, his papers, and his book *Quantum Chemistry*, has taught quantum mechanics to a whole generation of chemists. Without these presentations, quantum chemistry could not have progressed as rapidly as it has; all quantum chemists are grateful to him. The Molecular Physics Group at the University of Texas has a special reason to be grateful to Henry Eyring. He has, since its creation, been scientific advisor to the Robert A. Welch Foundation of Houston, Texas, to which the Group owes its existence. As part of the tribute to Henry Eyring, I will review briefly some of the computations of the Molecular Physics Group on the molecular states of helium.

Molecular helium is an attractive subject to the *ab initio*-ist for the following reasons: (a) It has a rich spectrum; (b) its ground state provides an example of inert gas atom interaction; (c) it exhibits a delightful array of obligatory and nonobligatory maxima[1]; (d) the doubly metastable $He(1s2s; {}^3S)$, $\tau \sim 24$ hr, and its interaction with the ground state $He(1s^2; {}^1S)$ has been the subject of considerable experimental study; and (e) He_2 is diatomic and carries only four electrons.

In our early calculations we followed the procedure outlined in *Quantum Chemistry*. The wavefunctions were constructed from large basis sets of eigenfunctions of S^2, symmetry adapted to $D_{\infty h}$. More recently we have employed the spin-free formulation.[2] Most of the states studied are the lower states of a particular symmetry type, in which case we minimize with respect

* Supported by the Robert A. Welch Foundation of Houston, Texas.

1

to the electronic energy the lowest root of the secular equation. In other cases we minimize a higher root.

The basis functions are products of Slater and/or elliptical orbitals. They are chosen so as to contain both separated and united atom contributions. Extensive, but not necessarily total, optimization of orbital exponents is performed at each internuclear separation. Miller and Browne have discussed in detail our method of integral evaluation.[3] The Molecular Physics Group has published an extensive table of integrals.[4]

A variation calculation yields an upper bound to the true Born-Oppenheimer potential curve. Error limit calculations have not yet proved useful. Since the experimental separated and united atom energies are known, the error at the endpoints is known. An adjustment procedure which corrects at both the separated and united atom ends has been proposed.[5] This procedure has not been fully evaluated.

The vibration (and rotation) energies for a given potential curve are computed as follows.[6]

1. Diagonalize R in a harmonic oscillator basis set.
2. Construct a diagonal representative of $V(R)$ expressed as a power series in R.
3. Transform the diagonal representation back to the harmonic oscillator basis by using the inverse transformation of step 1.
4. Diagonalize the full Hamiltonian in this space. The results can be presented in terms of the intervals $\Delta G(v + \frac{1}{2})$ or can be analyzed into the spectroscopic constants. A modification of this method using a Morse function basis set has proved to be superior to the harmonic oscillator basis.[7]

The computed curves are shown in Fig. 1. The $^{1,3}\Pi$ calculations are the most reliable, since considerably more work was invested in the choice of basis set and in optimization. The shapes of the curves appear to be in good agreement with experiment. We have predicted maxima for those cases for which there is experimental evidence for maxima. We have predicted correctly, in a qualitative way, the heights of the observed maxima. For example, for $He_2\{1s^2; {}^1S, 1s2s; {}^3S; {}^3\Sigma_u{}^+\}$ we have predicted a very small barrier with a height of 0.11 eV. Experiments indicate that the barrier exists and is small and of the order ~ 0.06 eV.[8–10]

Some of the spectroscopic constants are listed in Table I, and the $\Delta G(v + \frac{1}{2})$ values are listed in Table II. The dissociation energies are about as reliable as the best spectroscopic values. The bond distances are generally too large. We have found that we can reduce the computed bond distance by

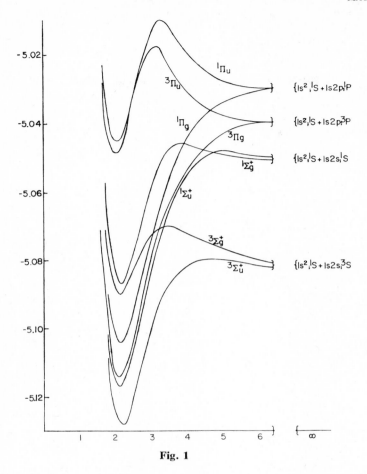

Fig. 1

introducing more united atom character into our wavefunctions. Accurate predictions of $\Delta G(v + \frac{1}{2})$, r_2, and ω_e are possible only if the wavefunction is carefully constructed as for $^{1,3}\Pi$. In general, the predictions are now accurate enough to be of considerable use to the experimentalists.

There is no question that at a rather high cost the accuracy of the computations on He_2 can be improved considerably, and there is also no doubt that the level of accuracy currently obtained for He_2 can also be obtained for the 10–20-electron diatomic molecules and for small polyatomic molecules. The open questions are (a) should such calculations be carried out, and (b) can they be adequately funded?

Table I[a]

Ref.		r_e (a$_0$) calc. ab initio	r_e (a$_0$) calc. adj.	r_e (a$_0$) expt.	D_e (a.u.) calc. ab initio	D_e (a.u.) calc. adj.	D_e (a.u.) expt.	ω_e (cm^{-1}) calc. ab initio	ω_e (cm^{-1}) calc. adj.	ω_e (cm^{-1}) expt.
($1s^2$; ^1S + $1s2s$; ^3S) 11, 5	$^3\Sigma_u^+$	2.14	2.12	1.985	0.044	0.044	—	1743	—	1811
12	$^3\Sigma_g^+$	2.15	—	2.072	0.006	—	—	—	—	—
($1s^2$; ^1S + $1s2s$; ^1S) 13	$^1\Sigma_u^+$	2.11	—	1.972	0.063	—	—	1665	—	1861
14	$^1\Sigma_g^+$	2.16	—	2.062	0.033	—	—	1655	—	1653
($1s^2$; ^1S + $1s2p$; ^3P) 15	$^3\Pi_u$	2.05	2.04	2.051	0.008	0.008	—	—	—	1662
16	$^3\Pi_g$	2.10	1.90	2.013	0.074	0.082	—	2195	1967	1769
($1s^2$; ^1S + $1s2p$; ^1P) 15	$^1\Pi_u$	2.07	2.05	2.051	0.014	0.014	—	—	—	1671
16	$^1\Pi_g$	2.15	2.10	2.014	0.073	0.075	—	2474	2749	1766

[a] The experimental data are from Herzberg[17] and Ginter[18].

Table II.   Vibrational Energy Intervals (cm^{-1})

	Experimental	*ab initio*	Adjusted
		$^3\Pi_g$	
$\Delta G_{1/2}$	1698.8	2103.5	2293.9
$\Delta G_{3/2}$	1628.4	2062.8	2252.9
$\Delta G_{5/2}$	1557.7	1925.1	2265.8
		$^1\Pi_g$	
$\Delta G_{1/2}$	1696.9	2096.2	2132.3
$\Delta G_{3/2}$	1627.9	1911.7	1928.2
$\Delta G_{5/2}$	1558.7	1781.0	1794.0
		$^3\Pi_u$	
$\Delta G_{1/2}$	1571.9	1573.5	1580.7
$\Delta G_{3/2}$	1482.3	1466.0	1474.8
		$^1\Pi_u$	
$\Delta G_{1/2}$	1590.5	1581.3	1537.4
$\Delta G_{3/2}$	1510.5	1488.2	1397.3
		$^3\Sigma_g^+$	
$\Delta G_{1/2}$	1480.0	1234.6	
$\Delta G_{3/2}$	1371.7	1005.9	

REFERENCES

1. R. S. Mulliken, *Phys. Rev.*, **136**, A962 (1964); *J. Am. Chem. Soc.*, **88**, 1849 (1966).
2. F. A. Matsen, *Advances in Quantum Chemistry*, Vol. 6, Academic, New York, 1964, p. 60.
3. J. Miller and J. C. Browne, Tech. Rept., Molecular Physics Group, The University of Texas, January, 1962.
4. J. Miller, J. M. Gerhauser, and F. A. Matsen, *Quantum Chemistry Integrals and Tables*, The University of Texas Press, Austin, 1958.
5. D. J. Klein, E. M. Greenawalt, and F. A. Matsen, *J. Chem. Phys.*, **47**, 4820 (1967).
6. D. F. Zetik and F. A. Matsen, *J. Mol. Spectry.*, **24**, 122 (1967).
7. E. M. Greenawalt and A. S. Dickenson, *J. Mol. Spectry.*, **30**, 427 (1969).
8. Y. Tanaka and K. Yoshino, *J. Chem. Phys.*, **50**, 3087 (1969).
9. F. D. Colegrove, L. D. Schearer, and G. K. Walters, *Phys. Rev.*, **135**, A353 (1964).
10. E. W. Rothe, R. H. Neynaber, and S. M. Trujillo, *J. Chem. Phys.*, **42**, 3310 (1965).
11. F. A. Matsen and R. D. Poshusta, *Phys. Rev.*, **132**, 307 (1963); F. A. Matsen and D. R. Scott, *Quantum Theory of Atoms Molecules and the Solid State*, Academic, New York, 1966, p. 133.
12. B. K. Gupta, unpublished results.
13. D. R. Scott, E. M. Greenawalt, J. C. Browne, and F. A. Matsen, *J. Chem. Phys.*, **44**, 2981 (1966).
14. E. M. Greenawalt, unpublished results.
15. B. K. Gupta and F. A. Matsen, *J. Chem. Phys.*, **50**, 3797 (1969).
16. M. N. Watson and D. J. Klein, unpublished results.
17. G. Herzberg, *Molecular Spectra and Molecular Structures*, Van Nostrand, Princeton, N.Y., 1950.
18. M. L. Ginter, *J. Chem. Phys.*, **42**, 561 (1965); **45**, 248 (1966); *J. Mol. Spectry.*, **17**, 224 (1965); **18**, 321 (1965).

A Free-Electron Study of Common Boron Frameworks[*]

JOEL A. HOUTMAN AND GEORGE H. DUFFEY

Department of Physics, South Dakota State University, Brookings

Abstract

In many boron compounds, the boron atoms form a fragment of the regular icosahedron, or a combination of such fragments. We have carried out free-electron calculations on such structures. The results have been checked by constructing the corresponding waves (except for $B_{18}H_{22}$). To explain the relative stabilities of various alternate structures, one has to make the delta-function potential at each nucleus positive.

[*] Supported by National Science Foundation Grant GP-7599.

A Personal Note

GEORGE H. DUFFEY

When I began my studies at Princeton in September 1942, I was shy and uncertain. Consequently, the human, friendly, unaffected personality of Henry Eyring appealed to me. He put me at ease and stimulated my development, without any conscious effort on his part.

Professor Eyring was not interested in polishing up what had been done in the past; he took little comfort in what was known. Instead, he was constantly involved in unraveling new problems—contributing to knowledge. This attitude dominated his classes, making them disorganized but interesting and useful.

Professor Eyring started me out on nerve-impulse research. He then felt that physicists and chemists could help the biologists solve their problems. But it was wartime, and we were asked to find out more about rate processes in detonation waves. Henry hesitated, because he was disturbed about the effects of war on civilization and humanity. He finally decided that any research necessary for the development of explosives would be carried out anyway. Our work would be essentially academic—directed toward understanding rather than development. Nevertheless, after the project got underway Professor Eyring let us carry the ball. He told me that one day I should explain to him what I was doing.

In 1945, as I was about to leave Princeton, Professor Eyring told me not to vegetate. I did not realize the significance of that advice then, but down through the years I have seen many scientists become spectators after getting their degrees.

INTRODUCTION

Most applications of free-electron theory[1-4] have been to planar conjugated systems. Nevertheless, the bonding electrons in various three-dimensional networks[5,6] are also delocalized. Treating them with the theory involves some additional assumptions[7,8] and is more approximate.

As in the planar systems, the bonds between neighboring atoms form roads along which the electrons move most readily. For simplicity we assume that the electrons are confined to the center line of each road and that they move at constant potential along the line. The net effect from branching at various junctions, occupied localized orbitals, and attraction to the nuclei is allowed for by adding a positive delta potential (a spike) at each junction (or joint) in the cage.[6,9,10]

Along each straight line joining nuclei the constant potential allows the eigenfunction ϕ to vary sinusoidally. At the joints the magnitudes from different branches are matched. The derivatives of ϕ are adjusted, however, to allow for the potential spike.

These approximations lead to the simultaneous equations

$$-[(C/k) \sin kd + m_j \cos kd]\phi(P_j) + \sum \phi(Q_j) = 0 \tag{1}$$

in which C is a measure of the delta potential spike strength, k the propagation constant of the de Broglie wave ϕ, and d the edge length, while P_j and Q_j label the jth nucleus and its m_j nearest neighbors, respectively. Letting F_j represent the expression in brackets reduces the form to

$$-F_j\phi(P_j) + \sum \phi(Q_j) = 0. \tag{2}$$

Note that the summation is over all the nearest neighbors of the jth nucleus.

A different equation (2) exists for each nucleus, or junction, in the framework; the number of relationships equals the number of different $\phi(P_k)$'s. Since these relationships are homogeneous and linear, the condition that they have a nontrivial solution is that the determinant of the coefficients be zero. The jth row of the determinant contains $-F_j$ on the principal diagonal,

9

1 in each jkth position for which a

$$Q_j = P_k \qquad (3)$$

and zero everywhere else.

Solving the resulting secular equation yields the F's. From these we obtain the kd's and the energies. The results are checked by actually plotting the waves.

ICOSAHEDRAL FRAGMENTS AND RELATED STRUCTURES

Boron frameworks not covered in previous papers[7,8] and described by Lipscomb,[5] Hoffmann and Lipscomb,[11] and Hawthorne[12] have been investigated. In these structures each kind of junction is labeled by the number of nearest neighbors, and so in the tables this number appears as the subscript on every F.

Most of the frameworks can be formed by successively removing boron atoms from the highly symmetric icosahedral $B_{12}H_{12}^{2-}$, adding appropriate edge hydrogens and, where necessary, removing electrons. The octadecaborane-22 is obtained on condensing two such fragments. The square-based pyramid is formed by removing hydrogens, then compressing and completing an edge bond, from the icosahedral fragment in B_5H_{11}. The triangular dodecahedron comes from symmetric condensation of two bent rhombi.

Except for the $B_{18}H_{22}$ structure, the pertinent waves were formulated and plotted.[7] The irreducible representation for which each wave is a basis was determined and labeled in the conventional manner[13] as a species. The corresponding energies were obtained, with each edge being 1.70 Å and with the parameter C being (a) 0 and (b) 1.77 Å^{-1}. The results appear in Tables I–XI. Information on the structures considered and the degree of each secular equation appear in Table XII.

From these energy levels and those calculated before[7,8] the results in Tables XIII and XIV were obtained. Table XIII compares the free-electron energies of various frameworks with those of possible cleavage products. On the second line of this table appear data for the square pyramid B_5H_9. Breaking a basal edge of the pyramid converts it to the C_{1v} icosahedral fragment in the first line. We see that 0.82 eV is required to break the basal B—B bond when $C = 1.77$ Å^{-1}. But when $C = 0$, energy is released. Although the calculations do not allow for a possible strengthening of the binding of an H, they do suggest that C should be positive, as we have assumed.

Decaborane-16 is the fourth entry in Table XIII. This molecule consists of two square pyramids linked apex to apex by a B—B bond. The fission products, two C_{4v} B_5H_8, appear as the third entry. On comparing the listed

Table I. Free-Electron Energy Levels for the Triangular Dodecahedron with 1.70-Å Edges

Species	Potential V Zero Everywhere ($C = 0$)				Potential Spike at Nuclei with $C = 1.77$ Å^{-1}			
	Root F_4 (extremal atom)	Root F_5 (midriff atom)	Angle kd	Energy (eV)	Root F_4 (extremal atom)	Root F_5 (midriff atom)	Angle kd	Energy (eV)
A_1	4	5	0	0	4.28	4.74	0.348π	1.58
B_2	1.278	1.597	0.397π	2.05	1.31	1.19	0.537π	3.75
E	$\dfrac{-5+\sqrt{105}}{10}$	$\dfrac{-5+\sqrt{105}}{8}$	0.458π	2.73	0.78	0.56	0.569π	4.21
A_1	-1.400	-1.750	0.614π	4.90	-1.31	-1.91	0.704π	6.45
E	$\dfrac{-5-\sqrt{105}}{10}$	$\dfrac{-5-\sqrt{105}}{8}$	0.624π	5.07	-1.47	-2.10	0.716π	6.68
B_2	-1.878	-2.347	0.656π	5.59	-1.70	-2.37	0.733π	6.99
A_1, A_2 B_1, B_2, E	Nodal solutions		π	13.0	Nodal solutions		π	13.0

Table II. Free-Electron Energy Levels for the Tetragonal Pyramid with 1.70-Å Edges

Species	Potential V Zero Everywhere ($C = 0$)				Potential Spike at Nuclei with $C = 1.77$ Å^{-1}			
	Root F_3 (basal atom)	Root F_4 (apical atom)	Angle kd	Energy (eV)	Root F_3 (basal atom)	Root F_4 (apical atom)	Angle kd	Energy (eV)
A_1	3	4	0	0	3.16	3.45	0.404π	2.12
E	0	0	0.500π	3.25	0	-0.442	0.646π	5.43
A_1	-1	$-\frac{4}{3}$	0.608π	4.81	-0.81	-1.43	0.713π	6.61
B_1	-2	$-\frac{8}{3}$	0.732π	6.98	-2	-2.87	0.831π	8.98
B_2, E	Nodal solutions		π	13.0	Nodal solutions		π	13.0

Table III. Free-Electron Energy Levels for the Decaborane-16 Structure (Two Square Pyramids Linked Apex to Apex) with 1.70-Å Edges

Species	Potential V Zero Everywhere ($C = 0$)				Potential Spike at Nuclei with $C = 1.77$ Å^{-1}			
	Root F_3 (extremal atom)	Root F_5 (apical atom)	Angle kd	Energy (eV)	Root F_3 (extremal atom)	Root F_5 (apical atom)	Angle kd	Energy (eV)
A_{1g}	3	5	0	0	3.33	4.00	0.390π	1.98
A_{2u}	2.722	4.537	0.138π	0.25	2.91	3.39	0.423π	2.33
E_g, E_u	0	0	0.500π	3.25	0	−0.88	0.646π	5.43
A_{1g}	−0.400	−0.667	0.543π	3.83	−0.05	−0.96	0.650π	5.50
A_{2u}	−1.322	−2.203	0.645π	5.42	−1.00	−2.33	0.730π	6.94
B_{1g}, B_{2u}	−2.000	−3.333	0.732π	6.98	−2.00	−3.72	0.831π	8.98
$A_{1g}, B_{1u}, B_{2g}, E_g, E_u$	Nodal solutions		π	13.0	Nodal solutions		π	13.0

Table IV. Free-Electron Energy Levels for the Tetraborane-10 Structure (Bent-Rhombus Fragment of Icosahedron) with 1.70-Å Edges

Species	Potential V Zero Everywhere ($C = 0$)				Potential Spike at Nuclei with $C = 1.77$ Å^{-1}			
	Root F_2 (extremal atom)	Root F_3 (midriff atom)	Angle kd	Energy (eV)	Root F_2 (extremal atom)	Root F_3 (midriff atom)	Angle kd	Energy (eV)
A_1	2	3	0	0	2.46	2.63	0.446π	2.58
B_1	0	0	0.500π	3.25	0	-0.57	0.693π	6.24
B_2	$-\frac{4}{3}$	-1	0.608π	4.81	-0.34	-1	0.730π	6.93
A_1	$-\frac{2}{3}$	-2	0.732π	6.98	-1.27	-2.16	0.849π	9.39
A_2	Nodal solution		π	13.0	Nodal solution		π	13.0

Table V. Free-Electron Energy Levels for the Pentaborane-11 Structure (Pentagonal Pyramid with One Base Corner Missing) with 1.70-Å Edges

Species	Potential Zero Everywhere			Potential Spike at Nuclei with $C = 1.77$ Å^{-1}				
	Root F_2	Angle kd	Energy (eV)	Root F_2 (extremal atom)	Root F_3 (basal atom)	Root F_4 (apical atom)	Angle kd	Energy (eV)
A_1	2	0	0	2.69	2.93	3.17	0.422π	2.32
A_2	$\dfrac{-1 + \sqrt{7}}{3}$	0.412π	2.20	0.73	0.37	0.01	0.617π	4.95
A_1	$-\tfrac{1}{3}$	0.553π	3.98	0.06	−0.49	−1.04	0.686π	6.12
A_1	−1	0.667π	5.78	−0.84	−1.64	−2.43	0.791π	8.14
A_2	$\dfrac{-1 - \sqrt{7}}{3}$	0.708π	6.52	−1.08	−1.92	−2.77	0.822π	8.79
A_1, A_2	Nodal solutions	π	13.0	Nodal solutions			π	13.0

Table VI. Free-Electron Energy Levels for the Pentagonal Pyramid with 1.70-Å Edges

Species	Potential V Zero Everywhere ($C = 0$)				Potential Spike at Nuclei with $C = 1.77$ Å^{-1}			
	Root F_3 (basal atom)	Root F_5 (apical atom)	Angle kd	Energy (eV)	Root F_3 (basal atom)	Root F_5 (apical atom)	Angle kd	Energy (eV)
A_1	3	5	0	0	3.28	3.93	0.394π	2.02
E_1	$\dfrac{-1+\sqrt{5}}{2}$	$\dfrac{5}{6}(-1+\sqrt{5})$	0.434π	2.45	0.62	0.01	0.598π	4.65
A_1	-1	$-\dfrac{5}{3}$	0.608π	4.81	-0.68	-1.87	0.702π	6.41
E_2	$\dfrac{-1-\sqrt{5}}{2}$	$\dfrac{5}{6}(-1-\sqrt{5})$	0.681π	6.04	-1.62	-3.19	0.789π	8.10
E_1, E_2	Nodal solutions		π	13.0	Nodal solutions		π	13.0

Table VII. Free-Electron Energy Levels for the Octaborane-12 Structure (Decaborane-14 Network with End Boron and Adjacent Side Boron Missing) Having 1.70-Å Edges

Species	Potential Zero Everywhere			Potential Spike at Nuclei with $C = 1.77$ Å^{-1}				
	Root F_3	Angle kd	Energy (eV)	Root F_3 (extremal atom)	Root F_4 (midriff atom)	Root F_5 (gable atom)	Angle kd	Energy (eV)
A_1	3	0	0	3.52	3.91	4.30	0.374π	1.82
B_1	1.556	0.326π	1.39	1.75	1.71	1.67	0.513π	3.42
B_2	$\dfrac{-1+\sqrt{7}}{2}$	0.412π	2.20	1.12	0.94	0.75	0.559π	4.07
A_1	−0.264	0.528π	3.63	−0.03	−0.48	−0.93	0.649π	5.47
A_1	−1.136	0.624π	5.06	−0.84	−1.46	−2.09	0.716π	6.66
B_1	−1.156	0.626π	5.10	−0.86	−1.49	−2.12	0.718π	6.70
A_2	−1	0.608π	4.81	−1	−1.66	−2.32	0.730π	6.93
B_2	$\dfrac{-1-\sqrt{7}}{2}$	0.708π	6.52	−1.77	−2.59	−3.41	0.805π	8.43
A_1, A_2, B_1, B_2	Nodal solutions	π	13.0	Nodal solutions			π	13.0

Table VIII. Free-Electron Energy Levels for the Symmetrical Enneaborane Ion (Decaborane-14 Network with One End Atom Missing) Having 1.70-Å Edges

Species	Potential Zero Everywhere			Potential Spike at Nuclei with $C = 1.77$ Å^{-1}				
	Root F_3	Angle kd	Energy (eV)	Root F_3 (upper edge atom)	Root F_4 (lower edge atom)	Root F_5 (bottom atom)	Angle kd	Energy (eV)
A_1	3	0	0	3.64	4.06	4.47	0.364π	1.72
E	1.391	0.347π	1.56	1.76	1.73	1.69	0.512π	3.40
A_1	0	0.500π	3.25	0.36	0	-0.36	0.618π	4.96
E	-0.851	0.592π	4.55	-0.60	-1.18	-1.76	0.695π	6.29
E	-1.140	0.624π	5.07	-0.89	-1.52	-2.16	0.720π	6.74
A_1	-1.800	0.705π	6.46	-1.70	-2.50	-3.30	0.797π	8.27
A_1, A_2, E	Nodal solutions	π	13.0	Nodal solutions			π	13.0

Table IX. Free-Electron Energy Levels for the Enneaborane-15 Structure (Decaborane-14 Network with One Side Atom Missing) Having 1.70-Å Edges

Species	Potential Zero Everywhere			Potential Spike at Nuclei with $C = 1.77$ Å^{-1}					
	Root F_2	Angle kd	Energy (eV)	Root F_2 (handle atom)	Root F_3 (corner atom)	Root F_4 (side atom)	Root F_5 (bottom atom)	Angle kd	Energy (eV)
A_1	2	0	0	3.17	3.56	3.96	4.35	0.370π	1.78
A_1	1.279	0.279π	1.01	2.06	2.11	2.15	2.20	0.485π	3.07
A_2	0.524	0.416π	2.25	1.33	1.15	0.97	0.79	0.557π	4.04
A_1	0.238	0.462π	2.78	0.83	0.50	0.17	−0.16	0.607π	4.79
A_1	−0.557	0.590π	4.53	−0.00	−0.57	−1.14	−1.71	0.693π	6.24
A_2	−0.565	0.591π	4.55	−0.07	−0.65	−1.24	−1.83	0.700π	6.37
A_1	−0.760	0.624π	5.07	−0.22	−0.85	−1.48	−2.12	0.717π	6.69
A_1	−1.033	0.673π	5.89	−0.81	−1.59	−2.37	−3.16	0.786π	8.04
A_2	−1.126	0.690π	6.20	−0.82	−1.60	−2.39	−3.17	0.787π	8.07
A_1, A_2	Nodal solutions	π	13.0	Nodal solutions				π	13.0

Table X. Free-Electron Energy levels for the Decaborane-14 Structure (Icosahedron with Two Adjacent Corners Missing) Having 1.70-Å Edges

Species	Potential Zero Everywhere			Potential Spike at Nuclei with $C = 1.77$ Å^{-1}				
	Root F_3	Angle kd	Energy (eV)	Root F_3 (handle atom)	Root F_4 (side atom)	Root F_5 (bottom atom)	Angle kd	Energy (eV)
A_1	3	0	0	3.72	4.16	4.60	0.356π	1.65
B_1	1.732	0.304π	1.20	2.18	2.24	2.30	0.480π	3.00
B_2	1.239	0.364π	1.73	1.77	1.74	1.70	0.511π	3.39
A_1	0.449	0.452π	2.66	0.85	0.60	0.35	0.581π	4.38
A_1	−0.735	0.579π	4.36	−0.37	−0.89	−1.42	0.676π	5.94
A_2, B_1	−0.750	0.580π	4.38	−0.46	−1	−1.54	0.683π	6.07
B_2	−1.089	0.618π	4.97	−0.69	−1.28	−1.88	0.703π	6.42
A_1	−1.364	0.650π	5.50	−1.17	−1.87	−2.56	0.745π	7.23
B_1	−1.732	0.696π	6.30	−1.59	−2.37	−3.15	0.786π	8.04
A_1, A_2, B_1, B_2	Nodal solutions	π	13.0	Nodal solutions			π	13.0

Table XI. Free-Electron Energy Levels for the Octadecaborane-22 Structure [Two Decaborane Cages Sharing a Common (5–6) Edge, but Opening up in Opposite Directions] with 1.70-Å Edges

	Potential Zero Everywhere		Potential Spike at Nuclei $(C = 1.77 \text{ Å}^{-1})$		
Root[a] F_3	Angle kd	Energy (eV)	Root[a] F_3 (extremal atom)	Angle kd	Energy (eV)
3	0	0	3.88	0.343π	1.53
2.680	0.148π	0.29	3.41	0.383π	1.91
1.629	0.317π	1.31	2.29	0.472π	2.90
1.463	0.338π	1.49	2.09	0.487π	3.08
1.013	0.390π	1.98	1.57	0.526π	3.60
0.918	0.401π	2.09	1.47	0.533π	3.70
0.292	0.469π	2.86	0.82	0.582π	4.41
−0.213	0.523π	3.55	0.22	0.629π	5.15
−0.596	0.564π	4.13	−0.06	0.651π	5.51
−0.742	0.580π	4.37	−0.40	0.678π	5.99
−0.742	0.580π	4.37	−0.41	0.679π	6.00
−0.895	0.596π	4.63	−0.51	0.687π	6.15
−1.031	0.612π	4.87	−0.64	0.699π	6.35
−1.061	0.615π	4.92	−0.66	0.700π	6.38
−1.198	0.631π	5.18	−0.81	0.713π	6.62
−1.264	0.638π	5.30	−0.93	0.724π	6.82
−1.569	0.675π	5.93	−1.39	0.766π	7.64
−1.684	0.690π	6.19	−1.49	0.776π	7.83
Nodal solutions	π	13.0	Nodal solutions	π	13.0

[a] Roots F_4, F_5, and F_6 are not listed.

energies, we see that the strength of the broken bond is not very sensitive to the choice of C, varying from 1.46 eV for $C = 0$ to 2.08 eV for $C = 1.77 \text{ Å}^{-1}$.

Again, parameter C must be positive to explain the bonding of the two C_{3v} B_9H_{11}'s in octadecaborane-22. (Compare the ninth and tenth entries in the table.) Furthermore, the two cleavage fragments C_{2v} B_8H_{10} and C_{2v} $B_{10}H_{12}$ (on the eighth line) are unstable by only 0.06 eV when $C = 0$, but reasonably unstable when $C = 1.77 \text{ Å}^{-1}$.

From the remaining entries in the table we see that the energy required to split decaborane-14 when $C = 1.77 \text{ Å}^{-1}$ is comparable to that needed to break the same number of bonds in splitting octadecaborane-22. In each case the calculated value is around 12 eV.

The free-electron energies of some observed boron cages are compared

Table XII. Description of the Configurations Treated and the Governing Secular Equations

Configuration	Geometric Figure	Example	Point Group	Degree of Secular Equation
Tetraborane-10	Bent rhombus	B_4H_{10}	C_{2v}	4
Tetragonal pyramid	Square-based pyramid	B_5H_9	C_{4v}	5
Pentaborane-11	Fragment of icosahedron	B_5H_{11}	C_{1v}	5
Hexaborane-10	Pentagonal pyramid	B_6H_{10}	C_{5v}	6
Triangular dodecahedron	Distortion of tetragonal antiprism	B_8Cl_8	D_{2d}	8
Octaborane-12	Fragment of icosahedron	B_8H_{12}	C_{2v}	8
Enneaborane ion	Fragment of icosahedron	$B_9H_{14}^-$	C_{3v}	9
Enneaborane-15	Fragment of icosahedron	B_9H_{15}	C_{1v}	9
Decaborane-16	Square pyramids linked apex to apex	$B_{10}H_{16}$	D_{4h}	10
Decaborane-14	Fragment of icosahedron	$B_{10}H_{14}$	C_{2v}	10
Octadecaborane-22	Condensed decaborane-14 cages	$B_{18}H_{22}$	S_2	18

with those of possible alternatives in Table XIV. Each framework is characterized by the number of boron atoms, their arrangement in space, and the number of levels that are filled.

When four levels are fully occupied in the four-boron system, the tetrahedron is most stable. When six levels are filled in the six-boron system, the octahedron is most stable. Adding two more electrons to form $B_6H_6^{2-}$ leaves this result unchanged.

Table XIII. Comparison of Various Networks with Possible Cleavage Products that are Fragments of Icosahedra

Number of B's	Bonds Broken	Configuration[a]	Energy (eV) $C = 0$	$C = 1.77$ Å^{-1}
5	1	B_5H_9 (C_{1v})	12.36	26.78
5	0	B_5H_9 (C_{4v})	13.00	25.96
10	1	$2B_5H_8$ (C_{4v})	35.62	65.14
10	0	$B_{10}H_{16}$ (D_{4h})	34.16	63.06
10	7	$2B_5H_7$ (C_{1v})	47.84	86.12
10	0	$B_{11}H_{14}$ (C_{2v})	47.36	73.84
18	7	$2B_9H_{11}$ (C_{1v})	104.32	164.08
18	5	B_8H_{10} (C_{2v}) + $B_{10}H_{12}$ (C_{2v})	102.74	158.44
18	5	$2B_9H_{11}$ (C_{3v})	102.44	158.16
18	0	$B_{18}H_{22}$ (S_2)	102.68	152.20

[a] Symmetry group of each boron cage is given.

Table XIV.   Free-Electron Energies of Alternative Boron Frameworks

Filled Levels in Cage	Formula	Structure	Energy (eV) $C = 0$	Energy (eV) $C = 1.77\ \text{Å}^{-1}$
4	B_4Cl_4	Bent rhombus	30.08	50.28
		Tetrahedron[a]	28.86	46.06
4	B_6H_{10}	Octahedron	19.50	33.34
		Pentagonal pyramid	19.42	35.46
6	B_6H_6	Pentagonal pyramid	43.58	67.86
		Octahedron[a]	42.66	63.10
8	B_8Cl_8	Octaborane part of icosahedron	57.42	87.00
		Bicapped trigonal antiprism[a]	56.14	81.68
		Triangular dodecahedron	56.28	81.10
8	$B_{10}H_{14}$	Bicapped tetragonal antiprism	48.76	72.22
		Decaborane part of icosahedron	47.36	73.84
9	$B_{10}H_{14}{}^{2-}$	Bicapped tetragonal antiprism	59.12	85.34
		Decaborane part of icosahedron	58.36	88.30
10	$B_{10}H_{10}$	Decaborane part of icosahedron	70.96	104.38
		Bicapped tetragonal antiprism[a]	70.04	99.28

[a] Best from Ref. 7.

For B_8Cl_8 the triangular dodecahedron is calculated to be the most stable as long as $C = 1.77\ \text{Å}^{-1}$, but when $C = 0$ the bicapped trigonal antiprism is more stable. Since natural B_8Cl_8 is dodecahedral, we have here more evidence that C should be positive. The dodecahedron can be considered a distortion of the square (Archimedean) antiprism.[14]

Table XV.   Correlation Between the Icosahedron and the $B_{10}H_{14}$ Icosahedral Fragment Energy Levels

Icosahedral Boron Cage (I_h)			Decaborane-14 Cage (C_{2v})		
Species	Analogous H orbitals	Energy (eV) for $C = 0$	Species	Analogous H orbital	Energy (eV) for $C = 0$
A_g	s	0	A_1	s	0
			B_1	p_x	1.20
F_{1u}	p's	1.62	B_2	p_y	1.73
			A_1	p_z	2.66
			A_1	$d_{x^2-y^2}$	4.36
			B_1	d_{zx}	4.38
H_g	d's	4.14	A_2	d_{xy}	4.38
			B_2	d_{yz}	4.97
			A_1	$d_{3z^2-r^2}$	5.50
F_{2u}	f_ξ's[a]	5.46	B_1	$f_{x(3y^2-x^2)}$	6.30

[a] J. H. Macek and G. H. Duffey, *J. Chem. Phys.*, **34**, 288 (1961).

For $B_{10}H_{10}$ and $B_{10}H_{10}^{2-}$, the bicapped tetragonal antiprism is calculated to be the most stable. The bicapped cube and the pentagonal antiprism[8] are less stable than the structures listed.

When $C = 1.77$ Å^{-1}, the calculated order is incorrect for B_6H_{10}, $B_{10}H_{14}$, and $B_{10}H_{14}^{2-}$. These complexes do stabilize as fragments of the isosahedron. However, the edge borons in such structures can bond the four excess hydrogens more strongly than the borons can in the corresponding compact structures. This effect is not allowed for in the calculations.

On comparing basis functions for irreducible representations, one can correlate many of the levels of related structures. Examining the plotted waves for the boron icosahedron and for the decaborane-14 led us to the results presented in Table XV. Note that three of the H_g levels remain very close together as A_1, B_1, A_2.

REFERENCES

1. J. R. Platt, in *Encyclopedia of Physics*, S. Flügge, Ed., Springer, Berlin, 1961, Vol. 37, Pt. 2, pp. 173–281.
2. J. R. Platt, K. Ruedenberg, C. W. Scherr, N. S. Ham, H. Labhart, and W. Lichten, *Free-Electron Theory of Conjugated Molecules—A Source Book*, Wiley, New York, 1964, Papers 1–21.
3. H. Kuhn, W. Huber, G. Handschig, H. Martin, F. Schäfer, and F. Bär, *J. Chem, Phys.*, **32**, 467 (1960); F. Bär, W. Huber, G. Handschig, H. Martin, and H. Kuhn, *ibid.*, **32**, 470 (1960).
4. N. S. Bayliss, *Quart. Rev. (London)*, **6**, 319 (1952).
5. W. N. Lipscomb, in *Advances in Inorganic Chemistry and Radiochemistry*, Vol. 1, H. J. Emeleus and A. G. Sharpe, Eds., Academic, New York, 1959, pp. 117–156.
6. J. A. Hoerni, *J. Chem. Phys.*, **34**, 508 (1961).
7. W. P. Rahilly, G. H. Duffey, and J. A. Schmidt, *J. Chem. Phys.*, **42**, 1330 (1965).
8. G. H. Duffey, J. A. Houtman, and J. Riashy, *J. Chem. Phys.*, **48**, 273 (1968).
9. H. Kuhn, *J. Chem. Phys.*, **25**, 293 (1956).
10. A. A. Frost, *J. Chem. Phys.*, **25**, 1150 (1956).
11. R. Hoffmann and W. N. Lipscomb, *J. Chem. Phys.*, **37**, 2872 (1962).
12. M. F. Hawthorne, in *Advances in Inorganic Chemistry and Radiochemistry*, Vol. 5, H. J. Emeleus and A. G. Sharpe, Eds., Academic, New York, 1963, pp. 307–345.
13. E. B. Wilson, Jr., J. C. Decius, and P. C. Cross, *Molecular Vibrations*, McGraw-Hill, New York, 1955, pp. 322–330.
14. S. J. Lippard and B. J. Russ, *Inorg, Chem.* **7**, 1686 (1968).

Basis of Extended Hückel Formalism[*]

G. BLYHOLDER† AND C. A. COULSON

Mathematical Institute, Oxford University, England

Abstract

It is shown that the LCAO molecular Hartree-Fock equations for a closed-shell configuration can be reduced to a form identical with that of the Hoffmann extended Hückel approximation if (i) we accept the Mulliken approximation for overlap charge distributions and (ii) we assume a uniform charge distribution in calculating two-electron integrals over molecular orbitals. Numerical comparisons indicate that this approximation leads to results which, while unsuitable for high accuracy calculations, should be reasonably satisfactory for molecules that cannot at present be handled with facility by standard LCAO molecular Hartree-Fock methods.

[*] Reprinted by permission from *Theoretica chimica Acta* (*Berlin*), **10**, 316–324 (1968).
† Present address: Department of Chemistry, University of Arkansas, Fayetteville, Arkansas 72701.

INTRODUCTION

Molecular-orbital theory has taken many forms and has been dealt with by many approximations. In 1963 Hoffmann[11] presented a formalism which he referred to as "extended Hückel" (EH). In the 1930's, however, this formalism would simply have been called "molecular-orbital," since it is a straight-forward application of molecular-orbital (MO) theory, using a one-electron Hamiltonian. Hoffmann referred to it as "extended Hückel" because it did not limit itself to π-electron systems and was able to deal with saturated molecules by including all overlap integrals. In these respects it did "extend" the usual, or "simple Hückel," method, which was customarily applied to π-electrons, and assumed complete $\pi - \sigma$ separability.

In the Hoffmann formalism each MO is assumed to be of the form

$$\phi_j = \sum_p C_{jp}\chi_p,\tag{1}$$

where the C_{jp} are LCAO coefficients, and the χ_p are normalized atomic orbitals. If we use the variation method with an effective one-electron Hamiltonian h, these coefficients are given by the secular equations

$$\sum_q (h_{pq} - \epsilon_j S_{pq})C_{jq} = 0, \qquad j = 1, 2, \ldots; \quad p = 1, 2, \ldots,\tag{2}$$

where the ϵ_j are the orbital energies, the S_{pq} are atomic overlap integrals, and the h_{pq} are the matrix components of h, defined by

$$h_{pq} = \int \chi_p^* h \chi_q \, d\tau.\tag{3}$$

Throughout this paper, subscripts i, j will refer to molecular orbitals, and p, q, r, s to atomic orbitals. These latter orbitals may be of π or of σ type.

Elimination of the C_{jq} in (2) leads to the typical secular determinantal equation

$$\|h_{pq} - \epsilon S_{pq}\| = 0\tag{4}$$

from which the orbital energies ϵ_j are found, if the S_{pq} and h_{pq} are known. When Slater orbitals are used it is easy to compute the S_{pq}. But since the

27

precise form of h cannot be written down in a one-electron model, we make no attempt to compute the h_{pq} directly. Instead, Hoffmann argues that h_{pp} should be identified with the valence-state ionization potential of χ_p and that h_{pq} should be proportional to S_{pq}, so that

$$h_{pq} = k \left(\frac{h_{pp} + h_{qq}}{2} \right) S_{pq}. \tag{5}$$

In (5) k is a constant subsequently to be chosen in such a way as to give best results for the total energy.* In any one-electron model such as this, the total energy is taken to be the sum of the orbital energies.

An alternative way of evaluating the h_{pq} in the EH method has recently been proposed by Newton, Boer, Palke, and Lipscomb.[21,22] It is clear that h is itself the sum of a kinetic energy operator and a potential energy term (in which nuclear attractions, Coulomb repulsions, and exchange interactions are all involved). Newton et al. calculate the kinetic energy part of h_{pq} exactly, and use a Mulliken approximation for the potential energy part. Diagonal elements and proportionality constants for this part were estimated from SCF procedures for small, chemically similar, molecules. Their approach was designed to see how well they could reproduce the SCF secular determinant within the EH formalism. In this paper we proceed in the opposite direction; our starting points are the elements of the SCF secular determinant, which are examined to see what approximations must be made within the SCF scheme in order to reduce the SCF equations to EH form.

There are, of course, certain circumstances in which the SCF equations can easily be shown to reduce to Hückel-type equations. Thus in the standard excited-state equations of Hall[9] for π-electron aromatic hydrocarbons, it can be shown that the LCAO coefficients are often precisely the same as in the Hückel scheme. However, this particular state is not of direct value for our purposes. Further, Pople[23] has shown that if the Pariser-Parr approximation of zero differential overlap is accepted, then, apart from some (usually) small terms in inverse internuclear distances, the total π-electron energy is given by an expression identical in form to that of Hückel theory. Both writers adopt the approximation of neglecting all overlap integrals. This is particularly serious with σ-orbitals, in which an overlap integral may be as large as 0.8. We have therefore thought it worthwhile to make an independent study of the equivalence of SCF and EH equations, in which overlap is not neglected from the start.

---

* Our conclusions in this paper would not be significantly altered if, instead of the arithmetic mean in (5), we used the Wolfsberg-Helmholtz-Mulliken geometrical-mean formula $h_{pq} = k\sqrt{(h_{pp}h_{qq})}S_{pq}$ recently advocated by Allen and Russell.[1]

SIGNIFICANCE OF THE OVERLAP INTEGRALS S_{pq}

The EH method is designed to take advantage of the fact that modern computers can solve complete secular determinants very readily, so that there is no need to neglect any of the off-diagonal elements in the secular determinants. It aims to produce reasonable values for important electronic properties by semiempirical methods, and its success must depend on how well it can approximate the more accurate SCF Hamiltonian matrix elements. Central to the EH method is the approximation of making the off-diagonal matrix elements h_{pq} proportional to the corresponding overlap integrals S_{pq}.

This particular assumption was introduced as early as 1942, by Wheland,[28] who used it in π-electron calculations, but without any theoretical justification. He applied it to near-neighbor overlaps, neglecting all further interactions. Mulliken[19] reported an empirical observation that for the π-electrons of ethylene the calculated values of the ratio β/S varied by only about 10% when the carbon–carbon distance changed from 0.998 to 1.664 Å. He also noted, empirically, that this ratio did not undergo large variations for the cases of H_2 and benzene. Within a very short time thereafter the proportionality of matrix elements h_{pq} to overlap integrals S_{pq} developed from being a method of estimating small changes in resonance integrals between adjacent atoms to being a method for estimating all off-diagonal elements h_{pq}.[2,5,12,13,16–18,29] In this sense the method was already in use before it was referred to as the "extended" Hückel method. Nevertheless, it is an appropriate label for a method of sufficient generality which embodies in the MO framework the following features:

(1) Use of LCAO representation of molecular orbitals
(2) Use of a one-electron Hamiltonian
(3) Overlap integrals not neglected
(4) h_{pq} proportional to S_{pq}
(5) Usable for saturated as well as unsaturated π-electron molecules.

In order that this method may give reliable predictions it is necessary that the corresponding secular determinant (4) should be almost identical with the SCF secular determinant. For then both orbital energies and LCAO coefficients given by the two methods will be closely similar. It is the way in which this equivalence depends upon feature (4), above, that provides the main concern of this paper. We shall first deal with the general theory and shall then conclude with a numerical study of some illustrative examples.

GENERAL THEORY

In the Roothaan form[25] of the SCF MO equations, we suppose that the total electronic wavefunction is of single-determinant Slater form, in which the molecular spin orbitals are products of space orbitals ϕ_j, as in (1), and conventional spin functions. The basic equations are

$$\sum_q (F_{pq} - \epsilon_j S_{pq}) C_{jq} = 0, \tag{6}$$

so that the orbital energies ϵ_j are found from the determinantal equation

$$\|F_{pq} - \epsilon S_{pq}\| = 0. \tag{7}$$

In (6) and (7) F_{pq} is usually written as

$$F_{pq} = H_{pq} + G_{pq}, \tag{8}$$

where H_{pq} (to be distinguished from our previous Hückel h_{pq}) comes from the 1-electron part of the total Hamiltonian and G_{pq} from the 2-electron part. Thus, if V_α is the potential due to nucleus α

$$H_{pq} = \int \chi_p^* \left\{ -\tfrac{1}{2}\nabla^2 - \sum_\alpha V_\alpha \right\} \chi_q \, d\tau, \tag{9a}$$

$$G_{pq} = \sum_i \{2(pq \mid ii) - (pi \mid iq)\}, \tag{9b}$$

where the i summation is over all doubly occupied orbitals ϕ_i, and

$$(pq \mid ij) = \int \chi_p^*(1)\chi_q(1) \frac{1}{r_{12}} \phi_i^*(2)\phi_j(2) \, d\tau_1 \, d\tau_2. \tag{10}$$

Our question is: Under what conditions will the F_{pq} of (7) and (8) be the same as the h_{pq} of (4)?

Now Mulliken[19] has shown that, to a good approximation, multicenter electron repulsion integrals involving atomic orbitals p, q, r, s may be simplified by putting

$$(pq \mid rs) \approx \tfrac{1}{4} S_{pq} S_{rs} \{(pp \mid rr) + (pp \mid ss) + (qq \mid rr) + (qq \mid ss)\}. \tag{11}$$

With this approximation the first term in (9b) becomes

$$\sum_i 2(pq \mid ii) = \sum_i 2 \cdot \frac{S_{pq}}{2} \cdot \{(pp \mid ii) + (qq \mid ii)\}.$$

The term $pp \mid ii$ is the Coulomb repulsion between an electron in the atomic orbital χ_p and an electron in the molecular orbital ϕ_i. But molecules have reasonably uniform charge distributions, and in particular, as the Coulson-Rushbrooke theorem proves, alternant hydrocarbons have exactly uniform atomic charge densities in their ground states. We may therefore expect that the sum $\sum_i (pp/ii)$ will be approximately the same, regardless of which atom p we have chosen. Thus we find that

$$\sum_i 2(pq \mid ii) \approx \text{constant} \times S_{pq}. \tag{12}$$

Applying the Mulliken approximation to the second term of (9b), we have

$$\sum_i (pi \mid iq) \approx \sum_i \tfrac{1}{4} S_{pi} S_{iq} \{(pp \mid ii) + (pp \mid qq) + (ii \mid ii) + (ii \mid qq)\} \tag{13}$$

On the assumption of a reasonably uniform charge distribution as before, the summations over i in (pp/ii), (ii/ii), and (ii/qq) would be expected to give constants independent of p and q. But the presence of the additional term (pp/qq) and the factors $S_{pi} S_{iq}$ in the summation means that there is no strictly analogous formula to (12) which applies to the exchange term (13). Fortunately, however (see later), it appears from numerical calculations for linear H_6, ethylene, butadiene, and benzene that the first term in G_{pq} is much larger than the second. If this is true generally, it will be approximately true to say that $G_{pq} \propto S_{pq}$. The numerical justification for this is given later.

We now turn to H_{pq}. Equation 9a shows this to be the sum of a kinetic energy term and a core-attraction term. Numerical study of the kinetic energy term shows it to be large if $p \equiv q$, small if p and q are adjacent atoms, and negligible otherwise. This is illustrated for s and p_π orbitals in Table I.

This still leaves the core-attraction terms, to which Barker and Eyring[3] have applied the Mulliken approximation, and found inaccuracies of about

Table I.   Values of Kinetic Energy Integral, $\int \chi_p^* (-\tfrac{1}{2} \nabla^2) \chi_q \, d\tau$ as a Function of the p–q Separation, Using Slater Orbitals

1s orbitals (orbital exponent 1)			p_π orbitals (orbital exponent 3.18)		
Separation (a.u.)	Kinetic Energy (a.u.)	Overlap Integral	Separation (a.u.)	Kinetic Energy (a.u.)	Overlap Integral
0	0.5000	1.000	0	0.5000	1.000
2	0.1128	0.586	2.64	0.0373	0.258
4	−0.0030	0.189	5.28	−0.0068	0.017
6	−0.0060	0.047	7.92	−0.0006	0.0017
8	−0.0021	0.010			
10	−0.0005	0.002			

10–20%. Some of the deviations were positive, others were negative. In general, however, this term gives

$$\int \chi_p^* \left(\sum_\alpha \frac{1}{r_\alpha} \right) \chi_q \, d\tau \approx \tfrac{1}{2} S_{pq} \left[\int \chi_p^* \left(\sum_\alpha \frac{1}{r_\alpha} \right) \chi_p \, d\tau + \int \chi_q^* \left(\sum_\alpha \frac{1}{r_\alpha} \right) \chi_q \, d\tau \right]. \quad (14)$$

Now $\sum_\alpha (1/r_\alpha)$ will have approximately the same value at all nuclei of the molecule, except for the end atoms of a chain, for which the number of close neighbor nuclei is less than for "internal" atoms. Thus, except possibly for such edge atoms, the potential energy part of H_{pq} is approximately proportional to S_{pq}.

Combining the various results just described we may therefore conclude that (1) if p and q are no closer than second neighbors, F_{pq} should be closely proportional to S_{pq}, (2) if p and q are first neighbors there should be a small variation from proportionality due chiefly to the kinetic energy operator, and (3) if p and q are the same atom, the kinetic energy part of F_{pq} shifts the numerical value right out of the sequence of other F_{pq} matrix elements.

Thus the EH formalism of putting $h_{pq} \propto S_{pq}$ $(p \neq q)$ is reasonably justified, and the necessity of the constant k in (5), with k not having the expected value of 1, is seen to arise from the kinetic-energy part of the Hamiltonian, which affects the diagonal terms differently from the off-diagonal ones. The importance of the kinetic-energy contribution to these off-diagonal elements has recently been recognized by Jørgensen[14] and by Radtke and Fenske[24] in some studies of transition metal complexes. These latter authors wrote: "terms of considerable magnitude which do not vary as functions of overlap integrals, make substantial contributions to off-diagonal elements." Our discussion has shown just what these are and why it is not possible to choose $k = 1.75$ in (5), as originally suggested by Hoffmann, nor to put $k = 2 - S_{ij}$ as urged by Carroll and McGlynn.[4] We can also see why Newton et al.[21,22] found it necessary to use a different value of k for each type of bond, and why Cusachs,[6] who did not consider any two-electron terms, adopted different variations for the kinetic- and potential-energy matrices, putting*

$$T_{ij} \approx S_{ij} |S_{ij}| \left(\frac{T_{ii} + T_{jj}}{2} \right), \qquad U_{ij} \approx S_{ij} \left(\frac{U_{ii} + U_{jj}}{2} \right).$$

TOTAL ENERGY EXPRESSIONS, AND ELECTRONIC SPECTRA

The discussion in the preceeding section dealt with the equivalence of orbital energies and LCAO coefficients in SCF and EH methods. This equivalence, however, does not imply equivalence of total energy E. In the SCF scheme

---

* Further comment on the Cusachs approximation can be found in M. D. Newton, *J. Chem. Physics*, **45**, 2716 (1966) and L. C. Cusachs, *ibid.*, 2717 (1966).

there are the relations[25]

$$E = 2 \sum_i H_i + \sum_{i,j} (2J_{ij} - K_{ij}), \qquad (15)$$

$$E = 2 \sum_i \epsilon_i - \sum_{i,j} (2J_{ij} - K_{ij}), \qquad (16)$$

$$E = \sum_i \epsilon_i + \sum_i H_i. \qquad (17)$$

The single summations are over all doubly occupied molecular orbitals ϕ_i, and the double ones are over all ϕ_i and ϕ_j including $i = j$. These equations show that although the Hückel energy is a good approximation for the various ionization potentials, it is bad for the total energy; for example, for linear H_6 (see the next section for an explanation of the calculation), the simple electronic energy is $2 \sum_i \epsilon_i = -3.23$ a.u., whereas in the SCF scheme the total energy from (17) is -7.42 a.u. If we are interested in total energies there seems no alternative to the calculation of the exchange and Coulomb integrals; if we have used the EH procedure to get the ϵ_i and LCAO coefficients, then we can determine the F matrix in terms of various atomic integrals. All those that involve the kinetic energy, and the diagonal elements for the nuclear attractions and r_{12}-integrals, can readily be evaluated numerically; and it is in the spirit of this approximation to put the remaining off-diagonal elements proportional to overlap, or—as suggested by Newton et al.[21,22]—estimate their values by comparison with suitable reference compounds. We shall show in the next section that both procedures give a reasonable reproduction of the F matrices calculated by direct evaluation of all integrals.

In the evaluation of E (17) is preferred to either (15) or (16), since the ϵ_i are already determined in solving the secular determinant; and only one-electron terms are involved in computing the H_i.

It is natural to estimate electronic excitation energies from the ground state, as suggested by Roothaan,[25] by using the virtual orbitals of the ground-state calculation as excited-state orbitals. However, an improvement of 5–20% has been indicated[7,15] if new orbitals are calculated for each excited state. But in view of the approximations inherent in the EH method, it is doubtful if this extra labor is justified. Moreover, most UV excitations lead to a displacement of charge, so that, except for situations such as the p-band of alternant hydrocarbons, in which no gross redistribution of charge occurs,[20] the assumption of a uniform charge distribution, required in our justification of the EH formalism, is in general less valid for excited states than for the ground state.

NUMERICAL COMPARISONS

We turn to some numerical illustrations of our previous general conclusions. First we have considered a linear complex H_6, using Slater $1s$

Table II.   Comparison of Matrix Elements Calculated from Roothaan's Equations with Elements Which Are Proportional to Overlap for the Case of Linear H_6

q	G'_{1q}	$\dfrac{G'_{12}}{S_{12}}\cdot S_{1q}$	G''_{1q}	$\dfrac{G''_{12}}{S_{12}}\cdot S_{1q}$	G_{1q}	$\dfrac{G_{12}}{S_{12}}\cdot S_{1q}$	H_{1q}	$\dfrac{H_{12}}{S_{12}}\cdot S_{1q}$	F_{1q}	S_{1q}
1	1.69	1.86	−0.46	−0.66	1.23	1.19				
2	1.09	(1.09)	−0.39	(−0.39)	0.696	(0.696)	−1.164	(−1.164)	−0.465	0.586
3	0.36	0.35	−0.14	−0.13	0.224	0.224	−0.409	−0.375	−0.185	0.189
4	0.090	0.87	−0.030	−0.031	0.060	0.056	−0.130	−0.093	−0.069	0.047
5	0.019	0.019	−0.0065	−0.0068	0.012	0.012	−0.023	−0.019	−0.010	0.010
6	0.0034	0.0037	−0.0052	−0.0001	−0.0018	0.0023	−0.0035	−0.003	−0.005	0.002

orbitals on each atom. We have made a SCF–LCAO–MO calculation for the system. Using Roothaan's equations, and an internuclear spacing of 2 a.u. in this calculation, all one- and two-center integrals were obtained from standard tables[27]; the three- and four-center electron-repulsion integrals were obtained with the Mulliken approximation[19]; the three-center nuclear attraction integrals were taken from Hirschfelder and Weygandt.[10] We found agreement, both as regards energies and LCAO coefficients, with previous calculations using similar methods for H_4 and H_6.[8,26] Table II gives comparisons of matrix elements calculated in this way and with the assumption of proportionality to overlap. In this table $G_{pq} = G'_{pq} + G''_{pq}$, where G'_{pq} is the first (i.e., Coulomb) part of G_{pq} given by (9b) and (12); and G''_{pq} is the second (i.e., exchange) part of G_{pq} as in (9b) and (13). A comparison of entries, particularly in the last two columns of the table, shows that for linear H_6 off-diagonal matrix elements of F are reasonably well represented by terms proportional to overlap.

Linear H_6 is a rather theoretical system. So we have made various similar calculations for the more realistic systems ethylene, *trans*-butadiene, and benzene. The results are shown in Tables 3, 4, and 5. Again there is a reasonable (but far from perfect) correspondence for the off-diagonal elements.

Some confirmation of our general conclusions can be found from the work of Newton et al.[21] on methane. These writers found a close correspondence between matrix elements calculated from Roothaan's formulas and elements calculated by adding together (1) a purely theoretical kinetic energy term and

Table III.   Comparison of Matrix Elements Calculated from Roothaan's Equations with Elements Which Are Proportional to Overlap for the Case of Ethylene

G'_{12}	$S_{12}G'_{11}$	G''_{12}	$S_{12}G''_{11}$	G_{12}	$S_{12}G_{11}$	$V_{12}{}^a$	$S_{12}V_{11}$
0.268	0.268	−0.256	−0.106	0.012	0.162	0.95	1.05

^{a} V denotes the core attraction term in H_{pq} as in (14).

Table IV.   Comparison of Matrix Elements Calculated from Roothaan's Equations with Elements Which Are Proportional to Overlap for the Case of Trans-Butadiene

q	G'_{12}	$\dfrac{G'_{12}}{S_{12}}\cdot S_{1q}$	G''_{1q}	$\dfrac{G''_{12}}{S_{12}}\cdot S_{1q}$	G_{1q}	$\dfrac{G_{12}}{S_{12}}\cdot S_{1q}$	H_{1q}	$\dfrac{H_{12}}{S_{12}}\cdot S_{1q}$	F_{1q}	$\dfrac{F_{12}}{S_{12}}\cdot S_{1q}$
1	1.30	1.49	−0.35	−0.95	0.95	0.53				
2	0.39	(0.39)	−0.25	(−0.25)	0.14	(0.14)	−0.419	(−0.419)	−0.276	(−0.276)
3	0.047	0.044	−0.033	−0.028	0.014	0.016	−0.049	−0.047	−0.035	−0.024
4	0.0028	0.0107	+0.013	−0.007	0.016	0.004	−0.0107	−0.0115	−0.0054	−0.0059

(2) a potential energy term of which the diagonal elements were taken over from Roothaan-type calculations for ethane, and the off-diagonal elements were made proportional to overlap.

The following conclusions follow from our present study:

1. Molecular LCAO Hartree-Fock equations can be reduced to the extended-Hückel form if we adopt Mulliken's approximation for an overlap-charge distribution, and assume an approximately uniform charge distribution in the molecule.

2. There is an advantage in distinguishing between the kinetic energy terms (which are not proportional to overlap) and the other items (most of which are much more closely proportional to overlap).

3. The origin of the difficulty with the numerical scale factor k in the Wolfsberg-Helmholtz formula (5) lies chiefly in the variation of kinetic-energy matrix elements with distance apart of the two atoms involved.

4. Numerical calculation for systems with σ or with π electrons indicate that the approximations considered produce results which, while not appropriate for a requirement of high accuracy, should be reasonably satisfactory for molecules that cannot at the present time be conveniently handled by full molecular Hartree-Fock methods.

Table V.   Comparison of Matrix Elements Calculated from Roothaan's Equations with Elements Which Are Proportional to Overlap for the Case of Benzene

q	G'_{1q}	$\dfrac{G'_{12}}{S_{12}}\cdot S_{1q}$	G''_{1q}	$\dfrac{G''_{12}}{S_{12}}\cdot S_{1q}$	G_{1q}	$\dfrac{G_{12}}{S_{12}}\cdot S_{1q}$
1	1.95	1.95	−0.39	−0.76	1.56	1.19
2	0.508	(0.508)	−0.20	(−0.20)	0.309	(0.309)
3	0.076	0.076	−0.037	−0.030	0.038	0.046
4	0.035	0.035	+0.015	−0.014	0.050	0.021

ACKNOWLEDGMENTS

Acknowledgment is made to the donors of the Petroleum Research Fund, administered by the American Chemical Society, for partial support of this research, while one of the authors (G. B.) was in Oxford.

REFERENCES

1. L. C. Allen and J. D. Russell, *J. Chem. Phys.*, **46**, 1029 (1967).
2. C. J. Ballhausen and H. B. Gray, *Inorg. Chem.*, **1**, 111 (1962).
3. R. S. Barker and H. Eyring, *J. Chem. Phys.*, **22**, 1182 (1954).
4. D. G. Carroll and S. P. McGlynn, *J. Chem. Phys.*, **45**, 3827 (1966).
5. B. H. Chirgwin and C. A. Coulson, *Proc. Roy. Soc. (London)*, **A201**, 196 (1950).
6. L. C. Cusachs, *J. Chem. Phys.* **43**, S 157 (1965).
7. L. Goodman and J. R. Hoyland, *J. Chem. Phys.*, **39**, 1068 (1963).
8. V. Griffing and J. T. Vanderslice, *J. Chem. Phys.*, **23**, 1035 (1955).
9. G. G. Hall, *Proc. Roy. Soc. (London)*, **A213**, 102, 113 (1952).
10. J. O. Hirschfelder and C. N. Weygandt, *J. Chem. Phys.*, **6**, 806 (1938).
11. R. Hoffmann, *J. Chem. Phys.*, **39**, 1397 (1963).
12. R. Hoffmann and W. N. Lipscomb, *J. Chem. Phys.*, **36**, 2179, 3489 (1962).
13. R. Hoffmann and W. N. Lipscomb, *J. Chem. Phys.*, **37**, 2872 (1962).
14. C. K. Jørgensen, *Chem. Phys. Letters*, **1**, 11 (1967).
15. R. Lefébvre, *Modern Quantum Chemistry*, O. Sinanoğlu, Ed., Part 1, Academic, New York, 1965, p. 125.
16. L. L. Lohr, Jr. and W. N. Lipscomb, *J. Amer. Chem. Soc.*, **85**, 240 (1963).
17. H. C. Longuet-Higgins and M. de V. Roberts, *Proc. Roy. Soc. (London)*, **A224**, 336. (1954).
18. H. C. Longuet-Higgins and M. de V. Roberts *Proc. Roy. Soc. (London)*, **A230**, 110 (1955).
19. R. S. Mulliken, *J. Chim. Phys.*, **46**, 497, 675 (1949).
20. J. N. Murrell, *Theory of the Electronic Spectra of Organic Molecules*, Methuen, London 1963, p. 108.
21. M. D. Newton, F. P. Boer, W. E. Palke, and W. N. Lipscomb, *Proc. Natl. Acad. Sci, U.S.*, **53**, 1089 (1965).
22. M. D. Newton, F. P. Boer, and W. N. Lipscomb, *J. Amer. Chem. Soc.*, **88**, 2353, 2361, 2367 (1966).
23. J. A. Pople, *Trans. Faraday Soc.*, **49**, 1375 (1953).
24. D. D. Radtke, and R. F. Fenske, *J. Amer. Chem. Soc.*, **89**, 2292 (1967).
25. C. C. J. Roothaan, *Rev. Mod. Phys.*, **23**, 69 (1951).
26. A. R. Ruffa and V. Griffing, *J. Chem. Phys.*, **36**, 1389 (1962).
27. R. C. Sahni and J. W. Cooley, NASA TN D-146-1 (1959) and II (1960).
28. G. W. Wheland, *J. Amer. Chem. Soc.*, **64**, 900 (1942).
29. M. Wolfsberg and L. Helmholtz, *J. Chem. Phys.*, **20**, 837 (1952).

Ionization Potentials and Electron Affinities of Some Polyphenyl Molecules

SOOK-IL KWUN

Department of Physics, Seoul National University, Korea

TAIKYUE REE

Department of Chemistry, University of Utah, Salt Lake City

Abstract

The calculation of the ionization potentials and electron affinities of the polyphenyl molecules, biphenyl, *para*-terphenyl, *para*-quaterphenyl, and *meta*-terphenyl is made from the simplified SCF-LCAO-MO theory for unsaturated hydrocarbons in their ground states by including the σ-bond deformation effect upon ionization. The calculated ionization potentials of conjugated hydrocarbons using Koopman's theorem is consequently modified. The resulting calculated ionization potential of the polyphenyl molecule is not simply the orbital energy of the highest occupied molecular orbital but includes an energy correction associated with the σ-bond deformation effect that accompanies ionization. The polyphenyl molecules and ions are assumed to be planar. The ionization potentials of polyphenyls obtained are 8.29, 7.96, 7.76, and 8.07 eV for biphenyl, *para*-terphenyl, *para*-quaterphenyl, and *meta*-terphenyl, respectively. The electron affinities of these molecules are also calculated by a similar method.

A Personal Note

SOOK-IL KWUN

I was a graduate student of Professor Henry Eyring from 1961 to 1965. Most of the results in this paper were obtained during this period. I thank Professor Eyring for his guidance, advice, and encouragement.

Since graduating, I have begun to study strong dielectrics. I was inspired by Dr. Eyring into this research, while attending his lectures in quantum mechanics and theory of rate processes, which were both interesting and stimulating. I will never forget my happy research life with him. Among many pleasant memories is the "foot race" with Dr. Eyring and his associates, which was conducted annually on the campus lawn.

INTRODUCTION

The quantum mechanical procedure for calculating the electronic structure and physical properties of polyatomic molecules has been developed by the use of a powerful tool, the molecular orbital theory proposed by Hund[1] and Mulliken.[2] The theory has been applied to large molecules with considerable vigor. However, because of the difficulty of carrying out all the molecular integral calculations involved, various approximations have been used.

Particularly for the large alternant hydrocarbon system, such as the polyphenyls, a π-electron approximation method has been developed extensively by many authors.[3-6] In this approximation, the π electrons are treated separately from the σ electrons. The latter are considered to form a constant core and make their contribution to the core potential.

In the Pariser-Parr-Pople scheme,[5,6] the so-called zero differential overlap approximation is used, and the σ-electron system is treated as a nonpolarizable core. The interelectronic repulsions are explicitly taken into account in the total Hamiltonian. Resonance integrals,[7] core integrals, and electronic repulsion integrals are given empirically,[6] and Coulomb penetration integrals are neglected.*

The calculated results from this scheme agree well with the experimental values for the ground states. However, the calculated ionization potentials of the conjugated molecules agree somewhat less well with experiment. This lack of agreement is due to the mistaken use of the total ground-state Hamiltonian for the ionized states; that is, in this calculation the σ-electron system has been treated as a constant core even after one π electron is removed from the σ framework. According to Koopmans' theorem,[9] the first ionization potential, which hereafter we will speak of as the ionization potential, is simply given by the negative of the Hartree-Fock orbital energy (the highest occupied orbital energy) for the ground state.† In the case of the

---

* The effect of neglecting the Coulomb penetration integrals is considered in detail by J. R. Hoyland and L. Goodman.[8]

† This is simply obtained by subtracting the energy of the ground state from the energy of the ionized state. For detailed derivation, see ref. 10.

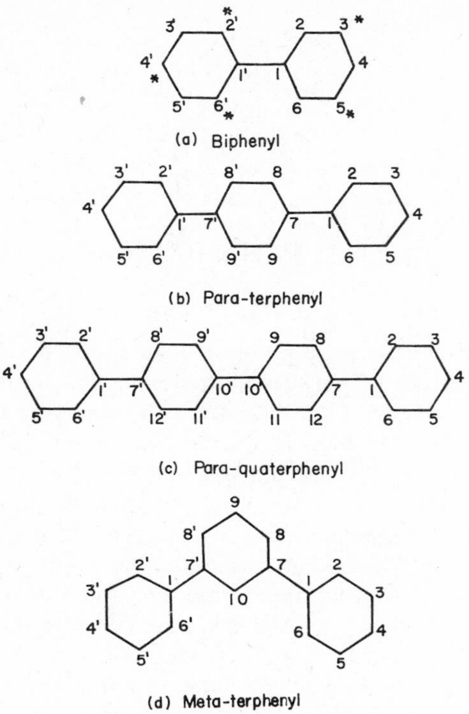

Fig. 1. Skeletal configuration of polyphenyls and numbering. (The starred and unstarred atoms are shown in the biphenyl molecule only.)

conjugated systems, ionization potentials calculated from Koopmans' theorem are usually higher than the experimental values.

Recently, Hoyland and Goodman[11] and I'Haya[12] have pointed out that the Hartree-Fock Hamiltonian matrix elements for the ionized states should be different from those calculated for the ground state. This implies that Koopmans' theorem should be modified to get rid of the deficiency mentioned above.

In this paper, in order to calculate the ionization potential, we considered the effect of the σ-bond deformation caused by the removal of one π electron from the highest occupied molecular orbital.

The skeletal configuration of polyphenyl molecules, that is, biphenyl, *para*-terphenyl, *para*-quaterphenyl, and *meta*-terphenyl, and the numbering of atoms in those molecules are shown in Fig. 1. In our previous papers[7] we calculated the mobile bond orders and discussed the related phenomena for the molecules. In this paper, we have assumed that all the polyphenyls are planar, according to the evidence of X-ray analysis.[13-15] (In general, the first four linear members of polyphenyl molecules have been proved to be planar in the crystalline state.[16]) Also, the regular hexagonal rings with C—C bond distances of 1.40 Å and a value of 1.50 Å for ring–ring bond distance are taken as a first trial. These values of bond distances are reasonable in light of recent studies[13] and are utilized in calculating the empirical values of resonance integrals and electronic repulsion integrals. Later, modified calculations using recent experimental bond lengths of biphenyl are carried out and compared with those of the regular hexagonal aromatic rings.

IONIZATION POTENTIALS AND ELECTRON AFFINITIES

A system with an even number of mobile electrons in a closed-shell ground state is considered.

The total wavefunction can be constructed from an antisymmetrized product of molecular spin orbitals and each of the molecular orbitals, ψ_i, expressed as a linear combination of atomic orbitals ϕ_μ,

$$\psi_i = \sum_\mu \phi_\mu C_{\mu i}, \tag{1}$$

where $C_{\mu i}$ is the LCAO coefficient which is the solution of the Roothaan equation

$$\sum_\nu F_{\mu\nu} C_{\nu i} = E_i C_{\mu i}. \tag{2}$$

E_i is the energy of the ith molecular orbital, and $F_{\mu\nu}$ is a matrix element of the Hartree-Fock Hamiltonian F, which is given by*

$$F_{\mu\nu} = \int \phi_\mu^* F \phi_\nu \, d\tau, \tag{3}$$

where $d\tau$ is a volume element including spatial coordinates only. The matrix elements of the Hartree-Fock Hamiltonian are obtained by a series of systematic approximations as follows:

$$F_{\mu\mu} = W_\mu + \left(\mu \left| \sum_{\nu(\neq\mu)} V_\nu \right| \mu \right) + \sum_\nu P_{\nu\nu}(\mu\mu \mid \nu\nu) - \tfrac{1}{2} P_{\mu\mu}(\mu\mu \mid \mu\mu), \tag{4}$$

and

$$F_{\mu\nu} = H_{\mu\nu} - \tfrac{1}{2} P_{\mu\nu}(\mu\mu \mid \nu\nu). \tag{5}$$

---

* The Hartree-Fock Hamiltonian operator can be obtained by the procedure in which the total energy of the system is varied to be minimum by an infinitesimal change of each molecular orbital. See Ref. 10.

In (4) and (5), W_μ is the atomic valence state ionization potential,[4,5] and

$$\left(\mu \left| \sum_{\nu(\neq\mu)} V_\nu \right| \mu\right) = \int \phi_\mu^* \sum_{\nu(\neq\mu)} V_\nu \Phi_\mu \, d\tau, \tag{6}$$

where $\sum_{\nu(\neq\mu)} V_\nu$ is the potential due to the core excluding the one in question (the μth). The quantity $P_{\mu\nu}$,

$$P_{\mu\nu} = 2 \sum_i^{(occ)} C_{\mu i}^* C_{\nu i}, \tag{7}$$

is called the mobile bond order[17] (π-bond order) in which the upper limit (occ) indicates the summation taken over all occupied molecular orbitals. The following,

$$(\mu\mu \mid \nu\nu) = \iint \left(\frac{e^2}{r_{12}}\right) \phi_\mu^*(1)\phi_\mu(1)\phi_\nu^*(2)\phi_\nu(2) \, d\tau_1 \, d\tau_2, \tag{8}$$

is called an electronic repulsion integral, and

$$H_{\mu\nu} = \int \phi_\mu^* H \phi_\nu \, d\tau \tag{9}$$

is called the resonance integral, where H is the core Hamiltonian operator of a given electron moving in the field of the nuclei alone.

The matrix elements (4) and (5) are valid for the ground state of the system. As mentioned in the preceding section, these matrix elements cannot be utilized directly for the ionized states. This is due to the fact that in the ionized states, a potential V_ν in (6) is not simply the potential due to one less electron than appears in (6), but there is an effect due to deforming the σ framework when one π electron is removed. To explore this situation (6) is further modified. By treating the charges at the nuclear centers as point charges, the approximation

$$(\mu| V_\nu |\mu) = \frac{Z_\nu}{R_{\mu\nu}}, \qquad (\mu \neq \nu) \tag{10}$$

is made, where Z_ν is the effective charge of the σ core of atom ν, and $R_{\mu\nu}$ is the distance between atoms μ and ν. However, after one π electron is removed to infinity from the σ framework, the effective charge of the σ core of atom ν should be changed. In other words, (10) becomes

$$(\mu| V_\nu' |\mu) = -\frac{Z_\nu'}{R_{\mu\nu}}, \qquad (\mu \neq \nu), \tag{11}$$

where Z_ν' is the new effective charge different from the ground state Z_ν. This Z_ν' gives some idea of the magnitude of the deformed potential. In this manner one can understand the σ-deformation effect in the polyphenyl molecules upon ionization.

To evaluate Z_ν' we use the semiempirical method. Let us equate Z_ν' to

$$Z_\nu' = Z_\nu - \Delta Z_\nu, \tag{12}$$

where ΔZ_v is an unknown quantity. Now (11) becomes

$$(\mu|\,V'_v\,|\mu) = -\frac{Z_v - \Delta Z_v}{R_{\mu v}} \qquad (\mu \neq v). \tag{13}$$

From the point charge approximation we can obtain the electronic repulsion integral (8) as

$$(\mu\mu\,|\,vv) = \frac{1}{R_{\mu v}} \quad \text{for} \quad R_{\mu v} > 2.88 \text{ Å}. \tag{14}$$

Using (13) and (14), (4) can be expressed by

$$F_{\mu\mu} = W_\mu + \tfrac{1}{2}P_{\mu\mu}(\mu\mu\,|\,\mu\mu) + \sum_{v(\neq\mu)} \frac{P_{vv} - Z_v + \Delta Z_v}{R_{\mu v}}, \tag{15}$$

where we neglect all two-electron integrals that depend on the overlapping of charge distributions of different orbitals. For hydrocarbons the values of $P_{\mu\mu} = 1$ and $Z_v = 1$ are used in (15).

Using (15) and (5), the ionization potential, I, of benzene is given by the following equation[6]:

$$-I = W_\mu + \tfrac{1}{2}(11\,|\,11) + H_{12} - \tfrac{5}{12}R_{12}{}^{-1} + \sum_{v(\neq\mu)} \frac{\Delta Z_v}{R_{\mu v}}. \tag{16}$$

Employing the experimental value[18] $I = 9.25$ eV for benzene in (16), we can fix the value of ΔZ_v, which turns out to be $\Delta Z_v = 0.04$. The physical interpretation of this value will be discussed in the following section. Since polyphenyls are comprised of benzene rings, a value of ΔZ_v for a molecule is the appropriate fraction of the value for benzene; for example, in the case of biphenyl, we take $\Delta Z_v = 0.02$ (just half of $\Delta Z_v = 0.04$) for benzene, because for the biphenyl the effect is spread over two benzene rings. In a similar way, $\Delta Z_v = 0.04/3$ and $0.04/4$ are taken for terphenyl and quaterphenyl, respectively.*

The quantity $\sum_{v(\neq\mu)} 1/R_{\mu v}$ in (14) should be different for different atoms, but for simplicity the mean value is taken for a molecule since it is within the error made from other sources. Therefore all the $F_{\mu\mu}$'s are the same for a particular molecule, but the value changes with the molecule. As we can see, the off-diagonal matrix elements $F_{\mu v}$'s are not affected by the potential in question. This indicates that the reference energy for each molecule changes with ΔZ_v.

The electron affinity of a molecule has been considered as the orbital energy of the lowest unoccupied molecular orbital of alternant hydrocarbons.

---

* The spreading of the σ-bond-deformation effect over the benzene rings involved in the molecules has not been theoretically proved yet. Since it yields good results, we may regard it as a reasonable assumption.

Thus, the σ deformation that occurs when an electron is brought from infinity to the lowest unoccupied molecular orbital was neglected in the past. However, by considerations analogous to those used in calculating ionization potential, the σ-deformation effect is taken into account in this paper. The effect in electron affinity is opposite in sign to what it was in the case of ionization potential.

CALCULATIONS

The molecular integrals are obtained in an empirical way using the following considerations. For the resonance integrals, we use the following interpolation formula:[7]

$$H_{\mu\nu} = -2490 \exp(-5R_{\mu\nu}), \tag{17}$$

where $R_{\mu\nu}$ is the distance (in Å) between carbon atoms μ and ν, and $H_{\mu\nu}$ is in units of eV. Equation 17 was obtained by using the values $H_{\mu\nu} = -2.92$, -2.39, and -1.68 eV, respectively, corresponding to the internuclear distances 1.35, 1.39, and 1.46 Å; all these values were recommended by Pariser and Parr[5].

For the electronic repulsion integrals the values $(\mu\mu \mid \nu\nu) = 7.30, 5.46$, and 4.90 eV, corresponding to the internuclear distances 1.40, 2.425, and 2.80 Å, respectively, are taken, and for the internuclear distances larger than 2.80 Å (14) is used. The value $\alpha = W_\mu + \frac{1}{2}(\mu\mu \mid \mu\mu)$ is taken as constant, namely -4.235 eV.

To calculate the Hamiltonian matrix elements shown in (5) and (16), the above-mentioned molecular integrals and the constant α are put into the equations. As a first trial input, mobile bond orders $P_{\mu\nu}$ are estimated from the Hückel approximation.*

The construction of symmetry orbitals makes use of the high symmetry of the polyphenyls. These symmetry orbitals are bases of the irreducible representations of the symmetry group. Group theory considerations simplify the problem considerably. Since biphenyl, *para*-terphenyl, and *para*-quaterphenyl molecules have a D_{2h} symmetry, symmetry orbitals of these molecules will transform as irreducible representations of b_{3u}, b_{2g}, b_{1g}, and a_u of the D_{2h} group. On the other hand, since the *meta*-terphenyl molecule has a C_{2v} symmetry, its symmetry orbitals will transform as irreducible representations of b_1 and a_2 of the C_{2v} group.

---

* For example, $F_{\mu\nu} = \alpha$, for all μ, and $F_{\mu\nu} = \beta$ when μ and ν are nearest neighbors, and $F_{\mu\nu} = 0$ otherwise.

After analyzing the symmetry orbitals, one can apply the well-known theorem[19]

$$\int \chi_\mu^* F \chi_\nu \, d\tau = 0, \tag{18}$$

where χ_μ and χ_ν belong to different irreducible representations. These procedures help to reduce the dimensions of the secular equation.[7] Using the pairing property of the conjugated hydrocarbons,[7] * we can reduce by half the number of secular equations which must be solved. The correlation between the symmetry orbitals and the atomic orbitals can be easily established, so that quantities expressed in terms of symmetry orbitals are now readily expressed in terms of atomic orbitals.

All orbital energies such as the E_i's and the LCAO coefficients $C_{\mu i}$ were calculated. The calculations were all carried out using an IBM 7040.

CONCLUSIONS

The calculated ionization potentials and electron affinities of the biphenyl, *para*-terphenyl, *para*-quaterphenyl, and *meta*-terphenyl molecules are shown in Table I, where they are compared with the values from other approaches and with the available experimental values. As shown in Table I, the present calculations yield the most satisfactory results. However, since some of the experimental values are not available in the literature, we cannot compare our results conclusively with the experimental values.

The present calculation convinces the writers that the ionization potential of the polyphenyls is not simply the highest occupied molecular orbital energy, but must include the σ-bond deformation effect that accompanies the ionization. Here ΔZ_ν is a measure of the deformation of the ground-state σ framework. Even though ΔZ_ν is a small fraction of Z_ν, still this small fraction makes an important contribution to the ionization potential.

From the above considerations it is hardly believable that the σ framework in the aromatic rings of the polyphenyl molecules is not deformed by removing a π electron from the highest occupied orbital to infinity.

---

* As we can see in Fig. 1a, the conjugated hydrocarbons are grouped into the starred atoms and unstarred atoms.

$$\psi_i = \sum_\mu^* \phi_\mu C_{\mu i} + \sum_\mu \phi_\mu C_{\mu i}$$

$$\psi_i' = \sum_\mu^* \phi_\mu C_{\mu i} - \sum_\mu \phi_\mu C_{\mu i}$$

are a pair of molecular orbitals, the orbital energies of which are E_i and $-E_i$, respectively. $\sum_\mu^*$ denotes the summation over the starred atoms while $\sum_\mu$ is over the unstarred atoms.

Table I.   Ionization Potentials and Electron Affinities of Polyphenyls

Polyphenyls	I (eV)		A (eV)	
	Present Work	Others	Present Work	Others
Biphenyl (regular)	7.99		0.48	
Biphenyl (modified)	8.29		0.19	
		8.30[a] (obs)		0.41[e] (obs)
		8.84[b]		−0.37[b,c]
		8.79[c]		
		8.53[d]		−0.78[d]
para-Terphenyl	7.96		0.51	
para-Quaterphenyl	7.76		0.71	
meta-Terphenyl	8.07	8.25[a]	0.40	−0.51[d]

[a] Y. K. Syrkin and M. E. Diatkine, *The Structure of Molecules*, Dover, New York, 1964, p. 265.
[b] N. S. Hush and J. A. Pople, see Ref. 20.
[c] S. Ehrensen, *J. Phys. Chem.*, **66**, 706 (1962).
[d] R. M. Hedges and F. A. Matsen, *J. Chem. Phys.*, **28**, 950 (1958).
[e] Quoted from Ref. 20.

In this calculation, we confirm the fact that the ionization potential I and the electron affinity A of alternant conjugated hydrocarbon molecules have the relationship[20]

$$I + A = \text{constant.} \tag{19}$$

In Table I "regular" indicates that the relevant values are obtained by the regular hexagonal configuration of biphenyl, and "modified" indicates that the relevant values are obtained by the experimental bond lengths of biphenyl.[14] The ionization potential of the biphenyl calculated from the experimental bond lengths gives the best results. This suggests that a calculation of the exact configurations may be necessary.

ACKNOWLEDGMENTS

This research was supported in part by the National Science Foundation, Grant No. GP-3698. S.-I.K. wishes to express sincere appreciation to the University of Utah Research Committee for the fellowship granted to him and also to Dr. Juergen A. Hinze of the Laboratory of Molecular Structure and Spectra, Department of Physics, The University of Chicago, for reading and commenting on this manuscript.

REFERENCES

1. F. Hund, *Z. Physik*, **51**, 759 (1928).
2. R. S. Mulliken, *Phys. Rev.*, **32**, 186, 761 (1928). For review articles, see also R. S. Mulliken, *J. Chim. Phys.*, **46**, 497, 675 (1949).
3. E. Hückel, *Z. Physik*, **70**, 204 (1931); **76**, 628 (1932).
4. M. Goeppert-Mayer and A. L. Sklar, *J. Chem. Phys.*, **6**, 645 (1938).
5. R. Pariser and R. G. Parr, *J. Chem. Phys.*, **21**, 466, 767 (1953).
6. J. A. Pople, *Trans. Faraday Soc.*, **49**, 1375 (1953).
7. (a) S.-I. Kwun, T. Ree, and H. Eyring, *J. Chem. Phys.*, **40**, 3320 (1960). (b) T. Nakamura, S.-I. Kwun, and H. Eyring, in *Molecular Orbitals in Chemistry Physics and Biology*, P.-O. Löwdin and B. Pullman, Eds., Academic, New York, 1964. p. 421.
8. J. R. Hoyland and L. Goodman, *J. Chem. Phys.*, **36**, 12 (1962).
9. T. Koopmans, *Physica*, **1**, 104 (1934).
10. D. D. J. Roothaan, *Rev. Mod. Phys.*, **23**, 69 (1951).
11. J. R. Hoyland and L. Goodman, *J. Chem. Phys.*, **33**, 946 (1960).
12. Y. I'Haya, *Mol. Phys.*, **3**, 513, 521 (1960).
13. J. Trotter, *Acta Cryst.*, **14**, 1135 (1961).
14. A. Hargreaves and S. H. Rizvi, *Acta Cryst.*, **15**, 365 (1962).
15. L. W. Pickett, *J. Amer. Chem. Soc.*, **58**, 2299 (1936).
16. J. Dale, *Acta Chem. Scand.*, **11**, 640, 650 (1957).
17. C. A. Coulson, *Proc. Roy. Soc. (London)*, **A169**, 413 (1939).
18. M. El-Sayed, M. Kasha, and Y. Tanaka, *J. Chem. Phys.*, **34**, 334 (1961).
19. H. Eyring, J. Walter, and G. E. Kimball, *Quantum Chemistry*, Wiley, New York, 1944, Chap. X.
20. N. S. Hush and J. A. Pople, *Trans. Faraday Soc.*, **51**, 600 (1955).

A Linear Sum-over-Points Approach for Computing Electronic Energies: Application to He and H₂ Using Conroy Diophantine Points

CHARLES M. CARLSON

Boeing Scientific Research Laboratories, Seattle, Washington

A Personal Note

Of all the many outstanding scientific lectures and seminars of Henry Eyring's that I attended, none affected me more than an early morning inspirational talk I once heard him deliver at the LDS Institute in Seattle. His strong faith in a supreme being, his fervent devotion to youth, and his dedication to provide young people with a clear perspective on how to view their own lives came across resonantly in this talk in which he likened the conduct of one's own daily affairs during his lifetime to the weaving of a fine Persian carpet to be displayed upon completion.

His endless energy and glowing personality have carried worldwide his scientific genius to perceive and formulate the most reasonable and simple physical models with which to interpret important characteristics of complex systems and processes. The theory of absolute reaction rates and the significant structure model of liquids are classic examples.

Countless others have been inspired to follow his pioneering path of model conception or to carry out more detailed analyses which refine or explore the limits of applicability of a fundamental concept.

An appreciable part of Eyring's total contribution to the scientific profession can consequently be measured in terms of the influence he has had in stimulating progress in the fields of chemistry, physics, and biology. Since the path he has taken requires the highest degree of intuitiveness and imagination, a student or colleague influenced by Eyring is more apt to be an innovator than a follower of a set course.

INTRODUCTION

As one of his numerous interests, Henry Eyring has inspired many of his students and colleagues to work along with him on the problem of the accurate computation of electronic energies for small molecules. It is natural for H_3 to become the focal point for such studies, and many outstanding contributions to the understanding and development of techniques for computing and approximating multicenter integrals developed from these studies.[1] My active participation in this project during my graduate career prompted an interest in this area that eventually led to the work described herein.

Vying with the elegant analytical techniques developed largely during and before the period of Eyring's active interest in this area, methods relying heavily on numerical summation or integration[2-5] have gained increasing prominence in recent years among the more popular and successful approaches for computing electronic energies. This is due entirely to the development of the high-speed electronic computer.

The major advantage of such approaches lies in their capacity to produce rapidly converging approximate energies utilizing wavefunctions of greater analytical complexity than that of a configuration interaction expansion. A common procedure required for all of them constitutes the use of an efficient numerical integration algorithm for approximating integrals sufficiently well to obtain an acceptable degree of accuracy in the least amount of computer time.

In certain of these approaches, such as the least squares local energy method,[3] the degree of accuracy to which the integrals need be approximated in order to obtain a given accuracy in the computed energy is significantly lowered via the use of a specific procedure for applying the numerical integration algorithm and utilizing the resultant matrix components to obtain the approximate eigenvalue and eigenvector.

In the least squares local energy method the criterion that serves as a basis for this procedure applies equally as well to sums over finite sets of points as to integrals. Theoretically, the local energy method can be applied using only a very small set of points, but in practice the best results have been obtained when the sums are reasonable approximations to their corresponding integral matrix components.[3] The use of an efficient procedure for approximating integration is thus an important factor in this approach which has been investigated extensively using Gauss quadrature points.[3b–3f] A further advantage offered by this approach lies in its ease of formulation for a given atom or molecule, requiring only a minimum of algebraic manipulation and programming effort.

The purpose of the present article is to describe the further application of a recently proposed linear sum-over-points procedure,[6] which, like the local energy method, in theory applies equally well for sums as for integrals, being identical to the Rayleigh-Ritz variation method in the latter case. The potential advantages of such an approach are the same as that described above for the local energy method, with the further asset of requiring only the solution of a linear eigenvalue problem. This can be carried out quite rapidly on a computer using such techniques as Parlett's method[7] or the QR algorithm.[8]

In comparison with the local energy method this procedure might be viewed as a global or volume average energy approach, for instead of the local energy $\epsilon_p = H(p)\phi(p)/\phi(p)$, the ratio utilized is

$$\epsilon_i = \frac{\sum_p \phi^{i*}(p)H(p)\phi(p)\omega_p}{\sum_p \phi^{i*}(p)\phi(p)\omega_p},$$

where $\phi = \sum_{i=1}^{n} c_i \phi^i$ is the trial wavefunction and ω_p is an appropriate weight factor for the point p. The quantity ϵ_i might be defined most appropriately as the global energy for the functions ϕ^{i*} and ϕ. If the exact eigenfunction ψ is inserted in place of ϕ in the expressions for ϵ_i corresponding to each term in the trial wavefunction, it is evident that the resulting ϵ_i values must all be identical and equal to the exact eigenvalue E corresponding to ψ. Consequently, if for the approximate wavefunction ϕ the criterion is used that all the ϵ_i be equal to one another or $\epsilon_i = \epsilon$, there result the linear eigenvalue equations

$$\sum_{j=1}^{n} c_j(H_{ij} - \epsilon S_{ij}) = 0, \tag{1}$$

which must be satisfied by the approximate eigenvalue ϵ and eigenvector c_j for the equality of the ϵ_i to hold. In these equations

$$H_{ij} = \sum_p \phi^{i*}(p)H(p)\phi^j(p)\omega_p \quad \text{and} \quad S_{ij} = \sum_p \phi^{i*}(p)\phi^j(p)\omega_p.$$

For a general set of points these equations are nonsymmetric due to the non-Hermiticity of H_{ij}, although in certain cases in which Gauss quadrature points are utilized, Goodisman[5a] has shown this quantity to be Hermitian.

The expression for the mean local energy has recently been utilized by Harriss and Carlson[6b] to derive (1) and prove the upper bound relationship existing between the eigenvalue solutions to these equations and the corresponding equations in which H_{ij} is replaced by the symmetric $H'_{ij} = \frac{1}{2}(H_{ij} + H_{ji})$.

Previous calculations[6a] utilizing these equations for $H_2{}^+$ have been found to yield approximate eigenvalues lying close to those obtained by the least squares local energy method for the same sets of points and trial wavefunctions.

For either the accurate or approximate computation of the high dimensional integrals occurring in many electron atomic and molecular problems, the most efficient numerical integration scheme developed hitherto is Conroy's recently reported closed Diophantine method.[9] This procedure, an improvement over Haselgrove's open method[10] shares the advantage with Monte Carlo methods of not suffering from the "dimensional effect." [11] Moreover, its associated error ideally decreases with the inverse square of the number of sample points, whereas that associated with Monte Carlo methods shows at most an inverse square root[11] dependence upon this number.

Ellis[12] has recently shown how this procedure can be applied effectively for evaluating the matrix components needed in unrestricted Hartree-Fock and configuration interaction calculations, while Boys and Handy[2f] have employed a similar method in the application of the transcorrelated method to lithium hydride.

Despite the relatively high efficiency of the closed Diophantine procedure, it is unlikely that this method would yield approximate energies for helium and perhaps hydrogen superior to those obtained using the Gauss quadrature method for the same number of summation points. This is due to the low dimensionality (three for helium and five for hydrogen) of these problems when expressed in terms of interparticle coordinates,[13] the reducibility to forms convenient for the application of Gauss quadrature formulas,[3b–3f,5c] and the applicability of symmetry reduction.[3b–3f]

Nevertheless, it is worthwhile to carry out calculations for these entities using Conroy's Diophantine points in order to develop an effective method of applying (1) to three electron species utilizing the Conroy method and to obtain an indication concerning the ease of applicability and the probable results obtainable in such future applications. The initial results obtained for He and H_2 are discussed herein.

In Section II the formulation of (1) in terms of coordinates suitable for the application of the Diophantine procedure is described, and a possible method for determining certain adjustable scale factors arising in the formulation is proposed. In succeeding sections the results obtained from the application of the approach to the computation of the ground-state electronic energies for helium and molecular hydrogen are reported and discussed.

FORMULATION FOR USE OF DIOPHANTINE POINTS IN (1)

In order to transform (1) into a form suitable for the application of Conroy's integration procedure, each of the independent coordinates must be mapped onto the 0–1 coordinate range via a coordinate transformation.[9] In order that the Diophantine sums be proper approximations to the integral matrix components entering into the Rayleigh-Ritz equations, the weight factor ω_p must be chosen equal to the nondifferential part of the new volume element arising from the transformation of both the differential and nondifferential parts of the volume element for the original coordinate system.

Designating the original coordinates via x's and the new coordinates via ω's, the transformation of the ith coordinate is given by

$$\omega_i = f_i(x_i) \tag{2}$$

where f_i is the transformation function.

The choice of $f_i(x_i)$ is governed by the form it imparts to the transformed integrand in the 0–1 coordinate range. Features which improve the efficiency of the resultant sum for approximating the corresponding integral are that the integrand expressed in terms of the new coordinates be slowly varying over this interval and its first few derivatives small in value at the endpoints.[14] It can then be reasonably well approximated by a low term Fourier series.[14] The optimum functions to use for more complex entities may be chosen by carrying out test calculations on atoms and molecules containing one or two electrons.[2f]

In certain types of transformation, such as those utilized for the radial coordinates of atoms, it is possible to introduce an arbitrary scale factor α which can be adjusted to make the Diophantine sum a better approximation to its corresponding integral. A possible method for determining such a parameter is proposed below.

For the helium radial coordinates r_i ranging from 0 to ∞, two types of transformations were used herein. They are, respectively, the transformation[14]

$$\omega_i = \frac{r_i}{r_i + \alpha} \tag{3}$$

and the transformation[4b]

$$\omega_i = 1 - \exp\left(\frac{-r_i}{\alpha}\right). \tag{4}$$

A transformation similar in form to (4) was utilized for the 1–∞ range confocal elliptical coordinates μ_i used for H_2. The transformations utilized for the remaining coordinates—angular, etc.—are described below.

For the determination of α, procedures for wavefunctions expressed in terms of atomic orbitals have been discussed by Boys and Handy.[2f] Since in the present calculations, two widely different types of wavefunctions (i.e., an expansion in terms of atomic orbitals[3b] for He and a Coolidge-James style expansion[5h] for H_2) were utilized, it was advantageous to formulate a numerical optimization procedure that could be applied to any type of wavefunction, thus eliminating the necessity of formulating a separate set of rules for each different case.

Since the matrix component sums in (1) are approximations to the integral matrix components of the Rayleigh-Ritz variation equations, a most plausible objective of such an optimization procedure would be to provide values of α that when inserted into (1) via the matrix component sums would result in computed eigenvalues lying closest to the corresponding variational eigenvalues. It would be a formidable task to determine a procedure that would fulfill this objective precisely. On the other hand, it should not be too difficult to develop one that would come close to fulfilling it by utilizing a property of the matrix component sums that arises as a result of their being inexact approximations to their corresponding integrals.

The procedure so developed must have the property of requiring for its application an amount of computer time sufficiently small to make its use advantageous. The only property of the matrix component sums from which such a procedure could be developed is the non-Hermiticity in the off-diagonal H_{ij}.

Since in the above Diophantine formulation of (1) only one value of α is utilized for the entire group of H_{ij} entering into these equations, in formulating the procedure it is appropriate to utilize all the off-diagonal H_{ij} in the form of a weighted sum over a set of functions V_{ij} that measure the amount of asymmetry for each individual H_{ij}. The first step in the development of this procedure, the most plausible and convenient that could be conceived utilizing the non-Hermiticity property, was the assumption of the ratio $V_{ij} = (H_{ij} - H_{ji})/(H_{ij} + H_{ji})$ for the measure of the amount of asymmetry in each H_{ij}. The factor $(H_{ij} + H_{ji})$ was included due to the fact that when the parameter α is varied appreciably from its optimum value, the value of the sum H_{ij} may decrease and approach zero as α is made larger or smaller. Consequently, the simple difference $(H_{ij} - H_{ji})$ might decrease due to this effect alone failing to serve as a true measure of the asymmetry associated with the deviation of H_{ij} from its corresponding integral value.

What was believed to be the most reasonable way of weighting each of the ratios V_{ij} in forming the final function, herein called the non-Hermiticity factor, was chosen according to the importance of the component H_{ij} in $\langle \phi | H | \phi \rangle$. This is given by the eigenvector coefficient product $c_i c_j$.

If the weighted sum over all the V_{ij} was formed without using magnitudes, terms corresponding to appreciable amounts of asymmetry but having different signs could cancel, and the resultant sum would not represent a meaningful average measure of the asymmetry for all the off-diagonal H_{ij}.

For this reason, the non-Hermiticity factor $V(\alpha, c_1, \ldots, c_n)$ was formed by multiplying each ratio V_{ij} by $c_i c_j$, taking the absolute value of the resultant term and summing over all i and j or

$$V(\alpha, c_1, \ldots, c_n) = \sum_{i=1}^{n} \sum_{j=1}^{n} |c_i c_j V_{ij}|. \tag{5}$$

Since this equation and (1) contain α and the coefficients c_j, the optimization procedure would consist in satisfying (1) and the minimization condition

$$\frac{\partial V}{\partial \alpha} = 0 \tag{6}$$

by these parameters. The procedure thus determines optimum values for α, the eigenvector coefficients c_j, and ϵ.

The solutions to these simultaneous equations were obtained by inserting various values of α into (1), substituting the resultant eigenvector solutions to these equations into (6), and plotting the resultant roots to these equations against the original values of α used in the (1). Such a plot is illustrated for helium in Fig. 1 in which the letter α is used to designate the original parameters used in (1) while α' designates the roots to (6). Since, as shown in Fig. 1,

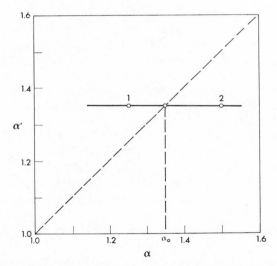

Fig. 1. Plot of α' versus α for helium. An adequate approximation, α_o, of the solution to the simultaneous equations (1) and (6) was given by the intersection of the line drawn through the computed points 1 and 2 with the diagonal $\alpha = \alpha'$.

the calculated values of α' were found to be nearly independent* of α, it was possible in practice to determine a self-consistent solution by carrying out the calculation using no more than two values of α.

In the calculations described below in which an exponential type of coordinate transformation such as that given by (4) was utilized, the optimization of the scale factor via this procedure was found to be accompanied by a substantially improved agreement between the eigenvalue solutions to (1) and those to the corresponding integrated form, the Rayleigh-Ritz variation equations. In all calculations the solution was carried out to an accuracy of no more than ±0.05, since the attainment of greater accuracy failed to bring about sufficient improvement in the computed eigenvalues to justify the additional computer time expended. For H_2 the optimum solution for a nine-term trial wavefunction was obtained utilizing a total of about 30 minutes of computer time on the IBM 360 model 44 computer, although improvements that could have substantially increased the efficiency of the computer program utilized had not been undertaken.

HELIUM

For the helium calculation, trial wavefunctions containing up to nine terms were used. These wavefunctions were composed of terms of the type[4d]

$$\phi_i = P\{r_1{}^l r_2{}^m \exp\{-(\rho_a r_1 + \rho_b r_2)\}\}r_{12}{}^n, \tag{7}$$

where P designates† the symmetrization operator; ρ_a and ρ_b are exponential parameters; and r_1, r_2, and r_{12} are interparticle coordinates identical to those used by Frost and co-workers.[4d,13] The untransformed coordinates utilized were r_1, r_2, and θ_{12}.[14d]

As stated in the preceding section, the transformations given by (3) and (4) were used to transform r_1 and r_2 onto the 0–1 coordinate range. For θ_{12} in all calculations the coordinate q defined through[14]

$$\theta_{12} = \pi(6q^5 - 15q^4 + 10q^3) \tag{8}$$

was used to map θ_{12} onto the 0–1 coordinate range. This formula has been proposed[14] for approximating the integrand closely by the first few terms of a Fourier series.

A total of 269 Diophantine summation points were used invoking Conroy's recommended Case II point selection algorithm.[9] The lesser number of 26

---

* When determined to an accuracy of no greater than ±0.05, α' actually showed no dependence upon α over the range of values covered.
† $P[\phi_i(1, 2)] = \phi_i(1, 2) + \phi_i(2, 1)$.

points whose generation parameters are given in Conroy's paper[9] were initially utilized but found to yield quite poor results. The 269-point calculation was then performed for the above-described coordinate transformations for r_1 and r_2, that given by (3) being designated 1 and that given by (4) designated 2. The results obtained using these two transformations are listed in Table I along with the corresponding variational eigenvalues.

Table I.   Comparison of Eigenvalue Roots, ϵ_1 and ϵ_2, Computed for Helium using Eqs. (1) with Eigenvalue Roots ϵ_{var} to Corresponding Integrated (Rayleigh-Ritz) Equations

Roots ϵ_1 were computed using coordinate transformation for r_i given by Eq. (3) with scale factor $\alpha = 1.05$, whereas roots ϵ_2 were computed using Eq. (4) with $\alpha = 1.35$. A total of 269 closed Diophantine points using Conroy's[a] recommended Case II point selection algorithm were used to compute the matrix component sums. Trial wavefunction terms, given by Eq. (7), are designated by integers l, m, n. A trial wavefunction, corresponding to a given row of the table contains terms designated in that row as well as those given in preceding rows. Accurate ground-state energy equals 2.90372 a.u.[b]

l	m	n	$-\epsilon_1$ (a.u.)	$-\epsilon_2$ (a.u.)	$-\epsilon_{var}$ (a.u.)
0	0	0	2.83557	2.81794	2.83150
0	0	1	2.89453	2.88708	2.89042
0	0	2	2.89343	2.88750	2.89053
1	0	0	2.89385	2.88757	2.89125
1	0	1	2.89303	2.88795	2.89149
1	0	2	2.89279	2.88831	2.89154
1	1	0	2.89652	2.89879	2.89916
1	1	1	2.89376	2.89987	2.89970
1	1	2	2.89304	2.89938	2.89973

[a] H. Conroy, *J. Chem. Phys.*, **47**, 5307 (1968).
[b] C. L. Pekeris, *Phys. Rev.*, **112**, 1649 (1958).

In discussing and evaluating these results and those tabulated in the succeeding section for hydrogen, the variationally determined eigenvalues have been used as the standards of judgment. The reason for this is that in all calculations carried out, the sum-over-points eigenvalues lay considerably closer to their variational counterparts than to the exact ground-state energy. The sum-over-points calculation is thus best viewed as an approximate solution to the Raleigh-Ritz equations with the computed eigenvalues approximations to variational upper bounds. This viewpoint is compatible with the fact that (1) must converge to the integral Rayleigh-Ritz equations as the number of summation points is steadily increased.

The initial results for helium listed in Table I were obtained using just one value of α for each type of transformation. This value was obtained by carrying out the above-described optimization procedure for the nine-term trial wavefunction. For transformation 1 an optimum value of $\alpha = 1.05$ was obtained, whereas that found for 2 was $\alpha = 1.35$.

As can be seen from the table, when transformation 2 was utilized, the eigenvalue corresponding to the nine-term trial wavefunction for which the

Table II. Eigenvalue Roots to Eqs. (1) Computed for Nine-Term Trial Wave-function of Helium Using Different Values of Scale Factor α to Generate Matrix Component sums

Roots ϵ_1 were computed using coordinate transformation for r_i given by Eq. (3), whereas roots ϵ_2 were computed using coordinate transformation given by Eq. (4). Optimum values of α determined for each transformation by self-consistant mini-mization of non-Hermiticity factor are enclosed in parentheses. Variationally determined root ϵ_{var} corresponding to this trial wavefunction is given at bottom of table.

Transformation 1		Transformation 2	
α	$-\epsilon_1$ (a.u.)	α	$-\epsilon_2$ (a.u.)
(1.05)	2.89304	1.25	2.90036
1.25	2.89433	(1.35)	2.89938
2.50	2.89834	1.50	2.89785
	$-\epsilon_{var} = 2.89973$ (a.u.)		

parameter α was optimized agreed closely (<0.0004 a.u.) with its variational counterpart, whereas in the case of transformation 2 this agreement was poor. Moreover, in the case of transformation 1 the agreement between the sum-over-points and variational eigenvalues was roughly the same for all the trial wavefunctions, whereas for transformation 2 it was found to be best for the 7-, 8-, and 9-term trial wavefunctions.

The results listed in Table II, in which eigenvalues computed for different values of α are tabulated, indicate that the above behavior may in general be related to the success of the optimization procedure as regards the value of α it determines. Thus producing an "optimum" value of α far removed from that which yields the eigenvalue closest to its variational counterpart, the optimization procedure for α in the case of transformation 1 was found to be quite ineffectual. In marked contrast the results listed for transformation 2 therein show the optimization procedure to have worked quite well in this case.

Using the observed agreement with the variational eigenvalues as the criterion of goodness, these results imply that for general use in this approach an exponential transformation of type 2 may be superior to that of type 1. Its utilization for the calculations described below on hydrogen was found to yield results equally encouraging to those described above.

HYDROGEN

For hydrogen, trial wavefunctions identical to those utilized by Goodisman[5h] were chosen. These, most conveniently expressed in terms of mixed inter-particle-confocal elliptical coordinates,[3f] are cusp corrected.[5h] They are composed of terms of the type[5h]

$$\phi_i = P\{\mu_1{}^p v_1{}^q \mu_2{}^w v_2{}^t \exp\left[-Z(\mu_1 + \mu_2)\right]\}g, \tag{9}$$

where

$$g = \frac{r_{12}{}^m + 2\partial_{im}}{1 + \partial_{im}}; \qquad m > 0.$$

The factor g is the cusp-correcting correlation factor which imparts the proper behavior to the wavefunction near the r_{12} singularity. The ranges for the original coordinates used are, respectively,

$$(1 \leq \mu_i \leq \infty); \qquad (-1 \leq v_i \leq +1); \qquad (0 \leq \phi = \phi_1 - \phi_2 \leq 2\pi)$$

The only type of coordinate transformation used for μ_1 and μ_2 is given by

$$\omega_i{}^u = 1 - \exp\left(1 - \mu_i/\alpha\right) \tag{10}$$

This is analogous to (4). A transformation analogous to (3) was not applied in the case of hydrogen, although this might have been beneficial in providing further knowledge concerning the questionability of its effectiveness as far as its use in the present approach is concerned.

For v_i and ϕ the transformations used,

$$\omega_i{}^v = (v_i + 1)/2 \tag{11}$$

and

$$\omega^\phi = \phi/2\pi \tag{12}$$

respectively, are algebraically the simplest that will map these coordinates onto the 0–1 coordinate range. Although more complex forms for these transformations may lead to improved results, this investigation was not undertaken.

The initial results obtained for hydrogen utilizing a total of 577 Diophantine summation points[9] and Conroy's Case II point selection algorithm[9] are listed in Table III along with Goodisman's variationally determined

Table III.   Comparison of Eigenvalue Roots ϵ Computed for Hydrogen Using Eqs. (1) with Eigenvalue Roots ϵ_{var} to Corresponding Rayleigh-Ritz Equations

Roots ϵ were computed using coordinate transformation for μ_i given by Eq. (10) with scale factor $\alpha = 1.00$. A total of 577 closed Diophantine points using Conroy's[a] recommended Case II point selection algorithm were used to compute the matrix component sums. Trial wavefunction terms, given by Eq. (9), are designated by integers p, q, w, t, m. The trial wavefunction corresponding to a given row of the table contains terms designated in that row as well as those given in preceding rows. Wavefunction exponential parameter $z = 0.95$; internuclear separation $R = 1.40$ bohrs. Accurate ground-state energy equals -1.88876 a.u.[b]

p	q	w	t	m	$-\epsilon$ (a.u.)	$-\epsilon_{var}$ (a.u.)
0	0	0	0	1	1.86502	1.86502
0	2	0	0	1	1.88411	1.88308
1	0	0	0	1	1.88598	1.88507
0	1	0	1	1	1.88608	1.88524
1	1	0	1	1	1.88620	1.88536
2	0	0	0	1	1.88674	1.88632
0	0	0	0	2	1.88793	1.88756
1	0	1	0	1	1.88808	1.88802
1	1	1	1	1	1.88821	1.88808

[a] H. Conroy, *J. Chem. Phys.*, **47**, 5307 (1968).
[b] W. Kolos and L. Wolniewiez, *Phys. Rev. Letters*, **20**, 243 (1968).

eigenvalues.[5h] The parameter α was again optimized only for the nine-term trial wavefunction. An optimum value of $\alpha = 1.00$ was found for this trial wavefunction.

As shown in Table III, the agreement obtained between the sum-over-points eigenvalues and their variational counterparts ($\sim$0.0001) for the eight- and nine-term trial wavefunctions is excellent, especially considering the relatively small number of 577 summation points utilized. In view of the fact that the parameter α was optimized for the nine-term trial wavefunction, the remarkable agreement shown in the case of the single-term function is somewhat puzzling, although that obtained for intermediate-sized wavefunctions is poorer (up to 0.001 a.u.). As observed for helium, the optimization procedure for determining α was found to work quite effectively for the exponential type of transformation.

A comparison of the best results obtained for helium and hydrogen shows the agreement between the sum-over-points and the corresponding variational eigenvalues to be slightly better for hydrogen. In one sense this is not surprising, since roughly twice as many summation points were utilized in the hydrogen calculation. However, the dimensionality of the hydrogen problem exceeds that of helium by a factor of 2. If a significant dimensional effect were present, one might expect this to offset the effect of increasing the number of points and lead to substantially poorer results.

CONCLUSIONS

Three results established are from these calculations:

1. The excellent agreement obtained for hydrogen and helium between sum-over-points and variational eigenvalues using relatively small numbers of Diophantine points for trial wavefunctions for which the scale factor was optimized.

2. The observed indication that, living up to expectations,[9] the dimensional dependence of the error for Conroy's Diophantine procedure is small.

3. The ease of applicability and effectiveness of the proposed optimization method for determining the scale factor α in the case where an exponential type of mapping transformation was utilized.

These factors serve as an indication that approximations to Rayleigh-Ritz upper bounds equally accurate to that obtained herein for hydrogen should be achievable for three-electron entities (at most an increase in dimensionality of four over that for hydrogen) using an easily manageable number of points.

Although the optimization of nonlinear trial wavefunction parameters was not carried out herein, the local energy method[3g] could have been utilized for one or two such parameters without invoking much additional effort. Determining nonlinear parameters in this manner and utilizing longer trial wavefunctions so that the variational upper bound lies close to the exact energy, the computation of accurate approximate energies via this procedure for such entities as lithium and H_3 should be feasible.

ACKNOWLEDGMENTS

I wish to thank the following professors for comments that were of considerable benefit toward the successful completion of this project: H. Conroy, D. E. Ellis, A. A. Frost, N. C. Handy, D. K. Harriss, R. T. Pack, and K. T. Tang.

Some of these discussions were carried out at the International Symposium on Atomic, Molecular, Solid-State Theory and Quantum Biology in Honour of Professor Henry Eyring held at Sanibel Island, Florida, January 1969.

Finally, gratitude is due to the Boeing Scientific Research Laboratories computing staff for their time in helping me carry out the above-described calculations.

REFERENCES

1. (a) R. S. Barker and H. Eyring, *J. Chem. Phys.*, **21**, 912 (1953); (b) R. S. Barker, H. Eyring, C. J. Thorne, and D. A. Baker, *ibid.*, **22**, 699 (1954); (c) R. S. Barker and H. Eyring, *ibid.*, **22**, 1182 (1954); (d) R. S. Barker and H. Eyring, *ibid.*, **22**, 2072 (1954); (e) R. S. Barker, J. C. Giddings, and H. Eyring, *ibid.*, **23**, 344 (1955); (f) R. S. Barker, H. Eyring, D. A. Baker, and C. J. Thorne, *ibid.*, **23**, 1381 (1955); (g) R. L. Snow, R. S. Barker, and H. Eyring, *ibid.*, **23**, 1686, (1955); (h) R. L. Snow and H. Eyring, *J. Phys. Chem.*, **61**, 1 (1957).
2. (a) S. F. Boys and P. Rajagopal, *Advan. Quantum Chem.*, **2**, 1 (1965); (b) S. F. Boys *Proc. Roy. Soc. (London)*, **A309**, 195 (1969); (c) S. F. Boys and N. C. Handy, *ibid.* **A309**, 209 (1969); (d) **A310**, 43 (1969); (e) **A310**, 63 (1969); (f) **A311**, 309 (1969).
3. (a) A. A. Frost, *J. Chem. Phys.*, **10**, 240 (1942); (b) A. A. Frost, R. E. Kellogg, and E. C. Curtis, *Rev. Mod. Phys.*, **32**, 313 (1960); (c) A. A. Frost, R. E. Kellogg, B. M. Gimarc, and J. D. Scargle, *J. Chem. Phys.*, **35**, 827 (1961); (d) B. M. Gimarc and A. A. Frost, *Theoret. Chim. Acta*, **1**, 87 (1962); (e) B. M. Gimarc and A. A. Frost, *J. Chem. Phys.*, **39**, 1698 (1963); (f) D. K. Harriss and A. A. Frost, *ibid.*, **40**, 204 (1964); (g) D. K. Harriss and R. K. Roubal, *Theoret. Chim. Acta*, **9**, 303 (1968).
4. (a) H. Conroy, *J. Chem. Phys.*, **41**, 1327 (1964); (b) **41**, 1331 (1964); (c) **41**, 1336 (1964); (d) **41**, 1341 (1964); (e) H. Conroy and B. L. Bruner, *ibid.*, **42**, 4047 (1965); (f) H. Conroy, *ibid.*, **47**, 912 (1967); (g) H. Conroy and B. L. Bruner, *ibid.*, **47**, 921 (1967);. (h) H. Conroy, *ibid.*, **47**, 930 (1967); (i) *ibid.*, **47**, 5307 (1967); (j) H. Conroy and G, Malli, *ibid.*, **50**, 5049 (1969).
5. (a) J. Goodisman, *J. Chem. Phys.*, **41**, 2365 (1964); (b) J. Goodisman and D. Secrest, *ibid.*, **41**, 3610 (1964); (c) J. Goodisman, *ibid.*, **41**, 3889 (1964); (d) *ibid.*, **43**, 2806 (1965); (e) *ibid.*, **43**, 3037 (1965); (f) J. Goodisman, *Theoret. Chim. Acta*, **4**, 343 (1966); (g) J. Goodisman and D. Secrest, *J. Chem. Phys.*, **45**, 1515 (1966); (h) J. Goodisman; *ibid.*, **45**, 3659 (1966); (i) *ibid.*, **47**, 5247 (1967).
6. (a) C. M. Carlson, *J. Chem. Phys.*, **47**, 862 (1967); (b) D. K. Harriss and C. M. Carlson, *J. Chem. Phys.*, **51**, 5458 (1969).
7. B. Parlett, *Math. Computation*, **18**, 464 (1964).
8. (a) J. G. F. Francis, *Computer J.*, **4**, 265 (1961); (b) **4**, 332 (1961); (c) J. H. Wilkinson, *The Algebraic Eigenvalue Problem*, Clarendon Press, Oxford, 1965.
9. H. Conroy, *J. Chem. Phys.*, **47**, 5307 (1968).
10. C. B. Haselgrove, *Math. Computation*, **15**, 323 (1961).
11. P. J. Davis and P. Rabinowitz, *Numerical Integration*, Blaisdell, Waltham, Mass., 1967, p. 129.
12. D. E. Ellis, *Intern. J. Quantum Chem.*, **2S**, 35 (1968).
13. A. A. Frost, *Theoret. Chim. Acta*, **1**, 36 (1962).
14. N. C. Handy, private communication.

The Potential Energy Surface of the H_3 System Using Floating Gaussian Orbitals

ARTHUR A. FROST

Department of Chemistry, Northwestern University, Evanston, Illinois

Abstract

Calculations of the H_3 potential energy surface have been carried out by an open shell valence-bond method using a localized orbital basis of spherical Gaussians. Results are presented for both single Gaussians as the localized orbitals and for linear combinations of two spherical Gaussians for each localized orbital.

INTRODUCTION

Henry Eyring arrived at Princeton in the fall of 1931 shortly after the appearance of the important paper by Eyring and Polanyi[1] which described semiempirical quantum mechanical calculations of the potential energy surface for the $H + H_2$ reaction. Eyring's enthusiasm immediately infected a group of students, some of whom extended the procedure to other reactions while others were interested in more fundamental calculations such as would now be referred to as *ab initio*.

Among this latter group was Joseph Hirschfelder, who together with Nathan Rosen and Henry Eyring,[2] and later H. Diamond[3] pushed through the difficult calculations of the H_3 potential energy surface using a basis of hydrogen atomic $1s$ orbitals. The difficulty arose from the three-center integrals for electron repulsion.

The present contribution relates to that of Hirschfelder, Eyring, and coworkers in that Gaussian orbitals are used in place of hydrogen orbitals. The corresponding integrals are much less difficult, as first shown by Boys,[4] and therefore the whole calculation is simplified.

RECENT H_3 CALCULATIONS

Since the early work mentioned above, H_3 calculations have proceeded along the two lines, semiempirical or *ab initio*. Semiempirical potential energy surfaces have been produced by Sato,[5] Porter and Karplus,[6] Cashion and Herschbach,[7] and Salomon.[8]

There have been many *ab initio* calculations, especially since the advent of high-speed digital computers. The more recent calculations involving a basis of Slater-type orbitals are by Shavitt, Stevens, Minn and Karplus,[9] by Gianinetti, Majorino, Rusconi, and Simonetta,[10] by Riera and Linnett,[11] and by Ladner and Goddard.[12] Of these the calculation by Shavitt et al. is the most accurate as judged by the low energy, which by the variation method is an upper bound to the exact solution. Conroy and Bruner[13] obtained an even lower energy using a wavefunction with explicit electron correlation and numerical integration. However, this has been criticized[9] as probably being inaccurate. Hayes and Parr[14] have made a single center calculation, but it does not lead to the desired accuracy. Gaussian orbitals have been used by Schwartz and Schaad[15] and by Edmiston and Krauss.[16] The latter calculation, which makes use of pseudo-natural orbitals, approaches the accuracy of Shavitt et al.

THE FLOATING SPHERICAL GAUSSIAN ORBITAL MODEL (FSGO)

A simplified model[17] (FSGO) using Gaussian orbitals has been found to reproduce quite accurately the geometry of a large class of molecules. It is of interest to investigate similar calculations of the H_3 potential energy surface.

The original FSGO calculations applied only to systems with an even number of electrons in a closed-shell singlet state represented by a single Slater determinantal wavefunction. Since the H_3 system has three electrons it was necessary to expand the computer program to allow for open shells and for linear combinations of Slater determinants. Furthermore, in order to improve the energy, it was considered desirable to include in the program the possibility of occupied orbitals formed as linear combinations of spherical Gaussians.

Of all the previous calculations on H_3, only that of Schwartz and Schaad[15] has any direct relation to what is reported here. Their calculation, however, is restricted to a single Slater determinant with one closed shell and therefore does not bring in sufficient electron correlation.

FLOATING GAUSSIAN ORBITAL CALCULATIONS ON H_3

These calculations were limited to linear configurations of the three nuclei since it is universally believed that this includes the transition state in the $H + H_2$ reaction. Furthermore, the calculations were first made for symmetrical nuclear configurations followed by asymmetrical configurations in the neighborhood of the transition state, which was found to be the symmetrical state of minimum energy. With similar sets of orbitals the energies of separated H atom and H_2 molecule were calculated so as to have a suitable reference for determining the activation energy.

Three localized orbitals were defined, one each associated with each of the nuclei, respectively. In the simplest case these were single floating spherical Gaussian orbitals of the form

$$\phi(r) = \left(\frac{2}{\pi\rho^2}\right)^{3/4} \exp\left(\frac{-r^2}{\rho^2}\right),$$

where r is the radial distance from the orbital center and ρ is the orbital "radius." Calculations were also performed with double Gaussian localized orbitals formed by linear combinations of single Gaussians which may be at different centers and have different radii.

Determinantal wavefunctions were then defined according to various electron occupations of the localized orbitals. A first calculation made use of only singly occupied orbitals. Designating the orbitals as a, b, and c, respectively, the three possible independent determinantal functions (for $M_s = +\frac{1}{2}$) were $|a\,b\,\bar{c}|$, $|a\,\bar{b}\,c|$ and $|\bar{a}\,b\,c|$ where the bar indicates β spin, the others being α. The linear combination of these three determinants produces two doublets and one quartet state, a doublet having the lowest energy and corresponding to a resonance among covalent valence bond structures.

Full configuration interaction (CI) calculations were also made in which all possible ionic states were included. These were the six possible doublet state determinants each with one doubly occupied orbital: $|a\,\bar{a}\,b|$, $|a\,\bar{a}\,c|$, $|b\,\bar{b}\,a|$, $|b\,\bar{b}\,c|$, $|c\,\bar{c}\,a|$, $|c\,\bar{c}\,b|$.

For each given choice of nuclear configuration, type of localized orbitals, and set of determinantal functions, the energy was minimized. The energy minimization with respect to the Gaussian parameters was carried out by a form of direct search,[17] but the minimization with respect to the linear combination coefficients of the determinantal functions was handled in the standard way by solving the appropriate secular equation.

RESULTS

Table I summarizes the results for H_3. In order to calculate the activation energy E_a, the results shown in Table II for the H atom and H_2 molecule at the same level of approximation were used. The total energy, of course, improves as the wavefunction is allowed more variation. However, because the cusp in the wavefunction is poorly represented by one or two Gaussians the total energy is still far from the experimental value. In comparison with the Gaussian work of Schwartz and Schaad,[15] our best total energy, which involves full configuration interaction, is naturally better than their double Gaussian SCF calculation but not as good as their results with larger basis sets.

Table I. H_3 System Transition States Using Floating Gaussian Orbitals[a]

	R_e	$-E$	E_a
I. Single Gaussians			
A. VB Calculation	1.96	1.392	0.025
B. Full CI	1.96	1.419	0.003
II. Double Gaussians			
A. VB Calculation	1.96	1.557	0.045
B. Full CI	1.93	1.576	0.032
Experimental[b]	—	1.6580	0.0156
		(9.8 $\pm$ 0.2 kcal/mole)	

[a] Linear symmetric nuclear configurations. R_e = neighbor internuclear distance in atomic units (bohrs). Energies in atomic units (hartrees). E_a = chemical activation energy.
[b] I. Shavitt, *J. Chem. Phys.*, **49**, 4048 (1968).

The calculated activation energy varied erratically as the wavefunction was improved. Unfortunately, this quantity is a small difference between large numbers and the error is greatly magnified. It was hoped that by calculating the E_a with suitably simplified wavefunctions of similar forms applied to H_3 and to H_2 + H this difference would be close to the accurate value, but this has not proved to be the case. The dissociation energy of H_2, shown in Table II, is also a small difference of large numbers but is surprisingly much more consistent than the E_a values.

Table II. Gaussian Orbital Results for H and H_2[a]

	H atom		H_2 molecule	
	$-E$	R_e	$-E$	D_e
I. Single Gaussians				
A. VB	0.424[b]	1.56	0.993	0.144
B. Full CI	0.424	1.59	0.998[c]	0.149
II. Double Gaussian				
A. VB	0.486[b]	1.45	1.116	0.145
B. Full CI	0.486	1.45	1.122	0.151
Experimental	0.500	1.40	1.174	0.174

[a] Atomic units: energies in hartrees, distances in bohrs.
[b] H atom results first given by R. McWeeny, *Nature*, **166**, 21 (1950).
[c] H_2 molecule by method I-B first given by C. M. Reeves, *J. Chem. Phys.*, **39**, 1 (1963).

It can be concluded that success for H_3 can be had only by using some more appropriate wavefunction or one with many more parameters to be varied.

ACKNOWLEDGMENTS

This work was supported by a grant from the National Science Foundation. Mr. Steven Schulman kindly assisted with the calculations.

REFERENCES

1. H. Eyring and M. Polanyi, *Z. Physik. Chem.*, **B12**, 279 (1931).
2. J. Hirschfelder, H. Eyring, and N. Rosen, *J. Chem. Phys.*, **4**, 121 (1936).
3. J. Hirschfelder, H. Diamond, and H. Eyring, *J. Chem. Phys.*, **5**, 695 (1937).
4. S. F. Boys, *Proc. Roy. Soc. (London)*, **A200**, 542 (1950).
5. S. Sato, *J. Chem. Phys.*, **23**, 592, 2465 (1955).
6. R. N. Porter and M. Karplus, *J. Chem. Phys.*, **40**, 1105 (1964).
7. J. H. Cashion and D. R. Herschbach, *J. Chem. Phys.*, **40**, 2358 (1964).
8. M. Salomon, *J. Chem. Phys.*, **51**, 2406 (1969).
9. I. Shavitt, R. M. Stevens, F. L. Minn, and M. Karplus, *J. Chem. Phys.*, **48**, 2700 (1968).
10. E. Gianinetti, G. F. Majorino, E. Rusconi, and M. Simonetta, *Intern. J. Quantum Chem.*, **3**, 45 (1969).
11. A. Riera and J. W. Linnett, *Theoret. Chim. Acta*, **15**, 181 (1969); J. W. Linnett and A. Riera, *ibid.*, **15**, 196 (1969).
12. R. C. Ladner and W. A. Goddard III, *J. Chem. Phys.*, **51**, 1073 (1969).
13. H. Conroy and B. L. Bruner, *J. Chem. Phys.*, **47**, 921 (1967).
14. E. F. Hayes and R. G. Parr, *J. Chem. Phys.*, **47**, 3961 (1967).
15. M. E. Schwartz and L. J. Schaad, *J. Chem. Phys.*, **48**, 4709 (1968).
16. C. Edmiston and M. Krauss, *J. Chem. Phys.*, **49**, 192 (1968).
17. A. A. Frost, *J. Chem. Phys.*, **47**, 3707, 3714 (1967), *J. Phys. Chem.*, **72**, 1289 (1968); A. A. Frost and R. A. Rouse, *J. Amer. Chem. Soc.*, **90**, 1965 (1968); A. A. Frost, R. A. Rouse, and L. Vescelius, *Intern. J. Quantum. Chem., Symp.*, **2**, 43 (1968); R. A. Rouse and A. A. Frost, *J. Chem. Phys.*, **50**, 1705 (1969).

A Forecast for Theoretical Chemistry[*][†][‡]

JOSEPH O. HIRSCHFELDER

Theoretical Chemistry Institute, University of Wisconsin, Madison

Congratulations, Henry, on the occasion of your seventieth birthday! We wish you continued health, happiness, and success in solving all of our problems! The name *Eyring* brings to mind a large number of brilliant achievements in the theory of reaction rates, the structure of liquids, the mechanism of biological processes, . . . ; indeed, all of the subjects which are discussed in this book. Henry's work is always thorough and his ideas are profound and novel. Besides being a wonderful scientist, Eyring is a wonderfully nice person and a great teacher. His tremendous enthusiasm has been a source of inspiration to three generations of scientists.

How much my own research and philosophy has been influenced by Eyring became evident a few years ago when I was given the Peter Debye Award of the American Chemical Society. After a bit of soul-searching, I decided to dedicate my award speech to Henry. In it, I tried to recollect the Golden Age of Quantum Chemistry 1931–36, during which I was a graduate student, and I predicted a new Golden Age into which we are currently entering.

* Reprinted from *Journal of Chemical Education*, **43**, 457 (September, 1966). Copyright 1966, by Division of Chemical Education, American Chemical Society, and reprinted by permission of the copyright owner.

† This research was supported by National Aeronautics Space Administration grant NsG275 62.

‡ Award Address on the occasion of the presentation of the Peter Debye Award in Physical Chemistry to Professor Hirschfelder at the 151st Meeting of the ACS, Pittsburgh, Pa., March 1966.

I was told that I should make this a very personal speech. Thus, I will take the occasion to do a bit of philosophizing. Let us start with some of the words of wisdom that have meant a great deal to me.

I owe a great debt to Henry Eyring for having given me a start on my scientific career. The Eyring philosophy I best remember is:

Be nice to the guys on the way up so that they will be nice to you on the way down.

In nature, or complicated systems, there is usually only *one* slow step or bottleneck which determines behavior.

Always ask yourself, how *might* the phenomena occur. Make a GEDENKS-model. It will suggest the proper groupings of variables—a big help in semiempiri-cizing.

A scientist's accomplishments are equal to the integral of his ability integrated over the hours of his effort.

To Eyring, every problem in nature can be studied theoretically. The first step is to ask yourself what *might* the solution look like. I think that Albert Einstein had much the same viewpoint:*

It is easy, just hard to do!

Nature is simple, it is we who are complicated!

Einstein seldom sought direct solutions to his equations. His first question was, "under what other conditions could these equations arise?" He learned a great deal from the way in which other people had solved these equations in other types of problems.

Surely it pays to spend a considerable amount of time looking at a problem from all different directions. Think before you leap! In this case, leaping corresponds to committing yourself to a particular research technique.

My first research as a graduate student was under the direction of Edward Condon. From him I learned that a theoretician should have a broad background.

A theoretician should be well-versed in a wide range of experimental techniques and facts. The three functions of a theoretician are the following:

1. Suggest new types of experiments.
2. Suggest new interpretations of existing data.
3. Further develop the theory so that it may become a more powerful tool in the understanding of nature.

At the Bikini Atom Bomb Tests in 1946, my title was "Chief Phenom-enologist." John Magee and I had the job of predicting all of the different

---

* I learned a great deal about Einstein from his research assistant, Nathan Rosen. Rosen collaborated with me in our calculations of the energy of H_3 and H_3^+. He showed me that by systematizing the procedure and the format properly, I could reduce my computational effort by a factor of 10.

effects of the bomb so that the experiments could be properly set up. Enrico Fermi felt very strongly that *every* scientist should be trained as a "phenomenologist." He trained his students to get order of magnitude solutions to arbitrary questions. A typical Fermi question was, "Estimate the number of railroad locomotives in the United States." It is interesting to note that chemical engineers are regarded as the most flexible and as possessing the best background for Operational Research Analysis, where such ability is required.

I was, indeed, fortunate to be a graduate student at Princeton during the "First Golden Age of Quantum Chemistry," 1931–36. This was a wonderful period of great discoveries, and Princeton was the hub. It was a period during which many people had high hopes of explaining all of the physical and chemical properties of material and, indeed, all of natural phenomena in terms of the basic laws of physics. Even such erudite physicists as Dirac, Van Vleck, and London concocted simple approximation procedures which they hoped would help to explain molecular structure. It was a period of high hopes in which theoreticians took a fresh look at all sorts of natural phenomena and guessed at their mechanism. The methods which were used in those days were crude compared to our present techniques, but the men who used them were not afraid to stick their necks out. Of course, the theoretical predictions served as challenges to the experimental scientists and were especially valuable when the theory suggested a critical experiment. Thus a very stimulating rivalry developed between the experimental and the theoretical men. The result was an era of exciting discoveries.

In a nutshell, the message which I want to convey today is that for the last 30 years theoretical chemists have been preoccupied with developing their mathematical techniques—that is, developing molecular quantum mechanics and statistical mechanics on a firm *ab initio* basis resulting from accepted laws of physics. The importance of this theoretical development cannot be overemphasized. With the advent of giant high-speed computing machines, new types of mathematical methods, and the renewed help of the theoretical physicists, it appears that within five to ten years we will be able to make accurate theoretical predictions of most of the physical and chemical properties of matter. Then there will be a change of emphasis.

The theoretical chemists must start preparing themselves and their students *right now* for this change. At present, theoretical chemists are concentrating on high-powered mathematics. They are also concentrating on very specific problems and using only a narrow range of techniques. The emphasis will shift from developing new methods to applying existing methods. Then the theoretical chemists must have a broad knowledge of experimental physics and experimental chemistry as well as a broad knowledge of theoretical chemistry and theoretical physics.

In the development of basic *ab initio* approaches to molecular quantum mechanics and statistical mechanics *rigor is a virtue* and absolutely essential in order to establish definitive results. But in the applications of theory to the world around us, the theory is most useful when it is stretched beyond those things which we know for sure. Thus, it will be important that the theoreticians of the future not be afraid to stick their necks out and make "guestimates" which can serve to guide experiments.

THE GOLDEN AGE OF QUANTUM MECHANICS

During the period 1931–36, thanks to the genius of Linus Pauling and Henry Eyring, theoretical chemists did try to apply their meager theoretical knowledge as far as they could stretch it. Out of this came many exciting discoveries. I predict that the period 1973–80 will resemble this first golden age of quantum chemistry.

In 1931 quantum mechanics, in its present form, was only a few years old. No one knew its limitations, and important discoveries were made every few months.

Quantum mechanics was first applied to problems of atomic energy. The explanation of the relative spacings or the "flags" corresponding to the splitting of multiplets was one of the first triumphs. The determination of the absolute values of the splittings required the evaluation of a set of radial integrals. Professor Condon set a number of us young graduate students to work on these integrals, which subsequently became a part of Condon and Shortley's famous treatise.

Of course, modern quantum mechanics was designed to give the correct energy levels for atomic hydrogen. It was Egil Hylleraas who showed that it also worked for the two-electron helium atom. At one time or other, Hylleraas used or developed almost every basic technique which we use in quantum mechanics: different types of perturbation and variational principles, correlated orbitals, configurational interaction, etc. All of Hylleraas' calculations were made on a hand-cranked desk calculating machine. The next generation of quantum mechanicians had electric power to turn the crank.

The early treatment of many-electron atoms was very crude. Hartree expressed the wavefunction for an atom as the product of one-electron orbitals. These orbitals were supposed to be the solutions to a set of coupled differential equations which seemed hopelessly difficult to solve. Eventually Hartree was able to obtain approximate solutions on a mechanical differential analyzer which he built out of "Mechano" parts.

Mechanical Differential Analyzer

The early differential analyzers were very interesting.[1] Vannevar Bush's first model was completely electrical, using a house-type watt-hour meter for

Fig. 1. The differential analyzer, built in 1935 at the Moore School of Electrical Engineering, University of Pennsylvania. A similar machine was constructed for the Ballistics Research Laboratory at Aberdeen Proving Grounds for the U.S. Army. These differential analyzers were constructed by the Civil Works Administration as part of President Roosevelt's emergency relief program. The work was directed by Professors Vannevar Bush, S. H. Caldwell, L. S. Frost, and Irven Travis.

integration. However, he could not reduce the error in the watt-hour meter to less than 2%. Therefore he switched to an all-mechanical contrivance. I used the analyzer (see Fig. 1) at the Moore School of Electrical Engineering of the University of Pennsylvania[2] during the winter of 1938–39. The independent variable was represented by the rotation of a rod 40 feet long rotated by a 3-hp motor. Each of the dependent variables was represented by the rotation of another long rod which was rigidly connected to the independent variable rod by means of gear trains, "adders," "integraters," etc. in accordance with the constraints of the mathematical relations. Addition and subtraction was accomplished by an "adder" which embodied the principles of the differential gears of an automobile (see Fig. 2): the rotation of the drive shaft is proportional to the sum of the rotations of the two rear wheels. Multiplication by a constant factor was easily accomplished by two intermeshing gears of different numbers of teeth, but multiplication by a variable factor required integration by parts. The "integrator" looked like a phonograph turn-table with the pickup arm replaced by a steel disc perpendicular

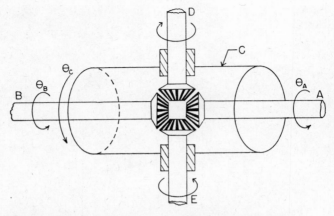

Fig. 2. Schematic diagram showing the principles of an "adder." Shafts D and E are "idlers" which are free to rotate and have no external connections. The four pinion gears in the center of the diagram are always engaged. Thus, when the two shafts A and B are rotated by θ_A and θ_B, respectively, the housing C is constrained to rotate by $\theta_C = (\theta_A + \theta_B)/2$.

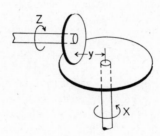

Fig. 3. Schematic diagram showing the principles of the "integrator." If the turntable is rotated through an angle x and the disk is moved in and out so that the point of contact of the disk is at the variable distance y from the center of the turntable, then the disk is constrained to rotate through an angle z where $z = a \int y \, dx$. Here a is the reciprocal of the radius of the disk, an instrumental constant.

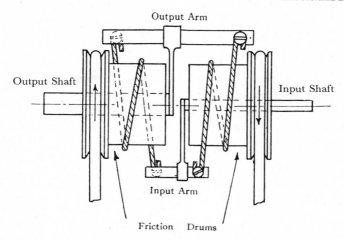

Fig. 4. Schematic diagram showing principles of the torque amplifier. The friction drums are attached by pulleys to an electric motor.

to the record (see Fig. 3). The rotation of the disc was equal to the integral of the distance of the disc from the center of the record integrated over the number of rotations of the record. The difficulty was that the disc rested only lightly on the record and therefore could not supply much torque without slipping. The torque amplification was accomplished with two stages of mechanical amplification known as a double motorized winch (see Fig. 4). This involved a fish line in the first stage and a shoe string in the second stage which had to be perfectly adjusted according to the humidity, etc. Computing machines have come a long way during the past 30 years!

Early Applications of Quantum Mechanics

A big breakthrough in the applications of quantum mechanics to atomic problems resulted from the development of screening constants—first by Schrödinger in 1921, then by Pauling in 1927, and finally by Slater in 1931. With the use of Slater screening constants, it was possible to estimate the energy or almost any physical property of an atom. Of course, the accuracy of these predicted values was not always very good, but the predicted values were seldom wrong by as much as a factor of 2, and they were very useful.

Wigner deserves the credit for bringing order out of chaos in atomic physics by introducing group theory. Since the Hamiltonian for an atom or molecule is invariant with respect to the interchange of two electrons, the spatial parts of the wavefunctions are bases of irreducible representations of a finite permutation group. Then, too, the Hamiltonian for an atom is invariant with respect to rotations of the coordinate system. Thus, the wavefunctions must also be bases of irreducible representations of the infinite three-dimensional rotation group. As part of my doctor's thesis, I helped Wigner to separate off the rotational coordinates from the electronic, vibrational, and rotational motions of a molecular system. This is still unfinished business—Charles Curtiss, Felix Adler, and I are still trying to separate off these rotational coordinates without ending up with messy coupled equations. The introduction of group theory was very much resented by most of the physicists and referred to as "the group pest." The problem was that group theory required learning a new mathematical language. I remember frequently hearing the statement that there was nothing you could do with group theory that you could not do without it. However, the group theoreticist could derive in one page what would otherwise require fifteen pages. Thus, the poor physicists and now the poor chemists must all learn this elegant bit of mathematical formalism.

In 1927, Burrau calculated the energy of H_2^+ and Heitler and London treated the hydrogen molecule. In 1928, the Heitler-London or valence bond method was applied to many electron systems, and simultaneously Hund and Mulliken started the development of the molecular orbital theory. In 1931, Slater expressed the wavefunctions of complex molecules in terms of Slater determinants made up of linear combinations of atomic orbitals. Thus, the Golden Age was born.

The Golden Age was marked by a feeling of great confidence in being able to solve *any* physical problem. The keynote was Dirac's often quoted statement in the preface of his 1930 book, "With the event of quantum mechanics, all of the basic physical laws which are required for the solution of chemical problems are now known." The actual difficulty of solving the Schrödinger equation for many electron problems was not fully realized. Even the crudest approximations were extremely difficult to test when all the calculations had to be made by desk computers. Thus, naively, the young physicists and theoretical chemists felt that they had the world by its tail. They generated the atmosphere of a Florida real estate boom with its get-rich-quick flamboyancy.

The physicists developed impressive equations which were supposed to represent the solutions to chemical problems, and they became discouraged because the chemists found it too difficult to test the equations numerically. In contrast, the chemists, such as Polanyi, Pauling, and Eyring, made whatever additional assumptions were required in order to get numerical solutions quickly and easily. The chemists' approach resembled engineering empiricism. They invented simple formulas which superficially agreed with the physicists' theoretical results on the one hand, and which possessed parameters which could be adjusted to agree with the known experimental data on the other hand. London told me that he was appalled at the way the chemists mangled his formula and still attached his name to the semi-empirical results. Whatever the justification of the semiempiricism, it gave very reasonable results. Pauling's notions of hybrid orbitals explained the angles between chemical bonds and the resonance stability of aromatics. Eyring and Polanyi's 14% coulombic approximation explained the activation energy for a large class of chemical reactions.

Henry Eyring: The Compleat Scientist

I should tell the following story about Eyring. In the spring of 1931, Eyring presented a paper at the Buffalo meeting of the ACS in which he claimed that the reaction $H_2 + F_2$ had a large activation energy since it required the breaking of two chemical bonds. Most people at the time thought that a homogeneous mixture of hydrogen and fluorine would explode instantaneously. Hugh S. Taylor, however, had been studying the reaction and found that the gas-phase reaction was indeed slow, the fast reaction taking place on a surface. Consequently, Hugh Taylor hired young Eyring to be an assistant professor at Princeton. And, to this day, Taylor has had the greatest confidence in Eyring's predictions.

After I passed my prelims in theoretical physics in 1933, I started to do my thesis work with Eyring. It was a tremendously stimulating experience. Eyring was young, vigorous, and even athletic. He prided himself on being able to beat his graduate students at the 100-yard dash. Eyring thought about his work from five or six o'clock in the morning, when he did his setting up exercises, until he went to sleep around midnight. He studied all of the theoretical physics literature and new techniques. He tried to apply everything he learned to the behavior of nature. Missing links in his arguments were temporarily filled by conjectures. Thus, he was prepared to attack any problem. In this sense he was the *compleat* scientist. Each morning he would come to work bubbling over with new ideas. Most of his ideas were wrong, and it was the responsibility of his graduate students to find the logical errors or the reasons why the ideas were not workable. However, there remained 5 to 10% of the ideas which were inherently interesting and provided a useful concept of the gross way in which the phenomena occurred. Thus each day we had a very practical demonstration of winnowing and sifting of ideas. The ability to recognize which approaches were blind alleys was most important.

Potential Energy Surfaces

For a time, all of us graduate students were engaged in a potential energy surface factory. We mechanized the work in the following manner[3]: I made

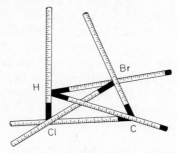

Fig. 5. A device used to simplify the calculation of potential energy surfaces. Here the steel rules correspond to a particular atomic configuration occurring in the reaction HCI + BrC → HBr + CIC. The carbon atom is bonded to three other atoms which do not enter into the reaction. The scales on the rods give both the interatomic separations in angstroms and the corresponding Morse curve energies for the "diatomic molecule" in kcal/mole.

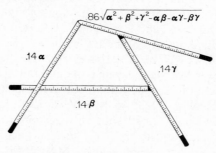

Fig. 6. Another device used in the calculation of the potential energy surfaces. Here, in accord with a suggestion by W. Altar, the angles between the α and the β and between the β and the γ are restricted to 60°. Double scales are given on the rods so that one can read in α, β, and γ and read out the 0.14α, 0.14β, and 0.14γ. The distance between the ends of the α and γ scales is the required square root. The potential energy for a particular atomic configuration is then the sum of four numbers as shown on the rulers.

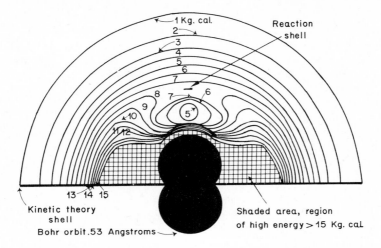

Fig. 7. Potential energy contours corresponding to a hydrogen atom approaching a hydrogen molecule.

up a set of scales giving the Morse curve energy of diatomic molecules as a function of the interatomic distance. These scales were glued onto steel rods. The steel rods were then fastened together so that their intersections corresponded to a configuration of the reacting atoms (see Fig. 5). The sum of the Morse curve energies on opposite sides of the quadrilateral were called α and β; the sum of the energies of the diagonals was called γ. A second gadget (see Fig. 6) consisting of steel rulers (with two of them constrained to a 60° angle with respect to a third) made it possible to read in α, β, and γ and read out 0.14α, 0.14β, 0.14γ, and $0.86(\alpha^2 + \beta^2 + \gamma^2 - \alpha\beta - \alpha\gamma - \beta\gamma)^{1/2}$. With such devices it was easy for a graduate student to construct a potential energy surface for a four-atom reaction with the expenditure of one afternoon's work. Figure 7 illustrates my favorite potential energy surface. It shows what a hydrogen molecule looks like to a hydrogen atom. The energy contours show that it is much easier for the hydrogen atom to penetrate deeply into the hydrogen molecule if it approaches along the internuclear axis rather than perpendicular to it.

A number of interesting principles emerged from these crude potential energy surface calculations:

At the *activated state*, the separations between the atoms is about 15% greater than their separations in normal molecules. This fact led Sherman and Eyring to explain why the atoms on the surface of an active catalyst should have unusually large separations.

The *activation energy* is generally very large for a reaction in which two chemical bonds are broken. Although this principle is usually true, there may be exceptions.

Many people, including myself, have tried to show errors in the Eyring theory of absolute reaction rate. The results have usually shown that the Eyring theory is surprisingly accurate. I think that the most important feature of this theory is that it defines an *entropy*, as well as an *enthalpy*, *of activation*. Thus it provides not merely one, but two significant parameters which may be determined from the experimentally observed temperature dependence of the rate constant. In biological reactions, the *entropy of activation* may play a dominant role. For example, in the cooking of egg white, the entropy of activation is extremely large and positive. This tends to compensate for the 140,000 calories per mole activation energy and make the boiling of eggs occur at a very sharp temperature.

So much for philosophizing. The Golden Age ran out of steam because the theoretical predictions were far beyond the known experimental facts. We were always ready and willing to add another adjustable constant to our equations to fit some new piece of experimental data. The theory of the Golden Age was an interesting mixture of engineering empiricism and basic physics. The Golden Age was sparked by the genius of two men: Henry Eyring and Linus Pauling.

THE PRESENT AND FUTURE OF QUANTUM MECHANICS

Now let us jump to 1966 and take stock of our situation. What is the current status of quantum chemistry and where do we go from here?

Molecular Quantum Mechanics

With the use of high-speed computing machines and many-configurational interactions, A. C. Wahl and others are calculating the energy of a large number of diatomic molecules with a precision comparable to, or better than, experiment. They have programmed the computers to make similar calculations for triatomic molecules. Clementi has now been able to make precise calculations of the energy of simple organic molecules. These calculations are *very, very* expensive: $10,000–$20,000 for each potential energy curve. We can afford sample calculations of every different kind. However, we should try to squeeze the most information from each calculation. Thus, every time a precise wavefunction is determined, the corresponding first- and second-order density matrices should also be calculated. These density matrices are easy to determine, and they provide the essential features of the electron density and the electron correlations.

Perturbation theory provides another approach to molecular quantum mechanics. From perturbation theory we are learning how to start with an approximate wavefunction and estimate good values for the energy or other physical property. The problem of calculating accurate first-order properties, such as dipole moments, and accurate second-order properties, such as polarizability, is very important. The dipole moment is very sensitive to the wavefunction. From perturbation theory, Peter Robinson was able to show that in order for the wavefunction to yield a good value for a dipole moment, it must satisfy a particular hypervirial theorem as a constraint. Thus we can provide a theoretical constraint that will insure that our wavefunction gives a good value for the dipole moment. Similarly, we need to find theoretical constraints which will insure our calculation of accurate values for many different types of physical properties. Indeed, we shall not be content until we can calculate good upper and lower bounds so that we can bracket each of our theoretical estimates of physical properties.

Since the earliest days of quantum chemistry, it has been hoped that at least some parts of the molecular wavefunctions would be interchangeable as one goes from one molecule to another. It now appears that the inner orbitals are characteristic of the atomic species and the outer orbitals are characteristic of the valence state of the chemical bond. The basic significance of the chemical bonds is evident from the geminal calculations of Shull. Simple regularities in the electron densities in atoms and molecules have been observed by Clementi, Allen, Sinanoglu, and others. Therefore we can hope that computers will ultimately be used to solve the steady state Schrödinger equation for an arbitrary molecular aggregate.

Prediction of Reaction Rates

One of the first applications of these molecular energies will be the large scale calculations of potential energy surfaces. Once we can construct potential energy surfaces, we can once more predict chemical reaction rates—the bread-and-butter problem of chemistry. The problem of constructing accurate energy surfaces may not be so difficult as we might suppose. In 1964, Kenneth Cashion and Dudley Herschbach[4] used the old Heitler-London first-order perturbation formulas to calculate the potential energy surface for the $H + H_2$ reaction. Here, the coulombic energy of a diatomic hydrogen molecule at an interatomic separation R_{ab} is

$$J_{ab} = \tfrac{1}{2}[^1E(R_{ab}) + {}^3E(R_{ab})]$$

and the exchange energy is

$$K_{ab} = \tfrac{1}{2}[^3E(R_{ab}) - {}^1E(R_{ab})]$$

The energy of the three atom system is then

$$E_{abc} = J_{ab} + J_{bc} + J_{ac} + 2^{-\frac{1}{2}}[(K_{ab} - K_{bc})^2$$
$$+ (K_{ab} - K_{ac})^2 + (K_{bc} - K_{ac})^2]^{\frac{1}{2}}$$

Eyring and Polanyi used the same formulas in 1931. The difference is that Cashion and Herschbach used the Kolos and Roothaan,[5] and Dalgarno and Lynn[6] accurately calculated values of the singlet and triplet sigma diatomic molecular energies, ${}^1E(R_{ab})$ and ${}^3E(R_{ab})$, whereas Eyring and Polanyi used Morse curves for the singlet sigma energy and the 14% coulombic approximation. The Cashion and Herschbach energy surface leads to an energy of activation within 1 kcal/mole of the experimental value. Vanderslice and Mason have used the same sort of treatment to calculate the interactions between oxygen atoms and nitrogen molecules, etc. Apparently this type of approximation is good and should be explored further.

Collision Phenomena in Molecular Beams

Molecular beams provide a much more precise way of studying molecular collision processes than do reaction rates. The collision cross sections as a function of energy are essentially the Laplace transforms of the reaction rate constant as a function of temperature.[7] Bernstein and others are studying collisions of molecules in specific electronic, vibrational, and rotational states and observing the quantum states of the product molecules as functions of scattering angles and translational energy. Experimental data are so precise that quantum mechanical interference effects play an important role in the observed cross sections. As a first step in explaining molecular beam behavior, one might use the Born-Oppenheimer separation of electronic and nuclear motions and construct a potential energy surface. As a second step, one might either calculate a manifold of classical trajectories by the methods of Karplus or Marcus and determine the semiclassical collision cross sections; or one might represent the molecular beam by a steady state wave packet moving across the potential energy surface, a technique which Eugene Wigner taught me in 1937. We have not estimated the magnitude of the errors which result from using the Born-Oppenheimer separation approximation. I think that the Born-Oppenheimer approximation is satisfactory if the collisions are truly adiabatic in the Ehrenfest sense. If, on the other hand, the collisions are frequently nonadiabatic in the sense of jumping from one potential energy surface to the other, then the Born-Oppenheimer approximation is not satisfactory. My guess is that *most* practical chemical reactions occur in the nonadiabatic fashion. Thus, it will be necessary for us to solve the full time-dependent Schrödinger equation including both the electronic *and* nuclear coordinates. This is an extremely difficult job. I doubt that this type of solution will be forthcoming for many-electron problems for another 5 to 8 years. Solution of the full time-dependent Schrödinger equation is definitely required in considering electron-transfer reactions.

The Need for Quantum Electrodynamics

Much of the present experimental data being obtained by physical chemists is very precise and sophisticated: nuclear magnetic resonance, electron spin resonance, laser optics, microwave line shapes, etc. These experiments require a much higher degree of theoretical precision. Indeed, they require the solution of quantum electrodynamic equations which are far more difficult than the Schrödinger equation. In considering nuclear magnetic resonance to the first approximation, it suffices to use the Breit-Pauli Hamiltonian, which is the Schrödinger equation with a set of relativistic-magnetic correction terms tacked on. For second-order effects, such as fine structure or line intensities, however, using the Breit-Pauli Hamiltonian does not correspond to considering all of the pertinent Feynman diagrams. In laser optics, the light is so intense that as many as five quantum transitions occur practically simultaneously. This compares with the two quantum transitions which characterize the Raman spectra. To consider such Auger effects, one needs quantum electrodynamics. Thus, to quote Bright Wilson, "It is no longer sufficient to teach our students how to solve the Schrödinger equation, they must also know quantum electrodynamics."

Solid State and the Many-Body Problem

I have concentrated on discussing quantum mechanics up to this point. However, it is clear that statistical mechanics and electricity and magnetism are equally important. From statistical mechanics we learn the behavior of macroscopic systems: thermodynamics, equations of state, transport properties, etc. Except for dilute gases, the molecules which we consider in the laboratory are not isolated, they are surrounded by interacting neighbors. Thus, after we learn to cope with two-body intermolecular forces, we must consider three-body, four-body, etc., interactions. Long-range forces between a very large number of ions are important in a plasma, strong electrolyte, or ionic crystal. Consequently, the quantum chemist is required to learn the techniques of solid state physics and the many-body problem.

Up to now quantum chemists have been preoccupied with trying to explain the structure and properties of simple molecules. As soon as they have simple molecules under control, they will have to learn the tricks of their physicist friends. Three-body forces determine the face-centered cubic packing of noble gas crystals. They probably determine the shapes of many high polymers. It is well known that crystals of aromatic substances have very interesting optical and physical properties. Indeed, some chemists are already studying these "exciton" effects which form an important class of many-body interactions. Howard Reiss tells me that transistors, photoelectric cells, etc., offer a very interesting challenge to chemists and involve the same sort of considerations which occur in strong electrolytes—an interesting mixture of statistical mechanics, electricity and magnetism, and quantum mechanics.

Solid state chemistry is needed to explain the physical and chemical properties of surfaces including surface catalysis. The activity of a catalyst depends strongly on traces of impurities and on the detailed history of its preparation which determines the metastable configuration of atoms on its surface. We are still puzzled over the strong correlation between para-magnetism of a surface and its catalytic activity. The growth of crystals and the theory of nucleation are other avenues for theoretical research.

Matter under Extreme Conditions

Perhaps the most important usage of quantum chemistry will be in predicting the behavior of materials under extreme conditions of tempera-ture, pressure, and electromagnetic fields. Chemical reactions in electrical discharges are extremely interesting and provide a real challenge to a theoretician. John Magee has shown that quantum chemistry is very useful in explaining radiochemical phenomena. Bates and Dalgarno have applied quantum chemistry to the chemistry of the upper atmosphere. In partially ionized plasmas and in flames, the excited states of molecules occur in large numbers. Theoretical chemists are needed to predict the physical and chemical properties of excited state molecules, since there is very little experimental data to serve as a guide. Then, there is the chemistry of extremely high pressures—making diamonds, polymerizing simple molecules, etc., where quantum mechanics should help in predicting the optimum conditions.

BE READY FOR THE NEW GOLDEN AGE

If time permitted, I could add other areas where quantum chemistry will be applied. The points which I have wanted most to stress are:

Thanks to high-speed computing machines and new techniques, quantum chemistry is making very, very rapid progress.

When quantum chemistry matures, it will be applicable to many types of experimental problems.

In order to prepare your students for this new Golden Age which is coming, give them a broad experimental and theoretical background.

Finally, let us stop being cynical conservatives afraid to stick our necks out. Let us have the fun of predicting useful and critical experiments, even if now and then we make a mistake!

REFERENCES

1. V. Bush, *J. Franklin Inst.*, **212**, 447 (1931).
2. I. Travis, *Machine Design*, **7**, 15 (1935).
3. J. O. Hirschfelder, *J. Chem. Phys.*, **9**, 645 (1941).
4. J. K. Cashion and D. R. Herschbach, *J. Chem. Phys.*, **40**, 2358 (1964); **41**, 2199 (1965).
5. W. Kolos and C. C. J. Roothaan, *Rev. Mod. Phys.*, **32**, 219 (1960).
6. A. Dalgarno and N. Lynn, *Proc. Phys. Soc.* (*London*), **A69**, 821 (1956).
7. M. A. Eliason and J. O. Hirschfelder, *J. Chem. Phys.*, **30**, 1426 (1959).

Studies on Rates of Nonequilibrium Processes

SHANG J. YAO AND BRUNO J. ZWOLINSKI

Thermodynamics Research Center, Department of Chemistry, Texas A&M University, College Station

Abstract

The Zwolinski-Eyring discrete energy model of microscopic rate processes, originally limited to four levels, is extended to a 10-level unimolecular kinetic system by appropriate solutions of the master equation. The distribution of energy levels for the reactants and products in the initial and final states, arranged in a variety of ways, was chosen to simulate real kinetic systems and the levels were accordingly spaced anharmonically, that is, $\epsilon = n(1 - n\alpha)h\nu$. Landau-Teller-Schuler-type anharmonic probabilities were used for excitational transitions with realistic values assumed for transitions defined as chemical transformations. The equilibrium assumption for the activated state was examined for kinetic systems defined in terms of zero heat of reaction, endothermicity, and exothermicity with appropriate choice of nonequilibrium concentrations, i.e., boundary conditions, including cases of simultaneous reactions defined by systems of three species which serve as a basis for nonadiabatic reaction systems.

91

INTRODUCTION

In recent years studies in nonequilibrium and nonadiabatic reaction processes have made significant contributions to chemical kinetics. Experiments employing modern mass spectrometers, laser spectrometers, and shock wave techniques have produced an enormous amount of data which have offered great opportunity for a deeper insight into the theory of chemical rate processes. Microscopic description of nonequilibrium chemical rate process, which dates back to the original study of Zwolinski and Eyring,[1] has played a very important role in the understanding of modern rate processes such as energy transfer, relaxation processes, and problems in intersystem crossing. Important attempts have been made to correlate microscopic descriptions to experimental observables; for example, the mean-first-passage-time studies[2] on the rate of decomposition of diatomic molecules and eigenvalue interpretation of reactions[3] have finally led to the identification of the lowest eigenvalue for the solution of the set of coupled linear rate equations as the rate constant of a unimolecular reaction.[4] This work is a continuation of the previous study on nonequilibrium theory of absolute rates of reaction.[1] As before, quantized molecular energy levels are considered. The system considered here is essentially a closed system, as is usually described in nonequilibrium statistical mechanics. Transitions and reactions among energy levels are assumed to be caused by collisions of reacting molecules with each other and with the large excess of inert particles contained in the closed heat reservoir characterized by a constant temperature T. Evolution or redistribution of molecular population among energy levels is described by the Pauli master equation. No attempt is made to correlate the microscopic transition parameters to experimental observables such as the equilibrium activation energy and specific rate constants[5]; instead, we shall examine the validity of the equilibrium rate hypothesis from nonequilibrium studies. The present model assumes that intramolecular relaxation and reaction (intermolecular relaxation) can take place at any individual energy level. Each intra- or intermolecular transition leading to reaction in either direction involves surmounting of a potential barrier of its own. Energy levels are designated and are classified as a set of reactant levels and sets of product levels. Physical meanings for the elementary processes are specified. The rate of the nonequilibrium reaction is expressed as the sum of rates of the elementary processes. Since the entire system is closed, the whole analysis is basically the same as the usual description for the approach to equilibrium in a closed system. The long-time nonequilibrium rate of reaction approaches a steady value as the entire closed system approaches a new equilibrium state.

METHOD OF CALCULATION

Consider a closed system characterized by a constant temperature T. The system is prepared in such a way that molecules in energy levels are distributed in departure from their equilibrium distribution. Transitions of molecules among energy levels take place by collisional excitation or deexcitation. The redistribution of molecular population is described by the rate equation or the Pauli master equation. The values for the microscopic transition probability k_{ij} for transition from ith level to jth level are, in principle, calculable from quantum theory of collisions.[6] Let the set of numbers ν_R be vibrational quantum numbers of the reactant molecule and ν_P be those of the product molecule. The collisional transitions or intermolecular relaxation processes will be described by:

$$R(\nu_R = 0) + M \to R^+ \to P(\nu_P = 0),$$
$$R(\nu_R = 0) + M \to R'^+ \to P(\nu_P = 1),$$
$$R(\nu_R = 1) + M \to R''^+ \to P(\nu_P = 3), \tag{1}$$

$$\begin{matrix} \cdot & & \cdot \\ \cdot & & \cdot \\ \cdot & & \cdot \end{matrix}$$

The master equation for describing the evolution among n energy levels of species A within the closed system is given by:

$$\frac{dA_i}{dt} = \sum_{j \neq i}^{n} (k_{ji} A_j - k_{ij} A_i) \tag{2}$$

The solution for the set of first-order linear differential equations (2) subject to the condition that

$$\sum_i^n A_i = D, \tag{3}$$

where D is the total concentration of species A, can readily be obtained by the method of determinants. The solution for A_i is expressed by

$$A_i = \sum_{k=1}^{n} B_{ik} \exp(-\lambda_k t) \tag{4}$$

where B_{ik} are elements of the eigenvector B and are constants, and λ_k is the kth eigenvalue. For multilevel problems, the solution can readily be obtained by means of numerical integration on a digital computer.

The actual rate of the forward reaction is determined by the sum of the instantaneous rates from reactant levels, i, to the final product levels, p; that is,

$$v_a = \sum_i \sum_p k_{ip} A_i, \tag{5}$$
$$\scriptstyle i \neq p$$

where A_i is the instantaneous distribution of species A at level i. The forward equilibrium reaction rate is expressed by

$$v_e = \sum_i \sum_p k_{ip} A_i^e \tag{6}$$
$$\scriptstyle i \neq p$$

with

$$A_i^e = \frac{\exp(-\beta\epsilon_i)}{\sum_j \exp(-\beta\epsilon_j)}$$

the constant equilibrium distribution of A as given by Maxwell-Boltzmann distribution for closed systems. Here, β is equal to $1/kT$ and statistical weights are taken to be unity for all levels.

The ratio of the actual rate to equilibrium rate is defined as:

$$\Gamma(t) = \frac{v_a}{v_e} = \frac{\displaystyle\sum_i \sum_p k_{ip} A_i}{\displaystyle\sum_i \sum_p k_{ip} A_i^e}, \tag{7}$$

where Γ, the quantity of prime interest to us, is analyzed in detail in the next section.

NONEQUILIBRIUM MODELS

In all calculations below, we shall consider only the cases when the closed system consists of ten energy levels, i.e., $n = 10$ in (2). These energy levels are arranged in specific manners to simulate the study of specific physical models of reactive systems.

The Anharmonic Models

Consider the initial reactant state, R, to consist of eight levels and the product state, P, to consist of two levels. The energy at the nth level of the reactant is given by

$$E_n = n(1 - \alpha n)h\nu, \tag{8}$$

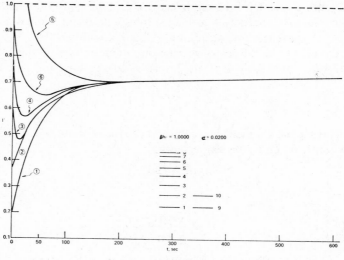

Fig. 1

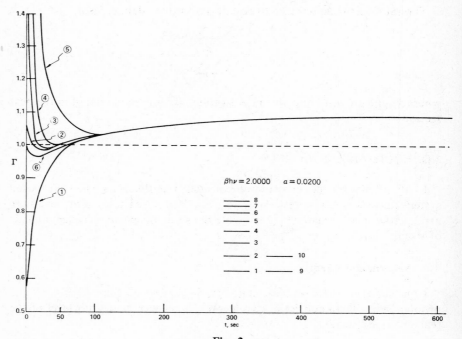

Fig. 2

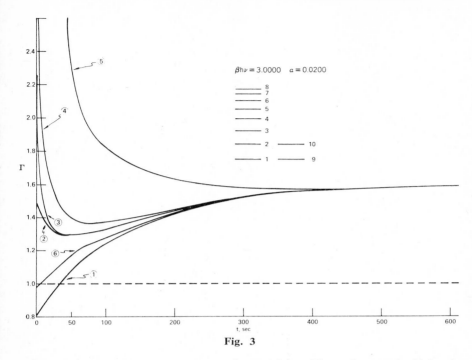

Fig. 3

where α is a constant to express anharmonicity. The anharmonic 10-level model is shown in Figs. 1–3. The selected nonequilibrium distributions of the initial reaction state are presented in Table I. The transition probabilities are given in Table II. These transition probabilities were evaluated using Bazley

Table I.   Initial Distributions for the Anharmonic Nonequilibrium Models[a]

Initial Distribution in Levels	System					$6 = A^e$		
	1	2	3	4	5	$\beta h\nu = 1$	$\beta h\nu = 2$	$\beta h\nu =$
$A_8^{(0)}$	0	0	0	0	0	0	0	0
$A_7^{(0)}$	0	0	0.0020	0.0050	0	0.0031	0	0
$A_6^{(0)}$	0	0	0.0180	0.0150	0.0500	0.0068	0.0001	0
$A_5^{(0)}$	0	0	0	0	0	0.0155	0.0005	0
$A_4^{(0)}$	0	0	0	0.0300	0.1500	0.0367	0.0030	0.0002
$A_3^{(0)}$	0	0.0500	0.0300	0.0700	0	0.0903	0.0184	0.0030
$A_2^{(0)}$	0	0.1000	0.1000	0.0800	0.8000	0.2313	0.1207	0.0501
$A_1^{(0)}$	1.0000	0.8500	0.8500	0.8000	0	0.6162	0.8571	0.9467

$$^a A_n{}^e = \frac{\exp(-\beta E_n)}{\displaystyle\sum_m \exp(-\beta e_m)} = \frac{\exp[-n(1 - \alpha n)\beta h\nu]}{\displaystyle\sum_m \exp[-m(1 - \alpha m)\beta h\nu]}$$

Table II. Values of Transition Probabilities for the Anharmonic Model

$\beta h\nu = 1.0000$, $\alpha = 0.0200$, and $\kappa = 0.0100$

k_{12}	k_{23}	k_{34}	k_{45}	k_{56}	k_{67}	k_{78}	k_{13}	k_{24}	k_{35}	k_{46}	k_{57}	k_{68}
0.0102	0.0208	0.0318	0.0432	0.0550	0.0672	0.0798	0.0002	0.0006	0.0012	0.0020	0.0030	0.0042

k_{21}	k_{32}	k_{43}	k_{54}	k_{65}	k_{76}	k_{87}	k_{31}	k_{42}	k_{53}	k_{64}	k_{75}	k_{86}
0.0272	0.0532	0.0782	0.1021	0.1249	0.1466	0.1673	0.0014	0.0038	0.0070	0.0107	0.0149	0.0192

$k_{19}=k_{1,10}$	$k_{29}=k_{2,10}$	$k_{39}=k_{3,10}$	$k_{49}=k_{4,10}$	$k_{59}=k_{5,10}$	$k_{69}=k_{6,10}$	$k_{79}=k_{7,10}$	$k_{89}=k_{8,10}$	
0.0010	0.0050	0.0100	0.0100	0.0200	0.0400	0.0650	0.1000	0.5000

$k_{91}=k_{92}=k_{93}=k_{94}=k_{95}=k_{96}=k_{97}=k_{98}=0.0010$

$k_{10,1}=k_{10,2}=k_{10,3}=k_{10,4}=k_{10,5}=k_{10,6}=k_{10,7}=k_{10,8}=0.0050$

$k_{9,10}=0.0101,\ k_{10,9}=0.0269$

$\beta h\nu = 2.0000$, $\alpha = 0.02000$, and $\kappa = 0.0100$. Except for the following changes, other transition probabilities are the same as those for $\beta h\nu = 1.0000$ cited above.

k_{21}	k_{32}	k_{43}	k_{54}	k_{65}	k_{76}	k_{87}	k_{31}	k_{42}	k_{53}	k_{64}	k_{75}	k_{86}
0.0724	0.1363	0.1924	0.2413	0.2835	0.3198	0.3506	0.0093	0.0238	0.0405	0.0576	0.0736	0.0878

$k_{9,10} = 0.0101$, $k_{10,9} = 0.0732$

$\beta h\nu = 3.0000$, $\alpha = 0.0200$, and $\kappa = 0.0100$. Except for the following changes, other transition probabilities are the same as those for $\beta h\nu = 1.0000$.

k_{12}	k_{23}	k_{34}	k_{45}	k_{56}	k_{67}	k_{78}	k_{13}	k_{24}	k_{35}	k_{46}	k_{57}	k_{68}
0.0010	0.0021	0.0032	0.0043	0.0055	0.0067	0.0080	0	0	0.0001	0.0002	0.0003	0.0004
k_{21}	k_{32}	k_{43}	k_{54}	k_{65}	k_{76}	k_{87}	k_{31}	k_{42}	k_{53}	k_{64}	k_{75}	k_{86}
0.0189	0.0352	0.0476	0.0567	0.0644	0.0695	0.0736	0.0063	0.0149	0.0196	0.0309	0.0364	0.0382

$k_{9,10} = 0.0101$, $k_{10,9} = 0.0189$

et al.'s results of the slightly anharmonic oscillator model,[7]

$$k_{n,n+1} = \kappa(n + 1)[1 + \alpha(n + 1)],$$
$$k_{n,n+2} = \tfrac{1}{2}\kappa\alpha(n + 1)(n + 2), \tag{9}$$

and

$$k_{n,n\pm m} = 0 \qquad (m = 3, 4, \ldots, N),$$

where κ defines the appropriate units. Furthermore, the microscopic reversibility condition

$$k_{ij} = k_{ji} \exp(-\beta\epsilon_{ij}), \tag{10}$$

with $\epsilon_{ij} = E_j - E_i$, is also employed as previously. In the estimation of transition probabilities, the usual assumption is made that a molecule in the activated state has the same probability for decomposition along the reaction coordinate to any level of the final state; that is, $k_{19} = k_{1,10}$, $k_{49} = k_{4,10}$, and similarly, for the reverse processes, $k_{10,1} = k_{10,5}$, $k_{92} = k_{94}, \ldots$.

The results of the calculations shown in Figs. 1–3 reveal that, in general, Γ will attain constant values within the first 400 seconds of the reaction. Since no source is available for continuously supplying the reactant molecules, Γ will not attain the value of unity even in the special case, system 6, in which the initial populations are arranged as equilibrium distribution, that is, $A_i^{(0)} = A_i^e$. It appears that the whole rate process is dominated by internal relaxation for the first 100 seconds. Major reaction or intermolecular relaxation becomes important only after the completion of the internal relaxation process. Nonequilibrium rates for all systems, that is, systems 1, 2, 3, 4, 5, and 6, approach each other with constant values as the major induction period or the internal relaxation period is over. All nonequilibrium rates, including the long-time steady value rates, are appreciably different from reaction rates of which the reactant molecules are maintained in constant equilibrium distribution among energy levels. It is also noted that for the first 100 seconds, the nonequilibrium rates, v_a, are much faster than the equilibrium rates, v_e, when the upper levels are *highly populated*. However, these v_a or their respective Γ values decrease to the leveling-off values much faster than those for other nonequilibrium cases investigated.

Table III. Initial Distribution for the Anharmonic Nonequilibrium Models With Different Anharmonicities. $\beta h\nu = 2.0000$

Initial Distribution in Levels	System			$4 = A^e$	
	1	2	3	$\alpha = 0.0080$	$\alpha = 0.0200$
$A_6^{(0)}$	0	0	0	0	0
$A_5^{(0)}$	0	0.0200	0.1000	0.0004	0.0005
$A_4^{(0)}$	0	0	0.9000	0.0025	0.0030
$A_3^{(0)}$	0	0	0	0.0168	0.0184
$A_2^{(0)}$	0	0.1800	0	0.1185	0.1207
$A_1^{(0)}$	1.0000	0.8000	0	0.8618	0.8572

The possibility of anharmonic effects has also been investigated. The initial nonequilibrium distributions are listed in Table III. Transition probabilities are tabulated in Table IV. The reaction model chosen is one where the reactant state consists of only six levels. However, two final or product states are assumed via parallel reaction paths. Each product state consists of only two levels. The two reactions are distinguished by two distinct sets of rate constants. For example, $k_{1,7} = k_{1,8} = 0.0010$ and $k_{2,7} = k_{2,8} = 0.0050 \ldots$, whereas $k_{19} = k_{1,10} = 0.0006$ and $k_{29} = k_{2,10} = 0.0030$. Results for cases $\beta h\nu = 2.0000$, $\alpha = 0.0080$ and $\beta h\nu = 2.0000$, $\alpha = 0.0200$ are shown in Figs. 4 and 5, respectively. These results indicate that no significant difference was found between the two cases of $\beta h\nu = 2.0000$ with different values of α, the anharmonicity parameter.

Table IV.   Transition Probabilities for the Anharmonic Models With Different Anharmonicities. $\beta h\nu = 2.0000$

$\alpha = 0.0080$

k_{12}	k_{23}	k_{34}	k_{45}	k_{56}	k_{13}	k_{24}	k_{35}	k_{46}
0.0101	0.0203	0.0307	0.0413	0.0520	0.0001	0.0002	0.0005	0.0008

k_{21}	k_{32}	k_{43}	k_{54}	k_{65}	k_{31}	k_{42}	k_{53}	k_{64}
0.0731	0.1427	0.2094	0.2718	0.3328	0.0051	0.0010	0.0225	0.0337

$\alpha = 0.0200$

k_{12}	k_{23}	k_{34}	k_{45}	k_{56}	k_{13}	k_{24}	k_{35}	k_{46}
0.0102	0.0208	0.0318	0.0432	0.0550	0.0002	0.0006	0.0012	0.0020

k_{21}	k_{32}	k_{43}	k_{54}	k_{65}	k_{31}	k_{42}	k_{53}	k_{64}
0.0724	0.1363	0.1924	0.2413	0.2835	0.0093	0.0238	0.0405	0.0576

Common inputs of transition probabilities:

$k_{17} = k_{18}$	$k_{27} = k_{28}$	$k_{37} = k_{38}$	$k_{47} = k_{48}$	$k_{57} = k_{58}$	$k_{67} = k_{68}$
0.0010	0.0050	0.0100	0.0500	0.1000	0.5000

$k_{19} = k_{1,10}$	$k_{29} = k_{2,10}$	$k_{39} = k_{3,10}$	$k_{49} = k_{4,10}$	$k_{59} = k_{5,10}$	$k_{69} = k_{6,10}$
0.0006	0.0030	0.0060	0.0300	0.0600	0.3000

$k_{78} = k_{9,10}$	$k_{87} = k_{10,9}$
0.0101	0.0732

$k_{71} = k_{72} = k_{73} = k_{74} = k_{75} = k_{76} = 0.0010$
$k_{81} = k_{82} = k_{83} = k_{84} = k_{85} = k_{86} = 0.0050$
$k_{91} = k_{92} = k_{93} = k_{94} = k_{95} = k_{96} = 0.0006$
$k_{10,1} = k_{10,2} = k_{10,3} = k_{10,4} = k_{10,5} = k_{10,6} = 0.0030$

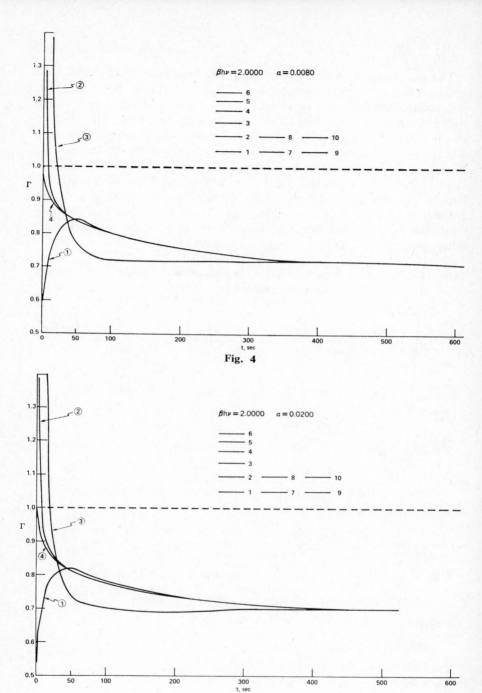

Fig. 4

Fig. 5

102

Harmonic Models for Parallel Exothermic and Endothermic Reactions

The anharmonic model for parallel reactions can be extended to the study of parallel exothermic and endothermic reactions. For the sake of simplicity, we shall assume that the energy levels are evenly spaced, i.e., harmonic. For example, consider the model described in Fig. 7, where the zero energy of product B is level 7, which is lower than the zero energy of the reactant by an amount of $2kT$. The reaction from reactant R to product B is an exothermic process. Conversely, the reaction from R to product C, which has its energy of zero at level 9, is an endothermic reaction. Various models of exothermic and endothermic reactions have been investigated. Results are shown in

Table V. Initial Distribution of the Evenly Spaced
Energy Levels for the Nonequilibrium Parallel
Reaction Models

Initial Distribution in Levels	System		
	1	2	3
$A_6^{(0)}$	0	0	0
$A_5^{(0)}$	0	0.0200	0.1000
$A_4^{(0)}$	0	0	0.0900
$A_3^{(0)}$	0	0.0500	0
$A_2^{(0)}$	0	0.0800	0
$A_1^{(0)}$	1.0000	0.8500	0

Figs. 6–15. Nonequilibrium distributions for initial populations are described in Table V. Transition probabilities are presented in Tables VI and VII. In all exothermic cases studied, Γ will eventually decrease to zero. With no source of supply for the reactant and with the product levels acting like a sink for the reaction, the decreasing of Γ with increasing time is not significant. Only for the slightly endothermic cases, such as the models described in Figs. 12 and 15, does one find the leveling-off of values of Γ at about unity.

With proper assignment for transition probabilities, these parallel reaction models can be modified to describe simultaneous adiabatic and nonadiabatic reactions. For example, if one considers the reactant R and product B in the same electronic state and the product C in an excited electronic state, and if the spin–orbit interaction is negligible, then the reaction R → B can be considered as adiabatic, whereas R → C is a nonadiabatic crossing reaction.

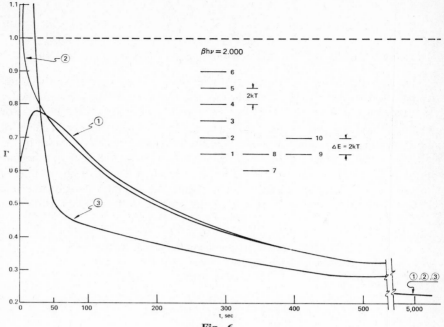

Fig. 6

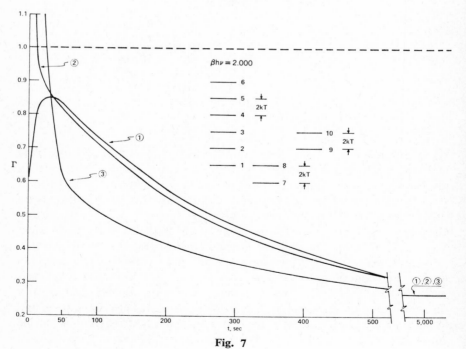

Fig. 7

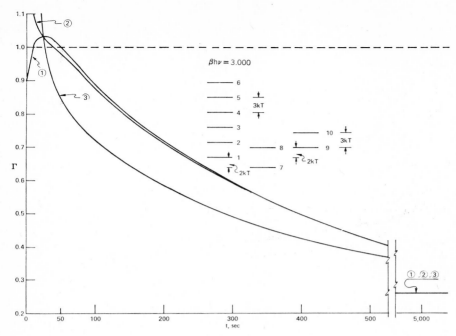

Fig. 8

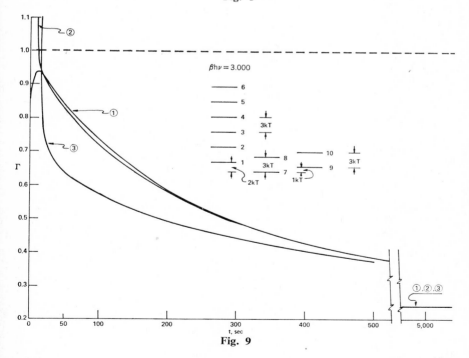

Fig. 9

105

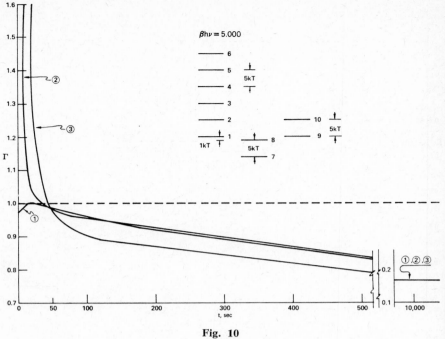

Fig. 10

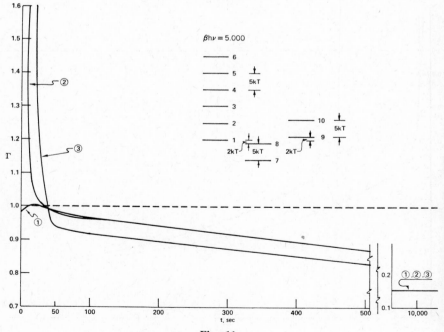

Fig. 11

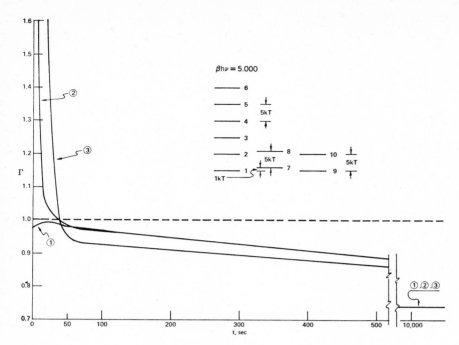

Fig. 12

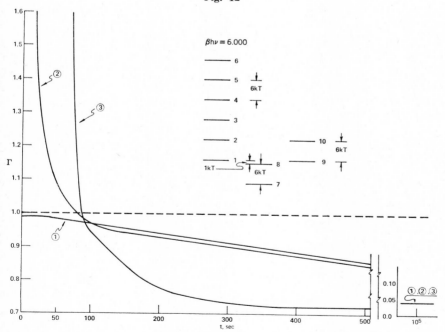

Fig. 13

107

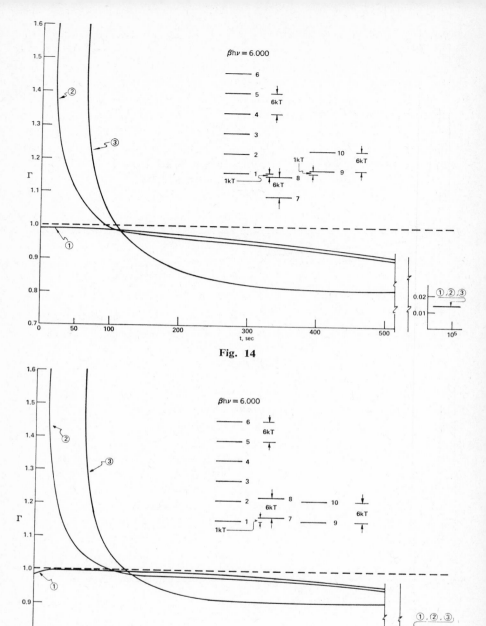

Fig. 14

Fig. 15

Table VI. Transition Probabilities for the Evenly Spaced Energy Levels Models, $\beta h\nu = 2.0000$ and $\beta h\nu = 3.0000$ Cases

$\beta h\nu = 2.0000$

$k_{12} = k_{78} = k_{9,10}$	k_{23}	k_{34}	k_{45}	k_{56}	k_{65}	k_{54}	k_{43}	k_{32}	$k_{21} = k_{87} = k_{10,9}$
0.0100	0.0200	0.0300	0.0400	0.0500	0.3695	0.2956	0.2217	0.1478	0.0739

$k_{18} = k_{17}$	$k_{28} = k_{27}$	$k_{38} = k_{37}$	$k_{48} = k_{47}$	$k_{58} = k_{57}$	$k_{68} = k_{67}$
0.0010	0.0050	0.0100	0.0500	0.1000	0.5000

$k_{1,10} = k_{19}$	$k_{2,10} = k_{29}$	$k_{3,10} = k_{39}$	$k_{4,10} = k_{49}$	$k_{5,10} = k_{59}$	$k_{6,10} = k_{69}$
0.0010	0.0050	0.0100	0.0500	0.1000	0.5000

Model, see Fig. 6. (For $\beta h\nu = 3.0000$, see Fig. 8)

$k_{81} = k_{82} = k_{83} = k_{84} = k_{85} = k_{86} = 0.0007$
$k_{71} = k_{72} = \quad \cdot \quad \cdot \quad = k_{76} = 0.0001$
$k_{10,1} = k_{10,2} = \quad \cdot \quad \cdot \quad = k_{10,6} = 0.0050$
$k_{91} = k_{92} = \quad \cdot \quad \cdot \quad = k_{96} = 0.0010$

Model, see Fig. 7. (For $\beta h\nu = 3.0000$, see Fig. 9)

$k_{81} = k_{82} = \quad \cdot \quad \cdot \quad = k_{86} = 0.0007$
$k_{71} = k_{72} = \quad \cdot \quad \cdot \quad = k_{76} = 0.0001$
$k_{10,1} = k_{10,2} = \quad \cdot \quad \cdot \quad = k_{10,6} = 0.0369$
$k_{91} = k_{92} = \quad \cdot \quad \cdot \quad = k_{96} = 0.0074$

$\beta h\nu = 3.0000$. Except for the following changes, transition probabilities for $\beta h\nu = 3.0000$ are the same as those for $\beta h\nu = 2.0000$.

k_{65}	k_{54}	k_{43}	k_{32}	$k_{12} = k_{87} = k_{10,9}$
1.0000	0.8034	0.6026	0.4017	0.2009

Table VII. Transition Probabilities for the Evenly Spaced Energy Levels Models, $\beta h\nu = 5.0000$ and $\beta h\nu = 6.0000$

$\beta h\nu = 5.0000$

$k_{12}=k_{78}=k_{9,10}$	k_{23}	k_{34}	k_{45}	k_{56}	k_{65}	k_{54}	k_{43}	k_{32}	$k_{21}=k_{87}=k_{10,9}$
0.0010	0.0020	0.0030	0.0040	0.0050	0.07420	0.5936	0.4452	0.2968	0.1484

$\begin{array}{c}k_{1,10}=k_{19}\\=k_{18}=k_{17}\end{array}$	$\begin{array}{c}k_{28}=k_{27}\\=k_{2,10}=k_{29}\end{array}$	$\begin{array}{c}k_{38}=k_{37}\\=k_{3,10}=k_{39}\end{array}$	$\begin{array}{c}k_{48}=k_{47}\\=k_{4,10}=k_{49}\end{array}$	$\begin{array}{c}k_{58}=k_{57}\\=k_{5,10}=k_{59}\end{array}$	$\begin{array}{c}k_{68}=k_{67}\\=k_{6,10}=k_{69}\end{array}$
0.0001	0.0005	0.0010	0.0050	0.0100	0.0500

Model, see Figs. 10 and 13

$k_{81} = k_{82} = k_{83} = k_{84} = k_{85} = k_{86} = 0.0002$
$k_{71} = k_{72} = \cdot \quad \cdot = k_{76} = 0.0000$
$k_{10,1} = k_{10,2} = \cdot \quad \cdot = k_{10,6} = 0.0005$
$k_{91} = k_{92} = \cdot \quad \cdot = k_{96} = 0.0001$

Model, see Figs. 11 and 14

$k_{81} = \cdot \quad \cdot = k_{86} = 0.0002$
$k_{71} = \cdot \quad \cdot = k_{76} = 0$
$k_{10,1} = \cdot \quad \cdot = k_{10,5} = 0.0014$
$k_{81} = \cdot \quad \cdot = k_{96} = 0.0003$

Model, see Figs. 12 and 15

$k_{81} = \cdot \quad \cdot = k_{86} = 0.0014$
$k_{71} = \cdot \quad \cdot = k_{76} = 0.0003$
$k_{10,1} = \cdot \quad \cdot = k_{10,6} = 0.0005$
$k_{91} = \cdot \quad \cdot = k_{96} = 0.0001$

$\beta h\nu = 6.0000$. Except for the following changes, transition probabilities for $\beta h\nu = 6.0000$ are the same as those for $\beta h\nu = 5.0000$.

$k_{12}=k_{78}=k_{9,10}$	k_{23}	k_{34}	k_{45}	k_{56}
0.0001	0.0002	0.0003	0.0004	0.0005

k_{65}	k_{54}	k_{43}	k_{32}	$k_{21}=k_{87}=k_{10,9}$
0.2015	0.1612	0.1209	0.0806	0.0403

SUMMARY

The results of the present 10-level model studies have shown that non-equilibrium rates are in general quite different from rates calculated based on constant equilibrium distributions of reactant molecules. In particular, for exothermic reactions, the nonequilibrium rates will decrease to a value of zero. It is possible that the nonequilibrium rates may approach the equilibrium rate within a reasonable amount of time, but only for some slightly endo-thermic cases or for cases where both reactants and products are almost in the same energy state and consist of the same number of energy levels such as the previous studies of the four-level model.[1] Further studies of this treatment for a larger number of energy levels and for complicated but realistic systems would be of interest.

ACKNOWLEDGMENTS

This work was supported by a grant from the Robert A. Welch Foundation. We wish to thank Mr. Wayne Hathaway and Mr. Charles O. Reed, Jr. for computational assistance.

REFERENCES

1. B. J. Zwolinski and H. Eyring, *J. Am. Chem. Soc.*, **69**, 2702 (1947).
2. E. W. Montroll and K. E. Shuler, *Advan. Chem. Phys.*, **1**, 361 (1958); S. K. Kim, *J. Chem. Phys.*, **28**, 1057 (1958); F. H. Ree, T. S. Ree, T. Ree, and H. Eyring, *Advan. Chem. Phys.*, **4**, 1 (1962); B. Widom, *ibid.*, **5**, 353 (1963).
3. F. A. Matsen and J. L. Franklin, *J. Amer. Chem. Soc.*, **72**, 3337 (1950); E. E. Nikitin, *Dokl. Akad. Nauk SSSR.*, **119**, 526 (1958); E. E. Nikitin, in *Theory of Thermally Induced Gas Phase Reactions*, E. W. Schlag, Ed., Indiana University Press, Bloomington, 1966; E. V. Stupochenko and A. I. Osipov, *Zhur. Fiz. Khim.*, **32**, 1673 (1958).
4. W. G. Valance and E. W. Schlag, *J. Chem. Phys.*, **45**, 216 (1966).
5. S. Arrhenius, *Z. Physik. Chem. (Leipzig)*, **4**, 226 (1889).
6. N. F. Mott and H. S. W. Massey, *The Theory of Atomic Collisions*, 3rd ed., Oxford, New York, 1965.
7. N. W. Bazley, E. W. Montroll, R. J. Rubin, and K. E. Shuler, *J. Chem. Phys.*, **28**, 700 (1958).

The Current Status of Eyring's Rate Theory

KEITH J. LAIDLER AND ALAN TWEEDALE

Department of Chemistry, University of Ottawa, Canada

Abstract

Henry Eyring's theory of reaction rates, formulated in 1935, focuses attention on the activated complexes, which are assumed to be in equilibrium with the reactants. Much recent work has shed light on the validity of this hypothesis. Provided that the reaction is not too fast, the evidence suggests that activated complexes are in equilibrium with reactants to a good approximation. The present authors have recently carried out dynamical calculations on the $H + H_2$ system in which account is taken of the zero-point energies of vibration throughout the course of reaction; the zero-point vibrational energy of the reactants is required to pass into the zero-point symmetric vibrational energy of the activated complex, the excess energy being utilized in crossing the barrier. These calculations lend some support to the view that activated-complex theory applies satisfactorily to this reaction.

INTRODUCTION

One of the most important contributions to the theory of reaction rates was Henry Eyring's formulation,[1] in 1935, of what is perhaps best referred to as *activated-complex theory*.* The essence of the theory is that it focuses attention on the activated complexes, which are the species present at the col or saddle point in the potential-energy surface. These activated complexes are taken to be in a special type of equilibrium with the reactants. By ignoring the manner in which the activated complexes are formed, and by dealing only with the concentration of complexes and with the speed with which they cross the barrier, Eyring's theory introduced an enormous simplification into the problem of calculating reaction rates. If a good experimental value is available for the activation energy, and a reliable guess can be made as to the structure of the activated complex, a rate-constant calculation can be made in an hour or two using a simple calculating machine. When attempts are made to improve the treatment, for example by making dynamical calculations, vastly more computational effort is required, and at the present time it is doubtful whether the final value will be more reliable.

THE EQUILIBRIUM HYPOTHESIS

It is important to appreciate exactly what is involved in the equilibrium hypothesis. Suppose that a reaction has proceeded completely to equilibrium; the activated complexes will then, like all other species, be in equilibrium with the reactants and products. Figure 1 shows a schematic energy diagram for a system in equilibrium, with half the activated complexes crossing the barrier in one direction and half in the other. A Maxwell demon observing the reaction would be able, if he had at least a short memory, to distinguish the two classes of activated complex: those that were reactants in the immediate past (i.e., those which in our diagram are therefore crossing the

* It is also commonly known as "absolute-rate" theory, "the theory of absolute reaction rates," and "transition-state theory." Our preference is for "activated-complex theory."

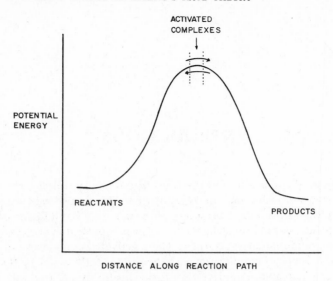

Fig. 1. A schematic energy diagram for a reaction, showing the flow of activated complexes in the two directions.

barrier from left to right), and those that were products in the immediate past (crossing the barrier from right to left). At equilibrium there must be, on the average, equal numbers of the two, crossing at the same rates.

If the product molecules are suddenly removed from the reaction system the flow from right to left in Fig. 1 will cease. The flow from left to right, however, will continue, and it is the essence of the equilibrium hypothesis that this flow will continue exactly as before. In other words, the hypothesis is that the concentration of complexes moving from left to right (i.e., those that were reactant molecules in the immediate past) is the same as if the product molecules were present at their equilibrium concentrations.

Looked at in this way, the equilibrium hypothesis becomes extremely plausible, because we expect the flows of complexes in the two directions to be completely independent of each other. The equilibrium hypothesis does not imply a classical type of equilibrium: addition of activated complexes moving from the initial to the final state would not disturb the equilibrium, as it would if the equilibrium were classical. The activated complexes are transient species, passing from the initial to the final state and unable to turn back. They are at equilibrium with the reactants not because they perform a number of vibrations and come to equilibrium, but because they are *created* in a state of equilibrium.

Support for the view that the concentrations of activated complexes moving from left to right may validly be calculated using equilibrium theory is provided by various lines of evidence:

1. The equilibrium hypothesis is certainly correct when the whole system is at equilibrium, so that rates calculated on its basis should be valid at equilibrium. If the hypothesis were in error before a reaction has come to equilibrium, rate constants would be expected to change as equilibrium is approached. There is no evidence that this ever happens.

2. A number of "stochastic" treatments of chemical reactions have been given;[2] these are based on the principles of probability and involve the concept of the random walk. These are rather formal treatments which have not yet been applied to particular reactions, but they do have the virtue of allowing estimates to be made of the kind of error that might be introduced by the assumption of equilibrium. Broadly speaking, the conclusion to which these treatments lead is that reactions are satisfactorily interpreted (rates within a few percent) on the equilibrium assumption provided that E/RT has a value of 5 or larger (E is the activation energy, R the gas constant, T the temperature in degrees Kelvin). The significance of this is that, if E/RT is too small, the reaction will occur so rapidly that there will be a depletion of the more energized reactant species and therefore of the activated complexes. For reactions under ordinary conditions E/RT is well above 5, so that the error in assuming equilibrium is extremely small.

3. During the last few years a number of calculations have been made of the dynamics of motion of systems over potential-energy surfaces, and these have provided some support for the activated-complex theory. The remainder of this paper is devoted to a consideration of these dynamical calculations.

MOLECULAR DYNAMICS

In the early 1930's, Eyring and his co-workers made some preliminary studies[3] of the trajectories of systems on potential-energy surfaces, but not much progress could be made until the development of high-speed computers. There has recently been a revival of interest in this field, now known as *molecular dynamics*. In particular, Karplus, Porter, and Sharma[4] have carried out calculations on the $H + H_2$ system, using what appears to be a very reliable potential-energy surface. The calculations are "quasi-classical" in nature; the vibrational and rotational states in the H_2 molecule are quantized, but the course of the collision is treated classically.

Several important conclusions have emerged from these calculations. One is that on the $H + H_2$ surface the collisions that lead to reaction are of very short duration—of the order of 10^{-14} sec. This, of course, is in line with Eyring's assumption about the activated state, through which the system passes smoothly. The calculations also show that a system that has reached the activated state rarely turns back and forms reactants again. Thus, the activated complex is not a species that undergoes a number of vibrations, thereby coming to equilibrium, and sometimes breaking down to give the reactants again. Rather, for a surface like that for $H + H_2$, the system passes straight through the activated state, which is a "point of no return," and the

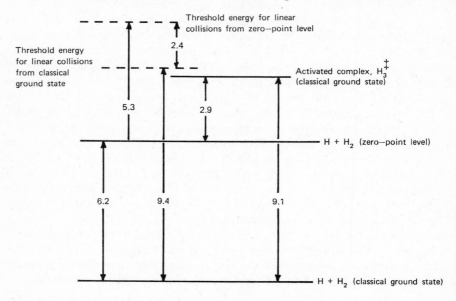

Fig. 2.   Energy diagram for the $H + H_2$ system, showing energies for linear collisions (classical calculations of Karplus et al.).

reaction is said to be *direct*. Calculations on special types of surfaces, having a basin, do indeed show that complexes may then undergo several vibrations; this, however, is probably an unusual situation, and it is better not to speak of activated complexes in such cases. Eyring's theory in its simplest form would not apply to such reactions.

Another interesting conclusion from the calculations of Karplus et al. relates to the energetics of collisions. For the potential-energy surface used, the barrier height for the linear $H \cdots H \cdots H^{\ddagger}$ system is 9.1 kcal with respect to the classical ground state of $H + H_2$, as shown in Fig. 2. The trajectory calculations showed, however, that for a linear collision between H and H_2, at least 9.4 kcal of relative translational energy is required for the system to cross the barrier; this is an energy of 0.3 kcal in excess of the barrier height. This small difference is due to the curvature of the reaction path, a little of the energy being forced into the symmetric vibrational mode of the H $\cdots$ H $\cdots$ H complex and not contributing to crossing the barrier.

The zero-point vibrational energy of the initial state is 6.2 kcal, so that the minimum energy required to cross the barrier from the zero-point level is 9.1–6.2 = 2.9 kcal. Karplus et al. made classical trajectory calculations for linear collisions in which the H_2 molecule is at its zero-point vibrational level and found that the threshold energy is now 5.3 kcal. In other words, although

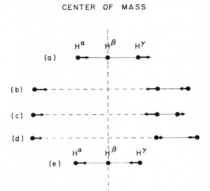

Fig. 3. A linear collision between H and H_2, showing the H_2 molecule in different vibrational phases (*b*), (*c*), and (*d*). The purely antisymmetrical vibration of the activated complex is shown in (*a*), the purely symmetric vibration in (*c*).

there is an additional 6.2 kcal of energy available, the barrier height relative to the zero-point level being only 2.9 kcal, the relative translational energy required is 5.3 kcal; that is, of the additional 6.2 kcal of vibrational energy only about 3.8 kcal can contribute to reaction. The rest, about 2.4 kcal, is forced into the symmetric stretching vibration of the activated complex. Some idea of why this is so is shown in Fig. 3. An activated complex in a purely antisymmetrical mode, as shown in (a), cannot arise from a collision between H and H_2, three possible phases for which are shown in (b) to (d). Atom H^γ will always have less motion to the right than shown in (a), which means that there is a contribution from the symmetric vibration (e). There is thus some adiabaticity in the vibrational energy, which cannot be freely converted into energy required to cross the barrier; some of it must remain as vibrational energy.

The essential feature of this type of calculation is that no consideration is given to the zero-point level in the activated state. The zero-point energy corresponding to the symmetrical stretching vibration of the H_3^+ complex is 3.1 kcal/mole, but the classical calculations permit systems to cross at a lower level than this (the threshold energy is 2.4 kcal), and therefore lead to somewhat higher values for the calculated rate constants than are given by activated-complex theory; the discrepancy is a factor of about 7 at 300°K and of about 1.25 at 1000°K.

The lifetime of an activated state is too short for one to be certain that there is full quantization throughout the trajectory, except where the reactants are well separated. However, complete neglect of quantization in the activated state does not satisfy the requirement that there must be full quantization in the products of reaction. The calculations of Karplus et al. do indeed lead to the result that the product hydrogen molecule has in general a different vibrational amplitude than corresponds to any quantum level. This is obviously a weakness of the treatment, and there is no reason to believe that the resulting rates are any more reliable than those calculated from activated-complex theory.

The most satisfactory way out of the difficulty is to make purely quantum mechanical calculations. However, the computational difficulties are very great, and it has always been necessary to introduce simplifications that lead to some doubt about the reliability of the results. Work of this kind has been done in particular by Child[5] and by Karplus and Tong,[6] and has again led to rates that are not inconsistent with activated-complex theory.

An alternative and simpler procedure has been explored by Tweedale and Laidler.[7] Their calculations were essentially classical, like those of Karplus et al.,[4] but they imposed the additional condition that at all stages along the reaction path the intermediate complex has its zero-point energy of vibration normal to the path. In other words, there is assumed to be full quantization at all stages of reaction, rather than no quantization in the intermediate stages. As previously indicated, there may only be partial quantization during reaction, so that the truth lies between the two extremes. The assumption of full quantization has the virtue of ensuring that the product molecules are in quantized states; some kind of partial quantization might ensure the same result, however.

A VIBRATIONALLY ADIABATIC MODEL FOR THE H + H₂ REACTION

The procedure of Tweedale and Laidler essentially involves imposing a "floor" on the potential-energy surface, corresponding to the zero-point energy at each point along the path. A rather abstract treatment of the problem has been given by Marcus.[8] Tweedale and Laidler made calculations based on the semiempirical potential-energy surface of Porter and Karplus.[9] The minimal reaction path was determined, and the vibrational frequencies normal to this path were calculated for a number of points along the path, including the col. From these frequencies the zero-point energies were calculated at each point. Figure 4 shows the energy profiles calculated in this way; curve P is the ordinary potential-energy profile for the Porter-Karplus surface, V shows the zero-point vibrational energies, and $P + V$ shows the resultant profile.

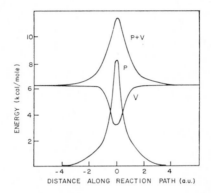

Fig. 4. Energy profiles for the $H + H_2$ system; curve P is the classical potential-energy profile; V shows the variation in the zero-point vibrational energy; $P + V$ is the resultant of the two.

One important point that emerges from these calculations is that the $P + V$ curve shows only a single maximum. It is quite possible, in principle, for a $P + V$ profile to have a double maximum, as shown schematically in Fig. 5. This would be the case if the shape of the surface were such that the V curve dipped rather sharply in the neighborhood of the activated state, rather than gradually as in Fig. 4. If the situation had been as in Fig. 5, an appreciable amount of energy would remain tied up as zero-point vibrational energy and would not be available for helping the system over the barrier. If this were the case, the true energy barrier would not lie at the col but would be at one side of it, and the activation energy would necessarily be higher than the height of the $P + V$ barrier at the col. In fact, however, for the $H + H_2$ reaction (and this is also true for $D + D_2$, $D + H_2$, and $H + D_2$),

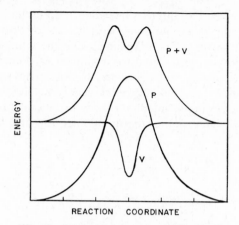

Fig. 5. Schematic energy profiles showing how $P + V$ can involve a double maximum.

there is a single maximum; the zero-point vibrational energy is therefore all available for helping the system over the barrier, and the threshold energy should therefore be the height of the $P + V$ barrier at the col.

This conclusion was confirmed by the trajectory calculations, which have been made so far only for linear activated complexes for the four reactions mentioned above. In the calculations of Karplus et al. the trajectories are determined by classical-dynamical considerations; in the Tweedale-Laidler calculations, on the other hand, the trajectories result from the condition that at all points along the reaction path the vibrational energy normal to the path must always be the zero-point energy; the rest of the energy is available for crossing the barrier. For high impact energies some additional energy will be required to cross the barrier, arising from the centrifugal effect, but this additional energy disappears as the energy of impact is reduced, and the threshold energy becomes exactly equal to the $P + V$ barrier height at the col. Table I shows these threshold energies for the four reactions.

Table I.   Threshold Energies

Reaction	Zero-Point Energies		Threshold Energies
	Reactants	Activated Complex	
$H + H_2 \rightarrow H_2 + H$	6.25	3.07	5.40
$D + D_2 \rightarrow D_2 + D$	4.42	2.17	6.33
$D + H_2 \rightarrow DH + H$	6.25	2.62	4.95
$H + D_2 \rightarrow HD + D$	4.42	2.65	6.81

Classical barrier height = 8.58 kcal/mole

The model on which these calculations are based can be referred to as *vibrationally adiabatic*, the zero-point vibrational energy of the reactant H_2 molecule passing smoothly into zero-point vibrational energy of all the intermediate complexes, including the activated complex. For moderate temperatures most reactant H_2 molecules will be in their ground vibrational states, and collisions in which this energy will pass nonadiabatically into other vibrational levels of the intermediate states will be improbable. The adiabatic assumption therefore seems very reasonable for the reactions under consideration, provided that the temperature is not too high. There is little evidence available as to how valid the adiabatic assumption would be for reactions in which higher vibrational states play an important role.

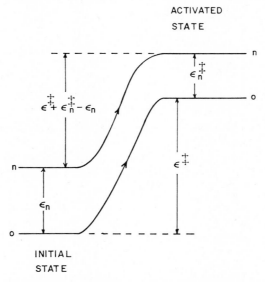

Fig. 6. Energy diagram showing two vibrational states for the reactants and activated complexes.

The results of the present calculations that the zero-point vibrational energy of the reactants can pass smoothly into that of the intermediate complexes is entirely consistent with the basic postulate of Eyring's theory that activated complexes are created from the reactants in equilibrium states. It is easy to show that the vibrationally adiabatic model, coupled with the assumption that collision cross sections are the same for all vibrational levels, leads to the conclusion that there is a Boltzmann distribution between the vibrational levels in the activated state. Thus, consider the situation represented by the energy diagram shown in Fig. 6; two levels are shown for the initial and activated states—the ground level and the nth vibrationally excited level—and it is assumed that the transitions occur adiabatically as represented by the arrows. The rate of formation of activated complexes in the ground vibrational state is then

$$v_0 = A[R_0] \exp\left(-\frac{\epsilon^{\ddagger}}{kT}\right) \tag{1}$$

where A is the preexponential factor (which depends on the reaction cross section) and $[R_0]$ is the concentration of reactant species in their ground vibrational states. Similarly, for the transition between the nth vibrational states,

$$v_n = A[R_n] \exp\left(-\frac{\epsilon^{\ddagger} + \epsilon_n^{\ddagger} - \epsilon_n}{kT}\right) \tag{2}$$

However, $[R_n]$ and $[R_0]$ are related by

$$\frac{[R_n]}{[R_0]} = \exp\left(-\frac{\epsilon_n}{kT}\right) \tag{3}$$

so that

$$v_n = A[R_0] \exp\left(-\frac{\epsilon^{\ddagger} + \epsilon_n^{\ddagger}}{kT}\right) \tag{4}$$

These rates of formation of activated states are equal to the rates with which they cross the barrier, namely,

$$\nu[X_0^{\ddagger}] = A[R_0] \exp\left(-\frac{\epsilon^{\ddagger}}{kT}\right) \tag{5}$$

$$\nu[X_n^{\ddagger}] = A[R_0] \exp\left(-\frac{\epsilon^{\ddagger} + \epsilon_n^{\ddagger}}{kT}\right) \tag{6}$$

where ν is the frequency with which systems cross. From (5) and (6) it follows that

$$\frac{[X_n^{\ddagger}]}{[X_0^{\ddagger}]} = \exp\left(-\frac{\epsilon_n^{\ddagger}}{kT}\right), \tag{7}$$

which means that there is an equilibrium distribution among the vibrational levels of the activated complex.

CONCLUSIONS

It is a striking fact that the more elaborate and detailed treatments of chemical reaction rates seem to be converging on conclusions that are the same as those given by activated-complex theory. It is a remarkable, and probably almost unique, feature of this theory that by trimming away details it has been able to yield such very satisfactory results. The more detailed treatments are, of course, of great value in providing information not given by activated-complex theory—such as information about energy distributions in product molecules, of importance in studies of molecular beams and chemiluminescence. It does not appear likely, however, that the more detailed treatments will show themselves to be much superior to activated-complex theory in giving reliable values for the rate constants of reactions or in providing insight into how chemical reactions occur.

However, a final word of caution should be added. The above conclusions have been drawn for $H + H_2$ and similar reactions for which the vibrational-energy separations are large. Deviations from the predictions of activated complex theory are more likely for very low energy barriers, or when the energy barrier is large in comparison with the vibrational quanta and is situated in the product valley. There is indeed theoretical evidence,[10] and evidence from molecular beam and chemiluminescence studies,[11] that such is the case for a number of reactions.

ACKNOWLEDGMENTS

We are indebted to Professor M. Karplus for valuable comments. One of us (K. J. L.) wishes to express his deep appreciation to Henry Eyring for his wise and inspiring teaching and constant encouragement. Eyring's influence has been enormous, not only on the whole field of theoretical chemistry, but also on the lives of those who have been fortunate enough to come into contact with him.

REFERENCES

1. H. Eyring, *J. Chem. Phys.*, **3**, 107 (1935); cf. also W. F. K. Wynne-Jones and H. Eyring, *ibid.*, **3**, 492 (1935); for general reviews see S. Glasstone, K. J. Laidler, and H. Eyring, *The Theory of Rate Processes*, McGraw-Hill, New York, 1941; K. J. Laidler, *Theories of Chemical Reaction Rates*, McGraw-Hill, New York, 1969.
2. H. A. Kramers, *Physika*, **7**, 284 (1940); B. Zwolinski and H. Eyring, *J. Amer. Chem. Soc.*, **69**, 2702 (1947); E. W. Montroll and K. E. Shuler, *Advan. Chem. Phys.*, **1**, 361 (1958); for a general review of stochastic theories see D. A. McQuarrie, *Stochastic Approach to Chemical Kinetics*, Methuen, London, 1967.
3. H. Eyring and M. Polanyi, *Z. Physik. Chem. (Leipzig)*, **B12**, 279 (1931); J. O. Hirschfelder, H. Eyring, and B. Topley, *J. Chem. Phys.*, **4**, 170 (1936).
4. M. Karplus, R. N. Porter, and R. D. Sharma, *J. Chem. Phys.*, **43**, 3259 (1965).
5. M. S. Child, *Proc. Roy. Soc. (London)*, **A292**, 272 (1966); *Disc. Faraday Soc.*, **44**, 68 (1967); *Mol. Phys.*, **12**, 401 (1967).
6. M. Karplus and K. T. Tang, *Disc. Faraday Soc.*, **44**, 56 (1967).
7. A. Tweedale and K. J. Laidler, *J. Chem. Phys.*, **53**, 2045 (1970).
8. R. A. Marcus, *J. Chem. Phys.*, **45**, 4500 (1966).
9. R. N. Porter and M. Karplus, *J. Chem. Phys.*, **44**, 1105 (1964).
10. J. C. Polanyi and W. H. Wong, *J. Chem. phys.*, **51**, 1439 (1969).
11. Cf. K. J. Laidler, *op. cit.*, ref. 1, Chap. 7.

Permeabilities for Reactions of the Type $H + H_2 = H_2 + H$ Treated as a Linear Encounter Using Variational and Distorted Wave Techniques*

EARL M. MORTENSEN†

Department of Chemistry, University of Massachusetts, Amherst, and Department of Chemistry, The Cleveland State University, Cleveland

* This work was supported under National Science Foundation Grant GP 8101.
† Present address Department of Chemistry, The Cleveland State University, Cleveland, Ohio, 44115.

A Personal Note

EARL M. MORTENSEN

One of Henry Eyring's most significant contributions has been the development of ideas and theories for investigating rate processes. This area has made great progress due to his fundamental work in kinetics, in both the construction of potential-energy surfaces and the formulation of absolute rate theory which enabled greater insights into kinetic processes. From this fundamental work Henry was able to apply his ideas to an ever-increasing number of problems in many different areas with considerable success. Beyond these and many other contributions, perhaps the most significant one has been the inspiration and guiding direction behind many colleagues and students who are now extending the work which one man alone could not do. His demands on his students and those under him were large, but this encouragement brought out the best in them. He expected that their full effort be devoted to furthering the work; they responded, and it was accomplished.

ANECDOTE

Because Henry traveled on a number of occasions during the school year, it was necessary to make up some missed class work in his statistical mechanics course. Several two-hour sessions were scheduled, and one of them was to be on Washington's Birthday, which was a university holiday. Perhaps to justify this action, he said that we could better honor George Washington by being at school studying and working than to take the day off. As a result, we honored George Washington that year as we devoted ourselves to higher things.

128

INTRODUCTION

Recently Mortensen and Gucwa[1] applied a modification of the Kohn[2,3] variational procedure to the reactive collision $H + H_2 = H_2 + H$ treated as a linear encounter. Their results were compared with the earlier calculations of Mortensen and Pitzer[4] and were found to be in good agreement. Because of the symmetry associated with this reaction at energies where no vibrationally excited states can occur, the problem can be completely formulated in terms of the phase of the waves associated with the relative translational motion, whereas the amplitudes can be determined by symmetry. Such a problem is ideally suited to variational techniques that correct the phases to first order. For symmetrical reactions such as that indicated above where excited vibrational states can be populated, and for unsymmetrical reactions, symmetry can no longer determine the amplitudes, and additional procedures such as a distorted wave technique must be used to correct these amplitudes. Using these procedures the reactions

$$H + H_2 = H_2 + H \tag{1}$$

$$D + D_2 = D_2 + D \tag{2}$$

$$D + H_2 = DH + H \tag{3}$$

were investigated and compared with the recent results of Mortensen.[5] The Sato[6] potential-energy surface was used as described by Weston[7] in order that the results reported here could be compared with those given in Ref. 5.

Although the linear collision cannot be expected to give a quantitative treatment of the above reactions, qualitative inferences can be made and the techniques that are used should be able to be extended to treat the more general collision problem associated with these reactions.

THEORY

The Hamiltonian[4] for the linear collision in reduced units may be written

$$\mathbf{H} = -\frac{\mu}{\mu_{12}}\frac{\partial^2}{\partial R_{12}^{\ 2}} + \frac{2\mu}{m_2}\frac{\partial_2}{\partial R_{12}\,\partial R_{23}} - \frac{\mu}{\mu_{23}}\frac{\partial^2}{\partial R_{23}^{\ 2}}$$
$$+ V_R(R_{12}, R_{23}) + E_\theta(R_{12}, R_{23}) - \frac{3\mu}{I} - \frac{2\mu}{m_2 R_{12} R_{23}} - \mathbf{K}', \qquad (4)$$

where R_{ij} is the distance between atoms i and j, m_i is the mass of atom i with atom 2 being the central one, V_R is the potential-energy surface for the linear motion, E_θ is an effective potential energy for bending motions of the complex, I is the moment of inertia associated with the three-atom system, $\mathbf{K}'$ is an operator containing coupling terms between the linear and bending motions which vanish in the reactant and product channels at infinite separation, and $\mu_{ij} = m_i m_j/(m_i + m_j)$. All distances in (4) are in units of the zero-point amplitude of a harmonic oscillator of frequency v and reduced mass μ, and the energies are expressed in terms of the zero-point energy of this oscillator. For the calculations reported here the frequency and reduced mass of the reference oscillator were taken to have the value of the hydrogen molecule. An alternate and often more convenient form of the above Hamiltonian may be written

$$\mathbf{H} = -\frac{\mu}{\mu_{ts}}\frac{\partial^2}{\partial R_{ts}^{\ 2}} - \frac{\mu}{\mu_{vs}}\frac{\partial^2}{\partial R_{vs}^{\ 2}} + V + \mathbf{K}', \qquad (5)$$

where $s = 1$ or 3 with

$$R_{t1} = R_{12} + \frac{m_3 R_{23}}{m_2 + m_3}, \qquad R_{v1} = R_{23}, \qquad (6a)$$

$$R_{t3} = R_{23} + \frac{m_1 R_{12}}{m_1 + m_2}, \qquad R_{v3} = R_{12}, \qquad (6b)$$

$$\mu_{t1} = \frac{m_1(m_2 + m_3)}{m_1 + m_2 + m_3}, \qquad \mu_{v1} = \mu_{23}, \qquad (7a)$$

$$\mu_{t3} = \frac{m_3(m_1 + m_2)}{m_1 + m_2 + m_3}, \qquad \mu_{v3} = \mu_{12}, \qquad (7b)$$

$$V = V_R + E_\theta - \frac{3\mu}{I} - \frac{2\mu}{m_2 R_{12} R_{23}} \qquad (8)$$

In (5) and subsequent equations, s indicates the channel or formulation relating to that channel, with the value of s referring to the atom which is separated from the diatomic molecule.

Since $V(R_{12}, R_{23}) \to V(R_{vs})$ as $R_{ts} \to \infty$ so that the Schrödinger equation separates, the asymptotic solutions may be written

$$\psi_{fi} \sim \phi_{v1i}(R_{v1}) \exp(-ik_{1i}R_{t1}) + \sum_j R'_{fij}\phi_{v1j}(R_{v1})$$

$$\times \exp(ik_{1j}R_{t1}), \qquad R_{t1} \to \infty \quad (9a)$$

$$\sim \sum_j T'_{fij}\phi_{v3j}(R_{v3}) \exp(ik_{3j}R_{t3}), \qquad R_{t3} \to \infty \quad (9b)$$

$$\psi_{bi} \sim \phi_{v3i}(R_{v3}) \exp(-ik_{3i}R_{t3}) + \sum_j R'_{bij}\phi_{v3j}(R_{v3})$$

$$\times \exp(ik_{3j}R_{t3}), \qquad R_{t3} \to \infty \quad (9c)$$

$$\sim \sum_j T'_{bij}\phi_{v1j}(R_{v1}) \exp(ik_{1j}R_{t1}), \qquad R_{t1} \to \infty \quad (9d)$$

where i refers to the internal state of the incoming molecule and j to the outgoing molecule, and the propagation vector or wavenumber k_{sj} is given by

$$k_{sj} = \left[\frac{\mu_{ts}(E - E_{vsj})}{\mu} \right]^{1/2}, \qquad (10)$$

where E is the total energy of the system and E_{vsj} is the vibrational energy of the diatomic molecule in vibrational state j. The phase angles associated with the reflected and transmitted waves are given by

$$R'_{\sigma ij} = R_{\sigma ij} \exp(i\epsilon_{r\sigma ij}), \qquad \sigma = f, b, \qquad (11a)$$

$$T'_{\sigma ij} = T_{\sigma ij} \exp(i\epsilon_{t\sigma ij}), \qquad \sigma = f, b, \qquad (11b)$$

where the phase angles are chosen such that $R_{\sigma ij}$ and $T_{\sigma ij}$ are real. The permeability of the reaction is given by[8]

$$G_{fij} = \frac{T_{fij}^2 v_{3j}}{v_{1i}}, \qquad G_{bij} = \frac{T_{bij}^2 v_{1j}}{v_{3i}} \qquad (12)$$

where v_{sk} is the magnitude of the relative velocity between atom s and the diatomic molecule in vibrational state k.

A set of real wavefunctions χ_i will be constructed such that upon taking appropriate linear combinations, the complex ψ_{fi} and ψ_{bi} having asymptotic forms given by (9a)–(9d) can be obtained. Following Kohn,[2,3] real wavefunctions may be constructed having the asymptotic form

$$\chi_i \sim \sum_j B_{sij}\phi_{vsj}(R_{vs})[\sin k_{sj}R_{ts} + \tan \epsilon_{sij} \cos k_{sj}R_{ts}], \qquad R_{ts} \to \infty \quad (13)$$

where

$$B_{sij} = A_{sij} \cos \epsilon_{sij} \qquad (14)$$

so that (13) may be written $\chi_i \sim \sum_j A_{sij}\phi_{vsj}(R_{vs}) \sin(k_{sj}R_{ts} + \epsilon_{sij})$ with A_{sij} being the amplitude and ϵ_{sij} the phase of the wave in channel s with the incoming molecule in vibrational state i and the outgoing molecule in vibrational state j. There are an infinite number of solutions of the above form to the Schrödinger equation; however, for a specific solution, the B_{sij} are taken to be constant and not subject to variation, such that each wavefunction χ_i is linearly independent of the others.

In Ref. 1 the variational functions $\chi_i = \chi_i(c_{ik}, \tan \epsilon_{sik})$ were constructed where the c_{ik} are variational parameters independent of the asymptotic regions. It was shown that the integral J_i, which is given by

$$J_i = I_i - \frac{\mu}{\mu_{t1}} \sum_j k_{1i} B_{1ij}^2 \tan \epsilon_{1ij} - \frac{\mu}{\mu_{t3}} \sum_j k_{3j} B_{3ij}^2 \tan \epsilon_{3ij}, \qquad (15)$$

is made invariant to the parameters c_{ik} and $\tan \epsilon_{sij}$, i.e.,

$$\delta J_i = 0, \qquad (16)$$

where

$$I_i = \langle \chi_i | \mathbf{H} - E | \chi_i \rangle, \qquad (17)$$

and that the phases may be corrected by the equation

$$\frac{\mu}{\mu_{t1}} \sum_j k_{1j} B_{1ij}^2 \tan \overset{0}{\epsilon}_{1ij} + \frac{\mu}{\mu_{t3}} \sum_j k_{3j} B_{3ij}^2 \tan \overset{0}{\epsilon}_{3ij}$$

$$= \frac{\mu}{\mu_{t1}} \sum_j k_{1j} B_{1ij}^2 \tan \epsilon_{1ij} + \frac{\mu}{\mu_{t3}} \sum_j k_{3j} B_{3ij}^2 \tan \epsilon_{3ij} - I_i, \qquad (18)$$

where the ϵ_{sij}^0 represent the corrected phases and the ϵ_{sij} the phases determined by solving (16).

In Ref. 1, reaction (1) was investigated using the above procedure where bending corrections in the collision region were not considered (i.e., V was taken to be equal to V_R in (8)) and the range of energies studied was such that only the ground vibrational state was populated. With these restrictions the χ_i can be constructed which are symmetric or antisymmetric with respect to interchange of R_{ij} so that the amplitudes of these functions are specified by symmetry alone and the phases can be corrected using (18). Where excited vibrational states can be populated, or for unsymmetrical reactions such as reaction (3), it is also necessary that the amplitudes be corrected. Such a correction may be applied using a distorted wave treatment.[9-12]

Let $\mathbf{H}_T$ be the kinetic energy portion of the Hamiltonian describing the collision process. If V is the potential-energy surface for the interaction, then the Schrödinger equation may be written $(\mathbf{H}_T + V)\psi_i = E\psi_i$, where ψ_i is the wavefunction associated with the collision having an asymptotic form given by (9a)–(9d). For a different surface V' the Schrödinger equation may be written $(\mathbf{H}_T + V')\psi'_{i'} = E\psi'_{i'}$. If the first Schrödinger equation is multiplied by $\psi_{i}^{'*}$, the complex conjugate of the second multiplied by ψ_i, and the two resulting equations subtracted and integrated over the range of the variables, the equation

$$\langle \psi'_{i'} \mid V - V' \mid \psi_i \rangle = \lim_{R_{t1} \to \infty} \frac{\mu}{\mu_{t1}} \int \left[\psi_{i'}^{'*} \frac{\partial \psi_i}{\partial R_{t1}} - \psi_i \frac{\partial \psi_{i'}^{'*}}{\partial R_{t1}} \right] dR_{v1}$$

$$+ \lim_{R_{t3} \to \infty} \frac{\mu}{\mu_{t3}} \int \left[\psi_{i'}^{'*} \frac{\partial \psi_i}{R_{t3}} - \psi_i \frac{\partial \psi_{i'}^{'*}}{\partial R_{t3}} \right] dR_{v3}, \quad (19)$$

is obtained after integrating by parts and requiring that the wavefunction vanish at $R_{ts} = 0$. Since R_{t1} or R_{t3} can become infinite by having either R_{12} or R_{23} become infinite, $R_{t1} \to \infty$ or $R_{t3} \to \infty$ will refer to channel 1 or 3 where $R_{12} \to \infty$ or $R_{23} \to \infty$, respectively. In order to evaluate the above integrals it will be necessary to specify the asymptotic forms of ψ_i and $\psi'_{i'}$. Now, ψ_i will be taken to be the wavefunction describing the forward reaction having potential-energy surface V with reactants in state i, and $\psi_{i'}^{'*}$, will be taken to be the reverse reaction having potential-energy surface V' with reactants in state i'. Thus, the equalities $\psi_i = \psi_{fi}$ and $\psi_{i'}^{'*} = \psi_{bi'}$ are made, where ψ_{fi} and ψ_{bi} have the asymptotic forms given by (9a)–(9d). Combining the above equations with (9a)–(9d) yields

$$\langle \psi_{bi'}^* \mid V_f - V_b \mid \psi_{fi} \rangle = 2i \left(\frac{\mu k_{3i'} T'_{fii'}}{\mu_{t3}} - \frac{\mu k_{1i} T'_{bi'i}}{\mu_{t1}} \right) \quad (20)$$

where the replacements $V = V_f$ and $V' = V_b$ have been made. For $V_f = V_b$, the integral on the left vanishes so that

$$\frac{k_{3i'} T'_{fii'}}{\mu_{t3}} = \frac{k_{1i} T'_{bi'i}}{\mu_{t1}} \qquad V_b = V_f. \quad (21)$$

Suppose that ξ_{fi} and $\xi_{bi'}$ are exact wavefunctions to the Schrödinger equation having the Hamiltonian

$$\mathbf{H}_1 = \mathbf{H}_T + U, \quad (22)$$

and have asymptotic solutions of the form of (9a)–(9d) with $R_{fij}^{(1)'}$, $R_{bij}^{(1)'}$, $T_{fij}^{(1)'}$, and $T_{bij}^{(1)'}$ replacing the corresponding constants in these equations. If $V_f = V$ is associated with wavefunction ψ_{fi} and if $V_b = U$ is associated with wavefunction $\xi_{bi'}$, then (20) becomes

$$T'_{fii'} = T_{fii'}^{(1)'} + \frac{\mu_{t3} \langle \xi_{bi'}^* \mid V - U \mid \psi_{fi} \rangle}{2i\mu k_{3i'}} \quad (23a)$$

using (21), and similarly for the reverse reaction

$$T'_{bi'i} = T_{bi'i}^{(1)'} + \frac{\mu_{t1} \langle \xi_{bi'}^* \mid V - U \mid \psi_{fi} \rangle}{2i\mu k_{1i}}. \quad (23b)$$

It should be noted that (23a) and (23b) are exact.

If ξ_{fi} and ψ_{fi} differ only slightly from one another, then the above equation may be approximated by replacing ψ_{fi} in the above two equations by ξ_{fi}. Using this approximation and $(V - U)\xi_{fi} = (V + H_T - E)\xi_{fi} = (H - E)\xi_{fi}$, (23a) and (23b) become

$$T'_{fii'} \simeq T^{(1)'}_{fii'} + \frac{\mu_{t3}\langle\xi^*_{bi'}| H - E |\xi_{fi}\rangle}{2i\mu k_{3i'}}. \tag{24a}$$

$$T'_{bi'i} \simeq T^{(1)'}_{bi'i} + \frac{\mu_{t1}\langle\xi^*_{bi'}| H - E |\xi_{fi}\rangle}{2i\mu k_{1i}} \tag{24b}$$

From (24a) and (24b) it should be noted that the potential-energy surface U need not be determined for a given ξ_{fi} which may be selected, and, as Miller[12] has pointed out, equations of this type are correct to first order in the residual interaction.

The ξ_i used in this work are those wavefunctions obtained by the variational procedures of Ref. 1 and are obtained by taking linear combinations of the real variational functions χ_i. Thus the ξ_i may be written

$$\xi_{fi} = \sum_k (\alpha_{fik} + i\beta_{fik})\chi_k \tag{25a}$$

$$\xi_{bj} = \sum_l (\alpha_{bjl} + i\beta_{bjl})\chi_l \tag{25b}$$

where the summation is over all independent χ_k needed for describing the reaction, and the coefficients α_{fik}, α_{bjl}, β_{fik}, and β_{bjl} are chosen to give the asymptotic form given by (9a) to (9d). Combining (25a) and (25b) with (24a) and (24b) and using (11b) along with the analogous equations for $T^{(1)'}_{fij}$ and $T^{(1)'}_{bij}$, the equations

$$T_{fij} \cos \epsilon_{tfij} = T^{(1)}_{fij} \cos \epsilon^{(1)}_{tfij} + \mu_{t3}\sum_k\sum_l \frac{(\alpha_{fik}\beta_{bjl} + \beta_{fik}\alpha_{bjl})I_{lk}}{2\mu k_{3j}}, \tag{26a}$$

$$T_{fij} \sin \epsilon_{tfij} = T^{(1)}_{fij} \sin \epsilon^{(1)}_{tfij} - \mu_{t3}\sum_k\sum_l \frac{(\alpha_{fik}\alpha_{bjl} - \beta_{fik}\beta_{bjl})I_{lk}}{2\mu k_{3j}}, \tag{26b}$$

and

$$T_{bji} \cos \epsilon_{tbji} = T^{(1)}_{bji} \cos \epsilon^{(1)}_{tbji} + \mu_{t1}\sum_k\sum_l \frac{(\alpha_{fik}\beta_{bjl} + \beta_{fik}\alpha_{bjl})I_{lk}}{2\mu k_{1i}}, \tag{27a}$$

$$T_{bji} \sin \epsilon_{tbji} = T^{(1)}_{bji} \sin \epsilon^{(1)}_{tbji} - \mu_{t1}\sum_k\sum_l \frac{(\alpha_{fik}\alpha_{bjl} - \beta_{fik}\beta_{bjl})I_{lk}}{2\mu k_{1i}}, \tag{27b}$$

are obtained, where

$$I_{lk} = \langle\chi_l| H - E |\chi_k\rangle. \tag{28}$$

From these equations the phases of the transmitted waves can be determined, and using (12) the permeabilities can be obtained. It should be noted by comparing (17) and (28) that $I_{ii} = I_i$, which is evaluated in the variational procedure. For symmetrical reactions such as reactions (1) and (2) where only ground vibrational states are occupied such that symmetrical and anti-symmetrical χ_i can be constructed, then $I_{ij} = 0$ for $i \neq j$ by symmetry, and if $I_i = 0$, the distorted wave treatment does not give additional improvement to the permeabilities.

If in (19) $\psi_i = \psi_{fi}$ and $\psi_i'^* = \psi_{i'} = \psi_{fi'}$ or if $\psi_i = \psi_{bi}$ and $\psi_i'^* = \psi_{i'} = \psi_{bi'}$, then equations analogous to (26) and (27) are obtained for R_{fij} and R_{bij}.

From the equations that relate the phases and amplitudes of the χ_i to those of the real and imaginary forms having asymptotic solutions given by (9a) and (9b) as outlined in Ref. 4, it can be shown, after some manipulation, that if no vibrational states are excited, then

$$T_f^2 = -\frac{4 \sin^2 (\epsilon_{31} - \epsilon_{32})}{D}, \tag{29a}$$

$$R_f^2 = 1 - \frac{4\rho_1\rho_2 \sin (\epsilon_{11} - \epsilon_{12}) \sin (\epsilon_{31} - \epsilon_{32})}{D}, \tag{29b}$$

where

$$\rho_1 = \frac{A_{11}}{A_{31}}, \qquad \rho_2 = \frac{A_{12}}{A_{32}}, \tag{29c}$$

and

$$D = 2\rho_1\rho_2 \cos (\epsilon_{11} - \epsilon_{12} - \epsilon_{31} + \epsilon_{32}) - \rho_1^2 - \rho_2^2. \tag{29d}$$

In these equations subscripts that refer to the vibrational states have been omitted to simplify the notation. Similar equations are obtained for T_b^2 and R_b^2.

In order to conserve particle flux, the equation

$$R_f^2 + \frac{T_f^2 v_3}{v_1} = 1 \tag{30}$$

is required, and upon combining it with (29a) and (29b) the equation

$$v_1 A_{11} A_{12} \sin (\epsilon_{11} - \epsilon_{12}) = -v_3 A_{31} A_{32} \sin (\epsilon_{31} - \epsilon_{32})$$

is obtained. From (14) and $B_{11} = B_{12} = B_{31} = -B_{32} = 1$, the above equation may be written

$$\frac{\mu k_1 \tan \epsilon_{11}}{\mu_{t1}} - \frac{\mu k_3 \tan \epsilon_{31}}{\mu_{t3}} = \frac{\mu k_1 \tan \epsilon_{12}}{\mu_{t1}} - \frac{\mu k_3 \tan \epsilon_{32}}{\mu_{t3}}, \tag{31}$$

where

$$v_s = \frac{k_s}{\mu_{ts}}. \tag{32}$$

Equation 31 is the condition on the phase angles of the two functions χ_i needed to conserve the total flux of particles. For reactions (1) and (2), where symmetrical and antisymmetrical χ_i can be constructed, $\epsilon_{11} = \epsilon_{31}$ and $\epsilon_{12} = \epsilon_{32}$ so that (31) is satisfied.

CALCULATIONS

To investigate the applicability of the variational and distorted wave treatments, reactions (1)–(3) have been studied to a linear collision approximation using the Hamiltonian of Refs. 4 and 5 given by (4) and (5). Using the computational procedures of these references, permeabilities were compared for reaction (1) where $\mathbf{K}'$ in (4) and (5) was retained and omitted at energies of 10 and 12.5 kcal/mole. At these energies the permeabilities obtained retaining $\mathbf{K}'$ were 0.00584 and 0.47851, and for those calculations taking $\mathbf{K}' = 0$ they were 0.00582 and 0.47855, respectively. These differences are within the uncertainties associated with the calculation. Since the retention of $\mathbf{K}'$ seemed to have little effect on the permeabilities calculated, all calculations in this work are with $\mathbf{K}' = 0$.

For the calculations reported here the total energy of the system is such that no vibrationally excited states are populated, so the variational function of Ref. 1 is applicable. Because the effective potential-energy surface given by (8) differs from the surface of Ref. 1, the one-dimensional effective potential-energy associated with the relative translation of the atom and diatomic molecule will be different and has been chosen to have the form

$$V_{ei} = A_0 \exp\left[-a_0\left(R_t \pm \frac{\pi}{4k}\right)\right] + \sum_{j=1}^{2} A_j \exp\left(-a_j R_t\right) + V_{eb}, \quad (33a)$$

$$V_{eb} = (B_0 + B_1 R_t) \exp\left(-b R_t\right), \qquad R_t \leq 2\,\text{Å}, \quad (33b)$$

$$V_{eb} = \sum_{j=1}^{8} \frac{C_j}{R_t^{\,j}}, \qquad R_t \geq 2\,\text{Å}, \quad (33c)$$

where $i = 1$ refers to the upper sign and $i = 2$ to the lower one in (33a). Equation 33a reduces to the corresponding potential of Ref. 1 by taking $V_{eb} = 0$. Similar potentials are needed for each of the two channels. Except for a_0 the parameters are applicable to both χ_1 and χ_2. The parameters in (33a), except those in V_{eb}, for channels having diatomic molecules D_2 and HD were obtained by fitting $\ln\left[A_0 \exp\left(-5.5R_t\right) + \sum_j A_j \exp\left(-a_j R_t\right)\right]$ to $\ln\left[\int(V_R - V_\infty)\phi_v^{\,2}\, dR_v\right]$ by a least squares procedure in the region $2 \leq R_t \leq 8\,\text{Å}$, where $V_R \to V_\infty$ as $R_t \to \infty$ and ϕ_v is the vibrational wavefunction associated with that channel. For channels associated with the diatomic molecule H_2, the corresponding parameters were taken from Ref. 1. The constant A_0 in making this least squares fit was taken to be 500 eV except for the channel associated with the diatomic molecule HD, where a better fit was obtained with 1000 eV. The permeabilities are quite insensitive to the choice of this parameter, whereas the other parameters used to fit the "near" asymptotic regions are much more important.

The parameters C_j in (33c) were obtained by fitting V_{eb} by a least squares fit to

$$\int \left(E_\theta - \frac{3\mu}{I} - \frac{2\mu}{m_2 R_{12} R_{13}} \right) \phi_v^2 \, dR_v$$

in the region $2 \leq R_t \leq 100$ Å. This potential-energy function will not give suitable solutions for the effective translational wavefunctions for small R_t, so the form given by (33b) was used at small R_t. The parameters in (33b) were

Table I.   Parameters used in the One-Dimensional Effective Potential Energy V_{ei} Given by (33a)–(33c), where the Unit of Energy is in Electron Volts and Distance in Angstroms

	Channel			
	H + H$_2$	D + D$_2$	D + H$_2$	H + HD
a_1	3.3983	3.1272	3.3983	2.7212
a_2	1.9403	1.9391	1.9403	1.9352
A_0	500	500	500	1000
A_1	11.8271	6.0485	11.8271	3.1660
A_2	2.9296	2.8870	2.9296	3.5962
b	1.9098762	1.4390671	1.8875599	1.4493748
B_0	3.1708467	1.1765805	2.9960965	1.9469287
B_1	−1.2861994	−0.46041077	−1.1987566	−0.77955506
C_1	−0.0055048165	−0.0027302928	−0.0054691311	−0.0054738808
C_2	−0.013248676	−0.0098263333	−0.012237669	−0.016367377
C_3	0.090490324	0.13522062	0.12794227	0.18757025
C_4	−1.4826342	−1.9156105	−1.8054652	−2.8337265
C_5	11.136670	13.518870	12.640669	20.999311
C_6	−42.712202	−50.346232	−46.676630	−81.106082
C_7	82.225140	95.551500	87.743624	156.65752
C_8	−56.955619	−66.885715	−60.115132	−110.33003

determined by requiring that the value of V_{eb} and its first two derivatives evaluated at $R_t = 2$ Å be the same as those obtained by (33c) at this distance. The values of the parameters in (33a)–(33c) are given in Table I.

The one-dimensional translational wavefunction is obtained by integrating a Schrödinger equation (see Ref. 1) having V_{ei} as a potential. Since V_{eb} decreases very slowly with distance, the Schrödinger equation must be integrated to large values of R_t. The integration is facilitated by using a WKB treatment to extend the solution to infinite separation. The numerical integration of the Schrödinger equation was carried to a value of R_t that satisfied $|V_{ei} - C_1/R_t| < 2.5 \times 10^{-4}$ (in reduced units). Beyond this distance the Schrödinger equation was integrated by the WKB method.

For reactions (1) and (2) the exponential constant a_0 in (33a) was determined by minimizing I_i^2, using 25 central functions. At each energy, a_0 was able to be selected such that $I_i = I_{ii} = 0$ and by symmetry $I_{ij} = 0$ for $i \neq j$. Thus, for these two reactions the variational procedure alone is sufficient, and the permeabilities obtained are given in Table II and compared

Table II. Permeability of the Sato Potential Energy Surface with Bending Corrections for the Linear Encounters $H + H_2 = H_2 + H$ and $D + D_2 = D_2 + D$ Using the Variational Procedures of ref. 1 with 25 Central Functions

E	Permeability			
	$H + H_2 = H_2 + H$		$D + D_2 = D_2 + D$	
(kcal/mole)	Variational	Ref. 5	Variational	Ref. 5
9.5			0.0074	0.0076
10.0	0.0058	0.0058	0.0278	0.0282
10.5			0.0873	0.0894
11.0	0.0557	0.0563	0.222	0.231
11.5	0.138		0.433	0.451
12.0	0.287	0.289	0.653	0.680
12.5	0.485	0.479	0.815	0.831
13.0	0.676	0.680	0.906	0.913
13.5	0.815	0.818	0.955	0.959
14.0	0.901	0.905	0.979	
15.0	0.974	0.975		

with the values reported in Ref. 5 where finite difference methods were used. For reaction (3), however, the integral I_{ii} in (26a)–(27b) may be made to vanish for suitably chosen a_0 for each channel, but in general the integrals I_{ij} ($i \neq j$) do not vanish so that the application of the distorted wave treatment to the variational functions can be used to further correct the permeabilities.

For reaction (3) the a_0 associated with both reactant and product channels for χ_1 were varied until both I_1^2 (before correction using (18)) and $R_1^2 = \int [(H - E)\chi_1]^2 \, d\tau$ (after correction) were minimized. Since many combinations of the a_0 gave $I_1 = 0$, it was important to also minimize R_1^2 to obtain the best solution where the phases associated with χ_1 are corrected to first order using (18). In actual practice I_1^2 is not minimized exactly, although its value is made very small for those cases in which I_1 can vanish, and (18) is used to make the necessary small corrections. Since (18) does not uniquely specify the values of the improved phases, the smallest displacement in the ϵ's were taken to obtain the ϵ^0's which cause I_1 to vanish to first order.

For χ_2 the phase angles are corrected using (18); however, it is also necessary to apply (31) to the corrected phases (i.e., where all ϵ_{ij} in (31) are replaced with $\epsilon_{ij}{}^0$) so that the total flux of particles is conserved. The left-hand side of (31) can be evaluated from the phases of χ_1, and (18) and (31) can be solved simultaneously to obtain the corrected phase angles associated with χ_2. This procedure corrects the phases and requires that the total flux of particles be conserved. From these corrected phases a value of $R_2{}^2$ can be calculated. This procedure is followed for each set of a_0, and the a_0 for both reactant and product channels are varied until $I_2{}^2$ (before correction) and $R_2{}^2$ (after correction) are minimized. The permeabilities calculated from both the variational calculations using 25 central functions and the additional

Table III.   Permeability of the Sato Potential Energy Surface with Bending Corrections for the Linear Encounter D + H$_2$ = DH + H Using Variational and Distorted Wave Treatments with 25 Central Functions

	Permeability		
E (kcal/mole)	Variational	Variational + Distorted Wave	Ref. 5
10.0	0.0109	0.0103	0.0102
11.0	0.0946	0.100	0.0959
12.0	0.417	0.415	
12.5	0.620	0.620	0.627
13.0	0.600	0.780	0.786
14.0	0.881	0.933	0.933

distorted wave calculations are given in Table III and compared with corresponding calculations in Ref. 5.

DISCUSSION

The permeabilities for reaction (1) that are given in Table II are in good agreement with the corresponding finite difference treatment reported in Ref. 5. The comparison reported here is better than that given in Ref. 1 where the original work of Mortensen and Pitzer[4] was compared. The finite-difference calculations of Ref. 5 were made over a larger range of the variables R_{12} and R_{23} and more care was taken in extrapolating phases to infinite separation of the atom and diatomic molecule than in Ref. 4. Since the variational calculations reported here and in Ref. 1 should be of comparable accuracy and since the variational and finite-difference calculations are in better agreement in this work, it should be concluded that the variational calculations of Ref. 1 are more reliable than the finite difference calculations in Ref. 4 except for the 17.5 kcal/mole collision which was discussed in Ref. 1.

The finite difference and variational calculations for reaction (2) are not in as good an agreement as those for reaction (1). The operator K', which has been neglected in these calculations, has the form:

$$K' = K_0 + K_{12} \frac{\partial}{\partial R_{12}} + K_{23} \frac{\partial}{\partial R_{23}}.$$

The coefficients K_0, K_{12}, and K_{23} are in general larger (often by a factor of 2) for reaction (1) than for reaction (2), although the magnitude of the derivatives $\partial \psi / \partial R_{ij}$ would be expected to be a little larger for the all-deuterium reaction. The net result of omitting K' in the Hamiltonian should not be appreciably different for the two reactions, and since K' had such a small effect on the permeabilities of reaction (1), it can only be concluded that the omission of K' for reaction (2) should not produce important changes in the permeabilities.

In order to obtain some indication of the reliability of these variational results, calculations were made with 16 central functions at those energies where the permeabilities by the two methods were in greatest disagreement. The permeability obtained for such a calculation for reaction (1) at 12.5 kcal/mole is 0.490, which is 0.005 larger than that obtained from the 25 central function calculation. This result is consistent with the permeabilities found in Table II for this reaction. For reaction (2) the permeability calculated using 16 central functions at 12.0 kcal/mole is 0.666, which is 0.013 larger than that listed in Table II. Even with a possible error of 0.01–0.02 in the variational calculations, there is still a small discrepancy between the two methods of calculation, which probably indicates that a somewhat similar error exists in the finite difference calculations for reaction (2).

By a comparison of the results reported in Table III with those of Ref. 5 it is clear that it is important to apply the distorted wave treatment to the variational calculations. The computational time for this additional refinement is certainly minimal, since only the one additional integral $I_{ij} = I_{ji}$ ($i \neq j$) need be calculated. The 11 kcal/mole permeability calculation is in greatest disagreement with Ref. 5. In this particular calculation the probability for reaction plus that for reflection is greater than unity, a disadvantage of the distorted wave treatment. This result indicates that the variational wavefunction calculated at this energy was not sufficiently close to the correct solution so that the replacement of ψ_{fi} by ξ_{fi} in (23a) and (23b) is not a very good approximation. Difficulties were also encountered with the 13 kcal/mole calculation where the phase angles associated with the χ_i were close to the poles $\pm \pi/2$ of the Kohn method. Again, for this calculation, the probability of transmission plus reflection was greater than unity. The calculations at other energies were fairly well behaved.

ACKNOWLEDGMENTS

I should like to express my appreciation to the computing centers at the University of Massachusetts and at Cleveland State University for the use of their facilities.

REFERENCES

1. E. M. Mortensen and L. D. Gucwa, *J. Chem. Phys.*, **51**, 5695 (1969).
2. W. Kohn, *Phys. Rev.*, **74**, 1763 (1948).
3. Y. N. Demkov, translated by N. Kemmer, *Variational Principles in the Theory of Collisions*, Macmillan, New York, 1963.
4. E. M. Mortensen and K. S. Pitzer, *Chem. Soc. (London) Spec. Publ.*, **16**, 57 (1962).
5. E. M. Mortensen, *J. Chem. Phys.*, **48**, 4029 (1968).
6. S. Sato, *J. Chem. Phys.*, **23**, 592, 2465 (1955).
7. R. E. Weston, Jr., *J. Chem. Phys.*, **31**, 892 (1959).
8. H. Eyring, J. Walter, and G. E. Kimball, *Quantum Chemistry*, Wiley, New York, 1944. 305.
9. A. Messiah, *Quantum Mechanics*, Vol. II (translated by J. Potter), Wiley, New York, 1962, Chapt. XIX.
10. L. Rosenberg, *Phys. Rev.*, **134**, B937 (1964).
11. R. D. Levine, *Quantum Mechanics of Molecular Rate Processes*, Oxford, London, 1969.
12. W. H. Miller, *J. Chem. Phys.*, **49**, 2373 (1968).

Calculation of Statistical Complexions of Polyatomic Molecules and Ions[*]

S. H. LIN AND C. Y. LIN MA

Department of Chemistry, Arizona State University, Tempe

Abstract

A method developed by Lin and Eyring for the calculation of the total number of states of a system, $W(E)$, based on the method of the steepest descent, has been applied to a system with a collection of harmonic oscillators, a collection of anharmonic oscillators, and a collection of harmonic oscillators coupled with rigid rotors. In the original approach the expression for $W(E)$ was derived only to the first-order approximation of the method of the steepest descent. In this study the second-order approximation has been included and it is shown that, in most cases, the agreement with the exact values is within 0.1%. The calculation of $\bar{\epsilon}_t$, the average translational energy in the reaction coordinate of the activated complex, which is important in discussing the distribution of the translational energies of reaction products, is also studied from the same viewpoint.

[*] Supported by the National Science Foundation and the A. P. Sloan Foundation.

143

A Personal Note

S. H. LIN AND C. Y. LIN MA

Professor Eyring's contribution to chemistry is well known. His other great contribution has been the training of the many young chemists from all over the world who have been fortunate enough to study with him. We both studied for our doctorates under Professor Eyring, and we are happy to join with the other contributors to this volume in expressing our deep appreciation for his guidance. We should like to add our sincere wishes for his continued good health and happiness in the years to come, which we hope will be many and pleasant.

The material presented here represents part of the work initiated when we were still graduate students working for Professor Eyring and it summarizes the developments since then.

INTRODUCTION

Theoretical studies of unimolecular reaction kinetics of polyatomic molecules in a mass spectrometer requires the knowledge of the number of states of a molecule with internal energy E, $W(E)$. It has been shown by Rosenstock et al.[1] and by Marcus and Rice[2] that, in general, the rate constant for the quasi-equilibrium dissociation of an isolated molecule can be expressed as

$$k(E) = \frac{1}{h} \int_0^{E-\epsilon_0} \frac{\rho^{\ddagger}(E, \epsilon_0, \epsilon_t)}{\rho(E)} \, d\epsilon_t, \tag{1}$$

where $\rho(E) \, dE$ represents the number of states of the system having energy between E and $E + dE$, and $\rho^{\ddagger}(E, \epsilon_0, \epsilon_t) \, d\epsilon_t$ is the number of states of the system in the activated complex configuration with the activation energy ϵ_0 and the translational energy between ϵ_t and $\epsilon_t + d\epsilon_t$ in the reaction coordinate, and with the total energy between E and $E + dE$. Thus, in terms of $W(E)$, (1) can be written as

$$k(E) = \frac{\sigma}{h} \cdot \frac{W^{\ddagger}(E - \epsilon_0)}{[dW(E)]/dE}, \tag{2}$$

where σ represents the symmetry number. Several methods[3-7] have been proposed to evaluate the number of states of the system with energy E, $W(E)$. In the previous investigations,[8,9] a method to calculate $W(E)$ based on the method of the steepest descent,[10] is developed by Lin and Eyring.

CALCULATION OF $W(E)$

First, let us consider a system of weakly coupled harmonic oscillators of frequencies $\nu_1, \nu_2, \ldots, \nu_m$ each having $g_1, g_2, \ldots, g_m$ degeneracies, respectively. If the total energy of the system is E, then

$$\sum_i n_i h \nu_i = E, \tag{3}$$

145

where the lowest energy level has been taken as zero. For convenience of calculation, (3) is made dimensionless by dividing both sides by $h\langle v \rangle$. Thus

$$\sum_i n_i \left(\frac{v_i}{\langle v \rangle} \right) = \frac{E}{h\langle v \rangle} = \langle n \rangle; \tag{4}$$

$\langle v \rangle$ is so chosen that $v_i/\langle v \rangle$, hence $\langle n \rangle$ are integers. The total number of quantum states with the energy values equal to or less than E is then[8,9]

$$W_s(E) = \sum_{\langle n \rangle = 0}^{\langle n \rangle} \sum_{n_i} \cdots \sum_{n_m} \prod_{i=1}^{m} \frac{(n_i + g_i - 1)!}{n_i!\,(g_i - 1)!}, \tag{5}$$

where the first summation is over the energy values and the remaining summations are taken over all values of the quantum numbers n_i, which are consistent with (4). By Cauchy's residue theorem we may express $W_s(E)$ by the following integrals:

$$W_s(E) = \sum_{\langle n \rangle = 0}^{\langle n \rangle} \frac{1}{2\pi i} \int_\gamma \frac{dz}{z^{\langle n \rangle + 1}} \prod_{i=1}^{m} (1 - z^{v_i/\langle v \rangle})^{-g_i}$$

$$= \frac{1}{2\pi i} \int_\gamma \frac{dz}{z} \frac{z^{-\langle n \rangle - 1} - 1}{z^{-1} - 1} \prod_{i=1}^{m} (1 - z^{v_i/\langle v \rangle})^{-g_i}$$

$$= \frac{1}{2\pi i} \int_\gamma \frac{dz}{z} [\phi(z)]^{\langle n \rangle}, \tag{6}$$

where

$$[\phi(z)]^{\langle n \rangle} = \frac{z^{-\langle n \rangle - 1} - 1}{z^{-1} - 1} \prod_{i=1}^{m} (1 - z^{v_i/\langle v \rangle})^{-g_i} \tag{7}$$

and γ is any contour lying within the circle of convergence of these power series enclosing the origin $z = 0$. In order that Cauchy's residue theorem be applicable, $\langle n \rangle$ must be an integer. For practical purposes one may choose $\langle n \rangle$ as close to an integer as possible.

By applying the method of steepest descent to (6), we obtain

$$W_c(E) = \frac{\theta^{-\langle n \rangle - 1}}{\theta^{-1} - 1} \prod_{i=1}^{n_i} \frac{(1 - \theta^{v_i/\langle v \rangle})^{-g_i}}{\{2\pi \langle n \rangle \theta^2 [\phi''(\theta)/\phi(\theta)]\}^{1/2}}, \tag{8}$$

where θ is the root of $\phi'(\theta) = 0$ and its value may be determined from the following equation:

$$\frac{\langle n \rangle + 1}{1 - \theta^{\langle n \rangle + 1}} = \frac{\theta^{-1}}{\theta^{-1} - 1} + \sum_{i=1}^{m} \frac{g_i(v_i/\langle v \rangle)}{\theta^{-v_i/\langle v \rangle} - 1}. \tag{9}$$

In general, $W_c(E)$ should be greater than or equal to $W_s(E)$.

It should be pointed out that $W_c(E)$ of (8) diverges when E approaches zero. This is because the method of steepest descent is an asymptotic expansion. But as is shown by our numerical calculation for various systems,[8,9] the divergence of $W_c(E)$ occurs only at extremely small values of E. So, for practical purposes, if $W_c(0) = 1$ is adopted, then (8) is applicable for any value of E. It is shown by our calculation[8,9] that for systems of coupled harmonic oscillators, the number of quantum states calculated from (8) is accurate within 5% for small energy values, and complete agreement with the exact values is obtained for large energy values.

Recently, the above approach has been applied to systems with a collection of harmonic oscillators coupled with rigid rotors and systems with a collection of anharmonic oscillators by Tou and Lin.[11] For a system with a collection of m harmonic oscillators $(\nu_1, g_1, \ldots, \nu_i, g_i; \ldots; \nu_m, g_m)$ coupled with k rigid rotors $(I_1, \ldots, I_j, I_k)$, where g_i is degeneracy, ν_i frequency, and I_j moment of inertia, if the internal energy is E, we have

$$\sum_{i=1}^{m} n_i h \nu_i + \sum_{j=1}^{k} \frac{l_j(l_j + 1)h^2}{8\pi^2 I_j} = E. \tag{10}$$

As before, (10) is divided by $h\langle \nu \rangle$ to make it dimensionless,

$$\sum_{i=1}^{m} n_i r_i + \sum_{k=1}^{k} l_j(l_j + 1)r_{j'} = \langle n \rangle, \tag{11}$$

where $\langle n \rangle = E/h\langle \nu \rangle$, $r_i = \nu_i/\langle \nu \rangle$, and $r_{j'} = h/8\pi^2 I_j \langle \nu \rangle$. The totality of states, $W_s(E)$, for this system can then be written as

$$W_s(E) = \sum_{\langle n \rangle = 0}^{\langle n \rangle} \sum_{n_i} \cdots \sum_{n_m} \sum_{l_j} \cdots \sum_{l_k} \prod_{i=1}^{m} \frac{(n_i + g_i - 1)!}{n_i!\,(g_i - 1)!} \prod_{j=1}^{k} (2l_j + 1), \tag{12}$$

where the summations are taken over all the possible vibrational and rotational quantum numbers, n_i and l_j, under the restriction given in (10) or (11). By using the generating functions G_{ν_i} and $G_{r_{j'}}$ for oscillators and rotors given by[11]

$$G_{\nu_i} = (1 - z^{r_i})^{-g_i}; \qquad G_{r_{j'}} = \sum_{l_j=0} (2l_j + 1)z^{l_j(l_j+1)r_{j'}} \tag{13}$$

we obtain $W_s(E)$ as[11]

$$\begin{aligned}
W_s(E) &= \sum_{\langle n \rangle = 0}^{\langle n \rangle} \frac{1}{2\pi i} \int_\gamma \frac{dz}{z^{\langle n \rangle + 1}} \prod_{i=1}^{m} G_{\nu_i} \prod_{j=1}^{k} G_{r_{j'}} \\
&= \frac{1}{2\pi i} \int_\gamma \frac{dz}{z} \cdot \frac{z^{-1-\langle n \rangle} - 1}{z^{-1} - 1} \prod_{i=1}^{m} G_{\nu_i} \prod_{j=1}^{k} G_{r_{j'}}.
\end{aligned} \tag{14}$$

Here we can introduce the method of steepest descent. The results are given in the original paper.[11]

Similarly, for systems with a collection of anharmonic oscillators

$(g_1, x_1, \nu_1, \ldots, g_i, x_i, \nu_i; \ldots, g_n, x_n, \nu_n)$, where x_i is the anharmonicity constant, we have

$$\sum_{i=1}^{n} [n_i - n_i(n_i + 1)x_i]h\nu_i = \sum_{i=1}^{n} N_{n_i}h\nu_i = E \tag{15}$$

or

$$\sum_{i=1}^{n} N_{n_i} \frac{\nu_i}{\langle \nu \rangle} = \frac{E}{h\langle \nu \rangle} = \langle n \rangle, \tag{16}$$

where $N_{n_i} = n_i - n_i(n_i + 1)x_i$. Now, since the quantum numbers n_i can take only the values from zero up to their maximum values,

$$n_{i,\max} = (2x_i)^{-1} - \tfrac{1}{2}, \tag{17}$$

the generating functions $G_{an,\nu_i}(z)$ can be expressed as[11]

$$G_{an,\nu_i}(z) = \left(\sum_{\langle n \rangle = 0}^{n_{i,\max}} z^{N_{n_i}\nu_i/\langle \nu \rangle} \right)^{g_i}. \tag{18}$$

By the Cauchy residue theorem we obtain[11]

$$W_s(E) = \sum_{\langle n \rangle = 0}^{\langle n \rangle} \frac{1}{2\pi i} \int_\gamma \frac{dz}{z^{\langle n \rangle + 1}} \prod_{i=1}^{n} G_{an,\nu_i}(z). \tag{19}$$

Again we can use the method of steepest descent to approximately evaluate the contour integral in (19). The results are given in the original paper.[11] The numerical results indicate that the total number of states for molecules or ions treated as a collection of anharmonic oscillators is larger than the number treated as collections of harmonic oscillators. Generally speaking, the calculated results for the numbers of states $W(E)$ for the molecular ions and models chosen, both being treated as systems of anharmonic oscillators or systems of harmonic oscillators coupled with rigid rotors, are in good agreement with exact values of $W(E)$ for small E values, and as E increases the agreement improves. Although in our previous calculations,[8,9,11] we investigated only the cases of systems with a collection of harmonic oscillators, a collection of harmonic oscillators coupled with rigid rotors, and a collection of anharmonic oscillators, the approach we developed can be extended to other cases, such as systems of a collection of anharmonic oscillators coupled with rotors.

It should be noticed that the above calculation for $W(E)$ is carried out only to the first-order approximation of the method of steepest descent.[10] In our recent investigation,[12] we extended the calculation of $W(E)$ to the second-order approximation. We shall give only the general results here and not the derivation. In general, the total number of states $W_s(E)$ of a system can be expressed in the contour integral as

$$W_s(E) = \frac{1}{2\pi i} \int_\gamma \frac{dz}{z} [\phi(z)]^{\langle n \rangle} \tag{20}$$

where $E = \langle n \rangle h \langle \nu \rangle$. Then to the second-order approximation of the method of steepest descent,[12] we have

$$W_c(E) = \frac{[\phi(\theta)]^{\langle n \rangle}}{\{2\pi\langle n\rangle\theta^2[\phi''(\theta)/\phi(\theta)]\}^{1/2}}$$

$$\times \left[1 + \frac{1}{8\langle n\rangle}\left\{\frac{B}{\{\theta^2[\phi''(\theta)/\phi(\theta)]\}^2} - \frac{\frac{5}{3}A^2}{\{\theta^2[\phi''(\theta)/\phi(\theta)]\}^3}\right\} + \cdots\right] \quad (21)$$

where θ is the root of $\phi'(\theta) = 0$ and B and A are defined as

$$B = 7\theta^2\frac{\phi''(\theta)}{\phi(\theta)} + 6\theta^3\frac{\phi'''(\theta)}{\phi(\theta)} + \theta^4\frac{\phi''''(\theta)}{\phi(\theta)} - 3\left[\theta^2\frac{\phi''(\theta)}{\phi(\theta)}\right]^2 \quad (22)$$

and

$$A = -3\theta^2\frac{\phi''(\theta)}{\phi(\theta)} - \theta^3\frac{\phi'''(\theta)}{\phi(\theta)}. \quad (23)$$

The preliminary numerical calculation[12] indicates that for a collection of harmonic oscillators, the total number of states calculated from (21), $W_c(E)$, shows agreement with the exact values $W_s(E)$ to within 0.1 % in the energy range of 0.01–10 eV, and better for E beyond 10 eV.

Recently we applied[13] the approach discussed above for the calculation of $W(E)$ to the calculation of rate constants of various hydrocarbons, and hence to the interpretation of the mass spectrum of the fragmentation of those compounds.[14–16] In general, from the rate constants and the characteristic transit time of the mass spectrometer, the breakdown curves—that is, the fraction of parent ions initially formed that are collected as such, as "metastable ions," and as product ions—can be calculated as a function of internal energy in the parent ion. For this purpose one has to adopt physically reasonable activated complexes and activation energies. The relative abundances of the major fragmentations, the metastable ions, and "missing metastable" ions are computed for different gas temperatures, which is important experimentally, because the fragmentation is significantly affected by the temperature of the gas. The calculated breakdown curves have been compared with the available experimental curves.[14–16] The equilibrium theory of unimolecular reactions gives satisfactory breakdown curves for the molecules chosen for calculation using physically reasonable activated complexes and activation energies. The good agreement between the calculated and experimental results seems to indicate the validity of the statistical approach to interpretation of the mass spectra, although the assignments of frequencies and activation energies are in many ways arbitrary.

TRANSLATIONAL ENERGIES OF REACTION PRODUCTS

To understand the microscopic details of reaction mechanisms a knowledge of the distribution of excess energy among the various degrees of freedom of reaction products is essential. Recently, the translational energies of the products of unimolecular dissociations of ions in the mass spectrometer have been studied experimentally.[17-20] To treat the correlation of excess energies of electron-impact dissociations with the translational energies of the products, one can use the theory of unimolecular reactions developed by Rosenstock et al.[1] Suppose we are concerned with the case in which the excitation energy is equal to the appearance potential. A major assumption is that the transfer of energy among the oscillators in the activated state is much faster than the dissociation into products, so that the oscillators reach energy equilibrium. With this assumption, the translational energy in the reaction coordinate in the activated complex becomes part of the translational energy observed for the dissociation products. If we let E^* represent the excess energy at the appearance potential, which is divided into the vibrational energy ϵ_v and translational energy ϵ_t, then the calculation of the average ϵ_t is essentially the problem of calculating the distribution of vibrational energy among a collection of oscillators whose total energy is E^*. Thus the probability that the system will have energy vibrational ϵ_v and total energy E^* is

$$P(E^*, \epsilon_v) = \frac{\rho(\epsilon_v)}{W(E^*)} \qquad (24)$$

where $\rho(\epsilon_v)$ and $W(E^*)$ represent the density of states and total number of states, respectively, as defined earlier. Notice that $\epsilon_v = E^* - \epsilon_t$. Hence

$$P(E^*, \epsilon_v) = \frac{\rho(E^* - \epsilon_t)}{W(E^*)}. \qquad (25)$$

Using (25), the average ϵ_t can be calculated by

$$\bar{\epsilon}_t = \int_0^{E^*} \epsilon_t P(E^*, \epsilon_t)\, d\epsilon_t. \qquad (26)$$

Equation 26 can equivalently be expressed as

$$\bar{\epsilon}_t = E^* - \frac{1}{W(E^*)} \int_0^{E^*} \epsilon_v \rho(\epsilon_v)\, d\epsilon_v \qquad (27)$$

$$= E^* - \frac{1}{W(E^*)} \int_0^{E^*} \epsilon_v \frac{dW}{d\epsilon_v}\, d\epsilon_v. \qquad (28)$$

As has been pointed out before,[1] the continuum approximation is not a good approximation for calculating either the density of states or the total number of states of a collection of harmonic oscillators. This can be avoided by using the following method. As in (1)–(22), let $\langle n \rangle = \epsilon_v / h\langle v \rangle$ and $\langle N \rangle = E^* / h\langle v \rangle$, where $\langle v \rangle$ is chosen to make all normal frequencies, $v_i / \langle v \rangle$, integers.

Then the integration in (28) can be replaced by summation.

$$\bar{\epsilon}_t = E^* - \frac{U(E^*)}{W(E^*)}\, h\langle v \rangle,$$ (29)

where

$$U(E^*) = \sum_{\langle n \rangle = 0}^{\langle N \rangle} \langle n \rangle\, \Delta W(\langle n \rangle),$$ (30)

with $\Delta W(\langle n \rangle) = W(\langle n \rangle + 1) - W(\langle n \rangle)$. By introducing (6) into (30) $U(E^*)$ can be evaluated as

$$U(E^*) = \frac{1}{2\pi i} \int_\gamma \frac{dz}{z}$$
$$\times \frac{\langle N \rangle z^{-\langle N \rangle - 3} + z^{-2} - (1 + \langle N \rangle) z^{-\langle N \rangle - 2}}{(1 - z^{-1})^2} \prod_{i=1}^{m} (1 - z^{v_i/\langle v \rangle})^{-g_i}.$$ (31)

The calculation of the total number of states, $W(E^*)$, based on the method of steepest descent, has been discussed before,[8,9,11] (2)–(22). Similarly, we can evaluate $U(E^*)$ by using the method of steepest descent. For this purpose, we rewrite (31) as

$$U(E^*) = \frac{1}{2\pi i} \int_\gamma \frac{dz}{z}\, [\Psi(z)]^{\langle N \rangle},$$ (32)

where

$$[\Psi(z)]^{\langle N \rangle} = \frac{\langle N \rangle z^{-\langle N \rangle - 3} + z^{-2} - (1 + \langle N \rangle) z^{-\langle N \rangle - 2}}{(1 - z^{-1})^2} \prod_{i=1}^{m} (1 - z^{v_i/\langle v \rangle})^{-g_i}.$$ (33)

Then, to the first-order approximation of the method of steepest descent, we obtain

$$U(E^*) = \frac{[\Psi(\theta)]^{\langle N \rangle}}{\{2\pi \langle N \rangle \theta^2 [\Psi''(\theta)/\Psi(\theta)]\}^{1/2}}.$$ (34)

The θ value in (34) is the root of $\Psi'(\theta) = 0$, or

$$\frac{(1 + \langle N \rangle)(2 + \langle N \rangle)\theta - 2\theta^{\langle N \rangle + 1} - \langle N \rangle(\langle N \rangle + 3)}{\langle N \rangle + \theta^{\langle N \rangle + 1} - (1 + \langle N \rangle)\theta}$$
$$= \frac{2}{\theta - 1} + \sum_{i=1}^{m} \frac{g_i \cdot v_i/\langle v \rangle}{\theta^{-v_i/\langle v \rangle} - 1}$$ (35)

Once θ is determined from (35), we can calculate $U(E^*)$ from (34) and ϵ_t from (29). The average translational energy, $\bar{\epsilon}_t$,[21] of the reaction coordinate has been calculated for various molecules by using the method developed above and the calculated results have been compared with the experimental data.[17–20] The agreement is satisfactory.

REFERENCES

1. H. M. Rosenstock, M. B. Wallenstein, A. L. Wahrhaftig, and H. Eyring, *Proc. Natl. Acad. Sci. U.S.*, **38**, 667 (1952).
2. R. A. Marcus and O. K. Rice, *J. Phys. Colloid Chem.*, **55**, 894 (1951).
3. B. S. Rabinovitch and J. H. Current, *J. Chem. Phys.*, **35**, 2250 (1961).
4. P. C. Haarhoff, *Mol. Phys.*, **7**, 101 (1963); **8**, 49 (1964).
5. E. Thiele, *J. Chem. Phys.*, **39**, 3258 (1963).
6. G. Z. Whitten and B. S. Rabinovitch, *J. Chem. Phys.*, **38**, 2466 (1963).
7. M. Vestal, A. L. Wahrhaftig, and W. H. Johnston, *J. Chem. Phys.*, **37**, 1276 (1962).
8. S. H. Lin and H. Eyring, *J. Chem. Phys.*, **39**, 1577 (1963).
9. S. H. Lin and H. Eyring, *J. Chem. Phys.*, **43**, 2153 (1965).
10. R. H. Fowler, *Statistical Mechanics*, Cambridge, New York, 1963, p. 16.
11. J. C. Tou and S. H. Lin, *J. Chem. Phys.*, **49**, 4187 (1968).
12. K. H. Lau, S. H. Lin, and C. Y. Lin Ma, *J. Phys. Chem.*, in press.
13. C. Y. Lin Ma and S. H. Lin, to be published.
14. B. Steiner, C. F. Giese, and M. G. Inghram, *J. Chem. Phys.*, **34**, 189 (1961); **40**, 3263 (1964).
15. W. A. Chupka, *J. Chem. Phys.*, **30**, 191 (1959).
16. M. Krauss and V. H. Dibeler, in *Mass Spectrometry of Organic Ions*, F. A. McLafferty, Ed., Academic Press, New York 1963; also, F. H. Field and J. L. Franklin, *Electron Impact Phenomenon*, Academic New York, 1959.
17. M. A. Haney and J. L. Franklin, *J. Chem. Phys.*, **48**, 4093 (1968).
18. C. E. Klots, *J. Chem. Phys.*, **41**, 117 (1964).
19. R. Taubert, *Z. Naturforsch.*, **19A**, 911 (1964).
20. J. L. Franklin, P. M. Hierl, and D. A. Whan, *J. Chem. Phys.*, **47**, 3148 (1967).
21. C. Y. Lin Ma and S. H. Lin, to be published.

Some Comments on the Theory of Photochemical Reactions

STUART A. RICE

Department of Chemistry and The James Franck Institute, The University of Chicago, Chicago

153

I

In the "science of science" it has been suggested that creative individuals may be classified on a scale spanning the range from "classicist" to "romantic." At the extremes, a "classicist" concentrates on only a few problems and rarely works with colleagues, whereas a "romantic" displays a passionate fondness for all of science, collaborates with many individuals, has considerable versatility and breadth of interest, and, along with great productivity, continuously introduces new interpretations and new ideas. Given this classification scheme Henry Eyring seems the model scientific "romantic." It is characteristic of "romantics" that they do not exhaust the possibilities opened by their own suggestions. Eyring's work in the field of chemical kinetics exhibits this characteristic. His contributions to chemical kinetics have been so extensive that a language has grown up based on the concepts he helped to introduce. Despite this evidence of widespread acceptance, the full consequences of the ideas advanced by Eyring are only now becoming apparent.

This paper deals with some aspects of the theory of photochemical reactions, and necessarily makes contact with the theory of unimolecular reactions and the theory of energy transfer. The indirect influence of Henry Eyring's work is manifest in much of what follows since, in part, we are concerned with the validity of the usual concepts of unimolecular reaction rate theory.

II

A photochemical reaction differs from a thermal reaction in at least one fundamental way: the absorption of a photon by a molecule "prepares" the molecule in a well-defined nonstationary state with a nonuniform distribution of energy, whereas a thermally excited molecule is almost always in a state in which the distribution of energy over possible modes of motion is uniform, or

155

nearly so. Thus, in the theory of thermal unimolecular reactions the rate of decomposition is related to two factors:

1. The collisional energy transfer leading to activation.
2. The breaking of a particular bond.

It is usually assumed that the collisions are so "strong" that the initial and final total energies are uncorrelated, so that the probability that a particular molecule will have an energy E after a collision is determined solely by the Boltzmann factor $g_E \exp(-E/kt)$, with g_E the degeneracy of the state with energy E. It is also assumed that the coupling between the internal degrees of freedom is so strong that the consequent energy rearrangement can be described statistically, that is, the probability that a particular molecule, with given total energy, will have a specified configuration after an internal energy exchange is calculated by describing the system in terms of the appropriate microcanonical ensemble. Finally, the bond which breaks is designated *a posteriori*, and is not prescribed by theoretical argumentation.

Our interest in the theory of photochemical reactions started with the observation that in a molecule "prepared" in a state with nonuniform distribution of the energy, the details of internal nuclear and electron dynamics should influence the rate and the products of the reaction. Now, in the upper excited states of a molecule it is often found that the vibrational levels corresponding to one Born-Oppenheimer (BO) manifold overlap the dense set of vibrational levels corresponding to a lower electronic Born-Oppenheimer state. If the energy separation between the overlapping levels is comparable with or less than the off-diagonal matrix elements of the nuclear kinetic energy operator in the basis of BO states, the BO representation of the states of the molecule is no longer adequate.[1] In this case the true molecular eigenstates can be thought of as mixtures of BO states. We have referred above to vibrational levels. In a highly excited molecule the harmonic approximation to the description of nuclear motion frequently is not adequate. We shall briefly return to this latter problem later; however, it is important to recognize at the outset that the theories which are familiar and satisfactory for the description of the ground state and the lower excited states of a molecule must be extended, or even replaced, when what is sought is a description of the upper excited states of a molecule.

Consider, as a first example, a molecule with an energy spectrum of the type shown in Figs. 1 and 2. This energy level scheme can be thought of as corresponding to the existence of a localized electronic excitation coupled to electronic ground-state vibrations, which in turn are coupled to a fragmentation continuum. Examples of energy spectra resembling this are found in many carbonyl compounds and in methyl-substituted benzenes. In both classes of

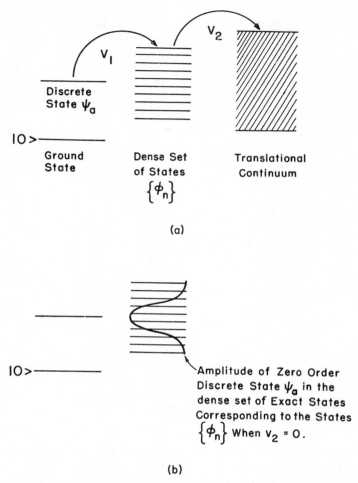

Fig. 1. Schematic energy-level diagram of (a) the zero-order states and (b) the exact eigenstates in the absence of coupling to the continuum.

compounds, localized excitation of a chromophore may lead to fragmentation at a distant chemical bond.[2]

The properties of the dissociative states of a molecule with the spectrum described above were studied by Rice, McLaughlin, and Jortner.[3] They represented the true eigenstates of the molecule in the absence of a radiation field as a superposition of zero-order states (which need not be BO states but which are the eigenstates of a zero-order separable Hamiltonian). Suppose

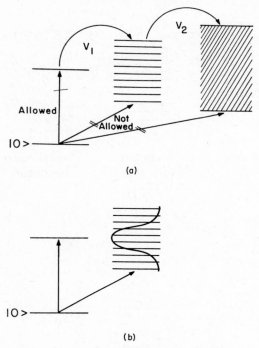

(a)

(b)

Fig. 2. Schematic diagram showing transitions between the sets of levels of the hypothetical spectrum: (*a*) allowed dipole transitions between zero-order states; (*b*) allowed dipole transitions between exact states in the absence of coupling to the translational continuum.

that, as shown in Fig. 2, transitions from the ground state to the dissociative continuum and to the dense manifold of bound zero-order vibronic states are forbidden, but that the transition to one discrete zero-order state is allowed. Then absorption of a photon prepares the molecule in a nonstationary state which can be described as a time-dependent superposition of exact eigenstates, each of which is itself a time-independent superposition of zero-order (say, BO) states.* To study photodecomposition it is imagined that the molecule is illuminated with a pulse of radiation, and the time evolution of the nonstationary state is followed. One of the interesting features of this description is the dependence of the probability of the molecule being in a given nonstationary state on the time correlations in the coupled radiation field. Finally, the probability of dissociation of the molecule is related formally to the matrix elements coupling the component manifolds of the spectrum, the densities of states in the dissociative continuum, and the manifold to which it is connected.

It is possible, using an approximation relating to the influence of the rate of decomposition on the distribution of amplitude of the discrete zero-order state in the dense set of exact vibrational states, to derive a simple expression for the decomposition rate. That expression is

$$k = \frac{2\rho_2 v_2{}^2}{h}\left[\frac{\pi}{2} - \tan^{-1}\left(\frac{E_m}{\pi(\rho_1 v_1{}^2)}\right)\right],$$

where E_m is the minimum energy required for fragmentation in the selected bond, ρ_1 and ρ_2 are the densities of states in the vibrational quasi-continuum and the fragmentation continuum, and v_1 and v_2 are the matrix elements connecting the discrete zero-order state to the zero-order vibrational quasi-continuum and the zero-order vibrational quasi-continuum to the fragmentation continuum, respectively. The limit $\rho_1 v_1{}^2 \ll \rho_2 v_2{}^2$ corresponds to the case where the molecule dissociates as rapidly as the nonstationary wavepacket can disperse into the vibrational quasi-continuum. This case is analogous to but not identical with that considered by Peters[6]—direct excitation to a nonbonding state without intermediate intramolecular energy transfer. In the limit $\rho_1 v_1{}^2 \gg \rho_2 v_2{}^2$, the rate of fragmentation is controlled by the rate of internal energy transfer.

Of what use is such a formal description? Can we learn from it anything not already known? I believe the answer to both these questions to be a qualified yes. Perhaps most important is the change in viewpoint put forward from that embodied in the classical theory of unimolecular reactions (thermal or photochemical). In the current theory the excited molecule about to dissociate is regarded as being in a resonant scattering state* and not a bound state. The process by which the molecule decomposes is then just a form of predissociation. Because the exact resonant scattering state is represented as a linear combination of the zero-order localized state, the dense set of zero-order vibrational states, and the zero-order continuum, and because the zero-order state that can be excited by absorption of a photon is spread over many exact resonant scattering states, absorption of a photon leads to the population of resonant scattering states spread over a range of energy. Predissociation occurs from this band of states, subject to appropriate selection rules. Thus, it is not necessary to ascribe to the activated molecule the properties of a stable molecule, and the difficulties associated with the

* Mies and Krauss have used a related method in their elegant theory of unimolecular reactions.[4,5] They explicitly employ a generalization of the Fano theory of autoionization, but they have given less emphasis than we to internal energy redistribution since they study, primarily, thermally excited molecules.

concepts of thermodynamic functions of the activated state, and the special role of the reaction coordinate, can be bypassed. Furthermore, by realizing that an excited molecule is in a resonant scattering state, the lifetime against decay into fragments appears naturally in the evaluation of the probability of dissociation—it need not be introduced as an ad hoc hypothesis defining irreversibility.

At a more detailed level the analysis raises the possibility that under suitable excitation conditions a coherently excited set of resonant scattering states can interfere[6]—perhaps with consequences of chemical interest. Also, depending on the magnitudes of the several coupling matrix elements and densities of states, the excited molecule in the resonant scattering state may behave like a long-lived vibrationally hot molecule, like a molecule that decomposes directly to fragments without internal energy transfer, or in an intermediate fashion not like either of the cited classical limits.

The model of the photodissociation reaction described above has been used by Heller[7] to elucidate the nature of hydrogen transfer and hydrogen abstraction reactions. Suppose that, starting from an initial nonuniform distribution of energy in the excited state of the molecule, electronic–vibrational and vibrational–vibrational coupling leads to the excitation of particular vibrations. It is known that high frequency anharmonic stretching vibrations are the preferred energy acceptors in molecule-conserving radiationless processes.[1] Then the abundance of hydrogen abstraction and transfer reactions is explained by preferential energy transfer to high frequency stretching vibrations involving hydrogen. As long as the rate of vibrational deexcitation of such modes is small relative to the rate of dissociation, the quantum yield of dissociated products should be large. The following rules are natural consequences of the properties built into the model considered:[7]

1. The most reactive atoms in a molecule are those that are involved in high frequency molecular motions.

2. Among bonds with nearly the same vibrational frequency, the more anharmonic the vibration and the lower the bond energy the greater the bond reactivity.

3. Among bonds with nearly the same vibrational frequency and dissociation energy, the closer the bond to the initial concentration of excitation energy the greater the bond reactivity.

Assuming that there are no changes in relative bond strengths in the excited states of a molecule (as compared to the bond strengths in the ground state), the above model and rules derived therefrom correlate the observed intramolecular abstraction of hydrogen by excited carbonyl compounds (in both singlet and triplet states) and the $Hg(^3P_1)$-sensitized fragmentation of alkanes. Of course, in some molecules there are changes in relative bond strengths in

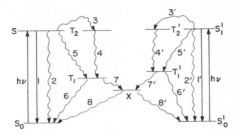

Fig. 3. A schematic plot of electronic energy versus angle of twist (about the double bond) in unsubstituted stilbene.

the excited state relative to the ground state; the above rules do not apply to the photochemistry of these compounds.

Further evidence of the potential value of representing a unimolecular reaction in terms of the time evolution of a resonant scattering state can be glimpsed from the as yet incomplete theory of photoisomerization.[8,9] Taking the *cis–trans* isomerization of stilbene as a prototype reaction, Gelbart and Rice[8] studied the properties of a model molecule in which a singlet vibronic level interacts (through spin-orbit coupling) with a nearly degenerate triplet state, which in turn is coupled (by vibronic interactions) to an isoenergetic set of closely spaced vibrational levels belonging to a lower triplet (see Fig. 3). Unlike the photodissociation reaction, the *cis–trans* isomerization does not involve a true continuum, since even in the limit of free internal rotation the moments of inertia of the molecular "ends" may be sufficiently small to lead to internal rotational spacings not grossly different from those characteristic of torsional motion. If the molecule is large enough to be in the limit where $\rho v^2 \gg 1$ (where ρ is the density of states in some manifold and v the matrix element coupling that manifold to a discrete state), and this condition holds for all discrete states and dense manifolds of interest, then it is shown that the radiative and nonradiative processes proceed independently of one another. The fluorescence quantum yield and the radiative lifetime predicted by this model depend explicitly on the (solvent-specific) energy difference between the discrete singlet and the triplet, and they may be conveniently parameterized in terms of the level density and coupling matrix elements of the zero-order states.

The role of hindered rotation in the breakdown of the BO approximation can, in fact, be analyzed in considerable detail.[9] It can be shown that the molecular mode whose equilibrium positions differ greatly in the initial and final electronic states (the strongly coupled degree of freedom) may be separated out from the other vibrations in the general expression for the rate of reaction. It is then found that, under conditions appropriate to the description of *cis–trans* isomerization in stilbene, most of the electronic energy change should go into the torsional degree of freedom rather than the usual vibrational modes which serve as acceptors if all modes are weakly coupled. The experimentally observable consequence of this prediction is the lack of (or great diminution of) an isotope effect in the deactivation of the lowest triplet to yield the isomer product. Sufficient experimental data are not now available to test these predictions.

In the classical theory of unimolecular reactions and in the theory of photochemical reactions recently developed, the internal transfer of energy plays an important role. Nevertheless, the mechanism of the internal energy transfer and its relationship to specific reaction product selection is not understood. Some insight into the energy transfer process and its coupling to molecular dissociation can be obtained from the study of models. Gelbart, Rice, and Freed[10] have examined the properties of a model in which all vibrations are coupled, some weakly and some strongly, to all others, and in which all bonds may break with specified dissociation probabilities. The model is sufficiently general that the nature of the vibrations need not be specified. Thus it is in principle possible to account for the breakdown of the usual harmonic oscillator description of a highly excited molecule—although only in a phenomenological fashion. A stochastic process defined by a set of transition probabilities that allows one to explicitly include the dynamics of the inter- and intramolecular energy exchange replaces the usual microcanonical ensemble and strong collision assumptions. By providing for the breaking of all bonds at all times and including the competing effects of energy transfer and bond dissociation, information can be obtained about which bond is most likely to break. The following cases of interest have been studied:

1. If the distribution of energy in the initial state is uniform, corresponding to collisional activation of a gas of polyatomic molecules originally in thermal equilibrium, it is shown that upon "heating up" the gas the probabilities for the breaking of the individual bonds are equal at all times regardless of the details of the vibrational interactions. This result is rendered plausible when it is noted that at sufficiently high pressures the rate of collisional excitation exceeds the rate of internal energy transfer. The lack of great specificity in collisional excitation results, in a collection of molecules, in almost equal probability of excitation of all degrees of freedom. Then the details of the vibrational interactions are clearly unimportant and the dissociation probabilities depend on the internal energy distribution. At very low pressures the reaction rate is determined by the rate of collisional activation. Thus, except for a small range of pressure where the rate of collisional excitation is comparable with the rate of internal energy transfer, it is to be expected that

individual bond dissociation probabilities will be sensibly independent of the nature of the internal coupling.

The result just cited provides a formal justification of the assumption implicit in classical unimolecular reaction rate theory, namely that the reaction rate depends only on the equilibrium internal energy distribution of the activated molecule.

2. If the distribution of energy in the initial state is nonuniform, corresponding to certain kinds of photochemical, hot atom, or photosensitization excitation, it is found that, all other things being equal (viz., all bond energies equal) the more strongly coupled vibrational motions will always be more likely to lead to fragmentation of the molecule than is the case for the weakly coupled vibrations. Furthermore, by virtue of the energy flow into strongly coupled degrees of freedom, a strong bond can break before a weak bond if that weak bond is more weakly coupled to the site where the energy is initially localized.

The result just cited is consistent with the rules deduced by Heller,[7] with the competing fragmentation reactions observed in the mass spectrometric studies of Rosenstock and Krauss,[11] and with the chemical activation studies of Rabinovitch and co-workers[12]; in all of these cases there are formed excited molecules, molecular ions, and/or radicals with strongly nonuniform internal energy distributions.

3. In case 2, each internal degree of freedom can exchange energy with every other degree of freedom. If, instead, interactions are restricted to "nearest neighbors," so that a linear chain of interactions exists, different results are obtained. As can be inferred from the work of Magee and co-workers,[13] in a linear chain of bonds one of which is weakly coupled, it is the weakly coupled bond which breaks, because the mean time that the energy spends localized in any of the individual bonds depends inversely on the strength of coupling of that bond to its neighbors. The weakly coupled bond then "traps" the otherwise delocalized energy for the longest time, and is therefore the most likely to break regardless of the dissociation mechanism.

4. The nature of the competition between energy transfer and fragmentation depends in a fundamental way on whether the molecule in question can be regarded as "effectively linear," is intrinsically nonlinear, or is intermediate between these two limits.

III

Although only deduced from a variety of phenomenological models, the preceding results show that the consideration of individual bond energies alone is not sufficient to develop a description of photochemical fragmentation.

In ordinary unimolecular reaction rate theory, the usual assumptions of "strong" collisions and random distribution of the internal energy simply serve to wash out precisely those features of the molecular dynamics that become of primary importance in the cases of photochemical, chemical, and electron impact excitation. Whereas evaluation of all the consequences is incomplete at present, it is already clear that the representation of an excited molecule in terms of the properties of resonant scattering states holds promise for the elucidation of those aspects of the internal dynamics that are important in photochemistry.

I have not mentioned at all in this set of comments the fascinating new studies that relate orbital symmetries and reaction pathways: these also have greatly increased our understanding of photochemical reactions. They are, however, concerned with aspects of reaction rate theory intrinsically different from those studied by Henry Eyring. For that reason the interested reader is referred to the literature.[14]

ACKNOWLEDGMENTS

This report includes unpublished material developed in collaboration with Dr. William Gelbart and Professor Karl Freed. As is evident from the textual citations, without their contributions the theory described would be much less complete and much less satisfactory.

The research reported herein has been supported by the Directorate of Chemical Sciences, Air Force Office of Scientific Research. We have also benefited from facilities provided by Advanced Research Progress Agency for materials research at the University of Chicago.

REFERENCES

1. A review dealing with the implications of the breakdown of the BO approximation for the description of radiative and nonradiative processes can be found in J. Jortner, S. A. Rice, and R. Hochstrasser, *Advan. Photochemistry*, **7**, 149 (1969).
2. See, for example, J. Calvert and J. Pitts, *Photochemistry*, John Wiley, New York, 1967. P. Johnson and S. A. Rice, *Chem. Phys. Letters*, **1**, 709 (1968).
3. S. A. Rice, I. McLaughlin, and J. Jortner, *J. Chem. Phys.*, **49**, 2756 (1968).
4. F. H. Mies and M. Krauss, *J. Chem. Phys.*, **45**, 4455 (1966).
5. F. H. Mies, *J. Chem. Phys.*, **51**, 787, 798 (1969).
6. D. Peters, *J. Chem. Phys.*, **41**, 1046 (1964).
7. A. Heller, *Mol. Photochem.*, **1**, 257 (1969).
8. W. Gelbart and S. A. Rice, *J. Chem. Phys.*, **50**, 4775 (1969).
9. W. Gelbart, K. Freed, and S. A. Rice, *J. Chem. Phys.*, **52**, 2460 (1970).

10. W. Gelbart, S. A. Rice, and K. Freed, *J. Chem. Phys.*, **52**, 5718 (1970). See, for example, H. Rosenstock, M. Wallenstein, A. Wahrhaftig, and H. Eyring, *Proc. Nat. Acad. Sci.*, **38**, 667 (1952).

11. H. Rosenstock and M. Krauss, in *Mass Spectrometry of Organic Ions*, F. W McLafferty, Ed., Academic Press, New York, 1963, Chapter 1.

12. C. W. Larson, B. S. Rabinovitch, and D. C. Tardy, *J. Chem. Phys.*, **47**, 4570 (1967); **45**, 1163 (1966); M. J. Pearson and B. S. Rabinovitch, *ibid.*, **42**, 1624 (1965); B. S. Rabinovitch, R. F. Kubin, and R. E. Harrington, *ibid.*, **38**, 405 (1963). R. E. Harrington, B. S. Rabinovitch, and R. W. Diesen, *ibid.*, **32**, 1245 (1960). B. S. Rabinovitch and R. W. Diesen, *ibid.*, **30**, 735 (1959).

13. J. Magee and K. Funabashi, *J. Chem. Phys.*, **34**, 1715 (1961); J. C. Lorquet, S. G. El Komoss, and J. Magee, *ibid.*, **35**, 1991 (1962).

14. See, for example, L. Salem, and J. Wright, *J. Amer. Chem. Soc.*, **91**, 5947 (1969) and the references contained therein.

Collisional Transfer of Triplet Excitations Between Helium Atoms*†

MITSUHIKO KODAIRA‡ AND TSUTOMU WATANABE

Department of Applied Physics, University of Tokyo

Abstract

The cross sections for 2^3S and 2^3P excitation transfer processes between He atoms are calculated by the perturbed stationary state method. The potential energy curves for $He(1^1S) + He^*(2^3P)$ and $He(1^1S) + He^*(2^3S)$ systems are obtained by the Heitler-London method taking the configuration interaction between them into account. The calculation involves the transfer, the diffusion, and the total elastic cross sections. The results show that the transfer cross section for 2^3P is 15–$23\pi a_0^2$ at thermal energy and larger than that for 2^3S. The transfer cross section for 2^3S does not so much differ from the calculation using rigorous potentials by Kolker and Michels [*J. Chem. Phys.*, **50**, 1762 (1969)].

*Reprinted from *Journal of the Physical Society of Japan*, **27**, 1301 (1969) by permission of the copyright owner.

† A part of this work is presented to the Sixth International Conference on the Physics of Electronic and Atomic Collisions, at Massachusetts Instutute of Technology, Massachusetts, U.S.A., July 28–August 2, 1969.

‡ Present address: Software Works, Hitachi Ltd., Totsuka, Yokohama.

A Personal Note

MITSUHIKO KODAIRA AND TSUTOMU WATANABE

Even in Japan the term chemical kinetics brings to mind the name Henry Eyring. Eyring's theory on rate processes in chemical kinetics as well as on general quantum chemistry have significantly influenced Japanese research in related fields. A few years ago we had a research meeting at the Institute for Solid State Physics, University of Tokyo, on "the present status of the theory of chemical kinetics." There we found that much of the current work in chemical kinetics is still based in principle on Professor Eyring's idea whereby the wave mechanical treatment of relative motion, accurate calculations for the potential energy surface, and so on, are carried out. In the field of atomic collision in which we are currently associated the method of perturbed stationary state is based on the idea of the adiavatic nuclear motion in many potential energy surfaces among which the electronic transitions are taken into account. In that sense this paper can be considered an extension of Eyring theory—the treatment of relative motion in a surface—to many electronic levels.

Henry Eyring's contributions not only to scientific research but also to under-graduate and graduate education in our country is enormous. The two books *Quantum Chemistry* (Eyring, Walter, and Kimball) and *Theory of Rate Processes* (Glasstone, Laidler, and Eyring) are most instructive and are familiar to Japanese students, for both have been translated into Japanese and have become standard textbooks. When Professor Eyring visited Japan, he gave many impressive lectures. The discussions at that time are still fresh in our memory.

We hope that Professor Eyring will keep his good health for many years to come and continue his outstanding contributions to the academic world. We also look forward to his visiting Japan again someday.

INTRODUCTION

Although the metastable excitation transfer by collision is an important process in gas-phase experiments, there are only a few theoretical papers on the subject.[1-4] Most theoretical papers on the collisional excitation transfer treat the cases of optically allowed excitation.[5,6] For the optically allowed transitions the impact parameter (IP) method can be applied and the cross section becomes large compared with the gas kinetical cross section. For metastable excitation the IP method cannot be applied because of shortness of the interaction range. Buckingham and Dalgarno[1] investigated the transfer of the singlet or triplet $2S$ excitation between helium atoms. They used the perturbed stationary state (PSS) method. Matsuzawa and Nakamura[2,3] treated the process of doublet $2p$ and $2s$ excitation transfers between hydrogen atoms, and compared the PSS method with the IP method. Furthermore, Kolker and Michels[4] performed a rigorous calculation of the interatomic potentials between $He(1^1S)$ and $He*(2^3S)$ and obtained the cross sections of excitation transfer, directly elastic scattering, and excitation diffusion.

Recently Schearer[7] measured the cross section of the excitation transfer of 2^3P_0 to 2^3P_1 and 2^3P_2 levels as a result of collisions with the ground-state helium atoms. For the 2^3S state, Colegrove et al.[8] measured the excitation transfer cross section and compared with the calculation by Buckingham and Dalgarno. Fitzsimmons et al.[9] and Pakhomov and Fugol'[10] discussed the transfer cross section obtained by the empirical potential from diffusion cross sections.

In this paper we calculate the cross section of the processes

$$He(1^1S) + He*((1s)(2p), 2^3P) \rightarrow He*((1s)(2p), 2^3P) + He(1^1S), \quad (1.1)$$

and also

$$He(1^1S) + He*((1s)(2s), 2^3S) \rightarrow He*((1s)(2s), 2^3S) + He(1^1S). \quad (1.2)$$

The interactions in the processes (1.1) and (1.2) are not of a long-range type. The excitation can be transferred only by exchange of electrons. As the incident energy of our interest is in the thermal range, the PSS method is applicable. We ignore the nonadiabatic effects for the determination of the phase shift. The potential energy curves are obtained by the Heitler-London

169

(valence bond) method using the single-term Slater-type orbitals (STO). In the calculation of potential energy curves, the coupling between molecular base energy levels of the $He(1^1S) + He^*(2^3S)$ and $He(1^1S) + He^*(2^3P)$ is taken into account. The purposes of this paper are (1) to examine the validity of the present simple calculation by comparison with more rigorous calculations for process (1.2), and (2) to calculate, for the first time, the cross section for process (1.1). We discuss the difference between ^3S and ^3P excited states in the excitation transfer cross section. From the comparison of the calculations with the experiments for the 2^3S case, a limitation on the adiabatic approximation will be discussed. We can conclude that the transfer cross section for the ^3P excitation is larger than that for the ^3S excitation.

CALCULATION FOR POTENTIAL ENERGY CURVES

In order to calculate the transfer cross sections of the processes by the PSS method, it is necessary to obtain the interaction potential curves of the system of $He + He^*$. The curves are calculated by the usual Heitler-London method. Considering the fact that the cross section will mainly be determined by the interaction at distant nuclear separation, we adopt the single term atomic base wave function of He and He*.

We consider the case where atom A with electrons 1, 2 is in the ground state and atom B with electrons 3, 4 is in an excited state. Let $\psi_A(1, 2)$, $\phi_B^\lambda(3, 4)$ and $\varphi_B(3, 4)$ be the atomic wave functions of 1^1S state, of 2^3P state and of 2^3S state, respectively. The superscript λ indicates the azimuthal quantum number $0, \pm 1$ with respect to the molecular axis of A—B. Since A and B are distant from each other, we express the total electronic wave function (Ψ_{AB}) of the system as the antisymmetrized product of respective atomic wavefunctions. We can write

$$\Psi_{AB}^\lambda = \psi_A(1, 2)\phi_B^\lambda(3, 4) - \psi_A(1, 3)\phi_B^\lambda(2, 4) - \psi_A(1, 4)\phi_B^\lambda(3, 2)$$

$$- \psi_A(3, 2)\phi_B^\lambda(1, 4) - \psi_A(4, 2)\phi_B^\lambda(3, 1) + \psi_A(3, 4)\phi_B^\lambda(1, 2) \quad (2.1)$$

for $He(1^1S) + He^*(2^3P)$, and can obtain Ψ_{AB} for $He(1^1S) + He^*(2^3S)$ by

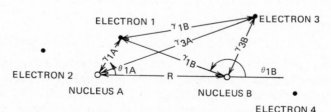

Fig. 1. Coordinate system of nuclei and electrons.

the replacement of $\phi_B{}^\lambda(i,j)$ by $\varphi_B(i,j)$. Further if we take χ_+ or χ_- as

$$\chi_\pm = \Psi_{AB} \pm \Psi_{BA}, \tag{2.2}$$

$\chi_\pm$ are the molecular wavefunctions symmetric or antisymmetric with respect to the interchange of nuclei. This sign is determined by the g or the u character of the state. In He(1^1S) + He*(2^3S), + and − correspond to g and u, respectively. In He(1^1S) + He*(2^3P), the signs should be reversed.

From combinations of atomic wavefunctions (2.1), we have six molecular states: $^3\Sigma_g$, $^3\Sigma_u$, $^3\Pi_g$, $^3\Pi_u$ for He(1^1S) + He*(2^3P) system and $^3\Sigma_g$, $^3\Sigma_u$ for He(1^1S) + He*(2^3S) system. When $\lambda = 0$, Ψ_{AB} belongs to the Σ molecular state and when $\lambda = \pm1$, Ψ_{AB} to the Π state. We shall put the number i to each of these molecular states as $i = 1, 2$ for the Π states, $i = 3$, 4 for the Σ state of He(1^1S) + He*(2^3P) and $i = 5, 6$ for the Σ states of He(1^1S) + He*(2^3S), respectively. Odd (even) i corresponds to + (−) sign in (2.2). Furthermore, we call a state I when $i = 1$ or 2, II when $i = 3$ or 4, and III when $i = 5$ or 6. In this approximation, we can obtain the electronic energy of the system by solving a secular equation

$$\det (H_{ij} - S_{ij}E) = 0, \tag{2.3}$$

where $H_{ij} = \langle \chi_i | \mathbf{H} | \chi_j \rangle$, $S_{ij} = \langle \chi_i | \chi_j \rangle$ and $\mathbf{H}$ is the total Hamiltonian. The interaction energy $v(j)$ of the system corresponding to each state is defined by

$$v(j) = E_j(R) - E_J{}^0. \tag{2.4}$$

Here, $E_J{}^0$ is the sum of atomic energies at infinite separation and $E_j(R)$ is the solution corresponding to the jth molecular state obtained from the secular equation (2.3).

For atomic wavefunctions, we employ the following single term STO:

$$\psi_A(1, 2) = a_{1s}(1)a_{1s}(2)\Theta^S(1, 2) \tag{2.5}$$

$$\phi_B{}^\lambda(3, 4) = \tfrac{1}{2}(b_{1s'}(3)b_{2p}(4) - b_{1s'}(4)b_{2p}(3))\Theta^T(3, 4), \tag{2.6}$$

$$\varphi(3, 4) = \tfrac{1}{2}(b_{1s''}(3)b_{2s}(4) - b_{1s''}(4)b_{2s}(3))\Theta^T(3, 4). \tag{2.7}$$

Here Θ^S and Θ^T are the singlet and the triplet spin eigenfunctions, respectively, of a two-electron system. The base functions employed are

$$a_{1s}(i) = (\alpha^3/\pi)^{\frac{1}{2}} \exp (-\alpha r_{iA}) \tag{2.8}$$

$$b_{1s'}(i) = (\beta^3/\pi)^{\frac{1}{2}} \exp (-\beta r_{iB}) \tag{2.9}$$

$$b_{2p\sigma}(i) = (\gamma^5/\pi)^{\frac{1}{2}}r_{iB} \exp (-\gamma r_{iB}) \cos \theta_{iB}, \tag{2.10}$$

$$b_{2p\pi}(i) = (\gamma^5/\pi)^{\frac{1}{2}}r_{iB} \exp (-\gamma r_{iB}) \sin \theta_{iB} \begin{cases} \cos \varphi_{iB} \\ \sin \varphi_{iB} \end{cases}, \tag{2.11}$$

$$b_{1s''}(i) = (\delta^3/\pi)^{\frac{1}{2}} \exp (-\delta r_{iB}), \tag{2.12}$$

$$b_{2s}(i) = (\mu^5/3\pi N)^{\frac{1}{2}}\{r_{iB} \exp (-\mu r_{iB}) - (3A/\mu) \exp (-\nu r_{iB})\}. \tag{2.13}$$

The orbital exponents are $\alpha = 1.6875$, $\beta = 1.9913$, $\gamma = 0.5441$, $\delta = 2.00$,

$\mu = 1.57$, and $\nu = 0.61$ in atomic units.[11,12] Coefficient A is determined in order to make $b_{1s''}$ orthogonal to $b_{2s'}$ and N is a normalization constant. Some procedures in the calculation of the relevant molecular integrals are shown in detail in the Appendix.

THE POTENTIAL ENERGY CURVES OF He + He*

The calculation described in the preceding paragraph is performed with aid of a HITAC 5020 E Computer at the Computer Center, University of

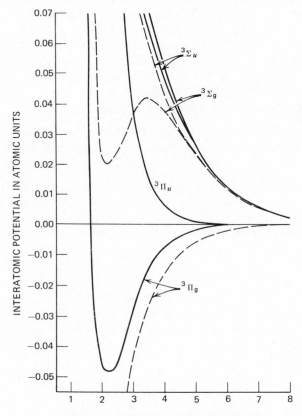

INTERNUCLEAR DISTANCE R IN ATOMIC UNITS

Fig. 2. Interatomic potentials between He(1^1S) and He*(2^3P). The full lines of Σ states include CI with Σ states He(1^1S) + He*(2^3S). The dotted lines of Σ states are without CI. The dotted lines of $^3\Pi_g$ state is the rigorous calculation by Kolker and Michels.[4]

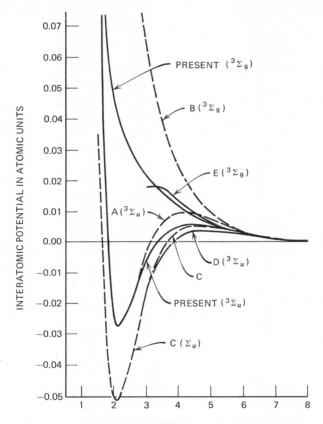

Fig. 3. Interatomic potentials between He(1^1S) and He*(2^3S). The full lines include CI with Σ states of He(1^1S) + H*(2^3P). The dotted lines A and B are without CI as in Buckingham and Dalgarno[1]. Curve C is the calculation by Mateson et al.[18,19] using a 21-term STO function. Curves D and E are by Kolker and Michels.[4]

Tokyo. The results are shown in Figs. 2 and 3.* Figure 2 illustrates the four states ($^3\Sigma_g$, $^3\Sigma_u$, $^3\Pi_g$, $^3\Pi_u$) in the He(1^1S) + He*(2^3P) system. Figure 3 illustrates the two states ($^3\Sigma_g$, $^3\Sigma_u$) in the He(1^1S) + He*(2^3S) systems. Dotted lines are those in the absence of configuration interaction (CI)

---

* The potential curves for He(1^1S) + He*(2^3P) (Fig. 1) in the abstract of the sixth International Conference on the Physics of Electronic and Atomic Collisions are not correct. Particularly, notations g and u should be interchanged as $^3\Pi_u \rightarrow {}^3\Pi_g$, $^3\Sigma_g \rightarrow {}^3\Sigma_u$, and so on.

between $He(1^1S) + He^*(2^3P)$ and $He(1^1S) + He^*(2^3S)$. With CI, the $^3\Sigma_g$ and the $^3\Sigma_u$ in Fig. 2 increase and the $^3\Sigma_u$ in Fig. 3 decrease as expected. Rigorous calculations by Kolker and Michels[4] for the $^3\Sigma_g$ and the $^3\Sigma_u$ state of $He(1^1S) + He^*(2^3S)$ and the $^3\Pi_g$ state of $He(1^1S) + He^*(2^3P)$ are also shown in the figures. There is a remarkable difference between the curves Σ and Π; the $^3\Pi_g$ state is monotonically attractive, whereas Σ states are repulsive at large distances. Furthermore, for the Π states the interactions are of shorter range than that for the Σ states. This feature comes from the direction of the $2p$ orbital. These situations lead to a large transfer cross section of $He^*(2^3P)$ in $He(1^1S)$ as compared with that of $He^*(2^3S)$, as mentioned later.

On the interaction potential for the $He(1^1S) + He^*(2^3S)$ system, there have been some published papers. Phelps and Molnar[13] indicated the existence of a maximum in the $^3\Sigma_u$ state experimentally. Burhop[14] estimated the height 0.0042 a.u. at $R = 4.2a_0$. The corresponding calculated value is 0.01 a.u. by Buckingham and Dalgarno[15] (identical with curve A ($^3\Sigma_u$) in Fig. 3). To reduce the discrepancy between the experimental barrier height and the theoretical one, Matsen's groups[16–19] made systematic theoretical investigations. Brigman et al.[16] calculated the $^3\Sigma_u$ potential using a four-term wavefunction with orbital exponents optimized for each internuclear distance. Poshuta and Matsen[17] and Matsen and Scott[18] calculated the potential by the use of 12- and 21-term wavefunctions with essentially constant orbital exponents, respectively. Klein et al.[19] extended semiempirical theory and made adjustments in the energy at infinite separation to the previous *ab initio* calculation. A rigorous *ab initio* calculation by Kolker and Michels[4] furnishes the potential curves of $^3\Sigma_g$ and $^3\Sigma_u$. They employed an extended elliptic function basis of 12 orbitals projected to form a 32×32 CI and also employed an STO basis of 20 orbitals projected to form a 108×108 CI for large internuclear distance. For comparison some of these potential curves are shown in Fig. 3. If we compare the interaction energy v (not the total electronic energy), a large difference appears only near the equilibrium internuclear distance. The comparison shows that the present calculation of $He(1^1S) + He^*(2^3S)$ (state III) for $R > 4$ a.u., which will be important for the excitation transfer process, is satisfactory as far as interaction energies are concerned. From this result and the comparison in Fig. 2, we may expect that the calculation for $He(1^1S) + He^*(2^3P)$ (states I, II) will not be very different from the true interaction potentials for large internuclear distance. We can conclude that the present interaction potential has sufficient accuracy for the calculation of phase shifts.

When R is sufficiently large, the interaction potential can be expanded in terms of R^{-1}. Since spin multiplicities of the two approaching atoms are different, the asymptotic behavior at large R is exponential, i.e., $e^{-\alpha R}$.

Asymptotic formulas are found to be

$$v(i) \simeq A_{\mathrm{I}}e^{-2\gamma R} \pm (B_{\mathrm{I}} + C_{\mathrm{I}} \ln R)R^4 e^{-(\alpha+\gamma)R} \qquad \text{for} \quad i = 1, 2 \qquad (3.1)$$

$$v(i) \simeq A_{\mathrm{II}}Re^{-2\gamma R} \pm (B_{\mathrm{II}} + C_{\mathrm{II}} \ln R)R^5 e^{-(\alpha+\gamma)R} \qquad \text{for} \quad i = 3, 4, \qquad (3.2)$$

and

$$v(i) \simeq A_{\mathrm{III}}R^2 e^{-2\mu R} \pm (B_{\mathrm{III}} + C_{\mathrm{III}} \ln R)R^5 e^{-(\alpha+\mu)R} \qquad \text{for} \quad i = 5, 6, \qquad (3.3)$$

respectively. Each coefficient is listed in Table I. The sign $\pm$ means $+$ or $-$ according to odd or even i, as mentioned before.

Table I. Coefficients A, B, C, in the Asymptotic Formulas of the Interaction Energies v for He + He*(2^3S) or (2^3P). Each Value is Given in Atomic Units

State	A	B	C
I ($i = 1, 2$)	-6.40×10^{-2}	1.77×10^{-1}	-5.56×10^{-2}
II ($i = 3, 4$)	2.66×10^{-1}	-1.62×10^{-2}	4.40×10^{-3}
III ($i = 5, 6$)	1.95×10^{-1}	3.30×10^{-2}	-1.02×10^{-2}

In the above discussion, the interaction energies are restricted to the first-order perturbation. When $R \gtrsim 6$ a.u., the term from the second-order perturbation, i.e., the van der Waals potential ($C \cdot R^{-6}$), will be important. The van der Waals potential is large between ground and excited atoms compared with that between ground and ground atoms. Although there are not many investigations on the subject, Buckingham[20] obtained a formula of van der Waals constant. Using Buckingham's expression, we calculated the van der Waals constant $C = 23.75$ a.u. for He(1^1S) + He*(2^3P). For He(1^1S) + He*(2^3S), Buckingham and Dalgarno[15] obtained two possibilities $C = 20$ and 30 a.u. from the static polarizability calculation and from some empirical modification to the van der Waals constant. We employed these values for $R \geq 6$ a.u. for the scattering calculation.*

SCATTERING CROSS SECTIONS

The method of perturbed stationary state (PSS) was applied to the heavy particle collision of symmetrical resonance by Buckingham and Dalgarno,[1] Matsuzawa and Nakamura,[2,3] and Kolker and Michels.[4] Although the essential idea of our method is the same as in these papers, except for the inclusion of four levels, we include a brief description of the PSS method for later discussion.

---

* There is a discontinuity of potential curves at $R = 6$, but this discontinuity has a negligible effect on the scattering calculation.

The basic idea of the PSS method is to start from the molecular wavefunction and to take the nonadiabaticity as a perturbation. However, in the case of symmetrical resonance the energy splitting of a g and corresponding u state gives the transition probability even when the nonadiabatic interaction is negligibly small. The Schrödinger equation describing the whole system composed of two atoms is

$$\left[-\frac{1}{2M} \Delta_R + H_0 + H' \right] \Psi(\mathbf{r}, \mathbf{R}) = E\Psi(\mathbf{r}, \mathbf{R}), \tag{4.1}$$

where M is a reduced mass of two atoms, $\mathbf{r}$ is a representative electron coordinate, and $\mathbf{R}$ is the relative coordinate of two nuclei. H_0 is the Hamiltonian of two separated atoms and H' is that of interaction. The molecular electronic wavefunction $\bar{\chi}_i$ (obtained from the solution of secular equation (2.3)) satisfies

$$(H_0 + H')\bar{\chi}_i = E(\mathbf{R})\bar{\chi}_i. \tag{4.2}$$

There are, in our problem, six molecular states which are orthogonal to each other. The total wavefunction Ψ including nuclear motion can be expanded in terms of $\bar{\chi}_i$ as

$$\Psi(\mathbf{r}, \mathbf{R}) = \sum_i u_i(\mathbf{r}, \mathbf{R})\bar{\chi}_i(\mathbf{r}, \mathbf{R}). \tag{4.3}$$

From (4.2) and (4.3) we can obtain the equations for u_i as

$$\{\Delta_R + 2\langle\bar{\chi}_i| \nabla_R |\bar{\chi}_i\rangle \nabla_R + \langle\bar{\chi}_i| \Delta_R |\bar{\chi}_i\rangle + 2M[E - E_i(\mathbf{R})]\}u_i$$
$$= -\sum_{j \neq i}\{2\langle\bar{\chi}_i| \nabla_R \bar{\chi}_j\rangle \nabla_R + \langle\bar{\chi}_i| \Delta_R |\bar{\chi}_j\rangle\}u_j. \tag{4.4}$$

When the incident velocity is very low, we can ignore the term involving the derivative of $\bar{\chi}_j$ with respect to R. We obtain

$$\Delta u_i + \{k^2 - 2Mv_i(\mathbf{R})\}u_i = 0, \tag{4.5}$$

where

$$k^2 = 2M(E - E_J{}^0), \tag{4.6}$$

$$v_j = E_j(\mathbf{R}) - E_J{}^0. \tag{4.7}$$

When R is large, the asymptotic form of u_i in (4.4) should be

$$u_i(R) \sim \alpha_i\left(e^{ikZ} + \frac{1}{R} e^{ikR}f_i(\theta)\right), \tag{4.8}$$

where Z, θ, φ are the polar coordinates of $\mathbf{R}$. Of the total wavefunctions given by (4.3), the solution of interest should correspond to the outgoing state (with respect to the direction of Z) of an excited helium. For this, if the nuclei are bosons, (4.8) should be

$$u_i(R) \sim \frac{1}{2}\left\{ e^{ikZ} + e^{-ikZ} + \frac{e^{ikR}}{R} [f_+(\theta) + f_+(\pi - \theta)] \right\}, \qquad (4.9)$$

for i odd, and

$$u_i(R) \sim \frac{1}{2}\left\{ e^{ikZ} - e^{ikZ} + \frac{e^{ikR}}{R} [f_-(\theta) - f_-(\pi - \theta)] \right\}, \qquad (4.10)$$

for i even. In the case of fermion nuclei, the suffixes of f will be reversed. The scattering amplitude $f(\theta)$ is given by

$$f(\theta) = \frac{1}{2}[f_+(\theta) + f_-(\theta) + f_+(\pi - \theta) - f_-(\pi - \theta)]. \qquad (4.11)$$

The scattering amplitudes $f_\pm(\theta)$ can be expanded in terms of Legendre polynomials and the derivation of the phase from that of u_i gives the phase shifts $\eta_l^\pm$ for each l as

$$f_\pm(\theta) = (2ik)^{-1} \sum_{l=0}^{\infty} (2l + 1)[\exp(2i\eta_l^\pm) - 1]P_l(\cos\theta). \qquad (4.12)$$

Various cross sections[21] can be expressed in terms of $\eta_l^\pm$ as

$$Q_{\text{tot}} = 2\pi \int_0^\pi |f(\theta)|^2 \sin\theta \, d\theta$$

$$= \frac{4\pi}{k^2} \sum_{l=0}^{\infty} (2l + 1) \sin^2 \delta_l, \qquad (4.13)$$

$$Q_D = 2\pi \int_0^\pi |f(\theta)|^2 (1 - \cos\theta) \sin\theta \, d\theta$$

$$= \frac{4\pi}{k^2} \sum_{l=0}^{\infty} (2l + 1) \sin^2(\delta_{l+1} - \delta_l), \qquad (4.14)$$

where $\delta_l = \eta_l^+$, $\delta_{l+1} = \eta_{l+1}^-$ for even l and $\delta = \eta_l^-$, $\delta_{l+1} = n_{l+1}^+$ for odd l. Since large-angle scattering seldom occurs in heavy-particle collisions, the scattering amplitude $f_\pm(\theta)$ is concentrated around $\theta = 0$ and $\theta = \pi$. Therefore, we can write the square of the total elastic scattering amplitude $|f(\theta)|^2$ as

$$|f(\theta)|^2 = \frac{1}{4}\{|f_+(\theta) + f_-(\theta)|^2 + |f_+(\pi - \theta) - f_-(\pi - \theta)|^2\}. \qquad (4.15)$$

The first term furnishes the differential cross section of direct elastic scattering, and the second term that of excitation transfer scattering. With this approximation

$$Q_{\text{tr}} = \frac{\pi}{k^2} \sum_{l=0}^{\infty} (2l + 1) \sin^2(\eta_l^+ - \eta_l^-), \qquad (4.16)$$

$$Q_{\text{tot}} = \frac{2\pi}{k^2} \sum_{l=0}^{\infty} (2l + 1)(\sin^2 \eta_l^+ + \sin^2 \eta_l^-), \qquad (4.17)$$

$$Q_D = \frac{2\pi}{k^2} \sum_{l=0}^{\infty} (2l + 1)[\sin^2(\eta_l^+ - \eta_{l+1}^-) + \sin^2(\eta_{l+1}^+ - \eta_l^-)], \qquad (4.18)$$

The phase shifts $\eta_l^\pm$ are calculated by the method of JWKB approximation with Langer's modification[1,22]

$$\eta_l^\pm = \int_{R_0'}^{\infty} \left(k^2 - 2Mv_\pm(R) - \frac{(l + \frac{1}{2})^2}{R^2} \right)^{1/2} dR - \int_{R_0}^{\infty} \left(k^2 - \frac{(l + \frac{1}{2})^2}{R^2} \right)^{1/2} dR, \quad (4.19)$$

where R_0 and R_0' are zeros of the respective integrands.

The formulas for various cross sections in (4.16) and (4.18) should take the sum over integer l to infinity. When l is large ($> l_0$), l can be considered a continuous argument and the summation can be replaced by an integration. Here the parameter l_0 is determined by the internuclear distance $R_0 = l_0/k$ which satisfies the condition $v_\pm(R_0)/k^2 \ll 1$ (of the order of 10^{-2}).

RESULTS AND DISCUSSION

$He(1^1S) + He(2^3P)$

The cross sections of transfer, diffusion, and total elastic scattering were obtained and are illustrated as functions of incident energy in Figs. 4–6.

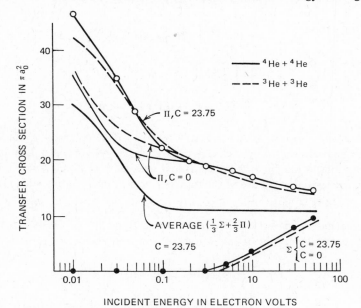

Fig. 4. Excitation transfer cross section for $He(1^1S) + He^*(2^3P)$ as a function of incident energy. $C = 23.75$ and $C = 0$ mean the cross sections with and without the van der Waals potential, respectively.

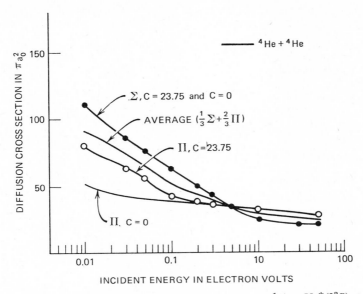

Fig. 5. Excitation diffusion cross section for $He(1^1S) + He^*(2^3P)$ as a function of incident energy. The meaning of $C = 23.75$ and $C = 0$ is the same as before. The isotope effect of $^3He + ^3He^*$ is negligibly small in this case.

It should be noticed that the cross section in the Π state increases with decreasing incident energy whereas the cross section in the Σ state almost disappears below 0.3 eV. This difference comes from the behavior of inter-atomic potentials for distant R. For Σ states, the interactions are both repulsive on $R \gtrsim 4$ a.u. Some amount of incident energy is needed for the colliding atoms to enter the region where the separation of $^3\Sigma_g$ and $^3\Sigma_u$ levels becomes significant. On the other hand, the interaction in the $^3\Pi_g$ state is attractive and no excess incident energy is needed for atoms to enter the effective region of excitation transfer. The transfer cross section for $He(1^1S) + He^*(2^3P)$ at thermal energy is about $23\pi a_0{}^2$ (in the case of $C = 23.75$ a.u.) and is larger than that for $He(1^1S) + He^*(2^3S)$. The process is attributable to the attractive character of the $^3\Pi_g$ state. A slight effect of van der Waals potential on the cross section is found only in low energy region (below 0.2 eV) as expected.

On the other hand, the cross section of excitation diffusion in Σ states is greater than in Π states (Fig. 5). The diffusion cross section corresponds to the subtraction of the contribution of small angle scattering from the total elastic scattering. The cross section is contributed not only from the transfer process but also from the large-angle elastic scattering. The scattering in the

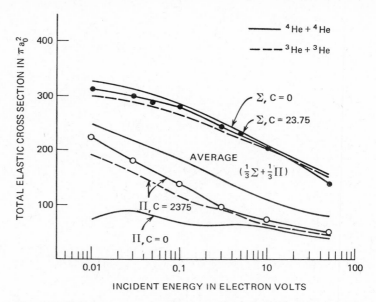

Fig. 6.   Total elastic cross section for $He(1^1S) + He^*(2^3P)$ as a function of incident energy. The meaning of $C = 23.75$ and $C = 0$ is the same as before.

two repulsive Σ states dominates the diffusion cross section through large-angle elastic process. In this case, the van der Waals effect is significant only for the Π states in low incident energy. The Σ scattering is little affected by the presence of the van der Waals potential, since the strong repulsive potential is dominant in this case.

The small-angle scattering dominates the total elastic scattering, and the total cross section is determined by the magnitude of each potential. Then the scattering cross section in the Π states is smaller than that in the Σ states. The van der Waals effect is significant only at low incident energy in the Π states as in the diffusion. The measurement of sublevel mixing among 2^3P_0, 2^3P_1, and 2^3P_2 states in collision with the ground-state He has been made by Schearer.[7] The cross section for collision-induced transition out of the 2^3P_0 level is $(68 \pm 3) \times 10^{-16}$ cm^2 $(77 \pm 3\pi a_0^2)$ at room temperature. A main part of the total cross section in our calculation comes from the scattering

$$He(1^1S) + He^*(2^3P) \rightarrow He(1^1S) + He^*(2^3P). \tag{5.1}$$

The total cross section $220\pi a_0^2$ (in the case of $C = 23.7$) at thermal energy is not inconsistent with experimental data. If we neglect the energy difference between the three sublevels, the relative probabilities of 2^3P_0: 2^3P_1:2^3P_2 out of 2^3P_0 would be $1:3:5$. However, it has been shown in the treatment of the IP method[23–25] that the resonance defect among sublevels is important. One should take the effect of resonance defect into account. One may expect that the fraction $\frac{8}{9}$ will be decreased by the effect of resonance defect.

$He(1^1S) + He(2^3S)$

Recently, Kolker and Michels[4] published the paper on the process using more rigorous interatomic potentials. At present, their calculation is conclusive within the limit of PSS and adiabatic approximation. Our calculation for the process is meaningful as a check of accuracy of the simple wavefunctions employed. Figures 7–9 illustrate the excitation transfer, excitation diffusion, and total elastic cross sections. For comparison, the data by Kolker and Michels and others are also shown. Essential features of the different calculation are the same. Accordingly, present calculations including the case of He*(2^3P) are considered to give the behavior and magnitude of the cross section properly. According to Kolker and Michels, the discrepancies with experiments[8–10,26] for the transfer and diffusion cannot be reduced by

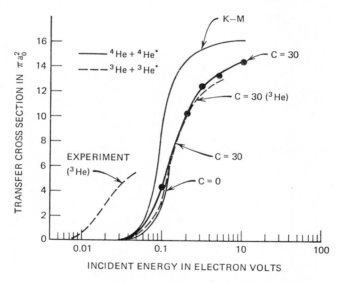

Fig. 7. Excitation transfer cross section for He(1^1S) + He*(2^3S) as a function of incident energy; $C = 30$ and $C = 0$ mean the cross sections with and without the van der Waals potential. In the case of $C = 20$ the cross section lies between the curves of $C = 30$ and $C = 0$. Curve K-M is by Kolker and Michels. "Experiment" refers to data by Colegrove, Schearer, and Walters.[8]

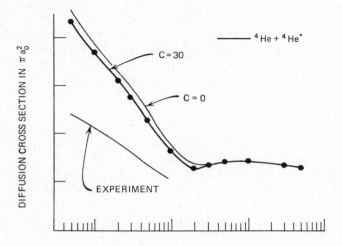

Fig. 8. Excitation diffusion cross section for He(1^1S) + He*(2^3S) as a function of incident energy. The meaning of $C = 30$ and $C = 0$ is the same as before. The curve for $C = 20$ lies between $C = 30$ and $C = 0$. "Experiment" is the data by Fitzsimmons, Lane, and Walters.[9] Isotope effect for ^{3}He is negligibly small in this case.

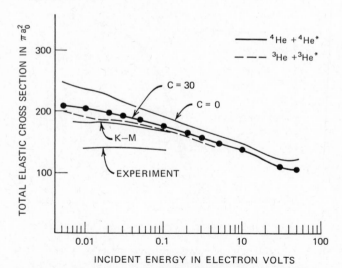

Fig. 9. Total elastic cross section for He(1^1S) + He*(2^3S) as a function of incident energy. The meaning of $C = 30$ and $C = 0$ is the same as before. The curve for $C = 20$ lies between $C = 30$ and $C = 0$. Curve K-M is by Kolker and Michels.[4] "Experiment" is the data by Rothe, Neynaber, and Trujillo.[26]

the employment of rigorous interatomic potentials. If experimental data are correct, we should evoke another mechanism.

In the incident energy region below 0.5 eV, the backward scattering of excitation hardly occurs. This is because the incident atom cannot enter the region where the separation between the g and u energy levels is significant. That is, the rapid decrease of the excitation transfer cross section (Fig. 7) with decreasing incident energy is due to the repulsive property of the interatomic potentials at $R > 4$ a.u. Coming back to the PSS method it is noted that we have ignored the nonadiabatic terms to determine u_i in (4.4). However, in this incident energy region (near the threshold of transfer cross section), there remains other possibilities of the excitation transfer through nonadiabatic term such as $\langle \chi_i | \nabla_R | \chi_j \rangle \nabla_R$ and $\langle \chi_i | \nabla_R^2 | \chi_j \rangle$ in (4.4).

Isotope Effect

In the case of ^{3}He nuclei, (4.11) should be replaced by

$$f(\theta) = \tfrac{1}{2}[f_+(\theta) + f_-(\theta) - f_+(\pi - \theta) + f_-(\pi - \theta)]. \qquad (5.2)$$

When the overlap between $f_\pm(\theta)$ and $f_\pm(\pi - \theta)$ is ignored, there is no difference between the expressions for $|f(\theta)|^2$ and those for $|f_+(\pi - \theta) - f_-(\pi - \theta)|^2$. Then isotope effect comes only from the difference of masses of nuclei. The isotope effects are shown in Figs. 4–9. The differences between the cross sections are small.

APPENDIX

Calculation of Two-Centered Molecular Integrals

By means of two-electron two-centered molecular integrals, Coulomb and ionic integrals can be reduced to one-electron integrals, and use can be made of Coulson's table.[27] * Since the exchange integrals cannot be reduced to

* In the table, $K_{49}(\alpha, \rho)$ should be corrected in his notation

$$K_{49}(\alpha, \rho) = \int \frac{1}{r_a} \cos^2 \theta_a \exp\left[-\alpha(r_a + r_b)\right] d\tau$$

$$= \frac{\pi \rho^2}{t^5}[-4(15 - 15t + 6t^2 - t^3)e^t E_i^*(2t)$$

$$- 4(15 + 15t + 6t^2 + t^3)e^{-t}(\gamma + \ln 2t)$$

$$+ t(120 + 60t + 15t^2 + t^3)e^{-t}],$$

where $t = \alpha\rho$, $\gamma = $ Euler's constant, and

$$E_i^*(x) = -E_i(-x) = \int_x^\infty (e^{-t}/t)\, dt.$$

one-electron integrals, we make use of Neumann expansion by elliptic coordinates,[28]

$$\frac{1}{r_{12}} = \frac{2}{R} \sum_{\tau=0}^{\infty} \sum_{\nu=0}^{\tau} D_{\tau\nu} Q_{\tau}^{\nu}(\lambda_>) P_{\tau}^{\nu}(\lambda_<) P_{\tau}^{\nu}(\mu_1) P_{\tau}^{\nu}(\mu_2) \cos \nu(\phi_1 - \phi_2), \quad (A.1)$$

where

$$\lambda = (r_A + r_B)/R, \qquad \mu = (r_A - r_B)/R,$$

and

$$D_{\tau 0} = 2\tau + 1,$$
$$D_{\tau\nu} = (-)^{\nu} 2(2\nu + 1)[(\tau - \nu)!/(\tau + \nu)!],$$

and $\lambda_>$ ($\lambda_<$) is the larger (smaller) of λ_1 and λ_2, and P_{τ}^{ν} and Q_{τ}^{ν} are the associate Legendre functions of 1st and 2nd kind, respectively. Each term can be expressed by $G_{\tau}^{\nu}(m, \alpha)$ and $W_{\tau}^{\nu}(m, n; \alpha, \beta)$ where

$$G_{\tau}^{\nu}(m, \alpha) = \int_{-1}^{1} e^{-\alpha\mu} P_{\tau}^{\nu}(\mu) \mu^{m} (1 - \mu^2)^{\nu/2} \, d\mu,$$

$$W_{\tau}^{\nu}(m, n; \alpha, \beta)$$
$$= \int_{1}^{\infty} \int_{1}^{\infty} Q_{\tau}^{\nu}(\lambda_>) P_{\tau}^{\nu}(\lambda_<) e^{-\alpha\lambda_1 - \beta\lambda_2} \lambda_1^{m} \lambda_2^{m} (\lambda_1^2 - 1)^{\nu/2} (\lambda_2^2 - 1)^{\nu/2} \, d\lambda_1 \, d\lambda_2.$$

Functions $G_{\tau}^{\nu}(m, \alpha)$, $W_{\tau}^{\nu}(m, n; \alpha, \beta)$ were obtained by the method described in Kotani et al.'s table.[28] The summation over τ should, in principle, be taken from 0 to ∞. In the present calculation, we took the first 10 terms for Π states, and the first 6 terms for Σ states. When R is extremely large the error from the finite sum may be significant. However, in the region $R < 15$ a.u., the results do not depend on the number of terms taken into account.

ACKNOWLEDGMENTS

This paper is dedicated to Professor Henry Eyring with sincere respect and gratitude, on his seventieth birthday, for his outstanding contribution to quantum chemistry.

The authors would like to express their thanks to Dr. M. Matsuzawa, Dr. H. Nakamura, and Dr. M. Natori of University of Tokyo for an offer of some programs in the calculation of interatomic potentials and for advice and checks in computational calculation. This work was partially supported by the Grant-in-Aid for Fundamental Scientific Research from the Ministry of Education. Computational calculations were carried out at the Computer Center, University of Tokyo.

REFERENCES

1. R. A. Buckingham and A. Dalgarno, *Proc. Roy. Soc. (London)*, **A213**, 506 (1952).
2. H. Nakamura and M. Matsuzawa, *J. Phys. Soc. Japan*, **22**, 248 (1967).
3. M. Matsuzawa and H. Nakamura, *J. Phys. Soc. Japan*, **22**, 312 (1967).
4. H. J. Kolker and H. H. Michels, *J. Chem. Phys.*, **50**, 1762 (1969).
5. T. Watanabe, *Radiation Chemistry*. II. *Advances in Chemistry Series*, R. F. Gould, Ed., American Chemical Society, Washington D.C., 1968, Vol. 82, p. 176, and references cited therein.
6. T. Watanabe, *J. Chem. Phys.*, **46**, 3741 (1967) and references cited in this paper.
7. L. D. Shearer, *Phys. Rev.*, **160**, 76 (1967).
8. F. D. Colegrove, L. D. Shearer, and G. K. Walters, *Phys. Rev.*, **135**, A353 (1964).
9. W. A. Fitzsimmons, N. Lane, and G. K. Walters, *Phys. Rev.*, **174**, 193 (1968).
10. P. L. Pakhomov and I. Y. Fugol', *Dokl. Akad. Nauk SSSR*, **179**, 813 (1968).
11. C. W. Scherr, F. C. Sanders, and R. E. Knight, *Perturbation Theory and Its Application in Quantumn Mechanics*, C. H. Wilcox, Ed., Wiley, New York, 1965, p. 97.
12. M. Morse, L. A. Young, and E. S. Hanwitz, *Phys. Rev.*, **48**, 948 (1935).
13. A. V. Phelps and J. B. Molnar, *Phys. Rev.*, **89**, 1203 (1953).
14. E. H. S. Burhop, *Proc. Roy. Soc. (London)*, **A67**, 276 (1954).
15. R. A. Buckingham and A. Dalgarno, *Proc. Roy. Soc. (London)*, **A213**, 237 (1952).
16. G. H. Brigman, S. J. Brient, and F. A. Matsen, *J. Chem. Phys.*, **34**, 958 (1961).
17. R. D. Poshuta and F. A. Matsen, *Phys. Rev.*, **132**, 307 (1963).
18. F. A. Matsen and R. D. Scott, *Quantum Theory of Atoms, Molecules and the Solid State*, Academic, New York, 1966.
19. D. J. Klein, E. M. Greenawalt, and F. A. Matsen, *J. Chem. Phys.*, **47**, 4820 (1967).
20. R. A. Buckingham, *Proc. Roy. Soc. (London)*, **160**, 113 (1937).
21. S. Chapman and T. G. Cowling, *The Mathematical Theory of Nonuniform Gases*, Cambridge University Press, Cambridge, 1952.
22. R. E. Langer, *Phys. Rev.*, **51**, 669 (1937); N. F. Mott and H. S. W. Massey, *The Theory of Atomic Collisions*, 2nd ed., Clarendon, Oxford, 1965, p. 98.
23. H. Nakamura, *J. Phys. Soc. Japan*, **20**, 2272 (1965).
24. J. Callaway and E. Bauer, *Phys. Rev.*, **140**, A1072 (1965).
25. J. Callaway and A. F. Dugan, *Phys. Rev.*, **163**, 26 (1967).
26. E. W. Rothe, R. H. Neynaber, and S. M. Trujillo, *J. Chem. Phys.*, **42**, 3310 (1965).
27. C. A. Coulson, *Proc. Cambridge Phil. Soc.*, **38**, 210 (1942).
28. M. Kotani, A. Amemiya, E. Ishiguro, and T. Kimura, *Tables of Molecular Integrals*, 2nd ed., Maruzen, Tokyo, 1963,

The Mechanism of Electronic Energy Transfer Between Excited Mercury ($^3P_1$) Atoms and Gaseous Paraffins*

KANG YANG

Central Research Division, Research and Development Department, Continental Oil Company, Ponca City, Oklahoma

Abstract

Excited mercury atoms in the $^3P_1$ state react with a paraffin in three different ways: (a) quenching to the metastable state ($^3P_0$), (b) quenching to the ground state ($^1S_0$) with the emission of light, and (c) quenching to the ground state with the rupture of a CH bond. The comparison of quenching rates determined by physical and chemical methods indicates that in the quenching by C_2H_6 and C_3H_8 process (c) is the major one, while in the quenching by $C(CH_3)_4$ and $CH_3CD_2CH_3$ process (a) predominates. The relative quenching rates between an isotopic pair, $CH_3CH_2CH_3$–$CH_3CD_2CH_3$, hardly change as the temperature is raised from 25 to 202°. The latter observation is used to draw two conclusions: (a) the resonance energy rule is not applicable, and (b) the potential barrier that can be surmounted by thermal motion is not responsible for the isotope effect. In the proposed model of quenching, an excited mercury atom with a definite value of J is assumed to form a planar complex with a paraffin which is approximated as a diatom RH. Symmetry arguments indicate that, while the quenching of $^3P_1$ to $^1S_0$ is allowed, the quenching of $^3P_0$ to

* Reprinted from *J. Am. Chem. Soc.*, **89**, 5344–5350 (October 1967). Copyright 1967 by the American Chemical Society. Reprinted by permission of the copyright owner.

187

$^1S_0$ is forbidden if the product R is in the S_g state. Available data support this conclusion quite well. Symmetry arguments also indicate that the quenching of $^3P_1$ to $^3P_0$ proceeds by the rotational excitation of RH. This is consistent with the observed increase in the metastable atom formation by D substitution. The mean life of the complex is assumed to be governed by two factors: (a) the polarizability of RH, which determines the rate of decomposition of the complex back to reactants, and (b) the CH bond strength, which determines the rate of the decomposition of the complex to HgH and R·. This model explains quite well the various differences observed among structurally similar paraffins in the quenching of $^3P_1$ atoms.

INTRODUCTION

In reaction kinetics two of the most often used of Eyring's many contributions are the concept of activated complex and the use of potential energy surface to follow the course of an elementary reaction. These are employed in the present paper to explore the mechanism of some photosensitized reactions.

Excited mercury atoms, $Hg^*(^3P_1)$, react with paraffins in three different ways: (1) quenching to the metastable state,[1] $Hg'(^3P_0)$; (2) quenching to the ground state with an emission of light[2]; and (3) quenching to the ground state with CH bond rupture.[3] Among structurally similar paraffins, the relative rates of these reactions are fascinatingly different. Thus, metastable atom formation is appreciable in the quenching by $C(CH_3)_4$ but not in quenching by any other undeuterated paraffins.[1,4,5] In general, deuteration increases the metastable atom formation. This increase is most drastic when the weakest bond, such as the secondary CH in propane, is deuterated.[1] Overall quenching efficiency increases in the order, $CH_4 < C_2H_6 < C_3H_8 < i\text{-}C_4H_{10}$, while the light emission from reaction (2) increases with the reverse order.[2] The efficiency of reaction (3) is usually reported in terms of the quantum yield, $\phi(H_2)$, of hydrogen measured at very low conversions and at very low intensities.[6,7] At atmospheric pressure and room temperature, $\phi(H_2)$ in methane quenching is nearly zero,[8] while in others it is close to unity.[7] In methane quenching, $\phi(H_2)$ increases markedly with increasing temperature and decreases with decreasing pressure. In quenching by other paraffins, $\phi(H_2)$ also decreases with decreasing pressure, this decrease being more pronounced in paraffins with stronger CH bonds.[7] To date, all these variations have not been consistently explained. The object of the present work is, then, to provide some experimental data which help to elucidate these variations and to propose an energy-transfer mechanism on the basis of which various observations can be rationalized.

The first part of the experiment investigates the temperature dependence of the isotope effect between a pair, $CH_3CH_2CH_3$–$CH_3CD_2CH_3$. This is

189

important for two reasons. Previously,[3b] a markedly lower quenching efficiency of $CH_3CD_2CH_3$ was explained by using absolute rate theory.[9] In this theory, the isotope effect mainly arises because of the difference in the height of the potential barrier that must be surmounted with thermal energy. This difference, $\Delta E_{1/2}$, originates from the difference in the zero-point energies of CH and CD bonds and contributes a factor, $\exp(\Delta E_{1/2}/RT)$, which governs the magnitude of the isotope effect. Hence, temperature should have a marked effect. It is tempting to suppose the existence of such a barrier, particularly in view of the quantitative agreement between theory and experiment at room temperature. Experimental data, however, show that the predicted temperature dependence is absent. The second reason for investigating the temperature dependence is to test the applicability of the resonance-energy rule in the quenching of $^3P_1$ atoms to the $^3P_0$ state. Arguments to be presented later show that the present experimental result is incompatible with this rule.

The second part of the experiment concerns the relative rates of the three reactions mentioned above. Here it is important to note the reported differences[1] in some quenching rates estimated by two different methods. In one method, called the physical method, the quenching cross section (σ^2_{phys}) is determined relative to the rate of fluorescence, $Hg^* \to Hg + h\nu$, while in the other method, called the chemical method, the quenching cross section (σ^2_{chem}) is determined relative to the reaction $Hg^* + N_2O \to Hg + O + N_2$. For some paraffins, such as C_3H_8, the two cross sections agree quite well, but for those paraffins which show appreciable quenching to the metastable state, σ^2_{phys} is found to be much larger than σ^2_{chem}. Using a recent modification of the physical method, we confirm this difference. It is then shown how the relative rates of the three reactions affect the magnitude of the above difference.

The last section treats a model of the collision complex which qualitatively explains why similar paraffins often behave so differently in these energy-transfer reactions.

EXPERIMENTAL SECTION

Material

Phillips' research grade hydrocarbons and Matheson's nitrous oxide were purified as described previously.[10] Deuteriopropane from Merck Sharp and Dohme Co. (Montreal, Canada) was passed through a H_2SO_4–P_2O_5 mixture to remove olefins and through a KOH trap to remove any acid spray. It was then thoroughly degassed.

Physical Method

The quenching of mercury fluorescence was investigated using equipment described previously[11] with a modification that the temperature of the fluorescence cell was kept within $\pm 1°$ at different temperatures, using a heater inserted in an aluminum block surrounding the cell. As before, the temperature of the mercury reservoir was kept within $\pm 0.1°$ using an ethyleneglycol bath. Table I shows an example of experimental data.

Table I. The Quenching of Mercury   Fluorescence   by $CH_3CD_2CH_3$[a]

p (torr)	Q_0'	Q'
0.504	101.0	86.5
0.576	100.8	85.0
0.684	100.7	81.8
0.720	100.0	80.6
0.772	99.1	76.4
0.792	101.0	80.8
1.33	101.0	71.5
1.66	100.2	66.3
2.45	100.0	59.9
3.49	99.2	51.5

[a] Cell temperature, 25°; mercury reservoir temperature, 20°; Q_0' and Q' denote photocurrents (arbitrary unit) in the absence and presence of $CH_3CD_2CH_3$.

Chemical Method

The equipment used in the determination of the quantum yield of nitrogen in the mercury-sensitized photolysis of N_2O–M systems was the same as before[10] except that the Vycor cell was placed in a temperature-controlling box equipped with a fan, heater, and quartz window. Nitrogen was analyzed using a Porapak Q column at room temperature with helium as a carrier gas. Both the stability and the sensitivity of this column were superior to the charcoal–molecular sieve column used in our earlier study. At room temperature, the column would not separate nitrogen from oxygen, but oxygen was absent in the present experiment. Nitrous oxide, which was not absorbed irreversibly, was backflushed as before. Both the plot of peak height against nitrogen pressure and the plot of peak height against irradiation time were

Table II.   Nitrogen Formed at Various $[C_3H_8]/[N_2O]^a$

$[C_3H_8]/[N_2O]$	$[N_2]$ (torr)	ϕ
1.00	0.232	0.914
2.08	0.214	0.843
3.00	0.209	0.823
5.15	0.175	0.690
7.00	0.159	0.626
8.10	0.148	0.583
9.00	0.138	0.544
10.40	0.137	0.540
12.30	0.133	0.524
15.00	0.119	0.469
25.70	0.0954	0.376
49.00	0.0695	0.274
65.70	0.0604	0.238
79.00	0.0563	0.222
99.00	0.0487	0.192

[a] Total pressure, 400 torr; irradiation time, 10 min; cell temperature, $28 \pm 2°$; mercury reservoir temperature, $20 \pm 0.2°$; intensity, 0.0254 torr/min.

good straight lines passing through the origin. Other details of similar experiments were described before.[7,10,11] Table II shows an example of experimental data.

RESULTS

Temperature Dependence of Isotope Effect

It is convenient to report the results of quenching experiments using the equation

$$\frac{Q_0'}{Q_0' - Q'} = \alpha + \beta[M]^{-1}, \tag{I}$$

where Q_0' and Q' are photocurrents in the absence and presence of M, while α and β are constants independent of [M]. As shown later, the ratio α/β is proportional to the rate constant of the quenching. According to (I), the plot of $Q_0'/(Q_0' - Q')$ against $[M]^{-1}$ should be a straight line. Such plots are shown in Fig. 1 for $CH_3CH_2CH_3$ and in Fig. 2 for $CH_3CD_2CH_3$. These data were obtained at a constant mercury reservoir temperature of $20°$ and at various

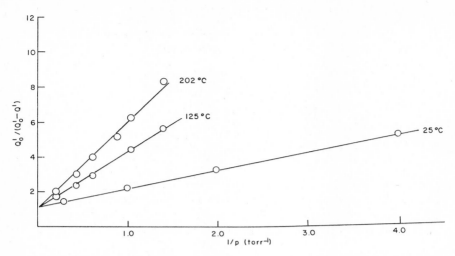

Fig. 1. The quenching of fluorescence by propane at t_{res} 20° and at various temperatures of the fluorescence cell.

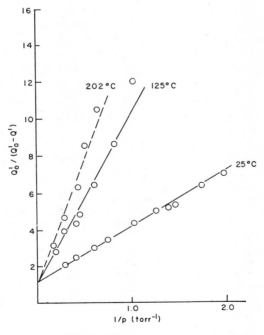

Fig. 2. The quenching of fluorescence by $CH_3CD_2CH_3$ at t_{res} 20° and at various temperatures of the fluorescence cell.

Table III.   The Relative Rate Constant of Quenching at Different Temperatures

Temperature ($^\circ$C)	α/β (torr^{-1})		$\dfrac{(\alpha/\beta)\mathrm{H}}{(\alpha/\beta)\mathrm{D}}$
	C_2H_8	$C_3H_6D_2$	
25	1.04	0.361	2.88
125	0.356	0.128	2.78
202	0.224	0.0835	2.68

temperatures of the fluorescence cell. Table III summarizes the ratio α/β obtained by a least-squares method. With increasing temperature, α/β sharply decreases (see Appendix), but the ratio of the two α/β values for the isotopic pair shows very little, if any, temperature dependence.

Absolute Quenching Rates

In the determination of quenching rates by the physical method, two opposing factors must be compromised. The pressure of M should be low so that the collision broadening of the absorption line can be neglected. Otherwise the amount of light absorbed by Hg atoms depends on the pressure of M, and a meaningful interpretation of the quenching data becomes very difficult.[7] On the other hand, the pressure of M must be high enough to quench the fluorescence appreciably, so that the difference $Q_0' - Q'$ can be measured accurately. At a given pressure of M, more light is quenched when the mercury vapor pressure is high.[11] Hence it is desirable to employ a higher

Table IV.   Constants in the Modified Stern-Volmer Formula[a]

Quencher	Reservoir Temperature ($^\circ$C)	α/β (torr^{-1})
C_2H_4	16.1	13.2
C_3H_8	16.1	0.489
C_3H_8	20.0	1.04
$CH_3CD_2CH_3$	20.0	0.361
$C(CH_3)_4$	20.0	0.734
C_3H_8	25.0	3.37
C_2H_6	25.0	0.693

[a] Equation (1) in text, at a constant cell temperature of 25° and at different mercury reservoir temperatures.

pressure of mercury in an investigation of a compound having a lower quenching rate. In the present work α/β for $CH_3CH_2CH_3$ is measured relative to that for C_2H_4 (whose absolute cross section is known to be 48.2 Å²) at a mercury reservoir temperature, t_{res}, of 16°. For $CH_3CD_2CH_3$ and $C(CH_3)_4$, α/β is determined relative to α/β for $CH_3CH_2CH_3$ at t_{res} 20°, whereas α/β for C_2H_6 and $C(CH_3)_4$ are compared at t_{res} 25°. Results are summarized in Table IV. The quenching rate follows the order: $C_2H_4 > C_3H_8 > C(CH_3)_4 > CH_3CD_2CH_3 > C_2H_6$.

Chemical Method

The results on the quantum yield ϕ of nitrogen are reported by plotting ϕ^{-1} against $X (= [M]/[N_2O])$, as is usually done. As shown in Figs. 3 and 4, the plot is linear if X is not too large. This linearity is expected if Hg^* undergoes a simple competitive reaction with N_2O and M. The slopes of the linear portion agreed well with literature values.[3,10] At high X values the plot curves down markedly both in $CH_3CH_2CH_3$ and $C(CH_3)_4$. It should be noted that if one considers the data only at high X values the plot is apparently linear, but the intercept is higher than unity and the slope is lower. In the C_2H_6–N_2O system the decrease in ϕ even at high X is very small. Because of this, even

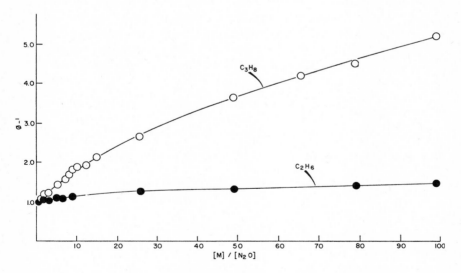

Fig. 3. Quantum yield ϕ of nitrogen at various $[M]/[N_2O]$ in $CH_3CH_2CH_3$–N_2O and CH_3CH_3–N_2O systems.

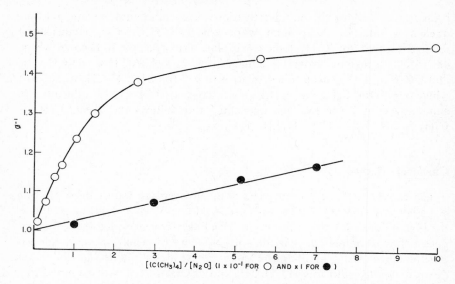

Fig. 4.   Quantum yield ϕ of nitrogen at various [M]/[N₂O] in C(CH₃)₄–N₂O systems.

though the slope at high X values was somewhat lower than at lower X, we could not reach a definite conclusion (see Fig. 3). Figure 5 shows a similar plot for the $CH_3CD_2CH_3$–N_2O system at 25 and 150°. The intercept here may be slightly higher than unity. For a definite conclusion, more experimental data are needed. It is clear, however, that the temperature dependence of σ^2_{chem} is negligible.

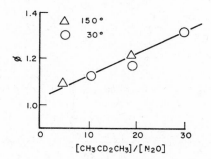

Fig. 5.   Quantum yield ϕ of nitrogen at various [M]/[N₂O] in $CH_3CD_2CH_3$–N₂O systems.

DISCUSSION

Physical Method

The results of quenching experiments are explainable with the following mechanism:

$$Hg + h\nu \rightarrow Hg^*,$$

$$Hg^* \rightarrow Hg + h\nu, \tag{1}$$

$$Hg^* + M \rightarrow (HgM)^*, \tag{2}$$

$$(HgM)^* \rightarrow Hg^* + M \tag{-2}$$

$$\rightarrow Hg + M + h\nu' \tag{3}$$

$$\rightarrow Hg' + M' \tag{4}$$

$$\rightarrow HgH + R, \quad \text{or} \quad H + Hg + R, \tag{5}$$

where ν' denotes the light other than the resonance line, and M' denotes rotationally or vibrationally excited species. The major modification of the previous mechanism is the inclusion of (3), which is shown to be important.[2] This modification considerably alters the rate equation. We suppose that the sensitivity of the photomultiplier tube is the same at ν and ν'; then the resulting rate equation is

$$\frac{Q_0}{Q} = \frac{1 + (k_2/k_1)\tau_c(k_3 + k_4 + k_5)[M]}{1 + (k_2/k_1)\tau_c k_3[M]}, \tag{II}$$

where Q_0 and Q denote the intensity of the light $(\nu + \nu')$ in the absence and presence of M, k_i is the rate constant of the ith reaction, and τ_c is the mean life of the collision complex, defined as

$$\frac{1}{\tau_c} = k_{-2} + k_3 + k_4 + k_5.$$

Most of the experimental data are obtained at less than 50% quenching; hence, $(k_2/k_1)\tau_c k_3[M] < 1$. We then expand the reciprocal of the denominator and retain the terms up to the first power of $[M]$:

$$\frac{Q_0}{Q} = 1 + \frac{k_2}{k_1}\tau_c(k_4 + k_5)[M]. \tag{III}$$

This is the Stern-Volmer formula in which k_2 is replaced by $k_2\tau_c(k_4 + k_5)$. Experimental photocurrents, Q_0' and Q', contain a contribution from stray light, Δ; hence $Q_0' = Q_0 + \Delta$ and $Q' = Q + \Delta$. To take into account the effect of imprisonment, k_1 is replaced by ck_1 $(c < 1)$, as before.[10] With these modifications (III) at once yields the modified Stern-Volmer formula (I), where $\alpha = [1 - (\Delta/Q_0)]^{-1}$ and

$$\frac{\alpha}{\beta} = \frac{k_2\tau_c(k_4 + k_5)}{ck_1}. \tag{IV}$$

The imprisonment correction factor, c, in (IV) depends on the geometry of the cell and the pressure of mercury but not on the nature of the quencher. Hence the ratio of the two α/β values between the pair $CH_3CH_2CH_3$–$CH_3CD_2CH_3$ in Table III represents the isotope effect in $k_2\tau_c(k_4 + k_5)$ values. It is now necessary to define σ^2_{phys} as

$$k_2\tau_c(k_4 + k_5) = \sigma^2_{phys}(8\pi RT/\mu)^{1/2}. \tag{V}$$

For ethylene $\sigma^2_{phys} = 48.2$ Å^2. The use of this value together with the data in Table IV gives various σ^2_{phys} as summarized in Table V. Although relative σ^2_{phys} values agree reasonably with published results,[3] the absolute values are much larger. This is due to the fact that although the published data are based on a theoretical c value the present work employs an experimentally determined c value.[11]

In quenching with nitrogen[12] the rate equation is much more complicated because Hg atoms contribute to photocurrent through the reactions

$$Hg' + N_2 \rightarrow Hg^* + N_2, \tag{a'}$$

$$Hg' \rightarrow Hg + h\nu. \tag{b'}$$

In this case the competing reaction

$$Hg' + N_2 \rightarrow Hg + N_2 \tag{c'}$$

is very slow, $k_c = 4.7$ mm^{-1} sec^{-1} at 25°. Available rate data[5] together with some reasonable approximations indicate that in the present experiments quenching reactions corresponding to reaction (c') are several orders of magnitude faster. Thus, under the experimental conditions in which (III) is valid, Hg' atoms are not likely to contribute significantly to photocurrent.

Table V. Quenching Cross Sections Estimated by Physical and Chemical Methods

Quencher	σ^2_{phys} (Å^2)	σ^2_{chem} (Å^2)
$C(CH_3)_4$	1.9	0.8
$CH_3CD_2CH_3$	0.8	0.3
CH_3CH_3	0.4	0.3
$CH_3CH_2CH_3$	2.2	2.4
$CH_2{=}CH_2$	48.2	48.2[a]

[a] Assumed value; see Ref. 11.

Chemical Method

In the presence of N_2O, in addition to reactions (1)–(5), reactions (6)–(8) should be considered:

$$Hg^* + N_2O \rightarrow N_2 + Hg + O, \tag{6}$$

$$Hg' + M \rightarrow \text{products}, \tag{7}$$

$$Hg' + N_2O \rightarrow Hg + N_2 + O. \tag{8}$$

The occurrence of reaction (6) is well known,[3a] but the occurrence of reaction (8) is at present only a reasonable presumption. In N_2O quenching, the quenching cross section is very high, indicating that $k_{-2} \simeq 0$. Metastable atoms are not detected here; thus $k_4 \simeq 0$. Also, k_3 is likely to be nearly zero because compounds having large cross sections, such as NO and C_2H_4, show no light emission. Hence the decomposition of the complex $(HgN_2O)^*$ is not considered explicitly. At high total pressure of quenchers the rate of reaction (1) is negligible, and the rate equation for the quantum yield of nitrogen is readily obtained (VI). Since $k_2/k_6 \ll 1$, and very likely $k_7/k_8 \ll 1$, the reciprocal of the denominator,

$$\phi^{-1} = \frac{1 + \{(k_7/k_8) + (k_2/k_6)\tau_c(k_3 + k_4 + k_5)\}X + (k_2/k_6)(k_7/k_8)\tau_c(k_3 + k_4 + k_5)X^2}{1 + \{(k_7/k_8) + (k_2/k_6)\tau_c k_4\}X}$$

$$X = \frac{[M]}{[N_2O]} \tag{VI}$$

can be expanded at not too large X. Retaining up to the first power of X, we obtain

$$\phi^{-1} = 1 + \frac{k_2}{k_6}\tau_c(k_3 + k_5)X. \tag{VII}$$

The form of this equation is the same as the one employed previously,[3] but the meaning of the cross section now becomes modified as

$$k_2\tau_c(k_3 + k_5) = \sigma^2_{\text{chem}}(8\pi RT/\mu)^{1/2}. \tag{VIII}$$

At low X values, (VII) is well satisfied, and the slopes together with the known σ^2_{chem} value of 27 Å^2 for N_2O provide σ^2_{chem} values for various paraffins as summarized in Table V.† Here again, relative values agree well with previous

† In estimating σ_{chem} for $CH_3CD_2CH_3$, complications arising from the fact that the intercept may be slightly higher than unity are not taken into account. True initial slope may give a somewhat higher σ^2_{chem} for this compound.

work, but absolute values are higher because a higher value of σ^2_{chem} for N_2O is employed.[11]

At high X values the ϕ^{-1} vs. X plot curves down markedly, indicating that some excess nitrogen other than that due to reaction (6) is formed. According to the present mechanism, reaction (8) supplies this nitrogen, but this cannot be the sole source because, in propane quenching where $k_4 \simeq 0$, the excess nitrogen is also appreciable at high X values. Probable reactions are

$$(\text{HgM})^* + N_2O \rightarrow N_2O^* + (\text{HgM}), \tag{9}$$

$$N_2O^* \rightarrow N_2 + 0, \tag{10}$$

$$(\text{HgM})^* + M \rightarrow (\text{HgM})^{*\prime} + M, \tag{11}$$

$$(\text{HgM})^{*\prime} \rightarrow \text{HgH} + R, \text{ etc.} \tag{12}$$

A basic assumption here is that the complex lives long enough to undergo collisional processes when $[M]$ and $[N_2O]$ are high. Reactions (11) and (12) are consistent with decreasing $\phi(H_2)$ with decreasing $[M]$. Much more extensive data than those given here are needed to prove or disprove the occurrence of these reactions. In the present work the possible occurrence of these reactions is neglected, hoping that the limiting equation (VII) remains substantially unaltered.

Comparison of the Two Cross Sections

Table V shows that the cross sections estimated by the two methods agree well in C_3H_8 quenching and probably also in C_2H_6 quenching but not in $C(CH_3)_4$ or in $CH_3CD_2CH_3$ quenching. This confirms previous work.[1] From (V) and (VIII)

$$\frac{\sigma^2_{\text{phys}}}{\sigma^2_{\text{chem}}} = \frac{k_4 + k_5}{k_3 + k_5}. \tag{IX}$$

If k_5 is much larger than k_4 or k_3, the two cross sections agree. Since $\sigma^2_{\text{phys}} > \sigma^2_{\text{chem}}$ in $C(CH_3)_4$ and $CH_3CD_2CH_3$ quenching, we conclude that here $k_4 > k_3$. The numerical values in Table V also indicate that $k_4 \geq k_5$. This is consistent with experimental observations that detectable Hg' atoms are formed in those quenchings where $\sigma^2_{\text{phys}} > \sigma^2_{\text{chem}}$.

The Collision Complex and Quenching Mechanism

In considering the quenching mechanism, it is important to note several facts which indicate that, in mercury, spin-orbit coupling is very strong and the total angular momentum quantum number J is the only reliable quantum number. The transition $^3P_1 \rightarrow {}^1S_0$ occurs readily even though the spin selection rule is violated, while the transition $^3P_0 \rightarrow {}^1S_0$, which violates the J selection rule, is forbidden. The quenching of Hg* atoms to the ground state by CO occurs[5] with high efficiency in spite of the fact that here the spin conservation rule is violated. Thus, a function of total angular momentum must be employed to represent the mercury orbital. Such an approach was first tested in the quenching of Hg* by N_2 with a satisfactory result.[13]

In the present work we assume the following model. An excited mercury atom, with a definite J value, forms a planar complex† with a paraffin which is approximated as a diatom, RH; the H atom to which R and mercury are bound is supposed to come from the weakest bond in the paraffin.‡ In the quenching of Hg*($^3P_1$) atoms, a total cross section for a molecule can often be estimated by simply adding the contribution from each bond.[3] This suggests as an approximation that, in the quenching, each bond behaves independently of the other; one is then justified to regard a paraffin as a perturbed diatom, CH, as is done in the above model. This is a crude approximation; it is, however, a useful one, and provides some illuminating results on the basis of symmetry arguments.[15,16] For the quenching reaction to proceed readily a state of the complex arising from the reactant side and a state of the complex arising from the product side must belong to the same symmetry species. In C_s symmetry, to which the present planar complex belongs, Hg*($^3P_1$) with a J value of unity gives three states of A', A'', and A' species, whereas Hg'($^3P_0$) with a J value of zero provides a single state of A'' species.[17] Since the Σ^+ ground state of RH becomes an A' state, we obtain

$$\mathrm{Hg}^*(J = 1) + \mathrm{RH}(\Sigma^+) \xrightarrow{C_s} (\mathrm{HgHR})^*(A', A'', A'), \qquad (13)$$

$$\mathrm{Hg}'(J = 0) + \mathrm{RH}(\Sigma^+) \xrightarrow{C_s} (\mathrm{HgHR})'(A''). \qquad (14)$$

To obtain the symmetry species involved in reaction (5), we use the result of theoretical work[16,18] on the linear Hg*–H_2 system, which indicates that, in reaction (5) with H_2, HgH is produced in the Σ^+ state and subsequently dissociates to give Hg and H. Since R may be approximated as having an electron in an s orbital or an electron in a p orbital, we have

$$\mathrm{HgH}(\Sigma^+) + \mathrm{R}(S_g) \xrightarrow{C_s} \text{complex } (A'), \qquad (15)$$

$$\mathrm{HgH}(\Sigma^+) + \mathrm{R}(P_u) \xrightarrow{C_s} \text{complex } (A', A'', A'). \qquad (16)$$

The comparison of (14) with (15) and (16) indicates that the quenching of metastable atoms to ground state is allowed if R has p character; but, as is

---

† The present planar complex involving a H_2 molecule is a generalization of a cyclic complex proposed in ref. 10. With paraffins, the present complex is essentially a bent triatomic molecule, because R–Hg interaction is expected to be much weaker than Hg–H interaction.
‡ Even though an excited mercury atom does not always complex at the most weakly bonded H, experimental data indicate that the interaction with the weakest CH is the most important process; see, for example, Refs. 3 and 14.

in the quenching by H_2, if R can only be in S state the transition is forbidden. On the other hand, the quenching of Hg* atoms is allowed, regardless of the state of R. In view of hybridization in the alkyl radical, R involved in paraffin quenching should be considered as an atom with a state between S_g and P_u. The quenching of Hg′ is thus partly allowed. One then expects that, compared with Hg*, Hg′ atoms should be less reactive but not negligibly so. Experimental data show Hg′ atoms to be about 20 times less reactive. In quenching by H_2, however, the quenching of Hg′ to the ground state is forbidden; hence here Hg′ atoms should be very much less reactive than Hg*. Experimental data indicate Hg′ atoms are about 600 times less reactive.[5] †

To discuss the quenching of Hg* to Hg′, it is necessary to consider the stability of electronic states. In one of the A' types, the positive lobe of the mercury orbital points toward the RH orbital (which has positive sign), while the negative lobe points away. This state should be attractive. In the other A' type, the negative lobe points to the RH orbital, and the state is repulsive. In A'' symmetry, however, both negative and positive lobes are about the same distance away from the RH orbital; this state cannot be bonding. Thus, it is evident that only one of the A' states leads to the formation of a strongly coupled collision complex. We assume that this state is involved in the quenching. The quenching of Hg* to Hg′ is then forbidden unless vibrational or rotational motion is excited. All the vibrational states in the present triatomic complex belong to the A' species[19] and cannot induce the $A'-A''$ transition. The rotation around the two axes in the symmetry plane, however, belongs to the A'' type; hence, the transition can be induced by this motion. When this rotationally excited complex decomposes, it should produce the rotationally excited RH. The recent phase space theory of reaction rate[20] indicates that, in the above quenching by rotational excitation, the cross section increases with decreasing spacing of rotational energy levels in RH. When H in RH is replaced by D, the rotational spacing decreases by nearly half, but deuteration of the alkyl group decreases the spacing only slightly. Recalling that H in RH comes from the weakest bond, we conclude that the deuteration at the weakest bond increases the rate of Hg′ formation, whereas this rate is affected very little by the deuteration at other bonds. Qualitatively this is in agreement with experimental observations.[1]

It has been customary to suppose that the quenching of Hg* to Hg′ proceeds by the vibrational excitation of RH.[21,22] This supposition is often used together with the resonance-energy rule according to which the quenching occurs most readily when the vibrational energy spacing is the closest to the energy difference (1768 cm^{-1}) between Hg* and Hg′. This view does not agree with experiments. The cross section for $CH_3CD_2CH_3$ contains a

† The quenching cross section for Hg* given in Ref. 5 is about two times too low; see Ref. 11.

large contribution from the quenching to the metastable state, but it shows negligible temperature dependence. Thus the quenching is either thermo-neutral or exothermic. In propane the vibrational frequencies assigned to the CH_2 group are[23] 940, 1278, 1338, and 1460 and doubly degenerate 2950 cm^{-1}. Hence one must assume that the 1460-cm^{-1} vibration is excited. Deuteration reduces this frequency, and the discrepancy from the resonance value of 1768 cm^{-1} increases. Thus deuteration should decrease the formation of Hg$'$ atoms. This is not true.

To explain the abnormal behavior of neopentane, it is now necessary to explore the factors which affect the lifetime, τ_c, of the complex. Since reaction (5) involves the breaking of a CH bond, motion along this reaction coordinate is likely to encounter a potential barrier. This situation is schematically shown in Fig. 6, in which the height of this barrier is denoted h'. In free-radical reactions involving a bond breakage, it is often found that activation energies increase with increasing bond strengths and also that a small fractional difference in bond strength leads to a large fractional

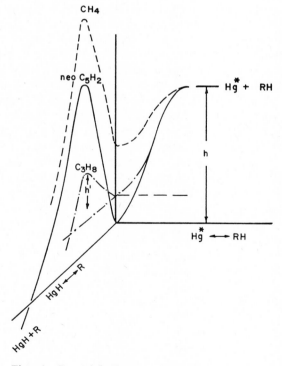

Fig. 6. Potential diagram for a planar complex between Hg* and RH.

difference in activation energy. A potential model used to explain this fact can also be used in the present case to demonstrate that h' increases with increasing bond strength, and also that a small difference in bond energy induces a large fractional change in h'. Since bond energy increases in the order[24] $i\text{-}C_4H_{10}$, C_3H_8, C_2H_6, $C(CH_3)_4$, CH_4, this must be the order with which k_5 sharply decreases. Another factor which is important in discussing τ_c is the energy difference between the complex and reactant state. This is denoted h in Fig. 6. Here h is likely to increase with increasing polarizability of R, which is essentially the same as the polarizability of a paraffin, because then a stronger bond is expected to form between R and Hg*. Hence k_{-2}, which increases with decreasing depth of h, must increase in the order $C(CH_3)_4$, $i\text{-}C_4H_{10}$, C_3H_8, C_2H_6, CH_4. On the basis of these arguments, potential surfaces involved in the quenching by C_3H_8, $C(CH_3)_4$, and CH_4 are schematically depicted as shown in Fig. 6. In CH_4, h is very shallow but h' is very high; hence, the maximum in the plane representing the plot of HgH–R against the energy plane may lie considerably higher than the energy of the reactants, as indicated in Fig. 6. If so, $\phi(H)$ at room temperature should be very small, but at high temperature $\phi(H)$ should become appreciable. This agrees with experimental observations. There are two other constants, k_3 and k_4, affecting τ_c. Nothing is known about these constants, but they are not likely to be affected by the change in bond strength or polarizability as much as k_5 and k_2 are affected. We hence assume that k_3 and k_4 for various RH are about the same. The difference in lifetime among paraffins is then governed by k_5 and k_{-2}. In all paraffins of present concern except neopentane, high h' is accompanied by a shallow h. Neopentane is unusual in that h' is high, while h is very deep. According to the present argument, then, neopentane should have the longest τ_c. Since the rates of reactions (3) and (4) are proportional to τ_c, one can thus rationalize at least qualitatively the fact that neopentane shows abnormally high light emission and also the fact that neopentane is the only undeuterated paraffin which produces detectable Hg′ atoms.

APPENDIX

Table III shows that α/β decreases sharply when temperature increases. At first sight, this may seem surprising. An attempt is hence made in this Appendix to treat this phenomenon in some detail. As shown in (IV), α/β is the product of $k_2\tau_c(k_4 + k_5)$ and $(ck_1)^{-1}$. When expressed in mm^{-1} sec^{-1}, $k_2\tau_c(k_4 + k_5)$ decreases with $T^{-1/2}$. The other factor, $(ck_1)^{-1}$, is the average time, τ, the photon spends (as Hg*) in the cell. In the present experiment, the pressure of mercury is high (1.2×10^{-3} torr), and the fluorescent light becomes absorbed several times before it escapes from the cell. Hence τ is much larger than τ_0, the mean life of an isolated atom (1.08×10^{-7} sec). Milne[25] derives an equation for τ which at high opacities ($= kp$) becomes

$$\frac{\tau}{\tau_0} = 1 + \frac{4}{\pi^2} (kp)^2, \tag{X}$$

where k is the absorption coefficient and p is the distance the photon traverses before escaping from the cell. Milne's theory has defects,[25] but it is simple and agrees well with experimental observations at low opacities.[11] For the moment we assume that k is the same as the absorption coefficient, k_0, at the center of the Doppler broadened absorption line. Then $k_0 = (2.59 \times 10^4) T^{-3/2}$ at a mercury vapor pressure of 1.2×10^{-3}. Hence, $k_2 \tau_c (k_4 + k_5)/ck_1$ decreases approximately with $T^{-3.5}$. A successful approximation for k at low opacities is[11]

$$e^{-kp} = T(\rho), \tag{XI}$$

where $T(\rho)$ is the properly averaged probability that the photon traverses a distance ρ without being absorbed. An approximate equation for $T(\rho)$ to be used at high opacities is[26]

$$T(\rho) = \frac{1}{k_{0\rho}(\pi \ln k_{0\rho})^{1/2}}, \tag{XII}$$

Table VI compares theoretical τ/τ_0, calculated with (X)–(XII), with experimental τ/τ_0's obtained from α/β values. At 202 and 125°, agreement is good; but at 25°, where $k_{0\rho}$ is the highest, the calculated value is too low. The above

Table VI.   Effect of Imprisonment on Mean Life[a]

Temperature (°C)	(τ/τ_0) obs.	(τ/τ_0) calc.
25	10.7	5.9
125	4.1	4.1
202	2.8	3.1

[a] τ = imprisonment lifetime, τ_0 = mean life of an isolated $Hg^*(^3P_1)$ atom.

argument nevertheless shows that Milne's theory provides at least a qualitative explanation for the sharp temperature dependence of the slope of the modified Stern-Volmer formula observed in Figs. 1 and 2.

ACKNOWLEDGMENTS

Mr. J. D. Reedy helped with the experiments, and Mr. C. L. Hassell offered valuable suggestions. Their assistance is gratefully acknowledged.

REFERENCES

1. S. Penzes, A. J. Yarwood, O. P. Strausz, and H. E. Gunning, *J. Chem. Phys.*, **43**, 4524 (1965).
2. S. Penzes, O. P. Strausz, and H. E. Gunning, *J. Chem. Phys.*, **45**, 2322 (1966).
3. For example, see these reviews: (a) R. J. Cvetanović, *Progr. Reaction Kinetics*, **2**, 39 (1964); (b) H. E. Gunning and O. P. Strausz, *Advan. Photochem.*, **1**, 209 (1964).
4. A. B. Callear and W. J. R. Tyerman, *Nature*, **202**, 1326 (1964).
5. A. B. Callear, *Appl. Opt., Suppl.*, **1965**, 145.
6. R. A. Back, *Can. J. Chem.*, **37**, 1834 (1959).
7. K. Yang, *J. Amer. Chem. Soc.*, **86**, 3941 (1964).
8. R. A. Back and D. Van der Auwer, *Can. J. Chem.*, **40**, 2339 (1962).
9. S. Glasston, K. Laidler, and H. Eyring, *The Theory of Rate Processes*, McGraw-Hill, New York, 1941.
10. K. Yang, *J. Amer. Chem. Soc.*, **87**, 5294 (1965).
11. K. Yang, *J. Amer. Chem. Soc.*, **88**, 4575 (1966).
12. J. E. McAlduff and D. J. LeRoy, *Can. J. Chem.*, **43**, 2279 (1965).
13. V. K. Bykhovskii and E. E. Nikitin, *Opt. Spectry.*, **16**, 111 (1964).
14. R. A. Holroyd and G. W. Klein, *J. Phys. Chem.*, **67**, 2273 (1963).
15. K. E. Shuler, *J. Chem. Phys.*, **21**, 624 (1953).
16. K. J. Laidler, *The Chemical Kinetics of Excited States*, Clarendon, Oxford, 1955, pp. 172–174.
17. G. Herzberg, *Molecular Spectra and Molecular Structure*, Vol. III, *Electronic Spectra and Electronic Structure of Polyatomic Molecules*, Van Nostrand, New York, 1966.
18. K. J. Laidler, *J. Chem. Phys.*, **10**, 43 (1942).
19. G. Herzberg, *Infrared and Raman Spectra*, Van Nostrand, New York, 1945, p. 134.
20. P. Pechukas, J. C. Light, and C. Rankin, *J. Chem. Phys.*, **44**, 794 (1966).
21. P. G. Dickens, J. W. Linnett, and O. Sovers, *Disc. Faraday Soc.*, **33**, 52 (1962).
22. A. C. G. Mitchell and M. W. Zemansky, *Resonance Radiation and Excited Atoms*, Cambridge University Press, Cambridge, 1961.
23. K. S. Pitzer, *J. Chem. Phys.*, **12**, 310 (1944).
24. J. A. Kerr, *Chem. Rev.*, **66**, 465 (1966).
25. E. A. Milne, *J. London Math. Soc.*, **1**, 1 (1926).
26. T. Holstein, *Phys. Rev.*, **72**, 1212 (1947).

Dynamics of Ion–Molecule Collisions

JOHN V. DUGAN, Jr.

NASA Lewis Research Center Cleveland, Ohio

JOHN L. MAGEE

Chemistry Department and the Radiation Laboratory* University of Notre Dame, Indiana

* The Radiation Laboratory of the University of Notre Dame is operated under contract with the U.S. Atomic Energy Commission. This is AEC Document No. COO-38-691.

207

A Personal Note

Professor Henry Eyring has been a leader in the field of reaction kinetics for almost forty years. His work has not only established a pattern for the expert in the field but it has changed the way scientists generally think about kinetics; his influence is ubiquitous. The work described here is a probing into the mechanism of ion–molecule reactions. We feel we owe much to the Eyring tradition and are honored to have this paper in a volume dedicated to him.

INTRODUCTION

Ion–molecule collisions have been of interest for a long time, starting with Langevin's treatment of the simplest possible case in 1905.[1] The recent interest in elementary reactions has inevitably led to experimental studies of ion–molecule collisions and these have raised questions of theory. For a number of years we have carried out theoretical investigations on collisions of ions with dipolar molecules. Although the collisions considered are relatively simple, they contain the essential features of many real systems. In fact, such collisions are prototypes for many of the ion–molecule collisions of small molecules occurring in experimental work of current interest. In this paper two important aspects of ion–molecule collisions are considered: collision cross section and collision time.

Langevin's treatment[1] was limited to the calculation of cross section. In this treatment of a collision between a singly charged ion and a spherical molecule, the potential V_P between the pair was taken as

$$V_P(r) = -\frac{1}{2}\frac{\alpha e^2}{r^4},\tag{1}$$

where α is the electronic polarizability of the neutral molecule, e is the charge on the electron, and r is the ion–molecule separation.

The phenomenon of the "capture collision"[2,3] can best be understood by consideration of the effective radial potential of a collision, V_{eff}. For a particular value of translational angular momentum L, this potential is

$$V_{\text{eff}}(r) = \frac{L^2}{2mr^2} - \frac{1}{2}\frac{\alpha e^2}{r^4}.\tag{2}$$

The angular momentum L is

$$L = mvb,\tag{3}$$

where m is the reduced mass, v is the relative velocity at infinite separation, and b is the impact parameter. The relative energy ϵ at infinite separation is

$$\epsilon = \tfrac{1}{2}mv^2\tag{4}$$

209

and substitution of (3) and (4) into (2) gives

$$V_{\text{eff}}(r) = \epsilon \left(\frac{b}{r}\right)^2 - \frac{1}{2} \frac{\alpha e^2}{r^4} \tag{2a}$$

as an alternative form of (2).

From (2a) the $V_{\text{eff}}(r)$ has a single maximum at

$$r^* = \left(\frac{\alpha e^2}{\epsilon}\right)^{1/2} \frac{1}{b}. \tag{5}$$

The maximum potential energy V_{eff}^* can be written in terms of the various parameters:

$$V_{\text{eff}}^*(r^*) = \tfrac{1}{2}\epsilon \left(\frac{b}{r^*}\right)^2 = \tfrac{1}{2}\epsilon^2 \frac{b^4}{\alpha e^2}, \tag{6}$$

or

$$V_{\text{eff}}^*(r^*) = \frac{1}{2} \frac{\alpha e^2}{r^{*4}}. \tag{6a}$$

The classical trajectories of this system have a very simple pattern. If the energy ϵ is less than the maximum potential energy given by (6); that is, if $b > \sqrt{2}\,r^*$, the pair turns at some distance larger than r^* and a simple scattering is effected. On the other hand, if ϵb^2 is larger than $[2V_{\text{eff}}^*(r^*)\alpha e^2]^{1/2}$, that is, $\epsilon < 2\alpha e^2/b^4$, the pair comes into separations smaller than r^*, and the molecule and ion can orbit about each other. It should be noted, however, that for impact parameters less than 10 Å, spiraling behavior (where the change in angle exceeds 2π) will only occur at separations ≤ 1 Å.[4] Thus, if ion–molecule reaction occurs at 1–2 Å, capture leading to reaction will occur before any spiraling is observed. (In our treatment we do not explicitly consider short-range interactions.)

For a particular energy ϵ all collisions with values of b less than that given by the relation

$$V_{\text{eff}}^*(r^*) = \epsilon, \tag{7}$$

which from (6) is

$$b^* = \left(\frac{2\alpha e^2}{\epsilon}\right)^{1/4}, \tag{7a}$$

lead to spiraling with the separation going to zero (i.e., capture collisions). The cross section for capture σ_L is, therefore, given by

$$\sigma_L(\epsilon) = \pi b^{*2} = \pi \left(\frac{2\alpha e^2}{\epsilon}\right)^{1/2}, \tag{8}$$

where b^* is given by (7a). This is the "Langevin cross section," which is frequently used in discussion of ion–molecule reactions.[1–8]

The energy-specific reaction rate constant is obtained from (9) by multiplication by v:

$$k'_L(\epsilon) = 2\pi e \left(\frac{\alpha}{m}\right)^{\frac{1}{2}}. \tag{9}$$

Here m is the reduced mass of the collision system. Clearly this formula is independent of energy ϵ and so the thermal-averaged rate constant is also given by (9). For the Langevin case the rate constant of an ion–molecule reaction just depends on the ratio α/m.

The first discussion of this problem, including a derivation of (9) by absolute reaction rate theory, was given by Eyring, Hirschfelder, and Taylor.[9]

The cross sections and trajectories for the Langevin case are completely determined by analytical techniques. On the other hand, no general analytical treatments are available for collisions between ions and dipolar molecules. We have investigated the use of numerical methods to obtain information on these collisions and most of the results presented here are obtained from computer calculations. First, however, we shall review some analytical considerations which yield useful approximate results.

CROSS SECTIONS

In this section we are considering only collision cross sections. Although many of the systems for which our models are prototypes have several channels (i.e., have several sets of reaction products), we will restrict ourselves to simple unreactive collisions.

Approximate Analytical Treatments of Cross Sections

Molecules of chemical interest are usually polyatomic and have more complicated potential functions of interaction with ions than that given in (1). Therefore a more general and realistic potential function of the form

$$V(r) = \frac{A}{r^2} + \frac{B}{r^3} + \frac{C}{r^4} + \cdots \tag{10}$$

must be used.

General expressions for the quantities A, B, and C of (10) are given by Hirschfelder, Curtiss, and Bird[10] in terms of the various moments of the molecule, the ion, and their mutual orientation. We chose the simplest potential which is compatible with the introduction of a dipole moment into the problem. The only nonvanishing constants in (10) are taken as

$$\begin{aligned} A &= -\mu e \cos \gamma, \\ C &= -\tfrac{1}{2}\alpha e^2, \end{aligned} \tag{11}$$

where μ is the dipole moment, γ is the angle it makes with $\mathbf{r}$, and α is the electronic polarizability. Literally, such a potential function means that only the neutral molecule has a dipole moment. Obviously, a molecule with a dipole moment cannot be treated as a point particle since it has a moment of inertia and rotational energy. These are essential complications which must be introduced into the problem.

Consider first the case of a polar molecule that is not rotating as it collides with an ion. For this case the angular momentum of the system is all in relative translational motion and depends only on the impact parameter b. If the dipole orientation can adjust sufficiently rapidly as the collision develops, the angle γ in the potential of (11) is always zero, and the largest possible cross section results. This maximum cross section is calculated according to the procedure outlined in the introduction with the effective potential

$$V_{\text{eff}} = \frac{L^2}{2mr^2} + V(r) = \frac{L^2}{2mr^2} + \frac{A}{r^2} + \frac{C}{r^4}. \tag{12}$$

The two terms in r^{-2} (i.e., A/r^2 and $L^2/2mr^2$) combine, and $\cos \gamma$ is set equal to 1.

$$V_{\text{eff}}(r) = \frac{L^2/2m - \mu e}{r^2} - \frac{1}{2}\frac{\alpha e^2}{r^2}. \tag{13}$$

A calculation parallel to that of (1)–(9) yields the cross section[4,5]

$$\sigma_M = \pi \left[\left(\frac{2\alpha e^2}{\epsilon} \right)^{1/2} + \frac{\mu e}{\epsilon} \right]. \tag{14}$$

The quantity σ_M is an estimate of the maximum possible cross section for an ion–molecule pair with the parameters α and μ.

This case can be described as "adiabatic" because of the ability of the dipole to adjust during the course of the collision. It does not apply to many real collisions, but is of interest in comparison with experimental results.

Capture Collisions for Rotating Molecules.   Molecules will generally be rotating before a collision. The rotational motion of a dipolar molecule will be hindered by the field of the ion at separations such that the relation[11]

$$E_R \approx \frac{\mu e}{r_h^2} \tag{15}$$

is approximately satisfied. Here E_R is the rotational energy of the molecule, and r_h is the hindering distance. If the separation r is considerably larger than r_h, the rotation is not hindered, and the influence on the rotational motion is described by the rotational Stark effect. Energy shifts resulting from the Stark effect are first or second order in the electric field strength.[12] The relative energy of the ion–molecule pair, therefore, has a dependence on the field strength given by

$$V_{\text{eff}}(r) = \frac{L^2}{2mr^2} + \Delta W_1 - \frac{1}{2}\frac{\alpha e^2}{r^4} + \Delta W_2. \tag{16}$$

The term ΔW_1 ($= A/r^2$) is the first-order energy shift, and ΔW_2 ($= C/r^4$) is the second-order shift. Equation 16 has the form of (2), and in the region of validity of the rotational Stark effect the simple Langevin theory applies. However, the effective angular momentum is determined by the first two terms of (16), and the effective polarizability is determined by the last two terms.

The previously described procedure gives, for the value of the potential energy at maximum,

$$V_{\text{eff}}^*(r^*) = \frac{1}{2}\frac{\alpha' e^2}{r^{*4}}, \tag{17}$$

where α' is the effective polarizability given by

$$\alpha' = \alpha - \frac{2r^4}{e^2}\Delta W_2. \tag{18}$$

Also, we can so define an effective angular momentum L' that

$$L'^2 = L^2 + 2mr^2\Delta W_1. \tag{19}$$

The position of the potential maximum is then given by

$$r^* = \left(\frac{\alpha' e^2}{L'^2/2m}\right)^{\frac{1}{2}}. \tag{20}$$

The cross section for capture collisions σ_T obtained by setting the right-hand side of (17) equal to ϵ is

$$\sigma_T = \pi b^{*2} = \pi\left(\frac{2\alpha' e^2}{\epsilon}\right)^{\frac{1}{2}} - \pi r^2\frac{\Delta W_1}{\epsilon}. \tag{21}$$

We may note again that $r^2\Delta W_1$ and $r^4\Delta W_2$ in (16), (18), and (19) are constants for a given ion–molecule pair at fixed ϵ.

The general cross section for the rotating molecule is a sum of two terms in $\epsilon^{-\frac{1}{2}}$ and ϵ^{-1} similar to σ (14). At small ϵ, therefore, symmetric top molecules should have cross sections that go over to the ϵ^{-1} dependence. The coefficient, however, depends directly on ΔW_1, the first-order Stark energy shift; and molecules that do not have a first-order Stark effect are expected to continue the $\epsilon^{-\frac{1}{2}}$ dependence even at small energies. Linear molecules do not have first-order energy shifts, for example. It is interesting to note that

Table I. Properties of Rotating Polar Molecules[12,14]

Type of Molecule	Rotational Energy in Absence of Field, E_R	First-Order Stark Energy Shift, ΔW_1	Second-Order Stark Energy Shift, ΔW_2
Linear	$\dfrac{J(J+1)\hbar^2}{2I}$	—	$\left[\dfrac{I\mu^2X^2}{J(J+1)\hbar^2}\right]$ $\times\left[\dfrac{J(J+1)-3M_J{}^2}{(2J-1)(2J+3)}\right]$
Symmetric top	$\dfrac{J(J+1)\hbar^2}{2I}$ $+\tfrac{1}{2}(K^2\hbar^2)(I_A{}^{-1}-I_B{}^{-1})$	$\dfrac{M_JKX}{J(J+1)}$	$\times\left(\dfrac{I_B\mu^2X^2}{\hbar^2}\right)$ $\times\left\{\dfrac{(J^2-K^2)(J^2-M_J{}^2)}{J^3(2J-1)(2J+1)}\right.$ $\left.-\dfrac{[(J+1)^2-K^2][(J+1)^2-M_J{}^2]}{(J+1)^3(2J+1)(2J+3)}\right\}$

J, K, and M_J are rotational quantum numbers; I, I_A, and I_B are moments of inertia; X is the field strength, which in this case is e/r^2.

the experimental cross section for linear molecules has only the $\epsilon^{-1/2}$ dependence,[13] whereas several symmetric-toplike molecules have been observed to make the transition to a ϵ^{-1} dependence at low energy.

A summary of the properties of rotating molecules is given in Table I.[12,14] It is noted that both ΔW_1 and ΔW_2 can be positive or negative. Thus the effective polarizability of a collision can be reduced to zero for certain states of rotation of a polar molecule; for such states, capture collisions do not occur.

One is generally interested in an average cross section for a rotating molecule that has a particular energy. Such a cross section involves an average over the various quantum numbers with proper weighting. This averaging is described in Appendix A along with the Stark effect and the capture cross sections for linear polar molecules.

A summary of formulas for the approximate analytical cross sections developed here is given in Table II.

Numerical Calculation of Cross Sections

The numerical calculations are based on a classical model for the ion–molecule collision. The potential energy, of course, includes both ion–dipole and ion-polarizability terms. It is necessary to use definite values of all physical constants in order to make a calculation. In all studies values of the reduced masses, moments of inertia, dipole moments, etc., have been chosen so that the calculations would correspond to some particular collision (e.g., HCl + HCl⁺). In all cases, however, the ion enters only through its charge and mass; it is a structureless particle.

The models and coordinate systems used for the numerical calculations are shown in Fig. 1. The Lagrangian formulation of the equations of motion was utilized. The coupled equations were programmed in FORTRAN IV for solution on the IBM 7094 digital computer using a variable order Runge-Kutta integration routine. The conservation of energy equation and behavior of trajectories with time reversal were used as checks on the accuracy of the integration technique. Trajectories were calculated for many impact parameters at two different rotational energies. For each impact parameter and fixed rotational energy 36 to 50 trajectories were calculated. Random initial conditions for the rotating molecule were chosen by a random number generation routine.

Treatment of the equations of motion and initial conditions are discussed in Appendix B.

The Langevin trajectories for ion–molecule collisions (described in the Introduction) have a very simple property: they either lead to a capture collision or they do not. It is a simple all-or-nothing situation. The treatment leading to the analytical "maximum cross section" for a dipolar molecule assumed that these collisions have the same property; it is also for an all-or-nothing situation. The calculated ion–dipolar molecule collisions do not have this property. There is no all-or-nothing answer to the question of capture in an ion–permanent-dipole collision for a fixed initial impact parameter. For

Table II.   Analytical Expressions for Ion–Molecule Capture Cross Sections[a]

$\sigma_L = \pi \left(\dfrac{2\alpha e^2}{\epsilon} \right)^{1/2}$	Langevin cross section
$\sigma_D = \pi \dfrac{\mu e}{\epsilon}$	Dipole cross section; rotation neglected
$\sigma_M = \sigma_L + \sigma_D$	Maximum cross section; rotation neglected
$\sigma_J = \pi \left(\dfrac{3\mu^2 e^2}{2E_R\epsilon} \right)^{1/2} \left[\dfrac{\sqrt{c+2}}{2\sqrt{3}} + \dfrac{(c-1)}{6} \ln \left(\dfrac{\sqrt{c+2} + \sqrt{3}}{\sqrt{c-1}} \right) \right]$	Dipole cross section for rotational state J $\qquad c > 1$
$\sigma_J = \pi \left(\dfrac{3\mu^2 e^2}{2E_R\epsilon} \right)^{1/2} \left[\dfrac{\sqrt{c+2}}{2\sqrt{3}} + \dfrac{(c-1)}{6} \ln \left(\dfrac{\sqrt{c+2} + \sqrt{3}}{\sqrt{1-c}} \right) \right]$	Dipole cross section for rotational state J $\qquad c < 1$

[a] $c = 4E_R\alpha/\mu^2$

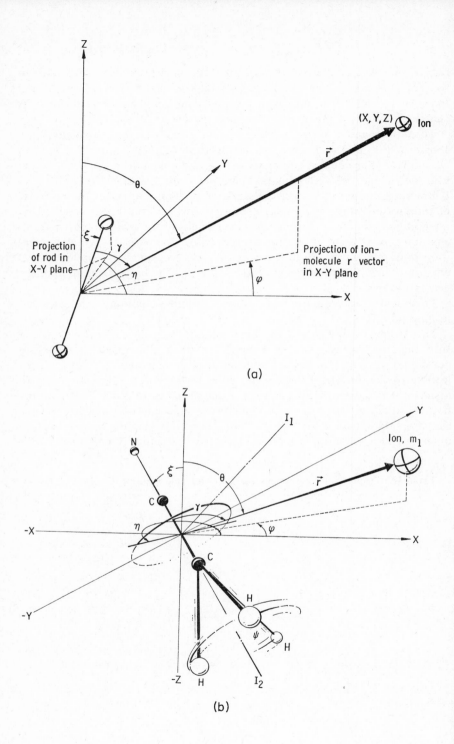

(a)

(b)

any initial value of energy and impact parameter, there is a probability that the system will arrive at an ion–molecule separation corresponding to capture. The fraction of collisions resulting in such a prescribed minimum separation we call the "capture ratio." Such a fraction is probably correlated with the probability for undergoing an ion–molecule reaction. Some calculated results are shown in Figs. 2–6. The ion energy, ϵ_1, in Fig. 5 equals $(m_1/m)\epsilon$, where m_1 and m are the ion and reduced masses. The maximum ion energy ϵ_m of Fig. 6 corresponds to an experimental voltage setting.[4,5]

The variation of the capture ratio for HCl–HCl (parent ion) collisions with impact parameter for two different relative velocities and two different temperatures is shown in Fig. 2. The plot of "all-or-nothing" type of capture ratio is a step function on this type of plot. Such functions are shown for Langevin ($b_L{}^2$ plot), permanent-dipole ($b_D{}^2$ plot), and the sum [$b_M{}^2$, (14)]. The Langevin case is, of course, for a molecule with the same polarizability as HCl but no dipole moment. It is clear that the dipole does not exert its maximum effect in bringing about capture. In fact, Fig. 2 shows that the dipole sometimes prevents capture which would occur if there were no dipole moment, since C_R falls below unity for values of b^2 smaller than $b_L{}^2$. The classical rigid rotor with dipole has an orientation changing from attractive to repulsive on each rotation. One might guess that the capture ratio at $b = b_L$ would be 0.5 because unfavorable orientation of the dipole would prevent half the captures. This is, however, not the case. Polarization of the rotor brings about an orientation with the field and the capture ratio is expected to increase as the rotational energy decreases. Clearly C_R is larger at the lower temperature. It is conceivable that as the rotational energy approaches zero the capture ratio would approach that predicted for the maximum cross section, but this case has not been studied.

A reasonable definition of the "capture collision" cross section is

$$\sigma_c = \pi b_c{}^2 = \pi \int_0^{b_0{}^2} C_R(b)\, d(b^2), \qquad (22)$$

where b_0 is the impact parameter at which C_R becomes zero. The integral is simply the area under the curve of C_R plotted against b^2, and b_c is defined by (22). The cross-section results are relatively insensitive to the choice of the cutoff ion–molecule separation (i.e., whether it is 1 or 2 Å).

As shown in Fig. 2, the shape of the C_R plots is sensitive to values of both rotational energy and ion velocity. Original calculations were performed with 36 cases for a fixed impact parameter. Then repeating the calculation with 50 new cases changed the cross section by only several percent.

Fig. 1. Coordinate systems used in computer studies of ion–molecule interaction: (a) coordinate system used in computer study of interaction between an ion and a linear polar molecule; (b) coordinate system used in computer study of interaction between an ion and symmetrical top polar molecules.

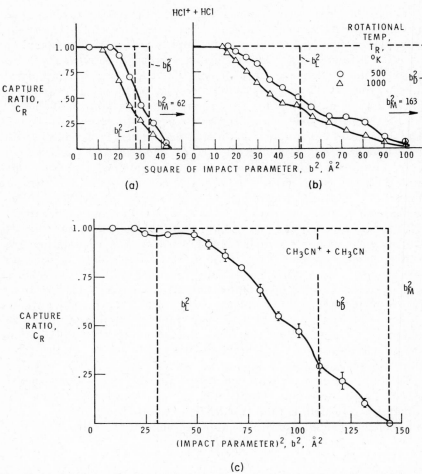

Fig. 2. Variation of capture ratio with square of impact parameter for linear HCl target rotators distributed at temperatures $T_R = 500$ and $1000°K$; 50 collisions were studied per point. (a) $v = 10^5$ cm/sec. (b) $v = 5.5 \times 10^4$ cm/sec. (c) $v = 10^5$ cm/sec.

Figure 3 gives the calculated capture cross sections for the HCl–parent ion system as a function of collision energy.

Figure 4 shows calculated plots of the capture ratio as a function of impact parameter for a dipolar symmetric top molecule in collision with an ion. The molecule is CH_3CN and the ion has the mass of the parent ion CH_3CN^+. The very large effect of the dipole on capture is clearly apparent when comparison is made with the Langevin prediction for a molecule with the polarizability of CH_3CN. On the other hand, as for HCl, the effectiveness of

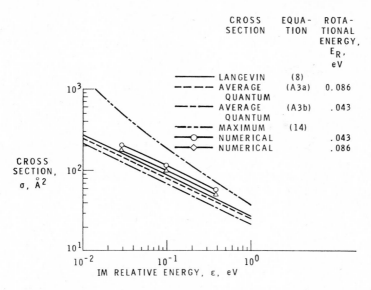

Fig. 3. Comparison of numerically calculated cross section σ_c for linear molecule HCl with various theoretical cross sections; electronic polarizability α, 2.63 Å³; dipole moment μ, 1.08 Debye units.

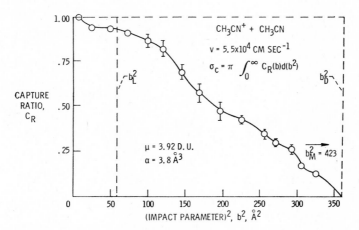

Fig. 4. Variation of capture ratio with square impact parameter for $CH_3CN^+ + CH_3CN$ system. The symmetric top target rotators are distributed at temperature $T_R = 500°K$. Bars indicate variation of calculation with different random number sets.

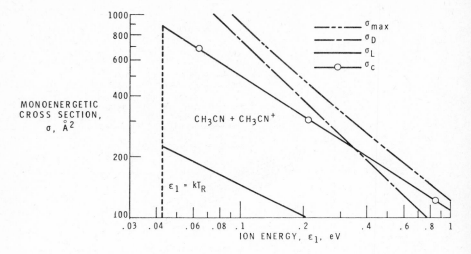

Fig. 5. Comparison of numerically calculated cross sections σ_c for methyl cyanide–parent ion collision with various theoretical cross sections. All cross sections are plotted as functions of ion translational energy for one rotational temperature $T_R = 500°K$; electronic polarizability α, 3.8 Å³; dipole moment μ, 3.92 Debye units.

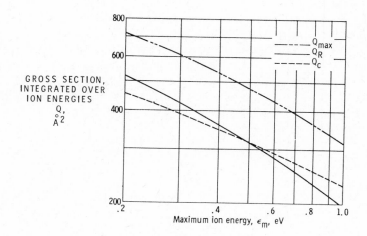

Fig. 6. Comparison of numerical capture cross section Q_c with experimentally assumed capture cross section Q_{max} and observed reaction cross section for methyl cyanide–parent ion collision as function of maximum ion energy ϵ_m (see Table II for definitions of cross sections).

the dipole is far less than the "adiabatic" maximum effect. Symmetric top molecules have first-order Stark effects and the larger capture ratio may be largely due to this effect.

In Fig. 5 the calculated cross section is shown as a function of energy and compared with various analytical estimates. The most obvious feature of this figure is how much larger the calculated cross section is than the Langevin cross section. The calculation also shows that the effect of the dipole is not as great at low energies as predicted on the basis of the adiabatic approximation. Although the Stark theory predicts an ϵ^{-1} contribution, it is not the dominant term in this range of energy.[11]

A comparison with experiment of Ref. 6 taken from Ref. 15 is presented in Fig. 6 and it shows that the calculation is satisfactory for a model of this degree of roughness.

COLLISION TRAJECTORIES AND COLLISION TIMES

The calculations described above yield trajectories as well as cross sections. If studies of chemical change in ion–molecule collisions are to be made by this method we must be interested in many features of the trajectories themselves. The most apparent difficulty in obtaining such details is the lack of information on the short-range interactions and the nature of the potential functions for the various chemical states of the systems. We make no contribution to the solution of these problems here.

Examination of various features of the calculated trajectories leads to the conclusion that there is very rapid interchange of energy between translational and rotational energy. The transfer of energy between internal degrees of freedom and translation leads to the possibility of long-lived collision complexes. Experimental evidence on the clustering of ions points to the existence of long-lived ion–molecule collision complexes.[16–18] If some of the longer lifetimes are to be understood, theory requires participation of several internal degrees of freedom (i.e., vibrations and internal rotations of the collision partners). Such long-lived complexes must, therefore, be formed by inelastic collisions. At the present time there is no theoretical technique powerful enough to describe such collisions rigorously. On the other hand, the "capture" collisions of ions with molecules depend only on long-range forces and are fairly well understood, as we have seen.[4,19]

In order to investigate the nature of trajectories adequately to understand lifetime of complexes, we must treat the short-range interactions in some fashion. We have used an extreme approximation: a hard inner core. This means that for some particular value of the separation, r_c, the potential energy is assumed to rise abruptly to infinity. In a trajectory at this separation the radial velocity reverses its sign. In this approximation the inward and outward trajectory times for the simple ion-polarizability potential are exactly the same. Specular reflection occurs at the hard core and the outward trajectory is symmetric with respect to the inward trajectory.

The situation is different for an ion–molecule collision involving a molecule with a dipole moment. In this case the rotating dipole alters the potential for the outward trajectory, outer turning points can be introduced, and multiple reflections occur.* The reflections can increase the collision time significantly. For parameters representing actual ion–molecule collisions of interest, as many as 2000 reflections have been observed in the numerical calculations, with collision times τ as large as 10^2 times the single reflection period.

Results of Calculations of Trajectories

Results of computer calculations on ion–molecule collisions are shown in Figs. 7–15. These time history plots have been made using the IBM 360/67 computer with the CDC DD280 plotter. Figures 7 and 8 show the variation of velocity and rotational energy as a function of separation. The collision shown in Fig. 7, for $HCl + NO_2^+$, involves a single reflection. The velocity increases as the separation decreases to the critical value where the potential energy is a minimum. The rotational energy changes drastically: there is hindered rotation at the small separations.

The trajectories in Fig. 8 for $CO + Ar^+$ involve multiple reflections. Variation of the outer turning point is shown in this figure in the rotational energy plot.

An ion–molecule collision between two monatomic entities would occur in a plane. The simplest systems on which our computations are based are essentially triatomic (i.e., diatomic rotors and monatomic ions). The trajectories do not lie in planes, as is shown in Fig. 9, which was obtained from calculations for $CO + Ar^+$. This figure shows a multiple-reflection collision.

Symmetric top dipolar molecules have one more degree of freedom than rigid rotors, and their collisions are essentially more complicated. Their trajectories involve more multiple reflections than rigid rotors. Figures 10 and 11 show trajectories for collisions of CH_3CN with its parent ion (i.e., as regards ion mass). These figures indicate a more effective transfer of energy between internal and translational degrees of freedom.

Figures 12–15 show variations of the polar and azimuthal angles for trajectories involving multiple reflections for $HCl + NO_2^+$, $CH_3CN + CH_3CN^+$, and $CO + Ar^+$. Figure 15 shows how distant some of the outer

---

* The position of the outer turning points varies because of dipole rotation, but the magnitude of the collision time τ is not very sensitive to this effect.

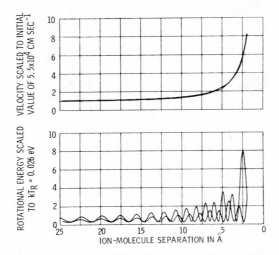

Fig. 7. Variation of ion velocity and polar molecule rotational energy during $NO_2^+ + HCl$ single reflection capture collision.

turning points in multiple reflection trajectories can be. One turning point for $CH_3CN + CH_3CN^+$ occurs at 22 Å separation.

It should be noted that multiple reflections are also to be expected for ion–molecule collisions involving polyatomic molecules without dipole moments since actual short-range potentials probably have sufficient angular dependence in most cases to produce the same effect in varying degrees.

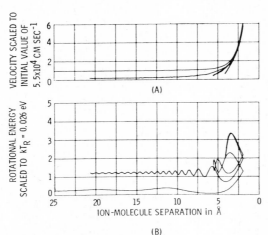

Fig. 8. Variation of ion velocity and polar molecule rotational energy during $Ar^+ + CO$ multiple reflection capture collision.

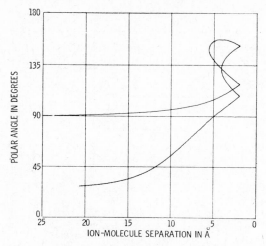

Fig. 9. Variations of polar angle θ for translational motion of Ar^+ relative to CO molecule during multiple reflection capture collision.

Many trajectories were calculated to investigate the numbers of reflections which would occur and thus to obtain estimates of the lifetimes of collision complexes. Summaries of the parameters of the three systems used are given in Table III. Summaries of the results of the calculated trajectories are shown in Table IV.[20]

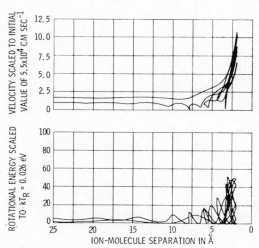

Fig. 10. Variation of ion velocity and polar molecule rotational energy during methyl cyanide–parent ion multiple reflection capture collision.

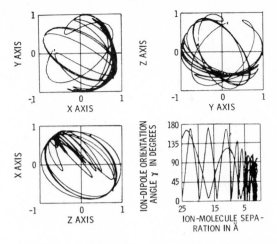

Fig. 11. Variations of dipole moment vector and ion–dipole orientation angle during CH_3CN–CH_3CN^+ multiple reflection capture collision with several turning points.

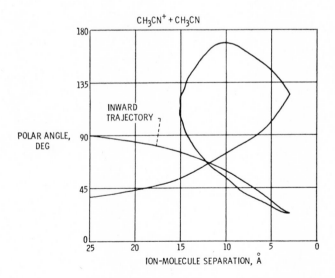

Fig. 12. Variation of polar angle for translational motion of CH_3CN^+ relative to CH_3CN molecule during multiple reflection capture collision.

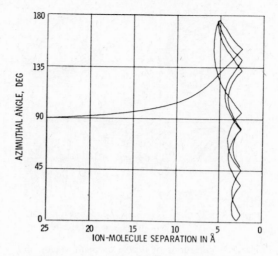

Fig. 13. Variations of azimuthal angle φ for translation motion of NO_2^+ relative to HCl molecule during multiple reflection capture collision.

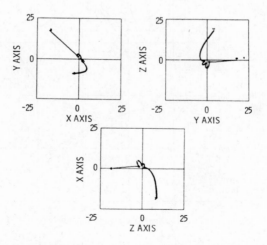

Fig. 14. Variation of ion projections tracing translational motion in $Ar^+ + CO$ capture collision with multiple reflections.

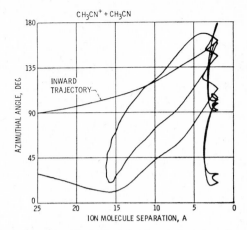

Fig. 15. Variation of azimuthal angle φ for translational motion of CH_3CN^+ relative to CH_3CN molecule during multiple reflection capture collision with maximum turning point of 16 Å.

Computer-Made Movies of Ion–Dipole Collisions

The computer-plotter studies have been extended to the making of motion pictures of ion–dipole single and multiple-reflection collisions. The collision movies provide instantaneous visual correlation of the relative translational motion of the pair and the dipole rotational motion. One of the most interesting phenomena observed in the movies is the hindering of the rotational motion of the dipole. The three polar rotors CO, HCl, and CH_3CN are seen to be hindered by the incident ion to varying degrees.[21] Each movie frame shows the projection of ion and polar molecules in a plane of the center-of-mass system. The ion is represented by a positive sign surrounded by a circle

Table III. Parameters Used in Trajectory Calculations[a,b]

Ion–molecule Pair	Electronic Polarizability ($Å^3$)	Dipole Moment (D.U.)	Reduced Mass m (kg ($\times 10^{26}$))	Moment of Inertia I (kg-m^2 ($\times 10^{46}$))
$Ar^+ + CO$	1.95	0.10	2.74	1.44
$NO_2^+ + HCl$	2.60	1.08	3.41	0.268
$CH_3CN^+ + CH_3CN$	3.80	3.92	3.41	9.12

[a] All collisions were studied for an initial ion–molecule relative velocity $v_0 = 5 \times 10^4$ cm/sec: the hard core (reflection barrier) was located at $r_c = 1, 1.5, 2,$ or 3 Å.

[b] Target rotors were chosen from a distribution at a rotational temperature $T_R = 300°K$.

Table IV. Results for Multiple-Reflection Collisions

Ion-molecule pair	Impact parameter, b, Å	Location of reflecting barrier, r_c, Å	Number of cases, n	Langevin potential, maximum, r^*, Å	Range of numerical interaction radius, defining collision time, Å	Fraction of cases resulting in multiple reflection, f_R	Maximum number of reflections, N_{max}	Average single-reflection time, $\bar{\tau}_o$, sec	Most probable number of multiple reflections, N_M	Average collision time for multiple reflection, $\bar{\tau}_R$, sec	Maximum collision time, τ_{max}, sec
$Ar^+ + CO$	2.0	1.0	33	16.5	5 to 8	0.63	15	1.5×10^{-12}	6	6.0×10^{-12}	1.7×10^{-11}
	3.0	1.5	21	11.0	4 to 8	.13	5	2.5	2	6×10^{-12} to 7×10^{-12}	8.2×10^{-12}
	3.0	3.0	12	11.0	5 to 8	.08	3	$\sim 10^{-12}$	3	1.7×10^{-11}	1.7×10^{-11}
	4.0	1.0	24	8.2	4 to 8	.33	18	1.5×10^{-12}	7	4×10^{-12}	8×10^{-12}
	6.0	1.0	49	5.5	4 to 8	.65	42	2.8	8	1.2×10^{-11}	3.5×10^{-11}
	6.0	2.0	23	5.5	5 to 8	.60	54	4	7	$\sim 10^{-11}$	2.7
$NO_2^+ + HCl$	4.0	1.0	10	8.5	~ 10	0.30	1115	1.5×10^{-12}	20	3.3×10^{-11}	7.5×10^{-11}
	4.0	2.0	40	8.5	8 to 12	.10	40	2	4	1.5	3.1
	6.0	1.0	8	5.7	~ 10	.50	192	2	7	8×10^{-12} to 9×10^{-12}	2
	6.0	2.0	56	5.7	~ 10	.14	530	1.7	3	5×10^{-12}	6.4
	8.0	1.0	6	4.3	~ 10	.67	715	4	6	1.6×10^{-11}	5.2
$CH_3CN^+ + CH_3CN$	3.0	3.0	8	13.7	10	0.50	192	2×10^{-12}	7	1.7×10^{-12}	1.7×10^{-11}
	5.0	2.0	33	8.2	15 to 20	.47	[a]25	4	25	1.3×10^{-11}	2
	5.0	3.0	12	8.2	15 to 20	.33	140	4	20	2	8
	6.0	2.0	24	6.5	15 to 20	.46	[a]25	5×10^{-12} to 6×10^{-12}	8	1.8	3.85
	7.0	2.0	24	5.6	15 to 20	.55	[a]25	4×10^{-12}	10	1.6	3.55
	8.0	2.0	12	5.2	15 to 20	.57	40	4.5	8	2	3.75
	10.0	2.0	12	4.1	15 to 20	.57	150	4.5	12	1.5	2.3
	10.0	3.0	12	4.1	15 to 20	.33	40	4.5	15	2	4.9

[a] Case limited to 25 reflections to minimize computer time.

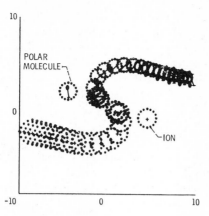

Fig. 16. Superposition of 15 movie frames for single reflection CH_3CN parent–ion collision. Ion–molecule pair are designated in pre-reflection positions; motion is traced during and after reflection.

of asterisks. The polar molecule is represented by a dumbbell with positive and negative signs attached to the ends. The dumbbell is also surrounded by a circle of asterisks. The radii of these two circles are variable. However, the sum of the radii is a constant equal to r_c. When either particle is above the viewing plane the radius increases; below the plane the radius shrinks. The maximum radius ($0.75\ r_c$) is reached when the particle is 10 Å above the plane; the minimum radius is $0.25\ r_c$ at 10 Å below the plane.

Figure 16 is a superposition of movie frames of a CH_3CN capture collision involving only one reflection. It provides a correlated history of ion–dipole interaction. The results are not presented stroboscopically since a variable step size is used. The postreflection hindering of the CH_3CN rotor at 5–15 Å is demonstrated in the movie. It has been suggested that preferential orientation of the negative end of the dipole toward the positive ion will favor a specific chemical reaction.[22]

Numerical cross sections for such reactions can be obtained by calculating the fraction of hindered collisions from plots of ion–dipole orientation angle and then multiplying this fraction by the capture ratio. The product, an effective reaction ratio, is then integrated over impact parameters to give the reaction cross section.

Theoretical Interpretations of Calculations

The calculations described here show in considerable detail the nature of "sticky" collisions arising in ion–polar molecule encounters. We have found that energy is transferred from relative translation to rotational energy of the dipole with great ease. We know that energy is transferred from translational to vibrational motion with less ease in collisions of neutral molecules.[23] The multiple reflections of the ion–molecule trajectories must vastly increase the probability of vibrational excitation, since multiple reflections are equivalent to many simple collisions. Furthermore, these collisions are relatively energetic, and the probability of vibrational excitation is enhanced for this reason.[23]

Theories of the lifetime of complexes that can dissociate have been developed[24,25] and can be applied to these cases. These theories assume that the energy of the complex is completely randomized over certain participating degrees of freedom, and the rather complicated nature of our calculated multiply reflected trajectories would suggest that such an assumption is reasonable for our cases. Application of any theory requires more knowledge of the potential functions of the complexes than we have at present, and the discussion presented here can be only fragmentary. The simplest "theoretical" expression for the lifetime of a complex with enough energy to decompose is[26]

$$\tau = \tau_0 \left(\frac{E}{E - E_b} \right)^s,$$ (23)

where τ_0 is a parameter having dimension of time made up of physical constants of the complex; E is the energy of the complex with its zero-point energy taken as the zero of energy; E_b is the magnitude of the binding energy of the complex; s is a parameter determined by the number of degrees of freedom of the system. We have that $E = E_b + \epsilon$, where ϵ is the relative energy of colliding pair. Thus

$$\tau \approx \tau_0 \left(\frac{E_b + \epsilon}{\epsilon} \right)^s,$$ (23a)

where τ_0 is a constant of the magnitude of a collision time without reflections; that is, $\tau_0 \approx 10^{-12}$ sec. E_b is in the range of 0.5 to several electron volts. For the various collisions we are considering here s varies from 0.5 to 1.0. Since $(E_b + \epsilon)/\epsilon \approx 10$–500, we expect that the collision time should be increased by a factor of 10 to several hundred by the reflection phenomenon.

Vibrational degrees of freedom have been entirely neglected in our studies, and the increased lifetime of the collisions arises from internal rotation of the dipole moment. As we have noted above, vibrations should be easily excited in ion–molecule collisions that involve multiple reflections. Inclusion of vibrations and additional internal rotational degrees of freedom (for complexes such as $CH_3CN^+ + CH_3CN$) increases the parameter s and the lifetime.*

* s increases one unit for every additional vibration and one-half for every internal rotation.

An ultimate objective of this work is the application to more complicated processes such as ion–molecule reactions and charge exchange. In such processes it is quite clear that relatively long-lived complexes must be involved. Systems of interest will usually involve more than one accessible potential surface. The rate-determining step in a given reaction is likely to be the transfer from one potential surface to another or the decomposition of a collision complex in competition with other possible modes (channels).

Although the complexing of gas-phase ions has been known for a long time, the probable importance of complexes in chemical reactions and charge exchange has not been appreciated. Most considerations of charge exchange involve simple scattering collisions[27]; such a mechanism applies if the relative energy is high enough, but not at ordinary thermal energies. In a study of a set of 17 charge-exchange reactions, Bohme, Hasted, and Ong[28] have found an interesting correlation between rate constant and asymptotic energy difference of initial and final states. The interpretation of these data in terms of collision complexes presents a challenge.

REFERENCES

1. P. Langevin, *Ann. Chem. Phys.*, **5**, 245 (1905); see also E. W. McDaniel, *Collision Phenomena in Ionized Gases*, Wiley, New York, 1964, p. 67 and App. II.
2. G. Gioumousis and D. P. Stevenson, *J. Chem. Phys.*, **29**, 294 (1958).
3. D. P. Stevenson and D. O. Schissler, *J. Chem. Phys.*, **29**, 282 (1958).
4. J. V. Dugan, Jr., and J. L. Magee, *J. Chem. Phys.*, **47**, 3103 (1967).
5. L. P. Theard and W. H. Hamill, *J. Amer. Chem. Soc.*, **84**, 1134 (1962).
6. T. F. Moran and W. H. Hamill, *J. Chem. Phys.*, **39**, 1413 (1963).
7. D. A. Kubose and W. H. Hamill, *J. Amer. Chem. Soc.*, **85**, 125 (1963).
8. E. E. Ferguson, "Thermal Energy Ion-Molecule Reactions," in *Advances in Electronic and Electron Physics*, Vol. 24, L. Marton, Ed., Academic, New York, 1968, pp. 1–50.
9. H. Eyring, J. O. Hirschfelder, and H. S. Taylor, *J. Chem. Phys.*, **4**, 479 (1936).
10. J. O. Hirschfelder, C. F. Curtiss, and R. B. Bird, *Molecular Theory of Gases and Liquids*, 2nd ed., Wiley, New York, 1964.
11. J. V. Dugan, Jr., and J. L. Magee, NASA TN D-3229, 1966.
12. C. H. Townes and A. L. Schawlow, *Microwave Spectroscopy*, McGraw-Hill, New York, 1955, Chs. 5 and 10.
13. L. P. Theard and W. H. Hamill, private communication.
14. G. Herzberg, *Infrared and Raman Spectra of Polyatomic Molecules*, Van Nostrand, Princeton, N.J., 1945.
15. J. V. Dugan, Jr., J. H. Rice, and J. L. Magee, NASA TM X-1586, 1968.
16. D. K. Bohme, D. B. Dunkin, F. C. Fehsenfeld, and E. E. Ferguson, to be published.
17. D. K. Bohme, private communication.
18. R. J. Munson and A. M. Tyndall, *Proc. Roy. Soc. (London)*, **A172**, 28 (1939).
19. J. V. Dugan, Jr., J. H. Rice, and J. L. Magee, *Chem. Phys. Letters*, **2**, 219 (1968).
20. J. V. Dugan, Jr., J. H. Rice, and J. L. Magee, *Chem. Phys. Letters*, **3**, 323 (1969).

21. J. V. Dugan, Jr., R. B. Canright, Jr., R. W. Palmer, and J. L. Magee, NASA TM X-52662, 1969; paper presented at Sixth Intern. Conf. Electronic and Atomic Collisions, Boston; MIT Press, July 1969, pp. 333–340.

22. L. J. Leger and G. G. Meisels, *Chem. Phys. Letters*, **2**, 661 (1968); L. J. Leger, G. G. Meisels, and T. O. Tiernan, paper presented at 17th Mass Spectrometry Conf., Dallas, Texas, 1969.

23. D. Rapp, *J. Chem. Phys.*, **43**, 317 (1965).

24. H. M. Rosenstock, M. B. Wallenstein, A. L. Wahrhaftig, and H. Eyring, *Proc. Natl. Acad. Sci. U.S.*, **38**, 667 (1952).

25. R. A. Marcus and O. K. Rice, *J. Phys. Chem.*, **55**, 894 (1951); R. A. Marcus, *J. Chem. Phys.*, **20**, 359 (1952).

26. M. Burton and J. L. Magee, *J. Phys. Chem.*, **56**, 852 (1952).

27. D. Rapp and W. E. Francis, *J. Chem. Phys.*, **37**, 263 (1962).

28. D. K. Bohme, J. B. Hasted, and P. P. Ong, *Chem. Phys. Letters*, **1**, 259 (1967).

APPENDIX A

The Linear Molecule, Stark Effect Cross Sections for High J

The total polarizability for the rigid rod for quantum numbers J and M_J is

$$\alpha_{tR} = \alpha + \frac{\mu^2}{4E_R J^2}(3M_J^2 - J^2), \tag{A1}$$

where α is the electronic polarizability. The partial cross section for this quantum state is

$$\sigma_J(M_J) = \pi P_J \left[\frac{2\alpha_{tR}(M_J)e^2}{\epsilon}\right]^{1/2}, \tag{A2}$$

where the probability the molecule is in the M_Jth state is $P_J = 2/(2J + 1) \approx 1/J$. Positive α_{tR} values correspond to net attraction, and only these trajectories can contribute to capture collisions. For constant J, (A2) must be averaged over M_J from the first integral value that gives a positive value of α_{tR} up to J. All J values of interest here are large enough so that M_J can be considered as a continuous variable. The average cross section is the properly weighted integral over the M_J values: that is,

$$\sigma_J = \frac{\pi}{J^2}\left(\frac{\mu^2 e^2}{2E_R \epsilon}\right)^{1/2} \int_{M_0}^{J}\left(\frac{4E_R J^2 \alpha}{\mu^2} + 3M_J^2 - J^2\right)^{1/2} dM_J. \tag{A3}$$

The limit M_0 is either

$$M_0 = \frac{J}{\sqrt{3}}\left(1 - \frac{4E_R \alpha}{\mu^2}\right) \qquad \text{for} \quad \alpha \leq \frac{\mu^2}{4E_R}$$

or

$$M_0 = 0 \quad \text{for} \quad \alpha \geq \frac{\mu^2}{4E_R}.$$

The average cross section is

$$\sigma_J = \pi \left(\frac{3\mu^2 e^2}{2E_R}\right)^{1/2} \left[\frac{\sqrt{c+2}}{2\sqrt{3}} + \frac{(c-1)}{6}\ln\left(\frac{\sqrt{c+2}+\sqrt{3}}{\sqrt{c-1}}\right)\right] \quad \text{for} \quad c > 1$$

(A3a)

and

$$\sigma_J = \pi \left(\frac{3\mu^2 e^2}{2E_R}\right)^{1/2} \left[\frac{\sqrt{c+2}}{2\sqrt{3}} + \frac{(c-1)}{6}\ln\left(\frac{\sqrt{c+2}+\sqrt{3}}{\sqrt{1-c}}\right)\right] \quad \text{for} \quad c < 1,$$

(A3b)

where

$$c = \frac{4E_R \alpha}{\mu^2}.$$

Experimentally, the quantity of interest is the cross section averaged over the rotational distribution in J values at gas temperature T_g

$$Q_R(T_g) = \int_0^\infty \sigma_J f(J)\, dJ,$$

(A4)

where, from Ref. 14,

$$f(J) = \left(\frac{hB}{kT}\right)(2J+1)\exp\left\{\frac{-[J(J+1)Bh]}{kT_g}\right\}$$

APPENDIX B

Equations of Motion and Initial Conditions

The Lagrangian expression for the ion–linear molecule system is

$$L = T - V = \frac{m}{2}(\dot{X}^2 + \dot{Y}^2 + \dot{Z}^2) + \frac{I}{2}\left[\dot{x}_1^2 + \dot{x}_2^2 + \frac{(x_1\dot{x}_1 + x_2\dot{x}_2)^2}{(1 - x_1^2 - x_2^2)}\right]$$
$$+ \frac{\mu e}{r^3}\{Xx_1 + Yx_2 + Z[\pm(1 - x_1^2 - x_2^2)^{1/2}]\} + \frac{\alpha e^2}{2r^4}, \quad (B1)$$

where $r = |(X^2 + Y^2 + Z^2)^{1/2}|$; x_1, x_2, and x_3 are the direction cosines of the dipole with respect to the fixed coordinate system. The coordinates are shown in Fig. 1.

The rotational term is written as a function of two coordinates and their velocities, because the rod has only two degrees of freedom. In this example of L, x_3 is the missing rotational coordinate. The program is written to calculate the largest coordinate of the polar molecule and then to omit its differential equation. The equations of motion follow from

$$\frac{d}{dt}\left(\frac{\partial L}{\partial \dot{q}}\right) - \left(\frac{\partial L}{\partial q}\right) = 0, \tag{B2}$$

where

$$q = X, Y, Z, x_i, \text{ and } x_j \qquad (i = 1, 2, \text{ or } 3; j = 1, 2, \text{ or } 3; i \neq j).$$

The translational coordinate equations to be solved for the respective accelerations have the form

$$\ddot{X} = \frac{\mu e}{mr^5}\,[r^2 x_1 - 3X(Xx_1 + Yx_2 + Zx_3)] - \frac{2\alpha e^2 X}{mr^6}\,, \tag{B3}$$

where again x_1 and x_2 are retained and are solved from the following differential equations

$$\ddot{x}_1 = -x_1[2F(E_R)] - \frac{\mu e}{Ir^3}\,\{x_1[\pm(1 - x_1^{\,2} - x_2^{\,2})^{1/2}]Z - X(1 - x_1^{\,2}) + x_1 x_2 Y\},$$

$$\tag{B4}$$

where

$$F(E_R) = \frac{1}{2}\left[\dot{x}_1^{\,2} + \dot{x}_2^{\,2} + \frac{(x_1\dot{x}_1 + x_2\dot{x}_2)^2}{(1 - x_1^{\,2} - x_2^{\,2})}\right] = \tfrac{1}{2}(\dot{x}_1^{\,2} + \dot{x}_2^{\,2} + \dot{x}_3^{\,2})$$

and

$$x_3 = \pm(1 - x_1^{\,2} - x_2^{\,2})^{1/2},$$

$$\ddot{x}_2 = -x_2[2F(E_R)] - \frac{\mu e}{Ir^3}\,\{x_2[\pm(1 - x_1^{\,2} - x_2^{\,2})^{1/2}]Z - Y(1 - x_2^{\,2}) + x_1 x_2 X\}.$$

$$\tag{B5}$$

Initial separations were taken as 50 Å to insure negligible interaction at $t = 0$. Sets of runs were made for fixed impact parameter, ion velocity, and rotational energy of target molecule. In each of these runs the orientation of the molecule was chosen using three random numbers. The first two numbers were used to locate the axis of the molecule on the surface of a sphere, and the third was used to locate the plane of rotation. This information was transformed to cartesian coordinates for solution of the equations of motion.

APPENDIX C

Translational Energy Assignment

In experimental studies of ion–molecule reactions the cross section as a function of energy is desired. The energy of the incident ion is determined by an applied voltage. If this energy is large compared with thermal energy, the latter can be taken as zero. This condition is fulfilled down to relatively low values of ϵ where the directed ion energy is nearly equal to the thermal energy of the target polar molecule. The ion energy $\epsilon_1 = m_1 v_1^2/2$ is simply referred to the center of mass to describe ion–molecule relative motion by replacing the ion mass m_1 by the reduced mass of the pair m. This m is equal to $m_1 m_0/(m_1 + m_0)$, where m_0 is the mass of the polar molecule.

Ion–molecule collision pairs were chosen to consist of equal mass molecules and molecular ions so that $m = m_1/2 = m_0/2$. This choice corresponds to the experimental "parent" ion case. The relative energy for such systems is merely $m v_r^2/2$ (initially, $m v_1^2/2$), and the center of mass moves with the constant velocity $m_1 v_1/(m_1 + m_0) = v_1/2$.

Activation Parameters and Deprotonation Kinetics of Intramolecularly Hydrogen Bonded Acids

EDWARD M. EYRING

in collaboration with

LARRY D. RICH, LAYTON L. McCOY,*
RICHARD C. GRAHAM, AND NEWELL TAYLOR

University of Utah, Salt Lake City

Activation enthalpies and activation entropies are widely used to elucidate reaction mechanisms in condensed phases. Indeed, solution chemists trained in the application of these parameters from textbooks or other secondary sources often do not know the origin of these now classic concepts (Wynne-Jones and Eyring[10]). Magee, Ri, and Eyring suggested[5] that positive entropies of activation in aqueous solution reactions are frequently integer multiples of a +5 e.u. that could be attributed to the "melting" of a mole of "frozen" water molecules in proceeding from reactant to activated complex configurations. This model provides an interesting interpretation of temperature jump relaxation method kinetic data obtained for several intramolecularly hydrogen bonded (H-bonded) organic acids in dilute aqueous solution.

* University of Missouri at Kansas City.

Before discussing the kinetic behavior of these acids I wish to relate the role Henry Eyring played in getting me and my co-workers interested in this type of fast reaction kinetics problem. His generosity with good-humored free advice to his colleagues and even casual acquaintances is well known, perhaps even legendary. It was he who counseled that I do a year of post-doctoral work (1960–1961) in Eigen's Goettingen laboratory. This suggestion

Table I. Spectrophotometrically Determined Mixed[a] Acid pK's for Azo Dye Anions in 0.1M Ionic Strength (Adjusted with KNO$_3$) Aqueous Solutions at 25°.

The pK's are for the intramolecularly hydrogen-bonded hydroxyl proton.

$$R_1 \text{—} \underbrace{\hspace{1cm}} \text{—} N \overset{\cdots HO \quad R_2}{\underset{N}{\diagdown}} \quad {}^-O_3S \cdots SO_3^-$$

R$_1$	R$_2$	p$K_a{}^M$	λ^b (nm)
H	NH$_2$	12.72 ± 0.04	527
NHCOCH$_3$	NH$_2$	12.49 ± 0.12	542
NH$_2$	NH$_2$	12.34 ± 0.04	560
CH$_3$	NH$_2$	12.64 ± 0.11	536
H	HNCOCH$_3$	10.65 ± 0.09	530
HNCOCH$_3$	HNCOCH$_3$	10.41 ± 0.05	524
CH$_3$	HNCOCH$_3$	10.43 ± 0.03	536

[a] Mixed acid dissociation constants are defined by $K_a{}^M = a_{H^+}[A^-]/[HA]$, where a_{H^+} denotes hydrogen ion activity and the bracketed concentrations are molar.[1]

[b] Maximum absorbance wavelength of deprotonated form of acid used in spectrophotometric determinations of pK_a.

was very timely, and my exposure to Manfred Eigen, the disciplined perfectionist, provided an interesting contrast to Henry Eyring, the buoyantly optimistic, free-wheeling generalist. Qualities I particularly admire in Henry Eyring's style of doing science are his complete lack of timidity in his choice of problems and his taboo against speaking critically in a personal way of the more modest achievements of others. To make a long story short, our kinetic measurements were made by Eigen's temperature jump relaxation method,[2] and our interpretation of these data borrows heavily from the now classic insights of Wynne-Jones and Eyring[10] and Magee, Ri, and Eyring.[5]

Table I lists seven similar azo dyes characterized by comparatively small acid dissociation constants of the di- or trianions. The dotted line in the formula of the dianion of 5-acetamido-3-(4'-acetamidophenylazo)-4-hydroxy-2,7-naphthalene disulfonic acid (denoted hereafter by I) depicts a postulated

(I)

intramolecular H-bond to which such high pK_a's are usually attributed. The observed differences in pK_a among these azo anions can be explained a posteriori largely on the basis of steric effects of the N-acetyl substituents.

In principle, it should be possible to measure the rate of the reaction

$$HA^{z-} + OH^- \underset{}{\overset{k_{23}'}{\rightleftharpoons}} A^{(z+1)-} + H_2O \tag{1}$$

for any of these acids by the Joule heating temperature jump relaxation method using spectrophotometric detection to follow the associated rapid color change (in the case of I: $\tau = 79$ μsec for pH $= 10.5$, $5 \times 10^{-5}M$ solutions). In fact, we have found it convenient to make such kinetic measurements only for compound I for which $k_{23'} = 2.2 \times 10^8$ M^{-1} sec^{-1} in aqueous $0.1M$ ionic strength solution (with KNO_3) at 25°. By way of comparison, the rate constant for the essentially diffusion-controlled deprotonation of the HPO_4^{2-} dianion by OH^- ion in $0.1M$ ionic strength aqueous solution at 25° is $\sim 2 \times 10^9$ M^{-1} sec^{-1}.[3] Eigen's explanation for the comparative slowness of deprotonation of an intramolecularly H-bonded anion is that H-bond bridging between the proton of the anion and surrounding solvent water molecules is disrupted by the involvement of the proton in a strong intramolecular H-bond. This structural discontinuity reduces the speed with which the proton of the anion can transfer by the Grotthuss mechanism[9] to the OH^- ion located two or more water molecules away in the bulk solvent.

Deprotonation kinetics of intramolecularly H-bonded dicarboxylic acid mono-anions are usually a good deal easier to study systematically by the temperature jump method because of their smaller pK_a values that do not require such high sample solution pH's as do the azo dyes of Table I. In Table II we have representative kinetic results for several dicarboxylic acids. The first nine acids in Table II are derivatives of malonic acid, $HOOCCH_2$-$COOH$, substituted at the number two carbon. Their specific rates, $k_{23'}$, for deprotonation by OH^- have been reported previously.[7,8] The kinetic data

Table II.   Acid Dissociation Constants and Rate Constants $k_{23'}$ for Monoanion Deprotonation of Several Dicarboxylic Acids in Aqueous 0.1M Ionic Strength Solution at 25°

Acid	pK_2	$K_1 K_2^{-1}$ $\times 10^4$	$k_{23'}$ $\times 10^7 \, M^{-1} \, sec^{-1}$
(1) Isopropyl malonic[a]	5.57	0.06	340
(2) Methyl (1-methyl butyl) malonic[a]	6.64	0.71	130
(3) Diethyl malonic[a]	7.06	9.3	24
(4) Ethylisoamyl malonic[a]	7.31	17	18
(5) Ethylphenyl malonic[a]	7.10	32	17
(6) Ethyl-n-butyl malonic[a]	7.25	13	16
(7) Di-n-propyl malonic[a]	7.34	16	3
(8) Ethylisopropyl malonic[a]	8.07	150	5.0
(9) Diisopropyl malonic[a]	8.58	330	4.6
(10) cis-1,2-Ethylene-dicarboxylic[b]	6.34	2.6	79
(11) Cyclopentene-1,2-dicarboxylic[b]	7.27	43	67
(12) Bicyclo[2.2.1]hepta-2,5-diene-2,3-dicarboxylic[b]	7.77	282	14
(13) Cyclobutene-1,2-dicarboxylic[b]	7.63	324	11
(14) Bicyclo[2.2.1]hepta-2-ene-2,3-dicarboxylic[b]	8.00	479	7.3
(15) Furan-3,4-dicarboxylic[b]	7.84	251	5.9

[a] Values of K_1 and K_2 are mixed acid ionization constants[1] for loss of first and second carboxylic acid protons, respectively. Values of $k_{23'}$ reported by Miles and co-workers.[7,8]
[b] K_1 and K_2 are thermodynamic dissociation constants reported by McCoy.[6] Values of $k_{23'}$ calculated from unpublished temperature jump data of L. D. Rich, E. M. Eyring, and L. L. McCoy obtained with added o-cresolsulfonphthalein indicator ($<10^{-5} \, M$) and KNO_3.

for acids 10–15 of Table II are unpublished but quite interesting because of the rigid skeletons of these acids

that make it possible to calculate spacing between the carboxyl groups: 1.68, 1.92, 2.07, 2.61, 2.11, and 2.23 Å for acids 10 through 15, respectively.[6]

Each of the two series of acids in Table II is arranged in order of decreasing rate constants $k_{23'}$, which means that the strongest intramolecular hydrogen bond is at the bottom of each series. The inverse correlation of $k_{23'}$ with $K_1K_2^{-1}$ is so good that much greater interest resides in comparisons of $k_{23'}$ for such acids in water and D_2O (see for example, Haslam et al.[4]) or in comparisons of activation parameters[7] than in the fairly predictable aqueous solution rate constants themselves. In Table III we see the activation

Table III. Activation Parameters for the Forward and Reverse Reactions of the Equilibrium $HA^- + OH^- \rightleftharpoons A^{2-} + H_2O$ in $0.1M$ Ionic Strength (Adjusted with $NaClO_4$) Aqueous Solution from Temperature Jump Relaxation Method Specific Rates Determined at 12 and 25° (Miles et al.[7])

	$HA^- + OH^- \xrightarrow{k_{23'}}$		$A^{2-} + H_2O \xrightarrow{k_{3'2}}$	
Malonic acid	$\Delta H^{\ddagger}$ (kcal/mole)	$\Delta S^{\ddagger}$ (e.u.)	$\Delta H^{\ddagger}$ (kcal/mole)	$\Delta S^{\ddagger}$ (e.u.)
Diethyl	4.5	−5	18.5	+3
Ethyl-n-butyl	5.5	−3	19.5	+6
Ethylisoamyl	5.5	−3	19.5	+6
Di-n-propyl	6	−1	19.5	+6
Ethylisopropyl	7	0	21	+12
Diisopropyl	7.5	+2	21.5	+16

parameters for both the forward and reverse reactions of equilibrium (1) for several of the malonic acids. The specific rate for the reverse reaction, $k_{3'2}$, (not shown) increases as one proceeds down the series of malonic acids whereas $k_{23'}$ decreases (see Table II) with increasing intramolecular H-bond strength. At 25° for the forward reaction we are below the isokinetic temperature (see Ref. 7), and the reaction of lowest $\Delta H^{\ddagger}$ has the highest specific rate. We note too that $\Delta H^{\ddagger}$ has the right magnitude to correspond roughly to the energy of an intramolecular H-bond. In the equally interesting case of the reverse reaction our kinetic measurements at 12 and 25° have been made above the isokinetic temperature in a region of control by $\Delta S^{\ddagger}$, a familiar situation when solvent influences reaction rate. The positive values of $\Delta S^{\ddagger}$ for the reverse reaction are attributable to melting of solvent molecules frozen about the A^{2-} ion in proceeding toward the activated complex configuration. For water,

$$\Delta S_{fusion} = \frac{\Delta H_{fusion}}{T_{fusion}} = \frac{1440 \text{ cal/mole}}{273 \text{ deg}} \cong 5 \text{ e.u.}$$

Thus, following Magee, Ri, and Eyring, we would guess that there are at least two more water molecules bound to the dianion of diisopropyl malonic acid ($\Delta S^{\ddagger} = +16$ e.u.) than to the diethyl malonic acid dianion ($\Delta S^{\ddagger} = +3$ e.u.) that must melt to reach the activated complex configuration.

Activation parameters are more and more frequently being reported as important features of temperature jump relaxation method kinetic studies. This is true in spite of the fact that the restricted temperature range accessible to a Joule heating temperature jump apparatus and the usual $\pm 10\%$ uncertainties in measured relaxation times are not favorable characteristics for achieving precise values of $\Delta H^{\ddagger}$ and $\Delta S^{\ddagger}$.

ACKNOWLEDGMENTS

Work supported in part by Grant AM 06231 from the National Institute of Arthritis and Metabolic Diseases. Azo dyes graciously provided by R. Walker, National College of Food Technology, Weybridge, Surrey, England.

REFERENCES

1. A. Albert and E. P. Serjeant, *Ionization Constants of Acids and Bases*, Methuen, London, 1962.
2. M. Eigen and L. DeMaeyer, in *Technique of Organic Chemistry*, Vol. VIII, Part II, S. L. Friess, E. S. Lewis, and A. Weissberger, Eds., Interscience, New York, 1963, p. 977 ff.
3. M. Eigen, W. Kruse, G. Maass, and L. DeMaeyer, in *Progress in Reaction Kinetics*, Vol. 2, G. Porter, Ed., Macmillan, New York, 1964, p. 300 ff.
4. J. L. Haslam, E. M. Eyring, W. W. Epstein, R. P. Jensen, and C. W. Jaget, *J. Amer. Chem. Soc.*, **87**, 4247 (1965).
5. J. L. Magee, T. Ri, and H. Eyring, *J. Chem. Phys.*, **9**, 419 (1941).
6. L. L. McCoy, *J. Amer. Chem. Soc.*, **89**, 1673 (1967).
7. M. H. Miles, E. M. Eyring, W. W. Epstein, and R. E. Ostlund, *J. Phys. Chem.*, **69**, 467 (1965).
8. M. H. Miles, E. M. Eyring, W. W. Epstein, and M. T. Anderson, *J. Phys. Chem.*, **70**, 3490 (1966).
9. W. J. Moore, *Physical Chemistry*, 3rd ed., Prentice-Hall, Englewood Cliffs, N.J., 1962, p. 338.
10. W. F. K. Wynne-Jones and H. Eyring, *J. Chem. Phys.*, **3**, 492 (1935).

The Brønsted α and the Primary Hydrogen Isotope Effect. A Test of the Marcus Theory

MAURICE M. KREEVOY AND DENNIS E. KONASEWICH

School of Chemistry, University of Minnesota, Minneapolis

Abstract

Rate constants, k_{HA}, have been evaluated for proton transfer from a variety of acids, HA, to diazoacetate ion. For a series of neutral oxygen acids including both carboxylic acids and phenols, $\log k_{HA}$ is a continuous but nonlinear function of $\log K_{HA}$ after both are corrected for small symmetry factors. This function can be reproduced by the Marcus theory of proton transfer, and a resulting parameter, W^r, which is the standard free energy of the associative step preceding proton transfer, is 8 kcal/mole, which is very reasonable. Other evidence is also presented that the associative step incorporated in the theory is necessary and must have a substantial free energy.

Rates and isotope effects on proton transfer from H^+ and H_2O to diazoacetate are in qualitiative agreement with those predicted by the theory, using the parameters derived from neutral acids, but some quantitative disagreement is evident. Data from other proton transfer reactions is concordant with the theory.

243

The theory of absolute reaction rates[1] has not and probably will not, in the foreseeable future, permit the satisfactory a priori calculation of most reaction rates, particularly in condensed phases. What it does do, instead, is to identify an attainable goal toward which speculation, semiempirical calculation, and experimental work may be directed. That is the elucidation of transition state structure. If transition state structure could be determined, by whatever means, as well as the structure of many stable molecules is now known, it *would* be possible to calculate reaction rates. Further, even well short of enough information to calculate rates, even very approximate and incomplete transition state structures may often suffice for the calculation of such rate-related quantities as isotope effects, Brønsted α values, and Hammett ρ values. When successful, such partial calculations tend to confirm the partial transition state structures on which they are based and, thus, to permit further progress on a more secure base. We cannot be certain how well pleased Henry Eyring is with this hybrid of theory and empiricism, but it doesn't seem likely that he is too disturbed. The pragmatist, pleased with what works and ready to take information from anywhere, has always seemed to be as well represented in his work as the rigorous theoretician.

In the present paper we describe an intermediate stage in this process. We show that a general formalism provided by Marcus fits data for proton transfer reactions of the type shown in (1), where HA is a carboxylic acid or a

$$N_2CHCOO^- + HA \rightarrow {}^+N_2CH_2COO^- + A^- \qquad (1)$$

p-substituted phenol. In this fitting, parameters are obtained which also permit prediction of the variation of the primary hydrogen isotope effect with the strength of the acid, HA. To the extent that these parameters have physical significance, they provide structural information about the activation processes and transition states for reactions of the type shown in (1) and also those for related reactions. The Marcus theory[2] of proton transfer reactions, which is a special elaboration of the general transition state theory of reaction rates, divides the activation process for such reactions into two parts. In the first part the reagents are brought together, and solvent, proton donor, and proton acceptor are reorganized so as to permit proton transfer. The second part is the proton transfer itself. In a third phase of the reaction, which does not directly influence the rate in the forward direction, the products undergo the reverse of the processes of the first part. This leads to (2) for the Gibbs free energy of activation.[2] The symbolism of (2) is that of Marcus. The standard free energy

$$\Delta F^* = W^r + \left(1 + \frac{\Delta F_R^{o\prime}}{\lambda}\right)^2 \frac{\lambda}{4} \qquad (2)$$

of the first, associative, step is W^r; that of the second, proton-transfer, step is

$\Delta F_R^{\circ\prime}$; and that of the third, dissociative, step is W^p. The average of λ_1 and λ_2 is λ. The free energy of activation for proton transfer within the properly oriented, structured, and solvated symmetrical donor–acceptor complex, AH·A$^-$, is $\lambda_1/4$. The analogous quantity for the other partner in the original proton-transfer reaction is $\lambda_2/4$.

Neglecting symmetry, the Brønsted α is $\delta(\log k_{HA})/\delta(\log K_{HA})$,[3] where k_{HA} is the rate constant for proton transfer from HA to substrate and K_{HA} is the acid dissociation constant of HA. To a good approximation λ is likely to be independent of the structure of A within a series of related acids. As variation in A makes HA a better proton donor, it simultaneously makes A$^-$ a poorer acceptor, so that λ_1 can remain largely unchanged. Since A has no role at all in determining λ_2, λ_2 will be accurately independent of the structure of A. It seems intuitively likely that W^r is also approximately independent of A in many systems, since most of the processes contributing to it do not depend acutely on the structure of A. This is true of the negative entropy associated with bringing the reagents together, and also of the free energy of reorganizing and resolvating the substrate. The process of desolvating the acidic proton of HA probably does depend on the acidity of HA, but not too strongly.[4] If both W^r and λ are independent of K_{HA}, then α is given by (3).

$$\alpha = \left(1 + \frac{\Delta F_R^{\circ\prime}}{\lambda}\right)^2 \tag{3}$$

For the same reasons as for W^r, W^p should be approximately independent of K_{HA}. Since ΔF_R°, the standard free energy of a proton transfer from HA to an acceptor, must vary with A in *exactly* the same way as the free energy of acid dissociation, ΔF_{HA}°, and since ΔF_R° is the sum of W^r, W^p, and $\Delta F_R^{\circ\prime}$ the latter must be given by ΔF_{HA}° plus a constant. Thus the variation in α as a function of the acidity of HA should be given by (4). In the present paper, (4) is shown to

$$\frac{\delta\alpha}{\delta\Delta F_{HA}^\circ} = \tfrac{1}{2}\lambda \tag{4}$$

fit a substantial body of data for the reaction shown in (1). The result is used to evaluate the other kinetically significant parameters.

Marcus' theory[1] also allows the primary hydrogen isotope effect to be calculated as a function of K_{HA}, using (5) or its equivalent (6). The parameters of these equations differ from those used previously only by the

$$\Delta F_H^* - \Delta F_T^* \simeq \tfrac{1}{4}(\lambda_H - \lambda_T)\left[1 - \left(\frac{\Delta F_R^{\circ\prime}}{\lambda_H}\right)^2\right] \tag{5}$$

$$\Delta F_H^* - \Delta F_T^* \simeq \tfrac{1}{4}(\lambda_H - \lambda_T)[1 - (2\alpha - 1)^2] \tag{6}$$

subscripts, which indicate the isotope to which they pertain. The only new assumption needed to obtain these equations from (2) and (3) is that W^r, $\Delta F_R^{\circ\prime}$, and W^p are not subject to hydrogen isotope effects. The only new parameter is $(\lambda_{\mathrm{H}} - \lambda_{\mathrm{T}})/4$, which is the isotope effect, expressed as a free energy, for the case that α is 0.5. That is the maximum isotope effect. In the present paper the ability of (5) and (6) to correlate the isotope effects for H^+, H_2O, and acetic acid is tested.

RESULTS

Using previously described techniques,[5] catalytic coefficients, k_{HA}, have been evaluated for the catalysis of diazoacetate hydrolysis (7) by four phenols, seven carboxylic acids, water, and the aquated proton—all at 25°, in aqueous solution, at an ionic strength of 0.105. The mechanism of this reaction involves several steps, and the rate law is somewhat complicated, but it is possible to

$$N_2CHCOO^- + H_2O \xrightarrow{\mathrm{HA}} HOCH_2COO^- \tag{7}$$

evaluate the catalytic coefficients, which are rate constants for reactions of the type shown in (1), with reasonable reliability, by the method which has been described. They are shown in Table I.

Separate plots of $\log k_{\mathrm{HA}}$ versus $\log K_{\mathrm{HA}}$ were made for phenols and for carboxylic acids. These plots are shown in Fig. 1. Neither one is visibly curved, but the slopes are obviously different. Each plot was assumed to be linear and its slope evaluated by the method of least squares. The slope for carboxylic acids, 0.51, is assumed to be the α pertinent to the midpoint of the carboxylic acid data at a K_{HA} of 6.7×10^{-4}. That for phenols, 0.74, was assumed to be α for the midpoint of the phenol data, at $K_{\mathrm{HA}} = 8 \times 10^{-10}$.

Before further calculation can be made, symmetry factors must be taken into account. In the present work K_{HA} is multiplied by q/p, where p is the number of acidic protons in the acid and q is the number of basic sites in the conjugate base. Similarly, k_{HA} will be multiplied by $1/p$. These symmetry corrections have been the subject of some disagreement,[6] but the original Brønsted proposals,[3] which have been used here, seem to us to be most nearly free of artifacts.

After making these symmetry corrections, $\Delta\Delta F_{\mathrm{HA}}^{\circ}$ is straightforwardly calculated from the midpoint K_{HA} values. It and $\Delta\alpha$ were substituted for $\delta\alpha$ and $\delta\Delta F_{\mathrm{HA}}^{\circ}$ in (4) and a value of 18 kcal/mole obtained for λ. This is probably uncertain by 2–3 kcal/mole, the uncertainty arising from the uncertainty in the "best" slopes of the plots of $\log k_{\mathrm{HA}}$ and also from uncertainty in the K_{HA} values in which the α's actually pertain. From (3) and (2), $\Delta F_R^{\circ\prime}$ and W^r

Table I.   Catalytic Coefficients for Diazoacetate Hydrolysis

No.	Acid	k_{HA}, M^{-1} sec^{-1}	K_{HA}, M[a]
1	H_2O	$6.9 \pm 0.1 \times 10^{-7}$ [b]	1.8×10^{-16} [b,c]
2	$p\text{-}CH_3C_4H_6OH$	$3.9 \pm 1.2 \times 10^{-2}$	5.6×10^{-11} [d]
3	C_6H_5OH	$7.0 \pm 1.0 \times 10^{-2}$	1.1×10^{-10} [e]
4	$p\text{-}BrC_4H_6OH$	$1.3 \pm 0.2 \times 10^{-1}$	4.5×10^{-10} [e]
5	$p\text{-}NCC_4H_6OH$	2.1 ± 0.1	1.1×10^{-8} [f]
6	$(CH_3)_3CCOOH$	$2.3 \pm 0.2 \times 10^2$	9.0×10^{-6}
7	CH_3COOH	$2.6 \pm 0.1 \times 10^2$	1.8×10^{-5}
8	$HCOOH$	$5.6 \pm 0.4 \times 10^2$	1.8×10^{-4}
9	$ClCH_2COOH$	$2.4 \pm 0.1 \times 10^3$	1.4×10^{-3}
10	$NCCH_2COOH$	$5.6 \pm 0.8 \times 10^3$	3.5×10^{-3}
11	$F_2CHCOOH$	$1.2 \pm 0.3 \times 10^4$	3.5×10^{-2} [g]
12	$Cl_2CHCOOH$	$1.6 \pm 0.2 \times 10^4$	5.6×10^{-2}
13	H^+	$6.5 \pm 0.1 \times 10^4$	5.5×10

[a] All acid dissociation constants were taken from G. Kortum, W. Vogel, and K. Andrussow, *Dissociation Constants of Organic Acids in Aqueous Solutions* (Butterworths, London, 1961) unless otherwise given.

[b] The conventional first-order rate constant for water[4] was divided by 55, the molar concentration of water, to get the rate constant shown, in order that it might be at least formally similar to the other values. The acid dissociation constant of water was treated similarly.

[c] The autoprotolysis constant of water was taken as 1.00×10^{-14} M^2 (L. Pentz and E. R. Thornton, *J. Amer. Chem. Soc.*, **89**, 6931 (1967), and references cited therein).

[d] P. D. Bolton, F. M. Hall, and I. H. Reece, *Spectrochim. Acta*, **22**, 1149 (1966).

[e] P. D. Bolton, F. M. Hall, and I. H. Reece, *Spectrochim. Acta*, **22**, 1153 (1966).

[f] G. W. Wheland, R. M. Brownell, and G. C. Mayo, *J. Amer. Chem. Soc.*, **70**, 2492 (1948).

[g] M. M. Kreevoy, T. S. Straub, W. V. Kayser, and J. L. Melquist, *J. Amer. Chem. Soc.*, **89**, 1201 (1967).

can then be successively evaluated. Equation (8) gives $\Delta F_R^{\circ\prime}$, and W^r is 8 kcal/mole. The latter has a statistical uncertainty of 1–2 kcal/mole. Using these

$$\Delta F_R^{\circ\prime} = \Delta F_{HA}^{\circ} - 3.6 \tag{8}$$

parameters ΔF^* was evaluated as a continuous function of ΔF_{HA}° for the entire range of acids represented in Table I. Equation (9) shows the relation between ΔF^* and k_{HA}.

$$\ln\left(\frac{kT}{h}\right) - \frac{\Delta F^*}{RT} = \ln\left(\frac{k_{HA}}{p}\right) \tag{9}$$

Figure 2 shows how well the experimental values are correlated by (2).

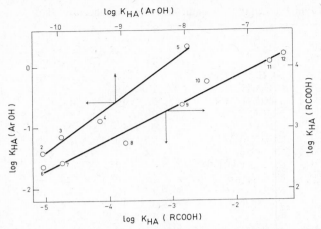

Fig. 1. Plots of $\log k_{HA}$ against $\log K_{HA}$ for carboxylic acids and phenols. The acids are numbered as in Table I. The lines were obtained by the method of least squares and have slopes of 0.51 for carboxylic acids and 0.74 for phenols.

In our earlier work[5] a quantity α_i, thought to be comparable with α,[7] was evaluated for H^+ and for H_2O. For H^+ it is 0.30 and for H_2O, 0.89.[5] These can now be compared with the α values given by (3): 0.36 for H^+ and 0.99 for H_2O.

The earlier work also provides a number of measures of the primary

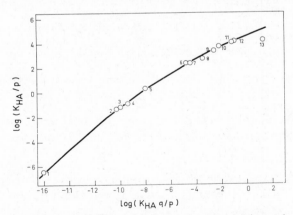

Fig. 2. Comparison of calculated and experimental correlations of k_{HA}/p with $K_{HA}q/p$. The solid line was generated with the ΔF^* values in (2), using the parameters described in the text. The circles are experimental points. The acids are numbered as in Table I.

hydrogen isotope effect for H^+, H_2O, and acetic acid. The one which most nearly meets the criteria that W^r, W^p, and $\Delta F_R^{0\prime}$ be free of isotope effects is κ_H/κ_T. This isotope effect compares the mobile proton of the transition state, in each case, with a solvent proton. From each of these isotope effects $(\lambda_H - \lambda_T)/4$ can readily be calculated, using (5) and the previously determined values of $\Delta F_R^{0\prime}$ and λ_H. The results are shown in Table II. Also given are the

Table II. Calculated Maximum Isotope Effects

Acid	α or α_i (exptl)	α (3)	$\lambda_T - \lambda_H{}^a$ (3)	$\lambda_T - \lambda_H{}^a$ (6)
H^+	0.26	0.36	1.35	1.5
CH_3COOH	0.57	0.57	1.52	1.5
H_2O	0.89	0.99	$\sim 10^2$	1.6

^{a} In kcal/mole.

values of $(\lambda_H - \lambda_T)/4$ obtained from (6). In that calculation, experimental α_i values were used for H^+ and H_2O, and the α given by (3) was used for acetic acid.

DISCUSSION

The carboxylic acid and phenol data are very well accommodated by the theory, as shown in Fig. 2, and the resulting derived quantity, W^r, is reasonable. The free energies associated with localizing HA in the vicinity of the substrate and breaking its hydrogen bond to its nearest-neighbor water molecule* are probably the principal contributors to W^r. The former process should have ΔF° of about 2.4 kcal/mole,[2] and the latter, about 6 kcal/mole.† The sum of these is in excellent agreement with W^r, 8 kcal/mole. This analysis of the value of W^r implies that it will be about the same for many, chemically different, reactions, a prediction which can be tested. It also implies, contrary to some previous thought,[10] that even an infinitely strong acid would give a

---

* It has recently been shown very likely that no water molecules intervene between the acid and the substrate in proton transfer to carbon.[8]

† The formation constant for the hydrogen bond between p-fluorophenol and aliphatic ethers is about 10 when all the constituents are solutes in CCl_4.[4] If pure liquid water is taken as the standard state for the acceptor and a comparable formation constant used a ΔF° of -3 kcal/mole is obtained. However, water gives a larger formation constant than ether as hydrogen bond acceptor by about 10^2.[9] When this is converted to a free energy it gives an additional -3 kcal/mole. These quantities appear to be approximately solvent-independent.[4]

rate several powers of 10 short of the diffusion limit, but otherwise having the characteristics of a diffusion-limited rate; i.e., a primary isotope $\sim 2^{1/2}$ and no selectivity among acids. These predictions are impossible to test in the present case because of the relative magnitudes of the rate constants. However, in two reactions of enolate ions with H^+, primary hydrogen isotope effects $\sim 2^{1/2}$ are observed with rate constants around 10^7 l mole^{-1} sec^{-1},[11,12] and a number of other carbon bases have similar rate constants for reactions with H^+, in spite of the fact that they are strongly spontaneous.[10]

Further evidence in favor of an associative step with a substantial standard free energy is provided by the observation that α is about 0.5 for an acid of K_{HA}, about $10^{-4}M$. If no such associative step is provided, this would imply a similar acidity for the protonated substrate.[10] However, aliphatic diazonium compounds, of which the protonated substrate is an example, are generally much stronger acids than this. For example, ethyl diazoacetate gives no evidence of significant protonation in HCl or HClO$_4$ up to $2.5M$.[13] The existence of such a step also permits the rationalization of α values larger than unity if W^r depends significantly on the structure of HA. Such a dependence might be anticipated in the case of nitroalkanes as acids, because of the extensive structural rearrangements that accompany dissociation, and it is in just such cases that α values greater than unity have been observed.[14]

There are indications[4,15] that free energies of formation of hydrogen bonds vary substantially with the structural type of the partners in ways that are not correlated with acid–base dissociation constants. If such a variation is incorporated in W^r it would be sufficient to account for the observation that acids covering a variety of structural types do not fit on a single Brønsted plot.[7]

When the theory is extended to H^+ and H_2O as acids, qualitative agreement but substantial quantitative discrepancies are observed for the present reaction. The α values are low and high, respectively, as predicted. The isotope effects, particularly that for H_2O, are reduced, again as predicted. However, the theory places the transition state with H_2O as the acid nearly at the products of the rate-determining transformation, with the result that α and κ_H/κ_T are both predicted to be close to unity. The experimental α, α_i, is significantly less than unity, and κ_H/κ_T, while reduced from its value for acetic acid, is still substantially above unity. When α_i is combined with κ_H/κ_T, using (6), for H^+ and H_2O, reasonable values of $(\lambda_T - \lambda_H)/4$ are obtained, in fair agreement with each other and with that from acetic acid. However, when κ_H/κ_T for H_2O is combined with the parameters calculated from phenol and carboxylic acid data, in (5), a value of $(\lambda_T - \lambda_H)/4$ is produced which is intuitively much too large and is also in striking disagreement with the reasonable values given by H^+ and acetic acid. These comparisons are all shown in Table II. In addition, k_H is almost a power of 10 below the value predicted by (2), as shown in Fig. 1. Parameters well outside of the apparent experimental uncertainties would be required to rectify these difficulties. At this time it is not known to what extent this problem may be unique to the solvent and an ion derived from the solvent.

Equation (4) predicts a continuous variation in α with changes in K_{HA}. However, the variation depends on λ. Assuming that a change of at least 0.2 is required for detectability, as suggested by Figs. 1 and 2, then either very small λ values, implying relatively basic substrates and fast reactions, or else studies spanning very long ranges of K_{HA}, are required to detect systematic variation in α. Most studies in this field to date have been limited to carboxylic acids and olefinic substrates,[7] a combination unlikely to meet these requirements, and no variation in α has generally been detected. There appear to be a few cases in which systematic variation in α with K_{HA} can be qualitatively observed. Two such reactions are shown in (10) and (11).[16,17]

$$ArCH_2^- + HA \rightarrow ArCH_3 + A^- \tag{10}$$

$$\text{(diagram)} + HA \longrightarrow \text{(diagram)} + TA \tag{11}$$

[In (10) Ar can be either the 2,4,6-trinitrophenyl group or the 2,6-dinitrophenyl group, so this equation represents two sets of results.] The results are qualitatively in accord with the theory, following a pattern similar to that of the present results, but quantitative fitting does not seem justified, because in two of these cases inhomogeneous series of acids are involved, and in the third a long extrapolation to higher temperatures is required to bring the k_{HA} values to the temperature at which the K_{HA} values are known.[16] In a fourth case, isotope exchange in trimethoxybenzene,[18] a long range of acid strengths but only a small number of acids were studied, and a number of different types of acids were involved, so that it is not possible to conclude anything about a possible systematic variation in α. In general, it can only be said that information previously available on $A-S_E2$ reactions is compatible with the Marcus theory.

In conclusion, then, the Marcus theory seems to provide a satisfactory framework for the rationalization of presently available results on $A-S_E2$ reactions. In particular, the introduction of an associative step with a significant standard free energy seems to be very attractive. In future work it is hoped that a more extensive test of the quantitative aspects of the theory can be provided.

ACKNOWLEDGMENTS

This work was supported, in part, by the National Science Foundation through GP-7915 and was written while M. M. Kreevoy was a participant in the exchange program between the U.S. National Academy of Sciences and the Council of Academies of the S.F.R. of Yugoslavia and a guest of Professor Dionis Sunko at the Institute "Ruder Bošković." He wishes to thank the Academies for their support, and the Institute and Professor Sunko for their warm hospitality. Summer fellowships from E. I. du Pont de Nemours, and Mobil Oil and a fellowship supported by General Mills are gratefully acknowledged by Dennis E. Konasewich.

REFERENCES

1. H. Eyring, *J. Chem. Phys.*, **3**, 107 (1935).
2. R. A. Marcus, *J. Phys. Chem.*, **72**, 891 (1968).
3. J. N. Brønsted and K. Pederson, *Z. Physik. Chem.*, **108**, 185 (1924); J. N. Brønsted, *Chem. Rev.*, **5**, 231 (1928).
4. R. W. Taft, D. Gurka, L. Joris, P. von R. Schleyer, and J. W. Rakshys, *J. Amer. Chem. Soc.*, **91**, 4801 (1969).
5. M. M. Kreevoy and D. E. Konasewich, *J. Phys. Chem.*, in press.
6. D. M. Bishop and K. J. Laidler, *J. Chem. Phys.*, **42**, 1688 (1965); other pertinent references are cited in this paper.
7. J. M. Williams, Jr., and M. M. Kreevoy, *Advan. Phys. Org. Chem.*, **6**, 63 (1968).
8. D. M. Goodall and F. A. Long, *J. Amer. Chem. Soc.*, **90**, 238 (1968), M. M. Kreevoy and J. M. Williams, Jr., *ibid.*, **90**, 6809 (1968).
9. I. M. Kolthoff and M. K. Chantooni, *J. Amer. Chem. Soc.*, **90**, 3320 (1968).
10. M. Eigen, *Angew. Chem.*, **75**, 489 (1963).
11. T. Riley and F. A. Long, *J. Amer. Chem. Soc.*, **84**, 522 (1962).
12. F. A. Long and D. Watson, *J. Chem. Soc.*, **1958**, 2019.
13. W. J. Albery and R. P. Bell, *Trans. Faraday Soc.*, **57**, 1942 (1961).
14. F. G. Bordwell, W. J. Boyle, Jr., J. A. Hautala, and K. C. Yee, *J. Amer. Chem. Soc.*, **91**, 4002 (1969).
15. D. Gurka and R. W. Taft, *J. Amer. Chem. Soc.*, **91**, 4801 (1969).
16. M. E. Langmuir, L. Dogliotti, E. D. Black, and G. Wettermark, *J. Amer. Chem. Soc.*, **91**, 2204 (1969).
17. R. J. Thomas and F. A. Long, *J. Amer. Chem. Soc.*, **86**, 4770 (1964).
18. A. J. Kresge and Y. Chiang, *J. Amer. Chem. Soc.*, **83**, 2877 (1961).

Self-Diffusion of Oxygen in Iota Phase Praseodymium Oxide

GARY R. WEBER* AND LEROY EYRING

Department of Chemistry, Arizona State University, Tempe

Abstract

The method of heterogeneous isotopic exchange was used to determine the self-diffusion coefficient of oxygen in Pr_7O_{12} (iota phase) as a function of temperature and pressure. The results indicated that there are two types of hyperstoichiometric material depending on the degree of oxygen excess. When the compound is close to stoichiometry, the diffusion coefficient can be written

$$D = 2.9 \pm 0.5 \times 10^{-6} P_{O_2}^{0.28 \pm 0.01} \exp\left(-19{,}050 \pm 375/RT\right).$$

When the compound has a larger deviation from stoichiometry,

$$D = 6 \pm 4 \times 10^{-9} P_{O_2}^{0.25 \pm 0.03} \exp\left(-8{,}210 \pm 750/RT\right).$$

Tensimetric data substantiated the existence of two types of Pr_7O_{12} and also yielded the expressions

$$f_{\mathrm{I}} = 4.7 \times 10^{-6} P_{O_2}^{1/2} \exp\left(10{,}000/RT\right)$$
$$f_{\mathrm{II}} = 2.2 \times 10^{-12} P_{O_2}^{1/2} \exp\left(37{,}700/RT\right)$$

for the mole fraction of oxygen in excess of the stoichiometric ratio in the nearly stoichiometric region and the region of larger deviation, respectively.

* Present address: Chemical Process Technology Department, Bell Telephone Laboratories Murray Hill, New Jersey.

It is proposed that intrinsic Frenkel disorder adds to the interstitial oxygen atoms resulting from the variation of oxygen pressure. From these assumptions, the diffusion coefficient in region I may be written

$$D = 1.28 \times 10^{-3} \exp\left(-19,050/RT\right)f_i$$

The data support an interstitialcy mechanism when the Frenkel contribution is taken into account.

These results are discussed against the increasing understanding of the nature of materials possessing extended defects.

INTRODUCTION

The iota phase ($Pr_7O_{12+\delta}$) is the most thermally stable[1] of the intermediate homologous series of praseodymium oxides having the general formula Pr_nO_{2n-2}.[2] Eight members of this series have been identified (those for $n = 4, 6, 7, 9, 10, 11, 12$, and ∞) including the end members, C-type (bcc) Pr_2O_3 and fluorite-type (fcc) PrO_2.[2,3] The praseodymium oxide system is the prototype of the fluorite-related oxides of the rare earths and actinides, there being striking analogies between it and the systems CeO_x, TbO_x, the A-type sesquioxides and fluorite dioxides of both series, UO_{2-x}, PuO_x, AmO_x, CmO_x and probably especially the higher members of the actinide oxide series.[4] This is the most extensive group of related oxides known.

The iota phase itself is specifically exhibited in the Ce,[5,6] Pr,[1,2] Tb[7,8] and Cm[9] pure binary oxides, and, in addition, oxides of this composition show a tendency for stability in the PuO_x[10,11] and AmO_x[12] systems. In addition, it is a very stable structure in the ternary oxides $MO_3 \cdot 3R_2O_3$, where M is W,[13] U,[14] or Mo,[14] and R is a rare earth element. The uranium-yttrium compound may have variable composition without phase transformation.[14] Closely related structures having the same space group have also been prepared and characterized in the zirconia-scandia system ($Zr_5Sc_2O_{13}$ and $Zr_3Sc_4O_{12}$).[15]

Except for the end members of the homologous praseodymium oxide series, the structure of only the iota phase is known.[14,16,17] The powder X-ray diffraction diagrams do, however, show the very close relationships among all the homologous series.[18] There seems little doubt that in all cases the metal positions shift very little from the fluorite PrO_2, but the oxygen sublattice is unique in each member of the series. The iota phase is a rhombohedral structure having the space group $R\bar{3}$ and the lattice parameters $a = 6.75$ Å, $\alpha = 99°23'$.[17] This structure may perhaps best be visualized as parallel infinite strings of 6-coordinated praseodymium atoms surrounded by sheaths of 7-coordinated praseodymium atoms.[8] This structure results when one-seventh of the oxygen atoms are removed in this regular way along the strings which are all in the $\langle 111 \rangle$ cubic directions.[18] Figure 1 illustrates a segment of the string showing the two vacant oxygen positions along the

255

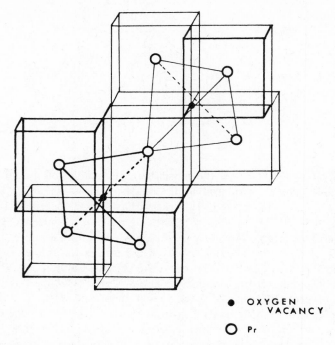

● OXYGEN
    VACANCY

○ Pr

Fig. 1. Ideal arrangement of a 6-coordinated praseodymium
atom surrounded by six 7-coordinated atoms.

$\langle 111 \rangle$ direction with the resulting one 6-coordinated and six 7-coordinated
metal atoms. It should be pointed out that these strings, in all four $\langle 111 \rangle$
directions such that they give all 6-coordinated metal atoms and do not
intersect, yield the C-type lower end member of the series and, as indicated
above, when all the vacant positions are filled the fluorite upper end member
of the series results. The exact relationships of the other intermediate phases
are not yet known, although some speculation has been expressed.[19]

Earlier isobaric studies[2] confirmed that at lower pressures the lower composition limit of the iota phase is near Pr_7O_{12} and that the phase does have a
small but easily demonstrated range of composition on the oxygen-rich side,
$Pr_7O_{12+\delta}$. At higher pressures complications arise and prominent pseudophase
behavior is observed.

The kinetic experiments to be described here are part of a concerted study
of these oxide systems as models for the behavior of materials having extended defects in contrast to the more commonly studied systems having low
concentrations of point defects.

It is already clear that the equilibrium behavior of materials having gross defects is not predicted by theories which are a simple extension of the classical point defect models. This is because strong interactions between defects in parent structures yield extended defects that become regular structural features of textured intermediate phases, which when ordered leave very low concentrations of point defects. The most studied of these structural features is the Wadsley defect resulting from crystallographic shear in typical 6-coordinated transition metal oxides (e.g., Ti_nO_{2n-1}) and the clustering as exhibited in, for example, FeO_{1-x}. The correct description of the structural feature in the complex praseodymium oxide system may involve the strings shown in Fig. 1, which, when parallel but shifted regularly in three ways with respect to each other, are the entire structure in Pr_7O_{12}. It is to be presumed that a full understanding of oxygen transport in this phase will form an important beginning in the description of the behavior of substances having extended defects.

The rare earth oxides share with calcia-stabilized zirconia the quality of having a high mobility in the oxygen sublattice coupled with a low mobility in the metal sublattice even at temperatures one-sixth the absolute melting temperature.[20] It has not been established whether this high mobility results from a high concentration or low energy of motion of the diffusing species.

The transport properties of oxygen in $Pr_7O_{12+\delta}$ were studied as a function of temperature and oxygen pressure. In principle, isobaric tensimetric runs could be treated to yield the enthalpy of formation of the defects, and the pressure dependence would indicate the charge carried by them. These results, combined with the activation energy for diffusion and the pressure dependence of the diffusion coefficient, might allow an enthalpy of motion to be calculated for the diffusing species and indicate the change in the diffusion coefficient with changes in stoichiometry. A partial report of this work was presented at the International Colloquy of the CNRS on the Rare Earth Elements at Paris-Grenoble, May 5–10, 1969.[21]

MATERIALS AND PROCEDURES

Preparation of the Diffusion Sample

Praseodymium oxide powder of 99.999% purity, obtained from the American Potash Company (Lindsay Division), was reduced to the sesquioxide in an evacuated platinum container at 850°C. Pressed pellets of the sesquioxide were sintered in vacuum by rf-induced radiative heating by a tantalum susceptor for 3 hr at 1800°C. The sintered oxide was crushed and dropped through an rf-induced argon plasma flame into a platinum crucible kept in air at 1000°C. The resulting oxide, in the form of smooth spheres of a range of sizes, was screened to obtain reaction samples of a small bracket of diameters. The spheres used for most of the runs were 0.105–0.088 mm in diameter. Other samples were also used, as specifically indicated in the section on experimental results. The starting material for each successive run was generated by repeated back-exchange with normal oxygen of the sample from the previous run.

The Diffusion Measurements

The method of heterogeneous isotopic exchange[22–24] was employed to determine the diffusion coefficients of oxygen in $Pr_7O_{12+\delta}$. The method consists of following the depletion of ^{18}O from enriched oxygen gas as isotopic exchange with the metal oxide proceeds at constant temperature. The reaction occurs in a volume such that the total number of moles of oxygen in each of the solid and the gas phases is constant and of comparable magnitude.

The solution of Fick's second law for the diffusion of solute from a limited and constant volume of well-stirred gas into a spherical solid has already been worked out[25] for a sphere of radius a when the mole fraction of ^{18}O in the gas is the same as that in the surface of the solid. The solution is

$$\frac{M_t}{M_\infty} = 1 - \sum_{n=1}^{\infty} \frac{6\lambda(1+\lambda)}{9(1+\lambda) + q_n^2\lambda^2} \exp\left(-\frac{Dq_n^2 t}{a^2}\right), \tag{1}$$

where M_t and M_∞ are the amount of ^{18}O taken up by the solid at time t and at infinite time, respectively, and γ is the ratio of the number of gram atoms of oxygen in the gas to that in the solid. The q_n's are the positive nonzero roots of the equation

$$\tan q_n = \frac{3q_n}{q_n^2\lambda + 3}. \tag{2}$$

Equation (1) cannot be solved for the diffusion coefficient in terms of M_t/M_∞; therefore, a graphical method of evaluating the term Dt/a^2 must be used. Tables of M_t/M_∞ versus Dt/a^2 were calculated for intervals of one percent in terms of the fractional uptake. The values of M_t/M_∞ were calculated from the equation

$$\frac{M_t}{M_\infty} = \frac{P_0 - P_t}{P_0 - P_\infty}, \tag{3}$$

where P_0 is the measured percentage of ^{18}O in the gas phase at $t = 0$, P_∞ the percentage when the exchange is complete, and P_t that observed at time t.

Values of Dt/a^2 were then obtained from the calculated tables and a plot of Dt/a^2 versus t was made. The slope of the curve is D/a^2. From the known radius of the spheres the diffusion coefficient was then calculated.

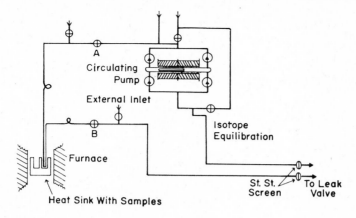

Fig. 2.   Heterogeneous isotope exchange apparatus.

Experimental Procedures

The apparatus has been discussed previously.[26,27] In the operating procedure the enriched oxygen was introduced over the sample without changing
the total oxygen pressure. Figure 2 shows the exchange section of the apparatus. The pressure was set initially using natural oxygen at equilibrium.
Valves A and B were then closed and the larger section of the exchange
volume evacuated. Enriched oxygen was admitted at the same pressure. The
exchange line was then isolated, the valves opened, and the timer and pump
started. The usual interval between opening the valves and starting the pump
was 5 seconds. The ^{18}O concentration in the gas was determined at selected
time intervals using a CEC 21-130 mass spectrometer. The length of a run
varied with temperature and pressure. At the lower temperatures about 1000
seconds was required for 90% exchange.

EXPERIMENTAL RESULTS AND DISCUSSION

A typical plot of Dt/a^2 versus t is shown in Fig. 3. At the lower temperatures
the straight-line portions do not extrapolate to the origin but intersect at a
finite time which becomes progressively shorter as the temperature is raised.
At temperatures near 900°C the lines go through the origin. The plots were
linear until the exchange was 95–98% complete. Some tailing at the end may
be due to the sphere size distribution.

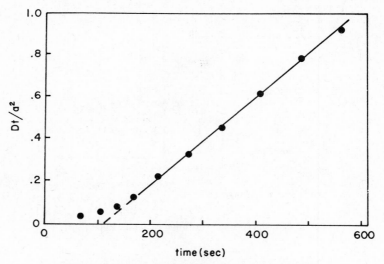

Fig. 3.   A typical plot for the determination of D/a^2.

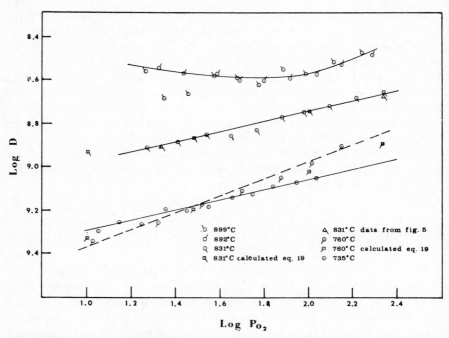

Fig. 4.   The narration of the diffusion coefficient with oxygen pressure at various temperatures.

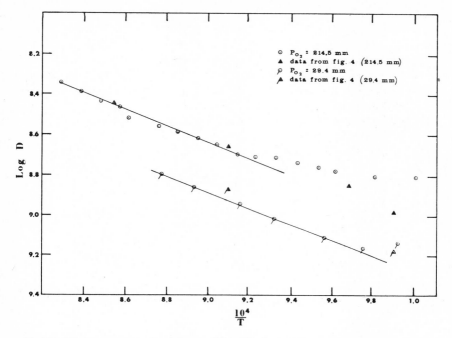

Fig. 5.   The temperature dependence of the diffusion coefficient at different pressures.

Diffusion Rate as a Function of Pressure and Temperature

The dependence of the diffusion coefficient on pressure at various temperatures is shown in Table I and Fig. 4. Table II and Fig. 5 display the temperature-dependence data from which the activation energy may be determined at two different pressures. The straight-line portions of the curves at high temperatures were fitted, yielding at 29.4 mm O_2

$$D = 5.5 \pm 0.6 \times 10^{-6} \exp\left(-18{,}440 \pm 250/RT\right), \tag{4}$$

an at 214.5 mm O_2

$$D = 1.3 \pm 0.3 \times 10^{-5} \exp\left(-19{,}095 \pm 470/RT\right), \tag{5}$$

or

$$D = 2.9 \pm 0.5 \times 10^{-6} P_{O_2}^{0.28 \pm 0.01} \exp\left(-19{,}050 \pm 375/RT\right) \tag{6}$$

when the pressures were taken into account.

The lower temperature region at high pressure was also fitted, yielding

$$D = 9 \pm 3 \times 10^{-8} \exp\left(-8{,}210 \pm 750/RT\right). \tag{7}$$

Since two activation energies are observed, two regions must be defined for the compound. In region I the deviation from stoichiometry is small, but it is larger in region II. The observations at 831°C (Fig. 4) indicated a $P_{O_2}^{0.26}$ dependence, in good agreement with that calculated from the temperature data (Table II and Fig. 5) for region I. These two points are shown on Fig. 4. At 760°C the slope is larger, but at this temperature there would be a mixture of results from both regions. The slope of the log D versus log P curve at 735°C indicated a $P_{O_2}^{0.25}$ dependence in region II. Figure 5 also displays results calculated from the pressure-dependence data.

The high temperature data suggest that diffusion is taking place by a vacancy mechanism at low pressures, as indicated by the decrease in the diffusion coefficient as the pressure is increased (Fig. 4).

Table I.   Pressure Dependence of the Diffusion Coefficient

735°C[a]		760°C[a]		831°C[a]	
$D \times 10^{10}$	P (mm)	$D \times 10^{10}$	P (mm)	$D \times 10^9$	P (mm)
5.03	11.21	4.57	10.50	1.23	18.61
5.57	13.93	5.52	20.83	1.29	25.53
5.48	17.65	6.70	33.01	1.41	34.04
6.36	22.41	7.72	50.04	1.39	44.41
6.29	28.01	8.91	74.25	1.46	57.59
6.50	35.08	10.30	103.88	1.68	75.31
7.17	44.80	12.40	139.16	1.77	95.30
7.42	55.59			1.89	123.90
8.12	69.06			2.08	162.59
8.26	87.23				
8.82	107.41				

891.6°C[a]		898.7°C[a]		902°C[b]	
$D \times 10^9$	P (mm)	$D \times 10^9$	P (mm)	$D \times 10^9$	P (mm)
2.78	21.03	2.75	18.61	3.66	14.41
2.67	27.51	2.06	22.23	3.42	18.38
2.67	36.27	2.16	28.20	3.30	22.55
2.48	48.78	2.59	37.28	3.01	27.59
2.50	62.55	2.56	47.43	3.12	36.61
2.54	82.38	2.35	59.82	3.10	48.19
2.63	107.93	2.84	75.97	2.70	59.24
2.95	140.15	2.66	96.92	2.62	72.90
3.26	191.19	3.00	129.12	2.46	89.80
		3.33	173.80	2.40	113.69

[a] Sphere size 0.105–0.088 mm.
[b] Sphere size 0.088–0.074 mm.

Table II.   Temperature Dependence of the Diffusion Coefficient

A. Spheres produced from plasma fusion

| $P_{O_2} = 29.4$ mm[a] | | $P_{O_2} = 214.5$ mm[a] | |
$D \times 10^{10}$	T (°C)	$D \times 10^9$	T (°C)
7.33	735.0	1.56	725.6
6.84	752.7	1.57	746.8
7.77	771.7	1.66	766.7
9.70	799.2	1.74	775.7
11.6	819.7	1.85	787.1
13.9	845.5	1.97	798.9
16.2	865.9	1.98	810.3
		2.02	820.1

| $P_{O_2} = 30.0$ mm[b] | | | |
$D \times 10^{10}$	T (°C)		
7.10	761.2	2.28	832.6
7.61	787.9	2.40	843.7
11.5	819.7	2.60	855.6
14.7	844.5	2.84	868.2
21.2	867.1	3.05	881.3
22.8	892.4	3.44	892.8
		3.71	904.9
		4.10	918.5
		4.56	932.3

B. Spheres produced by sol-gel technique

| $P_{O_2} = 30.25$ mm[a] | |
$D \times 10^9$	T (°C)
0.850	747.9
0.870	770.2
0.988	792.1
1.24	814.0
1.43	836.7
1.69	856.9
2.74	894.2

[a] Sphere size 0.105–0.088 mm.
[b] Sphere size 0.088–0.074 mm.

The Effect of Sample Preparation and Sphere Size

In order to determine the effect of sample preparation and sphere size on the transport properties of $Pr_7O_{12+\delta}$, two additional series of experiments were made. In the first, spheres prepared by the sol-gel technique were used. These were annealed at 875°C with frequent changes of oxygen atmosphere until CO_2 was no longer observed to be evolved; this required one week. In the second, the behavior of smaller spheres, screened from the plasma fused material, were compared with those of the main sequence of runs.

The results of the study of diffusion rate as a function of temperature for sol-gel spheres, together with results from fused spheres run at approximately the same temperature, are presented in Table II. It is apparent that the diffusion coefficients for the sol-gel material are consistently slightly greater at the same temperature. In addition to fundamental differences in the nature of the samples, the oxygen pressure was slightly higher (30.6 mm versus 29.4 mm) in the sol-gel runs, and there is no assurance that the size distributions are the same for the two processes for the same screening procedures. The activation energies are the same for the two runs if one ignores the two results at high temperature where the compound is near stoichiometry and Frenkel disorder may be playing a determining role.

Results of measurements of D versus T for different sphere sizes are given in Table II. The points virtually define the same curve. Table I gives the results of the pressure dependence runs on the small spheres. The observed effects are independent of sphere size to a first approximation. The specimen is apparently hypostoichiometric at 902°C, as shown by the negative pressure dependence over the entire pressure range.

The Composition of the Iota Phase as a Function of Temperature and Pressure

The weight of a 7.51-g sample of $PrO_{1.714}$ was determined as a function of pressure at several temperatures. The composition was calculated after corrections for buoyancy were made. (The buoyancy correction was determined from runs with a comparable sample of Nd_2O_3 in the apparatus.) The data shown in Table III were analyzed using a nonlinear least squares program[28] which fitted the data to the function $w - w_0 = KP_{O_2}{}^n$, where w signifies weight. Table IV lists values of w_0, K, and n together with their standard deviations for the temperatures used. The high temperature data are less reliable than those at lower temperatures. As can be seen, the run near 900°C indicates a much different pressure dependence than those at the lower temperatures. This is a result of the way in which the data were treated. In the equation $w - w_0 = KP_{O_2}{}^n$, the compound is assumed to be stoichiometric when the pressure is zero. This is a good approximation when the temperatures are low; however, it is not good at 900°C. The assumption that stoichiometry occurs at a finite pressure results in an increase in the pressure dependence calculated. This situation will be discussed in detail below. Thus a $P_{O_2}{}^{1/2}$ relationship seems most likely, as indicated at 760 and 831°C. This is in contrast to the $P_{O_2}{}^{1/4}$ dependence found for diffusion.

The temperature dependence of the composition was also measured; the

Table III.   Weight Change of the Iota Phase PrO_x as a Function of Pressure at Various Temperatures

727°C		760°C		828°C		890°C	
P_{O_2} (mm)	Corrected Weight (mg)	P_{O_2} (mm)	Corrected Weight (mg)	P_{O_2} (mm)	Corrected Weight (mg)	P_{O_2} (mm)	Corrected Weight (mg)
9.62	2.490	10.09	2.212	12.52	1.94	12.66	1.755
14.45	2.714	14.71	2.384	17.59	2.015	14.87	1.835
18.51	2.834	20.04	2.528	23.75	2.125	16.97	1.883
25.79	3.053	27.69	2.694	31.90	2.223	20.18	1.942
33.28	3.228	36.74	2.872	39.54	2.333	23.22	1.993
40.40	3.383	43.22	3.003	46.91	2.429	28.94	2.058
51.73	3.640	53.42	3.164	53.43	2.492	35.98	2.146
62.30	3.841	63.70	3.316	63.45	2.592	41.69	2.209
71.84	4.017	73.28	3.452	73.35	2.668	47.21	2.259
82.16	4.192	83.47	3.577	83.44	2.765	54.80	2.329
92.97	4.373	93.78	3.709	94.02	2.835	62.88	2.393
102.50	4.531	104.04	3.837	104.57	2.931	71.77	2.458
112.86	4.692	114.19	3.940	114.40	2.992	82.28	2.533
122.84	4.854	123.93	4.055	124.25	3.052	97.42	2.622
133.12	5.026	135.08	4.160	135.32	3.137	110.29	2.690
143.33	5.193	145.37	4.258	147.10	3.203	125.89	2.743
155.22	5.399	156.81	4.357	162.33	3.288	137.52	2.802
163.85	5.566	169.57	4.474	177.79	3.376	152.59	2.857
175.16	5.794	179.24	4.560			170.24	2.920
184.62	6.004					187.24	3.007
195.02	6.286						

data were not corrected for buoyancy. Typical differential plots of the data are shown in Figs. 6 and 7 for oxygen pressures of 10 and 212 torr, respectively. The logarithm of T^2 times the slope of the weight versus temperature curve is plotted against the reciprocal of the absolute temperature. The results of these computations were analyzed using a nonlinear least squares program by fitting them to the equation $w_0 - w = K \exp(-H/RT)$, where

Table IV.   Pressure Dependence of the Weight for a 7.51-g Sample of $PrO_{1.714}$

Parameters of the Equation $w - w_0 = KP_{O_2}{}^n$

T (°C)	w_0 (mg)	K	n
727	1.98 ± 0.02	0.135 ± 0.007	0.63 ± 0.01
760	1.54 ± 0.02	0.21 ± 0.01	0.51 ± 0.007
823	1.39 ± 0.03	0.15 ± 0.01	0.49 ± 0.01
890	0.68 ± 0.17	0.55 ± 0.11	0.27 ± 0.03

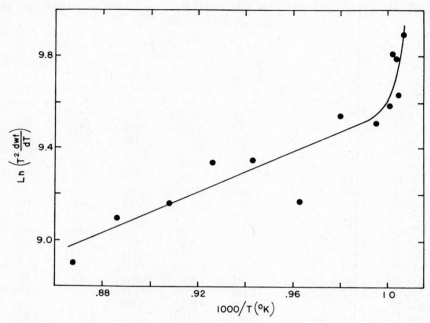

Fig. 6. The change in composition of iota phase with temperature at 10 torr.

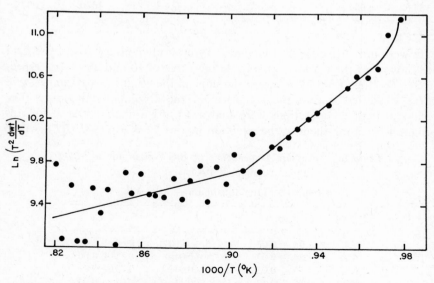

Fig. 7. The change in composition of iota phase with temperature at 212 torr.

Table V.   Parameters of the Equation $w - w_0 = K \exp(-H/RT)$

Sample weight $= 7.51$ g

Pressure	Temperature	K	H(kcal)	w_0
213 mm	$>825°$C	0.22 ± 0.18	$-7.0 \pm 1.5 \times 10^3$	0.43 ± 0.97
212 mm	$>825°$C	—	$-11.0 \pm 1 \times 10^3$	—
213 mm	$<825°$C	$4 \pm 2 \times 10^{-8}$	$-37.7 \pm 1.1 \times 10^3$	4.78 ± 0.05
212 mm	$<825°$C	—	$-37.1 \pm 1.1 \times 10^3$	—
10 mm	$<900°$C	$1.8 \pm 0.5 \times 10^{-2}$	$-10.0 \pm 0.5 \times 10^3$	4.3 ± 0.1

H is the activation energy. Values for the parameters of the equation are given in Table V.

Both the tensimetric and diffusion data clearly indicate two regimes for hyperstoichiometric Pr_7O_{12}. The change in slope for both sets of data (see Figs. 5 and 7) at about 214 mm oxygen pressure occurs at approximately the same temperature, thus indicating the close relationship between the defect concentration and the diffusion coefficient. Before this relationship can be fully determined, the pressure dependence of the composition must be decided.

The pressure dependence of the composition at 890°C appears to be anomalously low, as shown in Table IV. As was stated previously, the data were analyzed assuming the compound to be stoichiometric at zero pressure. If stoichiometry occurs at a finite pressure the analysis is in error and we must use the following derived expression.

If δ is the deviation per unit volume from stoichiometry, then

$$\delta = [O_i] - [V_O] = K_9 P_{O_2}^{1/2} - K_{10} P_{O_2}^{-1/2}, \tag{8}$$

where K_9 and K_{10} are the equilibrium constants for the reactions

$$\tfrac{1}{2}O_2 \rightarrow O_i \tag{9}$$

and

$$O_O \rightarrow V_O + \tfrac{1}{2}O_2. \tag{10}$$

Let P_0 be the pressure at which the compound is stoichiometric. At this pressure $\delta = 0$ and $K_9 P_0^{1/2} = K_{10} P_0^{-1/2}$. Further,

$$\delta = K_{10} P_0^{-1/2} \left(\frac{P_{O_2}}{P_0}\right)^{1/2} - K_9 P_0^{1/2} \left(\frac{P_{O_2}}{P_0}\right)^{-1/2} \tag{11}$$

or

$$\delta = K_D^{1/2}(\beta^{1/2} - \beta^{-1/2}), \tag{12}$$

where $\beta = P_{O_2}/P_0$ and $K_D = K_9 K_{10} = [O_i][V_O]$.

The agreement obtained between the experimental and the calculated points using this formula is shown in Table VI. This rationalizes the pressure

Table VI. Comparison of Experimental and Calculated Pressure Dependence of the Composition of the Iota Phase at 890°C

$P_0 = 21.0$ mm and $W_0 = 1.958$ (mg)

P_{O_2} (mm)	Wt (mg)	β	$\beta^{1/2} - \beta^{-1/2}$	δ (exptl.)	$\delta(K_0 = 39)$	$\delta(K_0 = 38)$
12.66	1.755	0.603	−0.511	−0.203	−0.199	−0.194
14.87	1.835	0.708	−0.347	−0.123	−0.135	−0.132
16.97	1.883	0.808	−0.213	−0.075	−0.083	−0.081
20.16	1.942	0.961	−0.040	−0.016	−0.016	−0.015
23.22	1.993	1.106	0.101	0.035	0.039	0.038
28.94	2.058	1.378	0.322	0.100	0.126	0.122
35.98	2.146	1.713	0.545	0.188	0.213	0.207
41.69	2.209	1.985	0.699	0.251	0.273	0.266
47.21	2.259	2.248	0.832	0.301	0.325	0.316
54.80	2.329	2.610	0.996	0.371	0.389	0.379
62.88	2.393	2.994	1.152	0.435	0.449	0.438
71.77	2.458	3.418	1.308	0.500	0.510	0.497
82.28	2.533	3.918	1.474	0.575	0.575	0.560
97.42	2.622	4.639	1.690	0.664	0.659	0.642
110.29	2.690	5.252	1.855	0.732	0.724	0.705
125.89	2.743	5.995	2.040	0.785	0.796	0.775
137.52	2.802	6.549	2.168	0.844	0.846	0.824
152.59	2.857	7.266	2.325	0.899	0.907	0.883
170.24	2.920	8.107	2.496	0.962	0.973	0.948
187.24	3.007	8.916	2.651	1.049	1.034	1.007

dependence at 890°C with that at lower temperatures. The pressure dependence at 727°C is measured in the edge of a two-phase region, as can be verified from the data in Figs. 6 and 7. Therefore, it should not be compared directly with the others at higher temperatures.

At 890°C, P_0 is near 21 mm, as may be seen from Table VI. The minimum in the diffusion curve, Fig. 4, occurs at about 50 mm, suggesting that oxygen vacancies are more mobile than oxygen interstitials.

The foregoing model for interpreting the data assumes uncharged oxygen atoms as the interstitials and uncharged vacancies. However, in an oxide such as this where the cations formally have two valences, other assumptions as to the charge on the defects lead to the same results. For example, if the interstitials were negatively charged, then

$$\tfrac{1}{2}O_2 \rightarrow O_i^- + h\cdot \tag{13}$$

This hole formally corresponds to the transformation of a Pr^{3+} to a Pr^{4+}. The concentration of both of these is very large in Pr_7O_{12}, hence

$$[O_i^-]p = KP_{O_2}^{1/2}, \tag{14}$$

$$[O_i^-] \simeq K'P_{O_2}^{1/2}. \tag{15}$$

The argument for charged vacancies is analogous. Thus the square root dependence of the composition upon pressure offers no definite information about the charge on the defects.

The established square root dependence of the composition in region I allows the mole fraction of excess oxygen to be expressed as

$$f_I = 4.7 \times 10^{-6} P_{O_2}^{1/2} \exp\left(\frac{10,000}{RT}\right). \tag{16}$$

As before, it has been assumed that the compound is stoichiometric at $P_{O_2} = 0$. For region II the data are not as unequivocable, since the boundaries have not been established, but assuming a square root dependence, they yield

$$f_{II} = 2.2 \times 10^{-12} P_{O_2}^{1/2} \exp\left(\frac{37,700}{RT}\right). \tag{17}$$

It has been observed that f_I is proportional to $P_{O_2}^{1/2}$ at any temperature (16) and D is proportional to $P_{O_2}^{1/4}$ (6) at the same temperature. The expression $D = D_i f_i$ should describe the relationship between the diffusion coefficient and the mole fraction of species responsible for diffusion; hence it is suggested that there is intrinsic Frenkel disorder in the iota phase over and above the extrinsic disorder f_I.

Table VII lists the mole fraction of extrinsic defects at 831°C calculated from (16) at the pressures listed. The tensimetric data yield information only on the extrinsic defects. To this calculated number a value has been added which is interpreted to be the mole fraction of interstitial oxygen atoms due to the Frenkel equilibrium

$$O_O \rightarrow V_O + O_i \tag{18}$$

having the equilibrium constant K_F.

Table VII. Computed Interstitial Concentrations and Diffusion Coefficients

P_{O_2} (mm)	f_I 831°C $\times 10^3$	f_I 760°C $\times 10^3$	$f_i = (f_I + c)$ $c_{831} = 4.12$ $\times 10^3$	$f_i = (f_I + c)$ $c_{760} = 2.20$ $\times 10^3$	D 831°C $\times 10^9$	D 760°C $\times 10^{10}$
10	1.42	1.94	5.54	4.14	1.15	4.71
30	2.47	3.36	6.59	5.56	1.37	6.34
100	4.49	6.12	8.61	8.32	1.79	9.47
214	6.58	8.96	10.70	11.16	2.22	12.70

The number of Frenkel interstitials was obtained from a fit of the tensimetric data to the observed pressure dependence behavior at 100 mm at 831°C. It has been assumed that the concentration of intrinsic interstitials is independent of pressure, whereas the concentration of extrinsic defects varies as the square root of the pressure. This is not strictly correct since they depend on the value of K_F. It should, however, be a good approximation at the lower temperatures.

A consequence of this suggestion is that the calculated pressure-dependence relationship has some curvature. In the region in which the experiments were performed, the curve approximates a straight line of slope $\frac{1}{4}$ since it is fit to the data at one point. Confirmation of this hypothesis could be obtained by performing further diffusion experiments at lower pressures where the curvature is larger. The experimental data of Fig. 4 seem to show a slight curvature in the direction expected.

Extensive Frenkel disorder in the Pr_7O_{12} structure is reasonable. The strings of missing oxygen in the structure contain holes that are easily capable of containing oxygen atoms. It is just these holes which are filled as the compound is oxidized.

Since the data were obtained in the transition region where intrinsic and extrinsic defects are contributing to the total defect concentration, the calculation of an enthalpy of motion cannot be made in a simple way because the temperature dependence of the Frenkel constant is not known. However, K_F probably increases with temperature while the extrinsic defect concentration decreases with temperature. If, to a first approximation, these two trends cancel, then the enthalpy of motion is just the experimentally determined activation energy. Using this value from (16) and the defect concentration shown in Table VII, the preexponential constant D_0 and hence the diffusion coefficient can be determined.

The tensiometric data are adjusted to the diffusion data at 100 mm where the diffusion coefficient is known from experiment to be 1.79×10^{-9} by calculating the intrinsic diffusion coefficient D_i. It is assumed that $f_i = (f_I + \text{constant})$ at any temperature

$$D_i = \frac{D}{f_i} = 2.13 \times 10^{-7} = D_0 \exp\left(-\frac{E}{RT}\right) = D_0 \exp\left[-\frac{19,050}{1.987(1104)}\right]$$

which yields $D_0 = 1.28 \times 10^{-3}$, in good agreement with expectation from the work of Stearn and Eyring.[29] It is then possible to calculate the diffusion coefficients from the equation

$$D = 1.28 \times 10^{-3} \exp\left(-\frac{19,050}{RT}\right) f_i \tag{19}$$

The calculated data are given in Table VII and are plotted together with the experimental data in Fig. 4.

The values of 4.12×10^{-3} and 2.20×10^{-3} for the concentration of intrinsic defects at 831 and 760°C, respectively, are in agreement with the expected decrease in the Frenkel constant with temperature.

In conclusion, the high oxygen mobility in the PrO_x system near the composition $PrO_{1.714}$ appears to be due to a low enthalpy of motion and to a high concentration of interstitial oxygens due to Frenkel-type disorder. All the data support an interstitialcy mechanism for diffusion. The region of large deviation from Pr_7O_{12} referred to as region II should be of great interest in that the larger number of defects in that region may lead to noticeable interaction between them.

ACKNOWLEDGMENT

It is a pleasure to express our indebtedness to Mr. S. R. Buxton and the Oak Ridge National Laboratory for supplying the sample of sol-gel-prepared PrO_x spheres used in one of the experiments; to our colleague, M. S. Jenkins, for help in obtaining some of the tensimetric composition data; and to Professors M. O'Keeffe and S. H. Lin for frequent discussions. The U.S. Atomic Energy Commission supports the ongoing studies of which this is a part. One of the authors (G.R.W.) enjoyed the honor of an NDEA fellowship during much of the time these experiments were done.

REFERENCES

1. R. E. Ferguson, E. D. Guth, and L. Eyring, *J. Amer. Chem. Soc.*, **76**, 3890 (1954).
2. B. G. Hyde, D. J. M. Bevan, and L. Eyring, *Proc. Roy. Soc. (London), Ser. A*, **259**, 583 (1966).
3. R. Turcotte, J. Warmkessel, and L. Eyring have seen a phase of upper composition limit $PrO_{1.67}$ at low temperatures (unpublished work).
4. L. Eyring, *Advances in Chemistry Series*, No. 71, ACS, 1967, p. 67. Lanthanide Actinide Chemistry, P. R. Fields and T. Moeller, Eds.
5. D. J. M. Bevan, *J. Inorg. Nucl. Chem.*, **1**, 49 (1955).
6. D. J. M. Bevan and J. Kordis, *J. Inorg. Nucl. Chem.*, **26**, 1509 (1964).
7. E. D. Guth and L. Eyring, *J. Amer. Chem. Soc.*, **76**, 5242 (1954).
8. B. G. Hyde and L. Eyring, *Rare Earth Research*, Vol. III, L. Eyring, Ed., 1965, p. 623.
9. T. D. Chikalla and L. Eyring, *J. Inorg. Nucl. Chem.*, **31**, 85 (1969).
10. T. L. Markin and M. H. Rand, *Thermodynamics* (AERE, Vienna), **1**, 145 (1965). Proceeding of a Symposium.
11. R. N. R. Mulford and C. E. Holley, Jr., USAEC LA-DC-8266 (1966).
12. T. D. Chikalla and L. Eyring, *J. Inorg. Nucl. Chem.*, **30**, 133 (1968).
13. H. Borchardt, *Inorg. Chem.*, **2**, 160 (1961).

14. S. F. Bartram, *Inorg. Chem.*, **5**, 749 (1966).
15. M. R. Thornber, D. J. M. Bevan, and J. Graham, *Acta Cryst.*, **B24**, 1183 (1968).
16. N. C. Baenziger, H. A. Eick, H. S. Schuldt, and L. Eyring, *J. Amer. Chem. Soc.*, **83**, 2219 (1961).
17. L. Eyring and N. C. Baenziger, *J. Appl. Phys.*, *Suppl.*, **33**, 428 (1962).
18. J. O. Sawyer, B. G. Hyde, and L. Eyring, *Bull. Soc. Chim. France*, **1965**, 1190.
19. B. G. Hyde, D. J. M. Bevan, and L. Eyring, International Conference on Electron Diffraction and Crystal Defects, Melbourne, II, C-4 (1965), published and distributed for the Australian Academy of Science by Pergamon Press (1966).
20. L. Eyring, Heterogeneous Kinetics at Elevated Temperatures Conference, Plenum Press, New York, 1970, p. 343.
21. G. Weber and L. Eyring, Proc. International Colloquy of the CNRS on the Rare Earth Elements, Paris-Grenoble, May 5–10, 1969, in press.
22. P. C. Carmen and R. Haul, *Proc. Roy. Soc. (London)*, **222**, 109 (1954).
23. R. Haul and D. Just, *J. Appl. Phys.*, *Suppl.*, **33**, 487 (1962).
24. R. Haul, D. Just, and G. Dumbgen, *Proc. 4th Intern. Symp. Reactivity of Solids Amsterdam*, 1960, p. 65.
25. J. Crank, *Mathematics of Diffusion*, Clarendon Press, Oxford, 1956.
26. G. D. Stone, Ph.D. Thesis, Arizona State University, Tempe, Arizona, 1968.
27. G. D. Stone, G. R. Weber, and L. Eyring, *Mass Transport in Oxides, Proceedings of a Symposium*, J. B. Wachtman, Jr. and A. Franklin, Eds, NBS Spec. Pub. No. 296, 1968.
28. R. H. Moore and R. K. Zeigler, *Los Alamos Scientific Laboratory Report LA-2367*, Los Alamos, N.M., March 1960.
29. A. E. Stearn and H. Eyring, *J. Phys. Chem.*, **44**, 955 (1940).

Application of the Absolute Reaction-Rate Theory to Non-Newtonian Flow

HIROSHI UTSUGI

Department of Applied Science, Faculty of Engineering, Tohoku University, Sendai, Japan

TAIKYUE REE

Department of Chemistry, University of Utah, Salt Lake City

Abstract

The Newtonian and non-Newtonian terms appearing in the Ree-Eyring equation were evaluated by applying this equation to experimental flow curves. The Newtonian viscosity number, η_{sp}/C, obtained in the low range of shear rates $\dot{s}$ of an Ostwald-type flow curve agrees with that obtained in the high range of $\dot{s}$. The values of η_{sp}/C approach the value of 2.5 in Einstein's equation. The values of η_{sp}/C were also obtained from pseudo-plastic flow curves. The value of η_{sp}/C obtained over the low range of $\dot{s}$ of pseudoplastic flow curves also approach 2.5. From these results, it was concluded that the flow units for the Newtonian term in the Ree-Eyring equation are rigid or semirigid spheres.

A Personal Note

HIROSHI UTSUGI

I was one of Professor Eyring's postdoctoral fellows from 1959 to 1961. A part of the present paper was written in this period and was presented in 1960 at the 137th annual meeting of the American Chemical Society at Cleveland, Ohio. I remember that Dean Eyring was very busy at that time because of his official activities, but he attended my presentation and helped me to answer questions in the discussions. After returning home I met him with the late Mrs. Eyring at Sendai when he was invited to visit our country by the Chemical Society of Japan. His talk at Sendai was extremely interesting and charmed the audience. After this seminar Dr. Izumi Higuchi guided the couple to Matsushima and the surrounding area. Both appeared to enjoy the scenery very much, frequently exclaiming "Beautiful! beautiful!"

Before going to Utah I was not familiar with rheology, but, through his kind guidance and stimulating discussions and because his approach was unique, I was strongly attracted to this field. I especially enjoyed the Saturday seminars at his office in which he taught me directly or indirectly how to study science. Now that I am deeply involved in the study of rheological phenomena I still retain his way of thinking and pursue his theories on viscosity and diffusion.

INTRODUCTION

There are two main streams in the study of rheological properties of suspensions and polymeric solutions. One is the hydrodynamic approach,[1] and the other is the molecular kinetic investigation.[2] The so-called Ree-Eyring equation[3] is a result from the latter category; it is written as follows:

$$f = \frac{\beta_0}{\alpha_0} X_0 \dot{s} + \frac{\beta_1}{\alpha_1} X_1 \dot{s} + \frac{X_2}{\alpha_2} \sinh^{-1} \beta_2 \dot{s} \tag{1}$$

Here f is the shear stress (force per area), $\dot{s}$ is the rate of shear, β_i $(i = 0, 1, 2)$ is a quantity proportional to the relaxation time, α_i is a constant proportional to the reciprocal of the shear modulus, X_i is the fraction of a shear plane occupied by the ith flow unit, and $i = 0, 1$, and 2 indicate, respectively, the solvent, the Newtonian, and the non-Newtonian flow units. It is a well-known fact that (1) was applied with great success to various cases including colloidal suspensions and polymeric solutions.[3-6] In this paper, we study the Newtonian terms [the first and second terms on the right of (1)] in more detail, and the nature of the Newtonian flow units of solutes or suspensoids will be considered.

THEORETICAL CONSIDERATIONS

In the application of (1) to experiments, the first two terms on the right-hand side appear together as one Newtonian term. Thus the two terms are defined as f^N; that is,

$$f^N = \frac{\beta_0}{\alpha_0} X_0 \dot{s} + \frac{\beta_1}{\alpha_1} X_1 \dot{s}, \tag{2}$$

and the viscosity arising from f^N is represented by η^N,

$$\eta^N = \frac{f^N}{\dot{s}} = \frac{\beta_0}{\alpha_0} X_0 + \frac{\beta_1}{\alpha_1} X_1. \tag{3}$$

275

The quantity X_i may be represented by

$$X_i = \frac{v_i}{V} N_i, \qquad (i = 0, 1, 2). \tag{4}$$

where v_i is the molecular volume of a flow unit belonging to the ith type and N_i is the number of the flow units in V, the total volume of the system. Accordingly, we express X_0, X_1, and X_2 by the following equations:

$$X_0 = \tfrac{4}{3}\pi R_0{}^3 \frac{N_0}{V}, \tag{5a}$$

$$X_1 = \tfrac{4}{3}\pi R_1{}^3 \frac{N_1}{V}, \tag{5b}$$

and

$$X_2 = v_2 \frac{N_2}{V}, \tag{5c}$$

where R_0 and R_1 are the radius of the solvent molecule and the solute molecule, respectively. By definition, the following relation holds:

$$X_0 + X_1 + X_2 = 1.$$

For a dilute solution, one may assume that $X_0 \simeq 1$ and $N_1 \gg N_2$; consequently, C_s (the concentration of the solute, molecules/cc) can be approximated by N_1/V. Thus (3) turns out to be

$$\eta^N = \eta_0\left(1 + \frac{\beta_1}{\alpha_1}\frac{1}{\eta_0} v_1 C_s\right), \tag{6}$$

where η_0 is the viscosity of the solvent (i.e., β_0/α_0), and (5b) was applied to X_1.

For a solution (or suspension) whose solute molecules are rigid spheres the viscosity is represented by the Einstein viscosity equation[7]:

$$\eta^N = \eta_0(1 + 2.5\varphi_1), \tag{7}$$

where φ_1 is the volume fraction of the solute, being equal to $v_1 C_s$. Comparing (6) and (7), we obtain the relation

$$K = \frac{\beta_1}{\alpha_1}\frac{1}{\eta_0} = 2.5;$$

that is, K is a constant irrespective of the nature of the system providing the solute molecules are rigid and the concentration is very dilute.

APPLICATIONS

We consider the following two systems: the solution of polyisobutylene in cyclohexane[8] and the suspension of Procain penicillin G in water.[9] Both systems exhibit the so-called Ostwald type of flow curves[10]; i.e., in these systems, structural breakdown occurs with the flow process, and the flow becomes a Newtonian type at high rates of shear. At low $\dot{s}$, however, we can apply (1) assuming that the breakdown is negligible. The results are shown in Tables I and II. The two values of $\eta_{sp}{}^N$ obtained from the low and high ranges of $\dot{s}$ agree very well (see the last column of Table I and the fourth column of Table II). One exception, however, is observed in the case of $C_s' = 2.0$ g/dl in Table I (see the Discussion).

Table I.   Solution of Polyisobutylene in Cyclohexane[a] at 30°C

Sample	Molecular Weight[b] $\times 10^{-6}$	C_s' (g/dl)	η^{N}[c] (poise)	X_2/α_2 (dyn/cm²)	$\beta \times 10^4$ (sec)	$\eta_{sp}{}^N/C_s'$[d] (dl/g)
Vistanex[e] L140	1.8–2.5 (2.2)	2.0	0.056	880	5.0	4.70 (2.92)
		0.51	0.027	110	2.0	4.75 (4.58)
Vistanex[e] L100	1.05–1.57 (1.3)	1.0	0.036	167	3.0	3.62 (3.39)
		0.5	0.022	65	2.0	3.62 (3.36)
Vistanex[e] LMMH	0.04–0.051 (0.046)	4.8	0.036	35	8.0	0.71 (0.71)
		2.65	0.023	45	2.0	0.70 (0.68)

[a] The viscosity of cyclohexane at 30° is 0.0082 poise.
[b] The molecular weight in parentheses is the averaged one.
[c] η^N represents the sum $X_0(\beta_0/\alpha_0) + X_1(\beta_1/\alpha_1)$, which was obtained in the low range of $\dot{s}$.
[d] η_{sp} is the specific viscosity, that is, $(\eta - \eta_0)/\eta_0$. The unparenthesized $\eta_{sp}{}^N/C_s'$ was obtained from the second Newtonian branch which appears in the Ostwald curve over the high range of $\dot{s}$, whereas the value in parentheses was calculated using η^N at low shear rates.
[e] Vistanex is a commercial name of polyisobutylene.

In Fig. 1 the flow curves for the solution of polyisobutylene in cyclohexane are shown. The curves calculated from (1) by using the parametric values in Table I are compared with the experimental curves. Figure 2 shows the flow curves for the suspension of Procain penicillin in water. In this case, the calculated flow curves were obtained by substituting the parametric values in Table II into (1). Figures 1 and 2 show that the agreement between the calculated and experimental curves is satisfactory in the low range of $\dot{s}$.

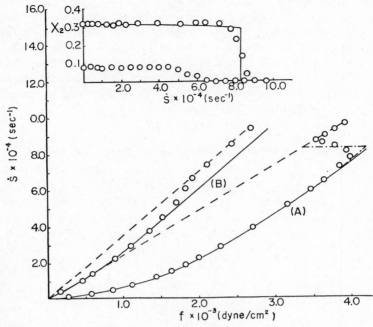

Fig. 1. Flow curves of polyisobutylene in cyclohexane. Circle experiment.[8] Curve A, calculated from (1): $f = 0.056\dot{s} + 880 \sinh^{-1} 5 \times 10^{-4}\dot{s}$, $C'_s = 2.0\,g/dl$. Curve B, calculated from (1): $f = 0.027\dot{s} + 110 \sinh^{-1} 2 \times 10^{-4}\dot{s}$, $C'_s = 0.5\,g/dl$. The inset diagram of X_2 versus $\dot{s}$ was calculated by applying the theory of thixotropy[10]. The broken curves are the extensions of the upper Newtonian curves.

In the following we consider the reason why the two values of $\eta_{sp}{}^N$ obtained over the low and high ranges of $\dot{s}$ should agree. In the low range of $\dot{s}$, the Newtonian viscosity η^N is given by (3), whereas the Newtonian viscosity $\eta_u{}^N$ in the high range of $\dot{s}$ is expressed by

$$\eta_u{}^N = X_0^\infty \frac{\beta_0}{\alpha_0} + X_1^\infty \frac{\beta_1}{\alpha_1}, \tag{8a}$$

where it is assumed that all of the non-Newtonian flow units of type 2 are transformed into those of type 1, that is, $X_1^\infty = X_1^0 + X_2^0$. The superscripts ∞ and 0 indicate that the shear rate is very high and very low, respectively. In (3) and (8a), one may consider that $X_0 \simeq 1$ and $X_0^\infty \simeq 1$; that is,

$$\eta^N = \eta_0 + X_1^0 \frac{\beta_1}{\alpha_1}, \qquad \eta_u{}^N = \eta_0 + X_1^\infty \frac{\beta_1}{\alpha_1}. \tag{8b}$$

Table II. Suspension of Procain Penicillin G in Water

φ	η^{N} [a] (poise)	η_u^{N} [b] (poise)	η_{sp}^{N} [c]	$X_2\beta_2/\alpha_2$ (poise)	β_2 (sec)
0.05		0.050	0.432		
0.1		0.073	1.091		
0.2	0.12	0.130	2.725 (2.438)	1.32	4.4
0.25	0.20	0.198	4.690 (4.701)	3.4	3.6
0.30	0.28	0.252	6.220 (7.023)	3.43	2.2
0.35	0.49	0.504	13.420 (13.040)	4.36	1.2

[a] The quantity η^{N} represents the sum $X_0(\beta_0/\alpha_0) + X_1(\beta_1/\alpha_1)$ obtained from the Ostwald flow curve in the low range of $\dot{s}$.
[b] η_u^{N} was obtained from the Newtonian branch appeared in the Ostwald curve over the high range of $\dot{s}$.
[c] The unparenthesized η_{sp}^{N} was calculated from η_u^{N}, whereas the value in parentheses was obtained from η^{N}.

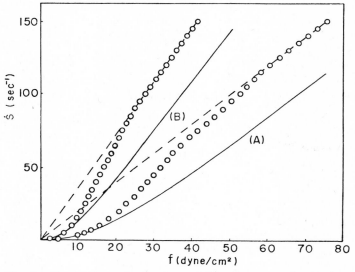

Fig. 2. Flow curves of procain penicillin G aqueous suspension. Circle experiment.[9] Curve A, calculated from (1): $f = 0.49\dot{s} + 4.36$ $\sinh^{-1} 1.2\dot{s}$, $\varphi = 0.35$. Curve B, calculated from (1): $f = 0.28\dot{s} + 3.43$ $\sinh^{-1} 2.2\dot{s}$, $\varphi = 0.30$. Broken curve, the upper Newtonian flow curve.

Table III. Solution of Polyisoprene in Benzene

Molecular Weight $\times 10^{-6}$	$C_s' \times 10^2$ (g/dl)	$X_0 + \dfrac{X_1\beta_1}{\alpha_1\eta_0}$	$\dfrac{X_2\beta_2}{\alpha_2\eta_0}$	$\beta_2 \times 10^4$ (sec)	$\eta_{\mathrm{sp}}{}^{\mathrm{N}}/C_s'$ (dl/g)
4.5	5.00	1.50	0.430	11.5	10.0
2.8	5.00	1.38	0.240	10.0	7.7
2.35	5.00	1.35	0.184	9.0	6.8
1.80	5.00	1.28	0.153	7.0	5.7
1.36	5.00	1.23	0.114	6.0	4.6
0.92	5.00	1.185	0.072	6.0	3.7
0.47	5.00	1.16			3.2
0.47	5.00	1.17			3.4
2.8	7.49	1.600	0.388	14.0	8.0
2.8	5.00	1.380	0.240	10.0	7.6
2.8	2.50	1.190	0.104	13.0	7.6
2.8	1.25	1.093	0.046	10.0	7.4
2.8	0.625	1.049	0.021	10.0	7.8
2.35	5.00	1.355	0.184	9.0	7.1
2.35	2.50	1.170	0.074	9.0	6.8
2.35	1.30	1.090	0.032	9.0	7.0
1.80	4.90	1.285	0.153	7.0	5.75
1.80	2.51	1.145	0.069	7.0	5.70
1.80	1.26	1.070	0.033	7.0	5.60

It may also be reasonable to assume that $X_1{}^0 \gg X_2{}^0$, that is, $X_1^\infty \simeq X_1{}^0$. Thus $\eta_u{}^{\mathrm{N}}$ becomes approximately equal to η^{N}, and as a result, the two values of η_{sp} calculated from $\eta_u{}^{\mathrm{N}}$ and η^{N} are equal. Returning to Tables I and II, it may be noted that η_{sp} from $\eta_u{}^{\mathrm{N}}$ is generally larger than η_{sp} from η^{N}. This fact is conceivable since $X_1^\infty = X_1{}^0 + X_2{}^0$ (that is, $X_1^\infty > X_1{}^0$).

The solution system, polyisoprene in benzene[11] does not show the Ostwald type of flow curves; that is, it is not a thixotropic system. The flow curves in this case are a general type of non-Newtonian or pseudoplastic flow. We applied (1) to this system[6,11]; the results are summarized in Table III. As shown in Table III, the values of $\eta_{\mathrm{sp}}{}^{\mathrm{N}}/C_s'$ are about constant irrespective of concentrations for the samples of molecular weight of $2.8 \times 10^6, 2.35 \times 10^6$, and 1.80×10^6. The values of $\eta_{\mathrm{sp}}{}^{\mathrm{N}}/C_s'$ in Tables I and III are used in the following discussions.

DISCUSSIONS

Calculations of K

From (6), we obtain

$$\frac{\eta_{\mathrm{sp}}{}^{\mathrm{N}}}{C_s} = \frac{\beta_1}{\alpha_1}\frac{1}{\eta_0} v_1 \equiv K v_1 \tag{9}$$

Table IV[a]. The Values of K

$M \times 10^6$	C_s' (g/dl)	Calculation 1		Calculation 2		Calculation 3	
		v_1'	K^b	v_1'	K^b	v_1'	K^b
(A) Solution of Polyisobutylene in Cyclohexane							
2.2	2.0	3.60	0.81	3.05	0.96	3.94	0.75
			(1.31)		(1.54)		(1.19)
	0.51		1.27		1.50		1.16
			(1.32)		(1.56)		(1.21)
1.3	1.0	2.78	1.22	2.35	1.45	3.03	1.12
			(1.30)		(1.61)		(1.20)
	0.5		1.21		1.43		1.11
			(1.30)		(1.61)		(1.20)
0.046	4.8	0.52	1.36	0.44	1.60	0.57	1.25
			(1.36)		(1.60)		(1.25)
	2.65		1.30		1.54		1.20
			(1.34)		(1.58)		(1.23)
(B) Solution of Polyisoprene in Benzene							
4.5	0.05	2.42	4.13	4.70	2.10	6.58	1.52
2.8	0.05	1.94	3.96	4.00	1.93	5.27	1.46
2.35	0.05	1.74	3.90	3.60	1.89	4.72	1.44
1.8	0.05	1.53	3.72	3.15	1.81	4.13	1.38
1.36	0.05	1.33	3.45	2.79	1.67	3.62	1.27
0.92	0.05	1.08	3.42	1.99	1.65	2.94	1.26
0.47	0.05	0.79	4.05	1.63	1.97	2.15	1.49
0.47	0.05	0.79	4.30	1.63	2.10	2.15	1.58
0.20	0.05	0.50	3.20	1.03	1.55	1.37	1.17
2.80	0.0749	1.94	4.12	4.00	2.00	5.27	1.52
2.80	0.0500	1.94	3.92	4.00	1.90	5.27	1.44
2.80	0.0250	1.94	3.92	4.00	1.90	5.27	1.44
2.80	0.0125	1.94	3.81	4.00	1.85	5.27	1.40
2.80	0.0063	1.94	4.03	4.00	1.95	5.27	1.48
2.35	0.0500	1.74	4.08	3.60	1.97	4.72	1.50
2.35	0.0250	1.74	3.90	3.60	1.89	4.72	1.44
2.35	0.0130	1.74	4.03	3.60	1.94	4.72	1.48
1.80	5.75	1.53	3.76	3.15	1.83	4.13	1.39
1.80	5.70	1.53	3.72	3.15	1.81	4.13	1.38
1.80	5.60	1.53	3.66	3.15	1.78	4.13	1.36

[a] The K values were calculated by using the following three methods: in calculation 1 v_1' and K were calculated by using the value of l obtained from the light-scattering experiment[12]; in calculations 2 and 3 the l values were estimated from the lengths of 20 and 25 carbon-atom chains,[13] respectively. For calculating K the values of η_{sp}^N/C_s' given in Tables I and III were used. The quantity v_1' is defined by (10).

[b] The values of K in parentheses were obtained from the η_{sp}^N values obtained from the Newtonian branch appeared in the Ostwald curve over the high range of $\dot{s}$, whereas all unparenthesized K values in this table were obtained in the low range of $\dot{s}$.

Next we calculate $C_s v_1$ as follows [see Stacey's book in Ref. (12)]:

$$C_s v_1 = 1.71 \times 20^{21} \left(\frac{l^2}{mN} \right)^{3/2} M^{1/2} C_s{}' \equiv v_1' C_s' \tag{10}$$

Here $v_1 = 4\pi R_1{}^3/3$; $R_1{}^2 = \langle r^2 \rangle/6 = nl^2/6$; l and m are the length and the mass of a segment, respectively; n is the number of segments in a molecule; $\langle r^2 \rangle$ is the mean square end-to-end distance of the polymer molecules with molecular weight M; and C_s' is the concentration expressed in units of g/dl whereas C_s is in number of molecules per cc. By substituting (10) into (9), K is readily obtained with the use of the experimental values of $\eta_{\mathrm{sp}}{}^{\mathrm{N}}$. For this purpose, however, we must know l. The value of l is obtainable from light-scattering experiments.[12] It may be also estimated to a 20 or 25 carbon-atom chain length according to Kauzman and Eyring.[13] The results are tabulated in Table IV.

The value of K varies between 1.2 and 4.0 depending on the method of estimating l or v_1'; in other words, K varies around 2.5. One may also notice that in a given method of estimation the K value is constant irrespective of the molecular weights and concentrations, in agreement with the theoretical requirement. Thus it is concluded that the flow unit for the Newtonian term in the Ree-Eyring equation behaves like a hard sphere.

Application of Robinson's Equation

Robinson's equation[14] is as follows:

$$\frac{\eta^{\mathrm{N}}}{\eta_0} - 1 = \frac{K \varphi_1 \varphi_c}{1 - \varphi_c \varphi_1} \tag{11}$$

In (11), φ_c is the critical volume fraction; it is the volume of the suspension at a critical concentration where the particles, whose total solid volume is 1 cc, make the suspension a complete agglomerate as a whole. Thus, if the volume fraction φ_1 is very small (that is $\varphi_c \to 1$), (11) becomes Einstein's equation with $K = 2.5$ for rigid spheres. Applying (11) to the suspension of procain penicillin in water,[9] the following values were obtained: $\varphi_c = 2.42$ cc and $K = 3.3$. The φ_c value, 2.42 cc, agrees very well with the sedimentation volume of 2.53 cc obtained by a centrifugal separation,[9] and the value of $K = 3.3$ is roughly equal to the theoretical value of 2.5. Thus one may conclude that in this case also the flow unit of the Newtonian term in the Ree-Eyring equation is a kind of hard sphere.

Hydrodynamic Consideration

Let W be the energy loss in unit volume per unit time for a pure solvent when the latter flows with $\dot{S}$. Then W is represented by

$$W = \eta_0 \dot{S}^2 = 2\delta^2 \eta_0 \tag{12}$$

Equation (12) is derived by combining Debye's relation[15] with the hydrodynamic equation.[7] In (12), δ^2 is a flow parameter appearing in the Einstein theory.[7] Next, let W^* be the energy loss when a system of suspension flows with $\dot{s}^*$. Then, the equation corresponding to (12) is written as

$$W^* = \eta^* \dot{s}^{*2} = 2\delta^{*2} \eta^* \tag{13}$$

The following relations are also well known[7]:

$$W^* = 2\delta^2 \eta_0 (1 + \tfrac{1}{2} v_1 C_s) \tag{14}$$

and

$$\delta^{*2} = \delta^2 (1 - v_1 C_s)^2 \tag{15}$$

From (12)–(15), we obtain the relations

$$f^* = F(1 + \tfrac{3}{2} v_1 C_s) \tag{16a}$$

and

$$\dot{s}^* = \dot{S}(1 - v_1 C_s) \tag{16b}$$

where F and f^* are the stress acting on the pure solvent and on the suspension, respectively. Since $\eta^* = f^*/\dot{s}^*$ and $\eta_0 = F/\dot{S}$, the following equation results from (16)

$$\eta^* = \frac{F}{\dot{S}(1 - v_1 C_s)}(1 + \tfrac{3}{2} v_1 C_s) = \eta_0 (1 + \tfrac{5}{2} v_1 C_s) \tag{17a}$$

which is the same as Einstein's equation, (7). By comparing (17) with (6), we obtain immediately the relation $K = \beta_1/\alpha_1 \eta_0 = 2.5$.

In order to help understand the above derivation, the following explanation will be useful: (a) a pure solvent flows with $\dot{S}$ when F is applied; (b) a suspensoid whose volume fraction is φ_1 is dispersed into the solvent; (c) with this mixing, $\dot{S}$ decreases to $\dot{s}^*$ whereas F increases to f^*; and (d) the relations between the corresponding quantities are given by (16a) and (16b).†

† In (12)–(17a) the stress and strain rate for *solution* are represented by f^* and $\dot{s}^*$, respectively, and the corresponding quantities for *pure solvent* are expressed by F and $\dot{S}$. Thus f and $\dot{s}$ in (1) and (2) are synonymous with f^* and $\dot{s}^*$, the asterisks emphasizing that these quantities are for solution. The capital letters F and $\dot{S}$ are used for pure solvent to stress the distinctivity between these quantities and $f_0, f_1, \dot{s}_0, \dot{s}_1, f^*,$ and $\dot{s}^*$ for solution.

Next we show that Einstein's equation (7) can be derived from the Ree-Eyring (R-E) theory (1). Here we assume that type 2 flow units are absent in dilute solutions. Then the Ree-Eyring equation is written as

$$f^* = X_0 f_0 + X_1 f_1 = (1 - X_1) f_0 + X_1 f_1 = (1 - \varphi_1) f_0 + \varphi_1 f_1 \quad (17b)$$

$$\dot{s}^* = \dot{s}_0 = \dot{s}_1, \quad (17c)$$

where f_0 and f_1 are the shear stress acting on the solvent molecules (in solution) and type 1 flow units (solute molecules or suspensoid particles), respectively; $\dot{s}_0$ and $\dot{s}_1$ are the shear rates of the respective component flow units. The relation (17c) is a fundamental assumption in the Ree-Eyring model. Here we note that the Ree-Eyring model corresponds to a mechanical model composed of Maxwell units of type 1 in parallel with other Maxwell units of type zero.

Equation 17b is a replica of (2), where f^N is replaced by f^*, since they are synonymous. The quantity f_0 in (17b) is represented in terms of F and $\dot{S}$ in the following way:

$$f_0 = F\left(\frac{\dot{S}}{\dot{s}^*}\right), \quad (17d)$$

since the system of the Ree-Eyring model flows with $\dot{s}^*$ but not with $\dot{S}$. Accordingly, f_1 in (17b) is written as

$$f_1 = \frac{[f^* - (1 - \varphi_1) f_0]}{\varphi_1} = \left[f^* - (1 - \varphi_1) F\left(\frac{\dot{S}}{\dot{s}^*}\right)\right] \varphi_1^{-1} \quad (17e)$$

We introduce (16a) and (16b) into (17d) and (17e), and obtain†

$$f_0 = F(1 - \varphi_1); \qquad f_1 = \tfrac{3}{2} F. \quad (17f)$$

By substituting (17f) into (17b) and making use of the relations $\eta^* = f^*/\dot{s}^*$ and $\eta_0 = F/\dot{S}$ in conjunction with (16b) we obtain the following equation:

$$\eta^* = \eta_0\left(1 + \frac{5}{2}\frac{\varphi_1}{1 - \varphi_1}\right), \quad (17g)$$

which becomes Einstein's (7) if $1 \gg \varphi_1$.

When the solution is fairly concentrated, the following equations hold[16]:

$$W^* = 2\delta^2 \eta_0(1 + \tfrac{1}{2}\varphi_1 - \tfrac{1}{2}\varphi_1{}^2)$$
$$= 2\delta^2 \eta^*(1 - 2\varphi_1 - 9.6\varphi_1{}^2). \quad (17h)$$

---

† The relation $f_1 = 1.5F$ is also obtained from another theoretical consideration not using (16a). The detailed report will appear elsewhere.

In this case (12) and (13) are also applicable. Thus from (17h) we obtain

$$f^* = F(1 + 1.5\varphi_1 + 6.3\varphi_1^2) \tag{17i}$$

$$\dot{s}^* = \dot{S}(1 - \varphi_1 - 5.3\varphi_1^2), \tag{17j}$$

$$\eta^* = \eta_0(1 + 2.5\varphi_1 + 14.1\varphi_1^2). \tag{17k}$$

Equation 17k is the Simha equation.

When the concentration of the system is considerably high, the Robinson equation (11) holds. Next we derive (11) from the Ree-Eyring model. Let us assume the following:

1. In the suspension system there are C_m micelles per unit volume.
2. Each of the micelles contains ν solute molecules; thus $C_m = C_1/\nu$.
3. Let v_m be the molecular volume of the micelle. Then φ_c which was previously defined, is given by the equation $\varphi_c = v_m/\nu v_1$.
4. The total volume of solvent molecules in micelles per unit volume is given by $C_m v_m - C_1 v_1 = C_1 v_1(\varphi_c - 1)$.
5. Since the solvent molecules inside micelles are bound, the volume of free solvent molecules per unit volume is given by $C_0 v_0 - C_1 v_1(\varphi_c - 1)$.
6. Φ_0 and Φ_m, which are the fraction of volume occupied by the free solvent molecules and by the micelles, respectively, are given by the following equations:

$$\Phi_0 = 1 - \varphi_1\varphi_c; \qquad \Phi_m = \varphi_1\varphi_c \tag{17l}$$

where the relations $C_0 v_0 = \varphi_0$, $C_1 v_1 = \varphi_1$, and $\varphi_0 + \varphi_1 = 1$, were used.

Equation 17g is applicable to this case also if φ_1 is replaced by Φ_m, that is,

$$\eta^* = \eta_0\left(1 + \frac{5}{2}\frac{\Phi_m}{1 - \Phi_m}\right) = \eta_0\left(1 + 2.5\frac{\varphi_1\varphi_c}{1 - \varphi_1\varphi_c}\right). \tag{17m}$$

Equation 17m becomes Robinson's equation if 2.5 is replaced by a constant K which takes the values around 2.5 in general. Thus we have derived (11) from the Ree-Eyring theory.

Correction in High Concentrations

In all our calculations of K we have assumed that

$$\frac{N_1}{V} \simeq C_s. \tag{18a}$$

Consequently

$$X_1 = \frac{v_1 N_1}{V} \simeq v_1 C_s. \tag{18b}$$

The above approximations, (18a) and (18b), do not hold at high concentrations, especially at low $\dot{s}$, because here $X_1 + X_2 = v_1 C_s$. Thus, some correction is needed.

From (8b), we obtain the following relations:

$$X_1^{\infty} = \frac{\eta_u^N - \eta_0}{\beta_1/\alpha_1}$$

and

$$X_1^0 = \frac{\eta^N - \eta_0}{\beta_1/\alpha_1}.$$

These two equations yield

$$X_1^0 = \frac{(\eta^N - \eta_0)X_1^{\infty}}{\eta_u^N - \eta_0} \tag{19a}$$

and

$$X_2^0 = \left(1 - \frac{\eta^N - \eta_0}{\eta_u^N - \eta_0}\right)X_1^{\infty}. \tag{19b}$$

In (19b) the relation $X_1^0 + X_2^0 = X_1^{\infty}$ was used. From (19a) and (19b) a factor α is derived:

$$\alpha = \frac{X_1^0}{X_1^0 + X_2^0} = \frac{\eta^N - \eta_0}{\eta_u^N - \eta_0} \sim \frac{\eta^N}{\eta_u^N}. \tag{20}$$

Using α, X_1 can be expressed by

$$X_1 = \frac{v_1 N_1}{V} = v_1 \alpha C_s. \tag{21}$$

Thus at high concentrations, (21) is used instead of (18b).

By applying (21) to the case of $C_s' = 2.0$ g/dl of the solution of polyisobutylene in cyclohexane the following result was obtained:

	Calculation 1 K	Calculation 2 K	Calculation 3 K
Uncorrected	0.81	0.96	0.75
Corrected	1.25	1.48	1.14

It can readily be noticed that the corrected K values agree well with those obtained in the high region of $\dot{s}$ (see Table IV). The corrected value of η_{sp}^N/C_s' ($C_s' = 2.0$ g/dl) is 4.50 whereas the uncorrected value is 2.92, and the former agrees better with the value (4.70 dl/g) obtained in the high region of $\dot{s}$ (see Table I).

REFERENCES

1. H. Lamb, *Hydrodynamics*, Dover, New York, 1932, p. 160; P. Debye, *J. Chem. Phys.*, 14, 636 (1946); P. Debey and A. M. Bueche, *ibid.*, 16, 573 (1948); J. G. Kirkwood and J. Risman, *ibid.*, 16, 565 (1948).

2. S. Glasstone, K. J. Laidler, and H. Eyring, *The Theory of Rate Processes*, McGraw-Hill, New York, 1941, p. 480.

3. T. Ree and H. Eyring, *J. Appl. Phys.*, **26**, 793, 800 (1955).

4. S. H. Maron and P. E. Pierce, *J. Colloid Sci.*, **11**, 80 (1956); S. H. Maron and A. W. Sisko, *ibid.*, **12**, 99 (1957); S. H. Maron, N. Nakajima, and I. M. Krieger, *J. Polymer Sci.*, **37**, 1 (1959).

5. H. Utsugi, S. Nishimura, and J. Takahira, *Nippon Kagaku Zasshi* (J. Chem. Soc. Japan), **91**, 702 (1970).

6. H. Utsugi, S. Nishimura, and M. Kato, *Nippon Kagaku Zasshi* (J. Chem. Soc., Japan), **91**, 792 (1970).

7. A. Einstein, *Ann. Phys.*, **19**, 289 (1906); **34**, 591 (1911); L. D. Landau and E. M. Lifshitz, *Fluid Mechanics*, Pergamon, London, 1959, p. 76.

8. E. W. Merrill, *Symp. Non-Newtonian Viscometry*, Am. Soc. Testing Materials, Washington, D.C., Oct. 11, 1960 (ASTM Special Technical Publication No. 299, p. 73).

9. H. Nogami and K. Umemura, *Yakugaku Zasshi* (Pharmeceutical Society of Japan), **82**, 273 (1962).

10. S. J. Hahn, T. Ree, and H. Eyring, *Spokesman* (J. Natl. Lub. Grease Inst.), **21**, No. 3 (1957); H. Utsugi, H. Iwasawa, and T. Ree, *Nippon Kagaku Zasshi* (J. Chem. Soc. Japan), **91**, 690 (1970).

11. M. A. Golub, *J. Polymer Sci.*, **18**, 27 (1955).

12. K. A. Stacey, *Light-Scattering in Physical Chemistry*, Butterworths, London, 1956, p. 124; S. Newman, W. R. Krigbaum, C. Langier, and P. Flory, *J. Polymer Sci.*, **14**, 451 (1954); W. R. Krigbaum and D. K. Carpenter, *J. Phys. Chem.*, **59**, 1166 (1955); P. Johnson and S. Bywater, *Trans. Faraday Soc.*, **47**, 195 (1951).

13. W. Kauzmann and H. Eyring, *J. Amer. Chem. Soc.*, **62**, 3113 (1940).

14. J. V. Robinson, *J. Phys. Colloid Chem.*, **53**, 1042 (1949); **55**, 455 (1951).

15. P. Debye, *J. Chem. Phys.*, **14**, 636 (1946).

16. V. E. Guth and R. Simha, *Kolloid-Z.*, **74**, 266 (1936); F. Eirich and R. Simha *Monatsh. Chem.*, **71**, 67 (1938).

Electronic States of Solid Explosives and Their Probable Role in Detonations*

FERD WILLIAMS

Physics Department, University of Delaware, Newark

Abstract

After a brief review of some experimental evidence for electronic processes having an effect on chemical reactions in crystals we consider theoretically the question of the influence of electronic states and electronic transport on the initiation and propagation of detonations in solid explosives. In an extension of the quantum mechanical adiabatic approximation the dynamics of atomic motion during detonation are shown to be determined approximately by an effective potential which is a function of the electronic eigenvalues. We analyze the general features of the electronic states and the electron and exciton transport in the region of the detonation front and conclude that the gradient in the energy bands arising from the large pressure gradient and the resulting anisotropic electronic transport may have, in some materials, appreciable effects on the detonation. Representative solid explosives are then considered as regards, for example, the effect of changes in the Fermi level on explosive characteristics. Initiation by electronic mechanisms is proposed, particularly by electron and positive hole injections. Special attention is given to initiation of amorphous explosives by high injection densities. Finally, inhomogeneous graded mixed crystals with nonsteady-state and unidirectional characteristics are considered.

* Based in part on a talk prepared for the Explosive Chemical Reaction Seminar at the Army Research Office, Durham in October 1968 and given limited circulation in a preliminary form in an AROD Report. Professor Eyring was instrumental in organizing the Seminar and in stimulating my interest in this subject.

289

INTRODUCTION

In this paper we consider the possible effects of the electronic structure and of electronic transport on the initiation and propagation of detonations in solid explosives. The analyses will be necessarily preliminary, rather than rigorous and thorough, and will be occasionally speculative. This is appropriate because of the complexity of the phenomena associated with initiation and detonation of solid explosives and because of the limited earlier consideration of electronic states and transport of solid explosives, evident by the lack of published works relating electronic properties to initiation and detonation.

The lack of published works on this subject is due to two causes. First, the theory of the electronic structure and electronic transport has only recently become anywhere near adequate to cope with materials as complex as lead azide on the one hand and 1,3,5,7-tetranitro 1,3,5,7-tetraza cyclooctane (HMX) on the other. Second, the hydrodynamic theory of detonations has been remarkably successful in explaining the velocities of detonations,[1] and initiation has been reasonably well explained in most cases as ultimately thermal in origin.[2]

EXPERIMENTAL EVIDENCE FOR THE INTERRELATION OF ELECTRONIC STRUCTURE AND CHEMICAL INSTABILITY

Nevertheless, there does appear to be experimental evidence that the electronic states, transitions, and occupational probabilities of these states may be involved in the chemical decomposition of inorganic and organic materials, including explosives.

A classical example of chemical changes arising from electronic processes is the radiation damage of alkali halide crystals. Ionizing radiation generates atomic defects, specifically halogen vacancies and halogen molecule negative ions, in these crystals, by electronic mechanisms.[3] These are genuine chemical changes in the sense of creating different molecular or structural entities. The generation of crystal defects by electronic mechanisms may occur in materials whose band gap E_g is large compared to the energy of formation of the defects E_d, because E_g is the maximum energy which is storable in electronic modes before atomic relaxation to form the defect. This requirement is not severe for explosives, in which E_d is negative for those defects involved in the decomposition. More immediately relevant to explosives than the radiation damage of alkali halides is the photodecomposition of alkali and other metallic azides to form defects such as N_4^- and N_2^- observed in electron spin resonance by King et al.,[4] Gilliam and coworkers,[5] Marinkas and Bartram,[6] and others. The photodecomposition of silver halides and of organic dyes also involves electronic transitions.

The electronic states of crystals are dependent on crystal structure. Some inorganic explosives (for example, lead azide) and many organic explosives (e.g., TNT and HMX) exist in several structures. The initiation characteristics for each of the polymorphs are found to be different. This may be explainable in terms of the differences in electronic states for the different crystal structures.

Finally, the effect of doping with charged impurities on the decomposition of explosives either through changes in the electronic energy levels or through changes in their occupational probability, that is, Fermi level, is of interest. Fair and Forsyth[7] have reported changes in the photolytic decomposition of lead azide by doping with iron. They have correlated the decomposition with exciton and impurity absorption and photoconductivity, all three of which are purely electronic processes.

We shall, therefore, be concerned with relating electronic structure of solids with chemical instability in the most general theoretical analyses, with energy transport by electronic mechanisms during detonation, with the effect of occupancy of electronic states (Fermi level) on initiation, and with the possibility of initiation by nonthermal, specifically electronic, mechanisms in both single-crystal and amorphous explosives.

GENERAL THEORY RELATING ELECTRONIC STRUCTURE
TO CHEMICAL INSTABILITY

Matter consists of electrons and atomic nuclei. We specify the spatial and spin coordinates of the i electron by ξ_i and the spatial coordinates of the j nucleus by R_j. According to quantum mechanics a system in a state is described as completely as is possible by a wavefunction, which is a function of the coordinates of all the particles. In general, the state is also time dependent; however, we are concerned in this analysis with stationary states and with the solutions of the time-independent Schrödinger equation. The time dependence of the chemical reaction will be taken care of through transitions between stationary states. For a system of n electrons and N

nuclei we have the many-particle wavefunction:

$$\psi(\xi_1, \xi_2, \ldots, \xi_n, \mathbf{R}_1, \mathbf{R}_2, \ldots, \mathbf{R}_N) \equiv \psi(\xi, \mathbf{R}), \tag{1}$$

where the dependence on ξ and $\mathbf{R}$ indicates the dependence on all electronic and nuclear coordinates, a notation we now follow. The many-particle wavefunction satisfies the Schrödinger equation:

$$\hat{H}\psi(\xi, \mathbf{R}) = E\psi(\xi, \mathbf{R}), \tag{2}$$

where $\hat{H}$ is the Hamiltonian operator involving terms for the kinetic energies of all particles and for all electrostatic interactions, and E is the energy eigenvalue.

From the difference in mass of the electrons m and of the nuclei M_j and from equipartition of energy it is evident that the electrons move rapidly compared to the nuclei and therefore will exist in separable, approximate stationary states that are smoothly modified by the motion of the nuclei. This is, of course, the well-known adiabatic approximation of Born and Oppenheimer.[8] In order to solve for these electronic stationary states we consider the nuclei fixed in the following Schrödinger equation:

$$\hat{H}_{el}\Phi_{\mathbf{R}}(\xi) = E_{el}(\mathbf{R})\Phi_{\mathbf{R}}(\xi), \tag{3}$$

where $\hat{H}_{el}$ is the part of $\hat{H}$ which involves electronic coordinates, $\mathbf{R}$ specifies the fixed $\mathbf{R}_1, \ldots, \mathbf{R}_N$, and $E_{el}(\mathbf{R})$ are the electronic eigenvalues or electronic structure, which are of course dependent on $\mathbf{R}$. Again, in accordance with the adiabatic approximation we take ψ to be the following form:

$$\psi(\xi, \mathbf{R}) = \Phi_{\mathbf{R}}(\xi)\,\chi(\mathbf{R}). \tag{4}$$

By substituting (4) in (2) and using (3) we obtain the following for the Schrödinger equation which governs the motion of the nuclei, if we neglect two terms to be discussed later:

$$\left\{ -\sum_j \frac{\hbar^2}{2M_j} \Delta_{\mathbf{R}_j} + V(\mathbf{R}) + E_{el}(\mathbf{R}) \right\}\chi(\mathbf{R}) = E\,\chi(\mathbf{R}), \tag{5}$$

where $\Delta_{\mathbf{R}}$ is the Laplacian and $V(\mathbf{R})$ is the direct interaction between nuclei. The important conclusion, however, is that the electronic structure $E_{el}(\mathbf{R})$ is a major term in the effective potential which governs the motion of the nuclei. In the usual application of (5) only displacements in $\mathbf{R}$ about equilibrium sites are involved. For the large excursions in $\mathbf{R}$ occurring during chemical reactions the approximations must be reexamined. It is expected, however, that $E_{el}(\mathbf{R})$ will persist in the effective potential for the motion of nuclei even during detonations. We believe that the initial assumption of the slow motion of nuclei versus rapid motion of electrons remains approximately valid for detonations. For typical electronic states of molecules and solids, $E_{el} \geq 1$ eV, and therefore the orbital time for electronic motion $\tau_0 \leq 10^{-15}$ sec, which is shorter than any process occurring in a detonation; for example, for shock velocities $v_s = 10^6$ cm/sec, the time to transverse an atomic layer $\tau \simeq 10^{-14}$ sec. The departure from stationary electronic states is actually one of the terms neglected in obtaining (5), and this term can be shown to be of order m/M compared to the terms retained.

The other term neglected in obtaining (5) is the following:

$$\sum_j \frac{\hbar^2}{M_j} \{\nabla_\mathbf{R} \Phi_\mathbf{R}(\xi) \nabla_\mathbf{R} \chi(\mathbf{R})\}. \tag{6}$$

The diagonal elements in the matrix formulation of this term vanish if the number of electrons is conserved, as is the case for non-nuclear reactions. The off-diagonal elements correspond to electron–phonon interaction. Further analysis is necessary to determine whether phonons remain a valid concept or whether (6) diverges in the chaos of the detonation.

We continue, assuming (5) to be approximately valid as the basis for Fig. 1. The effective potential

$$V_{\text{eff}} = V(\mathbf{R}) + E_{\text{el}}(\mathbf{R}) \tag{7}$$

is plotted versus the nuclear coordinates or reaction coordinates $\mathbf{R}$. Because

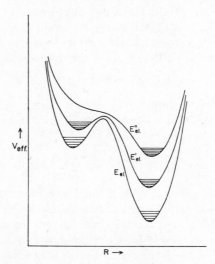

Fig. 1. Effective potentials for atomic motion during detonation versus generalized reaction coordinate. Different curves correspond to different electronic states; vibrational levels for initial and final configurations are also shown.

of the double minima, the eigenstates $\chi(\mathbf{R})$ for the system are not simple harmonic oscillator functions. They are, however, bound states because $V_{\text{eff}} \to \infty$ at the extrema of $\mathbf{R}$. The reaction is visualized as proceeding from an initial "clamped" state to the final equilibrium state by the perturbation from electron–phonon interaction. The initial and final states are describable as linear combinations of the complete set of exact $\chi(\mathbf{R})$ for the complex potential V_{eff}. The electronic states $E_{\text{el}}(\mathbf{R})$ at all intermediate configurations $\mathbf{R}$ between the initial and final configurations are, of course, involved in determining the $\chi(\mathbf{R})$. To reduce somewhat the complexity of the problem, the core electrons which survive the reaction unchanged can be grouped on their respective nuclei and $V(\mathbf{R})$ taken as the ion–ion interaction. In any case, it is evident from (5) and Fig. 1 that the electronic structure $E_{\text{el}}(\mathbf{R})$ plays an important role in chemical reactions, including detonations.

ELECTRONIC STATES AND ELECTRONIC TRANSPORT IN THE REGION OF THE SHOCK FRONT

The classic theory of detonations of Doering, von Neumann, and Zeldovich (see Ref. 1 for original references) considered the shock and chemical reaction zones separately, the pressure wave heating the unreacted explosive so that the reaction occurs thermally after an induction period. Hirschfelder and coworkers[9] and Wood[10] investigated the effects of coupling of the shock and reaction zones. We are concerned with the electronic structure in the region of the shock front and with the possibility of electronic transport getting ahead of the shock and contributing to initiation.

Therefore we look at the details fore and aft of the shock front in condensed explosives. Ilynkhin and coworkers[11] give the shock velocities v_s, the pressure behind the shock front P, and the ratio of specific volumes after and before shock compression V/V_0 for cast TNT, crystalline RDX, and nitromethane. These quantities are approximately 5×10^5 cm/sec, 10^5 atm, and 0.7, respectively, for these explosives. The important observation to be made is that the electronic structure $E_{\text{el}}(\mathbf{R})$ in the region of shock compression is in general quite different from that of the uncompressed explosive. As note earlier, the electronic states are approximately stationary at shock velocities. The initial predictions are that for most explosives the allowed electronic bands are wider, the forbidden energy gap is narrower, and all states are perturbed toward the vacuum level in the high pressure region. The first prediction is evident for organic explosives because the overlap of the molecular wavefunctions is increased by compression; the last, from consideration of the increase in kinetic energy of the electrons confined to the smaller volume.

The second feature of the region of the shock front with which we are concerned is the spatial dependence of the electronic structure. If the pressure gradient is only over one or two molecular distances, then we are dealing with an abrupt semiconductor heterojunction whose electronic transport properties are largely determined by space charge effects; if the pressure gradient at the shock front extends over at least 100 molecular distances, then, as was shown by Gora and Williams[12] for graded mixed crystals, the concept of a graded band gap is valid. Craig[13] estimated the reaction zone length for nitromethane at 10^{-5} cm from extrapolation of the equation of state for the unreacted material and from thermodynamic adiabatic explosion theory. The shock front is probably thinner; therefore, we take $l \approx 10^{-6}$ cm as an estimate of the length of the pressure gradient and consider the system as having graded band edges. More quantitative determinations of the width of the pressure gradient and of the electronic structure in the high pressure region are needed for typical solid explosives. Figure 2 includes these two features of the electronic states in the region of the shock front.

We now consider electronic transport in the region of the detonation front. The following analysis applies to conduction electrons or to positive holes in the valence band and with straightforward modifications to excitons. The component of the current density arising from ordinary diffusion is

$$J_D = n v_D = - D \frac{dn}{dx}, \tag{8}$$

where n is the carrier concentration, v_D is the diffusion velocity, and D is the diffusion constant. Our primary concern is with the order of magnitude of

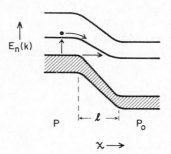

Fig. 2. Probable electronic band structure of unreacted crystalline explosive versus direction through the shock front. The pressure fore and aft the front are, respectively, P_0 and P; electronic transitions, transport, and tunneling are also shown.

velocities compared to shock velocities v_s; therefore

$$v_D \simeq -D\frac{n - n_0}{n} \cdot \frac{1}{l},\tag{9}$$

and for negligible charge carriers in front of the shock we have $n_0 \ll n$. Using the Einstein relation $D = (kT/e)\mu$, (9) can be solved in terms of the mobility μ. For an order of magnitude calculation we take $\mu \approx 1$ cm^2/V-sec, typical for a very poor inorganic semiconducting crystal, for an amorphous semiconductor, or for a good organic semiconductor; $l \approx 10^{-6}$ cm as estimated previously; and $T \approx 3000°$K; and obtain $v_D \sim 2 \times 10^5$ cm/sec, which is the magnitude of, but less than, v_s. Incidentally, μ is reduced by the high temperature but increased by the high pressure (10^5 atm); the latter effect is particularly pronounced for organic semiconductors. A value of μ for the conditions at the shock front of the magnitude observed at ordinary temperatures and pressures for good inorganic semiconductors, that is, 10^4 cm^2/V-sec, would obviously result in the electronic excitation outrunning the shock wave.

In addition to the ordinary diffusion current, there is an additional anisotropic current arising from diffusional effects in systems with graded band edges such as shown in Fig. 2. Van Ruyven and Williams[14] derived this anisotropic diffusional component of the current density:

$$J_A = nv_A = -\frac{D}{kT}n\frac{dE(x)}{dx}\tag{10}$$

where $E(x)$ is the position-dependent band edge for the n electronic carriers. For the same hypothetical system just analyzed for v_D and with $E - E_0$ estimated as 2 eV from considering the change in energy levels of an electron in a box of molecular dimension on compression ($V/V_0 \approx 0.7$), we find $v_A \sim 2 \times 10^6$ cm/sec. This exceeds v_s. In general, for $n_0 \ll n$, v_A exceeds v_D by the factor $(E - E_0)/kT$. In Fig. 2 electrons precede the shock; holes do not.

In order to have the electrons, positive holes, or excitons that are generated in the detonation region precede the shock front and conceivably contribute to initiation, it is also necessary that their lifetime τ_e exceed the threshold value $l/v_A \approx 10^{-12}$ sec. This requirement does not appear to be a problem.

The anisotropic diffusion of electrons or positive holes in advance of the shock front will yield a space charge and result in an electric field traveling with the shock front. This effect, which arises from the graded band edges, may explain the voltages measured in shocks; for example, the observations of Eichelberger and Hauver[15] of shock-induced electrical polarization in distilled water may originate from graded band edges, probably for the ionic states involved in protonic conduction.

In general, any theory of electronic effects in a detonation must include the effects of the graded band edges such as shown in Fig. 2. In addition to band edge transport, electron tunneling is also shown and can be included in a more complete analysis.

ELECTRONIC STATES AND THEIR OCCUPANCY, AND ELECTRONIC INITIATION

As a representative primary explosive, we consider lead azide; as a representative secondary, HMX. It is now possible to prepare single crystals of both materials and to measure their electronic properties. We shall consider in a preliminary way a few aspects of their electronic states and the occupational probability of these states, as described by the Fermi level, relevant to the decomposition of these materials.

Perhaps the most striking feature of a preliminary consideration of the electronic states of lead azide is the diversity of types of excitonic states that it may have. In addition to the charge-transfer excitons, well known in alkali halides, and effective mass excitons, well known in elemental semiconductors, one predicts for PbN_6: intra-cation excitons (describable in terms of excited states of Pb^{2+}, ^1P and ^3P states of the $1s^2 \cdots 5d^{10}6s6p$ configuration, modified by the complex crystal field of the PbN_6 structure) and intra-anion excitons (describable in terms of excited states of N_3^-). These offer possibilities for the transport of electronic energy.

Slow decomposition or initiation is predicted from (5) and illustrated in Fig. 1 to depend on occupied electronic states. The Fermi level is the energy for which the occupation probability is one-half. If we assume that during the early stages of decomposition, or during initiation, PbN_6 undergoes the reaction

$$2N_3^- \rightarrow 3N_2 + 2e^- \tag{11}$$

and that lead metal is not precipitated, the electrons ($2e^-$) of (11) will be in the conduction band, perturbed, however, by the field of the $[N_3^-]$ vacancies. The electrons will be in F-center bound states, the Fermi level will have risen as shown in Fig. 3, and the n-type electronic conductivity increased. The electronic states $E_{\text{el}}(\mathbf{R})$ in the effective potential V_{eff} of (7), which governs the motion of the nuclei in (5), will have changed. Higher $V_{\text{eff}}(\mathbf{R})$ functions on Fig. 1 should now dominate the chemical reaction. It is tempting to predict that initiation occurs rather generally by a rising Fermi level. The well known memory effect in explosives subjected to repeated partial decomposition may be explainable by accumulated changes in the Fermi level.

Similarly, if both positive holes p^+ and electrons are electrically injected into PbN_6 (double injection), the Fermi level will also rise if the positive

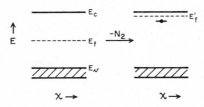

Fig. 3. Changes in Fermi level of lead azide accompanying the decomposition $2N_3^- \rightarrow 3N_2 + 2e^-$.

holes are removed by the reaction

$$2N_3^- + 2p^+ \rightarrow 3N_2 \qquad (12)$$

and the compensating electrons occupy perturbed conduction band states. In other words, appropriate electronic injection can in principle achieve the same Fermi level change which may be essential to initiation.

HMX or 1,3,5,7-tetranitro 1,3,5,7-tetrazacyclooctane exists in four polymorphic forms. These are molecular crystals. Both intramolecular excitons and intermolecular (charge transfer) excitons are predicted. Electronic charge transport depends on overlap of the molecular wave-functions and is therefore enhanced by the pressures (10^5 atm) in the shock wave during detonation. The mobilities of both types of excitons are also enhanced by pressure.

Surya Bulusu et al.[16] have determined with isotopes that the thermal degradation of HMX proceeds by breakage of C—N bonds. It is important to determine whether charged intermediates are formed during decomposition, and even during detonation because these would change the Fermi level.

In addition to the effects on initiation by changes in the Fermi level of solid explosives as a consequence of electronic double injection followed by chemical reactions involving electronic charge carriers of one sign, we also consider

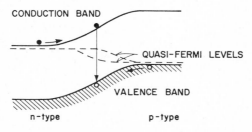

Fig. 4. Double injection of electrons and positive holes into a *p–n* junction biased in the forward direction. Electron–hole recombination in the junction is also shown.

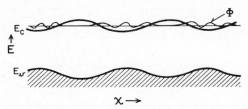

Fig. 5. Fluctuating band edges of amorphous semiconductors arising from compositional or structural inhomogeneities. A typical eigenstate near the conduction band edge is shown; transport of conduction electrons involves tunneling between neighboring minima or thermal activation.

effects arising from double injection only, which separates the Fermi level into two quasi-Fermi levels: one for electrons, the other for positive holes. A simple example is shown in Fig. 4 for a p–n junction biased in the forward direction. It is not yet clear whether either inorganic or organic explosive crystals can be made both p and n type; if not, injecting electrodes may provide the same separation of the quasi-Fermi levels.

In order to achieve initiation by purely electronic mechanisms, high injection densities are probably essential. Electronic switching and storage in amorphous semiconductors have recently received considerable attention.[17] The storage is accompanied by a phase transition, increase in crystallinity. The switching from a low to a high conductivity state involves high local current densities and has been reported for a diversity of amorphous materials, including amorphous films of elemental semiconductors, chalcogenide glasses, and organic solids.[18] Essential characteristics appear to be low electronic mobilities and ability to exist in more than one solid phase. Many solid explosives such as lead azide, HMX, and TNT have these characteristics. Some have been prepared as amorphous solids. The low mobilities of amorphous solids are understandable on the basis of the compositional or structural inhomogeneities which yield fluctuating band edges and tunneling such as shown in Fig. 5. If electronic switching to a high conductivity state is observed with solid explosives it will probably be followed by detonation.

HOMOGENEOUS AND GRADED MIXED CRYSTALS

It is well known in semiconductor physics that most homogeneous mixed crystals of semiconductors have electronic bands intermediate between those of the pure components. This can be explained theoretically on the basis of the virtual crystal approximation, in which the statistical distributions of unit cells in the random alloy are replaced by an average unit cell. Because the de Broglie wavelength for electron states within a few kT of band edges is large compared to unit cell dimensions, the electron responds to the field of the average cell and has the corresponding band structure. To the extent, therefore, that the electronic states within a few kT of band edges influence initiation or detonation, mixed crystal explosives may offer interesting intermediate characteristics; for example, within the constraints of solubility limitations due to differences in crystal structures of copper azide and sodium azide, $Cu_f Na_{1-f} N_3$ may warrant experimental studies.

In addition, mixed semiconductors with graded compositions have, as described earlier relevant to gradations due to pressure, graded electronic properties; for example, graded band gaps.[14] A graded mixed crystal explosive should have predictable, nonsteady-state, detonation characteristics. The nonsteady time dependence and partial directionality of the detonation may be of interest.

CONCLUDING REMARKS

In conclusion we emphasize that there are two points of view that can be taken regarding the impact that investigations of electronic structure and electronic transport may have on explosives research. The first, which is emphasized in this paper, concerns the understanding of phenomena well known for conventional explosives; the second, which is quite speculative, concerns the investigation of qualitatively different phenomena and materials relevant to initiation and detonation, as a consequence of detailed consideration of electronic processes.

ACKNOWLEDGMENTS

I am indebted to Professor H. Eyring and to Dr. R. S. Eichelberger for stimulating my interest in this subject; to Drs. J. Hershkowitz, S. J. Jacobs, and R. F. Walker for bringing to my attention many of the references used in preparing this paper; and to Drs. H. Fair and J. Hershkowitz for helpful and critical observations on an early version of the presentation and for their continued enthusiastic interest. Helpful discussions with Professor A. Halprin on the adiabatic approximation and with Mr. R. B. Hall on transport are acknowledged.

REFERENCES

1. M. W. Evans and C. M. Ablow, *Chem. Rev.*, **61**, 129 (1961).
2. A. Macek, *Chem. Rev.*, **62**, 41 (1962).

3. For a recent review see J. H. Crawford, *Advan. Phys.*, **17**, 93 (1968).
4. G. J. King, B. S. Miller, F. F. Carlson and R. C. McMillan, *J. Chem. Phys.*, **32**, 940 (1960); **34**, 1499 (1961); **35**, 1442 (1961).
5. O. R. Gilliam, A. J. Shuskus, C. G. Young, P. W. Levy and D. W. Wylie, *J. Chem. Phys.*, **33**, 622 (1960); *Phys. Rev.*, **125**, 451 (1961).
6. P. L. Marinkas and R. H. Bartram, *J. Chem. Phys.*, **48**, 927 (1968).
7. H. D. Fair, Jr., and A. C. Forsyth, *Sixth International Symposium on Reactivity of Solids* (August 1968, Schenectady), J. Mitchell et al. Eds., Wiley, New York, 1969.
8. M. Born and J. R. Oppenheimer, *Ann. Phys.*, **84**, 457 (1927).
9. J. Hirschfelder, C. F. Curtiss, B. Linder and M. P. Barnett, *J. Chem. Phys.*, **28**, 1130, 1147 (1958); **30**, 470 (1959).
10. W. W. Wood, *Phys. Fluids*, **4**, 46 (1961).
11. V. S. Ilynkhin, P. F. Pokhill, P. F. Rozanov and N. S. Shvedova, *Dokl. Akad. Nauk, SSSR* **131**, 793 (1960).
12. T. Gora and F. Williams, II–VI *Compound Semiconductors*, Benjamin, New York, 1967, p. 639; *Phys. Rev.*, **177**, 1179 (1969).
13. B. G. Craig, *Tenth Symposium on Combustion*, The Combustion Institute, Pittsburgh, Pa., 1965, p. 863.
14. L. J. Van Ruyven and F. E. Williams, *Am. J. Phys.*, **35**, 705 (1967).
15. R. J. Eichelberger and G. E. Hauver, *Les Ondes de Detonation*, CNRS, Paris, 1962, p. 363.
16. Surya Bulusu, J. R. Autera and R. J. Graybush, West Point Conference of the Army Material Command June, 1968.
17. S. R. Ovshinsky, *Phys. Rev. Letters*, **21**, 1450 (1968).
18. See *Proceedings of Symposium on Semiconductor Effects in Amorphous Solids*, Edited by W. Doremus, North-Holland, Amsterdam, 1970.

Notes on the Two-Particle Density Matrix in π-Electron Theory

C. A. COULSON AND R. G. A. R. MACLAGAN

Mathematical Institute, Oxford, England

Abstract

Since a normal chemical bond is often thought of in terms of the pairing of two electrons with opposed spins, it would appear plausible that indices of bond strength and reactivity in π-electron systems should be based on the two-particle density matrix rather than, as is more usual, on the one-particle density matrix. This paper examines this possibility, and obtains formulas for bond strength and Diels-Alder reactivity indices within the two-particle framework.

A Personal Note

It is a pleasure to dedicate this short paper to Professor Henry Eyring, whose versatility and originality in several distinct fields of science are matched only by his gaiety of spirit and dedication to his work. It is not given to many people to write a textbook which should remain standard reading for more than twenty-five years and be completely described by the names of its authors. Yet so well-known is his *Quantum Chemistry*, written in collaboration with John Walter and George Kimball, that as soon as it was published in 1944 no one bothered about the title: it was sufficient to refer simply to "Eyring, Walter, and Kimball."

Henry Eyring's contributions to the handling of spin functions in VB theory and to the quantitative theory of chemical reactivity are mere examples of his tremendous versatility. Some of his many friends have occasionally tried to devise a motto that should indicate this: it is by no means pure romance to suggest, "You describe it, I'll explain it."

304

INTRODUCTION

All the electronic properties of a molecule are given by its one- and two-particle density matrices. This means that it should be possible to discuss chemical bonding, as it occurs in conjugated and aromatic π-electron systems, and equally the important reactivity indices, in these terms. In traditional MO theory, which derives from the basic work of Hückel, this has usually been achieved by the one-particle matrix—the charge-bond-order matrix introduced by Coulson[1] and developed by Coulson and Longuet-Higgins,[2] but there is more information implied in a knowledge of the two-particle matrix, so that a better approach ought to start at this point. Moreover, ever since G. N. Lewis introduced the concept of an electron-pair bond, it has been natural to think of bonds in terms of two electrons. This also urges the desirability of considering the two-particle matrix.

This paper explores the use of the two-particle matrix in terms of two typical applications, both of which bear on Henry Eyring's work. They are (a) bond strength using a "molecule-in-molecules" approach, and (b) static indices for Diels-Alder reactions. We consider these separately. But in view of the different definitions of the density matrices, a preliminary section is desirable.

DENSITY MATRICES IN MOLECULAR THEORY

The majority of MO calculations involve the determination of certain orbital wavefunctions. However, the determinantal character of the total wavefunction implies that these individual orbitals have no objective significance, and all physically relevant information about the molecule (e.g., total energy, charge distribution) is obtained by appropriate summations, all of which are associated with one or other of the various density matrices. It remains a chief objective of theoretical chemistry to use the one- and two-particle matrices to avoid the ambiguity inherent in individual MO's, and find a simple physical picture not involving the many-dimensional total wavefunction. The foundations of this theory were laid by Landau,[3] von Neumann,[4] and Dirac,[5] but the introduction of reduced density matrices by Husimi,[6] followed by the work of McWeeny[7] and Löwdin[8] have led to their fairly widespread use. An example of this is Ruedenberg's review[9] of the physical nature of the chemical bond.

If $\psi(1, 2, \ldots, N)$ is the normalized wavefunction for an N-electron molecule, the second-order (or two-particle) reduced density matrix is defined by

$$\Gamma(1, 2 \mid 1', 2') = N(N - 1) \int \psi(1, 2, 3, N)\psi^*(1', 2', 3, N) \, d\tau_3 \ldots d\tau_N. \quad (1)$$

The first-order (or single particle) density matrix is similarly defined by

$$\gamma(1 \mid 1') = N \int \psi(1, 2, \ldots, N)\psi^*(1', 2, \ldots, N) \, d\tau_2 \, d\tau_3 \cdots d\tau_N$$

$$= \frac{1}{N - 1} \int \Gamma(1, 2 \mid 1', 2) \, d\tau_2. \quad (2)$$

For many purposes we need the spinless density matrices

$$\rho(1 \mid 1') = \int \gamma(1 \mid 1') \, ds_1 \quad (3)$$

$$\pi(1, 2 \mid 1', 2') = \int \Gamma(1, 2 \mid 1', 2') \, ds_1 \, ds_2. \quad (4)$$

Notice that with definitions (1) and (2) $\int \gamma(1 \mid 1) \, d\tau_1 = N$. The most obvious use of these matrices lies in the result that for any operator expressed symmetrically in the form of one- and two-particle terms,

$$H = \sum_i f(i) + \sum_{i < j} g(ij), \quad (5)$$

the expectation value of H is simply

$$\langle H \rangle = \text{tr} \, (f\gamma) + \tfrac{1}{2} \text{tr} \, (g\Gamma). \quad (6)$$

If H is the electronic Hamiltonian, (6) gives the energy immediately.

In the forms (1)–(4) the density matrices are expressed in terms of the continuous variables which define the positions of the electrons. But if, as often happens, our wavefunction is compounded out of atomic orbitals (e.g., the LCAO approximation), we may find it instructive to use an AO or an MO base. An example that we shall need later is the MO ground state of butadiene, where, in the π-electron approximation, we fill the two mutually orthogonal MO's a_u and b_g with two electrons each and the π-electron wave-function is

$$\psi = \sqrt{\frac{1}{24}}\, \det |a_u \bar{a}_u b_g \bar{b}_g|.$$

This leads to the following expressions for Γ and γ (in which we have dropped the suffixes u and g for simplicity):

$$\gamma(1 \mid 1') = aa + \bar{a}\bar{a} + bb + \bar{b}\bar{b}, \tag{7}$$

$$
\begin{aligned}
\Gamma(1, 2 \mid 1', 2') = {} & a\bar{a}a\bar{a} - a\bar{a}\bar{a}a - \bar{a}aa\bar{a} + \bar{a}a\bar{a}a \\
& + abab - baab - abba + baba \\
& + a\bar{b}a\bar{b} - \bar{b}aa\bar{b} - a\bar{b}\bar{b}a + \bar{b}a\bar{b}a \\
& + \bar{a}b\bar{a}b - \bar{b}\bar{a}\bar{a}b - \bar{a}b b\bar{a} + b\bar{a}b\bar{a} \\
& + \bar{a}b\bar{a}b - \bar{b}\bar{a}\bar{a}b - \bar{a}b\bar{b}\bar{a} + b\bar{a}b\bar{a} \\
& + b\bar{b}b\bar{b} - \bar{b}b b\bar{b} - \bar{b}b\bar{b}b + \bar{b}b\bar{b}b,
\end{aligned} \tag{8}
$$

where in (7) the first factor in each term refers to point 1 and the second to $1'$; in (8) the sequence in each term is 1, 2, $1'$, $2'$. Similarly,

$$\rho(1 \mid 1') = 2aa + 2bb, \tag{9}$$

$$\pi(1, 2 \mid 1', 2') = 2aaaa + 2bbbb + 4abab + 4baba - 2abba - 2baab. \tag{10}$$

Expressions (7)–(10) are given in a molecular-orbital basis. If we adopt the LCAO expansions of a_u and b_g, then we obtain similar types of expressions in an atomic-orbital basis. In particular, (9) then gives the familiar charge-bond-order matrix of Hückel theory.

Expressions (7)–(10) are easily generalized to deal with molecules containing any number of electrons in paired MO's, and to excited states and states of radicals.

BOND STRENGTH AND A MOLECULE-IN-MOLECULES APPROACH

We have already argued that discussions of bond strength should most naturally be given in terms of two-particle matrices. But before doing so it is worth looking briefly at earlier discussions in terms of the one-particle matrix. In Hückel theory, where in (5) there are no terms $g(ij)$, (6) shows that the energy involves only γ; hence a one-particle discussion is reasonable. The simplest example is the Coulson bond order, which is simply an off-diagonal element in the expansion of (9) in terms of atomic orbitals: it is also the differential coefficient of $\langle H \rangle$ with respect to the appropriate resonance integral, as shown by the relation

$$p_{rs} = \frac{1}{2} \frac{\partial \langle H \rangle}{\partial \beta_{rs}}. \tag{11}$$

Since the electron density is given by (7) or (9), any discussion of bond strength in terms of population analysis along the lines suggested by Mulliken[10] falls into this category. So also does the recent development by Polansky and Derflinger[11] and Sofer, Polansky, and Derflinger.[12] In effect they write each MO in terms of the MO's of smaller, simpler molecules, which are fragments of the complete molecule. This is equivalent to using a basis for the expansion of (9) which is intermediate between the AO and full MO forms. Let us consider a fragment L (e.g., an ethylenic unit of a dienoid or benzenoid unit). Then we define the character of this fragment in the complete molecule by

$$\rho_L = \frac{2}{N_L} \sum_{j < k \, \text{in} \, L} p_{jk}{}^L p_{jk}, \tag{12}$$

where $p_{jk}{}^L$ is the bond order of jk in the fragment molecule, p_{jh} is the corresponding bond order in the full molecule, and N_L is the number of atoms in the fragment. Thus for an ethylene fragment $N_L = 2$, $p_{jk}{}^L = 1$, and the ethylenic character ρ_L is equal to the conventional bond order. For a buta-dienoid fragment $N_L = 4$ and (labeling the atoms 1, 2, 3, 4 in order), the dienoid character of this part of the full molecule is

$$\rho_{\text{dienoid}} = \frac{1}{2\sqrt{5}} [2(p_{12} + p_{34}) + (p_{23} + p_{14})]. \tag{13}$$

It follows from (12) that these "characters" measure the extent to which a chosen fragment of a larger molecule resembles the smaller molecule when isolated from the rest of the larger molecule. A similar expression to (13) enabled Polansky et al. to justify the reasonableness of Clar's postulate[13] that certain of the six-membered rings in a polycyclic hydrocarbon have much stronger benzenelike character than the others. The following two examples illustrate how chemically interesting results can be obtained using these fragment characters.

Consider the zethrene molecular in Fig. 1. One may reasonably ask whether

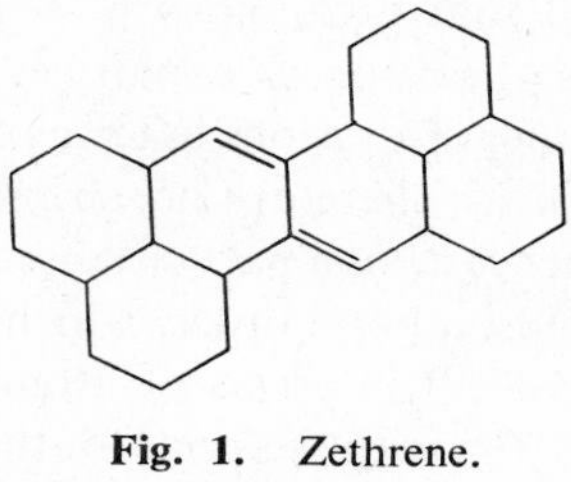

Fig. 1.   Zethrene.

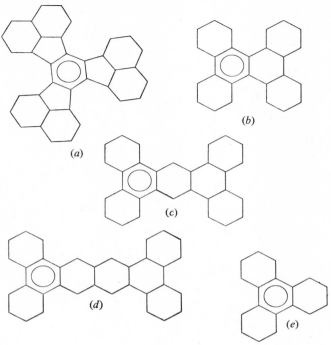

Fig. 2

or not it is reasonable to regard this molecule as essentially built up from two naphthalene units and a central dienoid region (marked as $= - =$ in Fig. 1). The π bond orders for the "double" bonds of the dienoid region are not unusually large, having a value 0.692,[14] but ρ_{dienoid} of (13) is equal to 0.767, considerably larger than its value in most similar structures, for example, tetracene, which is usually about 0.61. Thus the dienoid hypothesis is shown to be very plausible.

Another example of this type of analysis concerns the benzenoid character of a six-membered ring with three substituents. In particular, the question has been asked whether it is better to treat decacyclene (Fig. 2a) as three acenaphthylene units conjugated together, or as three naphthalene units surrounding a central benzene ring. Using the same kind of analysis as before, the benzenoid character of the central ring in decacyclene comes out to be 0.822, considerably larger than the values for the marked rings in tetrabenznaphthalene (0.749), tetrabenzanthracene (0.704), tetrabenztetracene (0.692), and triphenylene (0.714) shown in Fig. 2b–e.

At this stage we turn to the use of the two-particle density matrix. We shall find it convenient to introduce bonding and antibonding ethylenic orbitals for any chosen bond i–j. Let us call these φ_{ij} and φ_{ij^*} respectively. In the usual Hückel approximation,

$$\varphi_{ij} = [\varphi_i + \varphi_j]/2^{1/2},$$
$$\varphi_{ij^*} = [\varphi_i - \varphi_j]/2^{1/2}. \tag{14}$$

Any MO ψ_k can now be written

$$\psi_k = C_{ij}{}^k \varphi_{ij} + C_{ij^*}^k \varphi_{ij^*} + \text{terms from atoms other than } i \text{ and } j \tag{15}$$

where, in terms of the usual LCAO notation

$$C_{ij}{}^k = [C_{ki} + C_{kj}]/2^{1/2},$$
$$C_{ij^*}^k = [C_{ki} - C_{kj}]/2^{1/2}. \tag{16}$$

If the occupation member of ψ_k is n_k, the π bond order p_{ij} is given by

$$p_{ij} = \sum_k \tfrac{1}{2} n_k [(C_{ij}{}^k)^2 - (C_{ij^*}^k)^2]. \tag{17}$$

For the ground states of alternant hydrocarbons, for which the Coulson-Rushbrooke theorem tells us that $q_i = q_j = 1$, it follows that

$$\sum_k \tfrac{1}{2} n_k \{(C_{ij}{}^k)^2 + (C_{ij^*}^k)^2\} = 1. \tag{18}$$

We now imagine the two-particle matrix to be expanded in terms of φ_{ij}, φ_{ij^*} and other atomic or ethylenic orbitals: and we measure the coefficient with which the term $\varphi_{ij}(1)\bar{\varphi}_{ij}(2)\varphi_{ij}(1')\bar{\varphi}_{ij}(2')$ occurs. We choose this term because, for a pure ethylenic double bond, the coefficient of this term is unity. By seeking the coefficient of this term in the full molecular two-particle matrix, we are measuring some kind of double-bond character for the bond. For a closed-shell molecule it is not difficult to show that this coefficient is

$$\mathcal{C}_{ij} = \sum_k \sum_{k'} (C_{ij}{}^k)^2 (C_{ij}{}^{k'})^2. \tag{19}$$

For closed-shell alternant hydrocarbons it follows from (17) and (18) that

$$\mathcal{C}_{ij} = \left(\frac{1 + p_{ij}}{2}\right)^2. \tag{20}$$

For other types of molecule the formula for $\mathcal{C}_{ij}$ does not simplify so nicely.

We can similarly define the coefficient $\mathcal{C}_{ij^*}$ involving the antibonding orbitals. We find that

$$\mathcal{C}_{ij^*} = \left(\frac{1 - p_{ij}}{2}\right)^2. \tag{21}$$

Thus

$$p_{ij} = \mathcal{C}_{ij} - \mathcal{C}_{ij^*},$$

implying that the conventional bond order is a measure of the extent to which two electrons have bonding rather than antibonding character in this bond.

For large values of the bond order, $\mathcal{C}_{ij^*}$ is small, so that $\mathcal{C}_{ij}$ is nearly equal to p_{ij}, and the well-known correlation between bond length and p_{ij} will also apply to $\mathcal{C}_{ij}$. However, for smaller values of p_{ij} the length appears to correlate well with $\mathcal{C}_{ij}$, better in fact than with p_{ij}. Thus the point representing the central bond in butadiene for which $R = 1.464$ Å (Cole, Mohay, and Osborne[15]) lies on the straight line for $\mathcal{C}_{ij}$ as a function of R_{ij}, a situation not found for p_{ij}. It is interesting to note that in the theory which Coulson[1] proposed for a relation between bond order and bond length, viz.,

$$R = s - \frac{s - d}{1 + K(1 - p/p)}, \qquad (23)$$

where s and d are "natural" single- and double-bond lengths and K is a constant of magnitude about 0.765, the value of the quantity

$$\frac{1}{[1 + K(1 - p)/p]}$$

is very similar to the value of $\mathcal{C}$, provided that p is not too small. We may similarly expect that other quantities such as NMR coupling constants (e.g., Jonathan, Gordon, and Dailey, Ref. 16), vibrational stretching frequencies (e.g., Coulson and Longuet-Higgins, Ref. 17) whose observed correlation with bond order stems primarily from their dependence on bond length, should have similar correlations with $\mathcal{C}_{ij}$.

Before leaving this discussion of possible uses of the two-particle density matrix as a tool for studying bond strength, reference should be made to the Pauling superposition method in the theory of resonance among canonical structures. Pauling estimated bond order (which he called percentage double-bond character), by addition of the weights of all those structures in which a chosen bond appeared double. It is clear that this, effectively a process of integrating over all electrons except two in the many-electron density function, is closely related to our choice of the quantity $\mathcal{C}_{ij}$. The two quantities cannot be the same, however, because (if for no other reason) Pauling made no allowance in his addition of weights for cross terms in ψ^2 and in this way neglected overlap integrals between the various structures. Both Pauling's method and the spin-pairing theory of Penney,[18] however, are examples of the use of effective two-particle density matrices in the discussion of bond strength.

DIELS-ALDER REACTIVITY INDICES

Hückel MO theory has been successfully applied to a number of problems in chemical reactivity. One of these is the Diels-Alder diene reaction, where an attack, for example, by maleic anhydride, occurs across the 1–4 positions of an aromatic ring system. There are dynamic indices such as Brown's *para*-localization energy,[19] and there are static indices such as the sum of the two relevant free valences,[20] the 1–4 bond order[21] and dienoid character.[12] The good correlation between Brown's *para*-localization energy and experimental equilibrium constants has led to the use of the former rather than the latter, of which only scant data are yet available, as a test of the validity of any static index that may be proposed. The great advantage of a static index is that it can usually be obtained from existing tabulations of MO data without reference to the approaching reactant. The basic assumption in all theories of the Diels-Alder reaction (which has been shown to be valid in most cases) is that the attack on the diene occurs simultaneously at the two atoms involved, so that the energy required to localize two π electrons on these sites, with a view to the formation of two new σ bonds, would be expected to be a good measure of the activation energy for the reaction.

If we are considering the two-particle matrix the obvious thing to do is to expand in a basis consisting of at least the two atomic orbitals φ_A and φ_B on the two sites of attack, and look for the coefficient (C_{AB}) in the spinless density matrix of the form $\varphi_A(1)\ \varphi_B(2)\ \varphi_A(1')\ \varphi_B(2')$. The appropriate term in the one-particle matrix would then be $\varphi_A(1)\ \varphi_B(1')$, which is just the 1–4 bond order p_{AB}, studied by Epstein.[21] Just as in our earlier discussion of bond strength, we can show that

$$C_{AB} = q_A q_B - \tfrac{1}{2}p_{AB}^2. \tag{24}$$

For alternant hydrocarbons this becomes

$$C_{AB} = 1 - \tfrac{1}{2}p_{AB}^2. \tag{25}$$

Now it is known (see, e.g., Salem, Ref. 22) that p_{AB} measures the probability that if one π electron is on atom A with spin α, there is another π electron on atom B with spin β. If p_{AB} is large, this corresponds to a strong tendency for the two electrons, once localized on A and B, to pair together and form a bond, though admittedly it would be a long bond since it joins *para* positions in a ring. Thus we should expect that if p_{AB} were large it would be harder to disengage these electrons to permit the formation of the two new σ bonds, so that a large p_{AB} (i.e., small C_{AB}) should imply low reactivity. Epstein's work showed a fairly good correlation with p_{AB}, and so, from (25), with C_{AB}.

In Table I we show a variety of static indices for this reaction. In addition to Brown's localization energy L_{14}, these are (a) Epstein's p_{14}, (b) Sofer et al.'s $\bar{p}_{dienoid}$, (c) Brown's sum of the free valences $f_1 + f_4$, and (d) C_{14}. The agreement in all cases is good. A similar comparison for the polyenes led to the following values of the regression correlation coefficient for these four indices with Brown's *para* localization energy: (a) 0.972, (b) 0.984, (c) 0.987, and (d) 0.968. All four indices have a very high correlation, with $f_1 + f_4$ marginally better than the others. [It should be pointed out that in Brown's original paper (1951), due to an error in published bond orders, only recognized in a footnote added in proof, the conclusion was drawn that $f_1 + f_4$ was not so good as we now know it to be.] But because of its direct physical meaning C_{14} must have claims to be the best fairly reliable and easily calculated static

Table I.   Static Indices for Diels-Alder Reactivity—Polyacenes

Molecule	Positions	L_{14} (B)	$-P_{14}$	$\bar{p}_{dienoid}$	$f_1 + f_4$	C_{14}
Benzene	1–4	4.000	0.3333	0.820	0.7974	0.4444
Triphenylene	1–4	3.826	0.3340	0.787	0.8781	0.4442
3,4-Benzphenanthrene	1–4	3.826	0.3496	0.781	0.8953	0.4389
1,2-Benztetracene	1–4	3.796	0.3351	0.782	0.8907	0.4439
1,2-Benzanthracene	1–4	3.788	0.3379	0.782	0.8902	0.4429
Phenanthrene	1–4	3.765	0.3449	0.783	0.8906	0.4405
Pentaphene	1–4	3.762	0.3614	0.774	0.9141	0.4347
Chrysene	1–4	3.744	0.3495	0.782	0.8934	0.4389
3,4-Tetraphene	1–4	3.738	0.3507	0.782	0.8942	0.4385
3,4,6,7-Dibenzphenanthrene	1–4	3.733	0.3509	0.781	0.8962	0.4384
Naphthalene	1–4	3.683	0.3623	0.778	0.9056	0.4344
1,2-Benzanthracene	8–11	3.653	0.3639	0.773	0.9148	0.4338
3,4-Benztetraphene	8–11	3.647	0.3652	0.773	0.9157	0.4333
Anthracene	1–4	3.630	0.3687	0.772	0.9186	0.4320
1,2-Benztetracene	9–12	3.625	0.3690	0.770	0.9202	0.4319
Tetracene	1–4	3.618	0.3705	0.769	0.9214	0.4314
Pentacene	1–4	3.614	0.3710	0.769	0.9222	0.4312
3,4,6,7-Dibenzphenanthrene	10–13	3.560	0.3652	0.772	0.9080	0.4333
Pentaphene	5–14	3.450	0.3647	0.741	1.0148	0.4335
1,2-Benzanthracene	7–12	3.418	0.3767	0.743	1.0167	0.4290
3,4-Benztetraphene	7–12	3.391	0.3841	0.743	1.0214	0.4262
3,4,6,7-Dibenzphenanthrene	9–14	3.386	0.3839	0.741	1.0238	0.4263
1,2-Benztetracene	7–14	3.360	0.3836	0.736	1.0331	0.4264
Anthracene	9–10	3.313	0.4041	0.741	1.0397	0.4185
1,2-Benztetracene	8–13	3.278	0.4063	0.736	1.0524	0.4175
Tetracene	5–12	3.248	0.4144	0.739	1.0590	0.4141
Pentacene	5–14	3.231	0.4175	0.733	1.0636	0.4129
Pentacene	6–13	3.178	0.4262	0.729	1.0802	0.4092
Regression correlation coefficient			0.949	0.970	0.978	0.946

index. The situation with nonalternants is not so simple, since the atomic charges are not all unity, and the whole matter deserves more study than we have been able to give.

CONCLUSION

We have shown in this introductory study that the two-particle density matrix can be used as a basis for indices of bond order or strength, and of Diels-Alder reactivity. In the approximation used there was a close parallel between the indices thus obtained and those developed within the one-particle matrix. To some extent this may be a little fortuitous and depend on the fact that for single-determinant wavefunctions there is a simple formula relating the elements of the two matrices:

$$\Gamma(1, 2 \mid 1', 2') = \gamma(1 \mid 1')\gamma(2 \mid 2') - \gamma(1 \mid 2')\gamma(1' \mid 2) \qquad (26)$$

However, if we use more sophisticated wavefunctions such as those that result from the introduction of configuration interaction, (26) will no longer hold: but our definitions of C_{12} and C_{AB} will still hold well, so that the treatment is easily generalized. This point of view shows why, in Hückel theory, there is very little point in going beyond the single-particle density matrix. But it suggests that with more realistic wavefunctions the approach suggested in this paper may be worth pursuing further.

ACKNOWLEDGMENT

One of us (R. G. A. R. M.) wishes to thank the Council of the Australian National University for the award of a Travelling Postdoctoral Fellowship.

REFERENCES

1. C. A. Coulson, *Proc. Roy. Soc.* (*London*), **A169**, 413 (1939).
2. C. A. Coulson and H. C. Longuet-Higgins, *Proc. Roy. Soc.* (*London*), **A191**, 39 (1947); **192**, 16 (1947).
3. L. D. Landau, *Z. Physik*, **45**, 430 (1927).
4. J. von Neumann, *Mathematical Foundations of Quantum Mechanics*, Princeton University Press, Princeton, N.J., 1955.
5. P. A. M. Dirac, *Proc. Cambridge Phil. Soc.*, **25**, 62 (1929); **26**, 361, 376 (1930); **27**, 240 (1931).
6. K. Husimi, *Proc. Phys. Math. Soc. Japan*, **22**, 264 (1940).
7. R. McWeeny, *Proc. Roy. Soc.* (*London*), **A223**, 63 (1954); **A232**, 114 (1955); **A235**, 496 (1956).

8. P. O. Löwdin, *Phys. Rev.*, **97**, 1474 (1955).
9. K. Ruedenberg, *Rev. Mod. Phys.*, **34**, 326 (1962).
10. R. S. Mulliken, *J. Chem. Phys.*, **23**, 1833, 1841 (1955).
11. O. E. Polansky and G. Derflinger, *Intern. J. Quantum Chem.*, **1**, 379 (1967).
12. H. Sofer, O. E. Polansky, and G. Derflinger, *Monatsh.*, **99**, 1879 (1968).
13. E. Clar, *Polycyclic Hydrocarbons*, Academic Press, New York, 1964.
14. C. A. Coulson and C. M. Moser, *J. Chem. Soc.*, **1953**, 1341.
15. A. R. H. Cole, G. M. Mohay, and G. A. Osborne, *Spectrochim. Acta*, **23A**, 909 (1967).
16. N. Jonathan, S. Gordon, and B. P. Dailey, *J. Chem. Phys.*, **36**, 2443 (1962).
17. C. A. Coulson and H. C. Longuet-Higgins, *Proc. Roy. Soc. (London)*, **A193**, 456 (1948).
18. W. G. Penney, *Proc. Roy. Soc. (London)*, **A158**, 306 (1937).
19. R. D. Brown, *J. Chem. Soc.*, **1950**, 691.
20. R. D. Brown, *J. Chem. Soc.*, **1951**, 3129.
21. I. R. Epstein, *Trans. Faraday Soc.*, **63**, 2085 (1967).
22. L. Salem, *The Molecular Orbital Theory of Conjugated Systems*, Benjamin, New York, 1966.

T_n Frequency Functions as Energy Contours for Photon Absorbance in Condensed Systems[*]

JOHN R. MORREY AND LARRY G. MORGAN

Pacific Northwest Laboratories, Battelle Memorial Institute

Abstract

Spectroscopic absorption contours of condensed systems appear to be described by one of several frequency functions, i.e., Gaussian, Lorentzian (T_1), or T_n, where $1 < n \leq 3$. A phenomenological explanation is given in terms of a statistical model involving random perturbations on energy levels ($\langle \psi_m | H | \psi_n \rangle$, $\langle \psi_n | H | \psi_n \rangle$) and orthogonal coupling matrix elements ($\langle \psi_m | M_q | \psi_n \rangle$), where M_q is the light-coupling operator and n and m are the two states between which transitions occur. If the matrix elements are well shielded from perturbation or strong complexes exist such that only a few relaxations of the same order of duration can occur, Gaussian functions result. If, along with the energy levels, one of the orthogonal components is strongly perturbed by a statistically oriented environment, a Lorentzian function (T_1) results. If two such components are perturbed, the predicted function is T_2, and if all three components are significantly perturbed, then T_3 results.

[*] Reprinted from *Journal of Chemical Physics*, Vol. 51, 1969, p. 67, by permission of the copyright owner.

317

A Personal Note

JOHN R. MORREY

I fondly remember a number of humorous experiences with Henry. One of the most vivid is an incident in a statistical mechanics class. He was busy deriving a partition function which had in the denominator $N!$. I questioned the reasoning behind incorporation of $N!$, whereupon Henry patiently went through the derivation again. I still was not satisfied and responded with "intuitively, it does not make sense." Henry, with all the patience he could muster, turned to me and said: "John, $N!$ needs to be there and I'm going to tell you five more times. That will constitute a proof."

When I came into Henry's lecture room for my orals, only Henry was there. The other members of the committee had not yet arrived. He sensed my anxiety and in an attempt to relax me asked if I had ever seen him jump to the table from a standing position. I had never seen him do this so he made a mighty jump which didn't suffice. He cracked both shins on the edge of the table. For a few moments I thought the oral would have to be cancelled, but with pain and determination he backed off and tried it again, this time succeeding.

INTRODUCTION

Spectral line shapes have been of interest for over eighty years, gases in particular being the object of careful research. Two reviews by Breene[1] and Chen and Takeo[2] cover the development of the theory to 1957.

The shape of a transitional band can be attributed to three effects: (a) natural band widths caused by inherent uncertainties of energy levels, (b) Doppler broadening caused by translational motion, and (c) perturbational broadening caused by surrounding molecules or atoms. In principal, perturbational broadening can be classed with natural broadening, as will be seen below. (See Table I for Summary of equations.)

The natural line width was first derived from a classical radiation damping model* and was later derived on a quantum mechanical basis by Weisskopf and Wigner[3] and Hoyt.[4] It is due to uncertainty in the energy of the excited level having a finite lifetime. The natural line width is much smaller than will be of concern in this paper, of the order of 0.01 cm^{-1}.

Doppler broadening results from frequency shift due to translational motion and has no relationship to molecular interaction.[5] It also is negligible in comparison to widths treated in this paper.

The perturbational effects have been variously described as interruption broadening, resonance broadening, and statistical broadening. The perturbational line shape derived by Michelson[6] was not correct even for pure interruption broadening because he neglected to average over all times between collisions. To do so results in the simplified Lorentz model.[7]

In the more extended derivation Lorentz assumed that collisions were strong enough to destroy any memory of previous orientation and that the mean distribution of oscillators just after collision was completely random.

By assuming that an optical electron of momentum p natural frequency ω_0, and restoring force $k = m\omega_0^2 x$ is acted upon by a force due to the electric field of an electromagnetic wave of frequency ω, Van Vleck and Weisskopf[8]

---

* This model, which also can be applied to resonance radiation, requires several important approximations which render the form inappropriate for condensed systems (See Appendix I.)

Table I. Summary of Special Line Shapes Predicted by Various Theories Formulated from Motions and Interactions of Gases[a]

Model	Approximations	Absorption Line Shape $A(\sigma)$	Half-width (
Radiation damping or quantum mechanical uncertainty (natural line width)	No collisions or interactions	$\dfrac{2/(2.303 \cdot c \cdot \pi \cdot w)}{1 + 4[(\sigma_0 - \sigma)/w]^2}$	$\dfrac{4\pi e^2}{Mc^2} \sum_k \sigma_{ik}^2$
Doppler broadening	Classical translational motion—no interaction	$\dfrac{2l}{2.303cw}\left(\dfrac{\ln 2}{\pi}\right)^{1/2} \exp\left\{-4\ln 2\left(\dfrac{\sigma - \sigma_0}{w}\right)^2\right\}$	$2\sigma_0\sqrt{\dfrac{2RT \ln}{Mc^2}}$
Collisional interruption	No molecular interaction other than collision	$\text{const. } 4\pi^2\tau^2\left\{\left[1 + 4\left(\dfrac{\sigma - \sigma_0}{w}\right)^2\right]^{-1}\right.$ $\left. - \left[1 + 4\left(\dfrac{\sigma + \sigma_0}{w}\right)^2\right]^{-1}\right\}$ $\sim \dfrac{4\pi^2\tau^2}{1 + 4[(\sigma - \sigma_0)/w]^2}$	$\dfrac{1}{\tau c\tau}$ $\tau = \dfrac{1}{\pi\rho^2\sigma l}$
Semiclassical impact broadening quantum mechanical impact broadening	No intermolecular interactions except collision, thus interrupting natural frequency of optical electron	$\dfrac{2lN_1e^2}{2.303Mc^2w}\left(\dfrac{\sigma}{\sigma_0}\right)^2\left\{\left[1 + 4\left(\dfrac{\sigma_0 - \sigma}{w}\right)^2\right]^{-1}\right.$ $\left. + \left[1 + 4\left(\dfrac{\sigma_0 + \sigma}{w}\right)^2\right]^{-1}\right\}$	$\dfrac{1}{\pi c\tau}$
Resonance broadening a special case of radiation damping	Same as radiation damping	$\dfrac{8\pi^2N_1e^2l}{2.303c^2\eta_0M_e\sigma_0}\sum_i \dfrac{f_i}{\gamma_i\{1 + 4[(\sigma_{0i} - \sigma)/w]^2\}}$	$\dfrac{4\pi\gamma}{c}$
Stark broadening by electric fields of adjacent molecules	One kind of perturber only which occupied no volume. Intensities of Stark components constant within manifold	$\dfrac{4l\,\Delta\sigma_m}{\pi w}\left[1 + 4\left(\dfrac{\sigma_0 - \sigma}{w}\right)^2\right]^{-1}$	$\dfrac{9.08N\mu\,\Delta\sigma_m}{F}$
Time-dependent perturbations by a neighboring atom or molecule	Perturbation affects only the energy of the excited state of the absorber but not the charge distribution, i.e. matrix element between ground and excited states.	$\left[\displaystyle\int_{-\infty}^{\infty} \exp\dfrac{2\pi i}{h}\left\{\left[\displaystyle\int_0^t \Delta E(\tau)\,d\tau - \Delta E^0 t\right]\right\}dt\right]$	not explici defined
Statistical	van der Waal perturbation of upper level. Bimolecular collisions only	$\dfrac{l}{2.303(\sigma_0 - \sigma)^{3/2}c^{3/2}}\exp\left(\dfrac{-\pi\phi^2}{(\sigma - \sigma_0)c}\right)$ $\phi = \tfrac{2}{3}\pi\alpha^{1/2}\eta$	$\dfrac{1.85\pi\phi^2}{c}$

[a] Definitions

$I^0(\sigma)$ = intensity of incident radiation σ in ergs/cm-sec.

$I(\sigma)$ = intensity of exciting radiation σ in ergs/cm-sec.

$A(\sigma) = \log I^0(\sigma)/I(\sigma)$

$A^0 = A(\sigma_0)$

σ_0 = resonance energy in cm^{-1}

derived a slightly different form than Lorentz. However, all interruption broadening except that associated with microwave transitions still takes the form of a simplified Lorentzian distribution.

Karplus and Schwinger[9] have derived quantum mechanically the Van Vleck-Weisskopf formula, having extended it to nonthermal equilibrium conditions brought on by higher power levels of exciting electromagnetic energy.

For gases at normal pressures and temperatures (STP) impact broadening is also negligible when compared to broadening in condensed systems. However, if collisions could be approximated by crystal lattice vibrations in condensed systems, half-widths according to interruption broadening could range to 1500 cm^{-1}.

Resonance broadening arises from the radiationless transfer of energy from one atom or molecule to a like atom or molecule, thus diminishing the lifetime of the excited molecule or atom. The band width is thereby increased. The simplest classical formulation of this effect is made by replacing each molecule or atom by an oscillator of frequency ω_0 (see Appendix I), then introducing a coupling force between all pairs of oscillators. The quantum mechanical formulation was given by King and Van Vleck.[10]

The first attempt at a statistical model was made by Holtsmark,[11] who attributed broadening to Stark splitting of degenerate levels due to electric fields imposed by adjacent atoms. He made several simplifying assumptions:

1. Only one kind of perturber was present, that is, ions, dipoles, or quadrupoles, but not all three.

2. The intensities of the Stark components within the manifold were constant.

3. Atoms were considered as points in the integration over all space to obtain the frequency function for the field strength.

---

w	= width of band which encompasses the region where $A(\sigma) \geq A^0/2$
c	= speed of light in cm/sec
R	= gas constant
M	= molecular weight of absorbing-translating entity
T	= temperature in degrees absolute
σ_{2k}	= wave number corresponding to the energy of transition between states 2 and k.
f_{2k}	= oscillator strength for the transition $2 \to k$, i.e., probability that transition $2 \to k$ will take place given that the probability of all transitions is unity.
τ	= time between collisions
ρ	= average distance between centers of colliding atoms
n	= number of atoms per unit volume
$\bar{v}$	= root mean square velocity
l	= thickness of sample
σ_m	= width of Stark manifold
μ	= dipole moment of broadening molecule
F	= electric field produced by the broadener
α	= van der Waals attraction constant in cm^6 sec^{-1}
N_1	= population of lower state
η	= No. of foreign atoms or molecules in system

No attempt was made to factor in the effect of a field on the intensity probability of Stark components. Intensity probability is an important consideration that is treated in this paper.

After comparing the Stark broadening theory with experiment, Holtsmark concluded that Stark broadening was not the only significant contribution to line shapes. The very fact that nondegenerate levels still give rise to finite band widths, i.e., larger than natural or Doppler broadening, indicates the necessity of a theory of broader base.

Following Holtsmark, Margenau and Watson[12] formulated the time-dependent quantum mechanical theory which, in principle, includes all perturbational models. However, it still neglects perturbation of the lower of the states involved in the transition. More importantly, it neglects the shift of electronic distribution between the two states and assumes that simultaneous interaction of more than two particles is additive, a condition that will not apply to condensed systems.

Margenau has discussed the fact that the true distribution function contains two parts: (a) the statistical distribution giving rise to modified frequencies corresponding to perturbation values which the configuration of atoms has produced, and (b) additional frequencies which result when kinetic energy is lost or gained during an optical process. The formulation developed implicitly includes both effects.

The later statistical formulation by Margenau[13] reflects the assumption that van der Waals forces perturb the upper transitional energy state more than the lower state, that the perturbation is generally toward more stability, that is, lower energies, and that the approach distance between colliding molecules is not small enough to cause repulsion. This theory applies only to bimolecular collisions in the sense that the only interaction involved is that between the absorber or emitter and the perturbing molecule. It is important to recognize in this treatment that the absorbance (emission) at some frequency is proportional only to the Hilbert volume associated with this perturbational frequency. It does not take into account the fact that perturbation of one or both levels also causes a change in the matrix element $\langle \psi_n | M_q | \psi_m \rangle$ between the states n and m. (M_q is the operator which couples the electromagnetic radiation to the molecule.)

Recent work, notably by Kubo,[14] Huber and Van Vleck,[15] Fano and Cooper,[16] and Gordon[17] have formulated transition probability theory in the Heisenberg quantum mechanical representation wherein an observable A_h obeys the equation

$$ih \frac{dA_h}{dt} = [A_h, H_h] + i \frac{\partial A_h(t)}{\partial t},$$

where H_h is the Heisenberg Hamiltonian related to the Schrödinger Hamiltonian H_s by

$$H_h = \exp\left[\frac{iH_s(t - t_0)}{h}\right] H_s \exp\left[-\frac{iH_s(t - t_0)}{h}\right].$$

This formulation is particularly useful in examining relaxation processes which relate to band shapes in condensed systems.

Jørgensen[18] suggests that line shapes of electronic transitions should be Gaussian or skewed Gaussian. His reasoning is based on the shape of the eigenfunction of the ground vibrational states and the assumption that excited electronic potential surfaces might be approximately linear in the region of concern as predicted by the Franck-Condon principle.

In an attempt to resolve spectra of condensed systems into summations of frequency functions, several investigators[19–22] have reported that most often the frequency functions are neither Gaussian nor Lorentzian. Usually the Gaussian function does not provide absorbance in the wings of the peak, whereas the Lorentzian function often provides too much. They generally serve as opposite limits of the true function. Accordingly, linear combinations of Lorentzian and Gaussian frequency functions have been used to obtain empirical fits.

EXPERIMENTAL

After carefully examining the absorption spectra of a number of condensed systems, we have found that they can be satisfactorily resolved into Gaussian (g) or Student $T_n(t)$ line shapes as expressed in (1) and (2);

$$A_i^g(\sigma, \sigma_i^0, w_i, A_i^0) = A_i^0\left(\frac{\sigma}{\sigma_i^0}\right)\exp\left\{-\ln 2\left[\frac{(\sigma - \sigma_i^0)}{w_i}\right]^2\right\} \tag{1}$$

or

$$A_i^t(\sigma, \sigma_i^0, w_i, A_i^0) = \frac{A_i^0(\sigma/\sigma_i^0)}{\{1 + \alpha(n)[(\sigma - \sigma_i^0)/w]^2\}^{(n+1)/2}}, \tag{2}$$

where

$$A_T(\sigma) = \sum A_i^g \quad \text{or} \quad \sum A_i^t. \tag{3}$$

The symbols σ, w, and A^0 represent, respectively, the frequency, the width at half height, and the absorbance at the resonance frequency σ^0 (in wave numbers).

For the most part spectra were obtained on Cary 14, Cary 14H, or Perkin-Elmer 521 spectrophotometers. Spectra were recorded on tape for analysis with a digital computer.[23] Unless indicated in the text, the instrument resolution was good enough to provide true band shapes for the condensed systems reported in this paper.

The most difficult experimental problem was associated with the spectra of imperfect solid samples which scattered some light. Whenever possible, apparent absorption caused by scattered light was subtracted from the spectrum mathematically by use of Lagrangian interpolation between the regions in which the apparent absorption was due entirely to scattered light. It has been our experience that if the baselines are incorrect spectral convergence is difficult and the poor conditioning, evident.

Spectra of liquid solutions were corrected for solvent absorption by dual-beam measurements with carefully matched optical grade quartz cells.

Methods of spectral resolution have just recently been described by Fraser and Suzuki.[24] In our work two computer programs were used for the resolution of spectra. One was developed by Biggers and associates.[25] The other was developed by Duane[26] and modified by Kottwitz.[27] Both were modified further for our use. Although the program developed by Duane and Kottwitz is mathematically more rigorous, having in it the calculation of second partial derivatives for detection of saddle points and regression analysis, it is significantly slower than the program developed by Biggers, which employs only first partial derivatives. By using a DuPont 310 curve resolver to obtain approximate solutions, we can obtain rapid convergence with the modified Bigger's program, the mathematics of which are outlined as follows:

$$V = \sum_{i=1}^{n} [A_i(\sigma, p_j) - Y_i]^2, \qquad (4)$$

where V is to be minimized with respect to the parameters p_j. Using a Taylor's expansion and neglecting second and higher order terms, we can write

$$A_T(\sigma, p_j) = \sum_i \left[A_i(\sigma, p_j') + \sum_{j=1}^{m} \left(\frac{\partial A_i}{\partial p_j}\right)_{p_j'} (p_j - p_j') \right]. \qquad (5)$$

By substituting (5) into (4) and using the condition that $\partial V/\partial p_j = 0$ for the best fit, we obtain

$$C_{ik} = \sum_{i=1}^{n} \left(\frac{\partial A_i}{\partial p_j}\right)_{p_j} \left(\frac{\partial A_i}{\partial p_k}\right)_{p_k'}, \qquad (6)$$

$$d_k = \sum_{i=1}^{n} (A_i - Y_i) \left(\frac{\partial A_i}{\partial p_k}\right)_{p_k'}, \qquad (7)$$

$$(C)(P) = d, \qquad (8)$$

and

$$P = dC^{-1}, \qquad (9)$$

where C represents the square m-dimensional matrix of the coefficients C_{ik}, d the column matrix for the d_k's, and p the column matrix for the Δp_j's, where $\Delta p_j = p_j - p'_j$. Successive approximations are thus used to obtain the converged solution.

The question of uniqueness of fit is both important and troublesome. The contours of the m-dimensional hyperspace are complicated by experimental noise to the extent that multiple solutions as defined by $\partial V/\partial p_j = 0$ and $\partial^2 V/\partial p_j\,\partial p_k > 0$ will sometimes arise; it is difficult to determine whether the solution with smallest V actually has been obtained. Nevertheless, it has been experimentally verified in a number of complex cases (up to 10 strongly overlapping peaks) that if the data are sufficiently precise and the proper number of parameters is used, the solution is unique, that is, independent of the initial parametric estimates. This study was made on data generated analytically by the digital computer from overlapping distributions, the number and values of the parameters being accurately known. With typical experimental data this will not be the case.

Certainly as important as the precision of the data is the number of parameters and the type of function used in the fit. If they are not correct, multiple solutions will result.

Neptunium(V) in Perchloric Acid

The electronic band at 10,202 cm^{-1} of Np(V) in HClO$_4$ has been carefully studied because it is relatively unencumbered by overlapping transitions. It is undoubtedly a hypersensitive transition[28-31] having an ϵ_{max} of about 400 liters/mole-cm. Unfortunately, since hypersensitive transitions are not yet well understood, the assignment(s) of the band cannot be made. Our study suggests that the band may be a composite of two strong transitions as shown on the right of Fig. 1. Table II summarizes the analysis of this band. Examination will show that the two-band T_3 model fits to within experimental error. If the order n (T_n) is allowed to vary during the least-squares fitting, a one-band model produces an n of 3.13 ± 0.12 standard deviation. When two bands are used, n is 5.23 ± 0.19 for one peak and 1.96 ± 0.02 for the other. However, the fit is not significantly different from that of the T_3 model for two bands. After considerable investigation of this point we have concluded that unless data are extremely precise on the tails the statistical determination of n may be misleading. As shown here, the n for one peak decreases from 3 and the other increases. This is probably because of a very weak band on the low-energy side of the major bands.

Whether one adheres to either a one-band model or a two-band model, he can confidently conclude that T_3 fits the data much better than either T_1 (Lorentzian) or Gaussian.

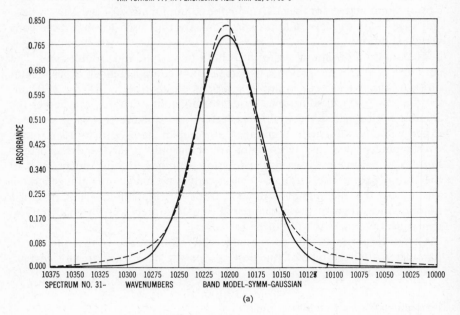

(a)

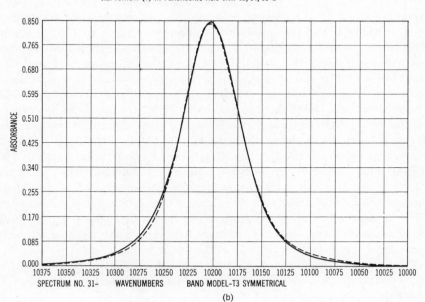

(b)

Fig. 1 Data: dotted line; fit: solid line.

NEPTUNIUM (V) IN PERCHLORIC ACID JRM 12/04/68-1

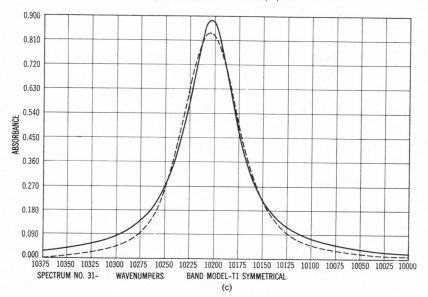

SPECTRUM NO. 31–   WAVENUMBERS   BAND MODEL–T1 SYMMETRICAL
(c)

PROGRAM TEST CASE ON NP (V), ALSO TWO PEAK TEST JRM 11/11/68-6

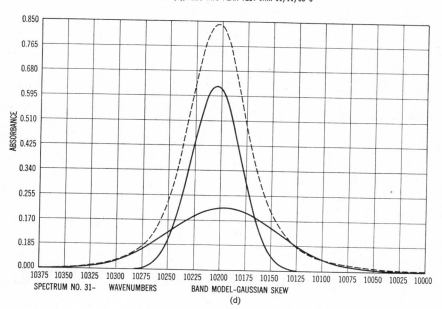

SPECTRUM NO. 31–   WAVENUMBERS   BAND MODEL–GAUSSIAN SKEW
(d)

Fig. 1.   (Continued).

PROGRAM TEST CASE ON NP (V), ALSO TWO-PEAK TEST JRM 11/11/68-6

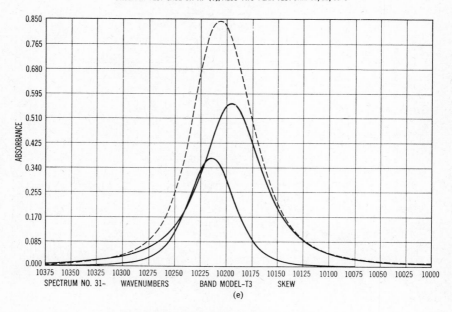

(e)

NEPTUNIUM (V) IN PERCHLORIC ACID JRM 12/04/68-5

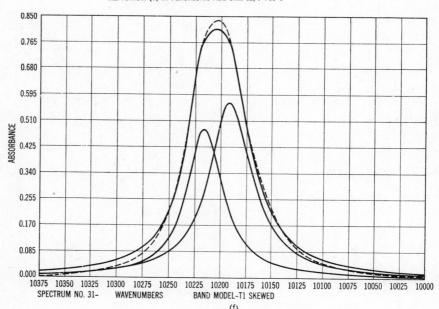

(f)

Fig. 1. (Continued).

Table II. Summary of Band Model Resolutions of the Np(V) Hypersensitive Band at 10203 cm^{-1}

Band Model	No. of Peaks	σ^0	w	A^0	n	Standard Deviation ($\times 10^3$)	Reference No.
Symmetrical G	1	10203	74.6	0.807	~24	26.71	JRM 12/04/68-3
Symmetrical T_3	1	10203	67.0	0.850	3	9.19	JRM 12/04/68-2
Symmetrical T_n	1	10203	67.3	0.849	3.14 ± 0.12	9.18	JRM 10/24/68-1
Symmetrical T_1	1	10203	59.7	0.889	1	29.56	JRM 12/04/68-1
Skewed T_3	1	10203	67.3	0.849	3	9.39	JRM 10/23/68-1
Skewed T_n	1	10203	67.3	0.849	3.13 ± 0.12	9.40	JRM 12/04/68-4
Symmetrical G	2	10196	136.3	0.188	24	4.70	JRM 11/11/68-2
		10204	59.6	0.654			
Symmetrical T_3	2	10199	68.7	0.667	3	4.12	JRM 11/11/68-5
		10216	48.1	0.225			JRM 11/26/68-1
Symmetrical T_n	2	10195	67.0	0.554	1.96 ± 0.02	1.35	JRM 11/09/68-2
		10215	51.9	0.369	5.23 ± 0.20		JRM 11/26/68-2
Symmetrical T_1	2	10191	50.8	0.561	1	15.27	JRM 11/11/68-3
		10217	43.3	0.499			JRM 12/04/68-6
Skewed G	2	10196	133.2	0.196	~24	4.74	JRM 11/11/68-1
		10204	59.2	0.646			
Skewed T_3	2	10194	67.2	0.556	3	1.39	JRM 11/11/68-6
		10215	51.8	0.367			
Skewed T_n	2	10195	67.0	0.554	1.98 ± 0.02	1.38	JRM 11/09/68-1
		10215	52.0	0.369	5.23 ± 0.19		JRM 11/26/68-2
Skewed T_1	2	10191	51.1	0.574	1	15.36	JRM 11/11/68-4
		10217	42.8	0.487			JRM 12/04/68-5

Atomic Mercury in Organic Media

Spectra of atomic mercury dissolved in cyclohexane, 1,2-dichloroethane, and absolute methanol have been examined as part of a search for simple condensed systems. The spectrum of the solvent was carefully subtracted from that of the mercury after the 10-cm paired cells were precisely balanced with solvent in both cells. Spectra of T_3-resolved distributions, Fig. 2, reveals that the spectra are much more complicated than *a priori* expectation.

These intriguing spectra cannot yet be explained completely. The bands between 38,000 and 39,000 cm^{-1} are derived from the $^3P_1$ level and have been presumably due to monatomic mercury. The splitting has been attributed to electric field effects and to an agglomeration effect where the $^3P_1$ level of the surface atoms loses its degeneracy. Since this splitting does not correlate with the dielectric of the solvent, Gunning and Vinogradov[32] prefer the agglomeration model. However, this makes it difficult to rationalize the existence of the rest of the spectrum. Furthermore, we have observed splitting

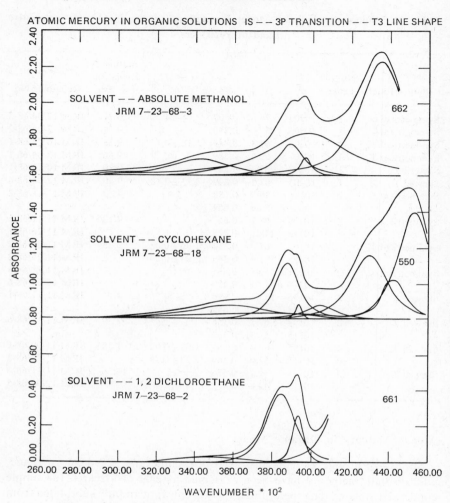

ATOMIC MERCURY IN ORGANIC SOLUTIONS   IS – – 3P TRANSITION – – T3 LINE SHAPE

SOLVENT – – ABSOLUTE METHANOL
JRM 7–23–68–3   662

SOLVENT – – CYCLOHEXANE
JRM 7–23–68–18   550

SOLVENT – – 1, 2 DICHLOROETHANE
JRM 7–23–68–2   661

WAVENUMBER * 10^2

Fig. 2. $^1S_0 \rightarrow {}^3P_1$ transition of atomic mercury in organic solvents.

of absorption spectra of mercury in a saturated vapor phase. To explain this we propose that Hg_2 molecules form weak bonds in the vapor phase (~4 Kcal) and that the induced Stark effect causes splitting. We also suggest that splitting in organics is caused by the same effect. The other wider bands may be due to trimers or tetramers.

It is not possible to unambiguously conclude the shapes of mercury bands in the organics we have studied because there is no guarantee that the true resolution of the spectral contour involves the least number of bands. It is possible to fit the contours to within experimental accuracy (0.002 absorbance units) with fewer T_3 transitions than Gaussian, Lorentzian, or T_4 transitions and on this basis we conclude that T_3 transitions are appropriate.

Table III.   Spectral Splittings of Hg $^3p_1$ in Various Media

	Std. Gas	Cyclohexane	1,2-Dichloroethane	MeOH
$\bar{\nu}_1$	39,434	39,356	39,357	39,673
$\bar{\nu}_2$	39,344	38,791	38,455	38,879
$\Delta\bar{\nu}$	90	565	902	794

Figure 3 illustrates our reasoning. This Gaussian resolution appears artificial since several bands are required for the fit which are not visible on the contour. By contrast, every required band for a T_3 resolution (Fig. 2, middle) is readily visible on the contour. Table IV further illustrates this point for the other solvents.

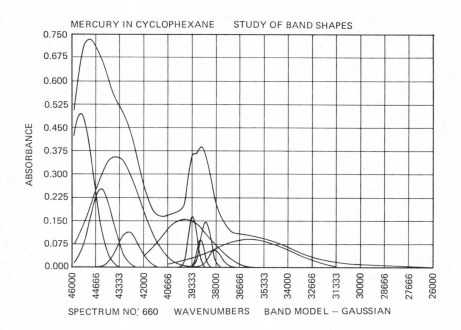

Fig. 3.   Example of the complexity needed to fit the Hg spectrum to Gaussian frequency functions. Compare with the more suitable result in Fig. 2 (*center*).

Table IV. Comparison of Band Shapes in Resolving Hg Spectra

		Cyclohexane	1,2-Dichloroethane	Absolute Methanol
Gaussian	Ref. No	JRM 7-20-68-12	JRM 7-20-68-14	No solution
	No. of parameters[a]	33	21	with 27
	Standard deviation	1.84×10^{-3}	1.61×10^{-3}	Parameters
Lorentzian	Ref. No.	JRM 7-20-68 2	JRM 7-20-68-4	JRM 7-20-68-7
	No. of parameters[a]	24	18	21
	Standard deviation	5.62×10^{-3}	3.20×10^{-3}	3.13×10^{-3}
T_3	Ref. NO.	JRM 7-20-68-18	JRM 7-23-68-2	JRM 7-23-68-3
	No. of parameters[a]	24	15	21
	Standard deviation	1.68×10^{-3}	1.53×10^{-3}	1.99×10^{-3}
T_4	Ref. No.	JRM 7-20-68-3	JRM 7-20-68-5	JRM 7-20-68-9
	No. of parameters[a]	24	18	21
	Standard deviation	3.42×10^{-3}	1.73×10^{-3}	2.00×10^{-3}

[a] The number of parameters necessary to fit contour to accuracy merited by data.

Molten Tetraphenylarsonium Hexachlorouranate (V)

Figure 4 shows the $\Gamma_7 \to \Gamma_7'$ electronic–vibronic transition of fused $\phi_4 AsUCl_6$. The resolution is unequivocally Gaussian. At lower temperatures in the solid state the distribution becomes Lorentzian.

Parenthetically, Fig. 4 illustrates what can be done by analyzing spectra once band shapes are understood. In octahedral symmetry there are six vibrational modes for the UCl_6^- entity. Some of these are infrared active; the others are Raman active. In this analysis we not only obtain all frequencies of the ground electronic state Γ_7 but also five of the six frequencies for the excited electronic state Γ_7' as shown in Table V.

Table V. Vibronic Transitions of $\phi_4 AsUC_6$

	Excited State Ferquencies (cm^{-1})	Ground State Frequencies (cm^{-1})
1	307.1	309.8
2	—	233.0
3	155.2	154.7
4	123.1	119.0
5	91.5	82.0
6	45.4	40.9

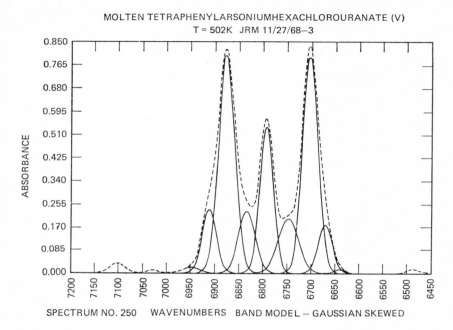

MOLTEN TETRAPHENYLARSONIUMHEXACHLOROURANATE (V)
T = 502K   JRM 11/27/68—3

SPECTRUM NO. 250   WAVENUMBERS   BAND MODEL – GAUSSIAN SKEWED

Fig. 4. Gaussian resolution of $\Gamma_7 \rightarrow \Gamma_7'$ transition, including vibronic levels.

Hexamminecobalti Chloride in Aqueous Solutions

Figure 5 illustrates another spectrum of Gaussian distribution. The transitions $t_2^6\,{}^1A_{1g} \rightarrow t_2^5e\,{}^1T_{1g}$ and $t_2^6\,{}^1A_{1g} \rightarrow t_2^5e\,{}^1T_{2g}$ have been assigned to these bands.[33–35] If the octahedral complex were distorted to D_{4h} or C_{4v} the ${}^1T_{1g}$ and ${}^1T_{2g}$ states would split into four states, that is, ${}^1A_{2g}(T_{1g})$, ${}^1E_g(T_{1g})$, ${}^1E_g(T_{2g})$, and ${}^1B_{2g}(T_{2g})$.[36] Figure 5 includes a four-band best fit of the data. If two very weak bands are included at 16,478 and 25,189 cm^{-1} the fit is within experimental error. The difficulty is that the ligand field theory does not predict distortion. Since the ground state is expected to be ${}^1A_{1g}$, neither dynamic nor static Jahn-Teller distortion is predicted. This point is being further investigated. Regardless of the outcome, however, the Gaussian function fits the data much better with either the two- or four-band model than any of the T_n distributions.

One would hope to determine whether skewing is important with spectra like those in Fig. 5, since they are broad enough to reflect skewing if it exists. Table VI summarizes these results. If the two-band model is used, it is quite

Table VI. Parametric Values and Fits of Skewed and Symmetric Gaussian Frequency Functions for d–d Transitions of $Co(NH_3)_6^{3+}$ Ions

Band Model	No. of Peaks	δ^0	ω	A^0	Standard Deviation	Reference No.
Skewed Gaussian	3	29,486	4137	44.58	1.956	JRM 11/30/68-1
		21,146	3544	54.85		
		16,503	1195	0.28		
Symmetrical Gaussian	5	30,323	4598	22.09	0.234	JRM 12/02/68-4
		29,113	3229	26.30		
		22,394	3692	22.80		
		20,778	2738	42.00		
		16,503	1195	0.28		
Skewed Gaussian	5	30,224	4579	21.79	0.243	JRM 12/02/68-5
		29,042	3244	26.68		
		22,304	3713	22.33		
		20,721	2754	52.29		
		16,478	1116	0.28		

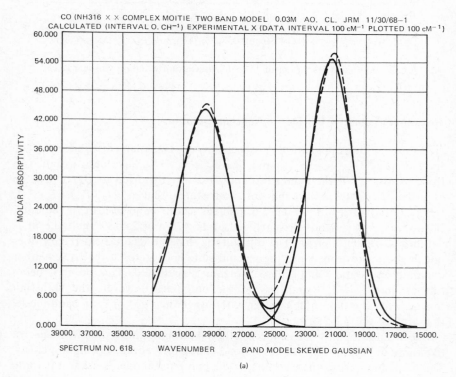

Fig. 5

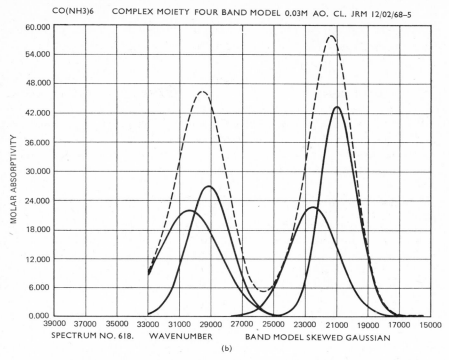

SPECTRUM NO. 618.   WAVENUMBER   BAND MODEL SKEWED GAUSSIAN

(b)

CO(NH3)6^{+++} COMPLEX MOIETY FOUR BAND MODEL 0.03M AU CL JRM 12/02/68—5

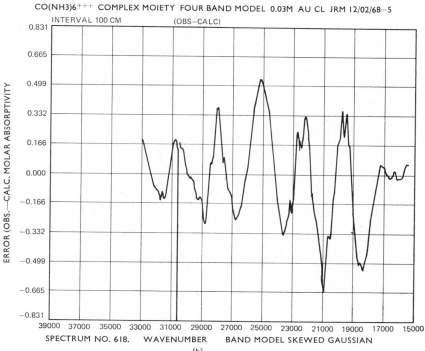

SPECTRUM NO. 618.   WAVENUMBER   BAND MODEL SKEWED GAUSSIAN

(b)

Fig. 5.   (Continued).

clear that skewed bands fit better than symmetrical bands. The situation is less clear with the four-band model because the high energy skewing of the low energy band tends to affect the increased attenuation of the low energy side of the high energy band. Thus a valid test of the skewed distribution has not yet been made although no data contradict the skewed model predicted below.

The error plot associated with this analysis is worth noting. The periodic error over two bands suggests very weak superimposed frequencies of $2750 \pm 80 \text{ cm}^{-1}$. The OH stretching frequency in water is 2750 cm^{-1}, suggesting the possibility that the periodicity is due to very weak vibronic coupling of the cobalt complex with the OH vibrator.

Crystalline UCl₄ at Low Temperatures

Figure 6 summarizes the resolution of the $^3H_4 \rightarrow {}^3F_2$ or $^3H_5$ electronic transition. At all temperatures studied (nominally 9.2, 22.5, and 77.5°K) the resolution is clearly Lorentzian (T_1). The fit in this case is not accurate at the peaks because the resolution of the instrument in not quite sufficient. No other model will fit without the inclusion of numerous small peaks to fill in the tails.

Results as shown in Fig. 6 suggest the possibility of determining the

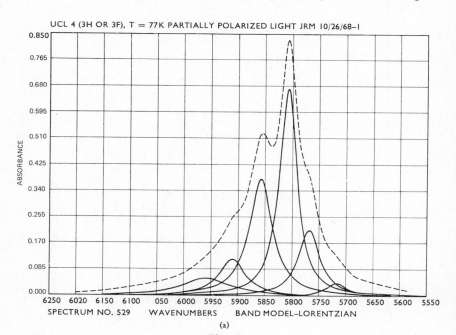

(a)

Fig. 6

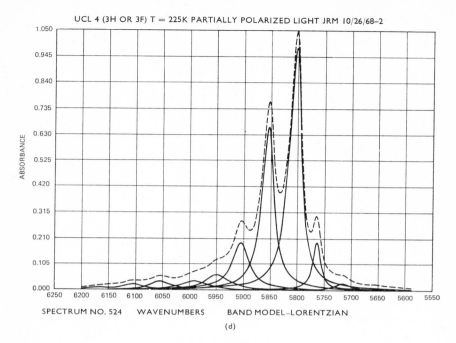

UCL 4 (3H OR 3F) T = 225K PARTIALLY POLARIZED LIGHT JRM 10/26/68–2

SPECTRUM NO. 524   WAVENUMBERS   BAND MODEL–LORENTZIAN

(d)

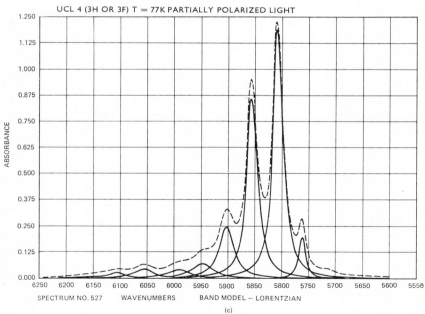

UCL 4 (3H OR 3F) T = 77K PARTIALLY POLARIZED LIGHT

SPECTRUM NO. 527   WAVENUMBERS   BAND MODEL – LORENTZIAN

(c)

Fig. 6.   (Continued).

temperature dependence on A_0, w, σ_0, and oscillator strengths. Data for this system are still too sparse for a meaningful study since cross correlations must also be considered.

Isomorphically Substituted Cr^{3+} in Al_2O_3

The absorption spectrum of Cr^{3+} in Al_2O_3 has been extensively studied by Sugano and Tanabe.[37] We have examined with polarized light the R_1 and R_2 transitions of a single Al_2O_3 crystal isomorphically substituted with 15% Cr^{3+}. The spectrum is shown in Fig. 7. This system was chosen because its optical anisotropy was of interest to us. From the crystal structure[38] it is determined that the Cr^{3+} ion is surrounded almost octahedrally by six nearest

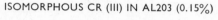

ISOMORPHOUS CR (III) IN AL203 (0.15%)

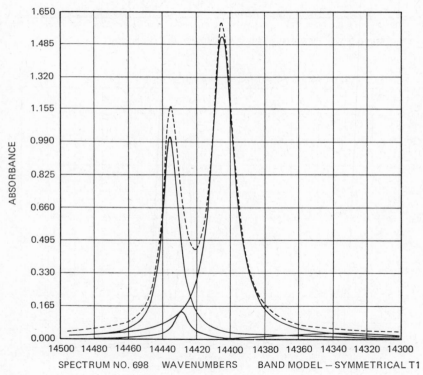

Fig. 7. $^4A_{2g} \rightarrow {}^2E_g$ transitions of Cr^{3+} isomorphously substituted into a single crystal of Al_2O_3. Calculated: solid line; experimental: dotted line.

O^{2-} ions and is, therefore, subjected to a strong cubic field. In addition, there is a relatively strong trigonal field whose optic axis is C_3. The structure at about 14,400 cm^{-1} has been attributed to the $t_2{}^3\,{}^4A_{2g} \to t_{2g}{}^3\,{}^2E_g$ transition where ${}^2E_g$ is split into two Kramer's doublets $2\bar{A}({}^2E)$ and $\bar{E}({}^2E)$ by the trigonal field. With some reservations the assignments

$$^4A_2 \to \bar{E}(^2E) \qquad (14{,}404 \text{ cm}^{-1})$$

$$^4A_2 \to 2\bar{A}(^2E) \qquad (14{,}434 \text{ cm}^{-1})$$

were made by Tanabe and Sugano.

Relatively weak vibronic modes are evident in the tails which broaden the bands. Even when these are corrected, however, the best fit is unequivocally Lorentzian. The inadequacy of the spectrophotometer to give correct contours at the tops of the peaks and between the two major peaks becomes evident in the resolutions.

MATHEMATICAL DERIVATIONS

We are unaware of any previously proposed model, statistical or otherwise, which will explain the frequency functions experimentally observed. The following discussion is an attempt to establish a consistent model.

The probability that a system having two quantum states n (the ground state) and m (the excited state) will be excited from n to m by radiation after unit time is given by the equation[39]

$$B_{n \to m}\,\rho(\nu_{mn}) = \frac{2\pi}{3\hbar^2}\,|R_{mn}|^2 \rho(\nu_{mn}). \tag{10}$$

Here $\rho(\nu_{mn})$ is the density of radiation in ergs/cm^3. $B_{n \to m}$ is the transition probability per unit of radiation density. $|R_{mn}|^2$ is defined by the equation

$$|R_{mn}|^2 = \sum_{q=1}^{3} [\langle \psi_m{}^0 | M_q | \psi_n{}^0 \rangle]^2 \tag{11}$$

where M_q is the appropriate operator for coupling to the electric or magnetic field.

The term $B_{n \to m}\,\rho(\nu_{mn})$ can be thought of as a rate constant for transitions between n and m. Under steady state conditions which generally apply, at least in conventional absorption spectrophotometry, the rate of absorption to state m must be equalled by the rate of emission. This emission may take place either by induced or spontaneous processes or by thermal degradation through coupling with the environment. Thus at steady state

$$B_{n \to m} \, \rho(\nu_{mn}) N_n = [B_{m \to n} \, \rho(\nu_{mn}) + D_{m \to n} + A_{m \to n}] N_m. \tag{12}$$

$D_{m \to n}$ and $A_{m \to n}$ are the respective rate constants for thermal degradation and spontaneous emission. From (10) it is apparent that $B_{n \to m} = B_{m \to n}$. Furthermore, from the Boltzmann Distribution Law, we can express N_m in terms of N_n:

$$\frac{N_m}{N_n} = \exp\left(\frac{-h\nu_{mn}}{kT}\right). \tag{13}$$

If the system were placed in a black box and thus allowed to attain complete equilibrium, the density of radiation ν_{mn} would be

$$\rho(\nu_{mn}) = \frac{8\pi h\nu_{mn}^{3}}{c^3} \frac{1}{\exp(h\nu_{mn}/kT) - 1}. \tag{14}$$

Substituting into Eq. (12), we obtain a relationship between $B_{n \to m}$ and $A_{m \to n} + D_{m \to n}$:

$$D_{m \to n} + A_{m \to n} = \frac{8\pi h\nu_{mn}^{3}}{c^3} B_{n \to m}. \tag{15}$$

The light flux of frequency ν_{mn} impinging on the surface of a sample normal to an arbitrary axis x can be represented by I_0 (ergs/cm²-sec). At steady state $\rho_{mn}^{\circ} = I_0/c$, where c is the velocity of light. Within the sample, the density can be expressed generally by I.

The net absorption within Δx length per unit area per unit time would be

$$I(X) - I(X + \Delta X)$$

$$= h\nu_{mn}\left\{B_{n \to m}\left(1 - \exp\left(-\frac{h\nu_{mn}}{kT}\right)\right)\frac{I}{c} - A_{m \to n}\exp - \frac{h\nu_{mn}}{kT}\right\} N_n \Delta x. \tag{16}$$

After the limiting process, this equation can be solved by integration:

$$A(\nu_{mn}, T) \equiv -\log\left(\frac{I}{I_0}\right) = -\log\left[\frac{a}{bI_0} + \left(1 - \frac{a}{bI_0}\right)e^{-bx}\right] \tag{17}$$

$$b = \frac{h\nu_{mn} B_{n \to m}[1 - \exp(-h\nu_{mn}/kT)]}{c} N_n,$$

$$a = h\nu_{mn} A_{m \to n} \exp\left(-\frac{h\nu_{mn}}{kT}\right) N_n. \tag{18}$$

When $h\nu_{mn}$ becomes much larger than kT, $\exp - [h\nu_{mn}/kT]$ goes to zero and $A(\nu_{mn}, T)$ is equal to $\log (I_0/I)$ or $(h\nu B_{n \to m}/2.303c)N_n X$.

The absorptivity $\epsilon(\nu_{mn}, T)$ is defined as

$$\epsilon(\nu_{mn}) = \frac{A(\nu_{mn})}{N_n X} = \frac{h\nu B_{n \to m}}{2.303c} = \frac{(2\pi)^2 \nu}{3\hbar c \, 2.303} |R_{mn}|^2 \tag{19}$$

If $A_{m \to n}$ is negligible but $\exp [-h\nu_{mn}/kT]$ is not, then

$$\epsilon(\nu_{mn}) = \frac{(2\pi)^2}{3\hbar c \, 2.303} |R_{mn}|^2 \left[1 - \exp\left(-\frac{h\nu_{mn}}{kT}\right)\right]\left[1 - \sum_{i=1} \exp\left(-\frac{h\nu_i}{kT}\right)\right]. \tag{20}$$

So far we have treated the levels m and n as though they were absolutely discrete. Due to thermal agitation and/or symmetry imperfections, the levels are perturbed to give a distribution in ϵ rather than a step function at $\nu = \nu_{mn}$.

Perturbations will affect the transition levels in two ways: (a) they will change the energy levels, and (b) they will change the off-diagonal matrix elements $\langle \psi_m{}^0| M_q |\psi_n{}^0\rangle$, hence transition probabilities.

The energy difference between two energy levels is represented by

$$\Delta E = h\nu_{mn} = \langle \psi_m{}^0| H |\psi_m{}^0\rangle - \langle \psi_n{}^0| H |\psi_n{}^0\rangle. \tag{21}$$

Because of the myriad of possible perturbations on an absorber in a condensed environment, it is not unreasonable to expect that the matrix elements $\langle \psi_n{}^0| H |\psi_m{}^0\rangle$ and $\langle \psi_n{}^0| H |\psi_n{}^0\rangle$ would be perturbed randomly from well-defined energies E_n and E_m; strong deviation from E_n or E_m would be caused by several successive perturbing events without possibility of restoration between events. (For the model to be correct there cannot be any net restoring force.)

We might also expect the matrix elements $\langle \psi_m{}^0| M_q |\psi_n{}^0\rangle$ of (11) to be randomly perturbed:

$$f\{\langle \psi_m| M_q |\psi_n\rangle\} = \frac{1}{\sqrt{2\pi}\,\sigma_q} \exp\left[-\frac{1}{2}\left(\frac{\langle \psi_m| M_q |\psi_n\rangle - \langle \psi_m{}^0 |M_q |\psi_n{}^0\rangle}{\sigma_q}\right)^2\right]. \tag{22}$$

It can be shown that if the frequency functions for the variables $X_1, X_2, \ldots, X_n$ are random, that is, Gaussian with mean zero and the same variance, then the frequency function for $|R_{mn}|^2 = \sum_{j=1}^n X_j'^2$ is chi squared (χ^2) with n degrees of freedom. Thus $|R_{mn}|^2$ will be represented by a χ^2 frequency function of nth degree, where n is 3 or less, depending on how many elements are significant. If we define

$$(\Delta X_j) = \langle \psi_m| M_j |\psi_n\rangle - \langle \psi_m{}^0| M_j |\psi_n{}^0\rangle \tag{23}$$

and

$$|\Delta R_{mn}|^2 = \sum_{j}^{n} (\Delta X_j)^2 \tag{24}$$

then

$$f_\alpha(|R_{mn}|^2) = \frac{(\Delta R_{mn})^{n-2}}{2^{n/2}\Gamma(n/2)\sigma_1{}^n} \exp\left\{-\left[\frac{\frac{1}{2}(|\Delta R_{mn}|)^2}{\sigma_1{}^2}\right]\right\} \tag{25}$$

(See Appendix II.)

It can also be shown that if the variations of $\langle \psi_m{}^0| H |\psi_m{}^0\rangle$ and $\langle \psi_n{}^0| H |\psi_n{}^0\rangle$ are random, then $\Delta E = \langle \psi_m{}^0| H |\psi_m{}^0\rangle - \langle \psi_m{}^0| H |\psi_n{}^0\rangle$ is also Gaussian with the relationships

$$\sigma_2 = \sigma_m{}^2 + \sigma_n{}^2 \tag{26}$$

and

$$\nu_0 = \nu_m{}^0 - \nu_n{}^0$$

where ν^0 represents a resultant mean frequency. Thus the frequency function for ΔE is

$$f_\beta(\Delta E) = \frac{1}{\sqrt{2\pi}\,\sigma_2} \exp\left\{-\left[\frac{1}{2}\left(\frac{h(\nu - \nu_0)}{\sigma_2}\right)^2\right]\right\} \tag{27}$$

The joint frequency function for ΔE, $|R_{mn}|^2$ becomes

$$f(\Delta E, |R_{mn}|^2) = f_\alpha f_\beta = \frac{|\Delta R_{mn}|^{n-2}}{2^{(n+1)/2}\Gamma(n/2)\pi^{1/2}\sigma_1{}^n\sigma_2}$$

$$\times \exp\left\{-\frac{1}{2}\left[\left(\frac{h(\nu - \nu_0)}{\sigma^2}\right)^2 + \frac{|\Delta R_{mn}|^2}{\sigma_1{}^2}\right]\right\} \tag{28}$$

We can now denote the marginal frequency function for $h(\nu - \nu_0)$ as

$$f_m[h(\nu - \nu_0)] = \int_0^\infty f[h(\nu - \nu_0), |R_{mn}|^2]\,d(|R_{mn}|^2). \tag{29}$$

Equation (28) represents the probability that an absorber in the system has the perturbed-energy relationship required to couple with electromagnetic radiation of frequency ν. Integration over $|R_{mn}|^2$ from 0 to ∞, however, renders

$$f_m[h(\nu - \nu_0)] = f_\alpha \tag{30}$$

unless the energy perturbation is correlated in some fashion with the perturbation on $|R_{mn}|$. This is not to infer that there is a functional dependence of R_{mn} on energy, because the joint frequency function can only be correct when ΔR_{mn} and ΔE are independent. It is only reasonable to assume at this point that the dispersion σ_1, being a perturbational parameter, is proportional to σ_2, which is also a perturbational parameter. Herein is the proposed correlation between ΔR_{mn} and ΔE. We then assume $\sigma_2 = \alpha\sigma_1$.

It can be shown that lifetimes of perturbed states in general are related to band widths through a Fourier transformation.[40] Figure 8 illustrates this for the band in Fig. 6. The ordinate of Fig. 8, $\langle u(0) \cdot u(t)\rangle$, is the component of the (in this case) dipole moment vector which remains in the direction of the dipole moment vector at time 0. In other words, as can be seen by Fig. 8, the width is inversely equal to the time required for molecular motion to disorient a molecule that is coupling with electromagnetic radiation. Thus the perturbation relates to the motion of the disorienting environmental cage surrounding the absorber.

The simplest way to take into account all possible perturbational motions of the cage is to integrate over all possible decay rates and, since the rates τ

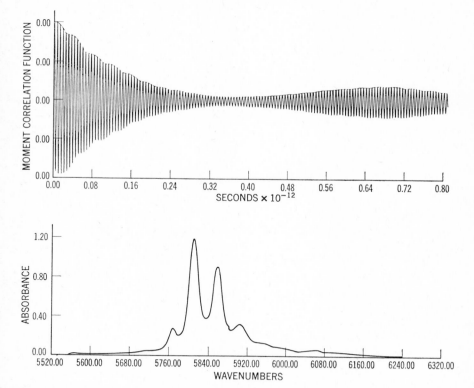

Fig. 8. The Fourier time transform of an electronic transition. The first decay period (neglecting the small undulations, which are not real) is about 3.5×10^{-13} sec, corresponding to a line width ($1/\pi\tau$) of about 30 cm^{-1}.

are related to $1/\sigma$, it follows that (28) can be written as follows:

$$f(\Delta E, |R_{mn}|^2) = \frac{|\Delta \bar{R}_{mn}|^{n-2} \delta^{n+1}}{2^{(n+1)/2} \Gamma(n/2) \pi^{1/2} \alpha}$$

$$\times \left[t^{n+1} \exp\left\{ -\frac{1}{2} \left[\left(\frac{h(\nu - \nu_0)}{\alpha} \right)^2 + |\Delta \bar{R}_{mn}|^2 \right] \right\} \delta^2 t^2 \right]. \quad (31)$$

When integrated over all possible values of t, (31) produces the marginal frequency function for ΔE; that is,

$$f_m(h(\nu - \nu_0)) = \frac{\Gamma[(n + 1)/2]}{2^{(n+1)/2} \Gamma(n/2) \pi^{1/2} \alpha |\Delta \bar{R}_{mn}|^3} \left[1 + \beta(n) \left(\frac{\nu - \nu_0}{w} \right)^2 \right]^{-(n+1)/2} \quad (32)$$

where the half-width of the band is w; that is,

$$w = \frac{\alpha \sqrt{\beta(n)} \langle \Delta R_{mn} \rangle_{ar}}{h} \quad (33)$$

and

$$\beta(n) = 4(2^{2/(n+1)} - 1). \quad (34)$$

It follows from (19) and (32) that

$$\epsilon_n(\nu) = \frac{(2\pi)^2 \nu |\bar{R}_{mn}{}^0|^2}{3\hbar(2.303)} [f_m(h(\nu - \nu_0))]$$

$$= \frac{\pi^{3/2} \nu |R_{mn}{}^0|^2 \Gamma[(n + 1)/2]}{3\hbar(2.303)2^{(n-3)/2} \Gamma(n/2) \alpha |\Delta \bar{R}_{mn}|^3 \{1 + \beta(n)[(\nu - \nu_0)/w]^2\}^{(n+1)/2}} \quad (35)$$

The operator M_q contains a scalar product $\boldsymbol{\epsilon} \cdot \mathbf{M}_q$, where $\boldsymbol{\epsilon}$ is a unit vector in the direction of the coupling electromagnetic vector. Thus, in a well-ordered optically perfect crystal, n must be less than or equal to 2, provided the dispersion on the elements $\langle \psi_m| \boldsymbol{\epsilon} \cdot \mathbf{M}_{q1} |\psi_n \rangle$ and $\langle \psi_m| \boldsymbol{\sigma} \cdot \mathbf{M}_{q2} |\psi_n \rangle$ is the same. If the transition happens to be permitted for only one orientation of $\boldsymbol{\epsilon}$ or if plane-polarized light is used, then n should be 1, thus producing a Lorentzian band contour.

In a liquid, provided the matrix elements $\langle \psi_m| \boldsymbol{\epsilon} \cdot \mathbf{M}_q |\psi_n \rangle$ have the same dispersion, one would expect n to be 3. This will only be the case, however, if the motions of the cage are random and would require integration over all t's as above. If the motion of the cage is highly coordinated, however, such as in strong simple symmetric complexes, and the vibrations have about the same frequencies, the dispersion on disorientation times will be small and integration will not be necessary. Under these conditions one would expect the band shape to approach Gaussian as in (30).

When dispersions of the elements $\langle \psi_m | M_q | \psi_n \rangle$ are not the same, (25) must be replaced by a much more complicated expression. However, an average n can be defined where $1 \leq n \leq 3$ that will approximate the complicated expression by (25).[41]

$$n = \frac{(\sum \sigma_i)^2}{\sum \sigma_i^2} .$$ (36)

This has not yet been required in our analysis.

SUMMARY

The stochastic formulation above has as a unique feature the inclusion of the perturbation on transition matrix elements. It promises to help classify the types of molecular motions that occur in condensed systems. Specifically, within the framework of the approximations, it predicts whether the shape will be Gaussian or T_n, depending on the entity interacting with light. If it is well shielded from a randomly perturbing environment by a strong complex sphere, the transitional band is expected to be Gaussian. If not, it should be T_3, T_2, or T_1 to the approximation that three, two, or one component of the dipole or quadrupole matrix elements can couple to the light and be perturbed. According to (36), if the dispersion of one element that can couple with the radiation is much larger than that of the other or if it can couple only with the radiation, then a Lorentzian band should result. If the dispersions of three components are about equal and all three components can couple, as they could if they were in randomly oriented solution, then T_3 is predicted.

These predictions are consistent with all of our experimental results. Np(V), being in a noncomplexing perchlorate medium, is expected to be perturbed quite randomly as is atomic mercury. Therefore one predicts T_3 frequency functions for these. By contrast, $Co(NH_3)^{3+}$ and UCl_6^- are relatively well shielded, suggesting Gaussian functions. The samples of UCl_4 and Cr^{3+} (Figs. 6 and 7) illustrate the last type where the radiation can couple to at most two elements in the case of UCl_4 and one in the case of Cr^{3+}, where polarized impinging light was used.

Needless to say, it should not be surprising to observe fractional values of n since integral values result only when the dispersions are equal to zero.

Furthermore, since T_n approaches G as n becomes large, n may not be limited to 3, although the concept of n vanishes at this point.

Finally, caution must be exercised in keeping the assumptions in mind. Although we have not seen such systems yet, it is anticipated that random energy perturbation will not be correct in some cases, as outlined by Margenau.[13]

APPENDIX I

The classical model for natural line widths and resonance radiation is based on several approximations and assumptions which are not valid for condensed systems.

Consider an nth electron that acts like an oscillator attached to a nucleus with a force constant k and resonant frequency ω_0. When an electromagnetic wave of frequency ω propagating in the z direction with amplitudes in the x direction acts on this electron, the forces are balanced as follows:

$$m\ddot{x}_i + s_i\dot{x}_i + k_ix = eF^0e^{i\omega t} \tag{1}$$

where m = mass of electron,

s = damping constant,

e = charge of electron,

F^0 = local electric field.

The solution to this equation is

$$x_i = \frac{e}{m} \frac{\omega_i^{02} - \omega^2 - i\gamma_i}{(\omega_i^{02} - \omega^2)^2 + \gamma_i^2\omega^2} F^0e^{i\omega t} \quad \left(\gamma_i = \frac{s_i}{m}, \omega_i^0 = \frac{\sqrt{k_i}}{m}\right). \tag{2}$$

The induced dipole moment is ex_i, which is also $\alpha_i F^0e^{i\omega t}$, where α_i is the polarizability of the electron. It follows that

$$\alpha_i = a_i - ib_i, \tag{3}$$

where

$$b_i = \frac{e^2}{m} \frac{\gamma_i}{(\omega_i^{02} - \omega^2)^2 + \gamma_i^2\omega^2}. \tag{3'}$$

By definition the dielectric constant, ϵ is given by

$$\epsilon = 1 + \frac{4\pi P}{E}, \tag{4}$$

where P is the polarization and E is the external field.

Also by definition the polarization is given by

$$P = N\left(\sum_i f_i\alpha_i\right)F, \tag{5}$$

where N is the number of electrons per unit volume and f_i is the "oscillator strength" representing the probability that the ith electron will contribute to the polarization.

From (4) and (5) we obtain

$$\epsilon - 1 = 4\pi N \sum_i f_i\alpha_i \frac{F}{E}, \tag{6}$$

and with the assumption that no adjacent electrons will alter the field around the ith electron, that is, where $F = E$ and the frequency ω is high enough that permanent dipoles cannot orient preferentially or that the Lorentz approximation holds, $\epsilon = \eta^2$, where η is the index of refraction, then

$$I_m\{\eta^2 - 1\} = I_m\{4\pi N \sum f_i \alpha_i\}. \tag{7}$$

It follows that

$$\Phi(N, \omega_i, \omega, \gamma_i) = \eta_0 k = \frac{2\pi N e^2}{\eta_0 m} \sum_i \frac{f_i \gamma_i}{(\omega_i^{02} - \omega^2)^2 + \gamma_i^2 \omega^2}, \tag{7'}$$

where

$$\eta = \eta_0(1 - ik). \tag{7''}$$

Now the electromagnetic wave can be expressed by the following equation as it passes through the medium of index η.

$$E(x, z, t) = E(x_0, 0, 0) \exp\left[i\omega\left(t - \frac{\eta z}{c}\right)\right] \tag{8}$$

where $\eta z \omega / c$ is the time-phase shift due to the change in wavelength within the medium. From (7'') and (8), recognizing that the light intensity is proportional to the square of the amplitude,

$$\frac{I^0(\omega, 0)}{I(\omega, z)} = \frac{E(x_0, 0, 0)^2}{E(x_0, 0, 0)^2 \exp(-2\eta_0 kz\omega)/c} = \exp\left[\frac{2\eta_0 kz\omega}{c}\right] \tag{9}$$

Since $A(\omega) = \log[I^0(\omega, 0)/I(\omega, z)]$, we have

$$A(\omega) = \frac{2\eta_0 kz\omega}{2.303c} = \frac{2\Phi z\omega}{2.303c} \tag{10}$$

and from (7')

$$A(\omega) = \frac{4\pi N e^2 z\omega}{2.303c\eta_0 m} \sum_i \frac{f_i \gamma_i}{(\omega_i^{02} - \omega^2)^2 + \gamma_i^2 \omega^2}.$$

Assuming $\omega \sim \omega_0$, we can write

$$A(\omega) = \frac{4\pi N e^2 z}{2.303c\eta_0 m\omega} \sum_i \frac{f_i}{\gamma_i[1 + 4[(\omega_i^0 - \omega)/2\gamma_i]^2]}. \tag{11}$$

It has been stressed that this equation is a direct result of the following approximations and assumptions:

1. The loss of energy due to damping, which can also include resonance energy loss, is proportional to the transverse velocity of the electron.
2. The internal and external fields are the same.
3. The frequency ω_0/ω is large enough so that it can be approximated by unity.

4. The temperature is high enough to insure completely random orientation of permanent dipole moments, if they exist. This is required to use the relationship $\eta^2 = \epsilon$.

If assumption 2 is replaced by the assumption that there are no permanent dipoles but that local fields are altered, we can use the relationship $F = E + 4\pi P/3$.[1]

This simple and necessary change for condensed systems renders $A(\omega)$ unobtainable in closed form. It must be expressed by (10) coupled with the following polynomial in $y = k\eta_0$:

$$y^4 + 5\eta_0^2 y^2 + \frac{6\eta_0 y}{\Phi(N, \omega_i^0, \omega, \gamma_i)} + (\eta_0^2 + 2)^4 = 0.$$

It is perhaps also important to emphasize that the oscillator strength f_i assumes a much more complicated form than normally realized, even for the first set of assumptions. In condensed systems it is formidable because simultaneous multiparticle interactions must be considered.

APPENDIX II

Derivation of x^2 with Nonunit Dispersion

Let $\Delta R^2 = \sum_{i=1}^n \Delta X_i^2$, where the frequency function for x_j is given by

$$f_j = \frac{1}{(2\pi)^{1/2}\sigma} \exp\left[-\frac{1}{2}\left(\frac{\Delta x_j}{\sigma}\right)^2\right].$$

The frequency function for the system of random variables $x_1, \ldots, x_n$ is the product of f_js:

$$f_n = \prod_j^n f_j.$$

The probability $f_n(\alpha)\,d\alpha$ that ΔR^2 be between α and $\alpha + d\alpha$ is obtained by integrating (3) between the limits $\alpha < x_1^2 + x_2^2 + \cdots + x_n^2 < \alpha + d\alpha$. We define

$$\Omega_n = \int dx_1\, dx_2 \cdots dx_n$$

extended to the domain $x_1^2 + x_2^2 + \cdots + x_n^2 < 1$. The value of this integral extended to $x_1^2 + x_2^2 + \cdots + x_n^2 < \alpha$ is $\Omega_n \alpha^{n/2}$. Extended to the domain

$\alpha < \sum_{i=1}^{n} x_i^2 < \alpha + d\alpha$ the value of the integral is

$$\Omega_n d(\alpha^{n/2}) = \Omega_n \left(\frac{n}{2}\right) \alpha^{n/2-1} \, d\alpha.$$

Since $\sum x_i^2 = \alpha$ in this region, we can write

$$f_n(\alpha) \, d\alpha = \frac{n}{2} \alpha^{n/2-1} \frac{n}{[(2\pi)^{1/2}\sigma]^n} \exp\left(-\frac{\alpha}{2\sigma^2}\right) d\alpha.$$

To obtain Ω_n we integrate from $\alpha = 0$ to $\alpha = +\infty$. The other form of this integral, extended to the entire n-dimensional space, is

$$\left[\frac{1}{(2\pi)^{1/2}\sigma}\right]^2 \int_{-\infty}^{\infty} \exp\left(-\frac{\sum \Delta x_i^2}{2\sigma^2}\right) d(\Delta x_i) \, d(\Delta x_2) \cdots d(\Delta x_n)$$

$$= \left[\frac{1}{(2\pi)^{1/2}\sigma}\right]^2 \left[\int_{-\infty}^{\infty} \exp\left(-\frac{\Delta x_1^2}{2\alpha^2}\right) d\Delta x_1\right]^n.$$

It follows that

$$\Omega_n = \frac{2\{\int_{-\infty}^{\infty} \exp\left[-\frac{1}{2}(\Delta x_1/\sigma)^2\right] d\Delta x_1\}^n}{\int n_0^{\infty} \exp\left[-\alpha/2\alpha^2\right]\alpha^{n/2-1} \, d\alpha} = \frac{2[\sigma\sqrt{2\pi}]^n}{n[\Gamma(n/2)2^{n/2}\sigma^n]} = \frac{2\pi^{n/2}}{n\Gamma(n/2)}$$

whereby

$$f_n(\alpha) = \frac{\alpha^{n/2-1}}{2^{n/2}\Gamma(n/2)\sigma^n} \exp\left(-\frac{\alpha}{2\alpha^2}\right).$$

It follows that

$$P\left(\alpha < \sum_{i=1}^{n} \frac{x^2}{\sigma^2} < \alpha + d\alpha\right) = f_n(\alpha) \, d\alpha.$$

ACKNOWLEDGMENT

We wish to acknowledge the contributions of W. L. Nicholson, whose comments initiated the research, D. G. Carter and P. Hoggard for experimental work, and B. Duane, D. Kottwitz, and R. E. Biggers, who supplied basic programs for resolution by computer. We also acknowledge the financial support of the Atomic Energy Commission Contract No. AT(45-1)-1830 for the development of experimental techniques and accumulation of certain experimental data, and Battelle Memorial Institute under AEC Use Permit AT(45-1)-1831, who supported the remainder of the effort.

REFERENCES

1. R. G. Breene, Jr., *Rev. Mod. Phys.*, **29**, 94 (1957).
2. Chen and Takeo, *Rev. Mod. Phys.*, **29**, 20 (1957).
3. V. Weisskopf and E. Wigner, *Z. Physik*, **63**, 54 (1930); **65**, 18 (1931).
4. F. Hoyt, *Phys. Rev.*, **36**, 860 (1931).

5. Lord Rayleigh, *Phil. Mag.*, **27**, 298 (1889).
6. A. Michelson, *J. Astrophys.*, **2**, 251 (1885).
7. H. A. Lorentz, *Proc. Roy. Acad. Amsterdam*, **8**, 591 (1906).
8. J. H. Van Vleck and V. F. Weisskopf, *Rev. Mod. Phys.*, **17**, 227 (1945).
9. R. Kraplus and J. Schwinger, *Phys. Rev.*, **73**, 1020 (1948).
10. G. W. King and J. H. Van Vleck, *Phys. Rev.*, **55**, 1165 (1939).
11. J. Holtsmark, *Ann. Physik.*, **58**, 577 (1919); *Z. Physik*, **20**, 162 (1919); **25**, 73 (1924); **34**, 722 (1925).
12. H. Margenau and W. Watson, *Rev. Mod. Phys.*, **8**, 22 (1936).
13. H. Margenau, *Phys. Rev.*, **48**, 755 (1935).
14. R. Kubo and K. Tomita, *J. Phys. Soc. Japan*, **9**, 888 (1954).
15. D. L. Huber and J. van Vleck, *Rev. Mod. Phys.*, **38**, 187 (1966).
16. U. Fano and J. W. Cooper, *Rev. Mod. Phys.*, **40**, 441 (1968).
17. R. Gordon, *J. Chem. Phys.*, **45**, 1635, 1643, 1649 (1966); **39**, 2788 (1963).
18. C. K. Jørgensen, *Absorption Spectra and Chemical Bonding in Complexes*, Pergamon Press, Oxford, 1962, pp. 85 et seq.
19. T. R. Lusebrinck, Ph.D. Thesis, University of California, Berkeley, 1965. Also Report UCRL-16344.
20. R. E. Biggers, Oak Ridge National Laboratory, private communication.
21. A. W. Baker and M. D. Yeoman, *Spectrochim. Acta*, **22**, 1773 (1966).
22. R. N. Jones, K. S. Seshadri, and J. W. Hopkins, *Can. J. Chem.*, **40**, 334 (1962).
23. J. R. Morrey, H. S. Gile, BNWL-343 (available from the Clearinghouse for Federal Scientific and Technical Information, National Bureau of Standards, U.S. Dept. of Commerce, Springfield, Va. 22151.
24. R. D. B. Fraser and E. Suzuki, in *Spectral Analysis II*, J. A. Blackburn, Ed., Marcel Dekker, New York, 1968.
25. R. E. Biggers, private communication.
26. E. Duane, AEC Research & Development Report, BNWL-390 (September 1967).
27. D. Kottwitz, private communication.
28. C. K. Jørgensen and B. R. Judd, *Mol. Phys.*, **8**, 281 (1964).
29. B. R. Judd, *J. Chem. Phys.*, **44**, 839 (1966).
30. D. E. Henrie and G. R. Choppin, *J. Chem. Phys.*, **49**, 477 (1968).
31. P. M. Hagan and J. M. Cleveland, *J. Inorg Nucl. Chem.*, **28**, 2905 (1966).
32. H. Gunning and S. Vinogradov, *J. Phys. Chem.*, **68**, 1962 (1964).
33. W. A. Yeranos, *Inorg. Chem.*, **7**, 1259 (1968).
34. R. A. P. Wentworth, *Chem. Commun.*, **1965**, 532.
35. J. S. Griffith, *The Theory of Transition-Metal Ions*, Cambridge Univ. Press, 1961, pp. 312 et seq.
36. A. Liehr, *J. Phys. Chem.*, **68**, 3629 (1964).
37. S. Sugano and Y. Tanabe, *J. Phys. Soc. Japan*, **13**, 880, 889 (1958).
38. O. Deutschbein, *Ann. Physik*, **14**, 712 (1932); **20**, 828 (1934).
39. H. Eyring, J. Walter, and G. Kimball, *Quantum Chemistry*, John Wiley, 1944, pp. 112 et seq.
40. R. G. Gordon, *J. Chem. Phys.*, **43**, 1307 (1965).
41. F. E. Satterthwaite, *Biometrics Bull.*, **2**, 110 (1946).

Covalency Effects in Octahedral $5f^1$ Complexes[*]

HARRY G. HECHT, W. BURTON LEWIS, AND MICHAEL P. EASTMAN[**]

University of California, Los Alamos Scientific Laboratory, New Mexico

Abstract

A molecular orbital theory for octahedral f^1 complexes is described. It is pointed out that the neglect of covalent bonding in the analysis of optical data for the actinide complexes is not justified, and that its inclusion leads to orbital reductions which are considerably greater than have usually been assumed.

[*] Work supported by the U.S. Atomic Energy Commission.

[**] Present address: Department of Chemistry, University of Texas at El Paso.

351

INTRODUCTION

The discussions of ligand field theory are generally divided into three limiting cases, depending on the relative sizes of the electronic repulsion, spin-orbit coupling, and crystal field terms. In the first case, spin-orbit $<$ crystal field $<$ electronic repulsion. This is generally referred to as the "weak-field" case and is typical of many transition metal ions. A second case is crystal field $<$ spin-orbit $<$ electronic repulsion, in which $J = L + S$ obviously becomes a meaningful quantum number. This situation is found in complexes of the lanthanide ions. In actinyl-type ions, AcO_2^{n+}, the "strong-field" case is found, in which spin-orbit $<$ electronic repulsion $<$ crystal field. This strong-field case is also found in some of the transition metal ions, particularly of the second and third series. A great deal of delocalization is usual in these ions, which are often referred to as "covalent complexes."

It is now recognized that yet another situation pertains to many compounds of the actinides. This may be termed the "intermediate-field" case, for which spin-orbit $\simeq$ crystal field $\simeq$ electronic repulsion. Obviously this situation is more complex theoretically, in the sense that none of the usual perturbation-type calculations are applicable here. Since the advent of high-speed digital computers, however, this need not be a serious limitation.

One might expect a considerable amount of covalent bonding in these intermediate-field actinide complexes, and there are several experimental results which do indeed indicate that such effects are significant; for example, Eisenstein and Pryce[1] have shown that the magnetic susceptibility data imply a high degree of covalency in UF_6, and more recently, Rigny and Plurien[2] have reported significant orbital reduction effects (which, as discussed later, also reflect the extent of covalent bonding), as inferred by EPR studies of some octahedral U(V) hexafluorides.

In the optical studies of actinide ions which have been reported, proper cognizance of covalency effects has not in general been shown, so that no reliable estimates of its extent are available. In this work we reexamine some of the optical data using a molecular orbital approach. For simplicity, we limit ourselves at the present time to octahedral $5f^1$ complexes. The most thoroughly studied optical spectrum for ions of this type is that of $CsUF_6$ reported by Reisfeld and Crosby.[3]

THE MODEL

Since the spin-orbit coupling and the crystal field interactions are regarded as comparable in magnitude, the choice of basis functions is completely arbitrary. We apply the crystal field interaction first, deriving the appropriate states and eigenfunctions within the $l_z s_z$ representation. The procedure is only briefly sketched here, since various other authors have described the calculation in detail.[1,5,10,12]

In an octahedral coordination we must consider the interaction represented by the crystal field Hamiltonian,

$$\mathcal{H}_{Xtl} = \frac{b_4}{60}(O_4^0 + 5O_4^4) + \frac{b_6}{180}(O_6^0 - 21O_6^4) \tag{1}$$

where $O_4^0 = 35l_z^4 - [30l(l+1) - 25]l_z^2 - 6l(l+1) + 3l^2(l+1)^2$,

$O_4^4 = \frac{1}{2}(l_+^4 + l_-^4)$,

$O_6^0 = 231l_z^6 - 105[3l(l+1) - 7]l_z^4 + [105l^2(l+1)^2 - 525l(l+1) + 294]l_z^2 - 5l^3(l+1)^3 + 40l^2(l+1)^2 - 60l(l+1)$,

$O_6^4 = \frac{1}{4}[11l_z^2 - l(l+1) - 38](l_+^4 + l_-^4) + \frac{1}{4}(l_+^4 + l_-^4)[11l_z^2 - l(l+1) - 38]$.

The coefficients b_4 and b_6 describe the strength of the crystal field. Considering only nearest neighbor interactions in a point charge model, they are

$$b_4 = \frac{Z_i e^2 \langle r^4 \rangle}{a_i^5} \frac{7}{66} = 8A_4^0 \frac{\langle r^4 \rangle}{33}$$

and

$$b_6 = -\frac{Z_i e^2 \langle r^6 \rangle}{a_i^7} \frac{5}{572} = -80A_6^0 \frac{\langle r^6 \rangle}{429}, \tag{2}$$

where Z_i is the charge on the ligands, located at a distance a_i from the metal ion, and the radial averages $\langle r^n \rangle$ are taken over the $5f$ electron wave function.

On a purely group theoretical basis, we know that the branching rule for an F state in an octahedral field is

$$\mathfrak{D}_3 \rightarrow \Gamma_2 + \Gamma_4 + \Gamma_5$$

so that one nondegenerate and two threefold degenerate states are expected. Direct calculation shows the energies and eigenfunctions, expressed in real form, to be

$$\Gamma_2: \quad W_{\Gamma_2} = -12b_4 - 48b_6 \qquad f_{\Gamma_2} = \left(\frac{105}{4\pi}\right)^{1/2} \frac{xyz}{r^3},$$

$$\Gamma_5: \quad W_{\Gamma_5} = -2b_4 + 36b_6 \qquad f_\xi = \left(\frac{105}{16\pi}\right)^{1/2} \frac{x(y^2 - z^2)}{r^3},$$

$$f_\eta = \left(\frac{105}{16\pi}\right)^{1/2} \frac{y(z^2 - x^2)}{r^3},$$

$$f_\zeta = \left(\frac{105}{16\pi}\right)^{1/2} \frac{z(x^2 - y^2)}{r^3}, \qquad (3)$$

$$\Gamma_4: \quad W_{\Gamma_4} = 6b_4 - 20b_6 \qquad f_x = \left(\frac{7}{16\pi}\right)^{1/2} \frac{x(5x^2 - 3r^2)}{r^3},$$

$$f_y = \left(\frac{7}{16\pi}\right)^{1/2} \frac{y(5y^2 - 3r^2)}{r^3},$$

$$f_z = \left(\frac{7}{16\pi}\right)^{1/2} \frac{z(5z^2 - 3r^2)}{r^3}.$$

It should be noted that the energy levels cannot be expressed in terms of a single parameter, as in the case of d electrons, due to the occurrence of both fourth- and sixth-order harmonics in this case. The ratio $\langle r^4 \rangle / \langle r^6 \rangle$ appears to be quite constant in a series of $5f^1$ ions,[3] however, so that the crystal field splitting can be described to a good degree of approximation in terms of a single parameter.

When spin-orbit coupling is also included, the energy states are found by decomposing the direct product of the above states with spin Γ_6 within the octahedral double group. The result is

$$\Gamma_2 \times \Gamma_6 \rightarrow \Gamma_7,$$
$$\Gamma_5 \times \Gamma_6 \rightarrow \Gamma_7 + \Gamma_8,$$
$$\Gamma_4 \times \Gamma_6 \rightarrow \Gamma_6 + \Gamma_8.$$

Thus one must solve a 2×2 secular equation for the Γ_7 states, a 2×2 for the Γ_8 states, and a 1×1 for the Γ_6 state. The secular determinants are of the form

$$\Gamma_7: \quad \begin{vmatrix} W_{\Gamma_7} & \sqrt{3}\zeta \\ \sqrt{3}\zeta & W_{\Gamma_5} - \dfrac{\zeta}{2} \end{vmatrix},$$

$$\Gamma_8: \quad \begin{vmatrix} W_{\Gamma_5} + \dfrac{\zeta}{4} & \dfrac{3\sqrt{5}}{4}\zeta \\ \dfrac{3\sqrt{5}}{4}\zeta & W_{\Gamma_4} - \dfrac{3\zeta}{4} \end{vmatrix}, \qquad (4)$$

$$\Gamma_6: \quad \begin{vmatrix} W_{\Gamma_4} - \dfrac{3\zeta}{2} \end{vmatrix},$$

where ζ is the appropriate spin-orbit coupling constant. It turns out that for any reasonable set of parameters the ground state is a Kramers doublet, which can be written in the form

$$|\pm\rangle = \cos\theta\, |\Gamma_7^\pm\rangle - \sin\theta\, |\Gamma_7'^\pm\rangle. \tag{5}$$

THE EFFECTS OF COVALENCY

The model just described, which is derived on the basis of pure crystal field theory, can be modified rather simply to include the effects of electron delocalization. Since the *ab initio* calculation of crystal field terms is at best totally unreliable, we follow the usual practice of treating these terms simply as adjustable parameters. Thus the question of paramount importance is how the spin-orbit coupling constant ζ is affected by covalent bond formation.

The electron spin is not affected by covalent bonding in any direct way, but the orbital moment may be considerably altered if the $5f$ electron spends a significant amount of its time in the ligand orbitals. It can readily be seen that the orbital moment, and thereby the spin-orbit coupling constant, since it is linear in the orbital moment, will be reduced by the factor

$$k_{ij} = \frac{\langle \psi_i |\, \mathbf{l}\, | \psi_j \rangle}{\langle f_i |\, \mathbf{l}\, | f_j \rangle}, \tag{6}$$

where the ψ's are the appropriate molecular orbitals. The k's thus defined were introduced by Stevens[4] and are known as *orbital reduction factors*.

A suitable set of antibonding molecular orbitals can be readily constructed by symmetry arguments. In this case[5] they are

$$
\begin{aligned}
\Gamma_2:\quad & \psi_{\Gamma_2} = f_{\Gamma_2} \\[4pt]
\Gamma_5:\quad & \psi_\xi = N'\{f_\xi - \tfrac{1}{2}\alpha_\pi'(-x_3 - x_6 + x_2 + x_5)\} \\
& \psi_\eta = N'\{f_\eta - \tfrac{1}{2}\alpha_\pi'(-y_1 - y_4 + y_3 + y_6)\} \\
& \psi_\zeta = N'\{f_\zeta - \tfrac{1}{2}\alpha_\pi'(-z_2 - z_5 + z_1 + z_4)\} \\[4pt]
\Gamma_4:\quad & \psi_x = N\{f_x - \tfrac{1}{2}\alpha_\pi(-x_3 - x_6 - x_2 - x_5) \\
& \qquad - \sqrt{\tfrac{1}{2}}\alpha_\sigma(-x_1 - x_4) - \sqrt{\tfrac{1}{2}}\alpha_s(s_1 - s_4)\}, \\
& \psi_y = N\{f_y - \tfrac{1}{2}\alpha_\pi(-y_1 - y_4 - y_3 - y_6) \\
& \qquad - \sqrt{\tfrac{1}{2}}\alpha_\sigma(-y_2 - y_5) - \sqrt{\tfrac{1}{2}}\alpha_s(s_2 - s_5)\}, \\
& \psi_z = N\{f_z - \tfrac{1}{2}\alpha_\pi(-z_2 - z_5 - z_1 - z_4) \\
& \qquad - \sqrt{\tfrac{1}{2}}\alpha_\sigma(-z_3 - z_6) - \sqrt{\tfrac{1}{2}}\alpha_s(s_3 - s_6)\},
\end{aligned}
\tag{7}
$$

where the functions x, y, z, and s are ligand p and s valence shell orbitals and the subscripts 1–6 indicate ligand positions of $(a_i, 0, 0)$, $(0, a_i, 0)$, $(0, 0, a_i)$, $(-a_i, 0, 0)$, $(0, -a_i, 0)$, and $(0, 0, -a_i)$, respectively. Thornley[5] has also given the resulting orbital reduction factors, which in this case are

$$k_{\Gamma_4\Gamma_4} = 1 - N^2\left\{\tfrac{4}{3}\alpha_\pi{}^2 + \alpha_\sigma{}^2 + \alpha_s{}^2 + \left(\frac{2\sqrt{2}}{3}\right)\alpha_\pi\left(\alpha_\sigma + a_i\alpha_s\langle x|\frac{\partial}{\partial x}|s\rangle\right)\right\},$$

$$k_{\Gamma_5\Gamma_5} = 1 - 2N'^2\alpha_\pi'^2,$$

$$k_{\Gamma_2\Gamma_5} = 1 - \tfrac{1}{2}N'\alpha_\pi'^2, \tag{8}$$

$$k_{\Gamma_4\Gamma_5} = 1 - NN'\left\{\tfrac{1}{2}(\alpha_\pi{}^2 + \alpha_\pi'^2 + \alpha_\sigma{}^2 + \alpha_s{}^2)\right.$$

$$\left. + \sqrt{\frac{2}{15}}\left(\alpha_\sigma\alpha_\pi' - \frac{\alpha_\pi\alpha_\pi'}{\sqrt{2}} + \alpha_s\alpha_\pi'a_i\langle x|\frac{\partial}{\partial x}|s\rangle\right)\right\}.$$

The normalization parameters are

$$N^2 = (1 - 4\alpha_\pi S_\pi - 2\sqrt{2}\alpha_\sigma S_\sigma - 2\sqrt{2}\alpha_s S_s + \alpha_\pi{}^2 + \alpha_\sigma{}^2 + \alpha_s{}^2)^{-1}$$

$$N'^2 = (1 - 4\alpha_\pi'S_\pi' + \alpha_\pi'^2)^{-1} \tag{9}$$

and the overlap integrals are given by

$$S_\pi = \langle f_x | -x_3\rangle, \qquad S_\sigma = \langle f_x | -x_1\rangle, \tag{10}$$
$$S_s = \langle f_x | s_1\rangle, \qquad S_\pi' = \langle f_\xi | -x_3\rangle.$$

The secular equations should of course now be modified to read

$$\Gamma_7: \begin{vmatrix} W_{\Gamma_2} & \sqrt{3}k_{\Gamma_2\Gamma_5}\zeta \\ \sqrt{3}k_{\Gamma_2\Gamma_5}\zeta & W_{\Gamma_5} - \tfrac{1}{2}k_{\Gamma_5\Gamma_5}\zeta \end{vmatrix},$$

$$\Gamma_8: \begin{vmatrix} W_{\Gamma_5} + \tfrac{1}{4}k_{\Gamma_5\Gamma_5}\zeta & \tfrac{3}{4}\sqrt{5}k_{\Gamma_4\Gamma_5}\zeta \\ \tfrac{3}{4}\sqrt{5}k_{\Gamma_4\Gamma_5}\zeta & W_{\Gamma_4} - \tfrac{3}{4}k_{\Gamma_4\Gamma_4}\zeta \end{vmatrix}, \tag{11}$$

$$\Gamma_6: \quad |\,W_{\Gamma_4} + \tfrac{3}{2}k_{\Gamma_4\Gamma_4}\zeta\,|.$$

It does not appear profitable to attempt an interpretation of the experimental data with so many parameters. Therefore, we assume, as do Eisenstein and Pryce,[1] that there are only two orbital reduction parameters k and k', which are characteristic of the Γ_5 and Γ_4 states, respectively. This is tantamount to setting $k_{\Gamma_4\Gamma_4} = k'$, $k_{\Gamma_5\Gamma_5} = k$, $k_{\Gamma_2\Gamma_5} = k^{1/2}$, and $k_{\Gamma_4\Gamma_5} = k^{1/2}k'^{1/2}$ in the above equations, in which case the g-factor for the Γ_7 ground state becomes

$$g = 2\cos^2\theta - 4\sqrt{k/3}\sin 2\theta - \tfrac{2}{3}(1 - k)\sin^2\theta. \tag{12}$$

APPLICATION TO CsUF$_6$

The analysis of the optical spectrum by Reisfeld and Crosby[3] shows that the levels may be assigned in this case as follows:

$$
\begin{array}{ll}
\Gamma_6 & 14{,}245 \text{ cm}^{-1}, \\
\Gamma_8' & 12{,}705 \text{ cm}^{-1}, \\
\Gamma_7' & 6928 \text{ cm}^{-1}, \\
\Gamma_8 & 4587 \text{ cm}^{-1}, \\
\Gamma_7 & 0 \text{ cm}^{-1}.
\end{array}
$$

Neglecting orbital reduction effects, they report a best fit to these levels for the parameters $A_4{}^0\langle r^4 \rangle = 2351$ cm^{-1}, $A_6{}^0\langle r^6 \rangle = 208.8$ cm^{-1}, $\zeta = 1955$ cm^{-1}. Assuming these parameters and the experimental g-factor of -0.709, Rigny and Plurien[2] conclude that $k = 0.78$.

The complete analysis of the optical data in terms of the model we have assumed requires five parameters: $A_4{}^0\langle r^4 \rangle$, $A_6{}^0\langle r^6 \rangle$, ζ_{5f}, k, and k', where ζ_{5f} is the *free ion* spin-orbit coupling constant. Unfortunately, ζ_{5f} has not been experimentally determined. Calculations using relativistic wavefunctions allow us to make an estimate of $\zeta_{5f} = 2172$ cm^{-1}, however.[6] This value is obtained by making a modified Hartree-Fock calculation of the $J = \frac{7}{2}$ and $J = \frac{5}{2}$ states, the difference being theoretically given by $\frac{7}{2}\zeta_{5f}$. By comparison with such experimental values as are available and making appropriate correction for changes in nuclear charge, the above value is obtained with an estimated reliability of ± 50 cm^{-1}.

Using this value of ζ_{5f}, an exact fit to the optical transitions can be obtained by adjustment of the remaining parameters. A set of numbers which is probably more meaningful is obtained if one requires an exact fit to the g-factor and an optimum fit to the optical levels, however. In this way we find $A_4{}^0\langle r^4 \rangle = 2602$ cm^{-1}, $A_6{}^0\langle r^6 \rangle = 224.9$ cm^{-1}, $k = 0.77$, and $k' = 0.53$. It will be observed that the value of k obtained in this way is very nearly the same as that obtained by Rigny and Plurien.[2]

In order to interpret these values further, it is necessary to have values for the overlap integrals (10). We obtained these as follows: An approximate *real* wavefunction which fits the outer regions of the $5f$ function well (within 0.4% for $a_0 \leq r \leq 5a_0$) was obtained by using a nonlinear least squares procedure. The square of a function composed of Slater-type orbitals was fit to the probability amplitude calculated from a relativistic SCF wavefunction. The result is

$$
\psi_{5f} = -0.41311\phi_1 - 0.44251\phi_2 - 0.24029\phi_3, \tag{13}
$$

where ϕ_1, ϕ_2, and ϕ_3 are Slater-type $5f$ functions with orbital exponents of 4.9584, 3.33482, and 2.3084, respectively. The overlap was calculated with fluorine atom $2s$ and $2p$ wavefunctions taken from Clementi,[7] using tables constructed by Brown and Fitzpatrick.[8]* For the internuclear separation of 1.98 Å appropriate for $CsUF_6$ we obtain $S_\pi = -0.028$, $S_\sigma = -0.063$, $S_s = -0.057$, and $S'_\pi = -0.036$.

From these overlap integrals and the experimental orbital reduction factors, it is possible to solve for the molecular orbital coefficients. For k we have

$$k \equiv k_{\Gamma_5\Gamma_5} = 1 - \frac{2\alpha'^2_\pi}{(1 - 4\alpha'_\pi S'_\pi + \alpha'^2_\pi)} = 0.77, \tag{14}$$

which leads to $\alpha'_\pi = 0.37$. Using $\langle 2x| \, \partial/\partial x \, |2s\rangle = -0.4$ a.u.,[9] we have for k'

$$k' \equiv k_{\Gamma_4\Gamma_4} = 1 - \frac{1.33\alpha_\pi^2 + \alpha_\sigma^2 + \alpha_s^2 + 0.943\alpha_\pi(\alpha_\sigma - 1.50\alpha_s)}{1 + 0.112\alpha_\pi + 0.178\alpha_\sigma + 0.161\alpha_s + \alpha_\pi^2 + \alpha_\sigma^2 + \alpha_s^2} = 0.53. \tag{15}$$

Unfortunately, it is not possible to solve for the other individual molecular orbital coefficients from this complex expression. We must content ourselves to point out that a considerable degree of covalency in the Γ_4 as well as in the Γ_5 molecular orbitals is implied.

CONCLUSIONS

It has unfortunately not been possible to make a thorough analysis of covalent bonding in actinide metal ions because of the paucity of experimental data in most instances. Attempts to extend the present analysis to NpF_6 and $PaCl_6^{2-}$, for example, are complicated by the fact that the optical assignment is ambiguous in the former case,[1] and not all the transitions can be observed in the latter.[10]

It seems evident from the work that we have presented here that although the overlaps are very small, the effect of covalency is certainly not negligible. Unfortunately the present analysis allows only one MO coefficient to be assigned a specific numerical value, but the orbital reduction factors we have derived do in fact show the trend that one would expect on the basis of physical arguments (i.e., $k' < k$). The actual magnitude of orbital reduction is, of course, dependent upon the value of ζ_{5f} taken for the free ion. The values

---

* The $1s-4f$ ($1s-5f$ according to Slater's rules) table is not given by Brown and Fitzpatrick but was constructed by us together with $1s-5f$ and $1s-6f$ tables. These tables are available upon request.

quoted here are not much affected for variations of this parameter within the stated limits of uncertainty.

The extension of this analysis to oxide systems where even more covalency might be expected should be of considerable interest. Some optical data for octahedral oxides have been reported by Kemmler-Sack.[11] In her analysis she argues that crystal field effects should be stronger in the case of the oxides, but in fact reports weaker crystal fields than are found in the fluorides, which makes her analysis rather suspect.

It would also be of interest to explore the relationship between the parameters for octahedral coordination and those for eightfold cubic coordination. Some compounds of the latter type are being investigated in this laboratory at the present time.

ACKNOWLEDGMENT

We wish to thank Mr. R. C. Kerns for assistance with calculation of the overlap integrals.

REFERENCES

1. J. C. Eisenstein and M. H. L. Pryce, *Proc. Roy. Soc.* (*London*), *A*255, 181 (1960).
2. P. Rigny and P. Plurien, *J. Phys. Chem. Solids*, 28, 2589 (1967).
3. M. J. Reisfeld and G. A. Crosby, *Inorg. Chem.*, 4, 65 (1965).
4. K. W. H. Stevens, *Proc. Roy. Soc.* (*London*), *A*219, 542 (1953).
5. J. H. M. Thornley, *Proc. Phys. Soc.*, 88, 325 (1966).
6. W. B. Lewis, J. B. Mann, D. A. Liberman, and D. T. Cromer, *J. Chem. Phys.*, 53, 809 (1970).
7. E. Clementi, "Tables of Atomic Functions," IBM Corporation, 1965.
8. D. A. Brown and N. J. Fitzpatrick, *J. Chem. Phys.*, 46, 2005 (1967).
9. J. Owen and J. H. M. Thornley, *Rept. Progr. Phys.* 29, 675 (1966).
10. J. D. Axe, "The Electronic Structure of Octahedrally Coordinated Protactinium(IV)," University of California Radiation Laboratory Report UCRL-9293 (1960).
11. S. Kemmler-Sack, *Z. Anorg. Allgem. Chem.*, 363, 295 (1968).
12. B. R. Judd, *Operator Techniques in Atomic Spectroscopy*, McGraw-Hill, New York. 1963.

The Calculated Heat of Adsorption of Water on Mercury, Silver, Gold, and Platinum*

DONALD D. BODÉ, Jr.

Lawrence Radiation Laboratory, University of California, Livermore

* This work was performed under the auspices of the U.S. Atomic Energy Commission.

A Personal Note

REMINISCENCE OF THE DEAN

Sometimes the Dean would let his pearls of wisdom fall softly in the quiet of his book-bedecked office. I particularly remember the occasion when he was writing at the board and fumbled the chalk. Quickly he stooped and snatched the piece in flight. Without a word he continued his paragraph. At the end, almost as a footnote, he added, "In life you may drop the chalk. Grab it and continue as though nothing happened. You'll amaze yourself and your audience and be marked as an accomplished man." I have since forgotten the subject of that particular lecture, but I'll not forget the twinkle in his eye.

INTRODUCTION

The heat of adsorption of water on a metal surface is needed in calculations explaining ion adsorption on metals.[1] A major contribution to the free energy of adsorption of an ion is the enthalpy lost by displacement of adsorbed water and by the reorientation of neighboring adsorbed water molecules in the hydration sphere when they have rotated their dipoles due to the influence of the adsorbed ion. Other contributions depend on changes in total hydration of the ion. This is a result of the angular relationship of the water and the ion and their respective distances from the metal surface.

This paper presents a calculation of the heat of adsorption (ΔH_{wm}) for water on mercury, silver, gold, and platinum as a function of the distance from the center of the molecule to the surface and as a function of the orientation of the water dipole. In a later paper the results will be used in the calculation of the free energy of adsorption of anions on those metals. The ultimate aim is to explain potentials of zero charge of hydroxyl and halide ions obtained by the open-circuit scrape method.[2,3]

CONTRIBUTIONS TO ΔH_{wm}

For two close atoms the dispersion energy is

$$U = \frac{-6mc^2\alpha_w\alpha_m}{\alpha_w/\chi_w + \alpha_m/\chi_m} \cdot \frac{1}{R^6}, \tag{1}$$

where m is the electronic mass, c is the speed of light, and R is the internuclear distance.[4] α_m and α_w are the polarizabilities and χ_m and χ_w are the diamagnetic susceptibilities of the metal and adsorbate atoms, respectively. When the dispersion energy is summed over the entire adsorbent, the result is

$$U_{\mathrm{disp}} = \frac{-Nmc^2\alpha_w\alpha_m}{\alpha_w/\chi_w + \alpha_m/\chi_m} \cdot \frac{1}{r^3}, \tag{2}$$

363

where N is the number of adsorbent atoms per cubic centimeter assuming 90% coverage by water,[1,5] and r is the normal distance from the atom to the metal surface. Variations in U_{disp} arise from the direction-dependent polarizabilities of water as reported by Le Fèvre.[6] Thus, as water rotates from an orientation where the oxygen is nearest the metal and the hydrogens furthest from the metal to a position where one hydrogen becomes closer to the metal, α_w changes from 1.32×10^{-24} cm^{-3} to 1.56×10^{-24} cm^{-3}. In this work, as the water rotates, α_w is taken as the sined contribution of the directional polarizabilities.

The polarizabilities of Ag and Au are found after the method of Bockris and Swinkels,[7] where $\alpha_m = 3V_m/4\pi N_{Av}$. V_m is the molar volume and N_{Av} is Avagadro's number. They found for mercury that the calculated α_m differs from the experimental value by 14% so they correct the calculated Pt value by this amount; the same was done here.

For the repulsion energy one can use the scheme of Kemball.[8] He employed the expression of Pollard for molecules possessing $1s$ electrons near a metal surface.[9]

$$U_{rep} = \frac{e^2[p + q(3p - 1)(1 + r/a)]}{(2a/3)e^{r/a} - (1 - r/a)},$$

$$p = \frac{(\rho/a)\cosh(\rho/a) - \sinh(\rho/a)}{(\rho/a)^3},$$

$$q = \frac{\cosh(\rho/a) - 1}{(\rho/a)^3}, \tag{3}$$

$$\rho = 2^{1/2}r_s,$$

$$a = \left(\frac{h^2\alpha_e}{32\pi me^2}\right)^{1/4},$$

where r_s is the radius of a sphere containing one metal electron of a given spin (assumed to be one electron per metal atom), r is the distance from the surface, and α_e is the polarizability for a $1s$ electron. He applied it to benzene by considering the 30 valence electrons as $1s$ electrons and dividing them into groups with the same α_e. With a water molecule there are eight bonding electrons, $(2s)^2(2p_x)^2(2p_y)^2(2p_z)^2$, contributing to α_e. The two p_x electrons project out of the oxygen toward the metal and would be expected to have a greater contribution to the polarizability and repulsive energy. However, in this work, α_e is taken as $\alpha_w/8$.

When the water molecule is rotated so that one hydrogen comes closer to the metal surface, there is more electron repulsion and there is a change in α_e (as discussed above with α_w and the dispersive energy). Thus the sined contribution of α_e and number of electrons are used in (3).

Rather than derive the image interaction of the water as a dipole,[10] consider the water molecule as three separate charges. Using partial charges of $0.98e$ for the oxygen and $0.48e$ for each of the hydrogens,[11,12] the image energy for each charge is[13,14]

$$U = \frac{-e^2}{4r} + \frac{e^2}{16W_a}\frac{e^2}{4r^2},\tag{4}$$

where W_a equals the sum of the electronic work function and the Fermi energy for adsorbent metal (with respect to $W = 0$ as $r \to \infty$) and r is the perpendicular distance from the center of the charge to the metal surface. As the water is rotated, however, the shielding of one charge by another in the molecule changes. The fraction of surface that the oxygen obscures is $\bar{\phi}/2\pi$ where $\bar{\phi}$ is the average projected angle of the tangents from the center of the hydrogen to the surface of the oxygen,

$$\bar{\phi} = 2\frac{\int_0^{s=(R_w^2-R_H^2\cos^2\theta_w)^{1/2}} \tan^{-1}(s/R_H\sin\theta_w)\,ds}{\int_0^{s=(R_w^2-R_H^2\cos^2\theta_w)^{1/2}}\,ds},\tag{5}$$

Table I

	Hg	Ag	Au	Pt	Refs.
W_a (10^{12} ergs)	1.54	1.53	1.54	1.44	1, 20–22
α_m (10^{24} cm^{-3})	5.05	3.51	3.48	3.10	1, 7
m (10^{29} cm^{-3})	−5.55	−4.25	−4.86	−3.30	20
N (10^{-22} cm^{-3})	4.08	2.33	2.35	2.64	23
r_s (Å)	2.27	2.01	2.09	1.93	

$$\alpha_w -2.13 \times 10^{-29} \text{ cm}^{-3} \text{ (ref. 18)}$$

where R_w is the radius of water, R_H is the O—H bond distance, and θ_w is the H—O—H bond angle. A similar geometrical relationship holds for the surface blocked by the hydrogens. In this case the oxygen will be able to see over or under the horizon of the hydrogen. To include this the limits of integration for summing the image energy over the surface must be altered. Equation (4) would have included the factor $1 - (\sin^2\psi_2 - \sin^2\psi_1)\cdot\bar{\phi}/2\pi$, where ψ_2 and ψ_1 are the angular limits formed by the normal to the surface and the tangents from the oxygen to the hydrogen at its uppermost and lowest points. Finally, the total image energy is taken as

$$U_{\text{image}} = -|U_{\text{oxygen}} - 2U_{\text{hydrogen}}|.\tag{6}$$

Values for some terms used in this work are given in Table I.

The heat of adsorption now is

$$\Delta H_{wm} = U_{\text{disp}}(r) + U_{\text{rep}}(r) + U_{\text{image}}(r).\tag{7}$$

Table II

	Hg	Ag	Au	Pt
U_{rep} (kcal/mole)	4.47	3.52	3.21	3.74
U_{disp} (kcal/mole)	−7.22	−3.54	−3.41	−3.57
U_{image} (kcal/mole)	−15.1	−16.0	−15.3	−16.7
ΔH_{wm} (kcal/mole)	−17.8	−16.0	−15.5	−16.4
r_e (Å)	2.09	1.99	2.03	1.93

Lateral water–water interactions are not included, so ΔH_{wm} is the heat of interaction of an H_2O molecule and the metal surface. The minimum ΔH_{wm} was determined by varying r between 1 and 4 Å using a CDC 6600 computer.

RESULTS

Table II reports the contributions to ΔH_{wm} for the most preferred orientation as shown in Fig. 1. The only experimentally measured value for ΔH_{wm} is

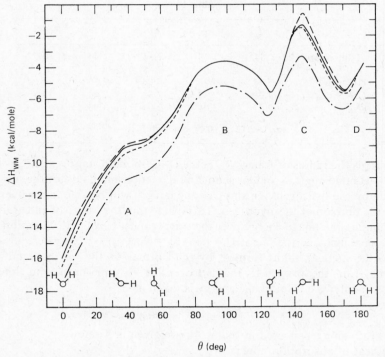

Fig. 1. Heat of adsorption of water on metals: (— —) Au; (—)Ag; (— − −)Pt; (− . −)Hg.

−17.6 kcal/mole for mercury.[15] The normal configuration has the hydrogens furthest from the surface.[10,16] This compares favorably with the work presented here.

For Ag, Au, and Pt no measurements of ΔH_{wm} are available. Indirectly the effects of water–metal interaction are evident in the dependence of the contact angle and the zeta potential on the metal. The contact angles for silver, gold, and platinum have been measured in hydrogen and nitrogen atmospheres.[17] In both, the order of decreasing angle is Au > Ag > Pt. Assuming that the surface tension between the solid and vapor and between the liquid and vapor is nearly equal among the metals, the order of increasing wetting of the metals and the decreasing water–metal interaction would be Au > Ag > Pt.

Looking at water–metal interactions as reflected by the zeta potential, one cannot draw any exact conclusions, though the ordering of the potentials follows the results here. In recent work, values of 49.0, 61.0, and 60.8 mV were reported for Ag, Au, and Pt, respectively, in distilled water.[18] The metals were allowed to remain in water for a considerable length of time, however, and oxide layers could have formed. The low value for Ag is explained by the author as related to the weaker metal–oxide bond and to differences in cleanliness of the metal prior to the experiment. In an older work the zeta potential is reported for suspended Ag, Au, and Pt in water as 34, 32, and 30 mV, respectively.[19]

The image energy is the main contributor to ΔH_{wm}. In Fig. 1 four positions of the water molecule are labeled A–D. The inflection at A occurs as the uppermost hydrogen passes through its maximum distance from the metal. The maximum at B comes just after the average distance of the hydrogens from the surface equals that for the oxygen and when the lower hydrogen passes through a position where it obscures the most surface area from the oxygen. At C the uppermost hydrogen has passed the point where it is equidistant with the oxygen from the surface and enters the region of its maximum shielding of the surface from the oxygen. Approaching D the shielding of the more advanced hydrogen is increasing while that for the second hydrogen is still in the region of maximum effect. These considerations of the image energy explain the variations in ΔH_{wm}. One atom reduces the ability of the others to induce charge on the surface. This effect combines with the changes in repulsive and dispersive energy due to the direction-dependent polarizability and the number of bonding electrons near the surface.

The equilibrium separation varies from A to D for mercury as 2.54, 2.73, 3.00, to 2.38 Å. The other metals show the same trend and differ from mercury by no more than $\frac{1}{2}$ Å.

Rather than let each valence electron contribute equally to α_e in U_{rep}, one can take the contribution of α_e for the p_x orbitals that project from the oxygen as twice the contribution of the other electrons. ΔH_{wm} then has minimum values of -16.7, -14.9, -14.5, and -15.2 for Hg, Ag, Au, and Pt when $\theta = 0°$.

The equations for the heat of adsorption of water on a metal as a function of the separation distance and of the angular orientation of the water will be used in future work to find the optimum free energy of adsorption for anions on these metals. In one theory of ionic adsorption on mercury the separation distance for water was assumed to be equal to the radius of the molecule and the energy difference in primary, adsorbed waters reorienting in the field of the adsorbed ion would be one-half the experimental ΔH_{wm}.[1] It is expected that this work will not only make it possible to extend this theory to other metals but it will improve the assumption of the theory by making more accurate predictions of the changes in number of molecules of water of hydration during adsorption.

REFERENCES

1. T. N. Andersen and J. O'M Bockris, *Electrochim. Acta*, **9**, 347 (1964).
2. D. D. Bodé, Jr., T. N. Andersen, and H. Eyring, *J. Phys. Chem.*, **71**, 792 (1967).
3. T. N. Andersen, J. L. Anderson, D. D. Bodé, Jr., and H. Eyring, *J. Res. Inst. Catalysis, Hokkaido University*, **16**, 449 (1968).
4. A. Muller, *Proc. Roy. Soc. (London)*, **A154**, 624 (1936).
5. A. I. Sarakhov, *Proc. Acad. Sci., Phys. Chem. Sect.*, **112**, 55 (1957).
6. C. G. LeFèvre, R. J. W. LeFèvre, B. Purnachandra Rao, and A. J. Williams, *J. Chem. Soc.*, **1960**, 123.
7. J.O'M. Bockris and D. A. J. Swinkels, *J. Electrochem. Soc.*, **111**, 737 (1964).
8. C. Kemball, *Proc. Roy. (London)*, **A187**, 73 (1946).
9. W. G. Pollard, *Phys. Rev.*, **60**, 578 (1941).
10. J. O'M. Bockris, M. A. V. Devanathan, and K. Muller, *Proc. Roy. Soc. (London)*, **A274**, 55 (1963).
11. D. D. Eley and M. G. Evans, *Trans. Faraday Soc.*, **34**, 1093 (1938).
12. J. D. Bernal and R. H. Fowler, *J. Chem. Phys.*, **1**, 515 (1933).
13. R. G. Sachs and D. L. Dexter, *J. Appl. Phys.*, **21**, 1304 (1950).
14. P. H. Cutler and J. J. Gibbons, *Phys. Rev.*, **111**, 394 (1958).
15. C. Kemball, *Proc. Roy. Soc. (London)*, **A190**, 117 (1947).
16. R. Parsons, "Equilibrium Properties of Electrified Interfaces," in *Modern Aspects of Electrochemistry*, Vol. 1, Butterworth, London, 1954, p. 169.
17. F. E. Bartell and M. A. Miller, *J. Phys. Chem.*, **40**, 895 (1936).
18. R. M. Hurd and N. Hackerman, *J. Electrochem. Soc.*, **103**, 316 (1956).
19. S. Procopiu, *Z. Physik. Chem.*, **A154**, 322 (1931).
20. *Handbook of Chemistry and Physics*, 49 ed., Chemical Rubber Publishing Co., Cleveland, 1968.
21. C. Kittel, *Introduction to Solid State Physics*, 2d ed., Wiley, New York, 1956, p. 250.
22. J. Friedel, P. Lenglart, and G. Leman, *J. Phys. Chem. Solids*, **25**, 781 (1964).
23. R. W. G. Wyckoff, *Crystal Structures*, 2d ed., Vol. 1, Interscience, New York, 1965, p. 10.

Intermolecular Potentials and Macroscopic Properties of Argon and Neon from Differential Collision Cross Sections

C. R. MUELLER, B. SMITH, P. McGUIRE, W. WILLIAMS, P. CHAKRABORTI, AND J. PENTA

Chemistry Department, Purdue University, Lafayette, Indiana

Abstract

The differential cross sections of argon and neon have been measured by using refinements of the modulated molecular-beam technique. From these measurements the intermolecular potentials were found. These potentials differ significantly from the Lennard-Jones potential. The neon and argon potentials have different shapes and are not related by any simple scaling factor. The macroscopic properties have been calculated and are in reasonable agreement with experiment. The face-centered cubic structure was found to be the most stable crystal lattice for neon. The effect of the argon potential on the critical properties and saturation pressures is also discussed.

INTRODUCTION

The molecular beam technique is a powerful tool for investigating the inter-molecular forces of atoms and molecules. The problem of measuring the differential cross section is somewhat formidable because the scattered beam intensities are a factor of 10^2–10^5 lower than in total cross-section measurements. In self-scattering measurements where both beams are of the same species, the scattered signal must be extracted from a background that is 10^4–10^8 greater. Absolute differential cross sections also require a knowledge of scattering particle density as well as the scattering path length, but these difficulties can be overcome by operation of the secondary beam well below the effusive limit.[1]

The background problem can be solved by a careful application of the modulated molecular beam technique.[2] Since electron bombardment detection is required for argon and neon, the relative inefficiency must be compensated for by sharp tuning of the ion current and by extended signal integration.

THE DETECTION SYSTEM

Apart from the detection system, the apparatus has been previously described in the literature.[1] A molecular beam was modulated at 41 cps by a tuning fork chopper. The driving current for this chopper was used to synchronize a lock-in amplifier and an enhancetron. The scattered beam was ionized by an E.A.I. quadrupole mass spectrometer. Mass selection was found to be useful in eliminating noise from the electron-bombardment source, even though it does not discriminate against the background gas.

After it emerged from an electron multiplier, the signal was fed into a pre-amplifier that consisted of four cascaded active bandpass filters (White Instrument Inc., Austin, Texas). Without these filters, the signal-to-noise ratio would have precluded any measurements of the differential cross

371

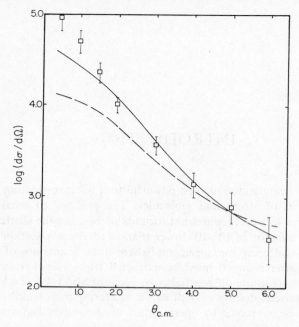

Fig. 1. The differential cross section of neon. The squares represent experimental data and the error bars, average deviations. The solid line represents the values calculated from the scattering potential of Fig. 3. The dashed curve was calculated using the Lennard-Jones potential obtained from virial coefficient data.

section. These filters increased the signal-to-noise ratio by a factor of approximately 10^2, reduced integration times, and eliminated some signals that were not averaged out by the lock-in amplifier.[3] At the lowest signal levels there appeared to be high frequency noise modulated at the beam frequency, which could be eliminated only after four stages of tuned amplification. These amplifiers must be carefully shielded with separate power supplies for each stage. The resultant bandwidth was approximately 0.2 cps.

The enchancetron was found to be an effective signal averager because its display of phase shifts allowed diagnosis of the background at low signal levels. We have used a four-cycle procedure (beam stops on both beams) with direct subtraction of the background, and this eliminates phase shifts in the scattered signal. Increasing the duration of signal averaging from 3 sec to 30 min improved the signal-to-noise ratio another factor of 10^2. Sharp tuning and long integration times were absolutely essential to the success of these experiments. Earlier experiments were done with an Aberth ion-gun with emission currents of 200 mA, but the noise from such a source completely vitiates its greater ionization efficiency.

The differential cross sections reported in this paper are absolute measurements for an apparatus having our geometry and resolution. The geometrical and resolution corrections are dependent, to some extent, on knowledge of an intermolecular potential and are included in the theoretical calculations. The largest corrections are for the finite resolution of the detector.

Any absolute measurements must include accurate knowledge of the number density of scattering centers. As a cross check on these figures, our measured total for argon, after resolution corrections, was 340 Å², which is just within the experimental limits of Rothe's measured value of 319 Å² ± 7%.[4]

The experimental data was reduced to the center-of-mass system by the method of Berkling[5] and Bernstein.[6] These reductions appear to be largely independent of an assumed intermolecular potential as long as the cross sections are calculated for relative velocities. The error bars in Figs. 1 and 2 indicate average deviations for a single cycle.

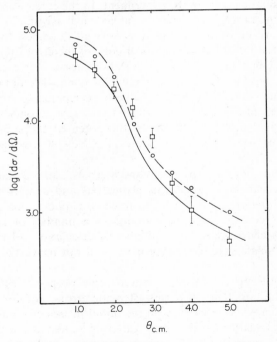

Fig. 2. The differential cross section of argon for various potentials. Symbols are those of Fig. 1 and the potentials are illustrated in Fig. 4.

The cycles were repeated until statistical scatter was reduced to a reasonable level. The statistical scatter of a single cycle was largely due to a variation of the mass spectrometer output with a roughly periodic variation of a few minutes duration. The amount of variation increased with decreased signal, and longer integration times are required to smooth out these irregularities. The most probable error at the poorest data point (neon at 6°) was ±15% (for groups of 10 cycles).

Most of the conclusions about scaling of the potentials and trends in potential shapes were based on factors of 2 or more in the magnitude of the differential cross section. On the basis of the statistical scatter and the use of trial potentials to calculate the differential cross sections, we would set limits on the most probable error of the potentials at about 7% for the forces around 5 Å, 10% in the region near the potential minimum, and 30% in the region near σ.

The actual error on the graphs of the potentials are roughly two to five times larger than the statistical errors would require. The increase in indicated error over statistical error was made to indicate lack of fit or the possible presence of systematic errors in the experiment. In the case of neon the error band on the long-range forces has been widened asymmetrically at the greater distances to indicate a lack of fit at very small angles. In the case of argon the potential indicated by the upper edge of the error and the Barker potential[7] almost bracket the average deviations for a single experimental cycle, and hence were generally outside the 90% confidence limit. The error bands on the potential extend almost to the Barker potential from 5 to 7 Å and elsewhere extend much lower. The asymmetric broadening of the error band is based on 15 variations of the potential shape and include, as in the case of neon, some account of the lack of uniqueness of the data.

Systematic error, both in experiment and analysis, can only be eliminated by using a wide variety of apparatus geometries and analytical methods. In the present situation, with respect to research support, this can only be done by a cross comparison of the results of a number of laboratories. As a preliminary check we report a number of calculations of macroscopic properties. Any single macroscopic property will not prove the absence of systematic error.

Reasonable agreement with many macroscopic properties will probably only eliminate the possibility of large systematic errors, and for this reason such calculations are very important in the initial phases of scattering work. Agreement of calculations, using our scattering potentials, with observed macroscopic properties lead us to believe that our error bands are on the conservative side.

DIFFERENTIAL CROSS SECTIONS AND INTERMOLECULAR POTENTIALS

The experimental differential cross section for neon, which is given in Fig. 1, was considerably higher at small angles, and at larger angles dropped below that of the Lennard-Jones potential. Qualitatively this requires an increase in the intermediate and long-range forces and some reduction of the potential in the neighborhood of the minimum.

The experimental results for argon, which are given in Fig. 2, indicate a quite different mode of deviation from the Lennard-Jones potential. Originally an attempt was made to fit the argon and neon data by a potential containing three segments. Two of these segments are Morse potentials joined at the minimum but with different curvatures. The third segment was a 6–8 potential which began at 4–5 Å. It soon became necessary to add an additional segment to cover the region just beyond the potential bowl and extending to about 6 Å.

The inversion of differential scattering cross section has been previously described.[8] To determine the potential parameters C_1 to C_j, a least squares criterion was used to minimize the sum of the squares of the difference between the experimental differential cross section $I(\theta_m)$ and the computed differential cross section $\sigma(\theta_m)$ at various angles, θ_m. With a given distribution of velocities, $\sigma(\theta_m)$ can be expanded in terms of the potential parameters and an initial estimate $\sigma'(\theta_m)$ from a starting potential with parameters C_k'.

$$\sigma(\theta_m) = \sigma'(\theta_m) + \sum_k (C_k - C_k')\left(\frac{\partial \sigma'(\theta_m)}{\partial C_k'}\right) \tag{1}$$

The difference between the sum of the squares of the differential between the cross section is minimized by the standard matrix least squares method. The partial derivatives are obtained by the method of partial waves.

In practice, the inversion procedure must be modified with Maxwellian beams to save computer time. Instead of determining all parameters at the same time, it is quicker to determine a smaller set of parameters that affect a given angular region. The potentials we have used are not quite flexible enough for an exact fit of the data, and it may be preferable to fit the more reliable points.[3] In the case of neon we could not fit the low-angle data exactly without increasing the number of parameters. We chose to fit the other data at the expense of these points, and this gave a potential that also fit the macroscopic potentials. A similar situation occurred fitting the argon data. The arbitrary selection of reliable points over a straight least squares fit resulted

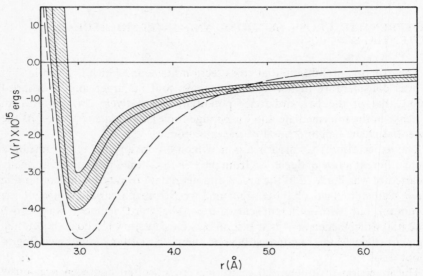

Fig. 3. Intermolecular potential of neon. The hatched area represents the error band. The solid line inside this area represents the scattering potential and the dashed curve is the Lennard-Jones potential.

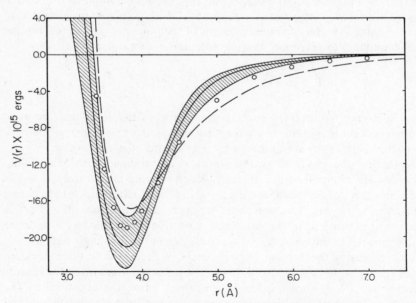

Fig. 4. Intermolecular potential of argon. Symbols are those of Fig. 3. The open circles represent the Barker potential.

in potential variations less than 10% and hence any differences are well within the error bands.

The potentials, inside the error bands of Figs. 3 and 4 indicated by solid lines gave the fits of the scattering data shown in Figs. 1 and 2. The neon potential is potential B of reference 3 and is the best fit of the most reliable points. The fit of the argon data is complicated by a region between $2\frac{1}{2}$ and $3°$ that was poorly described by every potential. Our scattering potential gives a slightly poorer fit in this region than the Barker and Lennard-Jones potentials, but is superior everywhere else.

The scattering measurements are quite sensitive to the region slightly outside the potential minimum and inside of 1.6 times the potential bowl separation. They become increasingly less sensitive as one moves into the repulsive part of the potential.

This variation in sensitivity is related to how uniquely one can specify the potential.[8a] Theoretically the potential is uniquely determined by exact data over the entire angular region and provided that the presence of quasi-bound states can be suitably taken into account. In practice, the data are over a limited angular region with some uncertainty. As a consequence, the intermediate-range forces are better determined than the repulsive forces. The data do, however, cover the crucial region between $0°$ and the first maximum in the differential cross section.

The use of segmented potentials would not appear to have any important effect on the reliability of the potential in view of the magnitude of experimental errors. It does appear, however, that spline fits are more suitable if the accuracy of the data improves, but this involves an increase in the number of parameters that must be handled and an increase in computer time.

The angular region in our experiments covers some structure in the differential cross section that is washed out by the absence of velocity selection. Energy control to permit exploration of the repulsive part of the potential and the extension of the measurements to cover higher angles would probably be more important than velocity selection in establishing the repulsive part of the potential.

Coupling the scattering measurements with the macroscopic properties provides a rather severe test of the potential. If, as with neon, the long-range and intermediate forces are increased to fit the scattering measurements, virial coefficient data can only be fitted by a severe reduction in the region of the potential bowl. In the case of argon, if the scattering data indicate a reduction in forces at larger distances, an increase in the depth of the potential is absolutely necessary to account for the macroscopic properties.

On the basis of scattering experiments alone, potential energies are probably uncertain to about 15%. If these potentials also fit macroscopic properties the uncertainty is greatly reduced.

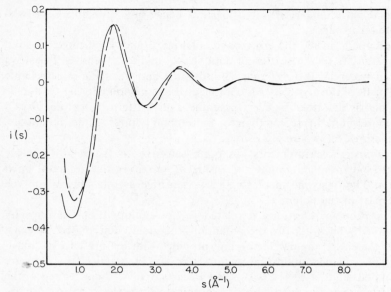

Fig. 5. X-ray scattering intensity function or Fourier intensity kernel $i(s)$ as a function of $s = (4\pi/\lambda) \sin (\theta/2)$; θ is the scattering angle for Ar at a density of 0.280 g/cm^3 and a temperature of $-125°$C.

In the neon data, the fit of the differential cross-section data at low angles can only be improved by breaking the potential beyond 6 Å into two regions. A similar situation may also apply to argon except that response in the $2\frac{1}{2}$–3° region seems to be confined to the potential inside the minimum. It is an outside possibility that quasi-bound states play a role, but probably a more flexible potential is required.

Argon has been intensively investigated and the changes which we have found are in the same direction as those found by Barker,[7] Guggenheim and McGlashan,[9] and Pings.[10] The minimum of the scattering potential is somewhat deeper and the long-range forces somewhat smaller than had previously been expected. It is perhaps not surprising that these differences do not have a dramatic effect on the macroscopic properties. The X-ray scattering data are also not very sensitive to the changes we have made in the potential. This is shown in Fig. 5.

Neon has been less intensively investigated, probably because of the difficulty of reconciling gas-phase data with the solid state. This may be due to the large vibrational amplitudes in the solid state. In the sections that follow, we have carried out more extensive calculations on neon because of the greater deviation of the scattering potential from the Lennard-Jones potential.

GAS-PHASE DATA

In all of the calculations that follow we have used the potentials of Figs. 3 and 4.

In Fig. 6 the fit of the second virial coefficient for neon is illustrated. The fit is good at low temperatures, but at high temperatures the coefficient is slightly high. On the supposition that this is due to the repulsive part of the potential curve, not accessible to low-energy scattering experiments, we have joined our low-energy potential at 2.55 Å with Amdur's[11] potential from high-energy scattering at 2.15 Å with a simple exponential function. This corresponds to energies of 0.04–2.1 eV. The effect of this is to greatly improve the high temperature properties. Figure 7 shows the fit of the second virial coefficient of argon. Figure 8 shows a comparison of the viscosity of neon from various potentials, and Fig. 9 shows a similar comparison of the calculated diffusion coefficients of neon. All of the calculations on neon are

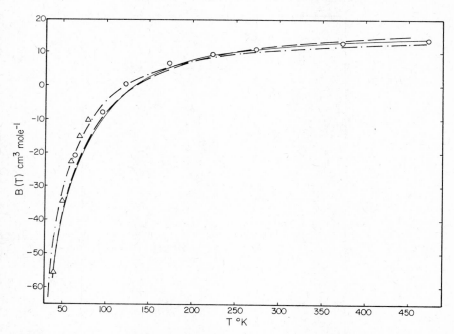

Fig. 6. The second virial coefficient of neon. The triangles are data from Ref. 15a and the open circles, from Ref. 15b. The dot–dashed curve was the Lennard-Jones fit of the virial coefficient data. The solid curve was calculated from the joining of the low energy scattering with Amdur's[11] high energy scattering potential. The dashed curve was from the scattering potential of Fig. 3.

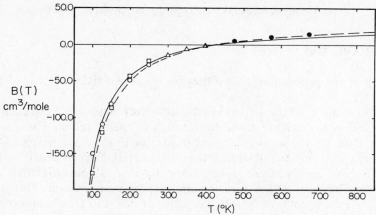

Fig. 7. The second virial coefficient of argon. The results calculated from the scattering potential of Fig. 4 are indicated by the solid line and those of the Lennard-Jones potential, by the dashed curve. The experimental data are represented by squares,[16a] open circles,[16b] triangles,[16c] and solid circles.[16d]

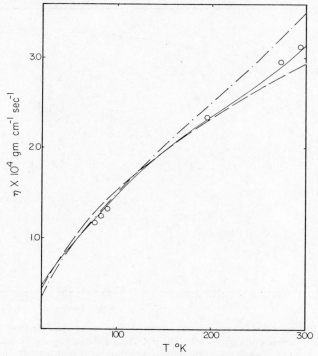

Fig. 8. The viscosity coefficient of neon as a function of temperature. The open circles represent the data of Ref. 17. The notation for the calculated curves is the same as in Fig. 6.

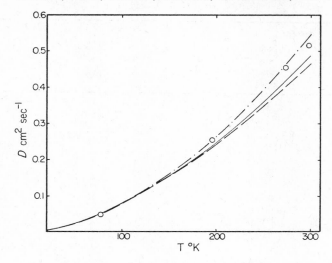

Fig. 9. The diffusion coefficient of neon as a function of temperature. The open circles are from the data of Ref. 18. The calculated curves have the same notation as in Fig. 6.

quantum mechanical except for the second virial coefficient of neon above 100°K. The 6–12 potential was obtained by a fit of the virial coefficient data and does a creditable job but at 30°K it becomes systematically too high for neon.

CRYSTALLINE PROPERTIES

At first sight crystal energy calculations should be relatively simple. In the case of neon, the lattice vibrations are so extensive that a rather elaborate procedure was required, which was similar to a self-consistent field calculation. We started first with the static lattice and found the cell potential, assuming a fixed position for neighboring atoms. This cell potential was then used to calculate the vibratory motion of neighboring atoms, which in turn had a large effect on a recalculation of the cell potential. This procedure was repeated until self-consistency was obtained.

The crystal properties of solid neon and argon at 0°K have been calculated both for the Lennard-Jones 12–6 potential and for the scattering potentials of Figs. 3 and 4.

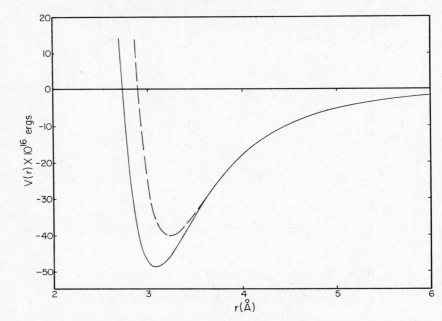

Fig. 10. Effective Lennard-Jones potential of neon in the solid state. The solid curve represents the potential before averaging over the vibrational motion. The dashed curve results from the averaging.

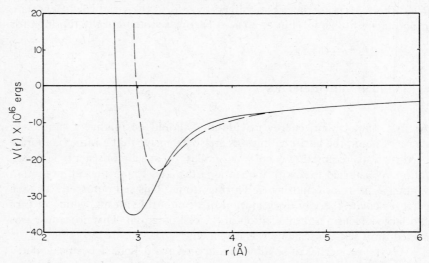

Fig. 11. Effective scattering potential of neon in the solid state. The notation is the same as in Fig. 10.

The static potential energy was calculated for the fcc structure and for the hcp by summing the potential interaction over a distance of about 8 Å from the central atom for both structures. The Lennard-Jones potential showed virtually no difference in the static potential energy of the two structures ($10^{-3}\%$), while the scattering potential for neon showed the fcc structure to be slightly more stable than the hcp (0.3%). In the case of argon, the hexagonal structure was found to be more stable by 0.05%, but a slight change in the scattering potential will reverse the stability of the two structures. Similar considerations should also apply to neon, but somewhat greater changes in the potential would be required.

The crystal properties of fcc neon at 0°K were first calculated from the averaged cell potential by a method similar to that used by Rice[12] and Corner.[13] In examining the calculations it became evident that the harmonic motion of the atoms was large enough to produce an effective or averaged potential which differed markedly from the two-body potentials obtained from scattering data.

We have attempted to take into account this oscillatory motion of the neighboring atoms, by averaging the potential over this motion using an uncoupled isotropic harmonic oscillator approximation. This determined the cell potential of the central atom. We assumed that the motion of the central atom can be described by the eigenfunction of a harmonic oscillator.

The effective potential $\langle u \rangle_d$ for two atoms, a distance d apart, is then obtained by averaging the pair potential $u(r)$ over the motion. We obtained the following expression for $\langle u \rangle_d$,

$$\langle u \rangle_d = \left[\frac{8\xi}{\pi}\right]^{1/2} \frac{e^{-2\xi d^2}}{d} \int_0^\infty u(r) r e^{-2\xi^2} \sinh(4d\xi r)\, dr, \tag{2}$$

$$\xi = \frac{2\pi^2 m v_0}{h}.$$

The results of this averaging procedure is illustrated in Fig. 10 for the Lennard-Jones and Fig. 11 for the scattering potential of neon, assuming a vibrational frequency of 1.2×10^{12} cps for the L-J and a frequency of 1.36×10^{12} cps for the scaled potential. The argon potential is also affected to a lesser extent than the neon potential by this averaging procedure, but the effect is still significant (5% reduction in depth, 1.3% outward shift in the minimum).

This effective potential was used in a self-consistent procedure to calculate the crystal properties. In this procedure, the atoms were at first considered to be fixed on their lattice sites. The frequency of vibration was then calculated by the method stated previously.[12,13] This frequency was then used to

Table I.   Properties of Crystalline Neon

	Sublimation Energy	Zero Point Energy	d_0	Cohesive Energy
Experimental[14]	+448	154	3.155	−602
L-J potential	+355	153	3.16	−508
Scattering potential	+381	185	3.16	−566

calculate an effective potential, which was then used to calculate a new frequency of vibration. This procedure was repeated until the starting frequency and the calculated frequency were the same. Seven to ten cycles are required for neon and three cycles for argon.

The results of these calculations for the L-J and scaled potentials are given in Table I along with the experimental values.[14] The energy of sublimation and zero point energy are given in cal/mole, d_0 is the first nearest neighbor distance in angstroms, and the cohesive energy is in cal/mole. It is to be remembered that this calculation does not take into account the fact that the oscillations of the atoms are actually coupled. This could lead to an over-estimation of the effect of the oscillations on the potential and improve the calculations. The results for argon are given in Table II.

These calculations indicate that a potential derived from crystal data without taking into account the oscillatory motion of the atoms would be seriously in error. Even for argon, sizable corrections need to be made. The cohesive energy listed in Table I is the energy in the absence of zero-point energy. For the scaled potential the error in cohesive energy was 6% low and 15% low for the Lennard-Jones potential. The Lennard-Jones potential had a slightly better zero-point energy so that the error in the energy of sublimation was 21% low, whereas the scaled potential was 15% low. Modification of the potential inside of 2.55 Å does not affect these conclusions. Both the scattering and L-J potentials give excellent agreement with the argon data.

Table II.   Properties of Crystalline Argon

	Sublimation Energy	Zero Point Energy	d_0
Experimental [14]	+1846	187	3.756
L-J potential	+1774	194	3.76
Scattering potential	+1808	189	3.76

LIQUID STATE PROPERTIES

Recently, Barker and Henderson[19] have made significant progress toward a realistic and practical theory of the liquid state by extending Zwanzig's perturbation method[20] in which the free energy is expanded about a rigid-sphere system. This results in an equation for the free energy as a function of the intermolecular potential, a rigid-sphere diameter, and the free energy of the rigid-sphere system. This equation for the free energy as derived by Barker and Henderson[19] is

$$\frac{F}{NkT} = \frac{F_0}{NkT} - 2\pi d^2 g_0(d)\left\{d - \int_0^\sigma (1 - e^{-\beta u(z)})\, dz\right\}$$
$$+ 2\pi\rho\beta \int_\sigma^\infty g_0(d)u(R)R^2\, dR - \pi\rho\beta\left(\frac{\partial p}{\partial P}\right)_0 \frac{\partial}{\partial\rho}\left[\rho\int_\sigma^\infty g_0(R)u^2(R)R^2\, dR\right], \quad (3)$$

where $\beta = 1/kT$, $\rho = N/V$ is the density, $g_0(R)$ is the radial distribution function of rigid spheres, $u(R)$ is the intermolecular potential, σ is defined such that $u(\sigma) = 0$, d is the rigid-sphere diameter, k is Boltzmann's constant, and T is the absolute temperature. The expression $(\partial\rho/\partial P)_0$ refers to the rigid-sphere system, where P is the pressure. F and F_0 are the free energies of the real and rigid-sphere systems, respectively. The equation of state follows from the above equation and is

$$\frac{PV}{NkT} = \left(\frac{PV}{NkT}\right)_0 + \sigma\frac{\partial}{\partial\rho}\left[\frac{F - F_0}{NkT}\right], \quad (4)$$

where $(PV/NkT)_0$ is the equation of state for a system of rigid spheres; $g_0(R)$ was calculated from the Percus-Yevick equation,[21,22] $(PV/NkT)_0$, from (3, 3) Padé approximant derived by Ree and Hoover,[23] and for $(\partial\rho/\partial P)_0$ the compressibility equation of state was derived from the Percus-Yevick equation.

The only parameter that has not yet been defined in (1) is the rigid-sphere diameter d. Barker and Henderson[19] define d such that

$$d - \int_0^\sigma (1 - e^{-\beta u(z)})\, dz = 0;$$

the diameter defined in this manner will be referred to as d_1. We have used an alternate definition of d (referred to as d_2) in an attempt to improve upon the choice of the diameter and have found that, although the two diameters

Table III

	Exp[24]	L-J (12, 6)[19] Potential $d = d_1$	Scattering Potential $d = d_1$	Scattering Potential $d = d_2$
ρ_c (g/cm³)	0.536	0.504	0.56	0.52
P_c (atm)	48.43	56.93	64.85	54.52
T_c (°K)	150.87	161.73	152.70	141.2
$P_c V_c / NkT_c$	0.293	0.34	0.37	0.36

differ by very little, there is a marked effect upon the properties of the system. However, this second choice for d does not lead to an overall improvement of the theory. d_2 is chosen such that $u(d_2) = \frac{3}{2}kT$, that is, to be that point on the repulsive part of the potential which corresponds to the mean kinetic energy of the system. This is the classical turning point.

The calculated critical constants of argon are compared with the experimental values for our scattering potential using both d_1 and d_2 in Table III. The critical constants calculated from the L-J (12–6) potential[19] are also included in this table. The parameters used for the L-J potential are $\epsilon/k = 119.8°$K and $\sigma = 3.405$ Å.

Using the scattering potential, the saturation pressures of the liquid have been calculated, using both d_1 and d_2, and they are compared with the experimental values in Table IV. Also included in this table are calculated and experimental pressures for the compressed liquid.

Using functional differentiation,[26,27] an attempt has been made to derive an equation for the radial distribution function from (3). The derivation will not be included here, but the final result will be given along with some calculations. The equation that has been derived is

$$g(r) = \begin{cases} \dfrac{d^2}{r^2 g_0(d) e^{-\beta u(r)}}, & r < \sigma, \\[2ex] g_0(r) - \left(\dfrac{\partial \rho}{\partial P}\right)_0 g_0(r) u(r) - \rho \left(\dfrac{\partial \rho}{\partial P}\right)_0 \dfrac{\partial g_0(R)}{\partial P} \bigg|_{R=r} u(r), & r > \sigma. \end{cases} \quad (5)$$

Table IV

T (°K)	ρ (g/cm³)	P_{exp} (atm)	Pd_2 (atm)	Pd_1 (atm)
100.94	1.3037	3.46[a]	30.49	−52.27
100.94	1.478	680.46	414.83	239.24
120.18	1.158	12.15[a]	85.74	−0.83
120.18	1.415	680.46	513.02	305.47
143.11	0.885	35.62[a]	96.17	44.99
143.11	1.317	680.12	604.91	381.0

[a] Saturation pressure.

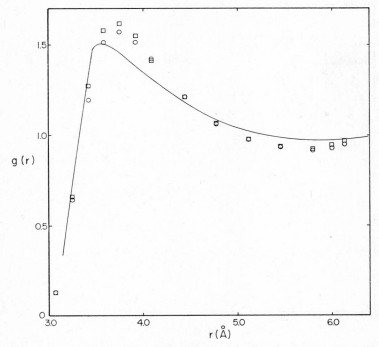

Fig. 12. The radial distribution function for argon calculated from (5) using the Lennard-Jones 12-6 potential and d_2 for the rigid-sphere diameter. The circles and squares are the results of Monte Carlo[28] calculations and calculations based on the Percus-Yevick[29] equation, respectively. All curves were calculated at a temperature of 327.5°K and a density of 0.67 g/cm³.

Two radial distribution functions, as calculated from (5) (using the L-J (12–6) potential and $d = d_2$), are shown in Figs. 12 and 13. In both figures the circles and squares are the results of Monte Carlo[28] calculations and calculations based upon the Percus-Yevick[29] equation, respectively. Both the radial distribution functions were calculated at a temperature of 327.5°K. In Figs. 12 and 13 the densities used in the calculation were 0.67 and 1.86 g/cm³, respectively. The pressures predicted by these radial distribution functions are given in Table V, where MC and PY refer to the Monte Carlo and Percus-Yevick calculations, respectively. Calc. refers to the calculations based on (5).

If we examine the agreement between calculations, based on the various potentials, and the experimental macroscopic properties, the question of how well these properties determine the intermolecular potential must surely arise. The relationship is not simple. Minor modifications of the potential do

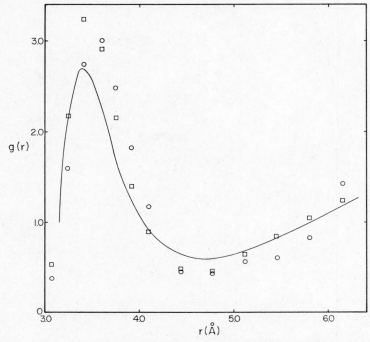

Fig. 13. The radial distribution function for argon calculated from (5). The conditions and notation are the same as in Fig. 12, except that the density is 1.86 g/cm³.

affect the macroscopic properties, sometimes drastically. At the same time, if these data are over a limited temperature range, they can be fitted by a variety of potentials.

The relationship between the transport properties and the scattering cross sections is quite straightforward. If we start from the differential cross section we find that the low-angle data have a greater sensitivity to long-range forces and back-scattering data to the short-range forces. Over the entire angular

Table V

ρ (g/cm³)	$(PV/RT)_{PY}$[a]	$(PV/RT)_{MC}$[b]	$(PV/RT)_{calc.}$
0.67	1.236	1.2–1.5	1.325
1.85	10.0	2.8	7.69

[a] Reference 29.
[b] Reference 28.

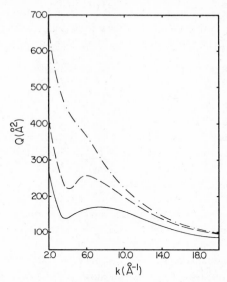

Fig. 14. The total collision cross section of neon as a function of the velocity ($k = 2\pi\mu v/h$). The solid line was calculated from the Lennard-Jones 6-12 potential, the dashed curve from the 6-8 potential, and the dot–dashed curve from the scattering potential of Fig. 3.

range we have reasonable sensitivity to the features of the potential that are explored by the collision.

When we pass to the total collision cross section we give these data a sin θ weighting; even before integration we lost a great deal of sensitivity to the long-range forces. This can be recovered by measurements at lower velocities. Total cross sections are probably effective when measured over at least an order of magnitude in the velocity. In terms of temperature, two orders of magnitude would be required. Figure 14 shows the divergence of various neon potentials at low velocities. By varying the forces beyond 6 Å we have found adequate sensitivity in the total collision cross section at velocities about one-fifth of those at room temperature or roughly at temperatures of 10°K. When we examine the viscosity and diffusion coefficients, the angular weighting deemphasizes even more the glancing type of collision, and this is augmented by velocity averaging. Since the total collision cross section over the complete energy range can uniquely determine the potential, it is quite possible that this uniqueness is carried over to the transport properties. The practical question of whether these measurements can be extended over a wide enough temperature range appears dubious because it appears that these measurements must be extended to temperatures at which vapor pressure is prohibitively low.

CONCLUSIONS

When the analysis of the neon data had been completed, the modification of the potential was so drastic that we expected some difficulty in the calculation of macroscopic properties. Considering the still primitive state of scattering measurements, the agreement with experiment seems reasonable for both argon and neon. The neon potential can and should be subject to further tests. Measurements of X-ray scattering and measurements of the total cross section at low velocities should be sensitive to the intermediate-range forces.

The most crucial measurements would probably be improved angular scattering measurements using the crossed supersonic beam technique. This method will simultaneously deliver an improvement of the signal-to-noise ratio and velocity selection. This will result in an extension of this work to wider angles and simplify the theoretical analysis. The emergence of quantum undulations will almost certainly increase the decisive character of the measurements.

ACKNOWLEDGMENT

This research was supported by the Office of Saline Water, U.S. Department of the Interior and by the Advanced Research Project Agency.

REFERENCES

1. R. W. Landorf and C. R. Mueller, *J. Chem. Phys.*, **45**, 240 (1966).
2. R. T. Brackmann and W. L. Fite, *J. Chem. Phys.*, **34**, 1572 (1961).
3. W. Williams, C. R. Mueller, P. McGuire, and B. Smith, *Phys. Rev. Letters*, **22**, 121 (1969).
4. E. Rothe and R. Neyhaber, *J. Chem. Phys.*, **43**, 4177 (1965).
5. K. Berkling, R. Helbing, K. Kramer, H. Pauly, C. Schlier, and P. Toschek, *Z. Physik*, **166**, 406 (1961).
6. F. Morse and R. B. Bernstein, *J. Chem. Phys.*, **37**, 2019 (1962).
7. J. A. Barker and A. Pompe, *Australian J. Chem.*, **21**, 1683 (1968).
8. (a) J. W. Brackett, C. R. Mueller, and W. A. Sanders, *J. Chem. Phys.*, **39**, 2564 (1963).
    (b) R. E. Olson and C. R. Mueller, *ibid.*, **46**, 3810 (1967).

9. E. A. Guggenheim and M. McGlashan, *Proc. Roy. Soc.* (*London*), **A225**, 456 (1960).
10. P. Mikoloj and C. Pings, *J. Chem. Phys.*, **46**, 1412 (1967).
11. I. Amdur and E. Mason, *J. Chem. Phys.*, **23**, 415 (1955).
12. O. K. Rice, *J. Amer. Chem. Soc.*, **63**, 3 (1941).
13. J. Corner, *Trans. Faraday Soc.*, **44**, 914 (1948).
14. G. Pollack, *Rev. Mod. Phys.*, **36**, 748 (1964).
15. (a) W. H. Keeson and J. Lammeren, *Physica*, **1**, 1161 (1934). (b) J. Otto, *Handbuck der Physik*, Vol. III, 132 Akademische Verlag-gesellshaft, Leipzig, 1929.
16. (a) A. Van Itterbeck and O. Van Paemel, *Physica*, **5**, 845 (1938), using Holborn's data. (b) A. Van Itterbeck and O. Van Paemel, *op cit.*, using Omne's data. (c) A. Michels and H. Wijker, *Physica*, **15**, 627 (1949), (d) L. Holborn and H. Schutze, *Ann. Physick*, **47**, 1089 (1915). L. Holborn and J. Otto, *Z. Physik*, **23**, 77 (1924); **30**, 320 (1924); **33**, 1 (1925).
17. H. L. Johnston and E. R. Gilly, *J. Phys. Chem.*, **46**, 948 (1942).
18. E. B. Winn, *Phys. Rev.*, **80**, 1024 (1950).
19. J. Barker and D. Henderson, *J. Chem. Phys.*, **47**, 4714 (1967).
20. J. Zwanzig, *J. Chem. Phys.*, **22**, 1420 (1954).
21. M. S. Wertheim, *J. Math. Phys.*, **5**, 643 (1964).
22. E. Thiele, *J. Chem. Phys.*, **39**, 474 (1963).
23. F. Ree and W. Hoover, *J. Chem. Phys.*, **40**, 939 (1964).
24. J. M. H. Levelt, *Physica*, **26**, 361 (1960).
25. W. Street and L. Stavely, *J. Chem. Phys.*, **50**, 2302 (1969).
26. S. A. Rice and P. Gray, *The Statistical Mechanics of Simple Liquids*, Interscience, New York, 1965, p. 49.
27. V. Volterra, *The Theory of Functionals and of Integral and Integro-Differential Equations*, Dover, New York, 1959.
28. W. W. Wood and F. R. Parker, *J. Chem. Phys.*, **27**, 220 (1957).
29. A. A. Broyles, *J. Chem. Phys.*, **35**, 493 (1961).

Distribution Function of Classical Fluids of Hard Spheres. I[*][†]

YONG-TEH LEE

Department of Physics, University of Utah, Salt Lake City

FRANCIS H. REE

Lawrence Radiation Laboratory, University of California, Livermore

TAIKYUE REE

Department of Chemistry, University of Utah, Salt Lake City

Abstract

The Born-Green-Yvon (BGY) hierarchy is truncated by introducing a super position approximation $g_{1234} \simeq g_{123}g_{124}g_{134}g_{234}/[g_{12}g_{13}g_{14}g_{23}g_{24}g_{34}]$ to the quadruplet correlation function g_{1234}. The resulting pair of two simultaneous integral equations including the triplet- and pair-correlation functions g_{123} and g_{12}, which hereafter will be called the BGY2 equations, is reformulated to give mathematically simpler forms. The BGY2 theory is checked for its internal consistency by using the hard-sphere potential. The following results have been obtained: (a) the fifth virial coefficient B_5 for the BGY2 theory is $(0.1090 \pm 0.0008) (B_2)^4$ or (0.1112 ± 0.0005)

* Reprinted from *The Journal of Chemical Physics*, **48**, No. 8, 15 April 1968, pp. 3506–3516.

† This work was supported by the National Science Foundation, Grant No. GP-5407. A part of the computorial work was done at the Lawrence Radiation Laboratory under the auspices of the U.S. Atomic Energy Commission.

393

$(B_2)^4$ depending on whether the virial theorem or the Ornstein-Zernike compressibility relation is used in the calculations of B_5; the exact Monte Carlo value, $B_5 = 0.1103(B_2)^4$, lies between the two values. (b) The third coefficient $g_{12}{}^3$ in the density series of g_{12} agrees exactly with the exact value given in the literature if $r/\sigma \geq 2$ (σ is the sphere diameter), and it differs only slightly from the exact value if $0 \leq r/\sigma \leq 2$, i.e., the deviation lies within 3 %. (c) The data for the second coefficient g_{123}^2 of the density series of g_{123} show that the Kirkwood superposition approximation is better for a symmetric configuration than an asymmetric one when particles 1, 2, and 3 lie close to each other. (d) It is shown that the pressure calculated from the virial theorem using the BGY $g(r)$ is exact for one-dimensional hard rods, and is independent of *any* approximation used to describe the g_{123}. A proof that the BGY2 $g(r)$ is exact for the hard-rod system is also given. These results indicate that the BGY2 theory compares favorably with any other theories for $g(r)$ and that it satisfactorily describes low and moderately dense fluids.

A Personal Note

TAIKYUE REE

It was in March of 1939 when I met Henry at Princeton University where I was a visiting scientist from Kyoto University. We started to work together in September of that year. At first we studied the orientation effect of substituents for benzene. Our first paper on this subject may have been the first to calculate successfully the charge distribution in substituted benzenes.

We had a lot of fun at Princeton. The "eraser game" which we invented was especially interesting. This game was an analog to horseshoes; we drew a circle a foot and a half in diameter on the cement floor of my office and standing five yards away threw erasers into the circle. The game was scored as follows: the person throwing his eraser completely inside the circle received the highest score (I forget the exact point); if he threw more than a half portion of the eraser inside the circle he received the second highest score, and so on. We played the game often—I believe at least twice a day. John L. Magee, a postdoctoral fellow at that time, was also one of the game's ardent players. Henry's visitors were often brought to my office to enjoy this game; Joe Hirschfelder and Allen Stearn were among the many who were introduced to it. Henry was the most skillful player among us; his throwing was stable and exact. We tried to beat him, but in vain. At last we discovered the reason why Henry was so skillful. It was very simple: because he did not drink *sake* or coffee at all, he had good equilibrium and steady hands.

In September of 1948 we met again in Salt Lake City. Since that time we have been cooperating in developing our theories of non-Newtonian flow and significant structures of liquids, and during this period of time Henry has had many Koreans and Japanese as his graduate students or postdoctorates and has given them fellowships from his research grants. After finishing their Ph.D. or postdoctoral work most of Henry's students return to their own countries and devote themselves to the development of science and engineering. In this respect his contributions to these countries are invaluable. I am especially indebted to him because my children, Francis and Teresa, received Ph.D. degrees under Henry's guidance; both are engaged in the study of liquids.

During this present work Henry gave us valuable advice and kind encouragement. Thus it is with extreme pleasure that we present our paper for the volume in honor of his seventieth birthday.

INTRODUCTION

In describing thermodynamic and equilibrium statistical-mechanical behaviors of a classical fluid, we often make use of a radial distribution function $g(r)$. The latter for a fluid of N particles in volume V expresses a local number density of particles situated at distance r from a fixed particle divided by an average number density ρ $(\equiv N/V)$, when the order of $1/N$ is negligible in comparison with 1. Various thermodynamic quantities are related to $g(r)$. For a single-component monatomic system of particles interacting with a pairwise additive potential $\Phi(r)$, the relationship connecting the pressure P to $g(r)$ is the virial theorem,[1]

$$\frac{P}{\rho k T} = 1 - \frac{2\pi\rho}{3kT} \int_0^\infty r^3 \frac{d\Phi(r)}{dr} \, g(r) \, dr, \tag{1}$$

where k and T denote, respectively, the Boltzmann constant and the absolute temperature.

There are various equations describing $g(r)$. The most frequently used integral equations for $g(r)$ can be grouped into the three categories: (a) BGY equation,[2] (b) PY equation,[3] and (c) CHNC equation,[4] where the symbols BGY, PY, and CHNC stand for, respectively, Born-Green-Yvon, Percus-Yevick, and Convolution-Hyper-Netted-Chain. These equations are the first equations appearing in each formally exact hierarchy of equations relating the distribution functions of different orders. To solve them, however, it is necessary in practice either to neglect certain subsets of graph integrals in the expression for $g(r)$, as in the case of the PY and the CHNC equations, or to truncate higher-order correlation functions appearing in the integral equation for $g(r)$ by using some appropriate combinations of $g(r)$ of lower orders at different distances, as is in the case of the BGY equation. For particles interacting with a hard-sphere potential, for example, the corresponding integral equations for $g(r)$ can be solved either exactly [PY $g(r)$],[5] or numerically [CHNC $g(r)$,[6] BGY $g(r)$[7]]. Pressure P then can be calculated from the virial theorem (1). At a half of the close-packed density $(\equiv \rho_0)$ of hard spheres, for example, the pressure calculated from the above theories give

$$P/(\rho kT) = 4.61 \; (BGY)^8$$
$$= 5.43 \; (PY)^8$$
$$= 6.24 \; (CHNC)^8$$
$$= 5.89 \; (MD),$$

where MD denotes the molecular dynamics (exact) data for a system of 108 hard spheres with periodic boundaries obtained by Alder and Wainwright.[9,10] The comparison of the above results with the experimental value (MD) shows that the pressures calculated from the BGY and the PY theories are low by 21.7% and 7.8%, respectively, while the CHNC value lies higher by 5.9%. Over the range of higher densities ($\rho/\rho_0 > 0.5$), especially in the neighborhood of the density ($\rho/\rho_0 \simeq 0.67$), where the hard-sphere fluid is predicted to undergo a phase transition to an ordered solid phase,[9] the discrepancies between the approximate theories and the experiments become more pronounced. At higher densities ($\rho/\rho_0 > 0.67$), moreover, the PY $g(r)$ becomes negative (thus, unphysical); i.e., as ρ increases, the first, the second, the third minimum of the PY $g(r)$ begin to be negative, respectively, at $(r/\sigma, \; \rho/\rho_0) = (1.30, 0.828)$,[11] $(2.188, 0.905)$,[8] and $(3.138, 1.215)$[8] (σ: hard-sphere diameter); and the PY pressure remains finite at the close-packed density.[5] Also, the BGY $g(r)$ and the CHNC $g(r)$ do not damp out exponentially fast at large distances when the densities exceed a certain critical density [$= 0.68 \; \rho_0$ for the BGY $g(r)$],[7,12] thus giving infinite compressibilities beyond this density (a flat top isotherm). By extending Kirkwood's three-dimensional results,[12] the following d-dimensional ($d = 1, 2,$ or 3) asymptotic formula of the BGY $g(r)$ for hard spheres is obtained:

$$g(r) = 1 + Ar^{-(d-1)/2}e^{-\beta r}\cos(\alpha r + \delta), \tag{2a}$$

where A and δ are constants, and α and β can be determined from the root ($z = \alpha - i\beta$) having the smallest $|\beta|$ of the equation

$$\rho g(\sigma)J_{d/2}(z) + (z/2\pi)^{d/2} = 0, \tag{2b}$$

where $J_{d/2}$ is the Bessel function of order $d/2$. For one-dimensional hard rods, β in (2a) is found to vanish at $P/\rho kT = 3.30$. Therefore, at higher values of pressure the one-dimensional BGY $g(r)$ shows a periodicity in particle positions as observed in crystals [cf. (2a)]. This cannot, however, be interpreted as a true limit of stability of the fluid phase since it contradicts the exact results[13,14] which show no such phase change. For the corresponding three-dimensional system of hard spheres, therefore, the critical density ($\rho/\rho_0 = 0.68$) of the BGY $g(r)$ might be due to the approximation inherent in the BGY $g(r)$, and it may not have a direct bearing to the observed density

$(\rho/\rho_0 = 0.67)$ at which the MD result indicates the initiation of an ordered phase.[9,15]

In order to refine these theories, we need to include the second equations in the corresponding hierarchy of the integral equations for the distribution functions. For the PY $g(r)$ and the CHNC $g(r)$, this procedure is equivalent to the inclusion of more complicated rooted graph integrals in calculating the $g(r)$.[16,17] For the BGY hierarchy, one possible way of the procedure is to truncate the quadruplet distribution function g_{1234} occurring in the second equation describing the triplet correlation function g_{123} [$\equiv g(r_{12}, r_{13}, r_{23})$; $r_{ij} = |r_i - r_j|$] by a proper combination of g_{ij} [$\equiv g(r_{ij})$] and g_{ijk}'s. The resulting equations are two simultaneous nonlinear integral equations relating both g_{12} and g_{123}. The solutions of the refined theories (PY2, CHNC2, and BGY2, where 2 denotes the refined theory) are not yet available. However, the first terms of the coefficients in the density expansion of the pressure P, the radial distribution functions g_{12} and g_{123} are obtained from the PY2 and CHNC2 theories. For the density series of the pressure, this procedure gives the exact virial coefficients B_n up to $n = 4$. However, because of the approximate nature of the theories, the expressions for B_5 and higher virial coefficients differ from the exact formula.[8,18,19] The values of B_5 have been calculated for hard spheres using the PY2 and the CHNC2 theories.[18,19] For example, the virial theorem (1) gives

$$B_5/(B_2)^4 = 0.1240 \text{ (PY2)},$$
$$= 0.0657 \text{ (CHNC2)}.$$

The above values differ from the exact Monte Carlo value[20] of $B_5/(B_2)^4 = 0.1103$; however, the value of PY2 B_5 has been significantly improved compared to the value [$= 0.0859$] calculated from the unrefined PY theory.[5,21]

Not too much attention seems to have been paid to the BGY2 theory. This may be partly due to the fact that the unrefined BGY theory generally tends to give larger errors in both the thermodynamic variables and $g(r)$ than those calculated from the unrefined PY theory. This discouraging finding might have, in turn, served as a weak stimulus to investigate further the BGY theory up to the present. From a computational view, there is an additional difficulty associated with the BGY2 equations, since they cannot be expressed in terms of the usual Mayer-Montroll rooted graph integrals,[22] as can be done in the case of the PY2 and the CHNC2 theories.

We shall consider the BGY2 equations in the present series of investigations. Section II of this report gives a mathematical formulation of the problem. In Section III, we expand the pressure P, the distribution functions g_{12} and g_{123} by density series and calculate the first several density coefficients for hard spheres. The final section is reserved for discussions of the results. From the present investigation, we find that the BGY2 $g(r)$ gives a superior value of the virial coefficients [for example, $B_5/(B_2)^4 = 0.1090$, calculated from the virial theorem] than those calculated from the PY2 or the CHNC2 theories. Furthermore, the density coefficient $g^n(r)$ of the BGY2 agrees closely with the exact result[8] for $n = 3$. Therefore, we intend to report in the future the complete results on the BGY2 $g(r)$ carried out by using the hard-sphere potential at several different densities. It is hoped that the present series of investigations will serve to decide whether the approach by the radial distribution function method to describe quantitatively the equilibrium state of a dense fluid is a practical and fruitful way or not. We believe that the present approach will give a partial answer to this question, since solving more refined theories such as the PY3, the CHNC3, and the BGY3 in a nontrivial case appears to be impractical, far exceeding the capability of the present-day numerical techniques.

MATHEMATICAL FORMULATION OF THE BGY2 THEORY

In the following discussions, we use the expressions for g_{12} and g_{123} derived independently by Born and Green,[2] Yvon,[2] Kirkwood,[12] and Bogoliubov[23] in various different forms. These basically equivalent hierarchy of equations (sometimes known as either the BGY or the BBGKY hierarchy) can be expressed as an infinite set of the following integrodifferential equations $(N \to \infty)$[24]:

$$-\nabla_1 g_{12} = g_{12}\nabla_1 u_{12} + \rho \int d\mathbf{r}_3 (\nabla_1 u_{13}) g_{123}, \tag{3a}$$

$$-\nabla_1 g_{123} = g_{123}[\nabla_1 u_{12} + \nabla_1 u_{13}] + \rho \int d\mathbf{r}_4 (\nabla_1 u_{14}) g_{1234}, \tag{3b}$$

$$-\nabla_1 g_{1234} = \text{equation involving five-particle correlation function}, \tag{3c}$$

where ∇_1 is the gradient operator acting on particle 1, and u_{12} is the reduced central potential energy [$\equiv \Phi_{12}/kT$] between particles 1 and 2. In the present report, we consider only the disordered (or fluid) state of a system under consideration. For a fluid, the singlet distribution function is the same as the average density ρ, and $g(\mathbf{r}_1, \mathbf{r}_2)$ is equal to $g(r_{12})$ ($\equiv g_{12}$), i.e., a scalar function of interparticle distance r_{12} only. Likewise, $g(\mathbf{r}_1, \mathbf{r}_2, \mathbf{r}_3)$ and $g(\mathbf{r}_1, \mathbf{r}_2, \mathbf{r}_3, \mathbf{r}_4)$ are, respectively, equal to $g(r_{12}, r_{13}, r_{23})$ ($\equiv g_{123}$) and $g(r_{12}, r_{13}, r_{14}, r_{23}, r_{24}, r_{34})$ ($\equiv g_{1234}$).

The BGY hierarchy (3) can be truncated by introducing a "closure approximation" for $g_{123\cdots n}$ which appears in the integrands of (3) in terms of lower-order correlation functions. In principle, by solving the resulting coupled $(n - 2)$ integrodifferential equations one obtains all the lower-order approximate correlation functions $g_{123\cdots i}$ $(i < n)$. Since the higher-order integrodifferential equations are neglected, no knowledge on the higher-order correlation function can, however, be deduced from the truncated equations.

The first equation (3a) of the BGY hierarchy can be truncated by a well-known Kirkwood superposition approximation[12] for g_{123}:

$$g_{123} = g_{12}g_{13}g_{23}. \tag{4}$$

In general, the correlation function $g_{123\cdots n}$ can be approximated by a proper combination of products of the lower-order correlation functions,[25] provided that it satisfies certain necessary consistency conditions for fluids. These conditions are a symmetry condition and an asymptotic condition. The symmetry condition requires the invariance of $g_{123\cdots n}$ under the exchange of particle positions; for example,

$$g_{123\cdots n} = g_{213\cdots n} = g_{132\cdots n}, \text{ etc.} \tag{5}$$

The asymptotic condition describes behaviors of the correlation function at large particle separations; for example:

$$g_{123\cdots n} = (g_{12\cdots k})(g_{k+1\cdots j}) \cdots (g_{l\cdots n}), \tag{6a}$$

$$g_{12} = 1 \quad \text{for} \quad r_{12} \to \infty. \tag{6b}$$

That is, if positions of n particles can be divided into an arbitrary number of small batches containing a smaller number of particles, all particles in a batch being separated sufficiently far away from the other particles in the different batches, the asymptotic condition (6) requires the joint correlation function $g_{123\cdots n}$ to be expressed as a product of the lower-order correlation functions corresponding to particles in each of these independent batches. One approximate expression for g_{1234} which satisfies both conditions is given by Fisher and Kopeliovich.[26] It has the following form:

$$g_{1234} = \frac{g_{123}g_{124}g_{234}g_{134}}{g_{12}g_{13}g_{14}g_{23}g_{24}g_{34}} \tag{7a}$$

$$= \chi_{1,23}\chi_{2,34}\chi_{3,14}\chi_{1,24}g_{12}g_{13}, \tag{7b}$$

where

$$\chi_{i,jk} \equiv \frac{g_{ijk}}{g_{ij}g_{ik}}. \tag{8}$$

Kirkwood's superposition approximation for g_{123} includes only three *independent* pair correlations and ignores the influence of any of these pair correlations when a third particle is in the neighborhood of a pair. Therefore, the BGY $g(r)$ obtained under the Kirkwood superposition approximation (4) is expected to be different from the exact values at small r.* For a repulsive system, the thermodynamic properties are largely influenced by $g(r)$ at small r. Therefore, as expected, this system exhibits some discrepancies in thermodynamic quantities at moderately dense states if the BGY $g(r)$ is used to calculate them. However, since the superposition approximation (7) for g_{1234} indirectly influences g_{12} through g_{123} in the BGY theory [see (3b)], it might improve the results obtained under the approximation (4). Calculations on one-dimensional system of hard rods supports this argument. It is proven in Appendix A that the use of the first two equations [(3a) and (3b)] in the BGY hierarchy gives a self-closed exact integral equation for $g(r)$ for the one-dimensional system of hard rods *independent* of the approximation (7b) for g_{1234}. In this respect, one may note that the use of the Kirkwood superposition approximation gives an incorrect integral equation for $g(r)$ for the same system.

We use the superposition approximation (7) in association with (3a) and (3b); however, these equations are not convenient to handle. Therefore, we modify these expressions slightly into the following more convenient expressions:

$$-\mathbf{\nabla}_1 \ln [g_{12} \exp (u_{12})] = \rho \int d\mathbf{r}_3 (\mathbf{\nabla}_1 u_{13}) g_{13} [\chi_{1,23} - 1], \tag{9}$$

$$-\mathbf{\nabla}_1 \ln [\chi_{1,23}] = \rho \int d\mathbf{r}_4 (\mathbf{\nabla}_1 u_{14}) g_{14} [\chi_{1,24} \chi_{2,34} \chi_{3,14} (g_{14})^{-1}$$

$$- \chi_{1,24} - \chi_{1,34} + 1]. \tag{10}$$

Expression (10) results from (3a), (3b), and (7b) when the quantity $-\mathbf{\nabla}_1 \ln [g_{12} g_{13} \exp (u_{12} + u_{13})]$ is subtracted from $-\mathbf{\nabla}_1 \ln [g_{123} \exp (u_{12} + u_{13})]$. The terms corresponding to -1 and 1 inside the brackets on the right-hand sides of (9) and (10) do not affect values of the integrals since

$$\int d\mathbf{r}_3 (\mathbf{\nabla}_1 u_{13}) g_{13} = 0, \tag{11}$$

where g_{13} is a symmetric function of r_{13}. Insertion of these terms guarantees the integrands in (9) and (10) to vanish when particles 1 and 2 become separated sufficiently far away. To illustrate this point further, we rewrite the

---

* In dense fluid and solid regions, the molecular-dynamics results indicate that, for three spheres in contact, the corresponding $g_{123} = g(1, 1, 1)$ is surprisingly well represented by the superposition approximation, i.e., $g_{123} = [g(1)]^3$. This is not, however, true for an asymmetric triplet configuration or at low densities (ref 27).

quantities inside of the brackets in the integrand of (10) as

$$\chi_{1,24}\chi_{2,34}\chi_{3,14}(g_{14})^{-1} - \chi_{1,24} - \chi_{1,34} + 1$$
$$= [\chi_{1,24} - 1][\chi_{1,34} - 1] + \chi_{1,24}\chi_{1,34}[(\chi_{4,23}/g_{23}) - 1], \quad (12)$$

where the following relations are used:

$$\frac{\chi_{i,jk}}{g_{jk}} = \frac{\chi_{j,ik}}{g_{ik}}, \tag{13}$$

$$\chi_{i,jk} = \chi_{i,kj}. \tag{14}$$

These relationships result from the fact that g_{ij} and g_{ijk} are symmetric functions of particle positions. The main contribution of the integral in (10) comes when particle 4 lies in the neighborhood of particle 1. If $r_{12} \gg 1$, $\chi_{1,24} \simeq 1$ and $\chi_{4,23} \simeq g_{23}$. Therefore, the integrand becomes very small. Any integral with small value whose integrand is not necessarily small can cause some difficulties in numerical integration, since the final answer in this case depends sensitively on the size of mesh chosen in the numerical integration. Therefore, the integrands in (9) and (10) are chosen to avoid this difficulty.

The gradient operator ∇_1 is a directional derivative when particle 1 makes an infinitesimal displacement $d\mathbf{r}_1$ from $\mathbf{r}_1$, while particles 2 and 3 are held fixed at $\mathbf{r}_2$ and $\mathbf{r}_3$. Next, the operator ∇_1 in (10) is expressed as a sum of two vectors, one directed along a vector $(\mathbf{r}_1 - \mathbf{r}_2)$ and the other along $(\mathbf{r}_1 - \mathbf{r}_3)$. Using this form for ∇_1, the component of (10) along the direction $\mathbf{r}_1 - \mathbf{r}_2$ can be shown, after several manipulations, to satisfy the equation,

$$-\frac{\partial}{\partial r_{12}} \ln \chi_{1,23} = \rho \csc^2 \theta_{213} \int d\mathbf{r}_4 \frac{du_{14}}{dr_{14}} [\cos \theta_{214} - \cos \theta_{314} \cos \theta_{213}]$$
$$\times [\chi_{1,24}\chi_{2,34}\chi_{3,14} - g_{14}\chi_{1,24} - g_{14}\chi_{1,34} + g_{14}], \quad (15)$$

where

$$\cos \theta_{ijk} \equiv \frac{r_{ij}^2 + r_{jk}^2 - r_{ik}^2}{2r_{ij}r_{jk}}. \tag{16}$$

A similar equation along the direction $\mathbf{r}_1 - \mathbf{r}_3$ is obtained from (15) by interchanging the subscripts 2 and 3.

Note that the partial derivative $\partial/\partial r_{12}$ in (15) is taken under the restrictions of fixed values of r_{13} and r_{23}. Therefore, the maximum possible value of r_{12} extends at most to $r_{13} + r_{23}$, which is not necessarily large. Therefore, the integration constant of the differential equation (15) is an unknown function of r_{13} and r_{23}. Moreover, if particles 1, 2, and 3 lie nearly along a straight line, the right-hand side of (15) tends to become indefinite. These difficulties can

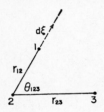

Fig. 1. Integration path appearing in the integral (18). Particles 2 and 3 are held fixed (r_{23} = constant). The integration path ξ measures the distance of particle 1 from particle 2 at a constant angle θ_{123}.

be avoided if a linear combination,

$$\left[\frac{\partial}{\partial r_{12}} + \cos\theta_{213}\frac{\partial}{\partial r_{13}}\right]\ln\chi_{1,23} = \frac{\partial}{\partial\xi}\ln\chi_{1,23}, \qquad (17)$$

of the two derivatives along the $(\mathbf{r}_1 - \mathbf{r}_2)$ and $(\mathbf{r}_1 - \mathbf{r}_3)$ directions is used. The derivative $\partial/\partial\xi$ is a directional derivative (see Fig. 1) along the $(\mathbf{r}_1 - \mathbf{r}_2)$ direction with fixed r_{23} and θ_{123}. The integration of $(\partial/\partial\xi)\ln\chi_{1,23}$ gives

$$\ln\chi_{1,23} = \ln g_{23} + \rho\int_{r_{12}}^{\infty}d\xi\int d\mathbf{r}_4\frac{du_{14}}{dr_{14}}\cos\theta_{214}$$

$$\times [\chi_{1,24}\chi_{2,34}\chi_{3,14} - g_{14}\chi_{1,24} - g_{14}\chi_{1,34} + g_{14}], \qquad (18)$$

where g_{23} is constant of integration for $\chi_{1,23}^{*}$ at $r_{12} \to \infty$. The distance between particles 1 and 2 in the integrand of (18) is represented by ξ with fixed values of r_{23} and θ_{123}. In an analogous manner, the integration of (9) yields

$$\ln[g_{12}\exp(u_{12})] = \rho\int_{r_{12}}^{\infty}dr_{12}\int d\mathbf{r}_3\frac{du_{13}}{dr_{13}}\cos\theta_{213}g_{13}[\chi_{1,23} - 1]. \qquad (19)$$

Equations (18) and (19) are two simultaneous equations in g_{12} and $\chi_{1,23}$. Therefore, an iterative scheme can be developed to obtain the solutions numerically. This work is now in progress for a hard-sphere system.

DENSITY EXPANSION OF P, g_{12}, AND $\chi_{1,23}$

The information on how much the BGY$_2$ equations (18) and (19) improve the results calculated from the unrefined BGY $g(r)$ can be obtained if these equations are used to calculate the first several coefficients in the density series of the following quantities:

$$P/(\rho kT) = 1 + B_2\rho + B_3\rho^2 + \cdots, \tag{20a}$$

$$g_{12}\exp(u_{12}) = 1 + g_{12}^1\rho + g_{12}^2\rho^2 + \cdots, \tag{20b}$$

$$g_{123}\exp(u_{12} + u_{13} + u_{23}) = 1 + g_{123}^1\rho + g_{123}^2\rho^2 + \cdots, \tag{20c}$$

$$\chi_{1,23}\exp(u_{23}) = 1 + \chi_{1,23}^1\rho + \chi_{1,23}^2\rho^2 + \cdots. \tag{20d}$$

The values obtained in this manner can be compared with the corresponding data for the exact, the PY2, and the CHNC2 theories. In order to carry this out, first substitute the density series (20b) and (20c) into (9) and (10), and then equate the coefficients of equal powers of ρ on both sides of the resulting equations. This procedure gives in a straightforward manner the BGY2 expressions for g_{12}^1, g_{12}^2, g_{123}^1, and $\chi_{1,23}^1$, which agree with the corresponding exact expressions.* This together with the virial theorem in turn implies that the BGY2 B_n is exact through $n = 4$. The first deviation $\epsilon_{1,23}$ occurs in $\chi_{1,23}^2$ (and g_{123}^2):

$$\text{BGY2 } \chi_{1,23}^2 = \text{exact } \chi_{1,23}^2 + \epsilon_{1,23}, \tag{21}$$

where $\epsilon_{1,23}$ is a solution of

$$\nabla_1\epsilon_{1,23} = -\int d\mathbf{r}_5\, d\mathbf{r}_4(\nabla_1 f_{14})(\tilde{f}_{24}\tilde{f}_{34} + f_{24}f_{34})f_{15}f_{25}f_{35}f_{45}, \tag{22a}$$

$$\tag{22b}$$

where the black dots for particles 4 and 5 represent integrations over all configurations of these particles. Various lines in the graph integral (22b) represent

$$\equiv f_{ij} \equiv \exp(-u_{ij}) - 1, \tag{23a}$$

$$\equiv \tilde{f}_{ij} \equiv \exp(-u_{ij}) \tag{23b}$$

$$ij \longrightarrow 1j = \nabla_1 f_{1j} \tag{23c}$$

* For the graphical expressions of these quantities, refer to a chapter by G. E. Uhlenbeck and G. W. Ford, in *Studies in Statistical Mechanics* (ref. 23).

The deviation $\epsilon_{1,23}$ in turn can alter values of the BGY2 $g_{12}{}^3$ and B_5 from the corresponding exact values.

In the following discussions, we use the hard-sphere potential to calculate the density coefficients in (20). The hard-sphere potential Φ_{ij} is given by

$$\Phi_{ij} = \infty \quad \text{if} \quad r_{ij} \le 1$$
$$= 0 \quad \text{if} \quad r_{ij} > 1, \tag{24}$$

where the hard sphere σ is taken as the unit of length ($\sigma \equiv 1$). For hard spheres,

$$f_{ij} = -1 \quad \text{and} \quad \tilde{f}_{ij} = 0 \quad \text{if} \quad r_{ij} \le 1 \tag{25a}$$

$$= 0 \quad \text{and} \quad = 1 \quad \text{if} \quad r_{ij} > 1. \tag{25b}$$

Furthermore,

$$\frac{df_{ij}}{dr_{ij}} = \delta(r_{ij} - 1). \tag{26}$$

For hard spheres, it is easy to prove that the deviation $\epsilon_{1,23}$ vanishes whenever any of the three distances r_{12}, r_{13}, and r_{23} exceeds 2 because of the property (25) of Mayer's function. Therefore, the BGY2 $\chi^2_{1,23}$ (also g^2_{123}) agrees with the exact result for those triplet configurations. Furthermore, the BGY2 $g_{12}{}^3$ becomes exact for $r_{12} \ge 2$, since $\chi^2_{1,23}$ is required in evaluating $g_{12}{}^3$ and agrees with the exact values for $r_{12} \ge 2$; that is,

$$\chi_{1,23}{}^2 = g_{23}{}^2 + \cdots \qquad ; r_{12} \ge 2, \tag{27a}$$

$$g_{12}{}^3 = \cdots + 3 \cdots + \cdots + \cdots \qquad ; r_{12} \ge 2, \tag{28a}$$

For $r_{12} \le 2$, $\chi^2_{1,13}$ and $g_{12}{}^3$ cannot be represented by the usual Mayer-Montroll graphs. Therefore, we obtain $\chi^2_{1,23}$ and $g_{12}{}^3$ from (18) and (19) by expanding $\chi_{i,jk}$ and g_{ij} in the density series and by using the hard-sphere properties (25) and (26) together with the fact that $\chi^2_{1,23}$ and $g_{12}{}^3$ become exact for $r_{12} \ge 2$.

After several simplifications, the final expressions for these quantities are

$$\chi^2_{1,23} = {g_{23}}^2 + \tfrac{1}{2}(\chi^1_{1,23})^2 - \tfrac{1}{2}(g^1_{23})^2 - H_{123}, \tag{27b}$$

$$
\begin{aligned}
g_{12}{}^3 = {}& -\tfrac{1}{3}(g_{12}{}^1)^3 + g_{12}{}^1 g_{12}{}^2 + g^2(1)g_{12}{}^1 + g^1(1) \\
& \times [g_{12}{}^2 - \tfrac{1}{2}(g_{12}{}^1)^2 - g^1(1)g_{12}{}^1 - g^2(2)] + g^3(2) + \eta_{12} - \eta(2) \\
& - 2\pi \int_{r_{12}}^2 dr_{12}(r_{12})^{-1} \int_{\max[|r_{12}-1|,1]}^{r_{12}+1} dr_{23} r_{23} \cos\theta_{213} \\
& \times [\tfrac{1}{2}(\chi^1_{1,23})^2 - \tfrac{1}{2}(g^1_{23})^2 - H_{123}]; \qquad r_{13} = 1, \tag{28b}
\end{aligned}
$$

where

$$\eta_{12} \equiv \eta(r_{12}) \equiv \int d\mathbf{r}_3 f_{13}\tilde{f}_{23} {g_{23}}^2, \tag{29}$$

$$
\begin{aligned}
H_{123} = {}& 2 \int_{r_{12}}^3 (d\xi/\xi) \int_{|r_{12}-1|}^{\min[r_{12}+1,2]} dr_{24} r_{24} \cos\theta_{214} \\
& \times \int_0^\pi d\phi [\tilde{f}_{24} f_{34}\chi^1_{1,24} + f_{24}\tilde{f}_{34}\chi^1_{1,34} + \tilde{f}_{24}\tilde{f}_{34}(\chi^1_{2,34} - g^1_{34}) + f_{24}f_{34}g^1(1)];
\end{aligned}
$$

$$r_{14}(=1); \; \theta_{123}, r_{23} \qquad \text{fixed.} \tag{30}$$

The function H_{123} is a symmetric function in the interchange of particles 1, 2, and 3. This can be easily proved from (27b) and the symmetry relations (13) and (14). In carrying out the numerical integration over the configuration of particle 4, the coordinate system (r_{14}, r_{24}, ϕ) rather than the tripolar coordinate system (r_{14}, r_{24}, r_{34}) is adopted to avoid possible divergence of the Jacobian in representing $d\mathbf{r}_4$, by the tripolar coordinate system, when particles 1, 2, and 3 lie on a straight line. The function η_{12} can be expressed in terms of the doubly rooted Mayer-Montroll graph integrals. Analytic expressions of most of the relevant graph integrals are given in Appendix B. In (30), $\chi^1_{1,23}$ is equal to

 $+ \; g_{23}{}^1,$

which can be easily proved from the definition of $\chi_{1,23}$, and the expression for

for hard spheres can be found in the literature.[8,28]

DISCUSSION

Values of $\chi^2_{12.3}$ and $g^3(r)$ are evaluated by numerical integrations of (27) and (28) using Simpson's rule and different mesh sizes for Δr. The numbers presented in Tables I and II are obtained by extrapolating these values to $\Delta r \to 0$. Table I presents, also, values of the hard-sphere $g^3(r)$ obtained from the exact and other approximate theories, partly using the analytic expressions for the relevant doubly rooted graph integrals obtained in Appendix B. Figure 2 shows behaviors of these $g^3(r)$'s. As mentioned previously, the BGY2 $g^3(r)$ is exact for $r \geq 2$. For $r \leq 2$, the comparison of the BGY2 $g^3(r)$ with the exact $g^3(r)$ shows that the difference between the two quantities

Table I.   Values of the Hard-Sphere $g^3(r)$ Calculated from the Exact, the Modified Born-Green-Yvon (BGY2), the Unmodified Born-Green-Yvon (BGY), the Percus-Yevick (PY), and the Convolution-Hypernetted-Chain (CHNC) Theories[a]

r	Exact g^3	BGY2 g^3	BGY g^3	PY g^3	CHNC g^3
0.0	32.990	32.918(6)	39.355	10.910	57.610
0.2	21.680(9)	21.933(6)	23.762	8.472	38.645
0.4	13.365(8)	13.522(6)	12.928	6.138	23.906
0.6	7.414(7)	7.462(6)	6.061	4.011	13.113
0.8	3.425(6)	3.424(6)	2.217	2.193	5.785
1.0	1.015(4)	1.005(6)	0.436	0.790	1.330
1.2	−0.210(4)	−0.214(4)	−0.144	−0.106	−0.879
1.4	−0.612(3)	−0.619(2)	−0.139	−0.485	−1.523
1.6	−0.531(4)	−0.531(2)	0.090	−0.443	−1.264
1.8	−0.175(3)	−0.173(1)	0.351	−0.123	−0.616
2.0	0.302	0.302	0.588	0.321	0.102
2.2	0.558	0.558	0.663	0.561	0.498
2.4	0.505	0.505	0.533	0.505	0.493
2.6	0.332	0.332	0.336	0.332	0.330
2.8	0.164	0.164	0.164	0.164	0.164
3.0	0.061	0.061	0.061	0.061	0.061
3.2	0.017	0.017	0.017	0.017	0.017
3.4	0.003	0.003	0.003	0.003	0.003
3.6	0.000	0.000	0.000	0.000	0.000

[a] The quantity $g^3(r)$ occurs in the density series

$$g(r) = \exp\left[-u(r)\right][1 + \rho g^1(r) + \rho^2 g^2(r) + \rho^3 g^3(r) + \cdots].$$

The hard-sphere diameter is taken as the unit of r. The numerals in the parentheses indicate the standard deviations in the last digits, i.e., $21.680(9) = 21.680 \pm 0.009$. The exact values of $g^3(r)$ in the second column are taken from Ref. 8.

Table II. Values[a] of $\chi^2_{1,23}$ and δ^2_{123} for Hard Spheres Calculated at Fixed Distances of r_{23} and r_{12} and at Different Angles θ_{123} between r_{12} and r_{23}

	$r_{23} = r_{12} = 1$			$r_{23} = 1, r_{12} = 1.6$			$r_{23} = 1, r_{12} = 2$		
θ_{123}	r_{13}[b]	$\chi^2_{1,23}$	δ^2_{123}	r_{13}[c]	$\chi^2_{1,23}$	δ^2_{123}	r_{13}	$\chi^2_{1,23}$	δ^2_{123}
0	0.000	−0.049	0.403	0.600	1.598	0.641	1.000	1.666	0.407
$\pi/6$	0.516	0.382	0.421	0.888	1.532	0.573	1.239	1.645	0.386
$\pi/3$	1.000	0.771	0.395	1.400	1.547	0.458	1.732	1.492	0.234
$\pi/2$	1.414	1.237	0.460	1.887	1.587	0.344	2.236	1.331	0.072
$2\pi/3$	1.732	1.701	0.583	2.272	1.372	0.113	2.646	1.266	0.007
$5\pi/6$	1.932	1.745	0.496	2.516	1.285	0.026	2.909	1.259	0.000
π	2.000	1.666	0.407	2.600	1.271	0.013	3.000	1.259	0.000

[a] The quantities $\chi^2_{1,23}$ and δ^2_{123} denote, respectively, the coefficients of ρ^2 terms in the density series of $\chi_{1,23}$ [$\equiv g_{123}/g_{12}g_{13}$] and δ_{123} [$\equiv g_{123}/(g_{12}g_{13}g_{23})$]. The hard-sphere diameter is taken as the unit of r.
[b] The values for $\chi^2_{1,23}$ and δ^2_{123} are uncertain in the last digit for $r_{13} \leq 1.732$.
[c] The values for $\chi^2_{1,23}$ and δ^2_{123} are uncertain in the last digit for $r_{13} \leq 1.400$.

lies within 3% from the exact values (see Fig. 3). The BGY2 $g(r)$ brings a marked improvement over the unrefined BGY $g(r)$ and is superior than the unrefined PY $g^3(r)$ and CHNC $g^3(r)$. Table II also presents the "correction" term δ^2_{123} in the density series of δ_{123} defined by the identity in the

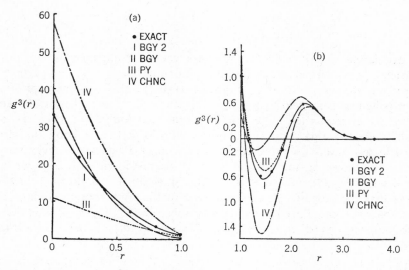

Fig. 2. Hard-sphere $g^3(r)$ calculated from the BGY2, the BGY, the PY, and the CHNC theories for (a) $r \leq 1$ and (b) $r \geq 1$. Note that the exact $g^3(r)$ [denoted by ●] is almost equal to the BGY2 $g^3(r)$.

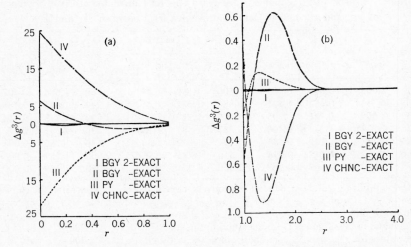

Fig. 3. Differences between the exact $g^3(r)$ and the $g^3(r)$ calculated from the various theories for hard spheres for (a) $r \le 1$ and (b) $r \ge 1$. Note that the difference of BGY2-exact is the smallest.

following equation:

$$\delta_{123} \equiv \frac{g_{123}}{g_{12}g_{13}g_{23}} = 1 + \delta_{123}^1\rho + \delta_{123}^2\rho^2 + \cdots . \tag{31}$$

In evaluating δ_{123}^2 of Table II, we have chosen configurations* of three particles, in which particles 2 and 3 are in contact ($r_{23} = 1$) and particle 1 is rotated at fixed values of r_{12} (see Fig. 4). If the Kirkwood superposition approximation is applied to $\chi_{1,23}$, $\chi_{1,23}^2$ is equal to $g_{23}^2 = g^2(1) = 1.259$, and all the correction terms δ_{123}^n ($n > 0$) drop out. The plots of $\chi_{1,23}^2$ (Fig. 5) show that, while the Kirkwood superposition approximation is good for large particle separations as expected, it is poor at small distances giving large deviations; however, the deviations do not decrease uniformly as angle θ_{123} increases. The plots of δ_{123}^2 (Fig. 6) also show this nonuniform behavior. Note that the plot of δ_{123}^2 corresponding to the configurations, where the two spheres 1 and 3 are in contact with sphere 2 (i.e., $r_{12} = r_{23} = 1$), has a minimum ($\theta_{123} \simeq 60°$ or $r_{13} \simeq 1$) and a maximum ($\theta_{123} \simeq 120°$) as the angle

---

* Rice and Lekner[29] obtained numerical solutions of δ_{123}^2 for two spatial configurations. The values of δ_{123}^2 corresponding to the other configurations are obtained through interpolation of the two solutions. They use the two-term (δ_{123}^1 and δ_{123}^2) Padé approximant to solve g_{12}, thus improving the results of ref. 7.

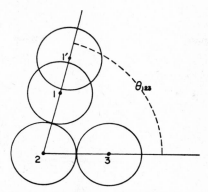

Fig. 4. Configuration of particles 1, 2, and 3: $r_{23} = 1$, $r_{12} = 1$ and $r_{1'2} = 1.6$. This configuration and the similar configurations at different values of θ_{123} are used to obtain the values of $\chi^2_{1,23}$ and δ^2_{123} given in Table II and Figs. 5 and 6.

θ_{123} changes. This indicates that Kirkwood's superposition approximation is better for symmetric three-particle configurations than for asymmetric ones. Alder reached the same conclusion based on an analysis of his molecular-dynamics data for hard spheres in dense fluid and solid phases.[27] From the present result, which is applicable to moderately dense states, it may be said

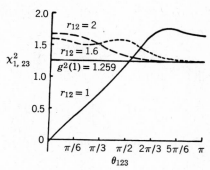

Fig. 5. The hard-sphere $\chi^2_{1,23}$ versus θ_{123} at $r_{23} = 1$, $r_{12} = 1$, 1.6, and 2. If the Kirkwood superposition approximation is used, then $\chi^2_{1,23} = g_{23}^2 = g^2(1) = 1.2587$.

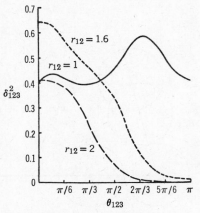

Fig. 6. Deviation δ_{123}^2 from the Kirkwood superposition approximation versus θ_{123} at $r_{23} = 1, r_{12} = 1, 1.6$, and 2.

that Alder's conclusion apparently is correct. In contrast to this, it is noted that the first correction term δ_{123}^1, which is equal to

is a uniformly decreasing function of particle separations.[8] At large particle separations, δ_{123}^2 also behaves uniformly, decreasing as the interparticle separations increase (see Fig. 6).

The fifth virial coefficients B_5 can be evaluated using either the virial theorem (1) which takes a simple form for hard spheres; i.e.,

$$\frac{P}{\rho kT} = 1 + B_2 \rho g(1), \tag{32}$$

or the Ornstein-Zernike compressibility relation,[1]

$$kT \frac{d\rho}{dP} = 1 + \rho \int d\mathbf{r}[g(r) - 1]. \tag{33}$$

The expressions for B_5 are obtained from these equations using the density series (20b) for $g(r)$:

$$B_5(\text{virial}) = B_2 g^3(1), \tag{34}$$

$$B_5 \text{ (O-Z)} = \tfrac{1}{5}[16(B_2)^4 - 36(B_2)^2 B_3 + 16 B_4 B_2 + 9(B_3)^2]$$

$$- \tfrac{6}{5} B_2 \int_1^4 g^3(r) r^2 \, dr. \tag{35}$$

If the BGY2 $g^3(r)$ is used, expressions (34) and (35) yield, respectively,

$$\frac{B_5 \text{ (virial)}}{(B_2)^4} = 0.1090 \pm 0.0008, \tag{36a}$$

$$\frac{B_5 \text{ (O-Z)}}{(B_2)^4} = 0.1112 \pm 0.0005. \tag{36b}$$

These values are, however, very close to the exact Monte Carlo value,[8,10,20] 0.1103 $\pm$ 0.0003. The difference between the two values of (36) is a measure of the accuracy of approximation (7) introduced in the BGY2 theory. Therefore, the present results show an internal consistency of the BGY2 theory. In Table III the BGY2 values of B_5 are compared with the B_5 values calculated from the other theories. It is rather remarkable that the BGY2 values of B_5 *agree even better* with the Monte Carlo B_5 than the values calculated from the refined Percus-Yevick theory.[18,19] Moreover, if the density series of the direct correlation function $c(r)$ is calculated from the known equations,[8] it can be easily shown that the BGY2 c_{12}^3 of the ρ^3 term is exact for $r \leq 1$, and that for $r \geq 1$, the c_{12}^3 (calculated using the BGY2 g_{12}^3 in Table I) and the exact c_{12}^3

Table III. Values[a] of the Hard-Sphere Virial Coefficients B_5 obtained from the Exact, the Percus-Yevick (PY), the Convolution-Hyper-netted-Chain (CHNC), and the Born-Green-Yvon (BGY) Theories[b]

	B_5	B_5 (virial)	B_5 (O-Z)
Exact[b]	0.1103 $\pm$ 0.0003		
PY[c]		0.0857	0.1211
CHNC[d]		0.1447	0.0493
BGY[e]		0.0475	0.1335
PY2[f]		0.1240	0.1074
CHNC2[f]		0.0657	0.1230
BGY2[g]		0.1090 $\pm$ 0.0008	0.1112 $\pm$ 0.0005

[a] Here, the unrefined (PY, CHNC, and BGY) and refined (PY2, CHNC2, and BGY2) theories are used in conjunction with the virial theorem or the Ornstein-Zernike (O-Z) relation in order to obtain B_5 in units of $B_2 \equiv 1$.
[b] See ref. 20.
[c] See refs. 5 and 21.
[d] G. S. Rushbrooke and P. Hutchinson, *Physica*, **27**, 647 (1961).
[e] B. R. A. Nijboer and R. Fieschi, *Physicsa*, **19**, 545 (1953).
[f] See refs. 18 and 19.
[g] Present work.
[h] Recently M. S. Wertheim [*J. Math. Phys.*, **8**, 927 (1967)] proposed the PYII theory which gives B_5 (O-Z) = 0.1098.

(calculated using the exact values[8] of $g_{12}{}^3$) agree approximately within the same errors observed in the case of $g_{12}{}^3$.

The primary objective of the present course of investigation is to find out how accurately an approximate radial distribution function can describe the equilibrium state of a dense fluid. As has been discussed earlier, the approximations for $g(r)$ can be in principle refined in a successive and systematic fashion to take into account higher-order corrections which are neglected in the lower-order calculations. In practice, however, it is difficult to do this, since the corrections usually involve considerations of complicated geometry of successively larger numbers of particles; in the present case, for example, it is required to consider various three-dimensional configurations of four particles with two of these hard spheres in contact. From this work we conclude, therefore, that the BGY hierarchy of equations under the closure approximation (7) for g_{1234} satisfactorily describes the low and moderately dense fluids. A further investigation of the BGY2 theory is required to check the theory over the range of denser fluids.

APPENDIX A

The BGY Hierarchy for One-Dimensional Hard Rods

Here we shall prove that the BGY $g(r)$ yields the exact pressure for one-dimensional hard rods if the virial theorem is used. This result is *independent of* any approximation on g_{123} such as the Kirkwood superposition approximation (4). We shall also prove that the BGY2 radial distribution function for one-dimensional hard rods is exact. This proof is *independent of* the approximation (7) used for the quadruplet correlation function g_{1234}.

To prove these, let v_{12}, v_{123}, and v_{1234} be defined as follows:

$$v_{12} \equiv v(r_{12})$$
$$\equiv g_{12} \exp (u_{12}), \tag{A1a}$$
$$v_{123} \equiv v(r_{12}, r_{13}, r_{23})$$
$$\equiv g_{123} \exp (u_{12} + u_{13} + u_{23}), \tag{A1b}$$
$$v_{1234} \equiv v(r_{12}, r_{13}, r_{14}, r_{23}, r_{24}, r_{34})$$
$$\equiv g_{1234} \exp (u_{12} + u_{13} + u_{14} + u_{23} + u_{24} + u_{34}). \tag{A1c}$$

When these relations are substituted into the first two equations ((3a) and (3b)) occurring in the BGY hierarchy, the integrals on the right-hand sides of these equations can be evaluated explicitly because of the restrictive conditions (25) and (26) on $f_{ij}, \tilde{f}_{ij}$, and df_{ij}/dr_{ij} for one-dimensional hard rods. The resulting equations are as follows:

$$\frac{dv_{12}}{dr_1} = -\rho v(r_{12}, 1, r_{12} + 1); \qquad 0 \le r_{12} < 2, \qquad (A2a)$$

$$= -\rho v(r_{12}, 1, r_{12} + 1) + \rho v(r_{12}, 1, r_{12} - 1); \qquad 2 \le r_{12}, \quad (A2b)$$

$$\frac{dv_{123}}{dr_1} = -\rho v(r_{12}, r_{13}, 1, r_{23}, r_{12} + 1, r_{13} + 1); \qquad 0 \le r_{12} < 2, \qquad (A3a)$$

$$= -\rho v(r_{12}, r_{13}, 1, r_{23}, r_{12} + 1, r_{13} + 1)$$
$$+ \rho v(r_{12}, r_{13}, 1, r_{23}, r_{12} - 1, r_{13} - 1)A(r_{13} - 2); \qquad 2 \le r_{12}, \quad (A3b)$$

where dr_1 represents the differential displacement of particle 1 when the other particles are held fixed. In (A3), only the ordered one-dimensional configuration (1–3–2) is considered in which particle 1 is always to the left of particle 3, and the latter lies to the left of particle 2. $A(x)$ in (A3b) denotes a step function; i.e., $A(x) = 0$ if $x < 0$ and $A(x) = 1$ if $x > 0$. The solutions corresponding to the other particle configurations can be obtained from this solution by simply reenumerating the particle numbers due to the symmetry property of v_{123}.

Let particle 2 be at the origin. Integration of (A2a) yields

$$v_{12} = 1 + \rho(2 - r_{12})v(1) - \rho \int_1^{r_{12}} dx[v(x, 1, x + 1) - v(1)]; \qquad r_{12} \le 2, \quad (A4a)$$

$$= 1 - \rho \int_{r_{12}-1}^{r_{12}} dx[v(x, 1, x + 1) - v(1)]; \qquad 2 \le r_{12}, \qquad (A4b)$$

where the boundary conditions on v_{12} is $v(\infty) = 1$ and the symmetry property (5) is applied, i.e., $v(x, 1, x + 1) = v(x + 1, 1, x)$. The solution of (A3) can be obtained in a similar fashion,

$$v_{123} = v_{23} + \rho(2 - r_{13})v_{23}v(1)$$
$$+ \rho \int_{r_{12}}^{1+r_{23}} dx[v(x, x - r_{23}, 1, r_{23}, x + 1, x - r_{23} + 1) - v(1)v_{23}];$$
$$r_{13} \le 2 \le r_{12} \quad \text{or} \quad r_{13} \le r_{12} \le 2, \quad (A5a)$$

$$= v_{23} - \rho \int_{r_{12}-1}^{r_{12}} dx[v(x, x - r_{23}, 1, r_{23}, x + 1, x - r_{23} + 1) - v(1)v_{23}];$$
$$2 \le r_{13} \le r_{12}, \quad (A5a)$$

where v_{123} is equal to v_{23} when r_{12} becomes large. Notice that the terms $v(1)$ and $v(1)v_{23}$ are inserted into the integrands of (A4) and (A5) in order to make the integrands vanish as r_{12} approaches to infinity.

Let us consider v_{123} at $r_{13} = 1$; i.e., $v_{123} = v(r_{12}, 1, r_{12} + 1)$. Only the triplet correlation function satisfying this configuration appears in the integral equation for v_{12}. When particles 1, 2, and 3 are ordered in this manner, both the upper and lower limits of the integration in (A5a) become equal. Therefore, the integration vanishes; i.e.,

$$v(r_{12}, 1, r_{12} + 1) = v_{12}[1 + \rho v(1)]. \tag{A6}$$

Introducing (A6) into (A4), we obtain

$$v_{12} = 1 + \rho v(1) - \rho[1 + \rho v(1)] \int_{\max(1, r_{12}-1)}^{r_{12}} dx v(x). \tag{A7}$$

Let $r_{12} = 1$, then

$$v(1) = 1 + \rho v(1) = (1 - \rho)^{-1}. \tag{A8}$$

By substituting (A8) into (A6) and (A7), we obtain

$$v(r_{12}, 1, r_{12} + 1) = v_{12} v(1), \tag{A9}$$

$$v_{12} = v(1) - \rho v(1) \int_{\max(1, r_{12}-1)}^{r_{12}} dx v(x). \tag{A10}$$

Several comments are in order for results (A8), (A9), and (A10). First, (A9) agrees with the exact one-dimensional triplet correlation function[14]:

$$g_{123} = g_{12} g_{13}, \qquad r_{23} = \max(r_{12}, r_{13}, r_{23}). \tag{A11}$$

The pair correlation function v_{23} corresponding to the two particles which are separated farthest away does not appear in v_{123}, unlike the case of Kirkwood's superposition approximation. Second, the integral equation (A10) for the BGY $g(r)$ is an exact integral equation for $g(r)$. This is true, since the triplet correlation function occurring in the integrand of the integral equation for the BGY2 $g(r)$ is correctly given by (A6). Third, $v(1)$ is equal to $(1 - p)^{-1}$ independent of the approximation for v_{123}, as can be seen from (A4). Therefore, from the virial theorem the unrefined BGY equation gives the exact pressure, $P/\rho kT = 1 + B_2 \rho g(1) = 1/(1 - \rho)$, for hard rods.[13]

APPENDIX B

Evaluation of the Rooted Graph Integrals

We used the method given in Ref. 8 to derive analytic expressions for various graph integrals appearing in $g^3(r)$ over the range $1 \leq r \leq 2$. The calculations are rather lengthy. Therefore, we present here only the results. For the sake of completeness, we also collected the known results for $r \geq 2$.

Here r represents the distance between two open circles in the following rooted graphs:

$$= -\pi^3\left[\frac{3}{453600}r^9 - \frac{3}{2520}r^7 + \frac{2}{945}r^6 + \frac{1}{30}r^5 - \frac{28}{225}r^4\right.$$
$$\left. + \frac{128}{315}r^2 - \frac{2176}{2835}\right]; \qquad 1 \le r \le 2,$$

$$= \pi^3\left[\frac{1}{453600}r^9 - \frac{1}{2520}r^7 + \frac{2}{945}r^6 + \frac{1}{90}r^5 - \frac{28}{225}r^4\right.$$
$$\left. + \frac{8}{27}r^3 + \frac{128}{315}r^2 - \frac{128}{45}r + \frac{11264}{2835} - \frac{2048}{1575}r^{-1}\right];$$
$$2 \le r \le 4,$$

$$= 0; \qquad 4 \le r. \tag{B1}$$

$$= = -\pi^3\left[-\frac{1}{453600}r^9 + \frac{1}{2520}r^7 - \frac{1}{90}r^5 + \frac{11}{1800}r^4\right.$$
$$\left. + \frac{47}{432}r^3 - \frac{23}{1260}r^2 - \frac{1219}{1440}r + \frac{27557}{22680} - \frac{4433}{25200}r^{-1}\right];$$
$$1 \le r \le 2,$$

$$= \pi^3\left[-\frac{1}{453600}r^9 + \frac{1}{2520}r^7 - \frac{2}{945}r^6 - \frac{1}{90}r^5 + \frac{71}{600}r^4\right.$$
$$\left. - \frac{37}{144}r^3 - \frac{163}{420}r^2 + \frac{363}{160}r - \frac{789}{280} + \frac{2313}{2800}r^{-1}\right];$$
$$2 \le r \le 3,$$

$$= 0; \qquad 3 \le r, \tag{B2}$$

$$= \pi^3\left[\frac{1}{226800}r^9 - \frac{1}{1260}r^7 + \frac{1}{945}r^6 + \frac{1}{45}r^5 - \frac{14}{225}r^4 - \frac{2}{27}r^3\right.$$
$$\left. + \frac{64}{315}r^2 + \frac{136}{315}r - \frac{2176}{2835}\right]; \qquad 1 \le r \le 2$$
$$= 0; \qquad 2 \le r, \tag{B3}$$

$$= = -+\pi^3\left[\frac{1093}{10080}r - \frac{6253}{22680} + \frac{1793}{25200}r^{-1}\right];$$
$$1 \le r \le 2,$$

$$= 0; \qquad 2 \le r, \tag{B4}$$

$$\text{(graph)} = \pi^3\left[\frac{1}{453600}r^9 - \frac{1}{2520}r^7 + \frac{2}{945}r^6 + \frac{1}{90}r^5 - \frac{101}{900}r^4 + \frac{47}{216}r^3\right.$$
$$\left. + \frac{233}{630}r^2 - \frac{1219}{720}r + \frac{18853}{11340} - \frac{1793}{6300}r^{-1}\right]; \qquad 1 \le r \le 2,$$
$$= 0; \qquad 2 \le r,$$

$$\text{(B5)}$$

$$\text{(graph)} = \pi^3\left[-\frac{1}{51840}r^9 + \frac{5}{2688}r^7 - \frac{17}{2520}r^6 - \frac{31}{960}r^5 + \frac{9}{40}r^4 - \frac{15}{64}r^3\right.$$
$$\left. - \frac{9}{8}r^2 + \frac{2187}{640}r - \frac{891}{280} + \frac{3159}{4480}r^{-1}\right]; \qquad 1 \le r \le 3$$
$$= 0; \qquad 3 \le r$$

$$\text{(B6)}$$

$$\text{(graph)} = -\text{(graph)} + \pi^3\left[\frac{251}{5760}r - \frac{2357}{22680} + \frac{87}{4480}r^{-1}\right]; \qquad 1 \le r \le 2,$$
$$= 0; \qquad 2 \le r,$$

$$\text{(B7)}$$

$$\text{(graph)} = -\text{(graph)} + \pi^2\left[\frac{1}{33600}r^7 + \frac{1}{16800}r^6 - \frac{7}{3600}r^5 + \frac{19}{5600}r^4\right.$$
$$\left. + \frac{5167}{756000}r^3 - \frac{23153}{756000}r^2 - \frac{3449}{25200}r - \frac{63769}{756000} + \frac{77821}{1521000}r^{-1}\right]$$
$$\times (2 + 2r - r^2)^{\frac12} + \pi^2\left[\frac{23}{450}r^4 - \frac{269}{315}r^2 + \frac{46}{45}r - \frac{51}{70} - \frac{758}{1575}r^{-1}\right]$$
$$\times \cos^{-1}\left\{\frac{r-1}{[3(3+2r-r^2)]^{\frac12}}\right\}$$
$$+ \pi^2\left[-\frac{1}{67200}r^9 + \frac{43}{40320}r^7 - \frac{1}{280}r^6 - \frac{29}{2880}r^5 + \frac{1}{40}r^4\right.$$
$$\left. + \frac{11}{192}r^3 - \frac{1}{8}r^2 + \frac{87}{128}r - \frac{123}{3200}r^{-1}\right]\cos^{-1}\left\{\frac{r^2-r-3}{[3(3+2r-r^2)]^{\frac12}}\right\}$$
$$+ \pi^2\left[-\frac{1}{67200}r^9 + \frac{43}{40320}r^7 - \frac{1}{280}r^6 - \frac{29}{2880}r^5 + \frac{61}{1800}r^4\right.$$
$$\left. + \frac{11}{192}r^3 - \frac{971}{2520}r^2 + \frac{4939}{5760}r - \frac{51}{140} - \frac{69701}{201600}r^{-1}\right]$$
$$\times \cos^{-1}\left\{\frac{-r^2+3r+1}{[3(3+2r-r^2)]^{\frac12}}\right\}; \qquad 2 \le r \le 1 + \sqrt{3},$$
$$= -\text{(graph)}; \qquad 1 + \sqrt{3} \le r \le 3,$$
$$= 0; \qquad 3 \le r.$$

$$\text{(B8)}$$

The above rooted-graph integrals are used in the text to calculate $g^3(r)$ and $c^3(r)$.

REFERENCES

1. J. de Boer, *Rept. Progr. Phys.*, **12**, 305 (1949).
2. J. Yvon, *Actualités Sci. et Ind.* No. 203 (1935); M. Born and H. S. Green, *Proc. Roy. Soc. (London)*, **A188**, 10 (1946).
3. J. K. Percus and G. J. Yevick, *Phys. Rev.*, **110**, 1 (1958).
4. J. M. J. Van Leeuwen, J. Groeneveld, and J. de Boer, *Physica*, **25**, 792 (1959); T. Morita and K. Hiroike, *Progr. Theoret. Phys. (Kyoto)*, **23**, 1003 (1960); E. J. Meeron, *J. Math. Phys.*, **1**, 192 (1960); G. S. Rushbrooke, *Physica*, **26**, 259 (1960); M. S. Green, *J. Chem. Phys.*, **33**, 1403 (1960); L. Verlet, *Nuovo Cimento* **18**, 77 (1960).
5. M. S. Wertheim, *Phys. Rev. Letters*, **10**, 321 (1963); E. Thiele, *J. Chem. Phys.*, **39**, 474 (1963).
6. M. Klein, *J. Chem. Phys.*, **39**, 1388 (1963).
7. J. G. Kirkwood, E. K. Maun, and B. J. Alder, *J. Chem. Phys.*, **18**, 1040 (1950).
8. F. H. Ree, R. N. Keeler, and S. L. McCarthy, *J. Chem. Phys.*, **44**, 3407 (1966).
9. B. J. Alder and T. E. Wainwright, *J. Chem. Phys.*, **33**, 1439 (1960).
10. F. H. Ree and W. G. Hoover, *J. Chem. Phys.*, **40**, 939 (1964).
11. B. J. Alder (private communication).
12. J. G. Kirkwood, *J. Chem. Phys.*, **3**, 300 (1935).
13. K. F. Herzfeld and M. G. Mayer, *J. Chem. Phys.*, **2**, 38 (1934).
14. Z. W. Salsburg, R. W. Zwanzig, and J. G. Kirkwood, *J. Chem. Phys.*, **21**, 1098 (1953).
15. F. H. Ree and W. G. Hoover, *J. Chem. Phys.*, **46**, 4181 (1967).
16. J. K. Percus, *Phys. Rev. Letters*, **8**, 462 (1962).
17. L. Verlet, *Physica*, **30**, 95 (1964).
18. G. S. Rushbrooke, in *Statistical Mechanics of Equilibrium and Non-Equilibrium*, J. Meixner, Ed., North-Holland Publ. Co., Amsterdam, 1965, p. 222.
19. L. Oden, D. Henderson and R. Chen, *Phys. Letters*, **21**, 420 (1966).
20. See refs. 8 and 10. The values of hard-sphere B_5 have also been obtained by: S. Katsura and Y. Abe, *J. Chem. Phys.*, **39**, 2068 (1963); J. S. Rowlinson, *Proc. Roy. Soc. (London)*, **A279**, 149 (1964); J. E. Kilpatrick and S. Katsura, *J. Chem. Phys.*, **45**, 1866 (1966).
21. G. S. Rushbrooke, *J. Chem. Phys.*, **38**, 1262 (1963).
22. J. E. Mayer and E. W. Montroll, *J. Chem. Phys.*, **9**, 2 (1941).
23. N. N. Bogoliubov, *J. Phys. (U.S.S.R.)*, **10**, 256, 265 (1946); English translation by E. K. Gora, in *Studies in Statistical Mechanics*, J. de Boer and G. E. Uhlenbeck, Eds., North-Holland Publ. Co., Amsterdam, 1962, Vol. 1, Pt. A.
24. For a comprehensive derivation of this hierarchy, refer to Chap. 6 in T. L. Hill, *Statistical Mechanics*, McGraw-Hill, New York, 1956.
25. See an article by G. Stell, in *The Equilibrium Theory of Classical Fluids*, H. L. Frisch and J. L. Lebowitz, Eds, W. A. Benjamin, New York, 1964.
26. I. Z. Fisher and B. L. Kopeliovich, *Dokl. Akad. Nauk USSR*, **133**, 81 (1960); [English transl.: *Soviet Phys.—Doklady*, **5**, 761 (1960–1961)].
27. B. J. Alder, Phys. Rev. Letters, **12**, 317 (1964).
28. H. L. Weissberg and S. Prager, *Phys. Fluids*, **5**, 1390 (1962); J. S. Rowlinson, *Mol. Phys.*, **6**, 517 (1963); M. J. Powell, *ibid.*, **7**, 591 (1964).
29. S. A. Rice and J. Lekner, *J. Chem. Phys.*, **42**, 3559 (1965).

Hard Spheres with Surface Adhesion: The Percus-Yevick Approximation and the Energy Equation*

R. O. WATTS†

Department of Applied Mathematics, University of Waterloo, Ontario

D. HENDERSON

IBM Research Laboratory, San Jose, California

R. J. BAXTER

Department of Mathematics, Massachusetts Institute of Technology Cambridge

Abstract

The equation of state and free energy of a system of hard spheres with surface adhesion are calculated from the "internal energy" of the fluid as given by the Percus-Yevick theory. A first-order phase change occurs. Further, the liquid–vapor coexistence curve, which cannot be found by the more usual routes to the equation of state, is calculated. It is found that the equation of state exhibits van der Waals type sigmoid isotherms in the region in which the Percus-Yevick theory has solutions.

* This research has been supported in part by grants from the National Research Council of Canada and the U.S. Department of the Interior, Office of Saline Water.
† Present Address: Diffusion Research Unit, Research School of Physical Sciences, Australian National University, Canberra, ACT, Australia.

421

A Personal Note

DOUGLAS HENDERSON

My first contact with Henry Eyring was at the University of British Columbia. Henry had been invited to speak by the metallurgy department. I was, at the time, an undergraduate in the mathematics department. Thus, a metallurgy seminar was not the usual place to find me. Presumably because Henry and I shared the same faith, I came along with some friends in metallurgical engineering to hear the talk. The talk was presented with the usual Eyring flair but I must confess that I can no longer recall the subject. My strongest recollections are of the question period following the talk. Unsurprisingly, Henry had attracted people with a wide range of interests and, as a result, the questions covered a wide range of topics. I was left with the impression, not entirely incorrect, that at any given moment only two people understood the question being considered: Henry and the poser of the question.

No doubt as a result of this experience, I found myself a few years later enrolled in a graduate program under his direction at the University of Utah. The two years immediately prior to my working under Henry had been spent completely away from science and mathematics. Because of this, some months were spent in less than full productivity. Despite his impatience with the slowness with which I commenced my research, Henry treated me with great consideration. His kindness also extended to my wife, freshly arrived from "darkest Africa," whom he made feel welcome to America.

His thoughtfulness was not confined to professionals. I recall one afternoon when he took nearly an hour out from his busy schedule to instruct a secretary in the mysteries of shifting gears in her new Volkswagen.

These, then, are my strongest impressions of Henry Eyring. Henry the scientist interested in any significant question and willing to offer an idea to assist in finding the answer to the question and Henry the man desiring and attempting to help his fellows.

INTRODUCTION

One of us has shown[1] that the Percus-Yevick[2] (PY) equation for the pair distribution function, $g(r)$, can be solved analytically for the intermolecular potential, $u(r)$, given by:

$$\frac{u(r)}{kT} = \psi(r) = \begin{cases} +\infty & r < R' \\ \ln\left[12\tau\dfrac{R-R'}{R}\right] & R' < r < R \\ 0 & r > R \end{cases} \qquad (1)$$

where $R - R'$ is allowed to become infinitesimally small. The parameter τ is a dimensionless measure of the temperature, T, and is such that, in the limit, the attractive well between R' and R gives a finite contribution to cluster integrals. The relation between τ and T is arbitrary. If desired, τ can be required to vary in such a way that $u(R)$ is independent of T. However, this restriction is quite unnecessary. From this solution, expressions for various thermodynamic properties of the fluid can be obtained.

In particular, for $r < R$:

$$g(r) = \begin{cases} 0, & r < R', \\ \dfrac{\lambda}{12}\dfrac{R}{R-R'} + O(1), & R' < r < R. \end{cases} \qquad (2)$$

The parameter γ depends both on τ and the density, $\rho = N/V$, being given by:

$$\tau = \frac{1+\eta/2}{(1-\eta)^2}\lambda^{-1} - \frac{\eta}{1-\eta} + \frac{\eta\lambda}{12}, \qquad (3)$$

where $\eta = (\pi/6)\rho R^3$. This is a quadratic equation having two solutions for λ for any given η and τ, as may be seen in Fig. 1. The larger solutions for λ

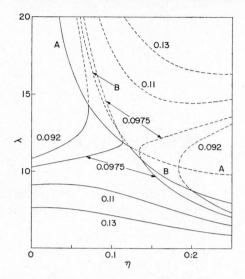

Fig. 1. Values of λ plotted against η for constant τ. Curve A separates the physical and unphysical values of λ. Curve B gives the values of λ for which $(\partial \rho / \partial p)_T$ diverges. Only the points below these two curves correspond to physically acceptable solutions of the PY equation.

can be rejected on thermodynamic grounds.[1] Thus

$$\lambda = \frac{6}{\eta} [\alpha - (\alpha^2 - \gamma)^{1/2}], \tag{4}$$

where

$$\alpha = \tau + \frac{\eta}{1 - \eta} \tag{5}$$

$$\gamma = \frac{\eta(1 + \eta/2)}{3(1 - \eta)^2}. \tag{6}$$

The curve separating the small (physical) from the larger (unphysical) values of λ is plotted in Fig. 1.

The equation of state may be calculated from the compressibility equation,

$$kT \left(\frac{\partial \rho}{\partial p} \right)_T = 1 + 4\pi\rho \int_0^\infty s^2 [g(s) - 1] \, ds, \tag{7}$$

which can be integrated[1] to give:

$$\frac{p}{\rho kT} = \frac{1 + \eta + \eta^2}{(1 - \eta)^3} - \mu \frac{1 + \eta/2}{(1 - \eta)^3} + \frac{\mu^3}{36\eta(1 - \eta)^3},$$ (8)

where

$$\mu = \lambda\eta(1 - \eta).$$ (9)

The equations above are physically meaningful only if $g(r) \to 1$ as $r \to \infty$ sufficiently fast for the integral

$$I = \int_0^\infty s^2[g(s) - 1]\,ds$$ (10)

to converge. The locus of points at which the integral becomes divergent, so that $(\partial\rho/\partial p)_T$ is infinite, is plotted in Fig. 1. All points lying above this curve and the curve separating the small and large values of γ are unphysical, corresponding either to divergent radial distribution functions or to the larger solutions for λ.

In Fig. 2 we have plotted these curves on a τ versus η graph. In this plot all the points below the solid curves are unphysical.

The equation of state, obtained from (8), predicts that the system undergoes a first-order phase transition. The critical values of the thermodynamic properties are given in Table I.

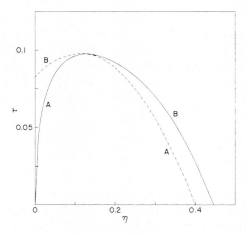

Fig. 2.   Curves A and B of Fig. 1 plotted on a τ versus η graph. The points below the solid portions of the two curves correspond to physically unacceptable solutions of the PY equation.

Table I.   Critical Values of the Thermodynamic Properties

	τ_c	η_c	$p_c/\rho_c k T_c$
Compressibility equation	0.0976	0.1213	0.4360
Energy equation	0.1185	0.32	0.32

The equation of state may also be calculated from the pressure equation:

$$\frac{p}{\rho k T} = 1 - \frac{2\pi\rho}{3kT} \int_0^\infty s^3 \frac{du}{ds} g(s)\, ds, \tag{11}$$

which can be integrated[1] to give

$$\frac{p}{\rho k T} = \frac{1 + 2\eta + 3\eta^2}{(1 - \eta)^2} - \lambda \frac{4\eta^2}{1 - \eta} + \lambda^2 \frac{\eta^2}{3}$$

$$- \frac{\eta}{3\tau} [3(1 + \eta)\delta + 4\epsilon + \tfrac{1}{6}\eta\lambda^2 + \tfrac{1}{24}\eta^2\lambda^3], \tag{12}$$

where

$$\delta = \frac{(1 + 2\eta - \mu)^2}{(1 - \eta)^4}, \tag{13}$$

$$\epsilon = \frac{-3\eta(2 + \eta)^2 + 2\mu(1 + 7\eta + \eta^2) - \mu^2(2 + \eta)}{2(1 - \eta)^4}. \tag{14}$$

Again (12) is meaningful only when the integral (10) converges. The isotherms calculated from (12) are similar to the compressibility isotherms at low densities and high temperatures. Near the critical point and at high densities however, the two sets of isotherms differ greatly. In particular, no critical point exists since the point of inflection, given by (12), corresponds to an unphysical point in Fig. 2.

ENERGY EQUATION

Another method[3] of obtaining the equation of state is to use $g(r)$ to calculate the internal energy, U, and then to integrate with respect to the inverse temperature to give the Helmholtz free energy, A. As usually stated, this procedure is applicable only to temperature-independent potentials. Although we could force the present system into this form, we prefer here to obtain the analogous results for the potential (1) by working directly with the parameter τ instead of the temperature.

Neglecting terms which are independent of density, the Helmholtz free energy, A, is given by

$$\beta A = -\ln \int \prod_{i<j} \exp\left(-\psi_{ij}\right) d\mathbf{r}_1 \cdots d\mathbf{r}_N, \tag{15}$$

where $\beta = 1/kT$, $\psi_{ij} = \psi(r_{ij})$ and $r_{ij} = |\mathbf{r}_i - \mathbf{r}_j|$ is the distance between molecules i and j. Thus differentiating (15) with respect to τ, gives

$$\frac{\partial(\beta A)}{\partial \tau} = -\tfrac{1}{2}N\rho \int \frac{\partial\psi}{\partial \tau} g(r)\, d\mathbf{r}. \tag{16}$$

Now

$$\frac{\partial\psi}{\partial \tau} = \frac{1}{\tau} \tag{17}$$

for $R' < r < R$ and zero elsewhere. Hence

$$\frac{\partial(A/NkT)}{\partial \tau} = 2\pi\rho\tau^{-1} \int_{R'}^{R} r^2 g(r)\, dr. \tag{18}$$

Substitution of (2) into (18) yields, in the limit $R' \to R$,

$$\frac{\partial(A/NkT)}{\partial \tau} = \tau^{-1}\eta\lambda, \tag{19}$$

which is the analog of the energy equation for this system.

When $\tau \to \infty$ the potential (1) becomes that of simple hard spheres of diameter R. Using the fact that λ is the smaller root of the quadratic equation (3), (20) can be integrated to give

$$\frac{A - A_0}{NkT} = -6 \int_{\tau}^{\infty} \left\{\alpha - (\alpha^2 - \gamma)^{1/2}\right\} \frac{d\tau'}{\tau'}, \tag{20}$$

where A_0 is the free energy of a system of hard spheres of diameter R and α is now given by (5) with τ replaced by τ'. The equation of state may be obtained from (20) by differentiating with respect to η. Thus

$$\begin{aligned}
\frac{p - p_0}{kT} &= -6\eta \frac{\partial}{\partial \eta} \int_{\tau}^{\infty} \left\{\alpha - (\alpha^2 - \gamma)^{1/2}\right\} \frac{d\tau'}{\tau'} \\
&= -6\eta \int_{\tau}^{\infty} \left\{\tau'(\alpha^2 - \gamma)^{1/2}\right\}^{-1} \left\{ [(\alpha^2 - \gamma)^{1/2} - \alpha] \frac{\partial\alpha}{\partial \eta} + \frac{1}{2}\frac{\partial\gamma}{\partial \eta} \right\} d\tau', \tag{21}
\end{aligned}$$

where p_0 is the pressure of a system of hard spheres of diameter R and

$$\frac{\partial\alpha}{\partial \eta} = \frac{1}{(1 - \eta)^2} \tag{22}$$

$$\frac{\partial\gamma}{\partial \eta} = \frac{1 + 2\eta}{3(1 - \eta)^3} \tag{23}$$

One virtue of this method is that, in contrast to the compressibility and pressure equations, an expression for the free energy is obtained.

The quantities A_0 and p_0 can be obtained from the hard-sphere isotherm of Carnahan and Starling,[4] which is the most accurate analytical expression available. Thus,

$$\frac{A_0}{NkT} = \ln \eta + \eta \, \frac{4 - 3\eta}{(1 - \eta)^2} + \text{terms independent of } \eta. \tag{24}$$

and

$$\frac{p_0}{\rho kT} = \frac{1 + \eta + \eta^2 - \eta^3}{(1 - \eta)^3} \tag{25}$$

NUMERICAL RESULTS

The integrals in (15) and (16) can be obtained analytically but are not given here because they are too lengthy. These expressions were checked by numerical integration.

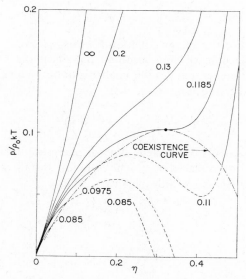

Fig. 3. Equation of state obtained from the energy equation. The curves are isotherms and are labeled with appropriate τ. The broken portion of the isotherms occur in the two-phase region which bounded by the liquid vapor coexistence curve. The solid dot is the critical point. The quantity $\rho_0 = \pi R^3/6$.

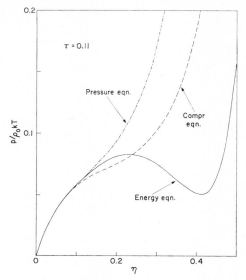

Fig. 4. Isotherms obtained from the compressibility, pressure, and energy equations for $\tau = 0.11$. The quantity $\rho_0 = \pi R^3/6$.

The resulting energy equation of state is plotted in Fig. 3. We do not plot values of the equation of state for $\eta > 0.5$ because the $\tau = \infty$ (hard-sphere) results from the Percus-Yevick theory are known to be in error for such densities. A typical van der Waals first-order phase change occurs. The point of inflection occurs in the region of convergent $g(r)$. The critical constants, which are very different from those obtained from the compressibility equation, are given in Table I. Below $\tau = 0.0976$ there are regions in which the energy equation of state does not exist.

The energy equation isotherm for $\tau = 0.11$ is compared with the corresponding compressibility and pressure equation isotherms in Fig. 4. The three isotherms are very different.

The Gibbs' free energy, $G = A + pV$, is plotted in Fig. 5 as a function of p/kT. Terms in G that are independent of η have been neglected. At temperatures above the critical temperature, G is a single-valued function of p. At temperatures below, but still near, the critical temperature (see $\tau = 0.11$, for example) there is a region in which there are three values of G for any given p. Thus, a first-order phase change occurs with the smallest value of G being the one appropriate to a given p. The crossover point gives the coexistence pressure. At lower temperatures (see $\tau = 0.0975$, for example) there is a region in which values of G and p do not exist. We have plotted the coexistence curve in Fig. 3.

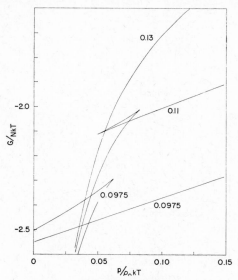

Fig. 5. Gibbs' free energy as a function of p. The curves are isotherms and are labeled with appropriate τ. The quantity $\rho_0 = \pi R^3/6$.

CONCLUSIONS

It has been shown that for a system of hard spheres with surface adhesion that, in the PY approximation, the equation of state can be obtained from the energy equation and that the resulting thermodynamic properties show typical van der Waals behavior. The critical temperature and density are considerably higher than those obtained from the compressibility equation.

The large discrepancies among the three methods of obtaining the equation of state indicate that the values of $g(r)$ given by the PY approximation are seriously in error for this system. It is still possible that one of the equations of state is moderately accurate, but in the absence of exact results no quantitative tests can be made of this hypothesis.

REFERENCES

1. R. J. Baxter, *J. Chem. Phys.*, **49**, 2770 (1968).
2. J. K. Percus and G. J. Yevick, *Phys. Rev.*, **110**, 1 (1958).
3. M. Chen, D. Henderson, and J. A. Barker, *Can. J. Phys.*, **47**, 2009 (1969).
4. N. F. Carnahan and K. E. Starling, *J. Chem. Phys.*, **51**, 635 (1969).

Inequalities for Critical Indices near Gas–Liquid Critical Point[*][†]

TERESA REE CHAY

Department of Chemistry, University of California, San Diego, La Jolla

FRANCIS H. REE

Lawrence Radiation Laboratory, University of California, Livermore

Abstract

By studying the behavior of the cell-pair correlation function $\chi(R)$ near the critical point we deduce the following new inequality under a plausible assumption on $\chi(R)$:

$$[\phi^{(n)}]^2[\phi^{(1)}]^{-n-1+(n-1)d/(2-\eta)} < \infty, \qquad n = 1, 2, \ldots$$

where d is the dimensionality of the system under study, and $\phi^{(n)}$ denotes $\rho^n[\partial^{n-1}\rho^{-1}(\partial\beta P/\partial\rho)_\beta/\partial\rho^{n-1}]_\beta$ [ρ = number density, P = pressure, $\beta = 1/(k_B T)$, T = temperature, k_B = Boltzmann constant]. The quantity η is a correction factor from the Ornstein–Zernike behavior of the leading term of $\chi(R)$ at the critical point at large distance $R[\chi(R) \sim R^{2-d-\eta}]$. The above inequality unifies further inequalities for the critical indices which can be generated from it by tracing different thermodynamic paths. Using Fisher's notations for the critical indices, we find the following relations from the above inequality:

$$\eta \geq 2 - \frac{d(\delta - 1)}{(\delta + 1)}, \qquad 2\Delta_n \leq \gamma + dv, \qquad 2\Delta'_n \leq \gamma' + dv', \quad \text{and} \quad 2\beta \leq dv' - \gamma'.$$

[*] Reprinted from *The Journal of Chemical Physics*, Vol. 49, No. 9, November 1968, 3977–3987.

[†] Work performed under the auspices of the U.S. Atomic Energy Commission. This work was started when one of the authors (T. R. C.) was under Contract AT(11-1) No. 34 Project 65.

431

INTRODUCTION

In this paper we present an approach to describe the equilibrium behavior of a classical fluid in the neighborhood of its critical point. The approach which we use here is based on a simplified version of work described earlier.[1,2] Some physical consequences, especially those which relate the critical indices of the various thermodynamic and statistical-mechanical quantities, can be drawn by properly interpreting the equations derived.

In the first section, we focus our attention on a quantity called *cell occupation probability* of a grand canonical ensemble with a given volume which is equally partitioned into a large number of identically shaped imaginary cells, where cell walls do not obstruct passage of particles. The cell occupation probability is defined as the probability of finding a given particle distribution in each cell. The cell occupation probability contains the particle-interaction potential indirectly through the thermodynamic quantities of the individual cell. In contrast to this, explicit knowledge on the particle-interaction potential is required in the usual particle-distribution functions. As we approach close to the critical point, the short-range ($\sim$ approximately equal to the order of range of potential) behavior of the particle-distribution functions do not change significantly, while the long-range behavior of the same functions starts to influence various equilibrium quantities. Therefore, as we shall see in the following discussions, the critical phenomena are more conveniently handled by means of the cell occupation probability than by use of the usual particle-distribution functions.

We next expand the probability so that the dominant contribution to the occupation probability comes from the Gaussian distribution centered around the most probable particle number in each cell and expand the remaining non-Gaussian corrections to the probability as a Taylor series. We examine in detail the non-Gaussian corrections which are usually neglected. A series expansion for the cell-pair correlation function can be generated from the series for the cell occupation probability. The first term for the cell-pair correlation function comes from the Gaussian distribution for the cell occupation probability. The first two non-Gaussian corrections are evaluated explicitly. We next make a main assertion in the present work which states that the series for the cell-pair correlation function is bounded in each term and is convergent. Their boundedness in turn can be utilized to obtain a new inequality. This inequality is used near the critical point, and we obtain from it various inequalities relating critical exponents and the dimensionality of system under study.

The inequalities somewhat relax the equalities of the corresponding relations obtainable from the Widom–Kadanoff's "scaling law,"[3,4] or "cluster" model introduced by Fisher.[5] Proper reviews on the approaches and assumptions employed in their works have already appeared, and we refer the reader to the literature[5-7] for the details. Experimentally, "the law of the corresponding states" near the critical point implied by the scaling law has been proved to hold within experimental accuracy by Kouvel and Rodbell[8] for the magnetization vs. magnetic field data for CrO_2 and Ni. Analogous scaling for the pressure–volume–temperature (PVT) data of a variety of gases is also confirmed by Green et al.[9] Furthermore, in the scaling-law theory, the equalities relating the critical indices which do not involve explicit dimensionalities are not inconsistent with the experimental data or theoretical values of the three-dimensional Ising model.[5-7] However, theoretical study of the Ising model in three dimensions indicates that the equality relations which contain both dimensionality and the critical indices appear to deviate slightly from the equalities. Use of the equalities in this case also contradicts a generally accepted view[5] that, when dimensionality becomes large, the corresponding critical indices should agree with the indices for the van der Waals gas. Instead of the equalities, if the inequalities are used for these relations such as one obtained by Josephson[10] for the heat-capacity exponent or the present works for the other exponents, the difficulties noted in the above discussions disappear.

SERIES EXPANSION OF THE CELL OCCUPATION PROBABILITY $P\{N\}$

Consider a system whose volume V is divided into M equal cells of volume Δ ($V = M\Delta$). In this section we present a series expansion of the occupation probability $P\{N\}$ of the cells 1, 2, ..., M, containing a set of particle numbers $\{N\} = N_1, N_2, \ldots, N_M$. The expansion parameter is chosen as the difference $\delta\rho_i$ between the number density ρ_i ($\equiv N_i/\Delta$) for the ith cell ($i = 1, 2, \ldots, M$) and the most probable density ρ_0 ($\equiv N_0/\Delta$).

For the grand canonical ensemble, with temperature T and volume V,

$P\{\mathbf{N}\}$ is written as

$$P\{\mathbf{N}\} = C \exp\left[\beta\left(\mu \sum_i N_i - A\{\mathbf{N}\}\right)\right], \qquad \beta \equiv 1/(k_B T), \tag{1}$$

$$C^{-1} \equiv \sum_{\{\mathbf{N}\}} \exp\left[\beta\left(\mu \sum_i N_i - A\{\mathbf{N}\}\right)\right], \qquad \sum_i \equiv \sum_{i=1}^{M}, \tag{2}$$

where k_B is the Boltzmann constant, $A\{\mathbf{N}\}$ being the Helmholtz free energy of the M closed systems with a given set of particle distributions denoted by $\{\mathbf{N}\}$ in each cell. The notation

$$\sum_{\{\mathbf{N}\}}$$

implies that the sum is taken over all distinct sets of $\{\mathbf{N}\}$. The quantity μ denotes the chemical potential and is related to the most probable particle number N_0 in Δ by

$$\mu = (\partial A\{\mathbf{N}\}/\partial N_i)_0, \qquad i = 1, 2, \ldots, M, \tag{3}$$

where the subscript 0 represents the quantity to be evaluated at N_0.

At a given temperature T below the critical temperature T_c of the system under consideration, there exists a chemical potential μ_t which can satisfy (3) by two different values of the most probable particle number N_0.[11] If the occupation probability $P\{\mathbf{N}\}$ is plotted against $\{\mathbf{N}\}$ at μ_t, $P\{\mathbf{N}\}$ will show double maxima, each centered around either one of the two possible values of N_0. The densities corresponding to these particle numbers are the liquid and gas densities of the coexistence line. In the following discussions applied to cases for $T < T_c$, we consider the one-phase regions for a given μ, which approaches close to μ_t from above (liquid side) and below (gas side) μ_t, thus avoiding the double maxima occurring at μ_t. For $T < T_c$, therefore, the present formalism is applicable to a study of the critical behavior along either the liquid or gas side of the coexistence line.

Let us impose the following two restrictions on the size of the cell Δ[1,4]: (1) Δ is chosen large enough to contain many particles, i.e., $N_0 \gg 1$; (2) however, it is much smaller than the correlation length of the pair correlation function. As the system approaches a critical point, the correlation length becomes large. Thus, both restrictions will probably not be difficult to satisfy by a system near its critical point. The second restriction is not necessary in the present section; however, it will be required in later sections, where the present formulation is applied to describe the behavior of a thermodynamic system near its critical point. The first restriction allows us to regard the number density ρ_i ($\equiv N_i/\Delta$, $i = 1, 2, \ldots, M$) of the ith cell as a continuous variable. Therefore the Helmholtz free energy $A\{\boldsymbol{\rho}\}$ can be expanded in

$\delta\boldsymbol{\rho}\ [\equiv (\delta\rho_1, \delta\rho_2, \ldots, \delta\rho_M), \delta\rho_i \equiv \rho_i - \rho_0]$ by using Taylor's formula around the most probable number density $\rho_0\ (\equiv N_0/\Delta)$; i.e.,

$$P\{\mathbf{N}\} \equiv P\{\boldsymbol{\rho}\} = C \exp \{-\beta A\{\mathbf{N}_0\} + \beta\mu N_0 M$$

$$- \tfrac{1}{2}\Delta^2 \sum_{ij} A_{ij}\, \delta\rho_i\, \delta\rho_j - \tfrac{1}{6}\Delta^3 \sum_{ijk} A_{ijk}\, \delta\rho_i\, \delta\rho_j\, \delta\rho_k \cdots$$

$$- [(n + 1)!]^{-1}\Delta^{n+1} \sum_{ij\cdots n+1} \gamma_{n+1}\, \delta\rho_i\, \delta\rho_j \cdots \delta\rho_{n+1}\}, \quad (4)$$

where

denotes

$$\sum_{ij\cdots n}$$

$$\sum_{i=1}^{M} \sum_{j=1}^{M} \cdots \sum_{n=1}^{M},$$

$\{\boldsymbol{\rho}\} \equiv \{\mathbf{N}\}/\Delta$; the remainder terms γ_{n+1} are $(n + 1)$th derivatives with respect to $\delta\rho_i$'s of $A\{\boldsymbol{\rho}\}$ evaluated at $\delta\boldsymbol{\rho} = \delta\boldsymbol{\rho}'$ which may lie slightly away from $\delta\boldsymbol{\rho} = 0$, and

$$A_{ij} \equiv \beta \left[\frac{\partial^2 A\{\mathbf{N}\}}{\partial N_i\, \partial N_j}\right]_0, \tag{5}$$

$$A_{ijk} \equiv \beta \left[\frac{\partial^3 A\{\mathbf{N}\}}{\partial N_i\, \partial N_j\, \partial N_k}\right]_0. \tag{6}$$

Finiteness of the series (4) for $A\{\boldsymbol{\rho}\}$ guarantees automatically its convergence, provided that conditions for the expansions are met. These are the conditions for existence of (a) the first n derivatives of $A\{\boldsymbol{\rho}\}$ at $\delta\boldsymbol{\rho} = 0$, plus (b) the similar condition for the remainder term γ_{n+1}; i.e., the $(n + 1)$th derivative of $A\{\boldsymbol{\rho}\}$ at $\delta\boldsymbol{\rho} = \delta\boldsymbol{\rho}'$. If a system is *at* the critical point or within the liquid–gas coexistence range, these conditions will *not* likely be satisfied if n is taken to be larger than 1. However, as long as we consider a system in the one-phase region *in the neighborhood of* the critical point and the coexistence range, the regularity of $A\{\boldsymbol{\rho}\}$ in the one-phase region will automatically fulfill the condition (a). Any clear evidence which might support or reject the condition (b) in the neighborhood of the critical point is lacking at present. Therefore, although breaking-down such an expansion is a possibility that is important to investigate, we take a contrary view in the present report and assume existence of the expansion, thus the fulfillment of the condition (b), and examine its consequences. In the last section of the present report, brief numerical results are quoted to partially support this assumption if dimension d of the system becomes large $(d \geqslant 3)$. Probably, a somewhat stronger condition than the condition (b) assumes the existence of a *convergent infinite* Taylor series of $A\{\boldsymbol{\rho}\}$ which can be generated from (4).

(However, it is possible that the two conditions may be equivalent.) For the sake of clarity of discussions in the following sections we assume this stronger condition; i.e., existence of the convergent series for $A\{\rho\}$, $P\{\rho\}$, and the cell-pair correlation function which can be obtained from $P\{\rho\}$. Note, however, that this assumption can always be replaced by a (possibly) weaker assumption [for example, the condition (b) for $A\{\rho\}$] in the ensuing discussion. Similar Taylor expansion for the normalization constant C in (4) can be carried out; however, we shall not discuss it here explicitly. Only the final result of the expansion will be presented at the end of the derivation.

Physical meaning of the coefficients A_{ij} and A_{ijk} becomes apparent if these quantities are separated into two parts by introducing the new quantities $\hat{A}_{ij}$ and $\hat{A}_{ijk}$,

$$A_{ij} = \hat{A}_{ij} + \Delta^{-1}A_2 \,\delta(i - j), \qquad A_2 \equiv \Delta \sum_j A_{ij}, \tag{7}$$

$$A_{ijk} = \hat{A}_{ijk} + \Delta^{-2}A_3 \,\delta(i - j)\,\delta(i - k), \qquad A_3 \equiv \Delta^2 \sum_{jk} A_{ijk}, \tag{8}$$

where $\delta(i - j)$ is the Kronecker delta ($=1$ if $i = j$; $=0$, otherwise). Note that the summation over the index j on $\hat{A}_{ij}$ and the summations over the double indices j, k on $\hat{A}_{ijk}$ yield zero. Thus, $\hat{A}_{ij}$ for $i \neq j$ can be regarded as proportional to the intercell force exerted on the ith cell by the jth cell, since the net force on the ith cell will vanish at equilibrium. The second term in the right-hand side of (7) is related to the inverse of the compressibility* $\beta(\partial P/\partial \rho)$ by the following reason: From the defining relation (3) for the chemical potential μ, the differential element $d\mu$ is given by†

$$d\mu = d\left[\frac{\partial A\{\mathbf{N}\}}{\partial N_i}\right]_0$$

$$= \sum_j \left[\frac{\partial^2 A\{\mathbf{N}\}}{\partial N_i \,\partial N_j}\right]_0 dN_0$$

$$= \beta^{-1}A_2 \,d\rho_0. \tag{9}$$

* The partial derivatives, $\partial/\partial\rho$ and $\partial/\partial\mu$ [$\mu \equiv$ chemical potential], are taken at constant temperature; i.e., $(\partial/\partial\rho)_\beta$ and $(\partial/\partial\mu)_\beta$, respectively. In the following discussions, the subscript β will be dropped when denoting these derivatives.

† The second equality of (9) is true from the following argument: For $i = 1, 2, \ldots, M$, let

$$\partial A\{\mathbf{N}\}/\partial N_i = \mu_i(N_1, N_2, \ldots, N_M),$$

$$(\partial A\{\mathbf{N}\}/\partial N_i)_0 = \mu.$$

Then,

$$d(\partial A\{\mathbf{N}\}/\partial N_i)_0 = d\mu$$

$$= \sum_j [\partial\mu_i(N_1, N_2, \ldots, N_M)/\partial N_j]_0 \,dN_0$$

$$= \sum_j (\partial^2 A\{\mathbf{N}\}/\partial N_i \,\partial N_j)_0 \,dN_0.$$

Since ρ_0 becomes independent of the cell size Δ for a sufficiently large N_0, (9) shows A_2 to be an intensive function of ρ_0. With the identities $\langle \delta \rho_i \rangle = \rho - \rho_0$ [$\langle \rho_i \rangle \equiv \rho$ = average density] and $(\partial \mu / \partial \rho) = \rho^{-1}(\partial P / \partial \rho)$, the above equation becomes

$$
\begin{aligned}
A_2 &= \beta \left(\frac{\partial \mu}{\partial \rho_0} \right) \\
&= \beta \left(\frac{\partial \mu}{\partial \rho} \right) \left(\frac{\delta \rho_0}{\delta \rho} \right)^{-1} \\
&= \rho^{-1} \phi^{(1)} \left(1 - \frac{\partial \langle \delta \rho_i \rangle}{\partial \rho} \right)^{-1},
\end{aligned}
\tag{10}
$$

where the dimensionless inverse compressibility $\phi^{(1)}$ comes from the *dimensionless* expression $\phi^{(n)}$ which occurs frequently in the subsequent discussions and is defined as

$$
\phi^{(n)} \equiv \rho^n \left[\frac{\partial^{n-1}(\rho^{-1} \, \partial \beta P / \partial \rho)}{\partial \rho^{n-1}} \right].
\tag{11}
$$

Similarly, from (8), (9), and (11), we obtain

$$
A_3 = \rho^{-2} \phi^{(2)} \left(1 - \frac{\partial \langle \delta \rho_i \rangle}{\partial \rho} \right)^{-2} + \rho^{-1} \phi^{(1)} \left(\frac{\partial^2 \langle \delta \rho_i \rangle}{\partial \rho^2} \right) \left(1 - \frac{\partial \langle \partial \rho_i \rangle}{\partial \rho} \right)^{-3}.
\tag{12}
$$

The quantities A_{ij} and A_{ijk} (therefore, A_2, A_3, $\hat{A}_{ij}$, and $\hat{A}_{ijk}$) appearing in the probability $P\{\rho\}$ are functions of the most probable number density ρ_0 instead of the average number density ρ. In the next two sections we are interested in evaluating the cell-pair correlation function and thermodynamic quantities, using ρ as an independent variable rather than using ρ_0. Hence it is convenient to transform A_{ij} and A_{ijk} as a function of ρ at this stage by expanding $\hat{A}_{ij}$, $\hat{A}_{ijk}$, A_2, and A_3 around ρ:

$$
A_{ij} = \hat{a}_{ij} - \Delta \sum_k \hat{a}_{ijk} \langle \delta \rho_k \rangle + \text{higher order terms in } \langle \delta \rho_i \rangle,
\tag{13}
$$

$$
\hat{A}_{ijk} = \hat{a}_{ijk} + \text{higher order terms in } \langle \delta \rho_i \rangle,
\tag{14}
$$

$$
A_2 = \rho^{-1} \phi^{(1)} \left(1 + \frac{\partial \langle \delta \rho_i \rangle}{\partial \rho} \right) + \text{higher order terms in } \langle \delta \rho_i \rangle,
\tag{15}
$$

$$
A_3 = \rho^{-2} \phi^{(2)} + \text{higher order terms in } \langle \delta \rho_i \rangle,
\tag{16}
$$

where $\hat{a}_{ij}$ and $\hat{a}_{ijk}$ are defined as $\hat{A}_{ij}$ and $\hat{A}_{ijk}$, evaluated not at ρ_0 but at ρ. The second term in the right-hand side of (13) follows from replacing $\partial A_{ij} / \partial \rho_0$ ($= \Delta \sum_k A_{ijk}$) evaluated at $\rho_0 = \rho$ by $\Delta \sum_k \hat{a}_{ijk}$.

If we introduce these expressions into (7) and (8), then use the resulting expressions for A_{ij} and A_{ijk} in (4) for the probability $P\{\rho\}$, and then carry out the similar series expansion in $\delta\rho_i$ for the normalization constant C, the resulting expression for $P\{\rho\}$ is composed of the quadratic terms in $\delta\rho_i$ (Gaussian probability) and the higher-order terms in $\delta\rho_i$ which are regarded as corrections to the quadratic terms. Expanding the non-Gaussian parts by Taylor series in $\delta\rho_i$, the straightforward but lengthy manipulations yield the following series for $P\{\rho\}$:

$$
\begin{aligned}
P\{\rho\} = P_G\{\rho\}\Bigg[&1 - \tfrac{1}{2}\rho^{-1}\phi^{(1)}\left(\frac{\partial\langle\delta\rho_l\rangle}{\partial\rho}\right) \sum_i (\delta\rho_i{}^2 - \langle\delta\rho_i{}^2\rangle_G) - \frac{\Delta}{6}\rho^{-2}\phi^{(2)}\sum_i \delta\rho_i{}^3 \\
&+ \frac{\Delta}{2}\sum_{ijk}\hat{a}_{ijk}\langle\delta\rho_k\rangle(\delta\rho_i\,\delta\rho_j - \langle\delta\rho_i\,\delta\rho_j\rangle_G) - \frac{\Delta^3}{6}\sum_{ijk}\hat{a}_{ijk}\,\delta\rho_i\,\delta\rho_j\,\delta\rho_k \\
&+ \frac{\Delta^2}{72}(\rho^{-2}\phi^{(2)})^2\sum_{ij}(\delta\rho_i{}^3\,\delta\rho_j{}^3 - \langle\delta\rho_i{}^3\,\delta\rho_j{}^3\rangle_G) \\
&+ \frac{\Delta^4}{36}\rho^{-2}\phi^{(2)}\sum_{ijkl}\hat{a}_{ijk}(\delta\rho_i\,\delta\rho_j\,\delta\rho_k\,\delta\rho_l{}^3 - \langle\delta\rho_i\,\delta\rho_j\,\delta\rho_k\,\delta\rho_l{}^3\rangle_G) + \cdots\Bigg].
\end{aligned}
$$

(17)

The subscript G appearing in (17) implies the Gaussian expectation of the quantities in brackets, using the Gaussian probability $P_G\{\rho\}$;

$$
P_G\{\rho\} = C_G \exp\left[-(\Delta/2)\rho^{-1}\phi^{(1)}\sum_i \delta\rho_i{}^2 - (\Delta^2/2)\sum_{ij}\hat{a}_{ij}\,\delta\rho_i\,\delta\rho_j \right], \quad (18)
$$

$$
C_G{}^{-1} \equiv \sum_{\{\rho\}} \exp\left[-(\Delta/2)\rho^{-1}\phi^{(1)}\sum_i \delta\rho_i{}^2 - (\Delta^2/2)\sum_{ij}\hat{a}_{ij}\,\delta\rho_i\,\delta\rho_j \right]. \quad (19)
$$

Note that (17) satisfies the normalization condition*

$$
\sum_{\{\rho\}} P\{\rho\} = 1.
$$

Equation 17 contains the Gaussian averages of the various moments of $\delta\rho_i$ such as $\langle\delta\rho_i{}^3\,\delta\rho_j{}^3\rangle_G$, which can be reduced into product of the Gaussian-pair fluctuations $\langle\delta\rho_i\,\delta\rho_j\rangle_G$ by using the following relationship:

$$
\langle\delta\rho_1{}^\alpha\,\delta\rho_2{}^\beta\cdots\delta\rho_M{}^\omega\rangle_G = 0, \qquad \text{if } \alpha+\beta+\cdots+\omega = \text{odd integer},
$$

$$
= \sum\prod \langle\delta\rho_i\,\delta\rho_j\rangle_G, \qquad \text{if } \alpha+\beta+\cdots+\omega = \text{even integer},
$$

(20)

---

* The Gaussian probability (18) differs from the expression given in (49b) in ref. 2. In ref. 2 the probability distribution by the quadratic terms in (4) of this paper is used to represent $P_G\{\rho\}$. When this form of $P_G\{\rho\}$ is taken to calulate the cell-pair correlation function (see the next section) as a function of ρ, the series for the cell-pair correlation function contains additional terms (ref. 2) which are not present in this work.

where $\sum$ represents the sum over all possible ways of arranging α $\delta\rho_1$'s, β $\delta\rho_2$'s, $\ldots$; ω $\delta\rho_M$'s into $(\alpha + \beta + \cdots + \omega)/2$—pairs of these quantities. The product $\prod$ represents product of these pairs. Using (20), $\langle\delta\rho_1{}^3 \delta\rho_2\rangle_G$, for example, is equal to $3\langle\delta\rho_1 \delta\rho_2\rangle_G\langle\delta\rho_1{}^2\rangle_G$.

In the same spirit the averages $\langle\delta\rho_i\rangle$ and $\partial\langle\delta\rho_i\rangle/\partial\rho$ occurring in (17) can be expressed as a series containing only the Gaussian averages of $\delta\rho_i$'s. The average $\langle\delta\rho_i\rangle$ is obtained by using (17) for $P\{\boldsymbol{\rho}\}$. Since the Gaussian averages of odd numbers of $\delta\rho_i$'s are identically zero because of the Gaussian property (20), the first several terms, including the terms containing $\langle\delta\rho_k\rangle$ and $\delta\langle\delta\rho_i\rangle/\partial\rho$, in (17) do not contribute to $\langle\delta\rho_i\rangle$. The first two nonvanishing terms of the expansion for $\langle\delta\rho_i\rangle$ follows from the higher terms in the expansion (17); i.e.,

$$
\begin{aligned}
\langle\delta\rho_i\rangle = {}& -\frac{\Delta}{6}\,\rho^{-2}\phi^{(2)}\sum_j \langle\delta\rho_i \,\delta\rho_j{}^3\rangle_G - \frac{\Delta^3}{6}\sum_{jkl} \hat{a}_{jkl}\langle\delta\rho_i \,\delta\rho_j \,\delta\rho_k \,\delta\rho_l\rangle_G + \cdots \\
= {}& -\frac{\Delta}{2}\,\rho^{-2}\phi^{(2)}\sum_j \langle\delta\rho_i \,\delta\rho_j\rangle_G\langle\delta\rho_j{}^2\rangle_G \\
& -\frac{\Delta^3}{2}\sum_{jkl} \hat{a}_{jkl}\langle\delta\rho_i \,\delta\rho_j\rangle_G\langle\delta\rho_k \,\delta\rho_l\rangle_G + \cdots \\
= {}& -\tfrac{1}{2}\rho^{-1}\phi^{(2)}(\phi^{(1)})^{-1}\langle\delta\rho_i{}^2\rangle_G - \frac{\Delta^2}{2}\,\rho(\phi^{(1)})^{-1}\sum_{kl} \hat{a}_{jkl}\langle\delta\rho_k \,\delta\rho_l\rangle_G + \cdots, \quad (21)
\end{aligned}
$$

where the expression following the second equality is obtained from the expression following the first equality using the Gaussian property (20). The quantity $\langle\delta\rho_j{}^2\rangle_G$ appearing after the second equality of the above equation is independent of the index j. For a fluid, $\hat{a}_{jkl}$ in (21) depends only on the intercell separations R_{jk} ($\equiv |\mathbf{R}_j - \mathbf{R}_k|$), R_{jl}, and R_{kl}; and $\langle\delta\rho_i \,\delta\rho_j\rangle_G$, appearing after the second equality of (21), is also a function of R_{ij}, where the vector $\mathbf{R}_j$ denotes the center of the jth cell. Therefore, the summations over the index j following the second equality can be performed independently. The third equality of (21) follows then immediately from the second equality by using the identity
$$\rho(\phi^{(1)})^{-1} = \Delta \sum_j \langle\delta\rho_i \,\delta\rho_j\rangle_G,$$
derived in Appendix A.

The quantity $\partial\langle\delta\rho_i\rangle/\partial\rho$ contained in (17) can be evaluated by differentiating $\langle\delta\rho_i\rangle$ with respect to ρ;

$$
\begin{aligned}
\rho^{-1}\phi^{(1)}\frac{\partial\langle\delta\rho_i\rangle}{\partial\rho} = {}& \frac{\Delta}{2}\,(\rho^{-2}\phi^{(2)})^2\sum_j \langle\delta\rho_i \,\delta\rho_j\rangle_G{}^2 + \tfrac{1}{2}\rho^{-3}(\phi^{(2)})^2(\phi^{(1)})^{-1}\langle\delta\rho_i{}^2\rangle_G \\
& + \Delta^3\rho^{-2}\phi^{(2)}\sum_{ijk} \hat{a}_{jkl}\langle\delta\rho_i \,\delta\rho_j\rangle_G\langle\delta\rho_i \,\delta\rho_k\rangle_G \\
& + \frac{\Delta^3}{2}\,\rho^{-2}\phi^{(2)}\sum_{jkl} \hat{a}_{jkl}\langle\delta\rho_i \,\delta\rho_j\rangle_G\langle\delta\rho_k \,\delta\rho_l\rangle_G + \cdots. \quad (22)
\end{aligned}
$$

The two terms,

$$A = \tfrac{1}{2} \sum_{jkl} \hat{a}_{jkl} \langle \delta\rho_i \, \delta\rho_j \rangle_G \langle \delta\rho_i \, \delta\rho_k \rangle_G$$

and

$$B = \tfrac{1}{2} \sum_{ijk} \hat{a}_{jkl} \langle \delta\rho_i \, \delta\rho_j \rangle_G \langle \delta\rho_i \, \delta\rho_k \rangle_G$$

occur in deriving (22). However, these terms are equal and combined in (22). Proof: Let $\tilde{G}_2(\mathbf{K})$ be defined by the relation

$$G_2(R_{ij}) = \int d\mathbf{K} \exp (i\mathbf{R}_{ij} \cdot \mathbf{K}) \tilde{G}_2(\mathbf{K}).$$

Using similar definition $\tilde{G}_2(\mathbf{K}')$ for $G_2(R_{ik})$, we obtain

$$
\begin{aligned}
A &= \iint d\mathbf{K} \, d\mathbf{K}' \tilde{G}_2(\mathbf{K}) \tilde{G}_2(\mathbf{K}') \Big[\sum_{kl} \hat{a}_{jkl} \exp (i\mathbf{R}_{jk} \cdot \mathbf{K}') \Big] \sum_j \exp [i\mathbf{R}_{ij} \cdot (\mathbf{K} + \mathbf{K}')] \\
&= \iint d\mathbf{K} \, d\mathbf{K}' \tilde{G}_2(\mathbf{K}) \tilde{G}_2(\mathbf{K}') \Big[\sum_{kl} \hat{a}_{jkl} \exp (i\mathbf{R}_{jk} \cdot \mathbf{K}') \Big] \delta(\mathbf{K} + \mathbf{K}') \\
&= \int d\mathbf{K} [\tilde{G}_2(\mathbf{K})]^2 \sum_{kl} \hat{a}_{jkl} \exp (i\mathbf{R}_{jk} \cdot \mathbf{K}), \\
B &= \iint d\mathbf{K} \, d\mathbf{K}' \tilde{G}_2(\mathbf{K}) \tilde{G}_2(\mathbf{K}') \\
&\quad \times \Big[\sum_{jk} a_{jkl} \exp (i\mathbf{R}_j \cdot \mathbf{K} + i\mathbf{R}_k \cdot \mathbf{K}') \Big] \sum_i \exp [i\mathbf{R}_i \cdot (\mathbf{K} + \mathbf{K}')] \\
&= \int d\mathbf{K} [\tilde{G}_2(\mathbf{K})]^2 \sum_{kl} \hat{a}_{jkl} \exp (i\mathbf{R}_{jk} \cdot \mathbf{K}),
\end{aligned}
$$

where $\sum_j$ in A was taken out of the bracket, since the quantity inside the bracket is just a function of R_{jk}. In deriving (22), the identity $\rho^2 \, \partial[\rho^{-1}\phi^{(1)}]/ \partial\rho = \phi^{(2)}$ and the following relationship have been used:

$$
\begin{aligned}
\frac{\partial \langle \delta\rho_i \, \delta\rho_j \rangle_G}{\partial\rho} &= \sum_{\{\rho\}} \delta\rho_i \, \delta\rho_j \frac{\partial P_G\{\boldsymbol{\rho}\}}{\partial\rho} \\
&= -\tfrac{1}{2} \sum_{\{\rho\}} P_G\{\boldsymbol{\rho}\} \, \delta\rho_i \, \delta\rho_j \Big\{ \Delta\rho^{-2}\phi^{(2)} \sum_k (\delta\rho_k^2 - \langle \delta\rho_k^2 \rangle_G) \\
&\quad + \Delta^3 \sum_{klm} \hat{a}_{klm} (\delta\rho_k \, \delta\rho_l - \langle \delta\rho_k \, \delta\rho_l \rangle_G) \Big\} \\
&= -\Delta\rho^{-2}\phi^{(2)} \sum_k \langle \delta\rho_i \, \delta\rho_k \rangle_G \langle \delta\rho_j \, \delta\rho_k \rangle_G \\
&\quad - \Delta^3 \sum_{klm} \hat{a}_{klm} \langle \delta\rho_i \, \delta\rho_k \rangle_G \langle \delta\rho_j \, \delta\rho_l \rangle_G. \quad (23)
\end{aligned}
$$

It is, therefore, possible to express $P\{\boldsymbol{\rho}\}$ entirely in terms of the Gaussian expectations. Performing this operation by using (21) and (22), we finally obtain the desired expression

$$
\begin{aligned}
P\{\boldsymbol{\rho}\} = P_G\{\boldsymbol{\rho}\}\Bigg\{ & 1 - \frac{\Delta}{6}\,\rho^{-2}\phi^{(2)}\sum_i \delta\rho_i{}^3 \\
& + \frac{\Delta^2}{72}\,(\rho^{-2}\phi^{(2)})^2 \sum_{ij} [\delta\rho_i{}^3\,\delta\rho_j{}^3 - \langle \delta\rho_i{}^3\,\delta\rho_j{}^3\rangle_G \\
& - 18(\langle\delta\rho_i\,\delta\rho_j\rangle_G{}^2 + \langle\delta\rho_i\,\delta\rho_j\rangle_G\langle\delta\rho_i{}^2\rangle_G)(\delta\rho_i{}^2 - \langle\delta\rho_i{}^2\rangle_G)] \\
& - \frac{\Delta^3}{6}\sum_{ijk}\hat{a}_{ijk}\,\delta\rho_i\,\delta\rho_j\,\delta\rho_k + \frac{\Delta^4}{36}\,\rho^{-2}\phi^{(2)}\sum_{ijkl}\hat{a}_{ijk}[\delta\rho_i\,\delta\rho_j\,\delta\rho_k\,\delta\rho_l{}^3 \\
& - \langle\delta\rho_i\,\delta\rho_j\,\delta\rho_k\,\delta\rho_l{}^3\rangle_G - 3\langle\delta\rho_k\,\delta\rho_l{}^3\rangle_G(\delta\rho_i\,\delta\rho_j - \langle\delta\rho_i\,\delta\rho_j\rangle_G) \\
& - 18\langle\delta\rho_i\,\delta\rho_l\rangle_G\langle\delta\rho_j\,\delta\rho_l\rangle_G(\delta\rho_k{}^2 - \langle\delta\rho_k{}^2\rangle_G) \\
& - 9\langle\delta\rho_i\,\delta\rho_j\rangle_G\langle\delta\rho_k\,\delta\rho_l\rangle_G(\delta\rho_i{}^2 - \langle\delta\rho_i{}^2\rangle_G)] + \cdots \Bigg\}.
\end{aligned}
\tag{24}
$$

As has been pointed out in the paragraph following (6), the present formalism will probably not be valid *at* the critical point, since the assumption involved in deriving it may break down. However, as long as we are interested in the behavior of a substance in either the liquid or gas phase (but not both) *in the neighborhood* of the critical point, we assume that the expansion (24) is valid. In the next section, the above result is used to find out the corresponding series for the cell-pair correlation function.

SERIES EXPANSION OF THE CELL-PAIR CORRELATION FUNCTION $\chi(R)$

Let us define the cell-pair correlation function $\chi(R)$ as the difference of the averages involving $\delta\rho_1$ and $\delta\rho_2$,

$$
\chi(R) = \frac{\langle\delta\rho_1\,\delta\rho_2\rangle - \langle\delta\rho_1\rangle\langle\delta\rho_2\rangle}{\rho^2},
\tag{25}
$$

where $R \equiv R_{12} = |\mathbf{R}_1 - \mathbf{R}_2|$.

When (25) is summed over all possible sites of cell 2 with respect to the fixed cell 1, the resulting quantity is equal to the compressibility $\partial\rho/\partial(\beta P)$ (see Appendix A). Therefore, the quantity $\chi(R)$ plays an analogous role as the total *particle*-pair correlation function occurring in the Ornstein–Zernike relationship. In this section, we derive the first three terms in the series expansion of $\chi(R)$, using expression (24) for the probability function $P\{\boldsymbol{\rho}\}$ obtained in the previous section.

Define the three terms by $G_2(R)$, $\omega_2(R)$, and $\hat{\omega}_2(R)$, that is,

$$
\chi(R) = G_2(R) + (\phi^{(2)})^2\omega_2(R) + \phi^{(2)}\hat{\omega}_2(R) + \cdots.
\tag{26}
$$

The first term $G_2(R)$ represents the contribution by the Gaussian probability $P_G\{\rho\}$,

$$\rho^2 G_2(R) = \langle \delta\rho_1 \, \delta\rho_2 \rangle_G - \langle \delta\rho_1 \rangle_G \langle \delta\rho_2 \rangle_G$$
$$= \langle \delta\rho_1 \, \delta\rho_2 \rangle_G. \tag{27}$$

The second and third terms in (26) are, respectively, the contributions by the terms containing factors like $\rho^{-2}\phi^{(2)}$ and $(\rho^{-2}\phi^{(2)})^2$, and by the term with $\hat{a}_{ijk}$ in (24). Use of the Gaussian property (20) reduces all higher fluctuations in $\delta\rho_i$'s occurring in $\omega_2(R)$ and $\hat{\omega}_2(R)$ as sums of products of the Gaussian pair fluctuations. Several arithmetic manipulations for the expressions occurring in $\omega_2(R)$ and $\hat{\omega}_2(R)$ simplify to give the final results,

$$\omega_2(R) = \frac{\Delta^2}{2} \rho^{-6} \sum_{ij} \langle \delta\rho_i \, \delta\rho_j \rangle_G^2 (\langle \delta\rho_1 \, \delta\rho_i \rangle_G \langle \delta\rho_2 \, \delta\rho_j \rangle_G$$
$$- \langle \delta\rho_1 \, \delta\rho_i \rangle_G \langle \delta\rho_2 \, \delta\rho_i \rangle_G), \tag{28}$$

$$\hat{\omega}_2(R) = \Delta^4 \rho^{-4} \sum_{ijkl} \hat{a}_{ijk} \langle \delta\rho_i \, \delta\rho_l \rangle_G \langle \delta\rho_j \, \delta\rho_l \rangle_G (\langle \delta\rho_1 \, \delta\rho_k \rangle_G \langle \delta\rho_2 \, \delta\rho_l \rangle_G$$
$$- \langle \delta\rho_1 \, \delta\rho_k \rangle_G \langle \delta\rho_2 \, \delta\rho_k \rangle_G). \tag{29}$$

Note that all the non-Gaussian corrections such as $\omega_2(R)$ and $\hat{\omega}_2(R)$ do not contribute to the compressibility. For example, the summations over all possible sites of the subscript 2 in (28) and (29) yield identically zero, giving the following exact relationship (proved in Appendix A), if the cell size Δ is chosen to include large number of particles (see Appendix A for the proof):

$$\frac{\partial\rho}{\partial(\beta P)} = \rho\Delta \sum_2 \chi(R_{12})$$

$$= \rho\Delta \sum_2 G_2(R_{12}). \tag{30}$$

If the cell size Δ is chosen small compared to the range of the pair-correlation function, the summations in (28)–(30) can be replaced by integrations; i.e.,

$$\omega_2(R) = \tfrac{1}{2}\rho^2 \iint d\mathbf{R}_i \, d\mathbf{R}_j [G_2(R_{ij})]^2 [G_2(R_{1i})G_2(R_{2j}) - G_2(R_{1i})G_2(R_{2i})], \tag{31}$$

$$\hat{\omega}_2(R) = \rho^4 \iiiint d\mathbf{R}_i \, d\mathbf{R}_j \, d\mathbf{R}_k \, d\mathbf{R}_l \hat{a}_{ijk} G_2(R_{il})G_2(R_{jl})$$
$$\times [G_2(R_{1k})G_2(R_{2l}) - G_2(R_{1k})G_2(R_{2k})], \tag{32}$$

$$\frac{\partial\rho}{\partial(\beta P)} = \rho \int d\mathbf{R}\chi(R) = \rho \int d\mathbf{R}G_2(R). \tag{33}$$

Next, we find a proper expansion parameter associated with the terms such as $\omega_2(R)$ in (26). An expansion parameter for the terms containing $\hat{a}_{ijk}$ such as $\hat{\omega}_2(R)$ in (26) is not discussed here, since any theoretically acceptable form for $\hat{a}_{ijk}$ has yet to be found. In Appendix B, a plausible argument is presented to show that, in the neighborhood of the critical point, the terms with $\hat{a}_{ijk}$ in the expansion (26) do not dominate over the other terms. Whether they dominate or not, the following discussions are valid, provided, of course, that such an expansion exists.

Let us assume that, at large R ($\gg \Delta^{1/d}$), the cell-pair correlation function takes the following form proposed by Fisher[12]:

$$G_2(R) = \frac{DG(\kappa R)}{R^{d-2+\eta}}, \tag{34}$$

where $G(x)$ is a decaying function of x, and κ^{-1} is known as the correlation length, d represents the dimensionality of the system, the exponent η is a correction factor to the Ornstein–Zernike form for $G_2(R)$, and it is less than one in most cases. In the neighborhood of the critical point, the proportionality constant D can be determined from (33) and (34) and is approximately equal to $\rho^{-1}\kappa^{2-\eta} \partial\rho/\partial(\beta P) = \rho^{-1}\kappa^{2-\eta}(\phi^{(1)})^{-1}$. This is so, since the dominant contribution to the compressibility comes from the long-range part of $G_2(R)$.

The asymptotic form (34) for $G_2(R)$ is probably valid for $\kappa R \approx 1$.[2] These two conditions, $R^d \gg \Delta$ and $R \approx \kappa^{-1}$, impose a restriction on the cell size to satisfy $\Delta \ll \kappa^{-d}$ (the second restriction given). This, together with the first restriction $\Delta \gg \rho^{-1}$, gives a bound[1,4] on Δ: $\rho \ll \Delta \ll \kappa^{-d}$. If an assumption identical to the one used to derive the Ornstein–Zernike form ($\eta = 0$) for the particle-pair correlation function is used here, this bound yields also the Ornstein–Zernike form for the cell-pair correlation function.[2] In the following discussions, we take the cell size to satisfy this bound but place no further restriction. In this respect, the present approach is different from the "scaling law" which requires an additional restriction. The additional restriction demands that the free energy of the whole system can be mapped onto that of the cell with the same functional form for any Δ (satisfying $\rho \ll \Delta \ll \kappa^{-d}$) by properly scaling temperature and magnetic field (or chemical potential for the liquid–gas case) between the system and its subsystems.

It is convenient to introduce a dimensionless distance $x \equiv \kappa R$ in describing equilibrium behaviors in the neighborhood ($\kappa \neq 0$) of the critical point. As we shall find out from the ensuing discussions, the convenience arises from the fact that κ^{-1} turns out to be the only nontrivial distance-reducing parameter for all terms occurring in the expansion of $\chi(R)$ which do not contain $\hat{a}_{ijk}$ or the higher terms such as $\hat{a}_{ijkl}$. In terms of this new variable x and a new dimensionless function defined by

$$f_1(x) \equiv \frac{G(x)}{x^{d-2+\eta}}, \tag{35}$$

the first term (34) of the expansion for $\chi(R)$ can be scaled as

$$G_2(R) = \xi\Phi_1 f_1(x),$$ (36)

where

$$\xi \equiv \rho^{-1}(\rho D)^{d/(2-\eta)},$$ (37)

$$\Phi_1 \equiv [\phi^{(1)}]^{-1+d/(2-\eta)}.$$ (38a)

The function Φ_n for the case $n > 1$ is defined here for future use:

$$[\Phi_n]^2 \equiv [\phi^{(n)}]^2[\phi^{(1)}]^{-n-1+(n-1)d/(2-\eta)}, \qquad n \geq 2.$$ (38b)

Several comments are in order as to the form (36) associated with $G_2(R)$, and these comments can apply also to the higher-order terms in the expansion of $\chi(R)$ as well. The factor ξ occurring in $G_2(R)$ is our expansion parameter in the present problem, so that the higher-order terms in the expansion of $\chi(R)$ contain higher powers of ξ. In the reduced form (36) for $G_2(R)$, the κ dependence (therefore, T and ρ dependence) occur only in the second factor Φ_1 as a proportionality constant. A detailed investigation we made shows that this linear occurrence of Φ_1 is a common feature associated with all terms involved in the expansion. Therefore this term can be factored out of the expansion and used to reduce the cell-pair correlation function $\chi^*(x)$ which is only a function of the reduced distance x, and its corresponding expansion has the following form:

$$\chi^*(x) \equiv \frac{\chi(R)}{\Phi_1} = \chi_1(x)\xi + \chi_2(x)\xi^2 + \cdots + \chi_n(x)\xi^n + \cdots$$

$$+ \text{ terms involving } \hat{a}_{ijk}, \text{ etc.,} \quad (39)$$

where $\chi_1(x) \equiv f_1(x)$ and $\chi_2(x)$ can be obtained from (26), (31), and (36); that is,

$$\chi_2(x) = [\Phi_2]^2 f_2(x),$$ (40)

$$f_2(x) \equiv \frac{1}{2}\iint dx_3\, dx_4 [f_1(x_{34})]^2 [f_1(x_{13})f_1(x_{24}) - f_1(x_{13})f_1(x_{23})].$$ (41)

Any explicit expression of the higher-order terms such as χ_3 and χ_4 is rather lengthy and is not at present directly of interest, since we are interested only in common features of each terms $\chi_n(x)$. This can be deduced directly by investigating the first several $\chi_n(x)$'s* and has the following expression:

$$\chi_n(x) = \sum_{m=1}^{m^{\ddagger}} H_n[\Phi_2, \Phi_3, \ldots, \Phi_m]h_{nm}(x),$$ (42)

* Using the analysis of this section, we obtain $\chi_3(x)$ and $\chi_4(x)$ as follows:

$\chi_3(x) = [\Phi_2]^4 h_{31}(x) + [\Phi_2]^2[\Phi_3]h_{32}(x) + [\Phi_3]^2 h_{33}(x),$

$\chi_4(x) = [\Phi_2]^6 h_{41}(x) + [\Phi_3]^3 h_{42}(x) + [\Phi_4]^2 h_{43}(x) + [\Phi_2]^2[\Phi_3]^2 h_{44}(x)$

$\qquad\qquad + [\Phi_2][\Phi_3][\Phi_4]h_{45}(x) + [\Phi_2]^4[\Phi_4]h_{46}(x).$

where $H_n(z_2, z_3, \ldots, z_m)$ is a certain well-defined product of its arguments, and $h_{nm}(x)$ is a bounded dimensionless function of x whose integral $\int h_{nm}(x)\, dx$ vanishes identically. The sum (42) extends to m^+ which is in turn determined by n.

In the case of Kac's model[13] with a finite long-range attraction, the proportionality constant D and the inverse range of the correlation κ appearing in the expression for $G_2(R)$ have been investigated by Hemmer.[14] From his investigation, therefore, it is possible to estimate the magnitude of the expansion parameter ξ in terms of shape of the intermolecular force, i.e.,

$$\xi \approx \left(\frac{\text{range of hard core}}{\text{range of attraction}}\right)^d.$$

Therefore, the quantity ξ is usually a small number. In the following section, the terms containing Φ_n's in (42) are investigated. Study of this term yields various inequalities relating the critical indices for both the thermodynamic and the statistical-mechanical quantities.

INEQUALITIES INVOLVING CRITICAL INDICES

In the neighborhood of the critical point, various thermodynamic and statistical-mechanical quantities show singular behaviors. We adopt the notations introduced by Fisher[15] to represent the critical behavior for the following quantities (these quantities applicable for a fluid can also be translated into thermodynamic quantities for a magnetic system; see Ref. 7):

$$P \sim |\rho - \rho_c|^\delta, \quad \text{at} \quad T = T_c, \tag{43}$$

$$\phi^{(1)} \sim [T - T_c]^\gamma \tag{44a}$$

$$\kappa \sim [T - T_c]^\nu, \tag{44b}$$

$$\left|\frac{\partial^3 P}{\partial \mu^3}\right| \sim [T - T_c]^{-\gamma - \Delta_3}, \tag{44c}$$

$$\left|\frac{\partial^4 P}{\partial \mu^4}\right| \sim [T - T_c]^{-\gamma - \Delta_3 - \Delta_4}, \text{ etc.}, \quad \text{at} \quad \rho = \rho_c,\, T > T_c, \tag{44d}$$

where the subscript c represents the critical point. For $T < T_c$, the primed exponents* γ' and ν' are used to distinguish them from the corresponding

---

* Note that the primed exponents for the liquid side of the coexistence curve may have different values from the corresponding ones for the gas side. In the following discussions, the relations involving the primed indices are applicable to either one (not both) of the two possibilities.

quantities for $T > T_c$. In addition, the critical behavior of the liquid (l)–gas(g) coexistence curve is represented by the index β;

$$|\rho_l - \rho_g| \sim [T_c - T]^\beta, \tag{45a}$$

and the gap exponents Δ'_n for $T < T_c$ are defined along the coexistence curve

$$\left|\frac{\partial^2 P}{\partial \mu^2}\right| \sim [T_c - T]^{\beta - \Delta_2'},$$

$$\left|\frac{\partial^3 P}{\partial \mu^3}\right| \sim [T_c - T]^{-\gamma' - \Delta_3'}, \text{ etc.} \tag{45b}$$

Relations connecting the critical indices γ, ν (also γ', ν'), and η follow immediately by using the compressibility relation (33) and (34) and (44):

$$\gamma = \nu(2 - \eta) \quad \text{and} \quad \gamma' = \nu'(2 - \eta). \tag{46}$$

Let us assume that the expansion (39) for $\chi^*(x)$ converges and each term $\chi_n(x)$ of the expansion is bounded near the critical point. (An alternative but perhaps somewhat stronger condition is an absolute convergence of the series.) This is the main assumption of the present work. If the dimensionality of the system under study becomes large, the above condition is probably satisfied. We return to this point in the next section, and present some numerical indications that a three-dimensional system may satisfy the condition. This assumption implies that the quantity Φ_n occurring in the expansion coefficient $\chi_n(x)$ (42) must be bounded;

$$(\Phi_n)^2 = [\phi^{(n)}]^2 [\phi^{(1)}]^{-n-1+(n-1)d/(2-\eta)} < \infty. \tag{47}$$

This new relationship can be used to obtain further inequalities relating the critical indices.

Along the critical isotherm ($T = T_c$), use of (11) and (43) leads to $\phi^{(n)}$ behaving as $|\rho - \rho_c|^{\delta-n}$. Therefore (47) yields the relationship among δ, d, and η,*

$$\eta \geq 2 - d(\delta - 1)/(\delta + 1). \tag{48}$$

Fisher[5] has shown that the equality relation $\eta = 2 - d(\delta + 1)/(\delta + 1)$ follows from the scaling-law argument. Although the equality is correct for the two-dimensional Ising model [$\eta = \frac{1}{4}$ and $\delta = 15$], it fails to predict the expected critical indices [$\eta = 0$, and $\delta = 3$] for the van der Waals gas ($d \to \infty$). Recently, Stell[16] obtained a new relationship,

$$\eta = \max\left[0, 2 - d\frac{(\delta - 1)}{(\delta + 1)}\right], \tag{49}$$

* If δ is an integer, $\phi^{(n)}$ (for $n \geq \delta$) becomes constant; then, (47) is trivially satisfied. Note that n occurring in the exponents of (47) does not contribute in (48), since it factors out.

by using the functional Taylor expansion of the direct correlation function which is assumed to be convergent at the critical point. Although Stell's relation circumvents the difficulty associated with the high dimensions, use of the numerical values [$\eta \approx \frac{1}{18}$,[17] $\delta = 5$ or 5.2[18]] for the three-dimensional Ising model studied both by Fisher and Burford[17] and by Gaunt[18] apparently do not satisfy this relation. This difficulty does not occur if the inequality is used in (48).

For $T > T_c$ we follow along the critical isochore ($\rho = \rho_c$). The inequality (47) relates the indices γ and ν with the gap exponents Δ_n's. To obtain the relationships, we use the definition $\partial P/\partial \mu = \rho$, and the identities, $\rho^{-1}\beta^{-1}(\partial^2 P/\partial \mu^2) = (\phi^{(1)})^{-1}$, $\rho^{-1}\beta^{-2}(\partial^3 P/\partial \mu^3) = -\phi^{(2)}/(\phi^{(1)})^3$, etc., together with (47) and obtain near the critical point, for example,

$$\rho^{-1}\beta^{-2}\left|\frac{\partial^3 P}{\partial \mu^3}\right| \leq [\phi^{(1)}]^{-3/2-d/(4-2\eta)}, \tag{50a}$$

$$\rho^{-1}\beta^{-3}\left|\frac{\partial^4 P}{\partial \mu^4}\right| \leq [\phi^{(1)}]^{-2-d/(2-\eta)}. \tag{50b}$$

For $T > T_c$, if the singular behavior of $\phi^{(1)}$ and $\partial^n P/\partial \mu^n$ described by (44) are inserted into (50), we obtain*

$$\Delta_3 \leq (\gamma + d\nu)/2 \quad \text{and} \quad \Delta_4 \leq (\gamma + d\nu)/2. \tag{51a}$$

The above relations can be generalized to an arbitrary Δ_n, i.e.,

$$\Delta_n \leq (\gamma + d\nu)/2, \qquad n = 3, 4, \ldots. \tag{51b}$$

Below the critical temperature ($T < T_c$), we follow the coexistence line.† The definition $\rho^{-1}\beta^{-1}(\partial^2 P/\partial \mu^2) = [\phi^{(1)}]^{-1}$ together with the critical behaviors of $(\partial^2 P/\partial \mu^2)$ (45b) and $\phi^{(1)}$ (44) immediately leads to the relation,

$$\Delta_2' = \beta + \gamma'. \tag{52}$$

The relationships connecting other Δ_n' with γ' and ν' can be found in a straightforward manner by using the same method employed in deriving the expressions relating Δ_n, γ, and ν,

$$\Delta_n' \leq (\gamma' + d\nu')/2, \qquad n = 3, 4, \ldots. \tag{53}$$

If all Δ_n''s are assumed to be equal, i.e., $\Delta' \equiv \Delta_2' = \Delta_3' = \cdots = \Delta_n'$, (25) and (53) reduce to

$$\beta \leq (d\nu' - \gamma')/2 \quad \text{and} \quad \Delta' \leq (d\nu' + \gamma')/2. \tag{54}$$

* The gap exponents Δ_n with odd n can exist for the liquid–gas case. For the magnetic problem, the gap exponents in (44) and (45) have similar meaning if P and μ are replaced by the free energy F and the magnetic field H, respectively. In this case, because of the symmetry of F for $H > 0$, and $H < 0$, the exponents Δ_n for odd n do not occur.

† See footnote *, p. 446.

If the equalities are used in the above equations, these equations become identical to the equations derived previously by Fisher. Although the equalities are correct for the two-dimensional Ising model, use of the equalities in (54) again fail to match the expected values ($\beta = \frac{1}{2}$, $\nu' = \frac{1}{2}$, $\gamma' = 1$, and $\Delta' = \frac{3}{2}$) for the limiting van der Waals behavior for large dimensions.

SUMMARY AND CONCLUDING REMARKS

Central assumption involved in the present work hinges upon bounded convergence of a series for the cell-pair correlation function $\chi(R)$ resulting from use of the series (24) for $P\{\rho\}$. The function $\chi(R)$ is assumed to converge in *single-phase* regions *near* the critical point. As has been noted in the paragraph following (6), this assumption can be replaced by possibly a weaker condition on $\chi(R)$. The assumption plus the leading asymptotic form (34) for $\chi(R)$ proposed by Fisher leads to the inequality (47) which is the main result of the present paper. The inequality can be regarded as unifying the inequalities (or equalities) which can be obtained by tracing different thermodynamic paths (constant temperature, constant density, or the coexistence curve) in the neighborhood of the critical point. When this relationship is used along the critical isotherm, for example, it yields the inequality (48) linking the critical indices η and δ. Recently, Gunton and Buckingham[19] derived the same inequality by a quasi-rigorous argument which relates the mean-square fluctuation for the magnetization with pair correlation length. Along the critical isochore ($\rho = \rho_o$), the inequality (47) gives the new inequalities (51), (53), and (54) relating the gap exponents Δ_n's with the other critical indices. Use of the inequalities circumvents the difficulties associated with the equalities for the same relations deduced from the scaling law when it is used in high dimension, $d \geqslant 3$.

Convergence of the expansion for the cell-pair correlation function near the critical point is still an open problem. However, the critical indices of a system whose dimension is large enough will eventually be expected to satisfy the relations (48), (51), (53), and (54) as inequalities rather than as equalities. Since the inequality relations imply that Φ_n (47) should become small near the critical point, the terms $\chi_n(x)$ for $n > 1$, which depend in turn on Φ_i's [see (40) and (42)], should also become smaller than the Gaussian contribution $\chi_1(x)$ to $\chi^*(x)$ near the critical point. The convergence of the series will then most likely be valid in this temperature and density region. That this might already appear in three dimensions is partly indicated by the numerical values of the critical indices[17,18] η [≈ 0.0555] and δ [≈ 5.2] for the Ising model in three dimensions. If these values were used in (48), we obtain the inequality relation, $\eta \approx 0.0555 > -0.03$.

Similarly, the cell-triplet correlation function

$$\rho^{-3}[\langle \delta\rho_1 \, \delta\rho_2 \, \delta\rho_3 \rangle - \langle \delta\rho_1 \rangle \langle \delta\rho_2 \, \delta\rho_3 \rangle - \langle \delta\rho_2 \rangle \langle \delta\rho_1 \, \delta\rho_3 \rangle$$
$$- \langle \delta\rho_3 \rangle \langle \delta\rho_1 \, \delta\rho_2 \rangle + 2\langle \delta\rho_1 \rangle \langle \delta\rho_2 \rangle \langle \delta\rho_3 \rangle]$$

can be expanded in a series by using the cell occupation probability (24). We have analyzed several terms in this series, using the form for the cell-triplet correlation function derived in Ref. 2.* The analysis is consistent with the result which is represented by (47). Finally, it is perhaps worthwhile to point out that any explicit use of interaction potential among particles does not occur in the present work. The interaction potential only indirectly affects the cell-occupation probability $P\{\boldsymbol{\rho}\}$ through the Helmholtz free energy $A\{\boldsymbol{\rho}\}$ appearing in the expression for $P\{\boldsymbol{\rho}\}$. It asserts the view that the interaction potential can govern the existence or nonexistence of a critical point as well as individual values of some critical indices to a lesser degree, yet it is presumably not required in *correlating* various critical indices.

ACKNOWLEDGMENTS

We would like to acknowledge our sincere thanks to Professor Joseph E. Mayer for the stimulating discussions, and to Professor Michael E. Fisher for the helpful correspondence.

APPENDIX A

Proof of $\rho[\Delta\partial(\beta P)/\partial\rho]^{-1} = \sum_j \langle \delta\rho_i \, \delta\rho_j \rangle_G = \sum_j \chi(R_{ij})$

Consider the identity,

$$\rho_0 = \langle \rho_l \rangle_G = C_G \sum_{\{\boldsymbol{\rho}\}} \rho_l \exp\left[-\tfrac{1}{2} \sum_{ij} v_{ij}(\rho_i - \rho_0)(\rho_j - \rho_0) \right], \qquad (A1)$$

where the Gaussian probability (18) is used explicitly in the right-hand side of the above equation, and the quantity v_{ij} stands for

$$v_{ij} \equiv \Delta^{-1}\rho^{-1}\phi^{(1)} \, \delta(i - j) + \hat{a}_{ij}, \qquad v_{ij} = v_{ji}. \qquad (A2)$$

The quantity $\hat{a}_{ij}$ is related to $\hat{A}_{ij}$ by (13). As has been described in the paragraph following (8), the quantities $\hat{A}_{ij}$, $\hat{A}_{ijk}$, etc., are evaluated at ρ_0

---

* The asymptotic form for the cell-triplet correlation function is given by (84b) in ref. 2.

corresponding to a given chemical potential, and satisfy the equilibrium conditions

$$\sum_j \hat{A}_{ij} = 0, \qquad \sum_{jk} \hat{A}_{ijk} = 0, \text{ etc.}$$

Since these relationships and (7) and (8) should hold at any density, if (7) and (8) are used at the chemical potential corresponding to the average density ρ (instead of ρ_0), we obtain similar conditions for $\hat{a}_{ij}$, $\hat{a}_{ijk}$, etc.,

$$\sum_j \hat{a}_{ij} = 0, \tag{A3a}$$

$$\sum_{jk} \hat{a}_{ijk} = 0. \tag{A3b}$$

Therefore, summing over the index j in (A2), we obtain

$$\sum_j v_{ij} = \Delta \rho^{-1} \phi^{(1)} = \Delta \rho^{-2} \frac{\partial(\beta P)}{\partial \rho}, \tag{A4}$$

If (A1) is differentiated with respect to ρ_0, we obtain

$$1 = \sum_j \left(\sum_k v_{jk} \right) \langle \delta \rho_i \, \delta \rho_j \rangle_G + \rho_0 \left(\frac{\partial \ln C_G}{\partial \rho_0} \right) - \frac{1}{2} \sum_{ij} \left(\frac{\partial v_{ij}}{\partial \rho_0} \right) \langle \rho_i \, \delta \rho_i \, \delta \rho_j \rangle_G. \tag{A5}$$

If the Gaussian property (20) is used, the last two terms in (A5) become zero, and the first equality of the desired relation follows immediately from (A4) and (A5).

In order to prove the second equality of the desired relation, we use

$$\langle N_i \rangle \equiv \sum_{\{N\}} N_i P\{N\}. \tag{A6}$$

Using the expression (1) for the distribution function $P\{N\}$, we obtain

$$\beta^{-1} d\langle N_i \rangle / d\mu = \sum_{\{N\}} N_i \{N_j - \langle N_j \rangle\} P\{N\}$$

$$= \Delta^2 \sum_j \chi(R_{ij}), \tag{A7}$$

where the definition (25) for the cell-pair correlation function has been used to obtain the second equality of the above equation. Since $d\langle N_i \rangle / d\mu = \Delta \, d\rho/d\mu = \rho \, \Delta(\partial \rho/\partial P)$, we immediately obtain the second equality of the desired relation from (A7).

The relation derived in this Appendix is an extension of the Ornstein–Zernike relation for the compressibility by using the cell-pair correlation function rather than the usual particle-pair correlation function.

APPENDIX B

Expansion Parameter Associated with $\hat{\omega}_2(R)$

In this Appendix, we find a proper expansion parameter associated with the terms such as $\hat{\omega}_2(R)$, defined by (32),

$$\hat{\omega}_2(R) = \rho^4 \iiiint d\mathbf{R}_i \, d\mathbf{R}_j \, d\mathbf{R}_k \, d\mathbf{R}_l \hat{a}_{ijk}\{G_2(R_{il})G_2(R_{jl})$$
$$\times [G_2(R_{1k})G_2(R_{2l}) - G_2(R_{1k})G_2(R_{2k})]\}, \quad (B1)$$

and we also discuss briefly inequalities for the critical indices arising from this function. The function $\hat{\omega}_2(R)$ contains an unknown function $\hat{a}_{ijk}$ which depends on the intercell separations, R_{ij}, R_{ik}, and R_{jk}. In the following, the functional form for $\hat{a}_{ijk}$ is considered to belong to either one of the following two rather general classes of functions:

Case (a): Class of functions which have the finite second moment,

$$\rho^2 \iint d\mathbf{R}_i \, d\mathbf{R}_k R_{ij}{}^2 \hat{a}_{ijk} \equiv -6(\rho\zeta)^{-2/d}; \quad (B2)$$

Case (b): Class of functions with the property,

$$\rho^2 \int \hat{a}_{ijk} \, d\mathbf{R}_k \sim -(\rho\bar{\zeta})^{(\bar{\eta}-2)/d} R_{ij}^{-d-2+\bar{\eta}}, \quad (B3)$$

where $\bar{\zeta}$ and $\bar{\eta}$ are temperature- and density-insensitive constants.

First, let us consider Case (a). In this case, the function $\hat{a}_{ijk}$ is strictly short ranged. Therefore the function,

$$G_2(R_{il})G_2(R_{jl})[G_2(R_{1k})G_2(R_{2l}) - G_2(R_{1k})G_2(R_{2k})],$$

occurring in the integrand of (B1) can be expanded by a Taylor series around both $\mathbf{R}_i$ and $\mathbf{R}_j$ at $\mathbf{R}_k$. The first term in the expansion does not contribute, since $\iint d\mathbf{R}_i \, d\mathbf{R}_j \hat{a}_{ijk} = 0$ [see (A3b) in Appendix A]. The second term which is linear in $(\mathbf{R}_j - \mathbf{R}_k)$ also does not contribute, since $\hat{a}_{ijk}$ is a symmetric function in R_{ij}, R_{ik}, and R_{jk}; that is, $\iint d\mathbf{R}_i \, d\mathbf{R}_j \hat{a}_{ijk} X_{jk} = 0$, $[X_{jk} \equiv x$ component of $\mathbf{R}_j - \mathbf{R}_k]$. The quadratic term of the expansion is assumed to have a nonzero moment, i.e.,

$$(\rho\zeta)^{-2/d} \equiv -\tfrac{1}{2}\rho^2 \iint d\mathbf{R}_i \, d\mathbf{R}_j X_{jk}{}^2 \hat{a}_{ijk}$$
$$= -\rho^2 \iint d\mathbf{R}_i \, d\mathbf{R}_j X_{ik} X_{jk} \hat{a}_{ijk}, \quad (B4)$$

where the identity, $X_{jk}{}^2 = -X_{jk}(X_{ij} + X_{ki})$, is used to derive the second equation. Let F represent the quantity inside $\{\ \}$ in (B1). The quadratic term in the Taylor expansion contains the factor,

$$\{\tfrac{1}{2}[\mathbf{R}_{ik} \cdot \mathbf{\nabla}_i]^2 + \tfrac{1}{2}[\mathbf{R}_{jk} \cdot \mathbf{\nabla}_j]^2 + [\mathbf{R}_{ik} \cdot \mathbf{\nabla}_i](\mathbf{R}_{jk} \cdot \mathbf{\nabla}_j)\}F,$$

where derivatives $\mathbf{\nabla}_i \ [\equiv \partial/\partial\mathbf{R}_i]$ and $\mathbf{\nabla}_j$ are evaluated at $\mathbf{R}_i$ and $\mathbf{R}_j = \mathbf{R}_k$. Now, the integrations over $\mathbf{R}_i$ and $\mathbf{R}_j$ in (B1) can be carried out first. Using the definition (B4) for $(\rho\zeta)^{-2/d}$, we obtain

$$\hat{\omega}_2(R) = (\rho\zeta)^{-2/d}\rho^2 \iint d\mathbf{R}_k \, d\mathbf{R}_l [\mathbf{\nabla}_i^2 + \mathbf{\nabla}_j^2 + \mathbf{\nabla}_i \cdot \mathbf{\nabla}_j]F$$

$$= \tfrac{1}{2}(\rho\zeta)^{-2/d}\rho^2 \iint d\mathbf{R}_k \, d\mathbf{R}_l [\mathbf{\nabla}_i^2 + \mathbf{\nabla}_j^2 + \mathbf{\nabla}_k^2]F, \qquad (B5)$$

where the derivatives are evaluated at $\mathbf{R}_i$ and $\mathbf{R}_j = \mathbf{R}_k$. The second equality follows by using the identity $\mathbf{\nabla}_i + \mathbf{\nabla}_j + \mathbf{\nabla}_k = 0$. We use the reduced distance $x \equiv \kappa R$ together with the definition for F and express $\hat{\omega}_2(R)$ in terms of the dimensionless function $\hat{f}_2(x)$; that is,

$$\phi^{(2)}\hat{\omega}_2(R) = -(\Phi_1)^{3/2}\Phi_2\xi^2(\xi\zeta^{-1})^{2/d}\hat{f}_2(x), \qquad (B6)$$

$$\hat{f}_2(x) \equiv \frac{1}{2} \iint d\mathbf{x}_k \, d\mathbf{x}_l \, \nabla^2\{[f_1(x_{kl})]^2[f_1(x_{1k})f_1(x_{2l}) - f_1(x_{1k})f_1(x_{2k})]\}, \qquad (B7)$$

where $\xi^2(\xi\zeta^{-1})^{2/d}$ in (B6) is the expansion parameter for those terms in $\chi(R)$ containing $\hat{a}_{ijk}$, and $x_{ij} \equiv \kappa \, |\mathbf{R}_i - \mathbf{R}_j|$ and $\mathbf{\nabla} \equiv \partial/\partial(\kappa\mathbf{R}_k)$. The quantity $f_1(x)$ is defined by (25), and the relation $\phi^{(1)} = [\kappa(\rho\xi)^{-1/d}]^{2-\eta}$ has been used to eliminate κ in (B6). Analysis of the higher-order terms involving $\hat{a}_{ijk}$ in the expansion (26) is essentially similar to the above discussions, and the proper expansion parameter for these terms are $\xi^2(\xi\zeta^{-1})^{2/d}$ in (B6).

Likewise, if $\hat{\omega}_2(R)$ belongs to Case (b), its reduced form is given in terms of a new function $f_2^+(x)$ which contains $f_1(x)$ (35);

$$\phi^{(2)}\hat{\omega}_2(R) = -(\Phi_1)^{3/2}\Phi_2[\phi^{(1)}]^{(\eta-\bar{\eta})/(2-\eta)}\xi^2(\xi\bar{\zeta}^{-1})^{(2-\bar{\eta})/d}f_2^+(x), \qquad (B8)$$

where

$$f_2^+(x) = \iiiint d\mathbf{x}_i \, d\mathbf{x}_j \, d\mathbf{x}_k \, d\mathbf{x}_l a_{ijk}^* f_1(x_{il})f_1(x_{jl})$$

$$\times [f_1(x_{1k})f_l(x_{2l}) - f_1(x_{1k})f_1(x_{2k})], \qquad (B9)$$

and a_{ijk}^* is given by (B3) and the following definition:

$$\int a_{ijk}^* \, d\mathbf{x}_k = x_{ij}^{-d-2+\bar{\eta}}. \qquad (B10)$$

The expansion parameter in this case is $\xi^2(\xi\bar{\zeta}^{-1})^{(2-\bar{\eta})/d}$.

If the series for the cell-pair correlation function $\chi^*(x)$ is assumed to be convergent, the term $\phi^{(2)}\hat{\omega}_2(R)/\Phi_1$ occurring in the expansion (39) of $\chi^*(x)$ must also be bounded. Near the critical point, it implies that the temperature- and density-sensitive part of $\phi^{(2)}\hat{\omega}_2(R)/\Phi_1$ must be bounded. Noting that this part for Case (a) given by (B6) is a special case ($\bar{\eta} = 0$) for Case (b), given by (B8), this requirement yields

$$|\phi^{(2)}[\phi^{(1)}]^{d/(2-\eta)-2+(\eta-\bar{\eta})/(2-\eta)}| < \infty. \tag{B11}$$

This inequality can be used to relate the critical indices η, $\bar{\eta}$, and δ. However, unless $\bar{\eta}$ is appreciably larger than η, the inequalities obtained in this way turn out to be weaker than the inequalities derived.

REFERENCES

1. T. R. Choy and J. E. Mayer, *J. Chem. Phys.*, **46**, 110 (1967).
2. T. R. Choy, *J. Chem. Phys.*, **47**, 4296 (1967).
3. B. Widom, *J. Chem. Phys.*, **43**, 3892, 3898 (1965).
4. L. P. Kadanoff, *Physics (N.Y.)*, **2**, 263 (1966).
5. M. E. Fisher, *J. Appl. Phys.*, **38**, 981 (1967).
6. L. P. Kadanoff, et al., *Rev. Mod. Phys.*, **39**, 395 (1967).
7. M. E. Fisher, *Rept. Progr. Phys.*, **30**, 615 (1967).
8. J. S. Kouvel and D. S. Rodbell, *Phys. Rev. Letters*, **18**, 215 (1967).
9. M. S. Green, M. Vincentini-Missoni, and J. M. H. Levelt Sengers, *Phys. Rev. Letters*, **18**, 113 (1967).
10. B. D. Josephson, *Proc. Phys. Soc. (London)*, **92**, 276 (1967).
11. See, for example, T. L. Hill, *Statistical Mechanics*, McGraw-Hill, New York, 1956, Appendix 9.
12. M. E. Fisher, *J. Math. Phys.*, **5**, 944 (1964).
13. M. Kac, G. E. Uhlenbeck, and P. C. Hemmer, *J. Math. Phys.*, **4**, 216 (1963).
14. P. C. Hemmer, *J. Math. Phys.* **5**, 75 (1964).
15. M. E. Fisher, *Natl. Bur. Std. U.S. Misc. Publ.*, **273**, 21 (1966).
16. G. Stell, *Phys. Rev. Letters*, **20**, 533 (1968).
17. M. E. Fisher and R. J. Burford, *Phys. Rev.*, **156**, 583 (1967).
18. D. S. Gaunt (unpublished); D. S. Gaunt, M. E. Fisher, M. F. Sykes, and J. W. Essam, *Phys. Rev. Letters*, **13**, 713 (1964).
19. J. D. Gunton and M. J. Buckingham, *Phys. Rev. Letters*, **20**, 143 (1967).

Application of a Short-Range Ordered Model to Strong Electrolytes[*]

FÉLIX CERNUSCHI[†] AND MARIO GIAMBIAGI[‡]

Department of Physics, Faculty of Engineering, Buenos Aires, Argentina

MYRIAM SEGRE[†]

Department of Physics, Faculty of Exact Sciences, Buenos Aires, Argentina

Abstract

The modifications introduced in the Debye-Hückel theory when considering the interaction between first neighbors are analyzed. Symmetrical electrolytes at low concentrations—that is, with small interaction among first neighbors—are studied. The expression obtained for κ (reciprocal to characteristic length) coincides with Debye-Hückel's in the limiting case of zero interaction. Fitting the theoretical activity coefficients to experimental data it is found that a (mean distance of closest approach between two ions) is not constant on varying concentration and temperature, and that U (interaction energy between a pair of first neighbors) is proportional to $m^{1/6}$ (m molality). These results are interpreted taking into account ionic hydration. We obtain a quite satisfactory agreement between the values calculated for dilution heats of NaCl and KCl, and the experiment up to $m = 0.2$.

* Reprinted from *Journal de Chemie Physique*, 9, 1966, pp. 1148–1155, by permission of the copyright owner.
† Present address: Department of Astronomy and Physics, Faculty of Humanities and Sciences, Montevideo, Uruguay.
‡ Work supported with the aid of Consejo Nacional de Investigaciones Cientificas y Técnicas and Centro Latino-Americano de Fisica.

455

INTRODUCTION

It has been noted[1] that the Debye-Hückel theory is a theory for long-range interactions. This theory is not applicable when short-range interactions are no longer negligible. We introduce some modifications to the Debye-Hückel theory by considering some short-range interactions.

One of us has presented a statistical theory for strong electrolytes according to this point of view.[2a] The model of order–disorder is similar to that applied by Bethe in his theory of binary alloys.[3] This model has been applied to adsorption phenomena of gases on surfaces,[2b] where it has explained the abnormal cases of isobaric adsorption, and it reproduces the isotherm of Langmuir. It has also been used for condensation phenomena,[2c] where we obtain from it the relations between physical quantities at the critical point.

We suppose that an electrolytic solution is, in a certain sense, a deformable crystalline structure made up of holes, and that these holes can be occupied by positive or negative ions or by solvent particles that can move from one hole to another. The possibility of empty cells has also been considered by other authors, but with a different formalism.[4,5]

In the Debye-Hückel theory it is assumed that the potential at a point is not modified by placing a positive or negative ion at that point. This is not correct, for the potential at one point of an electrolytic solution depends on the entire ionic distribution; in placing a positive or negative ion at a certain point, this ion will modify the ionic distribution about it, and also the potential at the given point. The model employed here, by permitting three possibilities for occupation, takes into account all the cases.

We consider, in this work, symmetrical electrolytes in weak concentrations (i.e., with a weak interaction between nearest neighbors); nevertheless, the complete development of the theory will permit the study of concentrated solutions.

We deduce for x an expression which coincides (in the limit case of negligible interaction between nearest neighbors) with the reciprocal of the characteristic Debye-Hückel length. Following current methods we establish the relation between x and the given experimental results. We study the variation of U (the energy of interaction between two nearest neighbors) and a (the average value of the minimum distance with which two ions can approach one another) with concentration (Ref. 6, p. 264).

We calculate also the values of $\partial a/\partial T$ that we need to determine the heats of dilution.

REVISION AND DEVELOPMENT OF THE THEORY

We consider spherical concentric shells of thickness a about a positive ion (Fig. 1a). We divide each spherical shell into cells of volume a^3, in such a way as to obtain ν neighboring cells, for each cell in the same shell. As the number of neighboring cells does not appreciably alter the results of our theory,[2c] we simplify the calculations by taking $\nu = 4$ (Fig. 1b). In calculating the electrical density ρ in a cell of any shell at a distance r from a positive ion (with Boltzmann's formula), we take into account the influence of neighboring ions, belonging to the same shell, on the ion which may be found in the considered cell. Thus, to each cell in the spherical shell there corresponds a ring of four neighboring cells; we give the designation "central" to each cell C surrounded by a ring (Fig. 1b). The central cell may be occupied by a positive or negative ion or may contain no ions. For each of these alternatives, each of the neighboring cells in its own turn may be occupied by a positive or negative ion or contain no ion.

A complete analysis would consider explicitly the interactions of the ion in the central cell, not only with its nearest neighbors, but also with its second

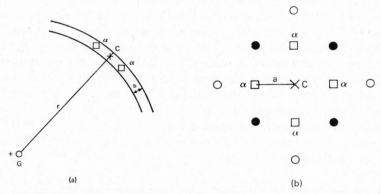

(a)   (b)

Fig. 1. (a) The intersection of a spherical shell at distance r from a positive ion G, in a plane passing through G. (b) The projection of the spherical shell on the tangent plane to the shell at the point C. (*) Central cell (C). (□) Nearest neighbors (α). (●) Second shell (average). (○) Second shell.

nearest neighbors, etc. As a first approximation, we limit ourselves to considering the interactions of each ion in the spherical shell with its nearest neighbors. The potential γ [see (1)], implicitly takes into account, however, the average interactions of all the exterior ions in the ring belonging to the same spherical shell.

In order to calculate the electrical density $\rho(\mathbf{r})$, we consider the following:

1. $\psi(\mathbf{r})$, the potential at a distance $\mathbf{r}$ from a positive ion, produced by the entire electrical distribution, with the exception of the spherical shell at the distance $\mathbf{r}$.

2. The interaction between an ion in the central cell and its nearest neighbors, in the same spherical shell. U is a measure of the energy of interaction.

3. The potential γ, which represents the average value of the potential produced by the ions outside the ring in the spherical shell, on each of the cells that form the ring.

We then define the following factors of Boltzmann:

$$\xi(\mathbf{r}) = \exp\left(-\frac{z\epsilon\psi(\mathbf{r})}{kT}\right), \qquad \eta = \exp\left(-\frac{U}{kT}\right), \qquad U = z^2\epsilon^2 U',$$

$$\zeta = \exp\left(-\frac{z\epsilon\gamma}{kT}\right); \qquad \delta = \xi\zeta, \tag{1}$$

where $\xi(\mathbf{r})$ corresponds to the potential ψ at the distance $\mathbf{r}$ from a positive ion on another positive ion; η represents the interaction between two nearest neighbor ions of the same sign, which are separated by the distance a; ζ the average interaction between the ions outside of the first ring and a positive ion which occupies a cell of the first ring; δ is a factor that we introduce to facilitate the calculations; k is the Boltzmann constant, T is the absolute temperature; $z\epsilon$ is the charge on a positive ion.

Let n_+ and n_- be, respectively, the average number of positive or negative ions per cubic centimeter of the solution and n_0 the average number of cells having no ions per cubic centimeter; we have then

$$n_+ + n_- + n_0 = n = \frac{1}{a^3},$$

where n is the number of cells per cubic centimeter of the solution. We can then attach a statistical weight a priori to each cell, according to whether or not it is occupied by a positive or negative ion or whether it contains no ions; that is to say,

$$\omega_+ = \frac{n_+}{n}; \qquad \omega_- = \frac{n_-}{n}: \qquad \omega_0 = \frac{n_0}{n}.$$

It follows from the definitions and hypotheses above that to a near factor the probabilities that the cell is occupied by a positive or negative ion or that it is vacant are, respectively,

$$f_+ = \omega_+ \xi (\omega_0 + \omega_+ \delta\eta + \omega_- \delta^{-1}\eta^{-1})^4,$$
$$f_- = \omega_- \xi^{-1} (\omega_0 + \omega_+ \delta\eta^{-1} + \omega_- \delta^{-1}\eta)^4,$$
$$f_0 = \omega_0 (\omega_0 + \omega_+ \delta + \omega_- \delta^{-1})^4, \tag{2}$$

We can then write

$$\rho(\mathbf{r}) = z\epsilon n \, \frac{f_+ - f_-}{f_+ + f_- + f_0}. \tag{3}$$

Equations 2 and 3 give

$$\rho(\mathbf{r}) = z\epsilon n$$

$$\times \left[\frac{\omega_+ \xi (\omega_0 + \omega_+ \delta\eta + \omega_- \delta^{-1}\eta^{-1})^4 - \omega_- \xi^{-1}(\omega_0 + \omega_+ \delta\eta^{-1} + \omega_- \delta^{-1}\eta)^4}{\Phi} \right] \tag{4}$$

in setting

$$\Phi = \omega_+ \xi (\omega_0 + \omega_+ \delta\eta + \omega_- \delta^{-1}\eta^{-1})^4 + \omega_- \xi^{-1}(\omega_0 + \omega_+ \delta\eta^{-1}$$
$$+ \omega_- \delta^{-1}\eta)^4 + \omega_0 (\omega_0 + \omega_+ \delta + \omega_- \delta^{-1})^4.$$

If we neglect the interaction between the neighboring ions, that is to say if $\eta = 1$, and if we set, for symmetrical electrolytes, $\omega_+ = \omega_-$,

$$\rho(\mathbf{r}) = z\epsilon n_+ (\xi - \xi^{-1}). \tag{5}$$

This expression is that of Debye.

We need to establish another relation between η, ξ, and δ. Due to the symmetry, every cell of the spherical shell can be considered as the central cell.[2a] In Fig. 1b, the cell designated C is the central cell of the ring formed by four other cells α; any one of the cells α can be the central cell of the ring to which cell C belongs. If we consider one of the cells α as the central cell according to whether cell C is occupied by a positive or negative ion, or contains no ions, the corresponding probabilities are proportional to

$$f'_+ = \omega_+ \delta (\omega_0 + \omega_+ \delta\eta + \omega_- \delta^{-1}\eta^{-1}),$$
$$f'_- = \omega_- \delta^{-1}(\omega_0 + \omega_+ \delta\eta^{-1} + \omega_- \delta^{-1}\eta), \tag{6}$$
$$f'_0 = \omega_0 (\omega_0 + \omega_+ \delta + \omega_- \delta^{-1}).$$

For weak concentrations $(\eta \simeq 1)\, f'_+ \ll f'_0$. By supposing that

$$\frac{f_+}{f'_+} = \frac{f_0}{f'_0},$$

we obtain,[2c]

$$\xi = \delta \frac{(\omega_0 + \omega_+ \delta + \omega_- \delta^{-1})^3}{(\omega_0 + \omega_+ \delta\eta + \omega_- \delta^{-1}\eta^{-1})^3}, \tag{7}$$

which is the new relation we are looking for.

The Debye-Hückel expression for ρ does not predict, even in its nonlinear form, a saturation effect for very large values of $ze\psi/kT$ as would be desirable.[7] Our equations show that for $ze\psi/kT \to \infty$, ξ and δ are of the same order and thus in the limit $\rho/zen_+ \to -1$. We encounter then the effect of saturation. In working with a linear approximation one cannot expect to find a saturation effect.

With the help of (7) and (4) we express $\rho(r)$ as a function of ξ and of η. Introducing the value of ρ thus calculated in the Poisson equation we will have a valid equation for all concentrations.

For $\eta = 1$, $\delta = \xi$. We can then develop δ about $\eta = 1$:

$$\delta = \xi + \left(\frac{\partial\delta}{\partial\eta}\right)_{\eta=1} (\eta - 1) + \frac{1}{2} \left(\frac{\partial^2\delta}{\partial\eta^2}\right)_{\eta=1} (\eta - 1)^2 + \cdots. \tag{8}$$

In taking the first two terms in the Taylor expansion of ξ

$$\xi \simeq 1 - \frac{ze\psi}{kT} = 1 - x. \tag{9}$$

With these approximations

$$\delta = c_1 + c_2 x; \qquad \delta^{-1} = \frac{1}{c_1} - \frac{c_2}{c_1^2} x, \tag{10}$$

where

$$c_1 = 1 + 3(\omega_+ - \omega_-)(\eta - 1),$$
$$c_2 = -[(4\omega_+ + 2\omega_-)(\eta - 1) + 1].$$

As a consequence, by always retaining the terms of the first order in x, we have, by replacing (10) in (2),

$$f_+ = \omega_+ E^4 + \omega_+ E^3(4F - E)x,$$
$$f_- = \omega_- E'^4 + \omega_- E'^3(4F' - E')x, \tag{11}$$
$$f_0 = \omega_0 E''^4 + 4\omega_0 E''^3 F''x,$$

where

$$E = \omega_0 + \eta\omega_+ c_1 + \omega_- c_1^{-1}\eta^{-1}; \qquad F = \omega_+ \eta c_2 - \omega_- c_2 c_1^{-2}\eta^{-1},$$
$$E' = \omega_0 + \omega_+ c_1\eta^{-1} + \omega_- c_1^{-1}\eta; \qquad F' = \omega_+ c_2\eta^{-1} - \omega_- c_2 c_1^{-2}\eta,$$
$$E'' = \omega_0 + \omega_+ c_1 + \omega_- c_1^{-1}; \qquad F'' = \omega_+ c_2 - \omega_- c_2 c_1^{-2}.$$

We thus arrive at the following electrical density:

$$\rho = \frac{zen}{S_1}\left[M_1 + \left(M_2 - M_1\frac{S_2}{S_1}\right)x\right], \tag{12}$$

where

$$M_1 = \omega_+ E^4 - \omega_- E'^4; \qquad M_2 = \omega_+ E^3(4F - E) - \omega_- E'^3(4F' + E')$$

$$S_1 = \omega_+ E^4 + \omega_- E'^4 + \omega_0 E''^4,$$

$$S_2 = \omega_+ E^3(4F - E) + \omega_- E'^3(4F + F') + 4\omega_0 E''^3 F^4.$$

In replacing these expressions in the Poisson equation (D being the dielectric constant of the solution)

$$\Delta\psi = -\frac{4\pi zen}{S_1}M_1 + \frac{4\pi zen}{DS_1}\left(M_2 - M_1\frac{S_2}{S_1}\right)x. \tag{13}$$

By a change of variable this equation becomes

$$\Delta\psi = -\kappa^2\psi$$

where

$$\kappa^2 = \frac{4\pi z^2\epsilon^2 n}{DkTS_1^2}[S_2M_1 - S_1M_2]. \tag{14}$$

This length has, in our theory, the same role as in the Debye-Hückel theory. In the limiting case where we neglect the interaction between nearest neighbors, $\eta = 1$, $U = 0$. We have, thus, setting $\omega_+ = \omega_-$,

$$c_1 = 1; \qquad c_2 = -1; \qquad E = E' = E'' = 1,$$

$$F = F' = F'' = 0; \qquad M_1 = S_2 = 0,$$

$$M_2 = -2\omega_+ = -\frac{2n_+}{n}; \qquad S_1 = 1.$$

Thus

$$\kappa^2 = \frac{8\pi z^2\epsilon^2 n}{DkT};$$

$1/\kappa$ being the characteristic Debye-Hückel length.

Following current methods[8] we establish the relation between κ and the given experimental data for the coefficients of activity.

In considering the potential (ψ_b) on a sphere of radius a, and the charge at a given moment of the process of $\lambda z\epsilon$ ($0 \leq \gamma \leq 1$), we calculate the work necessary to charge an ion. In repeating this procedure for all the ions, the potential energy of the solution will be

$$
W = (N_+ + N_-) \int_0^1 z\epsilon\psi_b(\lambda z\epsilon) \, d\lambda
$$

$$
= -\frac{N_+ + N_-}{D} \int_0^1 \frac{z^2\epsilon^2\lambda\kappa}{1 + a\kappa} \, d\lambda, \tag{15}
$$

where N_+ and N_- are the number of positive and negative ions in the solution, and κ depends on $\lambda z\epsilon$. The coefficients of activity f are related to W by

$$
\ln f_+ = \ln f_- = \frac{1}{kT} \frac{\partial W}{\partial N_+}, \tag{16}
$$

from which

$$
\ln f = -b_1 y_1 - b_2 y_2,
$$

$$
b_1 = -\frac{z^2\epsilon^2}{DkT} \left(1 - 2n_+ \frac{\partial D}{\partial n_+}\right),
$$

$$
y_1 = \int_0^1 \frac{\lambda\kappa}{1 + a\kappa} \, d\lambda,
$$

where

$$
b_2 = \frac{2n_+ z^2\epsilon^2}{DkT},
$$

$$
y_2 = \int_0^1 \frac{\lambda[\partial x/\partial n_+ - \kappa^2 \partial a/\partial n_+]}{(1 + a\kappa)^2} \, d\lambda.
$$

We can calculate these integrals if we know U and a. The other quantities can be determined experimentally.

In order to integrate we use Simpson's formula. We obtain U and a as we would like to see it, with the help of a four-point interpolation following Lagrange's method.

THE ENERGY OF INTERACTION U AND THE DISTANCE a

We will not attribute any one particular form to this function U. We limit ourselves to obtaining some numerical values for the alkali halides at some concentrations. We look next to express U as a function of concentration c.

The variation in the energy of interaction between nearest neighbors as a function of the concentration has already been studied in the case of solid solutions.[9] We believe that it is quite logical that U depends on the concentration, since the order at short range (related to U) is not completely independent of the long-range order (which must be related to c) even in solids.

The dielectric constant of the solution is calculated by basing it on the results of Hasted, Ritson, and Collie,[10] which are in agreement with the estimations of subsequent theories.[11] The formula used is $D = D_0 + 2\mu C$ (D_0, dielectric constant of water; μ, coefficient which depends on each ion). The analysis performed by Satoh[12] permits us to apply this linear formula even for studied concentrations. Satoh shows that the coefficient 2μ is valid since $c = 0$ to $c = 2\mu$ where he found a critical point. In using the given experimental value we will suppose that the given formula is still valid, μ remaining the same for different temperatures: we take for D_0 the dielectric constant of water at the considered temperature. The coefficients of activity are extracted from the usual tables.[13]

It is not possible to choose a constant a such that the experimental results are in agreement with formula.[16] The values of a that permit this agreement are obtained iteratively, starting with a preliminary calculation where we suppose for U a coulombic interaction. The coulombic interaction has already been used explicitly.[1,14] Nevertheless, some more detailed considerations of the interaction energy at short distances can take into account other terms.[15]

In Fig. 2 we see the variation of U with molality m for the alkali halides at 25°C. U is of the order of 10^{-14} erg; the ion–dipole interactions are of the order of 10^{-12} erg.[16] This reduction is reasonable since hydrate ions partially

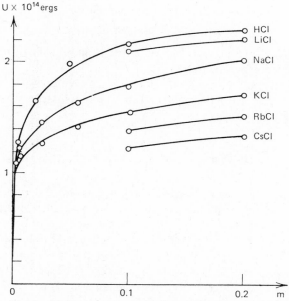

Fig. 2. Variation of the energy of interaction U with the molality m for the alkali halides at 25°C.

neutralize their interaction. Moreover, U is naturally smaller than the crystalline network of the alkali halides (10^{-11} erg).[17] We see that U decreases when the molecular mass increases; the same thing takes place for the crystalline network.

At a given concentration the interaction is strongest when the cations are smallest. This is in agreement with the decrease in the dielectric constant of the solution with the mass of the cations. When U increases, we deviate from the physical conditions of pure water (where $\eta = 1$); that is to say, the introduction of cations is thus the more disturbing when they are small.

The growth of U with concentration leads to a growth of the order at short range. We can expect that this growth is weaker than a linear growth, for otherwise it would not be able to lead to a saturation effect. This effect is due to a physical limitation of the order at short distance, which cannot increase indefinitely. The curves of Fig. 2 lead us to try

$$U = k'm^{1/n}. \tag{18}$$

If we represent the variation of log ($U/z^2\epsilon^2$) with log m we obtain straight lines, as indicated in Fig. 3 for NaCl. The exponent n of formula (18) is practically equal to 6 in all cases studied. Table I gives the values of $\dfrac{k'}{z^2\epsilon^2}$ for the considered alcalies halides. Let us point out that at $\sim 25°C$, kT is of the order of 4×10^{-14} ergs, therefore for $m = 0.1$, u is not negligible in front of the thermic energy kT.

As η and U are both time average values, it can happen that two ions act on one another for a long time at a certain distance a_1, then at another distance $a_1 < a_2$. Consequently, the growth of U with a is not surprising.

Let us return now to the values of a that have already been used for the calculation, and to their variation with concentration and temperature. As it is usual to adopt constant values of a for each salt, these values do not

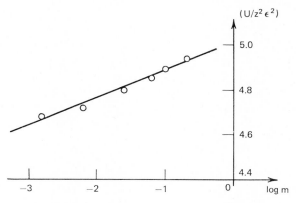

Fig. 3. Variation of log ($U/z^2\epsilon^2$) with log m, for NaCl.

always allow for a satisfactory agreement with the given experimental values.[18] It has been noted[19] that a depends on the concentration, and it has been thought that a could explain other short-range effects. Its variation with concentration has been suggested explicitly.[20] The length was introduced to take into account the volume of ions; the volume depends essentially on the degree of hydration.[21] We first take into account this ion–water interaction,[22,23] introducing the variation of the dielectric constant with concentration. Then we consider the interaction of the ions with the solvent in order to explain the variations in a.

When there are few ions in solution the tetrahedral structure typical of water is not appreciably affected. But if some ions succeed in breaking this structure, as when the ion–dipole interaction is stronger than the dipole–dipole,[16] the ions that are subsequently added to the solution will easily be able to capture molecules of water not belonging to a stable structure. As hydration increases with concentration so does a (Fig. 4).

With this interpretation it is evident that there must exist concentrations that we can consider with the first approximation of the theory where there exists only normal water.[24] After these concentrations, the ions disperse the molecules of water from their hydrated bands, and the numbers of hydration must necessarily diminish. However, it is possible that the critical concentration is weaker than $m = 3$ or even $m = 2$.[12]

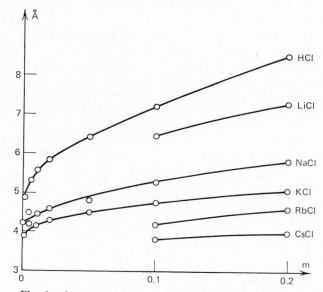

Fig. 4. Average minimum separation between two ions a as a function of concentration at 25°C.

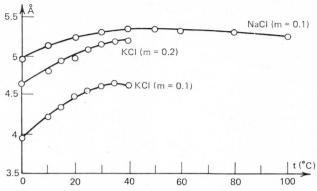

Fig. 5. Variation of the length a with temperature, for KCl ($m = 0.1$, and $m = 0.2$) and NaCl ($m = 0.1$).

This picture explains also the decrease of a for halides when the molecular weights increase with concentration and given temperature in agreement with experiment[25] (see Fig. 4). The ions possess the capacity to destroy the molecular order of water and to create a new order about themselves. This organizational capacity, calculated from entropy measures[26] decreases when the radii of considered monovalent cations increase. Consequently the smaller cations, like H^+ and Li^+, due to their large "organizational capacity" have larger effective radii.

In Fig. 5 we see the variation in a with temperature for NaCl at (from 0 to 100°C) and for KCl at $m = 0.1$ and $m = 0.2$ from 0 to 40°C. When the temperature increases, the increase in kinetic energy helps to break the tetrahedrons of water, thus producing ionic hydration. But when the majority of ions are completely hydrated and the temperature increases again, the ions will begin to dehydrate themselves, that is to say, above this temperature the hydration will diminish. The interaction between an ion and the first shell of water about it seems to be too large for the effects to be appreciably modified by the thermal agitation.[16,27] The process that we want to describe will take place most probably outside the first shell in the transition region.[28]

The coefficient of temperature $\partial a/\partial T$ has often been analyzed,[29] and it has even been proposed to give it an analytical expression.[30] Its importance comes from the fact that $\partial a/\partial T$ appears in the formulas of the heats of dilution. Bjerrum[29] obtained, as part of the experimental variation in the heats of dilution, positive values of $\partial a/\partial T$ for weak hydrated ions and negative values for very hydrated ions. Our curves suggest that even for a given salt, for example, NaCl, the coefficient may be positive or negative following the temperature (and possibly the concentration) considered. The values that we obtain are, respectively ($m = 0, 1$), 9.14×10^{-11} cm/°K and $8.13 \times$

Table I.   Values of $k'/z^2\epsilon^2$ for the Alkali Halides Expressed in Such a Way that U in Formula (18) is in ergs, m being the Molality

Substance	$k'/z^2\epsilon^2$
HCl	1.26×10^5
LiCl	1.10×10^5
NaCl	1.05×10^5
KCl	9.77×10^4
RbCl	8.71×10^4
CsCl	7.45×10^4

10^{-11} cm/°K for NaCl. Positive values of $\partial a/\partial T$ from 0° to 45°C for other salts of K have been found.[31]

Table II gives values of a for KCl with coulombic U and with the approximation of Levine and Bell.[15] They considered, outside of the coulombic term, another perturbation containing $(\partial D/\partial n_+)/a^3$. The values of a thus calculated are approximately equal and their variation entirely similar.

THE HEATS OF DILUTION

In applying current methods (Ref. 6, p. 161) we easily obtain the formulas that permit the calculation of heat integrals of dilution (1), in keeping with that expression. Since

$$L = W - T\left(\frac{\partial W}{\partial T}\right)_v,$$ (19)

Table II.   Values of a for KCl Obtained with Coulombic U and with the U of Levine[15] (a in $\mathring{A}$)

m	a with Coulombic U	a with U of Levine
0.001	3.87	3.87
0.005	4.21	4.21
0.02	4.24	4.26
0.05	4.40	4.43
0.1	4.61	4.67
0.2	5.07	5.16

by replacing W by the corresponding expression [see (15)], and referring the heat of dilution L to 1 mole we have

$$L = -b\left[\left(1 + \frac{T}{D}\frac{\partial D}{\partial T}\right)y_1' - Ty_2'\right], \tag{20}$$

where

$$b = \frac{2 \times 0.23889 \times 10^{-7}N'z^2\epsilon^2}{D},$$

$$y_1' = \int_0^1 \frac{\lambda\kappa}{1 + a\kappa}\,d\lambda; \qquad y_2' = \int_0^1 \frac{\left[\partial x/\partial T - \kappa\dfrac{2\partial a}{\partial T}\right]}{(1 + a\kappa)^2}\,d\lambda.$$

We consider only two types of salt: NaCl and KCl.

For a, U, and $\partial a/\partial T$ we use the values obtained previously and we suppose that $\partial U/\partial T = 0$.

In Fig. 5 we see that at 25°C, $\partial a/\partial T$ is nearly the same for KCl at $m = 0, 1$ and at $m = 0, 2$. Keeping the values of $\partial a/\partial T$ for $m < 0, 1$, we adopt then for all the concentrations the value obtained for $m = 0, 1$. For $\partial D/\partial T$ we take, as is usual, $\partial D_0/\partial T$ and choose the value -0.3557,[32] the most recent value reported by Hamer.[33]

The term $1 + (T/D)(\partial D/\partial T)$ in (20) is negative in cases which interest us. As y_1' is positive, the sign of L depends on the sign of y_2', L can thus be negative as is the case, for example, for NO_3K. In y_2', $\partial a/\partial T$ and $\partial x/\partial T$ appear and we note that (Ref. 6, p. 293) the variation of a with temperature would be able to produce a change of sign in L. Contrary to this the Debye-Hückel formula foresees only positive values of L. In Table III we see the theoretical and experimental results[34] for L. The agreement seems quite reasonable, taking into account the previously mentioned approximations,

Table III. **Theoretical Results and Experimental Values for the Heats of Dilution of KCl and NaCl at 25°C (L is given in cal/mole)**

m	KCl		NaCl	
	Theoretical	Experimental	Theoretical	Experimental
0.001	14.2	10.3	14.2	11.7
0.002	19.4	15.0		
0.005	28.7	25.8	28.9	27.0
0.01	37.9	37.3		
0.02	48.7	48.6	49.3	50.6
0.05	64.4	65.5	65.7	69.8
0.1	75.3	77	76.5	82
0.2	85.0	80	86.2	

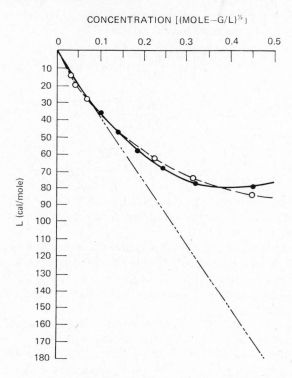

Fig. 6. Theoretical (– – –) and experimental (———) results for KCl, and the Debye-Hückel limiting law (— – – —) for the heats of dilution.

and considering that there are small uncertainties in the value of $\partial D/\partial T$ (8%) which affect greatly (30%) the theoretical estimations of L (Ref. 6, p. 161).[33]

The theoretical coefficient that we obtain 449 cal mole$^{-3/2}$ liter$^{1/2}$ does not differ from that of the right experimental limit 440 cal mole$^{-3/2}$ liter$^{1/2}$. We record in Fig. 6 the theoretical and experimental results for KCl and the Debye-Hückel limiting law.

It remains to note the influence of $\partial a/\partial T$ on the value of L. In fact, although $\partial a/\partial T$ is very small ($\sim 10^{-10}$ cm/°K), its omission can lead to some modifications in L of the order of 35%.

In summary the agreement of the experimental values for the heats of dilution with the values calculated with a, U, and $\partial a/\partial T$ demonstrates the possibility that our theory may be used to explain other effects.

ACKNOWLEDGMENT

We wish to thank Dr Jacques Danon for his discussions on this work.

REFERENCES

1. H. S. Frank and P. T. Thompson, *J. Chem. Phys.*, **31**, 1086 (1959).
2. (a) F. Cernuschi, *Rev. Mat. Fis. Teor. Univ. Tucuman*, **2**, (1941); (b) F. Cernuschi, *Proc. Cambridge Phil. Soc.*, **34**, 392 (1938), (c) F. Cernuschi and H. Eyring, *J. Chem. Phys.*, **7**, 547 (1939); F. Cernuschi and M. Segre, *ibid.*, **36**, 412 (1962).
3. H. Bethe, *Proc. Roy. Soc. (London)*, **A150**, 552 (1935).
4. J. Cohn, *Phys. Fluids*, **6**, 21 (1963).
5. M. Eigen and E. Wicke, *Naturwiss.*, **38**, 453 (1951).
6. H. Falkenhagen, *Electrolytes*, Alc.'n, Paris, 1934.
7. M. Eigen and E. Wicke, *J. Phys. Chem.*, **58**, 702 (1954).
8. H. S. Harned and B. B. Owen, *The Physical Chemistry of Electrolytic Solutions*, Reinhold, New York, 1954, Chapter III.
9. M. Hillert, *J. Phys. Radium*, **23**, 835 (1962).
10. J. B. Hasted, D. M. Riton, and C. H. Collie, *J. Chem. Phys.*, **16**, 1 (1948).
11. A. Takahashi, *Busseiron Kenkyn*, **96**, 1 (1956).
12. T. Satoh, *J. Phys. Soc. Japan*, **17**, 279 (1962).
13. (a) R. E. Conway, *Electrochemical Data*, Van Nostrand, New York, 1952. (b) T. Shedlovsky, *J. Amer. Chem. Soc.*, **72**, 3680 (1950).
14. J. F. Skinner and R. M. Fuoss, *J. Amer. Chem. Soc.*, **86**, 3423 (1964).
15. S. Levine and G. M. Bell, in *An International Electrolytes Symposium*, Pesce, Ed., Pergamon, New York, 1962, P. 77.
16. R. W. Gurney, *Ionic Processes in Solution*, McGraw-Hill, New York, 1940, p. 50.
17. F. Seitz, *The Modern Theory of Solids*, McGraw-Hill, New York, 1940, p. 80.
18. T. H. Gronwall, V. K. La Mer, and K. Sandved, *Physik. Z.*, **28**, 358 (1928).
19. M. A. V. Devanathan, *J. Sci. Ind. Res.*, **20B**, 256 (1961).
20. C. Dejak and I. Mazzei, *Ann. Chim.*, **47**, 1044 (1957).
21. M. Eigen and E. Wicke, *Z. Electrochem.*, **56**, 551 (1952).
22. E. Hückel, *Z. Physik*, **26**, 93 (1925).
23. P. S. Yastremskii, *Zh. Strukt. Khim.*, **2**, 269 (1961).
24. J. Satoh, *J. Phys. Soc. Japan*, **15**, 1134 (1960).
25. J. C. Hindman, *J. Chem. Phys.*, **36**, 1000 (1962).
26. H. S. Frank and M. W. Evans, *J. Chem. Phys.*, **13**, 507 (1945).
27. R. A. Robinson and R. H. Stokes, *Electrolyte Solutions*, Butterworth, London, 1959, p. 53.
28. H. S. Frank and Wen Yang Wen, *Discuss. Faraday Soc.*, **24**, 133 (1957).
29. See, for example, N. Bjerrum, *Trans. Faraday Soc.*, **23**, 445 (1927).
30. G. Scatchard and L. Epstein, *Chem. Rev.*, **30**, 211 (1942).
31. H. Brusset and M. Kikindal, *Bull. Soc. Chim.*, **1962**, 1150.
32. C. G. Malmberg and A. A. Maryott, *J. Res. Natl. Bar. Std. U.S.*, **56**, 1 (1956).
33. W. J. Hamer, *The Structure of Electrolytic Solutions*, Wiley, New York, 1959, Chapter IX.
34. Landolt-Bornstein, *Erg*, Bd. II/2, S. 1536.

Eyring's Theory of Viscosity of Dense Media and Nonequilibrium Statistical Mechanics

I. PRIGOGINE,* G. NICOLIS, AND P. M. ALLEN

Faculté des Sciences, Université Libre de Bruxelles, Belgium

* Also, Center for Statistical Mechanics and Thermodynamics, University of Texas at Austin, Texas.

INTRODUCTION

It is a great pleasure to contribute to this anniversary volume honoring Henry Eyring. As the subject of this small paper we have chosen the molecular theory of viscosity for dense systems. It is a field where the great originality of Eyring's approach is most conspicuous. For dilute gases we know how to calculate the viscosity from the Boltzmann kinetic equation. But for dense gases? It seemed impossible, especially thirty-five years ago, to extend Boltzmann's ideas to liquids. It is characteristic of Eyring's genius that he did not waste time in futile formal extensions of ideas valid only for dilute systems but introduced a new, simple, and direct approach[1-3] which immediately gave the correct order of magnitude for liquids. One of the authors (I.P.) was so impressed by these results that he devoted his first paper on the theory of liquids to a discussion of the concept of free volume as used in Eyring's theory.[4]

Since Eyring's work, much effort has been devoted to obtain a derivation of transport properties based on nonequilibrium statistical mechanics, which is the more general version of the classical kinetic theory.[5] It is very interesting that this development has led to ideas which largely validate Eyring's simple physical picture.

Perhaps the most interesting and at first paradoxical aspect of Eyring's theory is the lack of a kinetic equation. No Fokker–Planck or Boltzmann equation is introduced or solved to derive the viscosity. Now this feature is also characteristic for a simple model we have studied recently.[6] For this reason it seems to us appropriate to summarize this model here and to compare it in more detail with Eyring's theory of viscosity.

NONEQUILIBRIUM STATISTICAL MECHANICS

The starting point in our analysis is the Liouville equation, which describes correctly the complete evolution in time of an N-body system.[5] This equation is formally solved and the solution is expanded in (formal) perturbation series. The parameter involved in the perturbation expansion is roughly proportional to the ratio of the interaction energy between particles and of the energy of thermal motion of the particles. A rearrangement of this series leads, then, to an equation describing the time evolution of the product singlet distribution function

$$f = \prod_{i=1}^{N} f_1(\mathbf{x}_i, \mathbf{v}_i, t), \tag{1}$$

where $\mathbf{x}_i$, $\mathbf{v}_i$ are the position and velocity of the ith particle. In the limit of long times this equation becomes closed and takes a pseudo-Markovian form.[5] This form is referred to as a *kinetic equation.*

Within the same approximations the pair correlation function

$$g_{12}(1, 2) = f_{12}(1, 2) - f_1(1)f_1(2) \tag{2}$$

is expressed as a functional of f by the relation:

$$g_{12}(1, 2, t) = \int (d\mathbf{x}_i)^{N-2}(d\mathbf{v}_i)^{N-2} \mathcal{C}(1, 2 \mid \{N\})f(t) \tag{3}$$

where the operator $\mathcal{C}$ (the so-called creation operator) is a well-defined quantity in terms of the Liouville operator describing the system.

If attention is now focused on the linear domain of irreversible processes, the kinetic equation and (3) may be systematically expanded in powers of the external constraints (external fields, thermal gradients). In this way, closed equations are obtained for $f^{(1)}$ and $g_{12}^{(1)}$, the first-order deviations of f and g_{12} from their local equilibrium forms.[7,8] These equations still contain a number N of variables and therefore have been solved only in a few particular cases. The main complication arises from the fact that the solution of the kinetic equation requires the inversion of a complicated operator, the so-called collision operator whose spectrum is completely unknown for dense systems. In addition, it does not seem possible to develop a systematic approximation scheme to this operator, because of the lack of any evident smallness parameter which would permit a suitable expansion of the general equations. As a result the problem of deriving tractable expressions for liquid transport coefficients starting from the general kinetic equation and from (3) is still to a large extent an open question. Until recently, liquid transport has therefore been approached by appealing to physical arguments leading to approximate theories of the Kirkwood or the Rice–Allnatt type.[9]

THE LOCAL EQUILIBRIUM MODEL

An adequate treatment of liquid transport on the basis of the general kinetic equation and of (3) seems, however, to be suggested by Rahman's computer experiments,[10] which have shown that in a liquid the momentum relaxation time is very short (of the order of 10^{-13} sec) compared to its value in a moderately dense gas or in a dilute gas. It even appears to be shorter than the characteristic decay time of the pair correlation function toward a functional of f_1, that is, the collision time, although the separation between the two times at liquid densities does not seem to be really as large as an order of magnitude. However, one can reasonably expect that the study of transport processes could be considerably simplified in this respect because the singlet distribution function is practically always at local equilibrium. The only quantity that gives rise to the irreversible fluxes is therefore $g_{12}(1, 2)$, which is given by (3), where now $f(t)$ is known and equal to the product local equilibrium distribution function. In this way the problem of solution of the kinetic equation is avoided and one only has to express the creation fragment $\mathcal{C}$ in the linear domain of irreversible processes.

This observation constitutes the basic idea of the local equilibrium model of Prigogine, Nicolis, and Misguich (hereafter referred to as PNM).[6] One considers the case of a spatially nonuniform system and deduces from (3) an integral equation for the pair correlation function that is linear in the gradients. This equation is then approximated in a simple way that enables one to derive explicit expressions for all thermal transport coefficients (viscosities, thermal conductivity), both in simple liquids and in binary mixtures, excluding of course the diffusion coefficient. The latter is a purely kinetic quantity, which cannot be obtained from a local equilibrium hypothesis.

In the case of pure liquids numerical computations for the transport coefficients in argon, krypton, and xenon have been carried out by Palyvos et al.[11] using a modified Lennard–Jones potential and the radial distribution function of Kirkwood, Lewinson, and Alder.[12] The results, for instance for argon, represent percentages between 60 and 90% of the experimental values in a wide range of temperatures and densities. Besides, they agree with experiment better than the results derived from the Kirkwood of Rice–Allnatt types of theories.

CONNECTION WITH EYRING'S THEORY

The essential feature of the local equilibrium model is that the doublet distribution function in the system is distorted because of the thermal gradients around a local equilibrium form, and this asymmetry produces a nonvanishing flow. This clearly corresponds very closely to the Eyring ideas of liquid transport. In both theories the kinetic aspects of liquid transport are ignored. In the PNM model, however, the radial distribution function is used to express the average position of neighbors whereas Eyring's theory introduces a suitable description in terms of some packing arrangement.

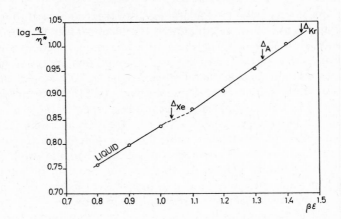

Fig. 1. Reduced shear viscosity ($n\sigma^3 = 0.818$). $\Delta =$ phase transition point.

Let us plot the logarithmic values of the shear viscosities calculated by the PNM model, divided by the dilute gas values (which vary as the square root of temperature T) as a function of ϵ/kT, ϵ being the depth of the Lennard–Jones potential well. The results are given in Fig. 1 for a reduced density of $n\sigma^3 = 0.818$, σ being the hard core diameter.[13] The experimental points are well fitted by two approximately straight lines. We observe that Eyring's formula for shear viscosity at constant density

$$\eta \sim \sqrt{kT} \exp\left(\frac{E_f}{kT}\right) \tag{4}$$

E_f being an activation energy, is in agreement with Fig. 1 and therefore also with the local equilibrium model. The value of the activation energy obtained from Fig. 1 agrees within 70% with the experimental values for argon, in a domain of temperatures between 90 and 110°K and within 50% with the values predicted by Eyring.[1–3]

In conclusion, it appears that the application of recent theories of non-equilibrium statistical mechanics to transport in dense media confirms Eyring's theory and provides in addition a convenient framework for possible extensions and refinements; for instance, Allen et al.[14] have recently combined the original PNM model with an approximate kinetic equation for the singlet distribution function and obtained a still better agreement with experiment.

REFERENCES

1. H. Eyring, *J. Chem. Phys.*, **4**, 283, 1936.
2. R. H. Ewell and H. Eyring, *J. Chem. Phys.*, **5**, 726, 1937.
3. For a recent presentation of Eyring's ideas, see H. Eyring, D. Henderson, B. J. Stover, and E. M. Eyring, *Statistical Mechanics and Dynamics*, Wiley, New York, 1964.
4. I. Prigogine, *Physica*, **9**, 405 (1942).
5. For a review of the theory see, for example, I. Prigogine, *Non-Equilibrium Statistical Mechanics*, Interscience, New York (1962); R. Balescu, *Statistical Mechanics of Charged Particles*, Interscience, New York, 1963; and P. Résibois, in *Physics of Many Particle Systems—Methods and Problems*, Vol. I, E. Meeron, Ed., Gordon and Breach, New York, 1967.
6. I. Prigogine, G. Nicolis, and J. Misguich, *J. Chem. Phys.*, **43**, 4516 (1965).
7. G. Severne, *Physica*, **30**, 1365 (1964); **31**, 877 (1965).
8. P. Résibois, *J. Chem. Phys.*, **41**, 2979 (1964).
9. S. A. Rice and P. Gray, *The Statistical Mechanics of Simple Liquids*, Interscience, New York, 1965.
10. A. Rahman, *Phys. Rev.*, **136**, A 405 (1964).
11. J. A. Palyvos, H. T. Davis, J. Misguich, and G. Nicolis, *J. Chem. Phys.*, **49**, 4088 (1968).
12. J. G. Kirkwood, V. A. Lewinson, and B. J. Alder, *J. Chem. Phys.*, **20**, 929 (1952).
13. J. Misguich, Ph.D. thesis, University of Brussels, 1968.
14. P. M. Allen and G. H. A. Cole, *Mol. Phys.*, **15**, 557 (1968); P. M. Allen, *Physica*, to be published.

Sound Velocity and van der Waals Force in Liquids According to Significant Structure Theory

MU SHIK JHON*

Department of Chemistry, University of Virginia, Charlottesville

* Present address: Dong Guk University, Seoul, Korea.

A Personal Note

I am greatly indebted to Professor Eyring for the special type of training in scientific research I received during the last five years of my academic and professional career.

It has been a real joy to meet him and to benefit from his great insight into the successful application of scientific models. Because of this experience I am now a "model man" rather than the man of "mathematical approach" I had formerly been.

Professor Eyring really loves science and spends all his time struggling to find scientific models that will contribute to scientific development. My observations during his three years as visiting lecturer at the University show that he is an everlasting friend of molecules. Each time he visited I had a chance to look into his carrying case which is full of books and research material and which he calls his "travelling library."

INTRODUCTION

The significant structure theory of liquid[1] has been one of the most widely applied of the various theories of liquids. It has been applied with success to predict the thermodynamic, dielectric, transport, and surface properties of many liquid systems ranging from simple liquids such as argon to complicated systems such as water or some mixtures. This model for liquids is based on several confirmed experimental facts and in this respect is the most acceptable of those currently offered.[2]

In this paper we discuss the further validity of the model by checking the velocities of sound and van der Waals forces in liquids. Since these properties are related to the second derivative of the partition function, their results will constitute a severe test of the model. According to the significant structure theory, the partition function of liquids can be expressed as follows:

$$f = f_s^{N(V_s/V)} \cdot f_g^{N(V-V_s)/V}, \tag{1}$$

where f_s is the partition function for the solidlike degree of freedom with positional degeneracy and f_g is the partition function for the gaslike degree of freedom; V and V_s are the molar volumes of the liquid and solid, respectively; and N is the number of molecules. In the recent literature[3] we see that for simple liquids the model works without any adjustable parameters.

SOUND VELOCITY IN LIQUIDS

According to well-known relations, the velocity of sound, C, obeys the equation

$$C = \left[\frac{\gamma}{\beta_T \rho}\right]^{1/2} = \left[\frac{C_p V}{C_v \beta_T M}\right]^{1/2}, \tag{2}$$

where γ is the ratio of C_p to C_v, β_T is the isothermal compressibility, ρ is the density, V is the molar volume, and M is the molecular weight. These

483

Table I. Sound Velocity of Liquid

	T (°K)	$C_{calc.}$ (cm/sec)	C_{obs} (cm/sec)	Δ (%)
Ar	84.76	916.7[a]	855.8[a]	7.1
	94.15	816.7	789.9	3.4
	105.31	704.4	707.0	−0.4
	114.94	613.2	629.2	−2.5
	124.09	529.6	546.8	−3.1
	133.11	447.7	454.9	−1.6
Kr	117.0	725.3	700.0[b]	3.6
	132.0	641.5	640.0	0.2
	142.0	586.1	597.0	−1.8
	152.0	532.6	551.0	−3.3
	172.0	429.9	455.0	−5.5
Xe	161.80	676.0	659.0[b]	2.6
	173.00	638.4	623.0	2.5
	203.00	529.4	533.0	−0.7
	223.00	460.2	471.0	−2.3
	243.00	392.9	402.0	−2.3
N_2	65.42	1024.3	968.1[a]	5.8
	70.03	960.9	923.5	4.0
	74.25	901.8	881.4	2.3
	79.34	832.7	830.5	0.3
	84.95	760.3	772.3	−1.6
	110.25	472.3	472.2	0.0
O_2	89.66	959.3	909.9[a]	5.4
	102.23	805.2	804.6	0.1
	113.69	683.1	700.5	−2.5
	131.90	512.1	517.1	−1.0
	140.74	434.9	412.6	5.4
CH_4	90.90	1724.2	1531.0[b]	12.6
	103.60	1533.5	1406.0	9.1
	121.00	1284.5	1236.0	3.9
	133.70	1123.2	1101.0	2.0
	147.40	965.4	954.0	1.2
	168.80	742.8	672.0	10.5
CCl_4	263.15	1080.5	1035.8[c]	4.3
	273.15	1046.8	1003.0	4.4
	283.15	1007.3	969.4	3.9
	293.15	966.5	938.1	3.0
	303.15	926.2	905.9	2.2
$TiCl_4$	273.15	1179.3	1064.2[c]	10.8
	283.15	1133.1	1037.8	9.2
	293.15	1089.2	1009.7	7.9
	303.15	1047.6	982.9	6.6

[a] W. Van Dael, A. Van Itterbeek, A. Cops, and J. Thoe, *Physica*, **1965**, 614.
[b] Yu. P. Blagoi, A. E. Butko, S. A. Mikhailenko, and V. V. Yajuba, *Russ. J. Phys. Chem.*, **41**, 908 (1967) [in English].
[c] H. Sackmann and A. Buczek, *Z. Physik. Chem.*, **29**, 329 (1961).

quantities may be evaluated in terms of the partition function as follows.

$$\beta_T^{-1} = -VkT\left(\frac{\partial^2 \ln f}{\partial V^2}\right)_T,$$

$$C_v = \left\{\frac{\partial}{\partial T}\left[kT^2\left(\frac{\partial \ln f}{\partial T}\right)_V\right]\right\}_V, \tag{3}$$

$$C_p = C_v + \frac{\alpha^2}{\beta_T}VT; \quad \text{here} \quad \alpha = \frac{[\partial^2(-kT \ln f)/\partial V\ \partial T]_{T\cdot V}}{VkT(\partial^3 \ln f/\partial V^2)_T}.$$

The velocity of sound in several liquids has been calculated and the data are shown in Table I.

VAN DER WAALS FORCE IN LIQUIDS

The van der Waals constant a in liquid can be evaluated from the relation,

$$\left(\frac{\partial E}{\partial V}\right)_T = kT^2\left(\frac{\partial^2 \ln f}{\partial V\ \partial T}\right)_{T\cdot V}, \tag{4}$$

together with the result derived from the van der Waals equation and the thermodynamic relation

$$\left(\frac{\partial E}{\partial V}\right)_T = T\left(\frac{\partial p}{\partial T}\right)_V - p,$$

$$a = V^2\left(\frac{\partial E}{\partial V}\right)_T. \tag{5}$$

Table II. Van der Waals Constant a of Liquid[a]

$T\,(^\circ\mathrm{K})$	a_{calc} (atm × liter^2)	a_{obs} (atm × liter^2)
	CCl$_4$	
293.55	28.620	31.18
293.78	28.619	31.23
298.99	28.575	31.18
304.40	28.533	31.23
310.24	28.489	31.22
	TiCl$_4$	
293.15	38.58	—
294.96	38.53	42.00
298.15	38.45	41.86
300.15	38.40	—
303.15	38.32	41.89

[a] a_{obs} values from J. H. Hildebrand and J. M. Carter, J. Amer. Chem. Soc., **54**, 3592 (1932).

The value of a was calculated for liquid CCl_4 and $TiCl_4$ at various temperatures: the results are shown in Table II along with the experimentally observed values at the same temperatures.

Data and partition functions used to calculate the velocity of sound and van der Waals constant were obtained from the following references: liquid Ar, Kr, Xe, and N_2, Ref. 3; liquid CCl_4, Ref. 4; liquid CH_4, Ref. 5; liquid O_2, Ref. 6; and liquid $TiCl_4$, Ref. 7.

DISCUSSION

The agreement between the calculated values and the experimentally observed values is satisfactory.

The results might be improved by modifying the ideal gas partition function[3] to allow for gas imperfections or by introducing the hindered rotation into the solidlike partition function. The success of significant structure theory in predicting the velocity of sound and the van der Waals constant a, which are dependent on the second derivatives of the partition function, is another piece of evidence for its general applicability.

ACKNOWLEDGMENT

I should like to express my appreciation to the donors of the Petroleum Research Funds, administered by the American Chemical Society, for supporting this research.

REFERENCES

1. (a) H. Eyring, T. Ree, and N. Hirai, *Proc. Natl. Acad. Sci. U.S.*, **44**, 683 (1958); (b) H. Eyring and T. Ree, *Proc. Natl. Acad. Sci. U.S.*, **47**, 526 (1961).
2. (a) H. Eyring and R. P. Marchi, *J. Chem. Educ.*, **40**, 562 (1963); (b) M. E. Zandler and M. S. Jhon, *Ann. Rev. Phys. Chem.*, **17**, 373 (1966); (c) H. Eyring and M. S. Jhon, *Significant Liquid Structures*, Wiley, New York (1969).
3. J. Grosh, M. S. Jhon, T. Ree, and H. Eyring, *Proc. Natl. Acad. Sci. U.S.*, **57**, 1566 (1967).
4. W. Paik and S. Chang, *J. Korean Chem. Soc.*, **8**, 29 (1964).
5. H. Pak and S. Chang, *J. Korean Chem. Soc.*, **7**, 174 (1963).
6. K. C. Kim, W. C. Lu, T. Ree, and H. Eyring, *Proc. Natl. Acad. Sci. U.S.*, **57**, 861 (1967).
7. D. S. Choi and M. S. Jhon, to be published.

Transient State Theory of Significant Liquid Structure Applied to Water[*]

HYUNGSUK PAK AND SEIHUN CHANG

Department of Chemistry, College of Liberal Arts and Sciences, Seoul National University, Korea

Abstract

The partition function for liquid water is developed according to the transient state theory of significant liquid structure proposed by Pak, Ahn, and Chang.[1] This theory assumes that the molecules may possess solid-like, transient, and gas-like degrees of freedom in liquid state.

Although liquid water has several special properties, for example, minimum molar volume at 4°C, the general theory of liquid can be applied successfully.

The theoretically calculated values for thermodynamic properties at the liquid temperature range and for the critical properties are in good agreement with the observed values.

[*] Reprinted by permission from *Journal of the Korean Chemical Society*, Volume 10, 1969, p. 91.

A Personal Note

SEIHUN CHANG

Professor Eyring developed two important theories; one is the theory of rate processes and the other the significant structure theory of liquids. Many liquid theories have been reported. I cannot think of any other however, which gives a clearer picture of liquid than the significant structure theory and is more applicable to so many types of liquid.

Professor Eyring is, I feel, a great scientist with a wealth of imagination. Whenever his students go to him for help with their research, he always presents a new idea immediately as if it were waiting in his desk drawer.

INTRODUCTION

The structure of liquid water has long been a subject of great interest because of the several unusual physical properties such as the high melting and boiling points, the high heat capacity, the high entropies of fusion and vaporization, the decrease of the molar volume on melting, and the subsequent contraction between 0 and 4°C, etc.

Various qualitative models for water have been proposed to explain its properties and to elucidate its structure. The earlier workers[2-15] attempted to explain its anomalous properties by postulating the existence of hydrogen bonds. However, none of these explanations are satisfactory.

Recently, Némethy and Scheraga[16] have developed a partition function for liquid water, based on the "flickering cluster" model proposed by Frank and Wen[13] assuming a distribution of the H_2O molecules over five energy levels, which represent from zero to four hydrogen bondings per molecule. The results from this model are in satisfactory agreement with the experimental values as to entropy, free energy, and internal energy, but the model fails to explain the behavior of the specific heat of water. In addition, Vand and Senior[18] have pointed out that the concentration of the species derived by Némethy and Scheraga do not show a satisfactory correlation with the experimental results of Buijs and Choppin[17] determined by infrared techniques. Buijs and Choppin have observed three absorption bands in the near-infrared spectrum of liquid water (in the region 1.1–1.3 μ), whose intensities vary with temperature and electrolyte concentration. Vand and Senior[19,20] have developed partition function for liquid water replacing Buijs and Choppin's concept of discrete energy level of water molecules by energy bands of water. They have assumed a Gaussian distribution of energy and expanded the partition function as a power series which involves many parameters to be determined using experimental data.

Marchi and Eyring[21] have developed a partition function based on significant liquid structure theory proposed by Eyring et al.[22,23] They assumed two solid-like species, a nonrotating hydrogen-bonded species and a rotating monomer. They further assumed that the monomer is able to rotate in the interstitial sites of the hydrogen-bonded species. But this is improbable because the interstitial site of hydrogen-bonded species does not have enough space for the free rotation of the monomer.

Pak and Chang[24] have previously developed a partition function for liquid water applying the modified theory of significant liquid structure proposed by Chang et al.[25] There, it is assumed that Ice-I-like, Ice-III-like, and gas-like molecules exist in liquid water and the molecules like Ice I and Ice III, both of which are oscillating torsionally, are in thermodynamic equilibrium. The equilibrium constant has been taken equal to the ratio of the partition functions of the two species. Various thermodynamic properties and the surface tension[26] of liquid water from the partition function were successfully calculated.

Recently, Jhon, Grosh, Ree, and Eyring[27] have modified the partition function proposed by Marchi and Eyring, assuming that there exist two species of solidlike molecules; one, a cagelike cluster of about 46 molecules with the density of Ice I, and the other, an Ice-III-like monomer. They also assumed that the cagelike cluster is dispersed in Ice III like a monomer. However, the partition function for the cagelike cluster is not given explicitly.

Though various quantitative models have been proposed and the thermodynamic properties have been calculated successfully by many workers, no general theory of liquid has yet been applied to water. Surely, the structure and the physical properties of real liquids depend on the chemical properties of the composing particles and the nature of the interacting forces among these particles. Yet a simple and clear model which might be taken as a "zeroth approximation" is absolutely essential for the further development of the liquid theory.

APPLICATION OF TRANSIENT STATE THEORY OF SIGNIFICANT LIQUID STRUCTURE TO WATER

Recently, Pak, Ahn, and Chang[1] have developed a theory of liquid state and called it transient state theory of significant liquid structure. (Hereafter, it is referred to as the transient state theory.) They have applied transient state theory to various liquids such as argon, nitrogen, benzene, chloroform, and carbon disulfide with success.

In the theory it is assumed that liquid molecules are partitioned by three kinds of degrees of freedom, that is, the solidlike, the transient, and the gaslike. And it is further assumed that the transition of the degrees of freedom from the solidlike to the gaslike does not occur directly, but the change takes place through a transient state in which the molecules are in different energy state by acquiring strain energy due to the neighboring molecular-sized holes in the liquid, supposing that the molar volumes for the solidlike portion and the transient portion are same. Since the holes are molecular size, $N(V - V_s)/V_s$ number of holes are introduced in liquid in random fashion, where N is Avogadro's number and V and V_s are molar volumes of liquid and solidlike molecules, respectively. Since the introduction of a molecular-sized hole requires an energy equal to heat of vaporization of a molecule, the molecules jumping into the holes should have the gas-like degrees of freedom. Then the fraction of the molecules having the gaslike degrees of freedom becomes

$$\frac{[(V - V_s)/V_s]N}{[(V - V_s)/V_s]N + N} = \frac{V - V_s}{V}.$$

The remaining fraction V_s/V are partitioned among solidlike and transient degrees of freedom by portions α and $1 - \alpha$, respectively. The strain energy, the energy difference between the transient state and the solidlike state are proportional to the heat of sublimation and inversely proportional to the number of the vacant sites around a molecule, $n(V - V_s)/V_s$, where n is given by $12(V_s/V_t)$, V_t being the molar volume of the liquid at the triple point. Einstein's solid model is applied for both of the solidlike and the transient state. The frequency of the lattice vibration is, in general, less in the transient state than in the solidlike state, since the former is at the higher energy state by the amount of strain energy, and it is less in both states than in the solid state. In other words, the molecules in the solidlike and transient states do not have the same degrees of freedom or the same structure, and in neither case must the molecular structure and the degrees of freedom be equal to solid. This is true especially in the molecules having molecular dissymmetry or high entropy of fusion as water molecules.

Therefore, in applying the transient state theory, the molecules in the solidlike and transient states may have different molar volumes, whose values are different from that of solid Ice I, 19.65 cc. The molar volumes at the solidlike and the transient state, V_σ and V_τ, as calculated by the trial and error method are 18.018 and 16.095 cc, respectively.

Then, according to the transient state theory, the partition function for liquid water is given as follows;

$$F = \frac{(N_s + N_t)!}{N_s! \, N_t!} f_s{}^{N_f} f_t{}^{N_g} \frac{1}{N_g!} = \frac{[(V_s/V)N]!}{[(V_s/V)N\alpha]! \, [(V_s/V)N(1 - \alpha)]!}$$

$$\times \left(\frac{e^{E_s/RT}}{(1 - e^{-\theta s/T})^6} \prod_{i=1}^{3} \frac{1}{1 - e^{-h\nu i/kT}} \right)^{(V_s/V)\alpha N}$$

$$\times \left(\frac{n[(V - V_s)/V_s]e^{(E_s-\epsilon)/RT}}{(1 - e^{-\theta t/T})^6} \prod_{i=1}^{3} \frac{1}{1 - e^{-h\nu i/kT}} \right)^{(V_s/V)(1-\alpha)N}$$

$$\times \left(\frac{(2\pi m kT)^{3/2}}{h^3} \frac{eV}{N} \frac{\pi^{1/2}(8\pi^2 kT)^{3/2}(I_A I_B I_C)^{1/2}}{2h^3} \prod_{i=1}^{3} \frac{1}{1 - e^{-h\nu i/kT}} \right)^{[(V-V_s)/V]N} \cdots$$

$$(1)$$

where θ is Einstein characteristic temperature; E_s, the sublimation energy of the solidlike molecule; V_s the mean molar volume of the molecules in the solidlike and in the transient state, $V_s = \alpha V_\sigma + (1 - \alpha)V_\tau$, here subscripts s, t, and g stand for the solidlike, the transient and the gaslike, respectively.

Because of the strong interactions among the molecules of liquid water, it is reasonable to assume that the molecules in the solidlike and transient states cannot rotate freely. In fact, Stevenson[28] has found that the rotating monomer concentration is very small. Therefore it is assumed that the molecules in the solidlike and in the transient state oscillate torsionally with the frequencies equal to those of the respective vibrations corresponding to θ_s and θ_t.

Here, it is assumed that the molecules in the gaslike state behave like an ideal gas. The three principal moments of inertia found in the literature[29] are 1.0243×10^{-40}, 1.9207×10^{-40}, and 2.9470×10^{-40} g cm² for I_A, I_B, and I_C, respectively.

The interatomic vibrational frequencies may be taken equal to those in the three states without appreciable error. They are also found in the literature,[29] that is, $\bar{\nu}_1 = 1595$, $\bar{\nu}_2 = 3652$, and $\bar{\nu}_3 = 3756$ cm^{-1}.

DETERMINATION OF PARAMETERS IN THE PARTITION FUNCTION

The values of α can be calculated using the thermodynamic equilibrium condition;

$$\left(\frac{\partial A}{\partial \alpha}\right)_{T,V,N} = -kT \left(\frac{\partial \ln F}{\partial \alpha}\right)_{T,V,N} = 0. \tag{2}$$

From the above relation α is obtained

$$\alpha = \frac{f_s}{f_s + f_t}. \tag{3}$$

Introducing (3), and using Stirling's approximation, (1) becomes

$$F = \left(\frac{e^{Es/RT}}{(1 - e^{-\theta s/T})^6}(1 + \lambda(x - 1)e^{-aEs/n(x-1)RT}) \prod_{i=1}^{3} \frac{1}{1 - e^{-h\nu i/kT}}\right)^{(1/x)N}$$

$$\times \left(\frac{(2\pi mkT)^{3/2}}{h^3} \frac{eV}{N} \frac{\pi^{1/2}(8\pi^2 kT)^{3/2}(I_A I_B I_C)^{1/2}}{2h^3} \prod_{i=1}^{3} \frac{1}{1 - e^{-h\nu i/kT}}\right)^{(1-(1/x))N}, \tag{4}$$

where

$$\lambda = n\left(\frac{1 - e^{-\theta s/T}}{1 - e^{-\theta t/T}}\right)^6 \tag{4a}$$

and

$$x = \frac{V}{V_s}. \tag{4b}$$

The parametric values, θ_s, θ_t, E_s, and a are determined at the triple point by the following method.

From the partition function the following equations can be derived.

$$\varphi \equiv \frac{A}{RT} = \frac{1}{N} \ln F = \frac{1}{x} [\sigma + y + (x - 1)(\gamma + \ln x)], \tag{5}$$

where

$$\sigma = \frac{E_s}{RT} - 6 \ln (1 - e^{-\theta_s/T}) - \sum_{i=1}^{3} \ln (1 - e^{-h\nu_i/kT}), \tag{5a}$$

$$y = \ln \{1 + \lambda(x - 1)e^{-\omega}\}, \tag{5b}$$

$$\omega = \frac{aE_s}{n(x - 1)RT}, \tag{5c}$$

and

$$\gamma = \ln \frac{(2\pi mkT)^{3/2}}{h^3} \frac{eV_s}{N} \frac{\pi^{1/2}(8\pi^2 kT)^{3/2}(I_A I_B I_C)^{1/2}}{2h^3} \prod_{i=1}^{3} \frac{1}{1 - e^{-h\nu_i/kT}}. \tag{5d}$$

$$\frac{PV}{RT} = \frac{\lambda e^{-\omega}(1 + \omega)}{1 + \lambda(x - 1)e^{-\omega}} - \frac{1}{x}(\sigma - \gamma + 1 + y - \ln x - x). \tag{6}$$

The equilibrium condition at constant temperature and pressure between the liquid and its vapor is

$$F_l = A_l + PV_l = F_g = A_g + PV_g.$$

By combining this with (5) and (6), and assuming the vapor behaves ideally, we obtain the following equation:

$$Z \equiv \frac{\lambda(x - 1)(1 + \omega)e^{-\omega}}{1 + \lambda(x - 1)e^{-\omega}}$$

$$= (x - 1)\left[\ln \frac{RT}{PV} - 2 + \frac{2PV}{RT} + \frac{1}{x}\right]. \tag{7}$$

The right-hand side of (7) is constant at a given temperature and pressure. It is independent on the values of ω and λ. Therefore by differentiating both sides with respect to ω at a constant temperature and pressure we obtain $(\partial Z/\partial \omega)_\lambda = 0$, and thence we get the following equation:

$$\omega = \lambda(x - 1)e^{-\omega}. \tag{8}$$

From (7) and (8), $\omega = z$, and $\lambda = \omega/(x - 1)e^{-\omega}$ are obtained. The calculated values of λ and ω at the triple point are 14.04 and 0.22435, respectively. By assigning appropriate value for θ_t, θ_s, E_s, a are obtained from (4a), (5a), and (5c), respectively. The obtained parametric values are given as follows:

$$E_s = 10940 \text{ cal/mole} \qquad a = 0.002618,$$
$$\theta_s = 234.6°K \qquad \theta_t = 224.0°K,$$
$$V_\sigma = 18.018 \text{ cc} \qquad V_\tau = 16.095 \text{ cc.}$$

The values of V_σ, V_τ, and θ_t are assigned so as the calculated vapor pressure and the molar volume at 10°C to coincide with the observed values.

CALCULATION OF THERMODYNAMIC PROPERTIES AND THE RESULTS

Vapor pressure and molar volume are obtained by plotting free energy, A versus V[22,23] as follows.

First, from (5), A is calculated at various V covering the liquid and vapor regions. Next, a common tangential line to the two regions is drawn. The slope of the line gives equilibrium vapor pressure and the tangential points give the molar volume of the liquid and the vapor.

The calculated and observed molar volumes of the liquid and its vapor pressures are given in Tables I and II, respectively.

Entropy is calculated from (5):

$$S = -\left(\frac{\partial A}{\partial T}\right)_v = \frac{R}{x}\left[-6\ln(1 - e^{-\theta_s/T}) + \frac{1}{e^y}\frac{6(\theta_s/T)e^{-\theta_s/T}}{1 - e^{-\theta_s/T}}\right.$$
$$+ \left(1 - \frac{1}{e^y}\right)\frac{6(\theta_t/T)e^{-\theta_t/T}}{1 - e^{-\theta_t/T}} + y + \left(1 - \frac{1}{e^y}\right)\omega$$
$$\left. + (x - 1)(\gamma + 3 + \ln x)\right]$$
$$+ R\left\{-\sum_{i=1}^{3}\ln(1 - e^{-h\nu_i/kT}) + \sum_{i=1}^{3}\frac{h\nu_i}{kT}\cdot\frac{e^{-h\nu_i/kT}}{1 - e^{-h\nu_i/kT}}\right\} \qquad (9)$$

and thence heat of vaporization from the following relation:

$$\Delta H_v = T(S_g - S_l), \qquad (10)$$

where S_g and S_l are entropies of the gas and the liquid at the temperature T. The calculated and observed values of liquid entropy and heat of vaporization are given in Tables III and IV, respectively.

Table I. Molar Volume of Liquid Water

$t\,^{\circ}C$	V_{calc} (cc)	V_{obs}^{30} (cc)	$\Delta\%$
0	18.019	18.019	0.00
4	18.018	18.017	0.01
10	18.021	18.021	0.00
20	18.035	18.048	−0.07
60	18.192	18.324	−0.72
100	18.498	18.799	−1.60
150	19.086	19.641	−2.83

Table II. Vapor Pressure of Liquid Water

$t\,^{\circ}C$	P_{calc} (atm.)	P_{obs}^{30} (atm.)	$\Delta\%$
0	0.006029	0.006029	0.00
4	0.008030	0.008028	0.02
10	0.01214	0.01212	0.17
20	0.02319	0.02307	0.52
60	0.1997	0.1966	1.58
100	1.020	1.000	2.00
150	4.766	4.698	1.45

Table III. Entropy of Liquid Water

$t\,^{\circ}C$	S_{calc} (e.u.)	S_{obs}* (e.u.)	$\Delta\%$
0	14.92	15.2	−1.9
10	15.57	15.8	−1.5
20	16.18	16.4	−1.4
60	18.44	18.7	−1.4
100	20.50	20.8	−1.4

* S_{obs} are the values obtained from $S_g - \Delta H_v / T$, where S_g is the entropy of gas obtained from the spectroscopic data[29] and ΔH_v is the observed heat of vaporization at the temperature T.

Table IV. Heat of Vaporization of Liquid Water

$t\,^{\circ}C$	$\Delta H_{v\,\text{calc}}$ (cal/mole)	$\Delta H_{v\,\text{obs}}$ (cal/mole)	$\Delta\%$
0	10820	10740	0.74
10	10720	10640	0.75
20	10620	10540	0.76
60	10230	10150	0.79
100	9800	9720	0.82

Table V. Heat Capacity at Constant Pressure of Liquid Water

$t°C$	C_p calc. (cal/mole.°K)	C_p obs. (cal/mole.°K)	$\Delta\%$
0	18.60	18.15	2.48
10	17.93	18.04	−0.61
20	17.57	17.99	−2.33
60	18.26	18.01	1.39
100	18.40	18.14	1.43

Heat capacity at constant pressure C_p is obtained from the following equation:

$$C_P = C_V + \frac{TV\alpha^2}{\beta} = C_V + \frac{TV[1/V(\partial V/\partial T)_p]^2}{-1/V(\partial V/\partial P)_T}$$

$$= T\left(\frac{\partial S}{\partial T}\right)_V - \frac{T(\partial P/\partial T)_V^2}{(\partial P/\partial V)_T}$$

$$= \frac{R}{x}\left\{U_s + \left(1 - \frac{1}{e^y}\right)(U_t - U_s + \omega)\right.$$

$$\times \left[\frac{1}{e^y}(U_t - U_s + \omega) + 1\right]$$

$$+ \frac{1}{e^y}U_s\left(\frac{\theta_s/Te^{-\theta_s/T}}{1 - e^{-\theta_s/T}} + \frac{\theta_s}{T} - 1\right)$$

$$+ \left(1 - \frac{1}{e^y}\right)U_t\left(\frac{\theta_t/Te^{-\theta_t/T}}{1 - e^{-\theta_t/T}} + \theta_t/T - 1\right)$$

$$\left. - \left(1 - \frac{1}{e^y}\right)\omega + 3(x - 1)\right\} + R\sum_{i=1}^{3}\frac{(h\nu_i/kT)^2 e^{-h\nu_i/kT}}{(1 - e^{-h\nu_i/kT})^2}$$

$$+ R\frac{(1/x^2)(U_t - U_s + \omega)\xi\{[(1 + \omega)x]/e^y - (x - 1)\}}{\xi^2[2\omega + 1 - \omega^2/(e^y - 1)]x + \xi(1 + \omega)}$$

$$\frac{+ x(1 + \xi) + 2 - [-6\ln(1 - e^{-\theta_s/T}) + U_s + y]^2}{+ \gamma + \ln x - (\sigma + y + 0.5)}, \quad (11)$$

where

$$U_s = \frac{6(\theta_s/T)e^{-\theta_s/T}}{1 - e^{-\theta_s/T}},$$

$$U_t = \frac{6(\theta_t/T)e^{-\theta_t/T}}{1 - e^{-\theta_t/T}}$$

Table VI. Critical Point Properties of Water

Property	Calculated	Observed	$\Delta\%$
$T_c \, ^\circ K$	745	647	15.1
P_c atm.	410	218	88.1
V_c cc	54.4	56.3	−3.37

and

$$\xi = \frac{\lambda e^{-\omega}}{1 + \lambda(x - 1)e^{-\omega}}$$

The calculated and observed values of C_p at various temperatures are compared in Table V.

Critical point properties are calculated using the conditions,

$$\left(\frac{\partial P}{\partial V}\right)_T = 0, \quad \text{and} \quad \left(\frac{\partial^2 P}{\partial V^2}\right)_T = 0.$$

which are given in Table VI.

DISCUSSION

The agreement between calculated and observed values are excellent in molar volume, vapor pressure, entropy, heat of vaporization and heat capacity at constant pressure.

It is known[31] that the calorimetrically measured entropy is less than the entropy obtained from the spectroscopic data by 0.81 e.u. This is due to the residual entropy. The calculated entropy of liquid is compared with the spectroscopic entropy which may be evaluated from the difference between the spectroscopic entropy of gas and the entropy of vaporization.

The calculated molar volume show the minimum at near 4°C and the heat capacity at constant pressure at near 35°C, which are in good agreement with the experimental observation.

The calculated critical point properties deviate from the observed values as might be expected since transient state theory is one of the zeroth approximation.

Nevertheless, we may say that the application of transient state theory to the complicated molecular states of liquid water is successful.

ACKNOWLEDGMENT

We are grateful to Dr, Kak-Choong Kim (Korea University, Seoul, Korea) for valuable discussions on the subject of this paper.

REFERENCES

1. H. Pak, W. Ahn, and S. Chang, *J. Korean Chem. Soc.*, **10,** 26 (1966).
2. H. M. Chadwell, *Chem. Rev.*, **4,** 375 (1927).
3. W. M. Latmer and W. H. Rodebush, *J. Amer. Chem. Soc.*, **42,** 1419 (1920).
4. W. H. Barares, *Proc. Roy. Soc. (London)*, *Ser.* A, **125,** 670 (1929).
5. J. D. Bernal and R. H. Fowler, *J. Chem. Phys.*, **1,** 515 (1933).
6. J. Morgan and B. E. Warren, *J. Chem. Phys.*, **6,** 666 (1938)
7. P. C. Cross, J. Burnham, and P. A. Leighton, *J. Amer. Chem. Soc.*, **59,** 1134 (1937).
8. K. Grjotheim and J. Krogh-Moe, *Acta Chem. Scand.*, **8,** 1193 (1954).
9. G. H. Haggis, J. B. Hasted, and T. J. Buchanon, *J. Chem. Phys.*, **20,** 1452 (1952).
10. L. Pauling, *in Hydrogen Bonding*, D. Hadzi Ed., Pergamon, London, 1959, p. 1.
11. J. Lennard-Jones and J. A. Pople, *Proc. Roy. Soc. (London)*, *Ser.* A, **205,** 155 (1951).
12. J. A. Pople, *Proc. Royl Soc. (London)*, *Ser.* A, **205,** 163 (1951).
13. H. S. Frank and W. Y. Wen, *Discussions Faraday Soc.*, **24,** 133 (1957).
14. H. S. Frank, *Proc. Roy. Soc. (London)*, *Ser.* A, **247,** 481 (1958).
15. H. S. Frank and A. S. Quist, *J. Chem. Phys.*, **34,** 604 (1961).
16. G. Némethy and H. A. Scheraga, *J. Chem. Phys.*, **36,** 3382 (1962).
17. K. Buijs and G. R. Choppin, *J. Chem. Phys.*, **39,** 2035 (1963).
18. V. Vand and W. A. Senior, *J. Chem. Phys.*, **43,** 1869 (1965).
19. W. A. Senior and V. Vand, *J. Chem. Phys.*, **43,** 1873 (1965).
20. V. Vand and W. A. Senior, *J. Chem. Phys.*, **43,** 1878 (1965).
21. R. P. Marchi and H. Eyring, *J. Phys. Chem.*, **68,** 221 (1964).
22. H. Eyring, T. Ree, and N. Hirai, *Proc. Natl. Acad. Sci. U.S.*, **44,** 683 (1958).
23. H. Eyring and T. Ree, *Proc. Natl. Acad. Sci. U.S.*, **47,** 526 (1962).
24. H. Pak and S. Chang, *J. Korean Chem. Soc.*, **8,** 68 (1964).
25. S. Chang, H. Pak, W. Paik, S. Park, M. Jhon, and W. Ahn, *J. Korean Chem. Soc.*, **8,** 33 (1964).
26. H. Pak and S. Chang, *J. Korean Chem. Soc.*, **8,** 121 (1964).
27. M. S. Jhon, J. Grogh, T. Ree, and H. Eyring, *J. Chem. Phys.*, **44,** 1465 (1966).
28. D. P. Stevenson, *J. Phys. Chem.*, **69,** 2145 (1965).
29. G. Herzberg, *Molecular Spectra and Molecular Structure*, Van Nostrand, Princeton, N.J., 1962.
30. *Handbook of Chemistry and Physics*, 42nd ed., Chemical Rubber Pub. Co.
31. Giaque and Stout, *J. Amer. Chem. Soc.*, **58,** 1144 (1936).

The Gaseous Fraction
in Liquid Metals

DAN McLACHLAN, Jr.

Department of Mineralogy, The Ohio State University, Columbus

Abstract

In this chapter of the Eyring Festschrift attention is called to some aspects of a subject which Dr. Henry Eyring originated; namely, the concept that a liquid metal is largely a crystalline mass thrown into long-range disorder by the presence of "fluidized vacancies" through which metal atoms can traverse as though they were gas atoms. No attempt is made in this chapter to regenerate the partition function of a liquid metal as Eyring has done or to reproduce the hundreds of computations with which he and his coworkers have had so much success in agreement with experiment.

This chapter outlines one possible interpretation of the nature of the fluidized vacancies. They are assumed to be thin layers of void space between the small crystalline domains in the liquid. Based on the Grüneisen potential function, the nature of the gap between crystalline domains is investigated and is found to be an interval of no barriers to atomic translation.

First some illustrations are given to support the concept of thin gas-filled layers between crystalline domains; and then several examples are given to show that the gaseous fraction which Eyring assumed to comprise a liquid metal agrees well with certain physical measurements which others have found by experiment. These examples include: heats of fusion, liquid coordination numbers, coefficients of thermal expansion of liquids, the effect of pressure on the melting point, and heat capacities of liquids.

INTRODUCTION

Although the name Henry Eyring had come to my attention many times before the autumn of 1947, it was only then that my intimate association with him began. It lasted for six inspiring years. Before that time he had already won fame in his application of reaction-rate theory to a myriad of problems, including chemical reactions, biological processes, and viscosity.

In 1947 Dr. Eyring was directing graduate students in projects under the general heading "creep and flow of metals" and was particularly intrigued by the sudden increase of viscosity of liquid metals as the temperature is lowered toward the melting point. This naturally enhanced his interest in a study of the liquid state of metals, where he found to his dismay that in liquids the vacancies or holes he expected to find were only a fraction of the volume of the space left after removing one atom from the inside of a crystal of metal and that, also, the energy required to produce this strange kind of vacancy was only a small fraction of the energy needed to remove an atom. But he accepted them as they were and named them "fluidized vacancies." He later discovered that during the time atoms are within these vacancies they behave as a perfect gas.

By applying his knowledge of the statistical mechanics of the perfect solid and of the perfect gas, Eyring was able to write a partition function for the liquid state on the assumption that a liquid consists of an intermixing of crystalline and gaseous fractions. This culminated in the theory of significant liquid structures.[1]

Although I am incapable of presenting Eyring's logic as clearly as he has already done in his various publications, I nevertheless have attempted to present, from a slightly different viewpoint, further evidence that the liquid state contains some atoms that behave as though they were gas atoms.

THE LIQUID METAL

A solid metal at any temperature above zero Kelvin can be thought of as an aggregate of perfect crystals any one of which shows perfect order

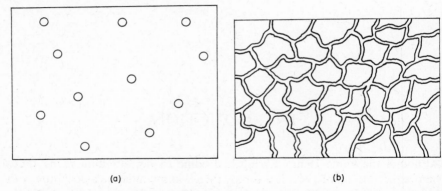

(a)   (b)

Fig. 1. (*a*) A continuous ordered region of a crystal: open circles represent holes or vacancies in the structure; (*b*) small ordered domains surrounded by a continuum of noncrystalline thin "shells." The nature of this noncrystalline portion is the topic of this paper. The thickness of the layers between the ordered domains is extremely thin.

throughout, with the exception of occasional sites where atoms are missing. Figure 1*a* shows, by open circles, the missing sites in a continuum of rigid ordered material. We propose that the liquid state (Fig. 1*b*) consists of small (subnuclear, for crystal growth) domains of ordered material "floating" about in a continuum of noncrystalline space. This is not very different from many theories previously proposed for liquids except with regard to the nature of the interglobular space. It has been conjectured to be amorphous by some theorists, and the distances between crystalline domains has been considered to be great. Here we consider that the globs are very close together so that an atom on the surface of any chosen glob is opposite an atom on an *adjacent glob* at a distance only about 20% greater than the distance between nearest neighbors *within a glob* or crystallite. An atom in flight between globs has the freedom of a perfect gas atom. As the temperature of the liquid is increased, the sizes of the globs (or ordered domains) become less; but the distance between globs stays fixed. This is developed from the Grüneisen potential function.[2]

THE POTENTIAL FUNCTION

A potential function used by many authors, including Mie, Grüneisen,[2] Lennard-Jones and Devonshire,[3] and London is of the form:

$$e(r) = \frac{e_0}{m-n}\left[-m\left(\frac{r_0}{r}\right)^n + n\left(\frac{r_0}{r}\right)^m\right], \qquad (1)$$

where r is the distance between two atoms, m and n are integers (determined for the elements by Fürth[4]), and e_0 is the energy required to separate two atoms by an infinite distance from one another. This function is shown as a graph in Fig. 2a. The function has the following properties: (a) at $r = r_0$ the potential is a minimum; (b) at the inflection point r_i, which is computed by

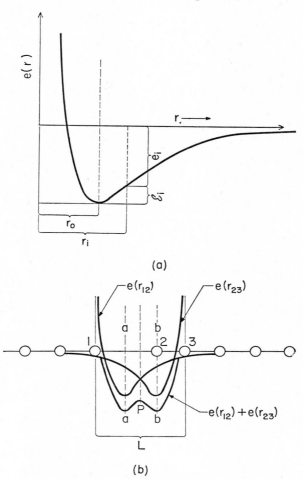

(a)

(b)

Fig. 2. (a) The potential energy of an atom as a function of its distance from a second atom according to the Grüneisen function; (b) the potential energy of an atom (2) as a function of its position between two other atoms (1 and 3). In the case shown for the particular value of L chosen there is a potential barrier at P that the atom must surmount to go from the positions of minimum energy at a and b.

equating to zero the second derivative of $e(r)$ with respect to r, is found to be

$$r_i = \left[\frac{m+1}{n+1}\right]^{1/m-n} r_0 \equiv B r_0, \tag{2}$$

and (c) the energy e_i, when $r = r_i$, is

$$e_i = \frac{e_0}{m-n} \left(-\frac{m}{B^n} + \frac{n}{B^m}\right) \equiv A e_0. \tag{3}$$

The change in distance in going from r_0 to r_i is $(B-1)r_0$, and the change in energy in going from r_0 to r_i is $\epsilon_i = (1-A)e_0$. For convenience Tables Ia and Ib give the $B-1$ and $1-A$ values for the various combinations of m and n.

Considering that n varies from 2 to 6 and m from 6 to 12, the values of $(B-1)$ and $(1-A)$ vary little within each table or between tables. When two atoms are separated to r_i we will say for convenience that the bond

Table Ia.   Values of $(B-1)$

			n		
m	2	3	4	5	6
6	0.236	0.205	0.183	0.167	—
7	0.217	0.186	0.170	0.160	0.148
8	0.202	0.176	0.158	0.145	0.135
9	0.187	0.165	0.149	0.136	0.126
10	0.176	0.157	0.141	0.129	0.120
11	0.167	0.147	0.134	0.123	0.114
12	0.158	0.137	0.127	0.117	0.109

Table Ib.   Values of $(1-A)$

			n		
m	2	3	4	5	6
6	0.1306	0.1837	0.1968	0.207	—
7	0.1454	0.1885	0.1994	0.213	0.225
8	0.1514	0.1765	0.1995	0.209	0.2185
9	0.1488	0.1778	0.1964	0.207	0.216
10	0.1451	0.1771	0.1948	0.207	0.217
11	0.1432	0.1717	0.1930	0.2095	0.2154
12	0.1394	0.1644	0.1980	0.2014	0.2138

between them is broken. The change in distance of separation is

$$\Delta r_i = r_0(B - 1) \tag{4}$$

and the energy required is

$$\epsilon_i = e_0(1 - A). \tag{5}$$

The dissociated energy e_0 has a physical meaning, since it is related to the heat of sublimation per mole,

$$e_0 = \frac{2\Delta H_S}{NW} \tag{6}$$

and r_0 has a physical meaning through the molar volume V,

$$r = (cV)^{1/3}, \tag{7}$$

where c is a structure constant. Thus (9) and (10) can be substituted into (1) to give $E(V)$, the potential energy of a mole of solid crystalline material as a function of molar volume V.

$$E(V) = \frac{\Delta H_S}{m - n} \left[-m\left(\frac{V_0}{V}\right)^{n/3} + n\left(\frac{V_0}{V}\right)^{m/3} \right]. \tag{8}$$

TRIUMPHS OF THE GRÜNEISEN EQUATION

Since it would be impossible to outline all the successful uses of (1), I shall discuss only a few that I have had something to do with.

Compressibility of Solids

When (8) is differentiated with respect to volume V, we get the negative of pressure, $P(V) = -dE(V)/dV$. A second differentiation gives

$$\frac{dP}{dV} = \frac{\Delta H_S mn}{3(m - n)} \left[-\frac{(n/3 + 1)V_0^{n/3}}{V^{n/3+2}} + \frac{(m/3 + 1)V_0^{m/3}}{V^{m/3+2}} \right]. \tag{9}$$

In the neighborhood of $V = V_0$, (12) becomes

$$\frac{dP}{dV} = \frac{\Delta H_S mn}{9V_0^2} \tag{10}$$

The bulk compressibility a of a solid is defined as

$$a = \frac{1}{V_0}\frac{dV}{dP}. \tag{11}$$

Table II.   Compressibility[a]

Element	m	n	V	ΔH_S	a, calc	a, exp. (ref. 6)
Al	7	4	10.5 cc	77,500 cal	10.52×10^{-7}	13.43×10^{-7}
Cr	7	5	7.23	95,000	4.73	5.19
Mn	10	3	7.62	66,730	8.30	7.91
Ca	6	4	26.0	42,200	56.0	56.97
Fe	7	4	7.7	99,830	6.01	5.87
Co	7	4	6.66	101,600	5.11	5.39
Ag	7	4.5	10.9	68,400	11.0	9.87
Cd	7	6	13.4	26,750	26.0	22.5
Ba	6	4	39.1	41,736	85.1	101.9
Li	6	1.5	13.3	38,439	84.0	86.92
Na	6	2	24.1	25,900	169.0	142.5
K	6	2	46.0	21,220	392.0	273.8
Rb	6	2	56.1	19,600	518.0	357.5
Cs	6	2	70.0	18,680	681.0	402.0
Mg	6	4	14.8	35,600	37.7	29.5

[a] The values of the heats of sublimation are taken from Ref. 5; m and n values are taken from Ref. 4

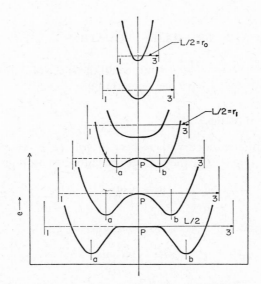

Fig. 3. As atoms 1 and 3 (see Fig. 2b) are separated to varying values of L the barrier at P changes and vanishes where $L/2 = r_i$, the inflection point on the Grüneisen curve (see Fig. 2a).

Application of (10) to (11) gives an equation for compressibility:

$$a = \frac{9V_0}{\Delta H_S mn} = \frac{0.218V_0}{mn\Delta H_S(\text{cal})}. \tag{12}$$

Table II[4,5] lists the values of a computed from (12) and compared to Bridgeman's[6] values in kilograms per sq cm.

The Debye Temperature

As has been described in two previous papers,[7,8] the potential of an atom in a row of atoms has a potential energy resulting from the presence of its nearest neighbors, as shown for atom 2 between atoms 1 and 3 in Fig. 2b. For each given distance that atoms 1 and 3 are apart, the potential of atom 2 is different, as shown in Fig. 3. When $L/2 = r_0$, atom 2 can vibrate harmonically with a characteristic frequency v which was ascribed[9] to the Debye frequency; the Debye temperature is computed from the derived equation:

$$\theta = \frac{h}{K}\frac{1}{2\pi r_0}\left(\frac{2\Delta H_S mn}{mW}\right)^{\frac{1}{2}} = \frac{218}{r_0}\left(\frac{\Delta H_S mn}{MW}\right)^{\frac{1}{2}}, \tag{13}$$

where h is Planck's constant, K is Boltzmann's constant, ΔH_S is the heat of sublimation, r_0 is in Å, M is molecular weight, and m and n are taken from Fürth.[4] Some results are shown in Table III.

The Critical Volume

It might be suspected that if a metal were thermally expanded under proper pressure conditions until all the bonds between nearest neighbors

Table III. Debye Temperature

Metal	θ_c	θ_D	Metal	θ_c	θ_D
Li	415	352.0 ± 17	Ba	89.3	110.5 ± 1.8
Na	154	157.0 ± 1	Ge	300	378 ± 22
K	90	99.4 ± 5	Sn	248	236 ± 24
Rb	49.1	54.0 ± 4	Al	446	423 ± 5
Cs	38.1	40 ± 5	Pb	103	102 ± 5
Cu	295	342 ± 2	Mo	427	459 ± 11
Ag	194.4	228 ± 3	W	349	388 ± 17
Au	193.0	165 ± 1	Fe	440	457 ± 12
Be	253	116.0	Co	348	452 ± 17
Mg	232	396 ± 5	Ni	354	427 ± 14
Ca	195.5	234 ± 5	Pt	251	234 ± 1
Sr	119	147 ± 1			

were r_i distance apart, then some interesting phenomenon might take place; that is, at the "critical point for the *solid–liquid* phases," where the densities of the solid and liquid are equal. This has been considered theoretically by Lennard-Jones and Devonshire,[10] who say that no such situation has been observed. If it were to be observed, the critical volume for the solid–liquid transition should be at VB^3. An inspection of Table Ia shows that over the range of commonly used m and n values this volume could be no higher than $1.86V_0$ and could be as low as $1.37V_0$. Also, one thing we are sure of is that this bond stretching to exactly r_i does not account for the critical volumes associated with the *liquid–gas* critical points.

The liquid–gas critical volume perhaps corresponds to a state where the atoms can pass freely between one another, in an almost amorphous state.

Table IV.   Critical Volumes

Substance	V_0	$4V_0$	V_c, exp
Elements			
Hg	14.65	58.60	63.7
Cu	7.96	31.84	31.4
Pb	19.42	77.68	68.3
Na	24.8	99.2	101.0
Cl	42.0	168.0	123.8
H_2	23.2	92.8	77.3[a]
Ne	13.98	55.92	41.74
Al	24.98	99.92	75.26
Kr	28.4	113.6	81.0[a]
Xe	36.5	146.0	113.8
Nonmetallic compounds			
Methane	30.94	123.76	99.02
Nitrogen	29.31	117.24	99.01
Benzene	77.00	308.00	256.0
Alkali halides[b]		$8V_0$	
NaCl	37.74	301.92	293[a]
KCl	48.80	390.4	431[a]
NaBr	44.08	352.64	342[a]
KBr	56.03	448.24	482[a]

[a] These are values calculated by Eyring's group.
[b] Note that the factor 8 is used instead of 4 for the alkali halides.

If, for example, a face-centered crystal having four atoms per unit cell were to lose its face atoms to become a simple cubic cell with only one atom per unit cell, the volume per mole would be four times V_0. Table IV shows a comparison of $4V_0$ with V_c from Eyring's papers.*

EVIDENCE FOR A GASEOUS FRACTION IN LIQUID METALS

An inspection of Fig. 3 shows that as atoms 1 and 3 are separated further apart, atom 2, which started with a potential minimum midway between 1 and 3, finally finds that it has two minima at a and b of Fig. 3. In between these two extremes there is a distance for $L/2$ at which the potential for atom 2 has a flat region over which it can traverse without a potential change much like a perfect gas atom. This occurs at $L/2 = r_i = r_0B$ [see (2)]. We postulate that this is the distance between the ordered globs in Fig. 2b, and that atoms can jump from one glob to the nearest glob across the gap without hindrance. The melting point is the temperature at which the ordered crystallites break up into these subnuclear domains leaving gaslike shells separating the globs. This naturally causes a sudden expansion of the material at the melting point.

We will now show some of the conclusions to be drawn from these assumptions regarding the liquid state of metals to supply further evidence that there is a gaseous fraction.

Phenomena at the Melting Point

Figure 2a shows that ϵ_i/e_0 relating the energy to break a bond is of the same order of magnitude as the fractional bond-length change $(r_i - r_0)/r_0$; and Tables Ia and Ib bear this out. Since ϵ_i and $r_i - r_0$ are measures of the heat of fusion and the volume change on melting, respectively, then we might expect that

$$\frac{\Delta V}{V} \approx \frac{\Delta H_M}{\Delta H_S}. \tag{14}$$

Table V verifies this relation† and Fig. 4 illustrates it graphically for a wide range of elements.

---

* Sources of critical volume data are: for gases, Eyring et al.[11]; for certain metals, Carlson et al.[12]; for hydrogen, Henderson et al.[13]; for chlorine, Thomson et al.[14]; and for the alkali halides, Carlson et al.[15]

† The values of V and $\Delta V/V$ for Table V were taken from Kaufman.[16]

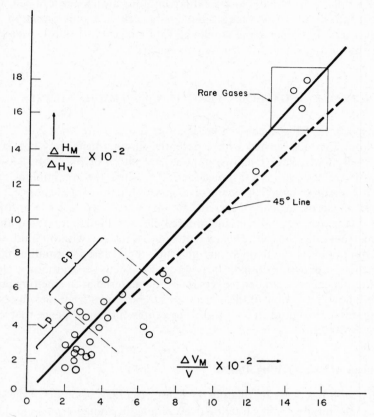

Fig. 4. The close correlation between $\Delta H_M/\Delta H_S$ and $\Delta V/V$ at the melting point. Some of these values were taken from Smithell.[16]

The Role of the Coordination Number

Barrett[17] has quoted some authors concerning the coordination number of certain metals in the liquid state. For example, liquid gold was given as $W_L = 11$; aluminum, 10–11; and lead, 8+. The last of these could be questioned because it disagrees with the values computed by another method[9] as well as the present work.

Here we attempt to compute the values of $\Delta H_M/\Delta H_S$ discussed in the last section and find that W_L is necessary for the final equation. Accepting that

$$\Delta H_S = \frac{e_0 N W_S}{2} \quad \text{or} \quad e_0 = \frac{2\Delta H_S}{N W_S} \tag{15}$$

Table V[a]

Element	$\Delta H_M/\Delta H_S$	$\Delta V/V$	Element	$\Delta H_M/\Delta H_S$	$\Delta V/V$
Ag fcc	4.1	3.8	Li	1.95	1.6
Al fcc	3.58	6.5	Mg hcp	5.85	4.12
Ar	15.0	12.7	Mn	5.0	1.7
Au fcc	3.39	5.1			
Cd hcp	5.70	4.75	N_2	12.3	7.3
Cs bcc	2.66	2.6	Na bcc	2.42	2.5
Cu fcc	3.82	4.2			
Fe bcc	3.81	2.6	O_2	6.52	7.4
Ba	2.0	−3.2	Pb fcc	2.54	3.5
H_2	12.9	12.3	Rb bcc	2.56	2.5
Hg	3.71	3.7	Sn	2.35	2.8
I_2	25.2	21.6			
In	1.41	2.5	Ti hcp	2.39	3.2
K bcc	2.6	2.55	Xe	16.6	15.1
Kr	17.8	15.1	Zn hcp	5.45	4.7

[a] All data in %.

and

$$e_i = e_0(1 - A),$$ (16)

we derive the equation for $\Delta H_M/\Delta H_S$:

$$\frac{\Delta H_M}{\Delta H_S} = 2(1 - A) \frac{W_S - W_L}{W_S}.$$ (17)

Table VI gives some values of $\Delta H_M/\Delta H_S$ on the rough assumption that $\frac{11}{12}$ of the bonds per atom are unbroken in the liquid state regardless of whether the structure is close-packed or body-centered.

Table VI. $(\Delta H_M/\Delta H_S)_{\text{calc}} = 2(1 - A)\frac{11}{12}$

Element	m	n	$(1 - A)$	W_S	W_L	$(\Delta H_M/\Delta H_S)_{\text{calc}}$	$(\Delta H_M/\Delta H_S)_{\text{exp}}$
Ag	7	4.5	0.20	12	11	3.34%	4.1%
Al	7	4	0.20	12	11	3.34	3.58
Au	8	5.5	0.21	12	11	3.5	3.39
Cd	7	6	0.225	12	11	3.75	5.70
Pb	12	3	0.164	12	11	2.73	2.54
Na	6	2	0.1306	8	7.3	2.16	2.42
K	6	2	0.1306	8	7.3	2.16	2.6
Rb	6	2	0.1306	8	7.3	2.16	2.50
Cs	6	2	0.1306	8	7.3	2.16	2.66
Li	6	1.5	0.110	8	7.3	1.83	1.95

The Coefficient of Thermal Expansion of Liquids

Referring back to Fig. 1b, we assume that the fraction of space between the globs just above the melting point is $\Delta V/V$, and this is the volume available for perfect gas atoms. However, since the thickness of the interglobular shell is assumed to maintain a constant value r_i, then the space is a two-dimensional gas. Since the bulk coefficient of expansion of a perfect gas at constant pressure is $1/T$, that of a two-dimensional gas is $\frac{2}{3}T$. So the bulk coefficient of thermal expansion of a liquid just above the melting point should be obtained by prorating the expansions due to the solid fraction $1 - \Delta V/V$ and the gaseous fraction $\Delta V/V$:

$$\alpha_L = \alpha_S\left(1 - \frac{\Delta V}{V}\right) + \frac{2}{3}\frac{\Delta V}{V}\frac{1}{T_M}. \tag{18}$$

A more sensitive means of testing the theory more rigorously than α_L is

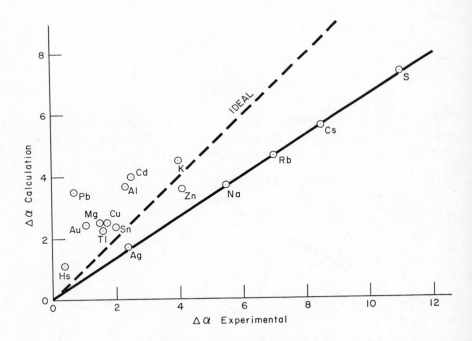

Fig. 5. The computed value of the change in bulk coefficient of expansion on melting using (19) plotted against the experimental differences for liquid and solid.

Table VII.   Liquid Expansion

Element	$\Delta V/V$	T_M	$\alpha_S \times 10^5$	$\alpha_L \times 10^5$	$(\alpha_S - \alpha_L)_{exp} \times 10^5$	$(\alpha_S - \alpha_L)_{calc}$
Ag	0.038	1234°K	8.1	10.5	2.4	1.71
Al	0.060	932.7	9.9	12.2	2.3	3.68
Au	0.051	1336.0	5.8	6.9	1.1	2.49
Cd	0.040	593.9	12.6	15.1	2.5	3.99
Cs	0.026	321.6	29.0	36.5	8.5	5.6
Cu	0.415	1361	7.0	9.5	2.5	1.72
Hg	0.037	234.13	17.1	18.2	1.1	0.40
In	0.020	429.4	12.4	12.0	−0.4	2.85
K	0.0255	335.3	25.0	29.0	4.0	4.45
Mg	0.041	934.0	11.0	12.5	1.5	2.49
Na	0.025	390.5	22.0	27.5	5.5	3.70
Pb	0.025	600.43	37.5	53.0	0.7	3.48
Rb	0.025	311.5	27.0	34.0	7.0	4.64
S	(0.055)	391.75	(36.0)	47.0	11.0	7.36
Se	(0.120)	491.0	18.0	40.0	22.0	8.96
Sn	0.028	704.89	9.5	11.5	2.0	2.37
Tl	(0.022)	575.0	12.4	14.0	1.6	2.24
Zn	0.042	692.0	11.3	15.4	4.1	3.57

the difference $\alpha_L - \alpha_S$ from the equation

$$\alpha_L - \alpha_S = \frac{\Delta V}{V}\left(\frac{2}{3}\frac{1}{T_M} - \alpha_S\right).\tag{19}$$

Table VII and Fig. 5 show the correlation between theory and experiment. The experimental values of coefficient of thermal expansion are taken from Schneider and Heymer.[18]

The Effect of Pressure on the Melting Point

The fact that the melting temperature of metals is influenced by applied pressure is of interest to people in many fields, particularly those geologists who are concerned with the core of the earth. In this paper we only go as far as computing the slope of the $T_M - P$ curve for relatively low pressures. The equation we present is a very simple one:

$$\frac{dT_M}{dP} = V\left(\frac{\Delta V}{V}\right)\frac{1000}{R} = \frac{V}{82.1}\frac{\Delta V}{V} \times 10^3,\tag{20}$$

in which the pressure is expressed in kilobars per sq cm. Table VIII and Fig. 6 show the fine correlation between the computed and measured values of dT_M/dP. These results are discouraging, however, when one considers the

Table VIII. $\dfrac{dT_M}{dP} = \dfrac{V(\Delta V/V)}{82.1} 1000$

Element	$\left(\dfrac{dT}{dP}\right)_{calc}$	$\left(\dfrac{dT}{dP}\right)_{obs}$
Li	3.30	3.3 and 1.5
Na	7.33	7.8 and 6.5
K	14.30	13.3 and 8.5
Rb	17.00	18.0
Cs	22.2	20.0
Al	7.70	6.4
Cu	3.84	4.2
Ag	5.51	5.5
Au	6.65	—
Ni	3.2	3.7
Pt	4.4	5.0
Rh	4.13	5.9
Pb	8.06	10.0 and 6.6
Fe	3.0	3.0
Tl	4.87	—
Mg	7.38	7.5
Zn	5.6	4.5 and 4.8
Cd	6.51	9.0 and 5.6
In	3.94	4.8 and 5.6
Sn	5.62	4.3 and 2.8

manner by which the equation came about. Starting with the equation for the gas in the liquid,

$$(P + P_i)V_g = N_g R T_M, \tag{21}$$

$$T_M = (P + P_i)\frac{V_g}{N_g R}, \tag{22}$$

where V_g is the volume of the gas in the liquid just above the melting point, P is the applied pressure, P_i is the internal pressure, and N_g is the number of moles of gas in the volume V_g. Differentiation gives

$$\frac{dT_M}{dP} = \frac{V_g}{N_g R} = \frac{V(\Delta V/V)}{N_g R}, \tag{23}$$

which becomes (20) only if N_g is assumed to be unity. The disturbing thing about $N_g = 1$ is that it implies that every atom in the liquid has access to the small volume, $V(\Delta V/V)$, and this is contrary to our starting picture of a liquid as depicted in Fig. 1b. The observed values of dT_M/dP are furnished by Larry Kaufman.[16]

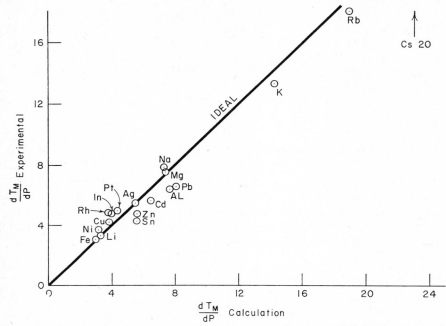

Fig. 6. The experimental slopes of the $T_M - P$ values compared with those computed from (20).

The Heat Capacity of Liquids

Like the coefficient of thermal expansion of liquid metals, the heat capacity of liquid metals should be expressible by an equation that prorates the crystalline fraction and the gaseous fraction:

$$C_{PL} = C_{PS}\left(1 - \frac{N_g}{N}\right) + \frac{3N_g}{N} + E_g \frac{dN_g}{dT} \qquad (24)$$

where N_g is the number of gas atoms in the liquid and the heat capacity of a perfect gas is 3 cal. Note that the last term in (24) accounts for the fact that as the temperature is increased, more bonds are broken and energy is required for breaking them. Since

$$\frac{n_g}{N} = \frac{e^{-e_g/kT}}{1 - e^{-e_g/kT}}, \qquad (25)$$

then

$$\frac{dn_g}{dT} = \frac{n_g e_g}{kT^2}\left(1 - \frac{n_g}{N}\right), \qquad (26)$$

where e_g is the energy to produce a gas atom in the way we prescribe.

Everything in (24) lends itself to computation, and the criterion as to whether the change in heat capacity $(C_{PL} - C_{PS})$ is positive or negative depends on whether the last term in (24) is sufficiently large to compensate for the fact that the heat capacity of gases is less than that of solids. Although this equation has been tried with partial success on a dozen or so metals, the procedure is not discussed further in this paper because of its length.

SUMMARY

On the basis of some very crude approximations, this paper reports some computations for (1) the phenomena at the melting point, (2) the role of coordination number in liquids, (3) the coefficient of thermal expansion, (4) the effect of pressure on the melting point, and (5) the heat capacity of liquids to corroborate Eyring's theory that liquids can be represented as a mixture of a solid and a gaseous fraction. The results appear to be rather encouraging considering the lack of refinement in the calculations.

REFERENCES

1. In addition to the many publications by Dr. Eyring we particularly wish to call attention to the book by Henry Eyring and Mu Shik Jhon, *Significant Liquid Structures*, Wiley, New York, 1969.
2. E. Grüneisen, *Ann. Phys. Leipsig*, **26**, 211 (1908).
3. J. E. Lennard-Jones and A. F. Devonshire, *Proc. Roy. Soc. (London)*, **A170**, 464 (1939).
4. R. Fürth, *Proc. Roy. Soc. (London)*, **A183**, 87 (1944).
5. K. A. Gschneidner, Jr., "Physical Properties and Interrelationships of Metallic and Semimetallic Elements," in *Solid State Physics*, Vol. 16, F. Seitz and D. Turnbull, Eds., Academic, New York, 1964, pp. 275–426.
6. P. W. Bridgeman, *The Physics of High Pressure*, Bell, London, 1958.
7. D. McLachlan and L. Chamberlain, *Acta Met.*, **12**, 571 (1964).
8. D. McLachlan, *Proc. Natl. Acad. Sci. U.S.*, **62**, 337 (1969).
9. D. McLachlan, *Acta Met.*, **15**, 153 (1967).
10. J. E. Lennard-Jones and A. F. Devonshire, *Proc. Roy. Soc. (London)*, **A169**, 464 (1939).
11. H. Eyring, T. Ree, and H. Hirai, *Proc. Natl. Acad. Sci. U.S.*, **44**, 683 (1958).
12. C. M. Carlson, H. Eyring, and T. Ree, *Proc. Natl. Acad. Sci. U.S.*, **46**, 649 (1960).
13. D. Henderson, H. Eyring, and D. Felix, *J. Phys. Chem.*, **16**, 1128 (1963).
14. T. R. Thomson, H. Eyring, and T. Ree, *Proc. Natl. Acad. Sci. U.S.*, **46**, 336 (1960).
15. C. M. Carlson, H. Eyring, and T. Ree, *Proc. Natl. Acad. Sci. U.S.*, **46**, 333 (1960).
16. L. Kaufman, "Phase Equilibria and Transitions in Metals," in *Solids Under Pressure*, W. Paul and D. M. Warschauer, Eds., McGraw-Hill, New York, 1963, Chap. 11, pp. 303–356.
17. C. S. Barrett, *Structure of Metals*, McGraw-Hill, New York, 1952, p. 265.
18. A. Schneider and G. Heymer, "Phenomena Accompanying Solid-Liquid Transformations of Metals and Alloys," in *The Physical Chemistry of Metallic Solutions and Intermetallic Compounds* (Symposium No. 9, Natl. Phys. Laboratory), Her Majesty's Stationery Office, London, 1959, pp. 4A.P2–4A.P18.
19. C. J. Smithell, *Metal Reference Book*, 3rd ed., Vol. II, Butterworth, London, 1962.

Application of the Significant Structures Theory to Plastic Crystals*

M. E. ZANDLER AND T. R. THOMSON

Department of Chemistry, University of Utah, Salt Lake City

Abstract

The significant structures theory as developed for liquids by Eyring, Ree, and collaborators is shown to be applicable also to the plastic crystal state. The properties of CBr_4 (in the plastic crystal state) were calculated and compared with the experimental data. The calculations were performed on a digital computer using, without alteration, a program developed for the liquid state.

A new development in the theory of states of matter is the recognition of a state which is called by Timmermans[1] a "plastic crystal" (review given in Ref. 2). This state normally occurs between the melting point and transition point(s) of substances composed of molecules with a high degree of symmetry. Substances in which this state appears are characterized by their low entropy of fusion, their relatively high entropy of transition, and their high plasticity (softness) in the plastic crystal range.

Many of the properties of this state suggest that the plastic crystal state is closely related to the liquid state. This viewpoint is in agreement with results

* Reprinted from *Solid State Communications*, Vol. 4, 1966, pp. 219–222, by permission of Pergamon Press Ltd.

of X-ray studies, optical properties, and dielectric constant measurements,[3,4] which indicate that rotation and/or movement of molecules takes place in the plastic crystal state to a degree comparable to the movement in liquids. Nevertheless, the motion cannot be free rotation.[5] Ubbelohde[6] suggests that the plastic property of this state must be associated with a comparatively large concentration of crystal defects such as lattice vacancies. He guesses that "marked plasticity would be expected for concentrations of vacancies of the order 0.1 to 1.0 per cent."

Since the significant structures theory as used by Eyring and collaborators[7] is able to predict quite accurately all the thermodynamic and many physical properties of a liquid using a model that assumes the volume expansion of a liquid to be due to the introduction of holes in the solid lattice, it seems quite likely that the theory might also be applied to the plastic crystal state. In order to test this idea, the theory was applied to the plastic crystal state of CBr_4, for which good experimental data are available.[5,8]

The significant structure theory of pure liquids is based on the recognition that the following three structures give the most important contributions to the partition function of a liquid: (a) molecules with oscillational degrees of freedom, (b) the positional degeneracy of these oscillatory molecules, and (c) molecules with translational degrees of freedom. A convincing argument is used to relate the relative proportions of molecules with oscillational and translational degrees of freedom to the excess volume of the liquid compared to the reference solid. More explicitly, the fraction of molecules having translational degrees of freedom is taken as $(V - V_0)/V$, where V and V_0 are the molar volumes of the liquid and the reference solid, respectively.

Based on these arguments the partition function is then written as a product of the partition functions written for each of the structures, weighted by the fraction of molecules in each structure,

$$Z_{liq} = \frac{(f_{sol} \cdot f_{deg})^{NV_0/V}(f_{gas})^{[N(V-V_0)]/V}}{N(V - V_0)/V!}$$

where the factorial term is due to the indistinguishability of the gaslike molecules (i.e., molecules undergoing translational motion). Using standard expressions for f_{sol} and f_{gas}, using Eyring's expression for f_{deg}, and applying Stirlings approximation for the factorial term the partition function for CBr_4 becomes

$$Z_{liq} = \left[\frac{e^{Es/RT}}{(1 - e^{-\theta/T})^6} \cdot \prod_{i=1}^{9} \frac{1}{1 - e^{-hv_i/kT}} \cdot \left(1 + 12 \cdot \frac{V - V_0}{V} \cdot e^{-\Phi}\right) \right]^{NV_0/V}$$

$$\times \left[\frac{(2\pi mkT)^{3/2}}{h^3} \cdot \frac{\pi^{1/2}(8\pi^2 kT)^{3/2}(I_a I_b I_c)^{1/2}}{\sigma h^3} \cdot \prod_{i=1}^{9} \frac{1}{1 - e^{-hv_i/kT}} \cdot \frac{eV}{N_0} \right]^{N(V-V_0)/V}$$

Table I. The Calculated and Observed Properties of Carbon Tetrabromide in the Plastic Crystal State

	T (°K)	P (mm Hg)	V (cc/mole)	S (cal deg^{-1} mole^{-1})	H^a (cal mole^{-1})	ΔH_{sub} (cal mole^{-1})	$\alpha \times 10^4$ (°K^{-1})	$\beta \times 10^6$ (atm^{-1})	C_v (cal deg^{-1} mole^{-1})	C_p (cal deg^{-1} mole^{-1})
Observed	320.00	3.14	102.91	—	—	11800	—	—	25.88	—
Calculated	(T_{tr})	(3.14)	(102.91)	59.59	−6941	12295	6.78	46.4	26.05	33.96
Observed	325.00	4.20	103.18	—	—	11800	6.10	38.0	25.81	33.6
Calculated		4.23	103.24	60.11	−6772	12238	6.43	45.3	26.11	33.54
Observed	330.00	5.62	103.50	—	—	11800	5.77	38.9	25.78	33.11
Calculated		5.63	103.57	50.62	−6605	12182	6.32	45.7	26.17	33.41
Observed	335.00	7.33	103.00	—	—	11800	5.82	—	25.74	33.03
Calculated		7.42	103.90	61.13	−6438	12127	6.30	46.7	26.22	33.40
Observed	340.00	9.57	104.14	—	—	—	6.02	—	25.71	33.09
Calculated		9.70	104.23	61.62	−6271	12071	6.35	48.2	26.28	33.45
Observed	345.00	12.3	104.46	—	—	—	—	—	25.68	33.25
Calculated		12.56	104.56	62.11	−6103	12016	6.42	49.9	26.33	33.54
Observed	350.00	15.9	104.77	—	—	—	—	—	25.65	33.38
Calculated		16.12	104.90	62.60	−5935	11961	6.51	51.8	26.38	33.65
Observed	355.00	20.3	105.10	—	—	—	—	—	25.62	33.56
Calculated		20.54	105.25	63.07	−5767	11905	6.62	53.9	26.43	33.77
Observed	360.00	25.7	—	—	—	—	—	—	25.60	33.90
Calculated		25.95	105.60	63.55	−5598	11149	6.73	56.2	26.47	23.90
Observed	363.25	30.1	—	—	—	—	—	—	—	—
Calculated	(T_{tp})	(30.1)	105.83	63.85	−5487	11183	6.81	57.7	26.50	33.99

^{a} Reference state ideal gas at 0.0°K

where $\Phi = [aE_S \cdot V_0/(V - V_0)RT]$, and the other notation is standard. The properties of the oscillatory molecules (the sublimation energy E_S, the molar volume V_0 and the Einstein characteristic temperature θ) as well as a are considered as parameters of the liquid state (or plastic crystal state). These are evaluated by fitting them to the following experimental data: the molar volume and vapor pressure of the liquid at the triple point (or of the plastic crystal at the transition temperature) and the vapor pressure at some other temperature. Knowing values for these parameters and the molecular constants, the thermodynamic and mechanical properties may be calculated using the standard formulas of statistical mechanics.

A computer program written in FORTRAN language was developed by one of the authors,[9] not only to make these computations, but also to find the parameters in a simplified consistent manner. This program was designed for calculations on the liquid state but could be used without alteration for the plastic crystal state.

For purposes of the program the properties of the plastic crystal state were calculated as though it were a liquid; that is, the transition point of the plastic crystal is used as equivalent to the melting point of a normal liquid. This means that in a substance in which the plastic crystal state occurs, crystal form I and crystal form II correspond, respectively, to the liquid and solid states for a normal substance.

The input data for CBr_4 was as follows: molecular weight = 331.647; the moments of inertia[10] $I_a = I_b = I_c = 1330 \times 10^{-40}$; the vibrational energies in units of cm^{-1} and their degeneracies[11] 269.0(1), 122.0(2), 183.0(3), and 667.0(3); $T_{trans} = 320.00°K$; $P_{trans} = 3.14$ mm; $V_{(I)trans} = 102.905$ cm^3; $T_{tp} = 363.25°K$; and $P_{tp} = 30.1$ mm. The values obtained for the parameters are: $V_0 = 101.08$ cm^3, $E_S = 13,648$ cal/mole, $a = 0.00015$ and $\theta = 25.17°K$. The results of the calculations for the plastic crystal state of CBr_4 are given in Table I along with the observed properties. The calculated concentration of vacancies at the transition point is 1.81%, agreeing well with the prediction made by Ubbelohde.

It is seen that the accuracy of the calculated properties is quite good—in most cases somewhat better than the accuracy expected for similar calculations of liquids. It must be concluded, therefore, that the significant structures theory yields an adequate description of the plastic crystal state as well as the liquid state.

REFERENCES

1. J. Timmermans, *J. Phys. Chem. Solids*, **18**, 1 (1961).
2. L. A. K. Staveley, *Ann. Rev. Phys. Chem.*, **13**, 351 (1962).
3. W. J. Dunning, *J. Phys. Chem. Solids*, **18**, 21 (1961).

4. C. P. Smyth, *J. Phys. Chem. Solids*, **18**, 40 (1961).
5. J. G. Marshall, L. A. K. Staveley, and K. R. Hart, *Trans. Faraday Soc.*, **52**, 19 (1956).
6. A. R. Ubbelohde, *J. Phys. Chem. Solids*, **18**, 90 (1961).
7. H. Eyring, T. Ree, and H. Hirai, *Proc. Natl. Acad. Sci. U.S.*, **44**, 683 (1958); H. Eyring and T. Ree, *ibid.*, **47**, 526 (1961); H. Eyring, D. Henderson, and T. Ree, *Progress in International Research on Thermodynamic and Transport Properties*, American Society of Mechanical Engineers, 88, New York, 1962; H. Eyring and R. P. Marchi, *J. Chem. Educ.*, **40**, 562 (1963); H. Eyring, T. S. Ree, and T. Ree, *Intern. J. Eng. Sci.*, **3**, 385 (1965).
8. R. S. Bradley and T. Drury, *Trans. Faraday Soc.*, **55**, 1844 (1959).
9. M. E. Zandler, Ph.D. Thesis, Arizona State University, 1965.
10. G. Gelles and K. S. Pitzer, *J. Amer. Chem. Soc.*, **75**, 5259 (1953).
11. G. Hertzberg, *Molecular Spectra and Molecular Structures*, Vol. I. *Spectra of Diatomic Molecules*, 2nd ed., Van Nostrand, Princeton, N.J., 1950.

On the Nature of Solutions of Organic Compounds in Fused Salts

LOUIS L. BURTON* AND E. R. VAN ARTSDALEN

Department of Chemistry, University of Alabama, Tuscaloosa

* Present address: E. I. Du Pont de Nemours and Company, Orange, Texas.

A Personal Note

ERVIN R. VAN ARTSDALEN

Henry Eyring is rightfully recognized and distinguished throughout the scientific world for his contributions to the theory of reaction kinetics and liquid structures. For some years my associates and I have investigated fused salts. Almost all of these studies have involved liquid structure problems and often kinetics as well. Our work has not only been influenced strongly by Eyring theories but often in a very personal way. Not infrequently involved and heated discussions with Henry have lasted into the late hours. Certainly no one has been more helpful and understanding than Henry Eyring.

My first meeting with Henry Eyring occurred at Princeton University while I was a graduate student at Harvard. I had made a trip to Princeton to attend a seminar being given by one of my former teachers and was graciously invited to join some of the faculty and the speaker at dinner and discussion in the Graduate House after the seminar. I recall how impressed I was to be sitting at dinner between Dr. (later Sir) Hugh Taylor and Dr. Henry Eyring. I was fascinated with the wit, charm, and scientific acumen of Henry Eyring.

After dinner we sat about and discussed scientific questions, largely arising from the seminar. In the course of this discussion Henry made the comment, "Give me some paper and pencils and I'll give you an answer to any question you pose. It may not mean anything much, it may even be wrong, but I'll give you an answer." I was greatly intrigued by this statement. To me it indicated Henry Eyring's interest in all matters of science, his confidence, yet humbleness, in his own capability, and a delightful sense of humor which permitted him to poke fun at himself.

In the years since then Henry and I have become close personal friends as well as scientific collaborators in both industry and university. I have come to realize how very right was my first judgment of him. Henry Eyring is one of the great minds and one of the great men in science. His energy, enthusiasm, and philosophy are boundless. It is a joy and inspiration to know him and to be associated with him.

There have been relatively few investigations of the use of fused inorganic salts as solvents for organic compounds. Of course, the high melting points of many salts are not conducive to such use and one is not accustomed to think of typical molecular organic substances as being particularly soluble in highly ionic liquids such as most fused salts. Nonetheless, a number of low melting inorganic salts, as well as some of their eutectic mixtures, can be used effectively as solvents for several classes of organic compounds, and instructive and useful studies can be made with them.

In general, it appears that many organic compounds which dissolve appreciably in molten inorganic salts, without extensive decomposition, do so by hydrogen bonding to solvent ions or in some cases by their own ionic dissociation.

Thus in a study of solubility in fused (Na, K)SCN eutectic Crowell and Hillery[1] found that pentaerythritol, ethyleneglycol, hydroquinone, sugars, and several other polyhydroxy compounds were quite soluble at 150°C. Their PMR spectrum of pentaerythritol in fused thiocyanate at 130°C taken with a Varian A-60 NMR spectrometer showed two singlet peaks characteristic of an alcohol. We have recently confirmed this finding in more detail using a Varian HA-100 spectrometer and will report on this elsewhere. These results suggest that the dissolved pentaerythritol is un-ionized in thiocyanate melts and that its solubility depends on hydrogen bonding, not ionization. Crowell and Hillery[1] confirmed this concept in the case of the indicator alizarin, which dissolves in fused thiocyanate to give the yellow color characteristic of its un-ionized form.[2] However, addition of a trace of sodium hydroxide produced the blue color of the ionic form and this could be reversed by addition of benzoic acid.

$$C_{14}H_8O_4 \text{ (yellow)} + B \rightleftharpoons C_{14}H_7O_4^- \text{ (blue)} + HB^+.$$

Similar solubility behavior of polyhydroxy compounds was shown in fused acetates by Burton and Crowell.[3] Burton[4] observed that solubility of several organic compounds at 200°C, in (Li, Na, K) acetate eutectic (mp 180°C) increases with the number of hydroxyl groups and with acidity. In weight percent the following approximate solubilities were determined: methanol, 0.05%; 2,4-dinitroaniline and 4-nitroaniline, ~0.5%; hydroquinone and resorcinol, ~1%; 2-amino-2-hydroxymethyl-1,3-propanediol and 2-amino-2-methyl-1,3-propanediol, ~10%; trimethylolethane and pentaerythritol were miscible in all proportions above their melting points.

In connection with a series of proton NMR studies in (Na, Rb, Cs) acetate eutectic (mp 95°C) we have observed similar solubilities. At 130°C pentaerythritol, trimethylolethane, 2,2-dimethyl-1,3-propanediol, *cis*- and *trans*-2,2,4,4-tetramethyl-1,3-cyclobutanediol, 2-amino-2-methyl-1,3-propanediol, 2-amino-2-hydroxymethyl-1,3-propanediol, glycerol, ethyleneglycol, phloroglucinol, pyrogallol, and pyrocatechol are soluble to 15% by weight or more. Lower solubility was observed for resorcinol, hydroquinone, and 2-methylhydroquinone (~10%) and for 2-*t*-butylhydroquinone and sucrose (~5%). PMR spectra were obtained for all but the 2-*t*-butylhydroquinone and sucrose solutions and will be reported in detail elsewhere. Suffice it to say here that the hydroxyl proton peak of these compounds in fused alkali metal acetates is shifted to lower field relative to many other solvents, possibly by rapid exchange with acetate ion. This bears out the concept of solubility resulting from strong interaction, between protons of the hydroxyl groups and the solvent anions.

We have recently extended investigations of organic materials in low melting salts to include studies of solubility, DTA, and proton NMR in molten alkali metal nitrates below 200°C, a temperature region in which there are few inexpensive ionic melts. Explosions have been reported on heating organic compounds in molten nitrates above 250°C[5]; therefore safety precautions were observed, but no violent reactions occurred in our experiments.

In a small-scale test behind a safety shield pentaerythritol (mp 260°C) dissolved readily in (Li, K)NO_3 eutectic (41.2–58.8 mole%, mp 125°C) at 130°C to form a 16% by weight solution. A 10% solution showed no visual evidence of reaction at 200°C. DTA runs made with small samples of 10% pentaerythritol–90% (Li, K)NO_3 eutectic with heating rates of 10° per minute showed endothermic peaks at 123–125°C, corresponding to melting, and broad exothermic peaks about 375°C attributed to decomposition of the solvent. Essentially, identical DTA's were observed for the pure eutectic solvent. Evidently, a 10% solution of pentaerythritol is reasonably stable in this nitrate melt and therefore we made NMR measurements. A PMR spectrum of pentaerythritol in (Li, K)NO_3 eutectic at 130°C, taken with a Varian HA-100 NMR spectrometer, was quite analogous to those obtained for pentaerythritol dissolved in thiocyanate or acetate eutectics. Two singlet peaks were observed, which we attribute to the methylene and hydroxyl protons. The hydroxyl peak was far downfield.

The weak acid 2,4-dinitroaniline ($pK_a \sim 15.5$ in alcoholic solution) dissolved in (Li, K)NO_3 at 140°C to form a yellow solution, which on heating gradually became orange (~180°C) and then red (~230°C), with the color changes being reversible on cooling. These color changes of this indicator suggest that it dissolves at the lower temperatures as the un-ionized molecule and gradually undergoes reversible acid–base reaction with the solvent as the temperature is raised, forming the resonance stabilized anion:

Although molten nitrate salts are powerful oxidizing agents at elevated temperatures, they may be useful solvents for selected organic substances below about 200°C. However, until data become available for other organic solutes, great caution and adequate safety measures are advised.

In a search for an aprotic solvent for acidity function studies with (Li, Na, K) acetate eutectic at 200°C (Na, K)SCN was suitably miscible; however, even small concentrations (1%) inhibited the known reaction of trimethylol-ethane,[3] which probably requires proton abstraction by the acetate anion to form an alkoxide intermediate. Since haloacetates are weaker bases than acetates, these salts were considered for diluents. Potassium trifluoroacetate (mp 135–137°C), reported[6] to be the most stable member of the series, was prepared from potassium hydroxide and excess trifluoroacetic acid with vacuum drying and fractional recrystallization from absolute alcohol. DTA of the white crystalline product detected decomposition as low as 125°C. Since the molten salt decomposed with bubbling at 145–150°C, this effort was discontinued.

Although ammonium salts are not typical ionic melts, in 1945 Khaish-bashev reported[7] the solubility in ammonium nitrate of many acidic organic compounds, including resorcinol, mannitol, cholesterol, acetamide, picric acid, and urea. Considerable work has been reported by Gordon[8] on fused quaternary ammonium salts.

As noted before, solubility of covalent organic compounds in molten salts seems to require proton interaction between acidic groups of the solute and anions of the solvent. For un-ionized solutes this may well involve structures analogous to the solvent-bridged ion-pair complexes postulated for concentrated aqueous solutions[9]:

$$\begin{matrix} R \\ | \\ M^+ \text{ιιιιιι} O\text{—}H \text{ιιιιιι} B^-. \end{matrix}$$

In the case of multifunctional acetate and nitrate anions, cyclic and polymeric structures can also be drawn. For molten salts containing the lithium cation, a hard acid,[10] solute molecules probably are more strongly polarized, increasing the tendency toward bridging and solubility; however, the melts containing lithium cations may not be 100% ionic.

Cryoscopic measurements in sodium nitrate[11] indicate that this is an ionic melt. The extension of studies with organic solutes to this solvent or (Na, K)NO$_3$ eutectic (mp 233°) could give information about the solvated species, role of the solvent, and at higher temperatures the mechanism of oxidation. Based on inorganic reactions,[12] currently there is disagreement about the existence in fused nitrates of oxide ion and nitronium ion. A fresh approach to the question with some organic reactions might help clarify the mechanism.

REFERENCES

1. T. I. Crowell and P. Hillery, *J. Org. Chem.*, **30**, 1339 (1965).
2. Also reported by B. J. Brough, D. H. Kerridge, and M. Moseley, *J. Chem. Soc., Ser. A*, 1556 (1966).
3. L. L. Burton and T. I. Crowell, *J. Amer. Chem. Soc.*, **90**, 5940 (1968).
4. L. L. Burton, Ph.D. Thesis, University of Virginia, 1968.
5. T. Greweling, *Anal. Chem.*, **41**, 540 (1969).
6. F. J. Hazelwood, E. Rhodes, and A. R. Ubbelohde, *Trans. Faraday Soc.*, **62**, 3101 (1966).
7. O. K. Khaishbashev, *Bull. Acad. Sci. URSS, Classe Sci. Chem.*, 587 (1945); *Chem. Abstr.*, **40**, 5981 (1946).
8. J. E. Gordon, *J. Amer. Chem. Soc.*, **87**, 1499 (1965).
9. R. A. Robinson and H. S. Harned, *Chem. Rev.*, **28**, 455 (1941).
10. R. G. Pearson, *J. Chem. Educ.*, **45**, 581, 643 (1968).
11. E. R. Van Artsdalen, *J. Tenn. Acad. Sci.*, **29**, 122 (1954).
12. R. F. Bartholomew and D. W. Donigian, *J. Phys. Chem.*, **72**, 3545 (1968).

Solid "Liquid-Crystalline" Films of Synthetic Polypeptides: A New State of Matter

A. V. TOBOLSKY AND E. T. SAMULSKI

Department of Chemistry, Princeton University and Textile Research Institute, Princeton, New Jersey

Abstract

Concentrated solutions of poly-γ-benzyl-L-glutamate (PBLG), while still in a fluid condition, have previously been shown to be in a liquid crystalline state, characterized as helicoidal or "cholesteric." We have found that this same structure persists in certain *solid* films of PBLG. This conclusion was reached from studies of swelling, X-ray, NMR, and optical properties. Orientation of the PBLG molecules could be achieved by casting films in a strong magnetic field. A new helical conformation for PBLG was observed in oriented films of this type which were cast from chloroform.

A Personal Note

ARTHUR V. TOBOLSKY

Professor Eyring has the qualities of a great teacher and missionary in that he attempts to reach and meet the soul and mind of every student he encounters. In my encounter with him twenty-five years ago he was not deterred by differences in religious and political viewpoints but strove to transmit to me his own understanding of all issues, scientific and otherwise. Even though I have not come to agree with him on all matters, I have moved very much toward his position on some, based on the life experience of twenty-five more years. What I can never forget was his concern for each individual, his attempt to join us to him in good fellowship, moral elevation, and inspirational thinking.

The liquid crystal state for low molecular weight compounds such as cholesterol esters has been known for more than 75 years. This is a meso-morphic state of matter, intermediate between the crystal state and the liquid state. In the liquid crystalline state the molecules retain preferential orientations relative to one another over large distances—a property normally associated with the crystalline state; yet liquid crystals from low molecular weight substances are highly fluid in their thermodynamically stable condition.

Recently liquid crystals have received much attention because of their optical properties that enable one to monitor minute changes in temperature and mechanical stress, for their ability to orient in electric and magnetic fields, and because of their implications on structures observed in biological systems.

In the 1930's it was shown by E. Vorlander [*Trans. Faraday Soc.*, **29**, 907 (1933)] that some liquid crystals could be quick frozen to a metastable brittle glassy state. By working with a polymeric molecule, a synthetic polypeptide called poly-γ-benzyl-L-glutamate, we have been able to obtain stable solid films with a liquid crystalline local structure. These films can be obtained in conditions describable as rubbery, leathery, or glassy, as is common for polymer films. The unusual local structure of the molecules in the liquid crystalline phase gives rise to magnetic and optical properties heretofore not obtainable with polymeric systems.

Fifteen years ago it was observed by Doty et al.[1] that synthetic polypeptides, $-(NH-CHR-CO)_{n}-$, in solution can exist in a rigid rodlike α-helical conformation, in contrast with the random coil shape assumed by most other synthetic polymers in solution. The observation has stimulated a large body of investigation of the dilute solution properties of this class of polymers. In more concentrated solutions (in the range of 10–50% polymer), Robinson[2] found that poly-γ-benzyl-L-glutamate (PBLG; $R = CH_2CH_2CO-O-CH_2C_6H_5$), a readily available synthetic polypeptide, forms a lyotropic liquid crystalline phase. This means that, locally, the PBLG molecules are arranged relative to one another in a very specific manner and this order extends to macroscopic dimensions throughout the solution. The liquid crystalline phase exists for melts of pure substances (thermotropic liquid crystals) as well as for solutions (lyotropic liquid crystals). The solutions that Robinson studied were quite fluid as are the liquid crystal phases of smaller molecules whether thermotropic or lyotropic. The molecular arrangement in the liquid crystalline phase of PBLG is similar to the helicoidal structure found in the liquid crystalline phase of many pure cholesterol esters and is easily recognized with a polarizing microscope. We refer to this type of structure found by

531

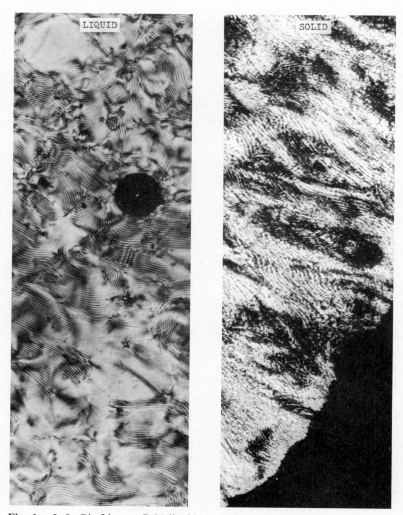

Fig. 1.   *Left:* Birefringent fluid liquid crystalline solution of poly-γ-benzyl-L-glutamate in *m*-cresol. *Right:* Birefringent solid film of poly-γ-benzyl-L-glutamate plasticized by 3,3'-dimethyl bisphenyl. Retardation lines characteristic of a helicoidal supramolecular structure are observed in the photomicrographs of both the liquid and solid states of this synthetic polypeptide.

Robinson as "cholesteric." When viewed between crossed polars, these birefringent solutions present an image very reminiscent of a fingerprint. The spacing between the alternating bright and dark retardation lines is equal to one-half of the pitch of the cholesteric structure (Fig. 1, liquid).

We have gone beyond the work of Robinson by casting *solid* films from various solvents and mixed solvents. These solid films are either pure PBLG or contain predetermined amounts of nonvolatile liquids, which act as plasticizers for the film.

Studies of X-ray diffraction patterns, specific volume, and mechanical properties of PBLG films led us to intuit that some of these solid films cast from certain solvents such as chloroform and methylene chloride retained the local structure of the liquid crystalline phase.[3] On the other hand, conventionally crystalline films of PBLG were obtained from the solvent dimethyl formamide.[4]

Consideration of the molecular structures of solvents which promoted the formation of the cholesteric liquid crystalline phase prompted the selection of the nonvolatile liquid 3,3'-dimethylbiphenyl as a plasticizer for the films. Using this plasticizer it is possible to obtain solid films even at relatively low concentrations of PBLG (less than 20% polymer). Controlled evaporation of chloroform solutions of PBLG + plasticizer resulted in solid films which retained the optical retardation lines characteristic of the cholesteric structure (Fig. 1, solid). This evidence is quite convincing that the unusual supramolecular arrangement of the liquid crystalline phase does exist in the solid state of mixtures of PBLG and plasticizer. When solid cholesteric films (with or without plasticizer) are cast from solvents such as $CHCl_3$ or CH_2Cl_2, X-ray evidence and anisotropic swelling characteristics clearly indicate that the PBLG rods lie in the plane of the film, but with no preferred direction in this plane.[6]

The fact that these films are solid with regard to mechanical properties but liquid crystalline in structure is different from the fluidity heretofore associated with liquid crystalline phases. This phenomenon probably results from the high molecular weight of the PBLG molecules.

The nuclear magnetic resonance (NMR) spectra of solute molecules in fluid nematic liquid crystal solvents show additional splittings due to direct dipole–dipole coupling. Three sets of workers (Samulski and Tobolsky,[6] Sobajima,[7] and Panar and Phillips[8]) found independently that in concentrated solutions of PBLG + CH_2Cl_2 the NMR spectrum of CH_2Cl_2 was split into a doublet. It appears that the cholesteric structure of the PBLG solution is untwisted by the magnetic field to form a nematic liquid crystalline structure in which the rodlike molecules are oriented parallel to the field. Splitting of the NMR absorption of CH_2Cl_2 molecules in this kind of environment is to be expected.

We found, in addition, that cast films of PBLG containing 5–20% CH_2Cl_2 (films that are solid) showed splitting of the CH_2Cl_2 absorption. The change in the NMR spectra of the swollen films with the orientation of the film in the field also indicates that the helical molecules in the solid film are randomly oriented in the plane of the film.[5]

It has been known for some time that magnetic fields of the order of

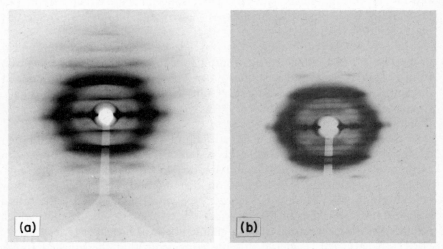

Fig. 2. CuK_α radiation, 200-μ collimator, fiber axis perpendicular to cylindrical camera axis and incident beam. Meridian is vertical direction: (*a*) PBLG cast from $CHCl_3$; (*b*) PBLG cast from CH_2Cl_2.

several hundred oersteds cause spontaneous ordering of liquid crystals. The local order present in the liquid crystalline phase allows the magnetic field to interact with the magnetic anisotropy of a large number of molecules in a cooperative manner, causing orientation of the direction along which the diamagnetism is the smallest parallel to the field. In fluid liquid crystals the orientation disappears rapidly when the magnetic field is removed.

Samulski and Tobolsky[6] and Sobajima[7] independently discovered that when films of PBLG are cast in the presence of a strong magnetic field they are highly oriented. Orientation occurs while the solutions are in the fluid liquid crystalline phase and becomes permanently locked in when the mixture of solvent plus PBLG becomes solid. We found, in fact, that the molecules could be oriented in any given direction in the plane of the film or even perpendicular to the plane of the film.[6] X-ray diffraction patterns from these films are very similar to the fiber patterns obtained from mechanically oriented fibers of PBLG. Figure 2a shows the diffraction pattern of an oriented film cast from chloroform in a magnetic field of 5000 Oe. The strong layer line at 10.4 Å and the absence of other layer lines between the equator and the turn layer line clearly demonstrates that the helical parameters are 7 residues in 2 turns (3.5 residues per turn) in contrast to the "normal" α-helix, 18 residues in 5 turns (3.6 residues per turn). The deformation of the normal α-helix to give a conformation with 3.5 residues per turn was postulated earlier by Tsuboi et al.[9] to permit sidechain–sidechain interactions between neighboring helices and thereby account for the diffraction pattern of a 50:50 mixture of D and L polybenzyl glutamate. The normal α-helical diffraction pattern shown in Fig. 2b with 3.6 residues per turn is observed in oriented films cast from methylene chloride. These observations suggest that small differences in molecular conformation present in different solvents might be detected by utilizing this new technique for orienting synthetic polypeptides.

Solid "liquid crystal" films of synthetic polypeptides present a new and very interesting phase of matter. It is quite possible that other rodlike molecules which exhibit a liquid crystal phase such as DNA can be induced to form in this same state, in vitro or in vivo.

ACKNOWLEDGMENT

We wish to thank Dr. Yukio Mitsui for his valuable discussions of the X-ray diffraction patterns.

REFERENCES

1. P. Doty, A. M. Holtzer, V. H. Bradbury, and E. R. Blout, *J. Amer. Chem. Soc.*, **76**, 4493 (1954).
2. C. Robinson, *Molecular Crystals*, **1**, 467 (1966) and references cited therein.
3. A. J. McKinnon and A. V. Tobolsky, *J. Phys. Chem.*, **70**, 1453 (1966); E. T. Samulski and A. V. Tobolsky, *Nature*, **216**, 997 (1967).
4. A. J. McKinnon and A. V. Tobolsky, *J. Phys. Chem.*, **72**, 1157 (1966).
5. E. T. Samulski and A. V. Tobolsky, "Liquid Crystals and Ordered Fluids," pp. 111–121, Plenum Press (1970).
6. E. T. Samulski and A. V. Tobolsky, *Macromolecules*, **1**, 555 (1968).
7. S. Sobajima, *J. Phys. Soc. Japan*, **23**, 1070 (1967).
8. M. Panar and W. D. Phillips, *J. Amer. Chem. Soc.*, **90**, 3880 (1968).
9. M. Tsuboi, A. Wada, and N. Nagashima, *J. Mol. Biol.*, **3**, 705 (1961).

The Thermal Stability of Collagen: Its Significance in Biology and Physiology

B. J. RIGBY

Division of Textile Physics, CSIRO, Ryde, Sydney, Australia

A Personal Note

My first contact with Henry Eyring came as a result of his interest during the forties and fifties in the mechanical properties of textile fibers at the Textile Research Institute, Princeton.

I had come from Australia, where I had been studying the mechanical properties of wool (and where there are about 10 times as many sheep as there are humans) to Utah (where there are about 10 times as many humans as there are sheep) to learn something about the theory of rate processes and its application to visco-elastic problems. Eyring and co-workers had developed the hyberbolic sine law of flow and applied it to many systems—both solid and liquid—and I had expected to find this kind of work going on full pace.

But this was not quite the case, for, as I was soon to learn, Henry Eyring was interested in everything from science to religion (and, in fact, applied rate theory to everything from science to religion), and all these things were being pursued actively.

So, noticing that Eyring included biology among his interests, I took the opportunity to expand my interests into biology, as I had always wanted to do, and decided to study the mechanical and physical properties of the biological fiber collagen. That is how Eyring started me off in the collagen field.

There are many fond memories of things Eyring said and did during my stay in Utah, but a remark which always comes to mind, because it illustrates the way in which he reacts with people and which I believe is the reason for his success, is the following: When assessing a person's worth we should measure not a scalar product as is usually the case but a vector product.

INTRODUCTION

Collagen is one of the most abundant and widespread proteins in the animal kingdom. It is the basic fibrous constituent of tendon, skin, cartilage, and bone, and occurs in practically every organ of the body. In its native state it is maintained in equilibrium with tissue fluids, the weight fraction of collagen being about 0.3. It is an extracellular protein and once formed is metabolically stable in most sites; this is in contrast to other body proteins which show a continuous turnover of molecules.[1]

Collagen is the supporting framework protein of the body, and since in humans it comprises about one-third of the total mass of protein, its importance in physiology and pathology is obvious.

THERMAL TRANSITION IN DILUTE SOLUTIONS

Vertebrate collagens, and perhaps all collagens, whether vertebrate or invertebrate which are manufactured by fibroblasts, have as their basic unit the tropocollagen molecule. This molecule, which consists of three polypeptide chains wound into a triple helix of approximate dimensions 2800 × 14 Å and molecular weight 300,000 packs into a quarter length staggered arrangement to give rise to the 600–700 Å banding pattern seen on fibrils in the electron microscope.[2] However, the cuticular collagens, which are apparently secreted by epidermal cells, have different molecular characteristics and do not give rise to the 600–700 Å pattern.[3,4]

Nevertheless, solutions of molecules can be prepared from both types of tissue by extraction with weak acids or neutral salts. Structural changes due to heat can be conveniently followed in these solutions by observing changes in viscosity and optical rotation. The underlying assumption in such studies is that the molecules are free of interactions with one another in contrast to the close interaction in the solid native state. For most soluble collagens the specific rotation $|\alpha|_D$ and intrinsic viscosity $|\eta|$, of dilute neutral solutions are almost temperature independent, until a particular value is reached at which they both decrease catastrophically over a range of a few degrees. For the majority of collagens $|\alpha|_D$ and $|\eta|$ change, during this transition, from about $-400°$ to $-125°$ and 13 dl/g to 0.4 dl/g, respectively.[2] These changes are consistent with the melting of the rodlike molecules to randomly coiled spherical structures. The reverse transition can be made to take place more or less completely depending upon the particular collagen system. The transition can thus be described as a first-order phase transition, or helix-coil melting. The melting point of the collagen molecule is defined as the temperature at the midpoint of the transition and is denoted by T_D.

MAJOR TRANSITIONS IN THE NATIVE STATE

If a specimen of collagen is taken directly from a warm-blooded animal (e.g., rat tail tendon) and slowly heated in 0.9% saline there is little evident change before 60°C, when the specimen quickly shrinks to about one-quarter of its original length. This is the most obvious and well known of the transitions in bulk collagen, and is denoted by T_S signifying shrinkage temperature. Confusion sometimes has arisen in the literature concerning the value of T_S for a given collagen. This is often due to excessive rates of heating or insensitive methods of detection. An apparatus which gives consistent results is shown in Fig. 1. The sample is clamped at its ends. One clamp is attached to the moving head of an extensometer, and the other is attached to a force transducer. At the beginning of a test the extensometer is adjusted so that the sample is just taut and the transducer output indicates a slight force. The liquid in which the sample is immersed is then heated at a constant rate. Figure 2 shows the result of such an experiment for kangaroo tail tendon in saline solution. The shrinkage at 58°C is very evident. As will be shown later, an adaption of this method also allows one to determine T_D.

As we have already mentioned, this shrinkage phenomenon is a manifestation of a first-order phase change in the collagen—it is the macroscopic result of crystalline aggregates of molecules melting and retracting into random coils. However, it is not the thermodynamic melting point of the collagen crystals. This temperature, defined by T_m, can only be measured when crystalline and melted regions coexist in the sample, and, as shown by the work of Oth et al.,[5] T_m is as much as 10°C lower than T_S for rat tail tendon. Flory and Garrett[6] have shown, using the system collagen–ethyleneglycol that T_m obeys the relation for the variation in melting point of a polymer with a volume fraction v_1 of added diluent, namely,

$$\frac{1}{T_m} - \frac{1}{T_m^0} = \frac{R}{\rho_2 V_1 \Delta H} (v_1 - X v_1{}^2),$$

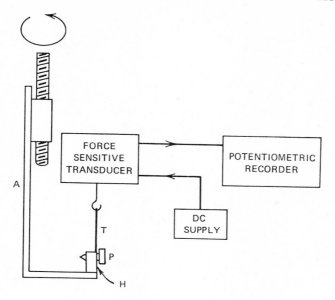

Fig. 1. Schematic apparatus used for controlling the length of tendon, T, or other collagen sample and for measuring the force developed when it contracts. Here A is a movable arm and H a metal holder into which the two free ends of the looped tendon are held by the plug P.

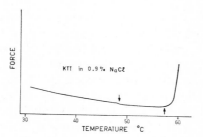

Fig. 2. A force–temperature curve for kangaroo tail tendon immersed in 0.9% saline. The tendon is mounted in the apparatus of Fig. 1. The arrow pointing down indicates a glass transition, and that pointing up indicates the shrinkage temperature T_S.

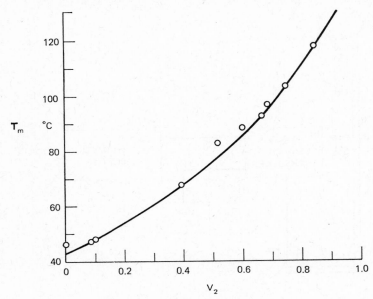

Fig. 3.   The melting temperature T_m plotted as a function of volume fraction of collagen V_2 for collagen–ethyleneglycol mixtures. The solid line is calculated from theory (after Flory and Garrett, Ref. 6).

where T_m^0 is the melting point of the undiluted polymer, ρ_2 the density of the amorphous polymer, v_1 the molar volume of the diluent, X the interaction constant, and ΔH the specific heat of fusion of crystalline polymer.

They used a wide range of concentrations extending right down to dilute solutions, and the relation between T_m and the volume proportion of collagen, V_2, is shown in Fig. 3. It is clear that there is a single melting process involved and that the actual value of T_m depends solely on V_2. Flory and Garrett[6] consider that the shrinkage temperature T_S is an artifact produced by "superheating." Nevertheless, although T_S cannot be considered a thermodynamic quantity, its reproducibility for collagens of quite different macrostructure (e.g., skin, tendon, and even cartilage taken from the same animal) is remarkable. This fact deserves more study.

OTHER TRANSITIONS IN THE NATIVE STATE

It is clear from the above discussion that the shrinkage of native collagen is due essentially to the melting of molecular collagen but that T_S is greater in value than T_D because both inter- and intramolecular forces must first be overcome. Further, as the work of Flory and Garrett[6] has shown, T_S (or T_m) may be lowered by "dilution" of the system, that is, by reducing molecular interactions.

One might expect that native collagenous structures, then, would exhibit no structural transition until either T_m or T_S is reached. However, there is evidence that native material does show small transitions at temperatures below T_S and T_m. These have usually been overlooked since they require sensitive techniques for their detection. We will now discuss this evidence.

Flory and Garrett[6] observed reversible glass transition behavior between 40 and 45° for beef achilles tendon collagen in water, and ascribed this to rotational motion in side groups. Mason and Rigby[7] have followed the thermal expansion of kangaroo tail tendon in 0.9% saline in both the native state and then after partial melting (Fig. 4). The main melting (corresponding to shrinkage) is clearly seen on both curves. On the curve for native material a small increase in expansivity occurs at about 40°C, but after melting has

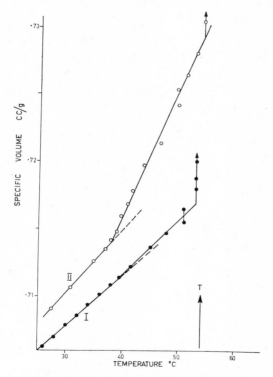

Fig. 4. Thermal expansion of kangaroo tail tendon in 0.9% saline. (I) Native material. (II) Partially melted material.

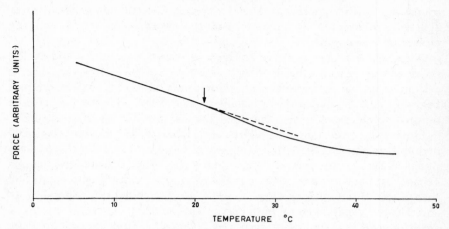

Fig. 5. The change in force, with temperature, of earthworm body wall collagen (*Digaster longmani*). In this experiment and that shown in Fig. 6 the sample was extended by about 1% and the force allowed to relax for 24 hr before heating was begun. The rate of heating for *Digaster longmani* was 6°C/hr. The arrow indicates structural transition.

taken place and the experiment repeated, a much larger increase occurs at the same temperature.

In the two experiments just described the sample was not subject to any mechanical stress. Characteristic temperatures have also been found at which abrupt changes occur in the mechanical behavior of collagen which *is* under stress. Thus Rigby et al.[8] found that the rate of force relaxation in rat tail

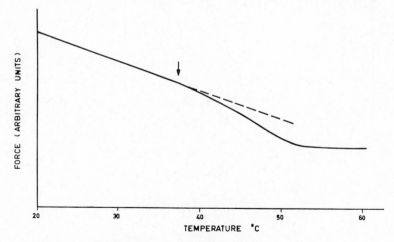

Fig. 6. Similar experiment to that described in Fig. 5 for *Ascaris* body wall collagen. The rate of heating was 15°C/hr.

tendon fibers extended by 1% in 0.9% saline at different temperatures increased rapidly at about 40°C. More recently (Rigby, unpublished data), transitions have been observed in extended samples in which the relaxation mechanisms have been allowed to reach equilibrium or to be eliminated before heating is begun. Two examples are shown in Figs. 5 and 6.

The interesting feature of all these results is that the lower transition temperature observed (T_g, after glass transition) is very close to T_D for dilute solutions of the same collagen where these have been measured. Thus, T_g and T_D for mammalian collagens (rat, beef) occur between 35 and 40°C.

Another example[7] of transitions in bulk material which have been observed using the apparatus of Fig. 1 is shown in Fig. 7. This simple experiment exhibits all three transitions so far discussed (i.e., T_s, T_m, and T_g). It is performed by noting the changes in force with temperature in a sample which

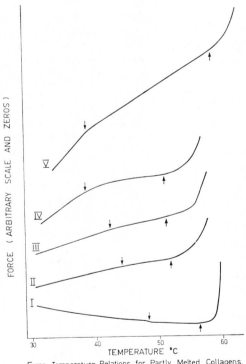

Force Temperature Relations for Partly Melted Collagens.
↓ Location of Tg ;   ↑ Location of Ts

Fig. 7. Force–temperature relations for partially melted tendon in 0.9% saline. Arrow pointing down indicates location of T_g and arrow pointing up indicates location of T_S.

is held at its natural length in an extensometer as described above. Curve I is for the native material, and the slight kink in the curve at 48°C represents T_g, whereas T_s is represented by the sudden large increase in force at about 57°C. Curves II, III, and IV were obtained on the same specimen in succession during the course of a day. It can be seen how both transition points are progressively reduced to approximately 39 and 51°C, respectively. After completing curve IV, the specimen was relaxed in the saline solution for 48 hr at room temperature and then retested. The resulting curve V shows that the shrinkage temperature had returned close to its original value, but that the lower transition point remained at 39°, the value produced by the successive heat treatments. In this experiment T_s, T_m, and T_g have the values 57, 51, and 39°C, respectively.

At this stage, it is appropriate to mention a method which determines a transition temperature for a wide range of collagens which in turn correlates closely with T_D measured on dilute neutral solutions. It is simply the force–temperature method already discussed except that the sample is first soaked in HCl at pH 1 and tested in this medium.[9] As is well known, the sample will swell reversibly in this medium at low temperatures. When the system is now heated a shrinkage phenomenon occurs that is quite analogous to that which takes place in saline at T_s.

We have called this temperature T_T. The method depends upon the large swelling of the system to break down interaction between molecules and groups of molecules. The system is apparently sufficiently close to dilute solution conditions to behave as such as far as molecular transitions are concerned. It should be pointed out that the pH is not crucial (pH 1–2) and that in fact any swelling agent which does not "melt" molecules serves equally well (e.g., $3M$ urea). This interpretation will be strengthened in the next section, where in fact it will be shown that T_T can also be equated to T_D.

Finally, we must offer some reason for the occurrence of the transition found by Flory and Garrett,[6] Mason and Rigby,[7] and Rigby which has been referred to as T_g. As we have said, for those collagen for which T_g has been determined, it is close to T_D for the same collagen, that is, to the value for the molecular melting temperature. For this, and other reasons that will become apparent later, we in fact suggest that T_g does indicate molecular melting. This situation could arise if there were a distribution of crystallite sizes, one end of the distribution containing single molecules or at least small groups of molecules that can melt at the dilute solution temperature.

CORRELATIONS BETWEEN THERMAL TRANSITIONS AND AMINO ACID COMPOSITION

So far we have discussed the thermal properties of collagen in a general way. We now give a list of the actual values of T_S, T_D, and T_T for a wide

Table I.   Proline + Hydroxyproline Unit in Residues per 1000 Total[a]

Sample	Proline + Hydroxyproline	T_S	T_D	T_T
1. RTT and skin	223	59	36	36
2. Calf skin	232	59	36	35
3. Maigre skin	200	48	30	26
4. Dogfish skin	166	41	16	19
5. Cod skin	155	37	16	12–16
6. Earthworm (*Lumbricus*) cuticle	170	40	22	22
7. *Ascaris* cuticle	295	59	52	40
8. Pike skin	199	45	27	24
9. Beef tendon	216	60	—	36
10. Human skin and tendon	218	60	—	37
11. *Helix aspersa*	204	52	—	29
12. Tunafish skin	188	49	—	26
13. *Ascaris* body collagen	219	59	—	38
14. *Melarapha unifasciata*	175	56	—	22
15. *Nodilittorina pyrimidalis*	176	59	—	22
16. *Helicella virgata*	199	59	—	29
17. Carp skin	197	49	—	30
18. Perch swim bladder	199	—	31	—
19. Shark skin	192	—	29	—
20. Cod swim bladder	160	—	16	—
21. Carp swim bladder	190	—	32	—
22. Seagull leg tendon	—	59	—	36
23. Octopus skin	—	47	—	26

[a] Temperatures are in °C.

range of collagen and the relationships that have been found between these and certain of the amino acid residues contained in the molecule, namely, proline, hydroxyproline (Table I).

The relation between the imino acid residues proline and hydroxyproline and thermal stability has been known for sometime. In the earliest studies of Gustavson[10] using vertebrate collagens, where the ratio of the content of proline and hydroxyproline is a constant (ca. 1.2), the correlation between T_S and hydroxyproline alone was considered, for it was assumed that extra stability associated with increased hydroxyproline was due to its hydrogen bonding capacity. It was natural to overlook the equally good correlation between T_S and proline, which would follow from the constancy of the ratio pro:hyp. However, as the range of collagen was extended to include invertebrate samples (especially earthworm and *Ascaris*, the latter a parasitic round worm) where the ratio of pro:hyp as well as the absolute amounts of each acid were found to vary greatly, the only relation that would now work for all collagens was a plot of T_S versus the sum of proline and hydroxyproline. (For comprehensive references, see Ref. 11.)

The underlying reason for this correlation appears to be the observation of Harrington and Sela[12] that poly-L-proline can exist in solution as a helical single chain, and that this structure is apparently stabilized by restriction on the rotation of the proline ring about the peptide backbone. Von Hippel and Harrington[13] suggested that the collagen would be stabilized in the same way in regions where there are contiguous imino acid residues.

Recent work, however,[11] has shown that a number of collagens do not fit even the relation T_S versus proline + hydroxyproline. In fact, the cuticle collagen of a number of worms has an amino acid composition quite different from that of their body collagen and yet both collagens have the same values of T_S and T_D. At this stage, it would seem that although proline and hydroxyproline are involved in the stabilization of the collagen molecule for the reasons mentioned many other factors must play a part also; for example, such factors could include (a) the order of amino acid residues along molecular

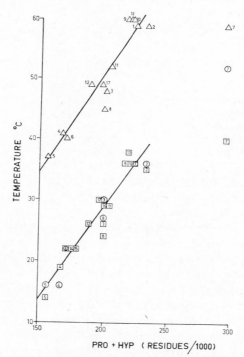

Fig. 8. Relation between T_S ($\triangle$), T_D ($\bigcirc$), T_T ($\square$), and the total imino acid content of various collagens. The numbers refer to samples listed in Table I. Sample 7 was not included when calculating regression lines. See Ref. 9 for details.

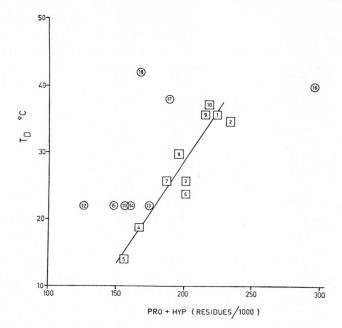

Fig. 9. Relation between T_D (the melting point of tropo-collagen) and the sum (pro + hyp). Data for vertebrate collagens are represented by □, and the line has been drawn for these points. Data for worm collagens are represented by ○. Sources of collagen were as follows: (1) Rattail tendon and skin. (2) Calf skin. (3) Maigre skin. (4) Dogfish skin. (5) Cod skin. (6) Pike skin. (7) Tuna skin. (8) Carp skin. (9) Beef tendon. (10) Human skin and tendon. (11) *Pheretima* cuticle. (12) *Digaster* cuticle. (13) *Digaster* body. (14) *Allolobophora* cuticle. (15) *Allolobophora* body. (16) *Ascaris* cuticle. (17) *Ascaris* body. (18) *Macracanthorpynchus* body.

chains,[14] (b) as yet unidentified crosslinks, and (c) the differences in composition among the three chains of the molecule,[15] since the separate chains would presumably differ in stability. Further, there is the curious negative correlation that has been found between thermal stability and serine.[16] This correlation is not the result of a correlation between serine and total imino acid, and even if it were it would not necessarily constitute an explanation for the correlation.

Figure 8 shows the relations between T_S, T_D, and T_T and total imino acid content for a range of collagen which obeys the correlation. It serves to illustrate the following points:

1. There is a constant difference between T_S and T_D of about 22°C, which allows the prediction of T_D from the easily measured T_S.

2. There is good agreement between T_D and T_T where these have been measured on the same collagen, and the agreement is strengthened when it is realized that the regression lines for each set of points (i.e., T_D and T_T versus pro + hyp) are identical within experimental error. In other words, T_T may be identified with T_D.

As stated above, there are exceptions to the linear plot of Fig. 8. For clarity these deviations are shown in Fig. 9.

CORRELATION BETWEEN ENVIRONMENTAL TEMPERATURE AND THERMAL STABILITY

It will be apparent from the preceding section that T_D of the collagen of the warm-blooded animals (human, beef, rat, etc.) is close to that of their normal body temperature, namely, $\sim38°C$. On the basis of shrinkage temperature measurements, Gustavson[10] and Leach[17] suggested some time ago that the thermal stability of various cold-blooded animals correlates with the temperature of their habitat. Thus cold-water fish had much lower values for T_S than warm-water fish, which were in turn lower than those for mammals. Because of the relation $T_S - T_D = $ const, it is clear that T_D for the collagen of all these animals would also correlate with the temperature of their habitat. As we have just stated, T_D for all mammalian collagens so far examined is close to that of their normal body temperature, and this leads to the question as to whether T_D of the collagens of cold-blooded creatures bears a similar relation to their environmental temperature. For the examination of such a relation the appropriate temperature would appear to be "the upper limit of the environmental temperature." By this definition, mammalian collagen environmental temperature remains fixed at body temperature.

Naturally, "the upper limit of the environmental temperature" or, for convenience, the environmental temperature, is unknown or at least difficult

Table II. Melting Temperature in Physiological Saline (°C) of the Collagen of Various Worms[a]

	P. mega-scolidiodes	Digaster longmani		A. caliginosa		As. lumbricoides		Macracan-thorynchus
	C	C	B	C	B	C	B	B
T_S	37	34	40	34	40	56	59	62
T_D	22	22	22	22	22	40	38	42

[a] T_S is the melting point of bulk, native sample, T_D the melting point of molecular unit. B, body. C, cuticle.

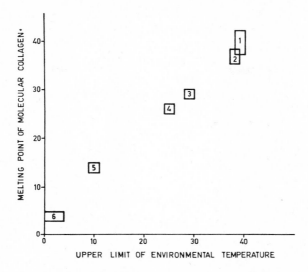

Fig. 10. The relation between the melting point (T_D) of the collagen molecule from various animals and the approximate temperature of the upper limit of their environment. (1) *Macracanthorynchus*, *Ascaris*, and hog; (2) rat, human, cow; (3) snail (*Helix aspersa*); (4) tuna skin; (5) cod skin; (6) Antarctic "ice-fish." See Ref. 18 for details.

to measure for many creatures. However, information in existence for a number of cold-blooded creatures does support the above notion. First we have the data on the collagens of two parasitic worms and their host: *Ascaris*, *Macracanthorynchus*, and the hog.[11] *Ascaris* contains two distinct collagens (cuticle and body wall) which are different in amino acid composition from each other and from that of *Macra* and the hog itself, yet each have similar values for both T_S and T_D and furthermore T_D is close to that of the body temperature of the hog.

A similar situation holds for a number of earthworms (see Table II), that is, each has two distinct collagenous proteins, yet as a group T_S and T_D are the same and T_D has a value (22°C) that can be shown to be consistent with the upper temperatures of the normal habitat of the animal.[18] Data culled from the literature for the environmental temperature of a diverse range of animals for which we have thermal data allows us to construct Fig. 10, which shows a good correlation. In particular, we have information for the skin collagen of the antarctic "ice-fish," which lives in the very narrow range of temperature −1 to +3°C and which falls on the straight line.

PHYSICAL BEHAVIOR AND THERMAL STABILITY

Now we discuss a number of alterations in physiological behavior of animals which occur suddenly at temperatures that coincide with the value of T_D for their collagen. One example is shown in Fig. 11 from the work of Kim[19] on the body temperature of the worm *Pheretima megascolidioides*. We see that there is an abrupt change in the body temperature of the worm at 23°C. From other evidence as well as this observation, Kim concluded that the upper limit of the worm's preferred temperature environment was 23–24°C. As Table II shows, T_D of the collagen of this worm is 22°C.

Second, Hatai[20] has shown that for several Japanese earthworms the rhythmic muscular contraction is normal and unaffected by heating again, up to a temperature of 23°C, beyond which the form of contraction alters suddenly.

Another physiological characteristic of animals that is frequently measured is the upper lethal temperature. The definition of this temperature is some-what arbitrary, and the values quoted for a given animal are thus widely spread. Assuming that the upper lethal temperature depends only upon temperature, it would seem that the upper lethal temperature could be defined by extrapolation of a time (for a given fraction of population to die) versus temperature relation from high temperatures, until time and temperature became independent. In other words, above the lethal temperature, time and temperature would be approximately inversely proportional; below, no relation would be evident.

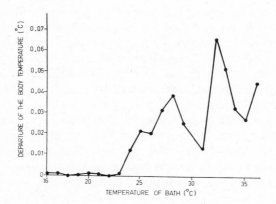

Fig. 11. Deviation in body temperature of worm from temperature of bath in which it is immersed. See Ref. 19.

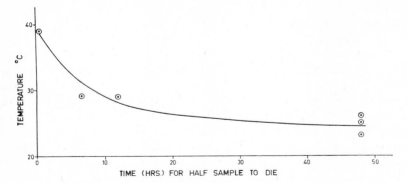

Fig. 12. The lethal temperature as defined by various authors (see text) plotted against time of exposure of animal population. It is suggested by the present author that the asymptotic value of temperature should be taken as the true lethal temperature.

The data from a number of authors (see Ref. 18) for such an experiment with earthworms have been plotted in Fig. 12. It can be seen that the curve is almost independent of time in the region of 24°C. All these worms have T_D values of 22°C. Our inference is that 22–24°C represents the true upper lethal temperature and that the high values of 29 and 39°C were obtained because the time of the experiment was too short. The results of Kim[19] and Hatai[20] discussed earlier are consistent with the conclusion that the upper lethal temperature is just above T_D. It appears that irreversible changes have begun to occur in the worm's tissues, but long times may still be required before the animal dies.

As a final example of physiological alterations with temperatures, we note the work of Newell.[21] He found that the oxygen uptake of various poikilotherms increases abruptly at temperatures that correspond to the upper limit of the environmental temperature and in one case for which collagen has been isolated (*Helix aspersa*) to the T_D of this collagen, namely, 27°C.[11]

Summarizing this section, we can say that for a number of animals, the melting point of their molecular collagen is very close to the upper limit of their preferred temperature environment, and in turn to the upper lethal temperature. Again, abrupt changes are noted in physiological and behavioral changes at this temperature. Whether or not collagen is uniquely involved in determining these limits cannot be said at this stage, although the fact that it is a protein with a very low rate of turnover (with certain exceptions[1]) and a sharp melting point is perhaps significant. It should be pointed out that probably other animal proteins begin to denature at this temperature also.

The results are direct evidence of a specific protein having easily measurable thermal properties that reflect thermal properties of the whole animal.

SOME SPECULATIONS

One would usually invoke natural selection as the explanation for the fact that T_D is coincident with the upper limit of environmental temperature. One would assume that the thermal stability of the molecule was genetically determined. But when we consider that many poikilotherms can be acclimatized with respect to temperature (within limits, of course) we should ask whether T_D can be influenced by the environmental temperature. Since, as we have pointed out, collagen is a protein with sharp melting points, this idea is worth testing. The coincidence of these two temperatures may be part of the biophysical requirements for the growth and remolding of collagen.

Turning to ourselves, we can think of some consequences of the fact that the melting temperature of the human collagen molecule is very near to normal body temperature. Special conditions may exist, for example, if body temperatures are abnormally high for any reason, for example, fever, or if the collagen is low in proline and hydroxyproline, and most particularly if the tissue is under a mechanical force. Under such conditions, structural transitions may take place with serious physiological and pathological consequences (vis-à-vis worms). These observations are relevant to rheumatic fever, where prolonged fevers are encountered. It may be significant that in this disease damage commonly occurs to the tendons that assist in controlling the position of heart valves, the implication being that the combination of elevated temperature and tension causes an irreversible change in the mechanical properties of the tendon. Similar considerations would apply to any part of the body where mechanical forces are prolonged or excessive, e.g., joints. Naturally these changes may be exceedingly small, but we must remember that collagen has a low or even negligible turnover in most parts of the body so that deleterious changes in structure will tend to be retained. Thus damage caused to connective tissue may only become manifest after accumulating for many years.[22]

In conclusion, we may become even more speculative and suggest that, by virtue of its property of undergoing a precise structural transition at a temperature very close to normal body temperature, degeneration of connective tissue over long periods of time is the basic cause of physiological and pathological conditions which have found no explanation in conventional pathology; for example, aging and the rheumatic diseases.

REFERENCES

1. R. D. Harkness, Biological Functions of Collagen. *Biol. Rev.*, **36**, 399 (1961).
2. W. F. Harrington and P. H. Von Hippel, *Advan. Protein Chem.*, **16**, 1 (1961).

3. M. D. Maser and R. W. Rice, *Biochem. Biophys. Acta*, **63**, 255 (1962).
4. O. W. McBride and W. F. Harrington, *Biochemistry*, **6**, 1484 (1967).
5. J. F. M. Oth, E. T. Dumitru, O. K. Spurr, Jr., and P. J. Flory, *J. Amer. Chem. Soc.*, **79**, 3288 (1957).
6. P. J. Flory and R. R. Garrett, *J. Amer. Chem. Soc.*, **80**, 4836 (1958).
7. P. Mason and B. J. Rigby, *Biochem. Biophys. Acta*, **66**, 448 (1963).
8. B. J. Rigby, N. Hirai, J. D. Spikes, and H. Eyring, *J. Gen. Phys.*, **43**, 265 (1959).
9. B. J. Rigby, *Biochem. Biophys. Acta*, **133**, 272 (1967).
10. K. H. Gustavson, *Chemistry and Reactivity of Collagen*, Academic, New York, 1956.
11. B. J. Rigby, *Symp. Fibrous Proteins*, Butterworth, Australia, p. 217.
12. W. F. Harrington and M. Sela, *Biochem. Biophys. Acta*, **27**, 24 (1958).
13. P. H. Von Hippel and W. F. Harrington, *Biochem. Biophys. Acta*, **36**, 427 (1959).
14. J. Josse and W. F. Harrington, *J. Mol. Biol.*, **9**, 269 (1964).
15. P. Bernstein and K. A. Piez, *Science*, **148**, 1353 (1965).
16. B. J. Rigby, *Nature*, **214**, 87 (1967).
17. A. A. Leach, *Biochem. J.*, **67**, 83 (1957).
18. B. J. Rigby, *Biol. Bull.*, **135**, 223 (1968).
19. H. Kim, *Tohoku Univ. Sci. Rept. Biol.*, **5**, 439 (1930).
20. H. Hatai, *Japan. J. Zool.*, **1**, 10 (1922).
21. R. C. Newell, *Nature*, **212**, 426 (1966).
22. B. J. Rigby, B. Bloch, and P. Mason, *Med. J. Australia*, **1**, 46 (1966).

Kinetics of the Interactions of Formaldehyde, Acetaldehyde, and Acrolein with Rattail Tendon*

DWAYNE H. CURTIS† AND JOHN D. SPIKES

Department of Biology, University of Utah, Salt Lake City

* This work was supported in part by Grant No. FR07092 from the U.S. Public Health Service and by Contract No. AT(11-1)-875 from the U.S. Atomic Energy Commission.
† Present address: Department of Biological Sciences, Chico State College, Chico, California 95926.

557

INTRODUCTION

Tendons are composed largely of white fibrous connective tissue with the fibers regularly arranged parallel to the long axis of the tendon. The white fibers consist primarily of the protein collagen, together with small amounts of mucopolysaccharides and other materials. Collagen has an unusually high content of glycine, proline, hydroxyproline, and hydroxylysine (see Ramachandran, Ref. 1). The present evidence indicates that a collagen molecule is made up of three polypeptide chains, each in the form of a left-handed helix. These, in turn, are wound around one another in the form of a right-handed coil.[2,3] The stability of the molecule is maintained by hydrogen bonds between the three peptide chains,[1-4] as well as by ionic bridges[5] and ester and imide linkages.[6,7]

Tendon contracts or shrinks at some definite temperature when heated in aqueous physiological solutions. This temperature, known as the shrinkage temperature (T_s), is approximately $61 \pm 1°C$ for mammalian tendons. Küntzel suggested[8] that the thermal shrinkage of tendon represents a fusion of its crystalline state. In 1950 Thies and Steinhardt postulated[9] that tendon shrinkage results from the rupture of interchain crosslinks between fibrils by heat or by agents that compete with the valency forces of the crosslinks. Another mechanism, proposed by Weir and Carter,[10] suggested that breakage of the interchain crosslinks results when a large number of bonds open at the same instant so that the chains can proceed to a more stable configuration (the shrunken state) by chain folding. Later, Flory and Spurr[11] applied the experimental and theoretical methods developed for the study of polymers to the examination of tendon. They compared hydrothermal shrinkage of rattail tendon crosslinked with p-benzoquinone to the melting of a polymer. According to their interpretation, the tertiary structure of the stabilized peptide chain is altered to a more random configuration during shrinkage.

The properties of collagen can be altered by treatment with so-called tanning agents, such as short-chain aldehydes; among other changes, the T_s of the collagen is increased.[12] Meyer first suggested[13] that tanning might involve a crosslinking of the polypeptide chains in collagen. Tanning agents

559

generally possess two reactive groups. Simple aldehydes are an exception; however, once they have reacted with a protein, a second reaction can take place at the same site.[12] In the formaldehyde–collagen system, the reactive sites in the protein are the ε-amino groups of lysine residues.[14,15]

As for most chemical processes, the rate of reaction between short-chain aldehydes and collagen increases with temperature. The effects of temperature on biological reaction rates can be treated most elegantly by Henry Eyring's absolute reaction rate theory (see Ref. 16). The present paper presents an analysis of the effects of temperature on the reaction of acrolein, formaldehyde, and acetaldehyde with the collagen of rattail tendon as examined using absolute rate theory. The change in the T_s of the tendon with the time of treatment was used as a measure of the aldehyde–collagen reaction rate. These measurements were carried out to provide comparative data for a study of the effects of photodynamic treatment (dye-sensitized photo-oxidation) on the mechanical properties of tendon.[17]

MATERIAL AND METHODS

The shrinkage temperature of rattail tendon was determined using an improved version of the stretching apparatus described by Rigby et al.[18] Single fibers were used in this arrangement in contrast to the looped fibers required for the earlier apparatus. Tendons were obtained from the tails of male albino rats (Sprague-Dawley strain) as described by Rigby et al.[18] Tendons with a diameter of 0.20–0.25 mm, as measured with a microscope provided with an ocular micrometer, were selected and stored in rat Ringer solution at 2–3°C until used.

The shrinkage temperatures were measured by a technique similar to the isometric method described by Verzár and Brocas.[19] One-inch lengths of tendon were clamped in the stretching machine and elongated, while immersed in Ringer solution, until a small load of approximately 20 mg was indicated on the recorder. The temperature of the Ringer solution was then increased at a relatively constant rate of 2.5°C/min starting at room temperature. The shrinkage temperature, T_s, was defined as the minimal temperature at which the stretched tendon showed a sudden increase in force. With this technique, the shrinkage temperature for native rattail tendon was found to be 61 ± 1°C. For the aldehyde studies, groups of tendons were immersed with stirring in 0.10M solutions of the compounds in distilled water at pH 7. The solutions were maintained at 2.5, 14.2, 30.2, 37.5, and 46.0°C. Then, at intervals, tendons were removed, washed for 5 min in Ringer solution, and the T_s determined as described above.

RESULTS

Native rattail tendon underwent a marked change in appearance during hydrothermal shrinkage. The normal, glistening white sheen disappeared and the tendon became transparent, jelly-like, and sticky. Typically the tendon became so weak that it broke under its own weight. In contrast, aldehyde-treated tendons did not break on heating, even at 90°C and became gray rather than transparent. Such tendons did not become sticky, and showed a rubberlike elasticity. In all cases, the T_s was increased by treatment with aldehydes. These observations agree well with the studies of Gustavson[12] on aldehyde-treated hide.

At any given temperature, the three aldehydes reacted with tendon at rather different rates as measured by the rate of increase of T_s, with acrolein reacting most rapidly and acetaldehyde most slowly. In all cases, the rate of increase of T_s with time was large when the tendons were first placed in the aldehyde solutions, and then decreased progressively with duration of treatment. Over the temperature range studied, the rate of increase of T_s with time of treatment increased with increasing temperature. The results of typical experiments over an extended time range are shown in Fig. 1 for acrolein, Fig. 2 for formaldehyde, and Fig. 3 for acetaldehyde. The duration

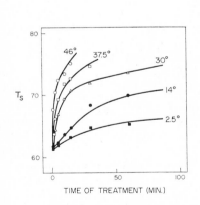

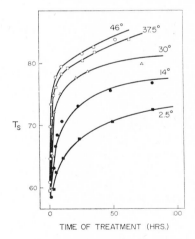

Fig. 1. Effect of temperature on the time course of the change in the shrinkage temperature (T_s) of rattail tendon in Ringer solution produced by treatment with 0.10M acrolein at pH 7.

Fig. 2. Effect of temperature on the time course of the change in the T_s of rattail tendon produced by 0.10M formaldehyde. Reaction conditions were the same as for Fig. 1.

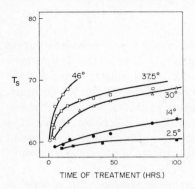

Fig. 3. Effect of temperature on the time course of the change in the T_s of rattail tendon produced by $0.10M$ acetaldehyde. Reaction conditions were the same as for Fig. 1.

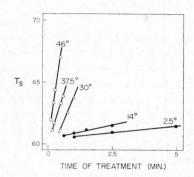

Fig. 4. Effect of temperature on the time course of the change in the T_s of rattail tendon produced by $0.10M$ acrolein during the initial, essentially linear, stage of the reaction. Reaction conditions were the same as for Fig. 1.

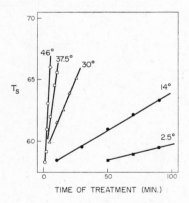

Fig. 5. Effect of temperature on the time course of the change in the T_s of rattail tendon produced by $0.10M$ formaldehyde during the initial, essentially linear, stage of the reaction. Reaction conditions were the same as for Fig. 1.

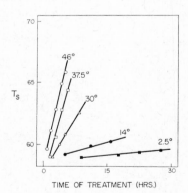

Fig. 6. Effect of temperature on the time course of the change in the T_s of rattail tendon produced by $0.10M$ acetaldehyde during the initial, essentially linear, stage of the reaction. Reaction conditions were the same as for Fig. 1.

of treatment was 100 min, 100 hr, and 100 hr respectively, in the three cases. It will be noted that the curves tend to plateau as the duration of treatment increases and that the T_s of the plateau level increases with the temperature at which the aldehyde treatment took place.

The curves for these long-term measurements, although they resemble a first-order process, deviate from such kinetics too much to permit derivation of a useful rate parameter for quantitative comparisons. It was found, however, that the initial change of T_s with time of aldehyde treatment was linear within experimental error. For this reason, the experiments described above were repeated, except that tendon samples were removed at very frequent intervals. The results of typical experiments of this type are shown in Fig. 4 for acrolein, Fig. 5 for formaldehyde, and Fig. 6 for acetaldehyde. The total duration of the experiments in this series was 5 min with acrolein, 100 min with formaldehyde, and 30 hr with acetaldehyde. As may be seen, the time course of increase in T_s is linear over these shorter periods. In this way the effects of temperature on the initial increase in T_s by aldehyde treatment could be compared quantitatively, even though rate constants for the overall reaction could not be obtained conveniently.

Rate constants, k_r, were calculated from the curves in Figs. 4–6 using a least-squares method and were expressed in terms of degrees increase in T_s per second of treatment with aldehyde. These rate data gave linear Arrhenius plots. Considerably more information can be obtained, however, by treating the data using absolute rate theory[16,20] according to the following equations:

$$k_r = \kappa \frac{kT}{h} e^{-\Delta F^{\ddagger}/RT}, \tag{1}$$

$$\Delta F^{\ddagger} = \Delta H^{\ddagger} - T\Delta S^{\ddagger}, \tag{2}$$

where k_r is the rate constant as defined above, κ is the transmission coefficient (which in reactions of this type can be taken as equal to unity), k is Boltzmann's constant, h is Planck's constant, R is the gas constant, T is the absolute temperature, $\Delta F^{\ddagger}$ is the free energy of activation, $\Delta H^{\ddagger}$ is the enthalpy of activation, and $\Delta S^{\ddagger}$ is the entropy of activation. The values of $\Delta F^{\ddagger}$ obtained with each aldehyde were plotted as a function of the absolute temperature as shown in Fig. 7. As may be seen, $\Delta F^{\ddagger}$ for the initial reaction of the aldehydes with tendon is a linear function of the absolute temperature in each case.

The values of $\Delta H^{\ddagger}$ and $\Delta S^{\ddagger}$, as defined in (2), were then calculated from the curves in Fig. 7. The slope of the curve is equal to $-\Delta S^{\ddagger}$, while $\Delta H^{\ddagger}$ was determined by extrapolating the curves to a temperature of $0°K$; $\Delta H^{\ddagger}$ is equal to the Y intercept at this point. A typical set of data as calculated from the curves in Fig. 7 is shown in Table I. As may be seen, $\Delta F^{\ddagger}$ is lowest for

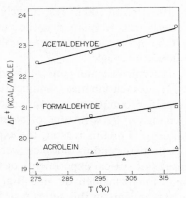

Fig. 7. Effect of temperature on the free energy of activation, $\Delta F^{\ddagger}$, of the reaction of $0.10M$ solutions of acrolein, formaldehyde, and acetaldehyde with rattail tendon as measured by effects on T_s.

acrolein treatment, intermediate for formaldehyde, and greatest for acetaldehyde. The differences, although small, are significant at the 1 % level. The values for $\Delta H^{\ddagger}$ go in the opposite direction, and the magnitudes of $\Delta S^{\ddagger}$ become progressively more negative in going from acrolein to acetaldehyde; the values of $\Delta S^{\ddagger}$ for the different compounds are significantly different at the 1 % level.

DISCUSSION

The thermal shrinkage of rattail tendon probably results from hydrogen bond breakage between the collagen fibrils during heating. Such a breakage results in a folding of the fibril chain, which is normally held in an extended

Table I. Thermodynamic Parameters^a for the Interaction of 0.10 M Solutions of Acrolein, Formaldehyde, and Acetaldehyde with Rattail Tendon as Measured by Changes in T_s

Compound	k_r	$\Delta F^{\ddagger}$	$\Delta H^{\ddagger}$	$\Delta S^{\ddagger}$	Standard Error of $\Delta S^{\ddagger}$
Acrolein	8.3×10^{-2}	19.3	17.1	-7.9	0.25
Formaldehyde	4.4×10^{-3}	20.8	16.6	-13.9	0.22
Acetaldehyde	1.6×10^{-4}	23.0	15.5	-25.1	0.16

^{a} $\Delta F^{\ddagger}$ and $\Delta H^{\ddagger}$ = kcal/mole; $\Delta S^{\ddagger}$ = e. u./mole; k_r = deg/sec.

state. When tendons are treated with crosslinking agents such as aldehydes, more heat energy is required to break the newly formed bonds, and thus T_s is increased.

Some observations may be made on the possible mechanisms of the interaction of aldehydes with tendons leading to the observed changes in T_s. The work of French and Edsall[21] indicates that aldehydes react with collagen primarily at the terminal amino groups of the lysine residues. Acrolein would be expected to react more rapidly with these groups than the other aldehydes, since the double bond has a negative inductive effect on the aldehyde functional group. Groups that attract electrons more strongly than hydrogen are said to exhibit negative inductive effects, while groups that are less powerful electron attractors than hydrogen are said to display positive inductive effects.[22] Acetaldehyde would be expected to react more slowly because of the positive inductive effect of its methyl group. Thus, based on a consideration of inductive effects, acrolein would be expected to react most rapidly with tendon, formaldehyde somewhat more slowly, and acetaldehyde most slowly. This agrees with the experimentally determined values for k_r and for $\Delta F^{\ddagger}$ (see Table I). It is assumed that the more negative the $\Delta S^{\ddagger}$ for a reaction, the more rigid the activated complex will be as compared to the ground state, and therefore the slower the reaction. Again, this is borne out in the present study since the values for $\Delta S^{\ddagger}$ with acrolein, formaldehyde, and acetaldehyde are -7.9, -13.9, and -25.1 e.u./mole, respectively (Table I). For all three aldehydes, the activation enthalpies are large, whereas the values for entropy are large in the negative direction. This suggests that the interaction of aldehydes with tendon is dominated by the chemical reaction of aldehyde with collagen, since, for most conformational processes, enthalpy and entropy both change positively or both change negatively.

ACKNOWLEDGMENT

We wish to thank Professor Rufus Lumry of the University of Minnesota for helpful discussions in connection with this work. As he pointed out, the present paper is typical of the kinds of studies which have been stimulated by the activities of Henry Eyring in the field of biology.

REFERENCES

1. G. N. Ramachandran, Ed., *Chemistry of Collagen*, Academic, New York, 1967.
2. A. Rich, and F. H. C. Crick, *J. Mol. Biol.*, **3**, 483 (1961).
3. G. N. Ramachandran, in *Aspects of Protein Structure*, G. N. Ramachandran, Ed., Academic, New York, 1963, p. 39.

4. W. F. Harrington, *J. Mol. Biol.*, **9**, 613 (1964).
5. K. H. Gustavson, *The Chemistry and Reactivity of Collagen*, Academic, New York, 1956.
6. P. M. Gallop and S. Seifter, in *Collagen*, N. Ramanathan, Ed., Interscience, New York, 1960, p. 249.
7. O. O. Blumenfeld and P. M. Gallop, *Biochemistry*, **1**, 947 (1962).
8. A. Küntzel, *Stiasny Festschrift*, Roether, Darmstadt, 1937, p. 191.
9. E. R. Theis and R. G. Steinhardt, Jr., *J. Am. Leath. Chem. Assoc.*, **45**, 591 (1950).
10. C. E. Weir and J. Carter, *J. Res. Natl. Bur. Std.*, **44**, RP 2106, 599 (1950).
11. P. J. Flory and O. K. Spurr, Jr., *J. Amer. Chem. Soc.*, **83**, 1308 (1961).
12. K. H. Gustavson, *The Chemistry of Tanning Processes*, Academic, New York, 1956.
13. K. H. Meyer, *Biochem. Z.*, **208**, 23 (1929).
14. J. H. Bowes and C. W. Cater, *Biochim. Biophys. Acta*, **168**, 341 (1968).
15. K. H. Gustavson, *Das Leder*, **13**, 233 (1962).
16. S. Glasstone, K. Laidler, and H. Eyring, *The Theory of Rate Processes*, McGraw-Hill, New York, 1941.
17. J. D. Spikes and R. Livingston, *Advan. Radiation Biol.*, **3**, 29 (1969).
18. B. J. Rigby, N. Hirai, J. D. Spikes, and H. Eyring. *J. Gen. Physiol.*, **43**, 265 (1959).
19. F. Verzár and J. Brocas, *Gerontol.*, **5**, 223 (1961).
20. F. H. Johnson, H. Eyring, and M. J. Polissar, *The Kinetic Basis of Molecular Biology*, Wiley, New York, 1954.
21. D. French and J. T. Edsall, *Advan. Protein Chem.*, **2**, 278 (1945).
22. E. S. Gould, *Mechanism and Structure in Organic Chemistry*, Holt, New York, 1959, p. 470.

Protein Conformations, "Rack" Mechanisms and Water

RUFUS LUMRY

Laboratory for Biophysical Chemistry, University of Minnesota, Minneapolis

The potential significance of changes in protein conformation for biological function was recognized relatively early by Haurowitz,[1] Wyman and Allen,[2] and Doherty and Vaslow,[3-5] all of whom presented evidence that such changes occur. Johnson, Eyring, and Polissar considered the possibilities of conformation change for a variety of physiological processes in their pioneering volume *The Kinetic Basis of Molecular Biology.*[6] The concept of strain and distortion was implicated in the model of Vaslow and Doherty for chymotryptic catalysis and in E. L. Smith's discussion of the role of trace metals in proteins and protein reactions.[7] The broad possibilities of strain and distortion of protein conformations were probably first recognized by Henry Eyring, who with his coworkers[8,9] introduced the unlovely but appropriate term "rack mechanism" to describe the mutual distortion of substrate and protein which could lead to enhanced catalytic efficiency. One can readily envision substrates twisted and bent on attachment to proteins into conformations approaching the geometry of difficultly formed activated complexes, and such special binding situations can be expected as the products of trial-and-error evolution like any other feature of organisms. One can also easily imagine that the (iron-porphyrin) heme groups of hemoglobin are distorted in the union of this coenzymic group with the protein parts of this molecule so as to induce electronic characteristics in the iron ions, where oxygen molecules bind, appropriate to the demands a given organism makes on its respiratory system.[9] Evolutionary modification of such electronic situations by substitutions of amino acids at various points in the protein is an attractive explanation for much of chemical evolution. In the same way, oxidation–reduction potentials of cytochromes can be established in a much more sensitive fashion than by simple exchange of ligands in the fifth and sixth positions of heme iron in these proteins.[10] Such protein systems appear to provide vehicles of nearly limitless flexibility for evolutionary experiments and in this way demonstrate the necessary characteristic of chemical mutability.[10,11]

The thermodynamic basis of "rack mechanisms" for hemoglobin and cytochrome c is the opposition of the tendency of the protein conformation to fold to a state having as low a free energy as possible and the tendency for the iron ion with its porphyrin partner to achieve the best directed-valency geometry so as to lower the free energy in this way.[9,11] This tug-of-war results in a compromise geometry of iron ion, ligands, and protein conformation with a desirable set of electronic properties. These examples can be generalized by noting that the major factor is the positive contribution to the free energy which arises in the proteins because neither iron complex ion nor protein conformation is in its own lowest free-energy state such as would be achieved if they did not each have to conform to the demands of the other. If no such compromise were required, the protein conformation would be acceptable only if it allowed complete satisfaction of the directed valency tendencies of the iron complex ion, but if this were possible in one state of the iron complex it would not be possible in the other; for example, oxy- versus deoxyhemoglobin, or reduced versus oxidized cytochrome. Hence, rack mechanisms are ubiquitous as a consequence of the intimate bonding situation which exists in protein–coenzyme–substrate systems. This type of bonding and its consequences are even more obvious in enzymes, such as lysozyme and carboxypeptidase A, which have polymeric substrates to which they are joined by many weak bonds of exactly the same kinds which hold polypeptides folded to form proteins. The substrate becomes part of the protein and the two must move together through the elementary steps of primary bond rearrangement. The mutual distortion accompanying these steps and primarily responsible for the catalytic process involves an increase in the total free energy of the system, but this positive contribution is apparently offset initially by the favorable features of the interaction between protein and substrate that occur in the binding process and make negative contributions to the free energy.[9,12] The term "mutual distortion" may be a bit too simple, since although the increase in free energy essential to the catalytic process may be stored and processed during catalysis in the form of strain, it may also be stored and utilized in the more sophisticated changes that accompany the replacement of solvent environment by protein environment. Not only must the substrate be so activated on binding but, as discussed by Jencks[13,14] and

Lovrien,[15] the protein must not restrict passage of the substrate system through activated complexes. Rack mechanisms of this type may receive some verification in X-ray diffraction studies of lysozyme. Tentative interpretation of such data have already suggested that there is considerable distortion of the susceptible primary bond toward the activated complex geometry expected for this hydrolytic process.[16,17] In carboxypeptidase A studies X-ray evidence[18] for a few large conformation changes, even though obtained with a poor substrate, has tended to emphasize the importance of conformational adjustments in facilitating electron rearrangements by direct attack of functional groups on substrate. Some adjustment of protein functional groups about the substrate would appear to be essential if catalysis is to follow the well-known mechanisms of organic chemistry, but the more subtle distortions are also undoubtedly important and, as a general consequence of the nature of protein conformations, are much more likely to be the source of the spectacular performance of proteins.

The rack concept grew out of the search for a chemically mutatable protein property that could support such apparently divergent phenomena as the interaction among the oxygen binding sites of hemoglobin, the so-called heme–heme interaction,[19] and peptide hydrolysis by carboxypeptidase. Thermodynamically the essential feature of rack mechanisms is the positive free-energy change generated at such a point and in such a way as to facilitate a biologically useful reaction. In the past we have put much emphasis on protein conformations acting as collections of springs able to undergo mechanical distortions so as to store free energy, to transport it from one place to another, as in heme–heme interaction, and to sum or subdivide free energy as mechanical energy to produce bundles of the sizes required for a subsequent chemical reaction.[20] Although many atoms of the protein may participate in these vital processes, the changes in their positions need not be large. Depending on the bonding, small displacements can have larger associated energy changes than large displacements.

It has been necessary to recognize the vehicle for manipulating free energy in order to describe protein processes. A particularly illuminating way to do this is shown in Fig. 1, which is an example of the usual greatly oversimplified but extremely useful potential-energy surfaces with which Eyring has made so much complex chemistry understandable. This particularly elaborate example is constructed for hemoglobin, and since conformational rearrangements were initially assumed to provide the means for free-energy manipulation, the vertical coordinate has become known as the "conformation" coordinate, but the term is no longer appropriate if it is considered to describe only changes in the conformation of the polypeptide (see below).

The "rules" for linkage systems such as that responsible for heme–heme interaction in hemoglobin and allosteric systems in general can be worked

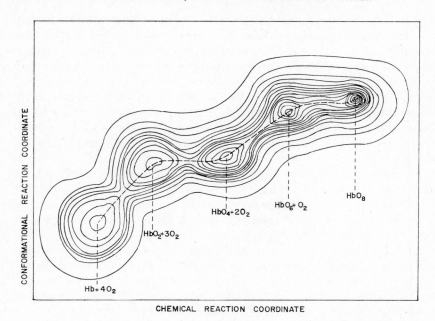

CHEMICAL REACTION COORDINATE

Fig. 1. Hypothetical potential energy surface for the oxygen-binding processes of hemoglobin to show relationship between chemical coordinate and conformation coordinate. See text.

out using Fig. 1 and are found[21] to be much less restrictive than has been generally suggested. The pattern and degree of cooperation among the four oxygen-binding sites of hemoglobin is determined by the relative depth (enthalpy) and area (entropy) of the potential wells for the several species, as becomes obvious after a few minutes examination of the figure. The dashed path roughly represents the average path for oxygenation on this arbitrarily drawn surface. Increasing oxygen pressure can be considered to increase the area of the wells in proportion to the first, second, third, and fourth powers of oxygen concentration as we move from Hb to HbO_8 along this line.

Hoard and coworkers[22,23] have proposed that there must be motion in the protein fabric generated by the movement of the iron ion from its initial position about 0.3 Å out of the average plane of the porphyrin nitrogen atoms on the fifth-ligand side (as determined with high-spin forms of myoglobin[24]) into the plane of the nitrogens on oxygen binding at the sixth-ligand position, since the iron ion goes into a low-spin state. Countryman, Collins, and Hoard[25] have recently demonstrated with a reasonable model, bis(imidazole)-$\alpha,\beta,\gamma,\delta$-tetraphenylporphinatoiron(III), that the iron is in fact in the plane in a small low-spin complex of this type. However, the X-ray studies of low-spin Fe^{III} derivatives of myoglobin indicate that the position of the iron ion is displaced well out of the plane.[26] It may be that the porphyrin moves over the iron ion if the latter is held too tightly in its lattice position by the fifth-position ligand to move. Such behavior has been demonstrated by X-ray studies of Love[27] on a single-subunit (single heme) worm hemoglobin. However, experiments of Albens et al.[28] and of Shulman and coworkers[29] suggest that there is no direct conformational pathway for coupling among the heme groups in human hemoglobin. If so, we will have to find a new mechanism for linkage. Such a vehicle is suggested in studies of Beetlestone and Irvine and their coworkers, some of whose major results are plotted in the main body of Fig. 2. They measured the standard enthalpy and entropy change in binding CN^-, N_3^-, SCN^-, and F^- to the Fe^{III} forms of human hemoglobins[30,31] and N_3^- binding to several hemoglobins,[32] myoglobin,[33] and the isolated α-chains of human hemoglobin[34] (which do not associate). Although there is no thermodynamic basis for a linear relationship between these quantities, they found good linear relationships between enthalpy and entropy points obtained by varying pH, as is seen in the figure. Furthermore, these enthalpy–entropy "compensation lines" are all roughly parallel. Their findings are particularly interesting because a number of small-solute phenomena have been found to give this same pattern of enthalpy compensation by entropy change with slopes of about the same value.[35,36] We will call the slope of such ΔH versus ΔS lines the compensation temperature, T_c, because it has the dimensions of temperature and because compensating enthalpy and entropy changes exactly cancel to give constant free energy changes when the experimental temperature equals T_c.

Linear enthalpy–entropy compensation is well known to physical organic chemists[36] and has been the subject of controversy[37] since the relationship was first discovered experimentally. We have discussed the complications elsewhere[35] and will only note here that the linearity found by Beetlestone et al. is statistically reliable for most of their examples. The most extensively studied set of small-solute compensation processes in water are the ionizations of weak acids. When acids such as acetic acid or benzoic acid are substituted in their nonpolar parts to form homologous series, the standard enthalpies and entropies of ionization are found to demonstrate compensation behavior with T_c values in the 280–290°K range but only after extraction of all the contributions to these quantities from the electronic rearrangements using methods developed by Hepler[38] and Ives[39] and their coworkers. The obvious conclusion is that this behavior in small-solute processes is due to solvation effects and thus a manifestation of some property of water. As a result of the comparison of their data with these small-solute examples, Beetlestone et al.[30] suggested that bulk water also plays an important role in the protein processes they studied.

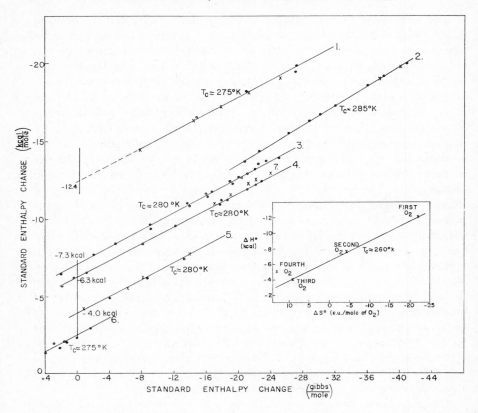

Fig. 2. Compensation plots of standard enthalpy and standard entropy changes in some reactions of hemeproteins. *Main figure:* (1) CN^- to Fe^{III} forms of human hemoglobins A and C (Ref. 29); (2) N_3^- to Fe^{III} forms of human hemoglobins A and C[30]; (3) N_3^- to Fe^{III} forms of dog, pigeon, and guinea pig hemoglobin[31]; (4) N_3^- to Fe^{III} form of α-chains from human hemoglobin A[33]; (5) SCN^- to Fe^{III} forms of human hemoglobins A and C[29]; (6) F^- to Fe^{III} forms of human hemoglobins A and C[29]; (7) N_3^- to Fe^{III} form of whale myoglobin.[32] Values of ΔH at $\Delta S° = 0$ are approximately equal to the chemical free-energy changes in these processes (see Ref. 44). *Insert:* Compensation plot of $\Delta H°$ and $\Delta S°$ values for the single oxygen-binding steps of sheep hemoglobin obtained by Roughton, Otis, and Lyster[58] and converted to $1 M$ standard states by George and Lyster.[59] Note that the variable used to produce compensation behavior is pH in the main figure but pH is constant for the insert.

Belleau[40] and Likhtenshtein[41] appear to have been the first to recognize that the pattern of linear enthalpy-entropy compensation when found in enzymic processes may indicate the participation of liquid water. Likhtenshtein[41] subsequently proposed that linear compensation is a ubiquitous feature of protein reactions. The data that Likhtenshtein and Sukhorukov[41,42] had available were not sufficient in number or precision to establish a strong case, but the aggregate of their analyses strongly suggests that these hypotheses deserve serious attention.

Linear enthalpy–entropy compensation has appeared in a number of other small-solute and protein processes in water.[35] The most comprehensive set of protein studies is that for α-chymotrypsin. Vaslow and Doherty[4] first observed linear compensation behavior in studies of inhibitor binding by chymotrypsin. They attributed it to changes in protein folding, which was quite reasonable in view of the cooperative nature of such folding. Yapel and Lumry,[35,43,44] in similar studies attempting to add detail to the early work of Vaslow and Doherty, found compensation behavior such that by varying the nature or optical geometry of the inhibitors or by varying pH sets of $(\Delta H^\circ, \Delta S^\circ)$ points could be generated which lie on very closely spaced compensation lines with T_c values in the range 265–285°K. A similar study of the binding of inhibitors of the tetramethylammonium-ion type to acetylcholinesterase has been carried out by Belleau and Lavoie.[45] Their regression line is identical to that obtained by Yapel and Lumry for indole binding to chymotrypsin. The identity of the slopes is no longer surprising, but the intercept values of ΔH° at $\Delta S^\circ = 0$ were also the same and this must be accidental. It is easy to show[35,46] that such intercepts are equal to the contribution to the total free energy from those part processes of the parent process, acid dissociation, inhibitor binding, etc., which do not manifest compensation. Such part processes can include inhibitor desolvation, electronic rearrangements, etc., and are unlikely to be the same with these two enzymes.

In further studies of α-chymotrypsin, Rajender, Han, and Lumry[35,47] have been able to estimate the standard enthalpies and entropies of formation of the intermediates in the catalytic mechanism given in (1) when applied to N-acetyl-L-tryptophanethylester, which is a good substrate, that is, a substrate rapidly hydrolyzed by this enzyme. The analysis, in addition to the assumption that (1) is valid, requires cancellation of pH dependencies of the individual rate constants. The latter is not quite complete so that the computed quantities are unreliable at pH values above 9. Nevertheless, they are adequate to show that the same compensation pattern found with inhibitors occurs in normal catalysis. These results are shown in Fig. 3, in which segments of the compensation line for indole binding obtained by Yapel and Lumry are given for reference. Rajender et al. also found linear compensation

$$E + S + H_2O \rightleftharpoons ES + H_2O \rightleftharpoons EA + P_1 + H_2O \rightleftharpoons$$
$$EP_2H + P_1 \rightleftharpoons E + P_1 + P_2 \quad (1)$$

behavior in the formation of the "on-acylation" activated complex in which the substrate Michaelis–Menten complex, ES, reacts with the hydroxyl group of Serine 195 leading to the acylated enzyme, EA, and releasing ethanol, P_1.

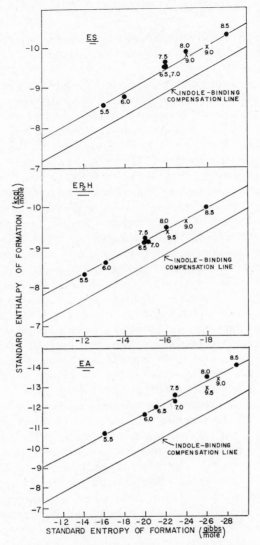

Fig. 3. Compensation plots of standard enthalpy of formation versus standard entropy of formation for the major intermediates in the chymotryptic hydrolysis of N-acetyl-L-tryptophanethylester.[34,46] ES and EP_2H are Michaelis-Menten complexes and EA is the acyl-enzyme. Compensation behavior was produced by varying pH (see text). The indole-binding compensation line was obtained by Yapel and Lumry.[43]

Similarly, the formation of the "off-acylation" activated complex, which takes EA to EP_2H, the Michaelis–Menten association complex between acid product and protein demonstrates the same compensation pattern. The T_c values are in the usual range, but the lines are much displaced along the vertical axis relative to those in Fig. 3.

The accumulated data for chymotrypsin show that inhibitors cause a large "motion" along the compensation line—as much as 100 e.u./mole and 29 kcal/mole.[43] The intermediates of the catalytic process, ES, EA, and EP_2H, on the other hand, occupy positions at a given pH considerably to the left of the corresponding equilibrium inhibitor–enzyme complexes. Thus the position of the equilibrium inhibitory complex between the enzyme and N-acetyl-L-tryptophan is considerably farther along the line to the right and thus toward more negative values of $\Delta H°$ and $\Delta S°$ than the position of the EP_2H species formed during the catalytic process. By such considerations it becomes clear that the conformation coordinate necessary in describing chymotryptic catalysis is identical with progress along the compensation line, so that the catalytic mechanism should be described by the pathway for catalysis shown in Fig. 4. The inhibited species, $[ES]_i$ and $[EP_2H]_i$ formed by large motions along the conformation coordinate are examples of non-productive substrate binding but probably not examples of so-called "wrong-way binding."

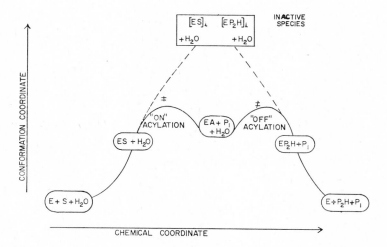

Fig. 4. Reaction pathway for chymotryptic catalysis showing relationship between chemical and conformation coordinates and self-inhibited species, $[ES]_i$ and $[EP_2H]_i$, formed between enzyme and substrate and between enzyme and acid product. See text.

The implications of the identification of the conformation coordinate for chymotrypsin with its compensation process are discussed elsewhere,[35,46,47] but the molecular description of motion along this coordinate remains a major puzzle. The accumulated data on linear compensation strongly suggest that we identify protein compensation behavior with that of small solutes in water and, as pointed out by Likhtenshtein and Sukhorukov,[41,42] Beetlestone et al.,[30] and Belleau and Lavoie,[45] this identification appears to implicate water as a direct participant in protein reactions. If we accept these hypotheses, it becomes necessary to determine how much of the enthalpy and entropy changes are due to changes in water and how much to polypeptide conformational rearrangements and what device couples the events involving directly the substrate or inhibitor to water. An examination of the properties of water necessary to produce linear compensation behavior shows either that the heat capacity changes in such reactions must be sensitive to the variables producing compensation (pH, chemical structure, and cosolvent concentration and type) but very weakly dependent on temperature, or that water must manifest two and only two phenomenological states, which interconvert cooperatively in a process with a weak temperature dependence.[35] There is evidence in favor of both possibilities. For the first alternative, despite some lack of quantitative agreement among measurements of the heat-capacity associated with solvation of nonpolar groups in water, there is general agreement that these quantities are not strongly temperature dependent.[48,49] Brandts has shown that the heat-capacity changes in exposure of nonpolar groups to water in protein unfolding are strongly dependent on cosolvent additions,[50] which also have been used to demonstrate compensation[35] and Shiao, Lumry, and Fahey have shown that such heat-capacity changes are very weakly dependent on temperature.[51] In support of the two-state case, a number of experts in the study of water claim that despite its complexity, liquid water does manifest the remarkably simple cooperative pattern attributable to the existence of only two important macroscopic (thermodynamic) states.[52,53] Jhon, Grosh, Ree, and Eyring have been very successful in quantitative computation of the properties of water, using what is essentially a two-state model built on the structures of ice I and ice III.[54] Applications of several experimental methods to determine the temperature and pressure dependencies of this two-state process have all given about the same relative values of enthalpy, entropy, and volume change: $\Delta H° \approx$ 5.0 cal/mole of cooperative units, $\Delta S° \approx 17$ e.u./mole, and $\Delta V° \approx -14$ ml/mole.[53,55] These numbers, if correct, demonstrate that with an input of 5 kcal of enthalpy the entropy of water increases by 18 e.u. and 14 ml of the free volume of the water is released. The ratio of $\Delta H°$ to $\Delta S°$ is 294°K and thus close to the experimental T_c values and if the free volume of the water released can be used to accommodate expanding solutes, we have a general explanation for the compensation behavior: Volume increases in expansion of solutes in water are accommodated primarily by free volume released from liquid water, and this process shifts the two-state equilibrium away from the lower-density state toward the higher-density state. This idea appears to have been first detailed by Eley,[56] and Ben-Naim[57] has recently proposed several important improvements.

Both of the alternatives are attractive and both are consistent with an important and quite rigorous example of the Gibbs–Duhem equation given in (2).[35]

$$A \, d\sigma = -V_w \, dW_w - V_p \, dW_p \qquad (2)$$

where V, T, P are constants, for convenience; A = interfacial area; σ = partial specific free energy of the interface per unit area; V_w = free volume of the water phase; W_w = partial specific free energy of the free volume in bulk water per unit volume; V_p = protein volume; and W_p = partial specific free energy of the protein phase per unit volume.

This equation provides a variety of ways to couple changes in water to processes that alter the extent of protein surface and volume or the partial specific free energies associated with these phases. Unfortunately, the experimental information necessary to estimate the magnitudes of the effects that might be produced and thus the practical importance of the equation is not yet established.

The various hypotheses about compensation behavior and its basis in water are far from established, but the experimental evidence is sufficient to suggest that there are major qualitative deficiencies in current descriptions of protein behavior. The original attractiveness of rack mechanisms persists, but molecular descriptions must be considered very unreliable as a result of the possibility that water sequestered inside such multiple-subunit proteins as hemoglobin or about protein surfaces will have to be included in these descriptions. It is quite possible that compensation behavior indicates only that proteins swell and shrink in their physiological processes, and we have elsewhere[35] shown that whatever the basis for these patterns the compensation phenomenon is not the basis of efficient catalysis with chymotrypsin. However, specificity for substrates in this process appears to be closely related to position along the compensation line. In the insert of Fig. 2 are plotted standard enthalpy and entropy changes for the single steps of oxygenation of ordinary (Fe^{II}) hemoglobin. The data due to Roughton, Otis, and Lyster[58] are the most reliable and probably the most precise that have been obtained for the temperature dependence of this process, but the demands made by the analysis into single-step quantities are so great that the apparent linear fit of these quantities for the first three steps of oxygenation, which was pointed out to us by John Beetlestone, cannot be taken as convincing evidence for compensation behavior. Nevertheless, the figure does suggest that the conformation coordinate in hemoglobin oxygenation is identical with the linear compensation process and thus closely related if not identical to the conformation coordinate in chymotryptic catalysis.

If Likhtenshtein[41] is correct in his proposal that compensation is ubiquitous in protein reactions, it may be a major vehicle for chemical evolution. It has begun to be possible to detect certain of the main themes of such evolution, and the most prominent among these themes appears to be the development of protein systems that allow chemical reactions to proceed ever closer and closer to thermodynamic reversibility without loss of speed.[44] We have proposed that this goal is approached by a mechanism we call "complementarity" or free energy complementarity to describe the efficient exchange of free energy between the substrate subsystem and the protein–water subsystem in such a way that deviations from an average, low-free-energy state for the total system are minimized.[44] Difficult chemical changes of the substrates and chemically reacting groups of the protein are facilitated by the protein–water subsystem, which is subsequently restored to high free-energy states at the expense of the chemical free energy the chemical subsystem releases. The existence of complex enzymic processes consisting of many elementary steps each involving only a small free-energy change and all difficult steps having about the same activation free energy suggests that there has already been considerable progress in evolution toward complementarity. In addition to providing a mechanism for linked processes, the compensation process, whether water based, rack based, or both, appears to have characteristics necessary to make such evolution possible. There are also interesting possible implications of compensation behavior for evolution, which are discussed in another place.[44]

ACKNOWLEDGMENTS

The preparation of this paper and much of the work on which it is based was supported by the National Science Foundation through Grant GB 7896. I am greatly indebted to Dr. Shyamala Rajender for her help in developing topics discussed in this paper. This is publication No. 62 from this Laboratory.

REFERENCES

1. F. Haurowitz, *Z. Physiol. Chem.*, **254**, 266 (1938).
2. J. Wyman and D. W. Allen, *J. Polymer Sci.*, **7**, 499 (1951).
3. D. Doherty and F. Vaslow, *J. Amer. Chem. Soc.*, **74**, 931 (1952).
4. F. Vaslow and D. Doherty, *J. Amer. Chem. Soc.*, **75**, 928 (1953).
5. F. Vaslow, *J. Phys. Chem.*, **70**(7), 2286 (1966).

6. F. H. Johnson, H. Eyring, and M. J. Polissar, *The Kinetic Basis of Molecular Biology*, Wiley, New York, 1954.

7. E. L. Smith, *Proc. Natl. Acad. Sci.*, *U.S.*, **35**, 80 (1959).

8. H. Eyring, R. Lumry, and J. D. Spikes, in *Mechanism of Enzyme Action*, W. D. McElroy and B. Glass, Eds., Johns Hopkins, Baltimore, 1956, p. 123.

9. R. Lumry and H. Eyring, *J. Phys. Chem.*, **53**, 110 (1954).

10. R. Lumry, A. Solbakken, J. Sullivan, and L. Reyerson, *J. Amer. Chem. Soc.*, **84**, 142 (1967).

11. R. Lumry, *Japan Biophys. Soc.*, **1**, (3), 1 (1961); *Un Seibutsu, Butsori*, **1**, 3 (1961).

12. R. Lumry and H. Eyring, in *The Present State of Physics*, American Association for the Advancement of Science, Washington, D.C. 1954, p. 189; R. Lumry, *Enzymes*, 2nd ed., **1**, 157 (1959).

13. W. P. Jencks, in *Current Aspects of Biochemical Energetics*, N. Kaplan and E. Kennedy, Eds., Academic, New York, 1966.

14. W. P. Jencks, *Catalysis in Chemistry and Enzymology*, McGraw-Hill, New York, 1969.

15. R. Lovrien, *J. Theoret. Biol.*, **24**, 247 (1969).

16. D. Phillips, *Proc. Natl. Acad. Sci.*, *U.S.*, **57**, 4811 (1967).

17. L. N. Johnson, D. C. Phillips, and J. A. Rupley, *Brookhaven Symp. Quant. Biol.*, **21**, 120 (1968).

18. W. Lipscomb, J. Hartsuck, G. Reeke, Jr., F. Quiocho, P. Bethge, M. Ludwig, T. Steitz, H. Muirhead, and J. Coppola, *Brookhaven Symp. Quant. Biol.*, **21**, 24 (1968).

19. J. Wyman, *J. Am. Chem. Soc.*, **89**, 2202 (1967); *Advan. Protein Chem.*, **19**, 223 (1964).

20. R. Lumry, in *Photosynthesis Mechanisms in Green Plants*, Publ. 1145 National Acad. Sci. and Natl. Res. Council, 1963, p. 625; R. Lumry, *Abhandl. Deut. Akad. Wiss. Berlin Kl. Chem. Geol. Biol.*, **1964**(6), 125.

21. R. Lumry, in *A Treatise on Electron and Coupled Energy Transfer in Biological Systems*, T. King and M. Klingenberg, Dekker, New York, 1970.

22. J. L. Hoard, M. J. Hamor, T. A. Hamor, and W. S. Caughey, *J. Amer. Chem. Soc.*, **87**, 2312 (1965).

23. J. L. Hoard, in *Hemes and Hemoproteins*, B. Chance, R. Estabrook, and T. Yonetani, Eds., Academic Press, New York, 1966.

24. J. C. Kendrew, R. E. Dickerson, B. E. Strandberg, R. G. Hart, D. R. Davies, D. C. Phillips, and V. C. Shore, *Nature*, **185**, 422 (1960); C. L. Nobbs, H. C. Watson, and J. C. Kendrew, *ibid.*, **209**, 339 (1966).

25. R. Countryman, D. Collins, and J. L. Hoard, *J. Amer. Chem. Soc.*, **91**, 5166 (1969).

26. L. Stryer, J. C. Kendrew, and H. C. Watson, *J. Mol. Biol.*, **8**, 96 (1964); C. L. Nobbs, H. C. Watson, and J. C. Kendrew, *Nature*, **209**, 339 (1966).

27. E. Padlan and W. Love, *Acta. Cryst.* **A25**, 5187 (1969).

28. J. O. Alben and W. S. Caughey, in *Hemes and Hemoproteins*, B. Chance, R. W. Estabrook, and T. Yonetani, Eds., Academic, New York, 1966, p. 139; *Biochemistry*, **7**(1), 175 (1968).

29. R. Shulman, S. Ogawa, K. Wüthrich, T. Yamane, J. Peisach, and W. Blumberg, *Science*, **165**, 251 (1969).

30. A. Anusiem, J. Beetlestone, and D. Irvine, *J. Chem. Soc. Ser. A.*, **1968**, 960.

31. J. Beetlestone and D. Irvine, *J. Chem. Soc. Ser. A*, **1968**, 951.

32. A. Anusiem, J. Beetlestone, and D. Irvine, *J. Chem. Soc., Ser. A*, **1968**, 1337.

33. J. Bailey, J. Beetlestone, and D. Irvine, *J. Chem. Soc., Ser. A*, **1969**, 241.

34. J. Bailey, J. Beetlestone, and D. Irvine, *J. Chem. Soc., Ser. A*, **1968**, 2778.

35. R. Lumry and S. Rajender, *Biopolymers*, **9**, 1125–1227 (1970).

36. J. Leffler and E. Grunwald, *Rates and Equilibria of Organic Reactions*, Wiley, New York, 1963.
37. O. Exner, *Coll. Czech. Chem. Commun.*, **29**, 1094 (1964).
38. L. G. Hepler and W. F. O'Hara, *J. Phys. Chem.*, **65**, 811 (1961).
39. D. Ives and P. Marsden, *J. Chem. Soc.* **1965**, 149.
40. B. Belleau, *Ann. New York Acad. Sci.*, **144**, 705 (1967).
41. G. Likhtenshtein, *Biofizika*, **11**, 23 (1966).
42. G. Likhtenshtein and B. Sukhorukov, *Zhur. Fiz. Khim.*, **38**, 747 (1963); *Biofizika*, **10**, 925 (1965).
43. A. Yapel and R. Lumry, *J. Amer. Chem. Soc.* in press (1971). A. Yapel, Dissertation; "A Kinetic Study of the Imidazole Groups of Chymotrypsin," University of Minnesota, 1967.
44. R. Lumry and R. Biltonen in *Structure and Stability of Biological Macromolecules*, S. Timasheff and G. Fasman, Eds., Dekker, New York, 1969, p. 65.
45. B. Belleau and J. Lavoie, *Can. J. Biochem.*, **46**, 1397 (1968).
46. R. Lumry, in *Proc. Fourth Johnson Foundation Symposium*, April 1969, Academic Press, 1970.
47. S. Rajender, M. H. Han, and R. Lumry, *J. Amer. Chem. Soc.*, **92**, 1378 (1970).
48. E. M. Arnett, W. B. Kover, and J. V. Carter, *J. Amer. Chem. Soc.* **91**, 4028 (1969).
49. F. Franks and D. I. G. Ives, *Quart. Revs.*, **20**, 1 (1966).
50. J. Brandts, in *Structure and Stability of Biological Macromolecules*, S. Timasheff and G. Fasman, Eds., Dekker, New York, 1969, p. 213.
51. D. Shiao, R. Lumry, and J. Fahey, *J. Amer. Chem. Soc.* **93**, (1970) in press.
52. J. F. Coetzee and C. D. Ritchie, Eds., *Solvent–Solute Interactions*, Dekker, New York, 1969.
53. G. Walrafen, in *Hydrogen-Bonded Solvent Systems*, A. Covington and P. Jones, Eds., Taylor and Frances, London, 1968, p. 9.
54. M. S. Jhon, J. Grosh, T. Ree, and H. Eyring, *J. Chem. Phys.*, **44**, 1465 (1966).
55. C. M. Davis, Jr., and J. Jarzynski, *Advan. Mol. Relaxation Proc.*, **1**(2), 155 (1969); D. I. G. Ives and T. H. Lemon, *Roy. Inst. Chem. Revs.*, **1**, 62 (1968).
56. D. D. Eley, *Trans. Faraday Soc.*, **35**, 1281, 1421 (1939).
57. A. Ben-Naim, *J. Phys. Chem.*, **69**, 1922, 3240 (1965).
58. F. Roughton, A. Otis, and R. Lyster, *Proc. Roy. Soc.* (*London*), **B144**, 29 (1955).
59. P. George and R. Lyster, in *Conference on Hemoglobin*, National Academy of Sciences (U.S.) and National Research Council publication 557, Washington, D.C. 1958, p. 39.

Biomolecular Conformation and Biological Activity

D. W. URRY

Institute for Biomedical Research, Education and Research Foundation, American Medical Association, Chicago, Illinois

INTRODUCTION

The study of biomolecules is entering a stage wherein the conceptualizations of the past are becoming realized in terms of three-dimensional descriptions of enzymes, enzyme–substrate complexes, ion-selective carriers, etc. One of the more formalized conceptualizations derives from the thermodynamic form of the absolute reaction rate equation, which, in addition to descriptions of enzyme reactions in terms of enthalpies, entropies, and volume changes on binding of substrate, allowed descriptions of the catalytic process in terms of the corresponding quantities for activation. Such treatments (which originated more than two decades ago[1]) represent in well-recognized terms the dynamics of enzyme catalysis. With the advances of X-ray diffraction techniques it has now become possible to observe changes in disposition of groups in enzymes on binding substrate.[2] One interesting example is that of carboxypeptidase A, which is being studied by the Lipscomb group.[3] The active site in carboxypeptidase A is seen as a pocket wherein the substrate is bound and identified by multiple, stereospecific interactions. On binding the glycyltyrosine substrate, a tyrosyl side chain is seen to swing over the substrate, bringing the acidic proton of the phenolic group into interaction with the peptide nitrogen; an arginyl side chain becomes visible as it interacts with and defines the carboxylate moiety; and the carboxylate side chain of a glutamic acid residue interacts with the positively charged amino group of the substrate. This beautiful work of the Lipscomb group puts in molecular terms an example of a dynamic enzyme–substrate binding process of the sort that had previously been conceptualized in thermodynamic terms. Although many workers have in one way or another described enzyme catalysis in dynamic terms, surely much credit for this concept should go to Eyring and his collaborators of two decades ago. The X-ray studies emphasize that the problem of biological activity is a problem in biomolecular conformation. It is now necessary to develop means of determining conformation of biomolecules in solution.

Another area of biological concern in which Eyring has provided concept, firmed by mathematical formalism, is membrane permeation. Again with great insight, his formalism provided for a carrier mechanism of selective ion transport across membranes.[4-7] This was at a time when the carrier mechanism was not in vogue. Recently, several ion-selective carriers have been identified and their conformations determined in the crystal[8-10] and in solution.[11-13] This communication reviews the determination of the solution conformation, by the nuclear magnetic resonance approach, of a biomolecule capable of selectively transporting potassium ion across membranes and of the ring moiety of a polypeptide hormone, and discusses these conformations in terms of biological activities. It then goes on to note other spectroscopic approaches for detailing the interaction between principles in the active sites of enzymes and for studying the conformation of molecules within membranes and at the solution–membrane interface.

SOLUTION CONFORMATION OF NATURALLY OCCURRING POLYPEPTIDES

Experimental approaches are presently reaching a state where the conformation of simpler naturally occurring polypeptides can be determined in solution. Specifically, proton magnetic resonance at 220 MHz provides sufficient resolution to determine the details of conformation. Combinations of chemical shifts, temperature dependences of chemical shifts, and coupling constants of amide proton resonances can be used to identify specific conformational features. The feature which will be reviewed below is the β-turn or β-fold.[14] The β-turn involves a hydrogen bond between the peptide oxygen of residue $i - 1$ and the peptide hydrogen of residue $i + 2$. This hydrogen bonding pattern is the same as for the 3_{10} helix,[15] but the β-turn is flattened in such a way that the entering and emerging polypeptide chains are in position for the antiparallel pleated sheet conformation.[16] (Looking ahead to Fig. 4, residue i is Ile and $i + 1$ is Gln.)

In general, in common solvent systems such as dimethylsulfoxide, methanol, and water, the β-turn is characterized in terms of amide proton magnetic resonances by a high-field shifted resonance with a decreased temperature dependence for residue $i + 2$, by a small αCH—NH coupling constant for residue i, and by a large αCH—NH coupling constant for residue $i + 1$. If the β-turn is part of an antiparallel pleated sheet, then $J_{\alpha CH—NH}$ will also be large, of the order of 7 cps, for the resonances of residues $i - 1$ and $i + 2$ and the temperature dependence of chemical shift of the amide proton resonance of residue $i - 1$ will be lowered.

The Solution Conformation of Valinomycin–Potassium Complex

Valinomycin is a dodecadepsipeptide, which is to say that it contains twelve residues with alternating peptide and ester groupings. Its sequence

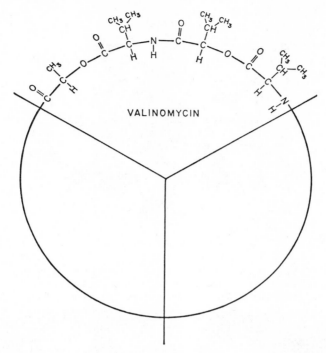

Fig. 1. Primary structure of valinomycin, with threefold symmetry.

contains four residues repeated three times,[17] [L-Lac-L-Val-D-HyV-D-Val]$_3$, as indicated in Fig. 1. An interesting feature of this antibiotic is that it effects a selective transport of potassium ions across membranes. The intriguing question is, of course, how can this sequence be folded into a unique conformation in order that it may impart such selective biological activity? The result of the NMR studies[12,14,18] is the complex in Fig. 2. When bound to potassium, valinomycin is essentially a series of β-turns in which the end group is an ester instead of a peptide and, most significantly, it is in a conformation in which the acyl oxygens of the ester groups point inward, forming a polar core of octahedral symmetry and of dimensions which result in favored binding of bare potassium ion over sodium ion. Thus a determination of the conformation of the valinomycin–potassium ion complex provides an understanding of its ion-selective activity in terms of a carrier mechanism.

In dimethyl sulfoxide in the absence of binding ion, valinomycin exhibits a conformation which actually appears to be an interconverting set of dish-shaped conformations. The back side contains the nonpolar side chains and the front side the polar peptide and ester moieties (see Fig. 3). This results in

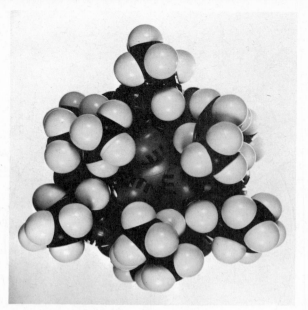

Fig. 2. The CPK molecular model of the valinomycin-potassium ion complex. The bare potassium ion is in an octahedral field made up of the polar acyl oxygens of the ester moieties. The side chains give a nonpolar exterior.

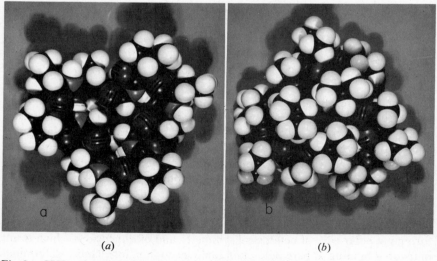

(a)   (b)

Fig. 3. CPK molecular models representing one of several nearly equivalent conformations in dimethyl sulfoxide-d_6: (a) polar side of the conformation with ester and peptide groupings available for interactions with a polar medium; (b) the nonpolar side containing the aliphatic side chains. This side would readily interact with nonpolar media. The polar side could be considered as a threefold symmetric bear trap in which the acyl oxygens of the esters are in a position to close readily on a potassium ion.

586

a structure which would explain the surfactant properties that are so striking for valinomycin. A small amount of valinomycin can immediately clear an emulsion. Furthermore, it allows a rather graphic description of the molecule in connection with its ion transport. One can imagine the valinomycin molecule at a membrane surface with its nonpolar side against the membrane and its polar side toward the aqueous solution. In this configuration it is poised like a threefold bear trap ready to close on a potassium ion as it passes by. Once enclosed on the potassium ion, a nonpolar exterior is formed and the complex can readily diffuse through the membrane.

The Secondary Structure of the Cyclic Moiety of Deamino Oxytocin

Again utilizing the NMR characterization of the β-turn, the secondary structure of deamino oxytocin has been determined in DMSO–d_6-methanol to be the β-type conformation[19] indicated in Fig. 4. The handedness of the disulfide bridge is assumed to be the same in DMSO–methanol as in the aqueous system where the optical rotation studies were carried out.[20,21]

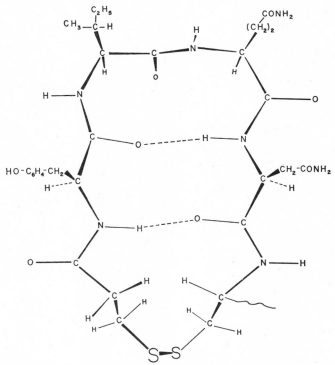

Fig. 4. Secondary structure of the ring moiety of deamino oxytocin in a dimethyl sulfoxide-d_6–methanol (3:7) solvent system. From Urry et al.[19]

While the conformation of the tail (-L-Pro-L-Leu-L-Gly-NH₂) has not yet been determined, some interesting points relating to biological activity are emerging.

A conspicuous aspect of the amide proton resonance spectrum for oxytocin and many of its biologically active analogs is the high-field shifted asparagine amide proton resonance. The resonance is high-field shifted due to shielding by the end peptide residue. Interestingly, in 5-valine-oxytocin, which is biologically inactive,[22] the high-field resonance is no longer apparent. This implies that the loss of a β-turn may relate to loss of activity. An argument can be presented suggesting that the orientation of the end peptide is a conformational feature discriminating between oxytocic and vasopressic activities. If future studies confirm this correlation it may be suggested that the β-turn provides a conformational constraint necessary for the formation of the active portion of the hormone. With neurohypophyseal hormones we are now entering a stage wherein biomolecular conformation can be related to biological activity.

RECIPROCAL RELATIONS IN OPTICAL ROTATION AND THE PROXIMITY OF AROMATIC GROUPS

Another means of achieving specific conformational information about

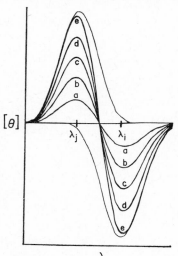

Fig. 5. Schematic representation of reciprocal relations as seen in circular dichroism spectra for an idealized case of close-lying bands. From Urry.[28]

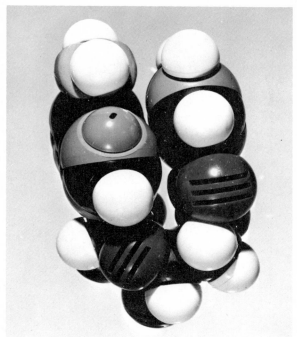

Fig. 6. A stacked conformation for adenosine-5′-mono-nicotinate, a model used for demonstrating reciprocal relations.

biologically active components, which would appear to be particularly promising in application to the active sites of enzymes, derives from the coupled oscillator source of optical activity. When dissymmetrically placed electric transition dipole moments are brought into juxtaposition they couple to give rise to rotational strength and they do so in a reciprocal manner; that is, as the rotational strength of one transition becomes more positive the rotational strength of the second transition becomes more negative.[23] The rotational strengths relate in an inverse or reciprocal way. This is schematically shown in Fig. 5 in terms of circular dichroism for the idealized case of two close-lying transitions.

From the standpoint of utilizing this effect it is fortunate that many coenzymes, prosthetic groups of enzymes, and substrates contain identifiable absorptions in the near ultraviolet as do the aromatic side chains of amino acids. Thus, if one could titrate a coenzyme or substrate onto an enzyme and note this type of reciprocal relations involving a transition in the enzyme, for example in a tryptophan or tyrosine group, and an absorption band in the group being added, then one would have identified an interaction at the active

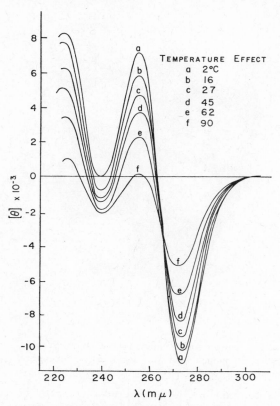

Fig. 7. Temperature effect on the circular dichroism of adenosine-5′-mononicotinate. As the temperature is decreased, more stacking is present. A reciprocal increase in negative ellipticity of the long-wavelength band and an increase in positive ellipticity of the first positive band are observed. From Miles and Urry.[24]

site. If the transition dipole moment directions were known, then relative orientations of the aromatic groups could be determined.

The model system that was used to demonstrate the reciprocal relations was adenosine-5′-mononicotinate,[24] which is given in the stacked or interacted conformation in Fig. 6. The temperature effect for this molecule as seen in the circular dichroism spectra is given in Fig. 7, where the reciprocal relations are beautifully apparent. This approach has been used to identify stacked conformations for flavin-adenine dinucleotide[25] and for the oxidized and reduced forms of α- and β-nicotinamide-adenine dinucleotide.[26]

CONFORMATIONS OF BIOMOLECULES IN PARTICULATE SYSTEMS

The living organism is replete with particulate structures, such as membranes, mitochondria, nuclei, and connective tissue containing insoluble elastin. These structures contain molecules which either are not soluble or when solubilized would not necessarily reflect the conformation of interest within the particle. Accordingly, one would like to develop techniques for assessing biomolecular conformations in particulate systems. Progress has been made, in this connection, with the optical rotation methods.[27-30]

Circular dichroism spectra of optically active, heterogeneous systems are distorted in several ways. There is an obscuring of chromophores due to light scattering; that is, there is a probability that the scattered photon would have been absorbed had the sample been molecularly dispersed. There is also an obscuring of chromophores due to the absorption flattening of Duysens[31]; that is, a particle casts a shadow and as the absorption by the particle increases the chromophores behind the particle in the light path are increasingly occluded from the beam. In addition in the circular dichroism measurement, which is the difference absorbance of left and right circularly polarized light, there is a differential scatter of the left and right circularly polarized beams by the optically active particle.[30] This is seen by the phototube as a difference absorbance.

The ellipticity of a sample $[\theta]$ is given by the expression

$$[\theta] = \frac{3300(A_L - A_R)}{C_0 l}, \tag{1}$$

where A_L and A_R are the absorbances for the left and right circularly polarized beams, C_0 is the analytical concentration in moles/liter, and l is the path length in centimeters. Neglecting differential scatter and under experimental conditions of short path length cells with relatively low concentrations, the distorted circular dichroism (CD) spectrum, $[\theta]_{\text{dist}}$, may be approximated from the correct spectrum,[28,29] i.e.,

$$[\theta]_{\text{dist}} = [\theta]_{\text{corr}}(Q_A - A_S), \tag{2}$$

where A_S is the absorption due to light scattering intensity losses, and Q_A is the flattening quotient, that is, the ratio of the suspension to the solution absorbance

$$Q_A = \frac{A(\text{susp})}{A(\text{soln})}. \tag{3}$$

In terms of the Rayliegh-Gans expression A_S takes on the form

$$A_S = -0.434 \ln \left[1 - \frac{K_D}{\lambda^v} (n_p^2 - n_s^2)^2 \right]. \tag{4}$$

With a knowledge of n_p and n_s, the particle and solvent refractive indices, a wavelength range devoid of true absorption can be fit with (4) to obtain the parameters K_D and v. A_S can then be calculated in wavelength ranges where true absorption occurs. If scattering particles are large it is best to use the Mie dependence.[29,32]

In order that an A_S relevant to the CD study may be experimentally obtained, it is important to have the same beam collimation and sample–phototube configuration for the absorption and CD measurements. This problem is well solved by simultaneously determining absorption and circular dichroism on the same phototube with a single beam and sample. A relatively simple modification of the Cary instrument allows this to be achieved.[29]

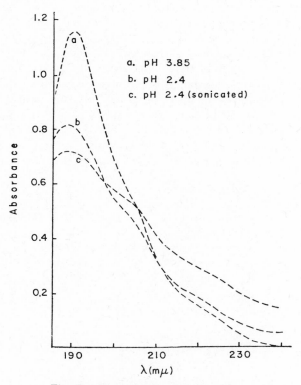

Fig. 8. Absorption curves for poly-L-glutamic acid in the reference state, pH 3.85, and as suspensions. The concentration is constant at 2 mg/ml; only the particle size is varied. These curves were obtained simultaneously on the same sample and phototube as the CD curves in Fig. 9. From Urry et al.[29]

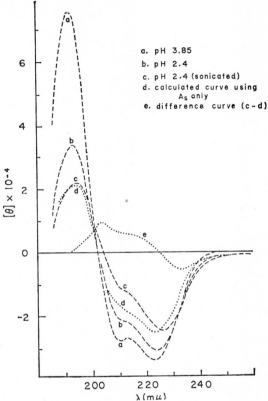

Fig. 9. Circular dichroism curves for poly-
L-glutamic acid in the reference state, pH 3.85,
and as suspensions. The concentration of PGA
is constant at 2 mg/ml; only the particle size
is varied. Curves *a*, *b*, and *c* were obtained
simultaneously on the same sample and phototube and with the same light beam as the absorption curves in Fig. 8. Curve *d* is the
calculated curve, using (2), and curve *e* is the
difference between the calculated and experimental curves. Note that curve *e* resembles the
ORD curve for an α-helix. From Urry et al.[29]

The adequacy and limitations of (2) can be determined with the poly-L-
glutamic acid (PGA) model system. At low pH, α-helical PGA aggregates in
a manner that can be controlled by pH and sonication.[33] Curves *a*, *b*, and *c*
of Figs. 8 and 9 are simultaneously determined absorptions and ellipticities
for the PGA system with increasing average particle size. Using the absorption

crossover points and the Duysens expression for Q_A as a function of the absorption along the diameter of a spherical particle,[31] A_p, that is,

$$Q_A = \frac{3}{2A_p} \left\{ 1 - \frac{2[1 - (1 + A_p) \exp(-A_p)]}{A_p^{2}} \right\} \tag{5}$$

it is possible to determine Q_A and A_S as a function of wavelength.[29] When these values are used in (2) the distorted curve d (Fig. 9) is calculated. The difference curve e of Fig. 9 approximates an optical rotatory dispersion curve and is in large part due to the differential scatter of the left and right circularly polarized beams.

On adding the differential scatter effect, (2) becomes

$$[\theta]_{\text{dist}} = \frac{3300}{C_0 l} [Q_A(A_L - A_R) - (A_{SL}A_L - A_{SR}A_R) + A_{SL} - A_{SR}], \tag{6}$$

where

$$A_{SL}(\lambda) = -0.434 \ln \left[1 - \frac{K_D}{\lambda^v} (n_{PL}^{2} - n_s^{2})^2 \right], \tag{7}$$

$$A_{SR}(\lambda) = -0.434 \ln \left[1 - \frac{K_D}{\lambda^v} (n_{PR}^{2} - n_s^{2})^2 \right], \tag{8}$$

$$n_{PL} = n_p + \frac{(n_L - n_R)_p}{2}, \tag{9}$$

$$n_{PR} = n_p - \frac{(n_L - n_R)_p}{2}, \tag{10}$$

$$(n_L - n_R)_p = \frac{[m]\lambda\rho_{\text{PGA}}}{18mw}, \tag{11}$$

where $[m]$ is the mean residue rotation for the α-helical polypeptide, ρ_{PGA} is the density of the PGA particles, λ is the wavelength in cm, and mw is the residue molecular weight. In order to utilize (6)–(10) the index of refraction of the particle and the particle density ρ_{PGA} are required. Vinograd and Hearst[34] report a density of 1.50 g/ml. Ifft et al.[35] carried out a thorough study of ρ_{PGA} as a function of pH using the density gradient technique, and in this laboratory we have confirmed the value of 1.50 g/ml on our samples.

The particle index of refraction was calculated in the following way:

$$n_p = 1 + n'(bkg) + n'(190)_{\text{PLA}} + n'(204)_{\text{PLA}} + n'(216)_{\text{PLA}}, \tag{12}$$

where

$$n'(bkg) = \rho'_{\text{PGA}} \left[\frac{n'(\text{HOAc})}{\rho'(\text{HOAc})} + \frac{n''(\text{DMF})}{\rho'(\text{DMF})} \right]; \tag{13}$$

ρ'_{PGA}, $\rho'(\text{HOAc})$, and $\rho'(\text{DMF})$ are the densities divided by the molecular weights. In the case of PGA it is the particle density divided by the mean residue molecular weight. The refractive indices of dimethyl formamide, $n(DMF)$, and acetic acid, $n(HOAc)$, were obtained by fitting the values in the International Critical Tables with the equation

$$n = 1 - \frac{a\lambda^2}{\lambda^2 - c}. \tag{14}$$

The local 196 mμ band of DMF was removed, and local bands for the α-helix of poly-L-alanine were inserted as indicated in (12) and below:

$$n'(HOAc) = n(HOAc) - 1. \tag{15}$$
$$n''(DMF) = n(DMF) - 1 - n'(196)_{DMF}. \tag{16}$$

The local absorption bands for the α-helix were taken from the resolved data of Quadrifoglio and Urry,[36] and the partial refractive indices were calculated using the Kronig-Kramers transforms.

The Kronig-Kramers transform[37,38] is the following:

$$n'(k) = \frac{\Delta_i K_{max}}{2\pi^{3/2}} \left[\frac{\Delta_i}{\lambda + \lambda_i} \frac{-2(\lambda - \lambda_i)}{\Delta_i} e^{-x^2} \int_0^x e^{x^2}\, dx \right]$$
$$+ \frac{\lambda_i K_{max}}{\pi^{3/2}} \left[e^{-x^2} \int_0^x e^{-x^2}\, dx - \frac{\Delta_i}{2(\lambda + \lambda_i)} \right] \tag{17}$$

and

$$x = \frac{\lambda - \lambda_i}{\Delta_i}, \tag{18}$$

$$K_{max} = \frac{2.303 \times 10^3 \varepsilon_i^0 \rho}{mw}, \tag{19}$$

where Δ_i is the half-bandwidth at ϵ_i/e, and Δ_i, λ, and λ_i are in centimeters. The resolved poly-L-alanine bands were Gaussian.[36] The integrations were achieved by a least-squares fitting of the table of the numerical values[39] of the integral for each value of x to a Fourier series. A sine series of 13 terms with a period of $x = 10$ was sufficient. The calculated partial refractive indices as well as $n'(HOAc)$, $n''(DMF)$, $n'(bkg)$, and n_p are given as a function of wavelength in Table I. And the quantities for calculating the differential scatter of left and right circularly polarized light are given in Table II. Preliminary comparative studies of red blood cell membranes and mitochondrial membranes indicate a greater differential scatter for the mitochondrial fragments.[40] This implies relatively more ordered protein at the surface of the mitochondrial membrane and would be consistent with the greater catalytic role of mitochondrial membranes. When the values in Table II and Q_A (see Ref. 29) are used, the experimental curve can be closely

Table I. Components of the Index of Refraction for the PGA Particle

λ(mu)	n'(190)	n'(204)	n'(216)	n'(HOAc)	n"(DMF)	n'(bkg)	n_p
240	.02522	.00925	.01766	.42951	.46236	.699745	1.75187
238	.02717	.00986	.01983	.43088	.46393	.70206	1.75892
236	.02937	.01071	.02213	.43228	.46555	.704451	1.76666
234	.03169	.01184	.02440	.43374	.46723	.70692	1.77484
232	.03407	.01310	.02636	.43524	.46897	.709471	1.78299
230	.03655	.01441	.02767	.43679	.47076	.712108	1.79073
228	.03929	.01590	.02797	.43839	.47261	.714836	1.79799
226	.04251	.01786	.02692	.44005	.47453	.717658	1.80495
224	.04644	.02056	.02427	.44177	.47652	.720579	1.81186
222	.05125	.02414	.01996	.44355	.47858	.723604	1.81895
220	.05704	.02878	.01409	.44539	.48071	.726739	1.82665
218	.06392	.03485	.00699	.4473	.48292	.72999	1.83574
216	.07205	.04249	.00087	.44928	.48522	.733361	1.84704
214	.08172	.05076	-.00886	.45133	.4876	.73686	1.86048
212	.09325	.05692	-.01639	.45346	.49007	.740493	1.87427
210	.10675	.05686	-.02288	.45567	.49265	.744269	1.88499
208	.12187	.04697	-.02794	.45797	.49532	.748194	1.8891
206	.13744	.02653	-.03134	.46036	.49811	.752278	1.88491
204	.15134	-.00103	-.03309	.46285	.50101	.756529	1.87375
202	.16059	-.02914	-.03335	.46544	.50403	.760958	1.85905
200	.16185	-.05095	-.03244	.46814	.50718	.765575	1.84403
198	.15210	-.06243	-.03072	.47095	.51047	.770392	1.82934
196	.12948	-.06371	-.02855	.47389	.5139	.77542	1.81263
194	.093995	-.05817	-.02625	.47696	.51749	.780675	1.79026
192	.047824	-.04999	-.02402	.48016	.52124	.78617	1.75998
190	.004889	-.04220	-.02199	.48351	.52518	.79192	1.72285

Table II. Quantities for Calculating the Differential Scatter of Left and Right Circularly Polarized Light

λ(mμ)	$\frac{(n_L-n_R)_p}{2}$ × 10⁴	n_{PL}	n_{PR}	n_s	A_{SL}	A_{SR}
240	−.96	1.75177	1.75197	1.38462	.158965	.15919
238	−1.10	1.75881	1.75903	1.38583	.168951	.169223
236	−1.23	1.76654	1.76678	1.38709	.164366	.164656
234	−1.31	1.77471	1.77497	1.38838	.182344	.182689
232	−1.30	1.78286	1.78312	1.38972	.179942	.180274
230	−1.15	1.79061	1.79085	1.3911	.17558	.175861
228	−.867	1.7979	1.79808	1.39253	.181825	.18043
226	−.452	1.8049	1.805	1.394	.185429	.185544
224	0	1.81186	1.81186	1.39553	.172048	.172048
222	.400	1.81899	1.81891	1.39711	.176406	.176313
220	.836	1.82673	1.82657	1.39875	.177082	.176889
218	1.13	1.83585	1.83563	1.40045	.18407	.183799
216	1.43	1.84718	1.8469	1.40221	.189003	.188658
214	1.63	1.86064	1.86032	1.40403	.178529	.178171
212	1.87	1.87446	1.87408	1.40593	.166721	.186297
210	2.10	1.8852	1.88478	1.4079	.190697	.190215
208	2.91	1.88939	1.88881	1.40994	.188498	.187842
206	3.59	1.88527	1.88455	1.41207	.191933	.191097
204	4.04	1.87415	1.87335	1.41428	.188578	.187634
202	4.45	1.85949	1.85861	1.41659	.177094	.176099
200	4.72	1.8445	1.84356	1.41899	.173146	.172082
198	4.68	1.82981	1.82887	1.42149	.186927	.185726
196	3.53	1.81298	1.81228	1.4241	.186025	.185082
194	2.33	1.79049	1.79003	1.42683	.200443	.199718
192	.768	1.76006	1.7599	1.42968	.205698	.205428
190	0	1.72285	1.72285	1.43265	.203327	.203327

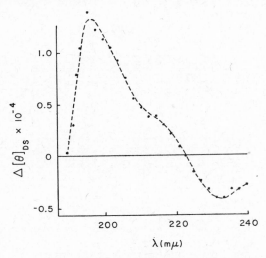

Fig. 10. The calculated contribution of the differential scatter of left and right circularly polarized light, $\Delta[\theta]_{DS}$, to the experimental ellipticity curve of a PGA suspension which exhibits an absorption due to light scattering of 0.16 to 0.20 absorption units in the 240- to 190-mμ wavelength range. This curve is due to the third term $(A_{SL} - A_{SR})$ in (6); that is, $\Delta[\theta]_{DS} = (3300/C_0 l)(A_{SL} - A_{SR})$. From Urry and Krivacic.[30]

calculated.[30] The pure differential scatter term is $A_{SL} - A_{SR}$. This is plotted in Fig. 10 for the PGA system that gave curve c of Figs. 8 and 9.

This work demonstrated (a) that there are distortions in the CD of particulate systems (this had not previously been appreciated and the distorted spectra had been interpreted in terms of unique conformational features), (b) that the distorted spectra can be calculated and hence corrected, and (c) that there is a measurable differential scatter of left and right circularly polarized light by optically active particles. Thus, in addition to correcting spectra for suspensions of particulate systems that may be interpreted in terms of biomolecular conformation, the third point makes it possible to obtain an optical rotatory dispersion spectrum for the particle surface. In the case of membranes this will allow determination of relative amounts of surface area which are covered by ordered protein. This information coupled with the CD spectrum for the whole membrane will provide considerable information on the structure of membranes.

REFERENCES

1. F. H. Johnson, H. Eyring, and M. J. Polissar, *The Kinetic Basis of Molecular Biology*, Wiley, New York, 1954, and references to literature therein.

2. M. F. Perutz, *European J. Biochem.*, **8**, 455 (1969).

3. W. N. Lipscomb, J. A. Hartsuck, G. N. Reeke, Jr., F. A. Quicho, P. H. Bethge, M. L. Ludwig, T. A. Steitz, H. Muirhead, and J. H. Coppola, *Brookhaven Symp. Quant. Biol.*, **21**, 24 (1968).

4. H. Eyring, D. Henderson, B. J. Stover, and E. M. Eyring, *Statistical Mechanics and Dynamics*, Wiley, New York, 1964, p. 454.

5. H. Eyring and D. W. Urry, in *Theoretical and Mathematical Biology*, T. H. Waterman and H. J. Morowitz, Eds., Blaisdell, New York, 1965, p. 57.

6. H. Eyring and D. W. Urry, *Ber. Bunsinges. Physik. Chem.*, **67**, 731 (1963).

7. B. J. Zwolinski, H. Eyring, and C. E. Reese, *J. Phys. Colloid Chem.*, **53**, 1426 (1949).

8. M. Dobler, J. D. Dunitz, and J. Krajewski, *J. Mol. Biol.*, **42**, 603 (1969).

9. B. T. Kilbourn, J. D. Dunitz, L. A. R. Pioda, and W. Simon, *J. Mol. Biol.*, **30**, 559 (1967).

10. M. Pinkerton, L. K. Steinrauf, and P. Dawkins, *Biochem. Biophys. Res. Commun.*, **35**, 512 (1969).

11. V. T. Ivanov, I. A. Laine, N. D. Abdulaev, L. B. Senyavina, E. M. Popov, Yu. A. Ovchinnikov, and M. M. Shemyakin, *Biochem. Biophys. Res. Commun.*, **34**, 803 (1969).

12. M. Ohnishi and D. W. Urry, *Biochem. Biophys. Res. Commun.*, **36**, 194 (1969).

13. Yu. A. Ovchinnikov, V. T. Ivanov, A. V. Evstratov, V. F. Bystrov, N. D. Abdulaev, E. M. Popov, G. M. Lipkind, S. F. Arkhipova, E. S. Efremov, and M. M. Shemyakin, *Biochem. Biophys. Res. Commun.*, **37**, 668 (1969).

14. D. W. Urry and M. Ohnishi, in *Spectroscopic Approaches to Biomolecular Conformation*, D. W. Urry, Ed., American Medical Association Press, Chicago, 263, 1970.

15. J. Donohue, *Proc. Natl. Acad. Sci. U.S.*, **43**, 213 (1957).

16. A. J. Geddes, K. D. Parker, E. D. T. Atkins, and E. J. Beighton, *J. Mol. Biol.*, **32**, 343 (1968).

17. M. M. Shemyakin, E. I. Vinogradova, M. Yu. Feigina, N. A. Aldanova, N. F. Loginova, I. D. Ryobova, and I. A. Pavlenko, *Experientia*, **21**, 548 (1965).

18. M. Ohnishi and D. W. Urry, *Science*, **168**, 1091 (1970).

19. D. W. Urry, M. Ohnishi, and R. Walter, *Proc. Natl. Acad. Sci. U.S.*, **66**, 111 (1970).

20. D. W. Urry, F. Quadrifoglio, R. Walter, and I. L. Schwartz, *Proc. Natl. Acad. Sci. U.S.*, **60**, 967 (1968).

21. R. Walter, W. Gordon, I. L. Schwartz, F. Quadrifoglio, and D. W. Urry, *Proc. Ninth European Peptide Symp.*, Orsay, France, North-Holland, Amsterdam, 1968, p. 50.

22. R. Walter, J. Rudinger, and I. L. Schwartz, "Conformation Studies on the Neurohypophyseal Hormone Oxytoxin and its Analogs." *Am. J. Med.*, **42**, 653 (1967).

23. D. W. Urry, *Proc. Natl. Acad. Sci. U.S.*, **54**, 640 (1965).

24. D. W. Miles and D. W. Urry, *J. Phys. Chem.*, **71**, 4448 (1967).

25. D. W. Miles and D. W. Urry, *Biochemistry*, **7**, 2791 (1968).

26. D. W. Miles and D. W. Urry, *J. Biol. Chem.*, **243**, 4181 (1968).

27. D. W. Urry and T. H. Ji, *Arch. Biochem. Biophys.*, **128**, 802 (1968).

28. D. W. Urry, in *Spectroscopic Approaches to Biomolecular Conformation*, D. W. Urry, Ed., American Medical Association Press, Chicago, p. 33, 1970.

29. D. W. Urry, T. A. Hinners, and L. Masotti, *Arch. Biochem. Biophys.*, **137**, 214 (1970).
30. D. W. Urry and J. Krivacic, *Proc. Natl. Acad. Sci. U.S.*, **65**, 845 (1970).
31. L. N. M. Duysens, *Biochim. Biophys. Acta*, **19**, 1 (1956).
32. D. W. Urry, T. A. Hinners, and J. Krivacic, *Anal. Biochem.*, **37**, 85 (1970).
33. T. H. Ji and D. W. Urry, *Biochem. Biophys. Res. Commun.*, **34**, 404 (1969).
34. J. Vinograd and J. E. Hearst, *Fortschr. Chem. Org. Naturstoffe*, **20**, 373 (1963).
35. J. B. Ifft, J. Zilius, and L. Lum, private communication.
36. F. Quadrifoglio and D. W. Urry, *J. Amer. Chem. Soc.*, **90**, 2755 (1968).
37. W. Moffitt and A. Moscowitz, *J. Chem. Phys.*, **30**, 648 (1959).
38. W. Kuhn and E. Braun, *Z. Physik. Chem.*, **8**, 281 (1931).
39. W. Lushmiller and A. R. Gordon, *J. Phys. Chem.*, **35**, 2785 (1931).
40. L. Masotti and D. W. Urry, unpublished results.

Eyring Rate Theory Model of the Current–Voltage Relationships of Ion Channels in Excitable Membranes[*]

J. W. WOODBURY

Department of Physiology and Biophysics, University of Washington School of Medicine, Seattle

Abstract

The bizarre properties of ion channels in excitable cell membranes are briefly described (Figs. 1 and 2). The movements of ions through ion-specific channels in a cell membrane are analyzed using Eyring rate theory. The Na^+ channels of squid nerve fiber membranes have a linear instantaneous current–voltage relation. This puzzling finding is explained by Eyring rate theory by assuming that there is a linear increase of potential energy barrier heights from inside to outside (Fig. 3). Linearity is obtained by making the total change in barrier height, $-FzV_1$, equal to $-FzV_{Na}$, where V_{Na} is the Na^+ equilibrium potential. This condition implies that there are about ten barriers in a Na^+ channel. The effects of changes in V_1 on the current–voltage relations are described (Fig. 4). Anomalous rectification is predicted for $V_1 > V_{Na}$, but the equation does not accurately describe the anomalous rectification found in the K^+ channels of skeletal muscle cell membranes. It is suggested that this discrepancy is due to interactions between K^+ ions in the channel (mean transit time greater than mean entry time). A challenge is given to reaction rate theorists to develop molecular-level explanations of three puzzling properties of exictable cell membranes: ion selectivity, voltage dependence, and nonindependance (ion interactions) in the membrane.

[*] This investigation was supported in part by PHS Research Grant number NSO1752 from the National Institute of Neurological Diseases and Stroke.

601

A Personal Note

Henry Eyring's courses in statistical mechanics and reaction rate theory opened a new world to me when I took them in 1947–48 while a graduate student in physiology. My most vivid memories of these courses are the clarity of his lectures, his enthusiasm for his subject, and the insight he imparted into the behavior of matter at the molecular level. The brilliance of his lectures was emphasized when Henry was out of town and his post-doctoral students had to substitute for him; they suffered in the inevitable comparison.

In going through my notes of the reaction rate theory course I found one of my test papers and was reminded forcefully of the learning pains of youth. The test consisted of three questions selected from a list of 21 given out in advance. At the end of my paper there is a plaintive note to the effect that one of the three was the one question out of the 21 that I was ill-prepared to answer (as the answer plainly shows). I was surprised to get a B + on the test. In retrospect, I suppose this grade was based more on my performance the previous three quarters than on that particular test.

The grade was generous and generosity is a word that helps to describe Henry Eyring. His easy, outgoing, gregarious, and gracious manner, combined with his generous praise of the efforts and attributes of students, aides, and peers and ready availability to all comers partially account for his exemplary impact as a scientist, teacher, and person.

Henry Eyring has had a twofold impact on my professional life: first as a superb teacher; second and more important, during an active apprenticeship with him I gained much insight into the nature of the creative process. My role in this process was to supply at need my imperfect knowledge of the phenomena of nerve impulse conduction. I learned much as I watched Henry build onto an embryonic theory of membrane function that fit most of the known facts. There resulted a paper[1] which outlines the application of rate theory to ion transport processes through membranes. On rereading our paper I found that it contains several ingenious proposals built on a sound theoretical framework. Although most of the proposals have been superceded by newer knowledge, the theoretical base is still useful; this paper goes on from where we left off.

SPECIAL PROPERTIES OF EXCITABLE CELLS

In their classic series of papers, Hodgkin and Huxley[2-6] gave a quantitative description of the unique electrical behavior of the giant nerve fibers (axons) of squid. This behavior is described in terms of permeability of the surface membrane (measured as conductance per unit area of membrane) to different ion species, particularly Na^+ and K^+. The current carried by an ion species through the membrane is then calculated from the product of conductance and driving force on them (transmembrane voltage, V, minus ion equilibrium potential). The specific ionic conductances have several unique properties which challenge explanation at the molecular level:

1. There are three or more separate systems carrying current through the membrane: a Na^+ system, a K^+ system, and at least one other system. These systems are usually called *channels*. Each system or type of channel is highly ion selective; e.g., Na^+ channels are 12 times more permeable to Na^+ than to K^+ (cf Ref. 7).

2. The conductances of the Na^+ and K^+ channels depend sensitively on the voltage across the membrane; the membrane molecules whose rearrangements are responsible for these changes in conductance have a minimum of about six charges per molecule, probably many times more.

3. The kinetics of the molecular rearrangements that cause conductance changes are complicated; the Na^+ channels obey fourth-order kinetics, whereas the K^+ channels have kinetics as high as twenty-fifth order (cf. Ref. 8). The reaction times are in the millisecond range.

4. The current–voltage relationships of both the Na^+ and K^+ channels are linear (constant conductance) when measured at times short compared with the times required for the molecular rearrangements that alter conductances to occur.

Although details and insights have been added in the years since 1952, the molecular bases of these properties remain largely unexplained. This paper gives an explanation in terms of Eyring rate theory of one puzzling membrane phenomenon, property (4) above. It is shown that a linear progression of potential energy barrier heights can be chosen to give (a) a linear current–voltage relation, (b) an inwardly directed rectification, or (c) an outwardly directed rectification.[9] This extension of our theory gives some insight into the nature of Na^+ ion channels because a linear current–voltage relation puts one constraint on the sequence of potential energy barrier heights in a channel. The theory also sheds some light on an even more

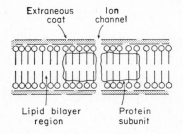

Fig. 1. Schematic drawing of a possible structure of an excitable cell membrane. Thickness of membrane is about 6 nm. See text.

puzzling phenomenon, anomalous rectification. The theoretical development is preceded by background material on the properties of cells and cell surface membranes.

Components of Cells

A nerve cell or fiber consists of three distinct components: a thin (ca. 10 nm), poorly conducting membrane (Fig. 1) separating two electrolyte solutions, the extracellular (outside) and intracellular (inside) fluids (cf. Fig. 2). The water activities and ionic strengths of the outside and inside solutions are equal, but in most other known properties these solutions are different. The extracellular fluid in comparison to the intracellular fluid has high concentrations of Na^+ and Cl^- and a low concentration of K^+. The absolute values of these concentrations vary from 0.1 to 0.5 M depending on the animal's environment. The ratios of external to internal concentrations of Na^+ and K^+ in nerve cells are less variable: $[Na^+]_o/[Na^+]_i \simeq 10$; $[K^+]_o/[K^+]_i \simeq 30$–40 regardless of species (Cl^- and other ions are not important in the present context and will not be further considered). Considerable concentration gradients of Na^+ and K^+ thus exist across the cell membrane.

Transmembrane Potential, V

The membrane is a poor conductor; the diffusion constants of Na^+ and K^+ in the membrane are only about 10^{-9} and 10^{-7} of their values in water, respectively. Hence there are relatively small but nonzero flows of these ions down their concentration gradients; Na^+ flows into and K^+ out of the cell. These fluxes are, on the average, balanced and the gradients maintained by active ion transport processes powered by the cell's metabolism. In addition, there is a steady potential difference, V_r, across the membrane of -60 to -90 mV, inside negative. This potential arises primarily because the membrane is 100 times more permeable to K^+ than to Na^+ and secondarily because of the active transport processes that maintain the concentration gradients.

Ion Penetration Through Membranes

The low values of ionic diffusion constants in the membrane could be due to (a) a high activation energy for ion entry and traversal of the membrane at any point; (b) a low transmission coefficient, e.g., widely scattered, hydrophilic channels or pores with low activation energy perforating an otherwise impenetrable membrane; or (c) a combination of both. There is convincing experimental evidence that the second alternative is correct. The most direct evidence is that the temperature coefficients of ion fluxes are low; Q_{10}'s are usually 1.3 to 1.5,[6,10] not much higher than the Q_{10}'s of electrolyte conductivity. Perhaps more convincing is the finding that the puffer fish poison, tetrodotoxin, in concentrations $\sim 10^{-8}$ M, selectively blocks Na^+ channels without affecting K^+ and other channels[11] and that less than 13 toxin molecules per μm^2 are required to block all Na^+ channels.[12]

Membrane Structure

There is considerable controversy about the structure of the cell surface membrane, but Fig. 1 shows a composite structure for the membrane that is reasonable, but unproven.[13] Most of the membrane surface is made of a lipid bilayer with coatings of mucoprotein (extraneous coat) in contact with

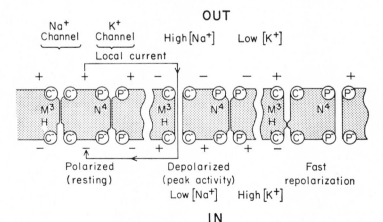

Fig. 2. Diagram of the properties of an excitable cell membrane. Voltages across each of the three segments of cell membrane are indicated by $+$ and $-$ signs. The ion selectivity properties of the Na^+ and K^+ channels are represented by negatively charged groups (C^- and P^-) guarding the entrances. See text for description of channel kinetics.

the intracellular and extracellular electrolyte solutions. The charged or dipolar "heads" of the lipid molecules face the mucoprotein and the hydrocarbon chains point inward. If cell surface membranes consisted entirely of the coated lipid bilayer, the ion permeability would be less than 10^{-6} of values found in cell membranes, assuming that resistivity measurements made on artificial lipid bilayer membranes are a reliable indication of the resistivities of coated lipid bilayers.

Ion Channels

There is good experimental evidence indicating that the lipid bilayer region of the membrane is interrupted at rare intervals (of the order of 1 μm) by channels that ions traverse easily ($\sim$ one-half the diffusion constant in water). The low ion permeability is ascribed to the paucity of channels. There are several kinds of channels named for the ion species that normally traverse them most readily as mentioned above. There are Na^+ channels, K^+ channels, and at least one other type of channel in excitable membranes. These channels are separable by their kinetics, ion selectivities, and pharmacology.[14]

Pore Proteins

The existence of specific ion channels through membranes is well established, but there is little knowledge of their structure. It is probable but not proven (cf. Ref. 13) that channels are built out of large protein molecules or groups of molecules embedded in the lipid bilayer as shown in Fig. 2 (protein subunits). The properties of each type of channel are conferred on it by the structure of the specific "pore" protein. Presumably this protein has a hydrophobic exterior for binding to the surrounding lipid and a hydrophilic channel for ion passage. In this picture, the dependence of Na^+ and K^+ permeabilities on the transmembrane potential difference, V, is attributable to conformation changes in the pore protein induced by changes in V.

MECHANISM OF THE ACTION POTENTIAL

The general nature of the excitation and propagation of the nerve impulse is illustrated in Fig. 2. The resting potential difference across the membrane is indicated by the + signs on the outside and − signs on the inside of the membrane at the left of Fig. 2. The Na^+ and K^+ channels are closed when V is near V_r. The processes determining the kinetics are represented by four first-order reactions,[5,15,16] three involving the reaction $M \rightleftharpoons M'$ and one the reaction $H \rightleftharpoons H'$ and K^+ channel kinetics by four reactions, $N \rightleftharpoons N'$.

The equilibria of these reactions depend sensitively on V. A Na^+ channel is open when 3 M molecules and one H molecule are in "place" (indicated by the unprimed letter). The whole process thus obeys fourth-order kinetics.

At the resting potential, the kinetics of the M, N, and H process are such that the Na^+ and K^+ channels are closed but the H portion of the Na^+ channel is open (Fig. 2, left). Depolarization (change of the negative V_r in the positive direction) changes the equilibria of all these reactions so that $M \rightleftharpoons M'$ and $N \rightleftharpoons N'$ are equilibrated left and $H \rightleftharpoons H'$ to the right. The M reaction is fast and so Na^+ channels open quickly (a few tenths of a millisecond) (Fig. 2, center). The H and N reactions are about 10 times slower; the H reaction closes the H region of the Na^+ channels and the N^4 reaction opens K^+ channels (Fig. 2, right).

When Na^+ channels are open and K^+ channels are still closed (Fig. 2, center), there is a greatly increased rate of Na^+ influx due to the high $[Na^+]_o$ and low $[Na^+]_i$. This flux quickly reverses the charge on the membrane, making the inside positive, that is, V approaches V_{Na}, the Na^+ equilibrium potential since $P_{Na} \gg P_K$ (P = permeability).

At a slightly later time, when the H process has closed many Na^+ channels and the N process has opened K^+ channels, V is quickly restored to near normal values by the increased efflux of K^+ and the decreased Na^+ influx. This depolarization due to Na^+ influx and the repolarization due to K^+ efflux is an action potential, lasting ~ 1 msec.

Since depolarization increases P_{Na} and an increase in P_{Na} increases Na^+ entry and further depolarizes the membrane, this is a regenerative process and is the main mechanism which generates the action potential. An action potential is propagated at constant speed along a cylindrical nerve fiber because of the depolarizing current flow from the region of high permeability to an undepolarized region (arrow in Fig. 2, left and center).

CURRENT–VOLTAGE RELATIONSHIP OF AN OPEN Na^+ CHANNEL

Experimental Current–Voltage Relation

Hodgkin and Huxley[3] studied the membranes of squid axons and found that the instantaneous current–voltage relations of both Na^+ and K^+ channels are linear when external concentrations are normal (property 4 above). "Instantaneous" means that measurements are made at times short compared with the kinetics of the M process. The Ohm's law statement of this relation for Na^+ channels is simply

$$I_{Na} = g_{Na}(V - V_{Na}), \tag{1}$$

where g_{Na} is the Na^+ conductance per cm^2, I_{Na} is the current density, and V is the voltage across the membrane (inside − outside). $(V − V_{Na})$ is the correct term for the driving force since, by the definition of V_{Na}, $I_{Na} = 0$ when $V = V_{Na}$. It must be emphasized that (1) holds only when $[Na^+]_o$ is normal. A change in $[Na^+]_o$ changes the I_{Na}–V curve in the manner predicted from the independence principle.[3]

Theory for a Homogeneous Membrane

The finding that current through Na^+ channels obeys (1) was unexpected and puzzling because theories of ion transport through homogeneous membranes predict a nonlinear current–voltage relation for any permeant ion, S, for $[S]_o \neq [S]_i$.

Three assumptions are usually made in calculating ion current density through a membrane: (a) an ion does not interact with any other ion while traversing the membrane (independence principle); (b) the membrane is homogeneous; and (c) the electric field in the membrane is constant in space. The linear transport equation then leads to the Goldman-Hodgkin-Katz equation[13,17] for the I_{Na}–V relation:

$$I_{Na} = P_{Na} \frac{F^2 z^2 V}{RT} \cdot \frac{[Na^+]_i e^{FzV/RT} − [Na^+]_o}{e^{FzV/RT} − 1}, \tag{2}$$

where P_{Na} is the membrane's permeability to Na^+ and F is the faraday. This relation is nonlinear unless $[Na^+]_o = [Na^+]_i$ and predicts an inwardly directed rectification when $[Na^+]_o > [Na^+]_i$, the normal case. Equation (2) also follows from Eyring rate theory using the same assumptions [Ref. 18, p. 454, (10)].

The linear I_{Na}–V curve is not always found in other excitable membranes. Dodge and Frankenhaeuser[19] studied the nodal membranes of toad myelinated nerve fibers and found that (2) accurately describes the I_{Na}–V relation. Frankenhaeuser[20] pointed out that a fixed surface charge density sufficient to reduce to zero the field in the membrane when $V = V_{Na}$ would give a linear current–voltage curve and was thus able to develop equations which permitted comparison of the Na^+ permeabilities of the membranes of squid and toad nerve fibers.

Thus a more general theory that can explain both linear and nonlinear experimental I–V curves would help in understanding these phenomena. Such a theory is presented here; it is shown that suitable choices of the sequences of potential energy barrier heights can lead to linear, inwardly rectifying, or outwardly rectifying I–V curves.

The discrepancy between homogeneous membrane theory and experimental results indicates that one or more of the assumptions are erroneous.

The independence assumption has been experimentally established for Na^+ channels.[2,3] Consideration of the other two assumptions indicates that there is no basis, except simplicity, for assuming that the membrane is homogeneous; the constant field assumption seems quite good in the sense that deviations are unlikely to account for linear I–V curves. In the picture of ion channels presented here the term "homogeneous" means that a hydrophilic ion channel has equal barrier heights throughout the membrane thickness. Since a channel is assumed to be through or between large protein molecules, barrier heights could vary considerably depending on local protein structure, particularly on the nature of charged or dipolar side groups.

Eyring Rate Theory for an Ion Channel

Figure 3a shows four barriers of a generalized potential energy diagram for an open Na^+ channel. Eyring's notation is used (cf Ref. 18, Fig. 16.1, p. 452). The distances between minima (λ) are assumed equal. Following

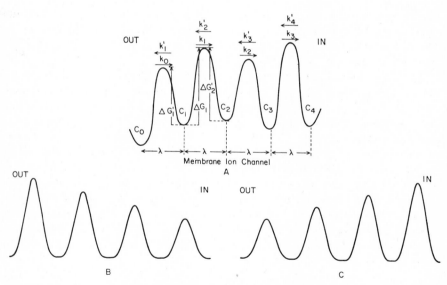

Fig. 3. A. Generalized potential energy profile of an ion channel through a membrane showing definitions of symbols used. Barriers are assumed to be symmetric and equally spaced. B. Linear decrease in barrier height from outside (OUT) to inside (IN). This can give rise to anomalous (outwardly directed) rectification in a Na^+ channel. Minima are all drawn at the same level for convenience, since their heights do not affect the current–voltage relationship. C. Linear increase in barrier heights from outside to inside. This accentuates normal, inwardly directed rectification in a Na^+ channel. See text.

Eyring, the fluxes of Na^+, M_{Na}, over successive barriers in the steady state are equal and of the form:

$$-M_{Na} = \lambda(C_{i-1}k_{i-1} - C_i k_i'),\tag{3}$$

where

$$k_i = \kappa_i \frac{kT}{h} e^{-\Delta G_i/RT}\tag{4}$$

and ΔG_i is the height of the ith barrier. The internal concentrations (C_1, $C_2, \ldots, C_{n-1}$) can be eliminated to give the flux in terms of C_0, C_n, and the forward and backward rate constants, k_i, k_i' (Ref. 18, p. 543):

$$-M_{Na} = \frac{k_0\lambda\left(C_0 - C_n \dfrac{k_1' \cdots k_n'}{k_0 \cdots k_{n-1}}\right)}{1 + \dfrac{k_1'}{k_1} + \dfrac{k_1'k_2'}{k_1k_2} + \cdots + \dfrac{k_1'k_2' \cdots k_{n-1}'}{k_1k_2 \cdots k_{n-1}}}.\tag{5}$$

The minus sign on M_{Na} is used because outward flux is defined as positive. This follows from the definition of V as the potential inside the cell minus the potential outside the cell. C_0 and C_n are defined as external and internal Na^+ ion concentrations (activities), respectively.

Since the internal and external solutions have nearly equal ionic strengths and water activities, the free energy difference between the two solutions is zero when $C_0 = C_n$ and $V = 0$, that is, the flux is zero under these conditions. This puts one constraint on the rate constants. The linearity requirement puts another constraint on the k_i's. Hence it is not meaningful to specify barrier heights by more than two arbitrary constants. Considerable mathematical experimentation showed that the simplest approach is adequate, that is, it is assumed that barrier heights change linearly through the membrane as shown in Figs. 3b and 3c. The minima are drawn at the same height since this results in considerable mathematical simplification without affecting the results; in the steady state and regardless of the depth, the C_i's take up the values necessary to maintain flux. The height of the ith barrier, ΔG_{i0}, when $V = 0$ is taken as

$$\Delta G_{i0} = \Delta G_{00} + i \cdot \delta G,\tag{6}$$

where δG is the difference in successive barrier heights and ΔG_{00} is the activation energy of the barrier immediately to the right of C_0 when $V = 0$. When $V \neq 0$, the field in the membrane is assumed to be constant and equal to $-V/n\lambda$, where n is the total number of barriers. This field adds an amount $Fz\lambda \cdot V/2n\lambda = FzV/2n$ to ΔG_i in the forward direction and subtracts an equal amount in the reverse direction (symmetrical barriers):

$$\Delta G_i = \Delta G_{00} + i \cdot \delta G + \frac{FzV}{2n},$$

$$\Delta G_i' = \Delta G_{00} + (i - 1)\,\delta G - \frac{FzV}{2n},$$

(7)

The term $(i - 1)$ is used for $\Delta G_i'$ because primed quantities refer to the $(i - 1)$th barrier (Fig. 3a).

The quantities k_i'/k_{i-1} and k_i'/k_i are needed in terms of δG and V to get flux as a function of voltage from (5). Using the definitions $g \equiv \delta G/RT$ and $v \equiv FzV/RT$,

$$\frac{k_i'}{k_{i-1}} = e^{v/n}, \qquad \frac{k_i'}{k_i} = e^{v/n+g}. \qquad (8)$$

Substitution of (8) into (5) and using $I_{Na} = FzM_{Na}$, gives the I_{Na}–V relation:

$$I_{Na} = \frac{Fzk_{00}\lambda[C_n e^v - C_0]}{1 + e^{(g+v/n)} + e^{2(g+v/n)} + \cdots + e^{(n-1)(g+v/n)}}$$

$$= Fzk_{00}\lambda[C_n e^v - C_0]\frac{[e^{(g+v/n)} - 1]}{[e^{n(g+v/n)} - 1]}, \qquad (9)$$

where $k = \kappa_0(kT/h)e^{-\Delta G_{00}/RT}$. For a homogeneous membrane $g = 0$ and $n \to \infty$. In this case (9) reduces to

$$I_{Na} = Fz\frac{k_{00}\lambda^2}{\lambda_n} v \frac{[C_n e^v - C_0]}{[e^v - 1]}, \qquad (10)$$

which is identical with (2), since $P_{Na} \equiv D_{Na}/d = k_{00}\lambda^2/\lambda n$ ($d = \lambda n =$ membrane thickness).

Linear I_{Na}–V Relation

Inspection of (9) shows that it can be made approximately linear by choosing g such that

$$ng = \ln\left(\frac{C_n}{C_0}\right) \equiv -v_{Na} \equiv -\frac{FzV_{Na}}{RT}, \qquad (11)$$

where V_{Na} is the equilibrium potential for Na$^+$, $V_{Na} = (RT/zF) \cdot \ln([Na^+]_o/[Na^+]_i)$. Equation (9) then becomes

$$I_{Na} = Fzk_{00}\lambda C_0[e^{(v-v_{Na})/n} - 1]$$

$$\simeq \frac{Fzk_{00}\lambda(v - v_{Na})}{n}. \qquad (12)$$

The larger n, the better the approximation. The I_{Na}–V curve is linear within experimental error over the range $-100 \text{ mV} < (V - V_{Na}) < 50 \text{ mV}$.[3] The experimental range was $-150 \text{ mV} < (V - V_{Na}) < 50 \text{ mV}$. The linear range puts a lower limit on the size of n: At the maximum value of $(V - V_{Na})$, $(V - V_{Na})/n < 25 \text{ mV}$. Thus, for $V - V_{Na} = 100 \text{ mV}$, $n > \frac{100}{25} = 4$. To get adequate linearity n must be two to three times larger (8 to 12), a not unreasonable range of values for a membrane 6–10 nm thick. Equation (12) is the same form as (1); hence, remembering that $v \equiv FzV/RT$,

$$g_{Na} = \frac{F^2 z^2 k_{00} \lambda C_0}{nRT} = \frac{F^2 z^2 P_{Na}[Na^+]_o}{RT}. \tag{13}$$

Hodgkin and Huxley[5] give a maximum value $g_{Na} = 0.13 \text{ mho/cm}^2$. $[Na^+]_o \simeq 0.5 \text{ mole/liter}$. Hence $P_{Na} \simeq 5 \times 10^{-5} \text{ cm/sec}$ (see Ref. 20). For comparison, the P_K of resting membrane is $\sim 10^{-6} \text{ cm/sec}$.

Woodbury et al.[13] studied a four-barrier membrane model and found that a slightly nonlinear sequence of barrier heights gives a satisfactorily linear $(2\%) I_{Na}$–V relation over the range $-150 \text{ mV} < (V - V_{Na}) < 150 \text{ mV}$. Their equation is more linear than (12) for $n = 4$. They found that all four barrier heights (not minima) were specified by using the three linearity constraints and the temperature coefficient of I_{Na}. For $[Na^+]_o = C_0 = 460 \text{ mmoles/liter}$ and $[Na^+]_i = C_4 = 60 \text{ mmoles/liter}$, the heights of the four barriers from inside to outside were 4330, 4390, 4650, and 5270 cal/mole, respectively. These are rather small changes in potential energy ($\sim RT$) and thus seem feasible.

Woodbury et al.[13] also used the theoretical equation for g_{Na}, measured values of g_{Na}, and the experimentally estimated number of channels per unit area[12] to compute the mean transmission coefficient $(\kappa_0 \kappa_4')^{1/4}$ interpreted as a constraint on an ion's angle of approach to a channel. Uncertainties limit the accuracy of this estimate, but the value is probably in the range 0.01 to 1: thus there appears to be little angular restriction on Na^+ entry into a membrane Na^+ channel.

NONLINEAR I_{Na}–V RELATIONS

The properties of (9) depend on the value chosen for the barrier height increment, g: At $g = 0$, the normal Goldman-Hodgkin-Katz relation, (2), is obtained and at $g = -v_{Na}/n$, a nearly linear relation results. This suggests that an even larger negative value for g would result in a reversed (outwardly directed) or "anomalous" rectification and a $g > 0$ would accentuate normal inward rectification. Figure 4 shows plots of I_{Na}–V relations calculated for $n = 4$, $V_{Na} = 58 \text{ mV}$ ($[Na^+]_o/[Na^+]_i = 10$) and for a number of values of

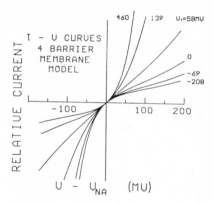

Fig. 4. Current–voltage relations of a Na$^+$ channel for various values of the parameter V_1. $V_{Na} = 58$ mV ([Na$^+$]$_o$/[Na$^+$]$_i$ = 10) and $n = 4$ (number of barriers). Abscissa is the difference between transmembrane potential V and the Na$^+$ equilibrium potential V_{Na}; $V \equiv$ potential of inside $-$ potential of outside. Outward current is positive. The slopes of the curves are normalized to one at $V - V_{Na} = 0$. All curves are tangent to one another at this point; none crosses. Curves for $V_1 = V_{Na} = 58$ mV is reasonably linear over the range -100 mV $< V - V_{Na} < 100$ mV. Anomalous rectification occurs for $V_1 > V_{Na}$ and accentuated normal rectification for $V_1 < 0$.

the parameter V_1 (mV) defined by $v_1 \equiv FV_1/RT \equiv -ng$; FzV_1 is thus the total decrement in barrier height from outside to inside. Equation (9) becomes

$$I_{Na} = Fzk_{00}\lambda C_0 \frac{[e^{(v-v_1)/4} - 1]}{[e^{(v-v_1)} - 1]} e^{(v-v_{Na})} - 1].$$ (14)

The curves in Fig. 4 are normalized to a slope of 1 at $V - V_{Na} = 0$. None of the curves cross; they are all tangent at $V = V_{Na}$. It can be seen that for $V_1 < 58$ mV ($= V_{Na}$), there is a normal (inward) rectification and for $V_1 > 58$ mV, there is an outward rectification. The curves for $V_1 = 460$ mV and -208 mV are close to the limiting curves for $V_1 \to \pm \infty$.

Accentuated Inward Rectification

Figure 3c shows a potential energy profile sequence that accentuates the normal inward rectification: the reason is that $[Na^+]_o$ is high and entry is easy (barrier is low) from the outside and vice versa on the inside. I know of no examples of this behavior in a cell membrane. The delayed rectification of a K^+ channel in membranes of squid nerve fibers is of this nature (N^4 process in Fig. 2) but this is undoubtedly due to the opening of more channels rather than the property of a single channel. As mentioned above, the instantaneous current–voltage relation of an open K^+ channel is linear.[3] The theory developed here, however, is not directly applicable to K^+ channels since the independence assumption does not hold for K^+ channels.[21]

Anomalous Rectification

A potential energy sequence like that in Fig. 3b can cause outward (anomalous) rectification. This rectification occurs because Na^+ ions can enter the membrane from the right (inside) quite easily but encounter a high barrier when entering from the left (outside). If the ease of entry from the inside more than compensates for the low $[Na^+]_i$, then there will be an outward rectification. This condition (cf Fig. 4) is that $V_1 > V_{Na}$, or $-ng >$ ln $([Na^+]_o/[Na^+])_i$. In Fig. 4, $[Na^+]_o/[Na^+]_i = 10$, $ng < -2.3$, and thus the total difference in ΔG from outside to inside is 1400 cal/mole or greater. For $n = 4$, $\delta G < -350$ cal/mole.

The extreme case shown in Fig. 4, $V_1 = 460$ mV requires a rather larger δG: $-ng = 18$, $g = -4.5$ and $\delta G = -2700$ cal/mole. The total $\Delta G = 10,800$ cal/mole. This predicts that current flowing through a membrane channel having a pronounced anomalous rectification will have a high temperature coefficient. I know of no pertinent experimental data.

Anomalous rectification for K^+ has been found in the membranes of skeletal muscle cells[22–24] and cardiac muscle cells.[25] It is therefore of interest to see if the I_K-V curves for muscle cell membranes are adequately described by (14) for appropriate values of C_0, C_n, n, and V_1. The shapes of the curves are similar, but (14) is not a good quantitative description primarily because the curvature of the experimental relation is too large. This can be shown by rewriting (14) for K^+ ions and taking the limit of I_K as V_1 approaches $+\infty$:

$$I_K = Fzk_{00}\lambda C_0[e^{(v-v_K)} - 1]. \tag{15}$$

For values of $(v - v_K) \geq 2$, I_K is approximately exponential with V. The voltage constant of (15) is thus $RT/Fz = 25$ mV. However, the corresponding range of the experimental data has a voltage constant of about 10 mV.[23] No adjustment of V_1 in (14) or of barrier heights or symmetry can circumvent this difficulty. The obvious (but not necessarily correct) explanation of this discrepancy is that $z \simeq 2$ ($RT/Fz \simeq 12$ mV), that is, K^+ ions might act as though they traverse an anomalous K^+ channel two or more at a time.

Such an assumption is not easy to evaluate quantitatively since the independence assumption is violated and special assumptions must be made about the nature of the K^+ ion interactions in the channels. The possibility that there are interactions between two or more K^+ within an anomalous rectification K^+ channel is supported by the experimental evidence that K^+ ions do interact in the K^+ channels of the squid nerve fiber membrane.[21] The interaction is the type expected if three or more K^+ were in single file in the channel. Thus, it seems possible that the anomalous rectification properties of the K^+ channels in muscle membranes have a potential energy profile similar to that in Fig. 3b but that there are usually two or three K^+ in any one channel. This implies that ion transit time is two to three times larger than the time between successful collisions of ions with the channel entrances. By contrast, Na^+ channels obey the independence principle implying that a Na^+ ion traverses a channel before there is another successful collision of an ion with the channel entrance. The longer transit time of K^+ ions could be due to deeper minima (tighter binding) in K^+ channels as compared with Na^+ channels.

CHALLENGE TO REACTION RATE THEORETICIANS

The foregoing analysis sheds some light on the molecular properties that determine the instantaneous current–voltage relations of ion channels through excitable cell membranes. The success of a simple sequence of potential energy barriers in predicting otherwise puzzling current–voltage behavior suggests that chemical kineticists could successfully tackle some of the more challenging aspects of membrane phenomena. Three problems seem particularly susceptible to attack:

1. The molecular basis of the ion selectivity sequences of membrane ion channels (cf. Ref. 7). It is not clear what molecular structures can preferentially select Na^+ over K^+ by a twelvefold ratio. Current theories attribute the selectivity to negative charge groups at the channel entrance (cf Fig. 2). These can account for ionic permeability sequences but not for the magnitudes of the selectivity.

2. The nature of the rearrangements in membrane molecules that give rise to the steep dependencies of sodium and potassium conductances on transmembrane voltage and their peculiar kinetics.[5] There are no satisfactory

molecular models of this behavior. Conformation changes in charged or dipolar proteins induced by changes in transmembrane voltage seems an attractive possibility but have not to my knowledge been investigated (cf Ref. 13).

3. The current–voltage relations of channels containing more than one ion at a time. This would be an extension of the theory given in this paper to see if the experimental properties of anomalous rectification channels can be explained in this manner.

REFERENCES

1. H. Eyring, R. Lumry, and J. W. Woodbury, "Some Applications of Modern Rate Theory to Physiological Systems," *Record Chem. Progr.*, **10**, 100 (1949).
2. A. L. Hodgkin and A. F. Huxley, "Currents Carried by Sodium and Potassium Ions Through the Membrane of the Giant Axon of *Loligo*," *J. Physiol.* (*London*), **116**, 449 (1952).
3. A. L. Hodgkin and A. F. Huxley, "The Components of Membrane Conductance in the Giant Axon of *Loligo*," *J. Physiol.* (*London*), **116**, 473 (1952).
4. A. L. Hodgkin and A. F. Huxley, "The Dual Effect of Membrane Potential on Sodium Conductance in the Giant Axon of *Loligo*," *J. Physiol.* (*London*), **116**, 497 (1952).
5. A. L. Hodgkin and A. F. Huxley, "A Quantitative Description of Membrane Current and Its Application to Conduction and Excitation in Nerve," *J. Physiol.* (*London*), **117**, 500 (1952).
6. A. L. Hodgkin, A. F. Huxley, and B. Katz, "Measurements of Current–Voltage Relations in the Membrane of the Giant Axon of *Loligo*," *J. Physiol.* (*London*), **116**, 424 (1952).
7. H. Meves, "Experiments on Internally Perfused Squid Giant Axons," *Ann. N.Y. Acad. Sci.*, **137**, 807 (1966).
8. K. S. Cole, *Membranes, Ions and Impulses*, University of California Press, Berkeley, 1968.
9. J. W. Woodbury, "Linear Current–Voltage Relation for Na^+ Channel from Eyring Rate Theory," *Abstr. Biophys. Soc. Ann. Meeting*, **9**, A250 (1969).
10. W. D. Stein, *The Movements of Molecules Across Cell Membranes*, Academic, New York, 1967.
11. T. Narahashi and J. W. Moore, "Neuroactive Agents and Nerve Membrane Conductances," *J. Gen. Physiol.*, **51**, 93S (1968).
12. J. W. Moore, T. Narahashi, and T. I. Shaw, "An Upper Limit to the Number of Sodium Channels in Nerve Membrane?" *J. Physiol.* (*London*), **188**, 99 (1967).
13. J. W. Woodbury, S. H. White. M. C. Mackey, W. L. Hardy, and D. B. Chang, "Bioelectrochemistry," in *Electrochemistry*, H. Eyring, W. Jost, and D. Henderson, Eds., Academic Press, New York, 1970, Chap. 9.
14. B. Hille, "Ionic Channels in Nerve Membranes," *Prog. Biophys. Molec. Biol.*, **21**, 1 (1970).
15. J. W. Woodbury, in *Handbook of Physiology*, Section 2, *Circulation*, W. F. Hamilton, Ed., Vol. I, pp. 237–286, American Physiological Society, Washington, D.C., 1962.
16. J. W. Woodbury, in *Physiology and Biophysics*, 19th ed., T. C. Ruch and H. D. Patton, Eds., W. B. Saunders, Philadelphia, 1965, Chapters 1 and 2, pp. 1–58.

17. A. L. Hodgkin and B. Katz, "The Effect of Sodium Ions on the Electrical Activity of the Giant Axon of the Squid," *J. Physiol. (London)*, **108**, 37 (1949).

18. H. Eyring, D. Henderson, B. J. Stover, and E. M. Eyring, *Statistical Mechanics and Dynamics*, Wiley, New York, 1964.

19. F. A. Dodge and B. Frankenhaeuser, "Sodium Currents in the Myelinated Nerve Fiber of *Xenopus laevis* Investigated with the Voltage Clamp Technique," *J. Physiol. (London)*, **148**, (1959).

20. B. Frankenhaeuser, "Sodium Permeability in Toad Nerve and in Squid Nerve," *J. Physiol. (London)*, **152**, 159 (1960).

21. A. L. Hodgkin and R. D. Keynes, "The Potassium Permeability of a Giant Nerve Fibre," *J. Physiol. (London)*, **128**, 61 (1955).

22. B. Katz, "The Electrical Properties of the Muscle Fiber Membrane," *Proc. Roy. Soc. (London)*, *B*, **135**, 506 (1948).

23. R. H. Adrian and W. H. Freygang, "Potassium Conductance of Frog Muscle Membrane under Controlled Voltage," *J. Physiol. (London)*, **163**, 104 (1962).

24. A. L. Hodgkin and P. Horowicz, "The Influence of Potassium and Chloride Ions on the Membrane Potential of Single Muscle Fibers," *J. Physiol. (London)*, **148**, 127 (1959).

25. A. E. Hall, O. F. Hutter, and D. Noble, "Current–Voltage Relations of Purkinje Fibres in Sodium-Deficient Solutions," *J. Physiol. (London)*, **166**, 225 (1963).

Theory of Rate Processes Applied to Release of Acetylcholine at Synapses

LUDVIK BASS AND WALTER J. MOORE*

Departments of Mathematics and Biochemistry, University of Queensland, Brisbane, Australia

* Permanent address: Indiana University, Bloomington.

A Personal Note

New graduate students in chemistry at Princeton 30 years ago soon came to accept the fact that Henry Eyring could see molecules. Henry spent eight to ten hours a day in his office talking with students and co-workers. After about an hour with him one would emerge into a new world in which the "little fellows," John and Joe and Sam, as Henry called them by name, were still dancing about in the friendliest way imaginable. Thus Henry showed how the ornery behavior of complex systems could often be tamed by a kindly approach to the molecules. With blackboard, chalk, and relentless discussion, many mysteries were unraveled. In this paper we shall try a similar method on a tantalizing neurophysiological problem, in which every experimental sign seems to point in two opposite directions and a perfectly lovely hypothesis now finds itself in a most precarious situation.

INTRODUCTION

When an action potential traveling down the axon of a motoneuron reaches the myoneural endplate, a process occurs that releases acetylcholine into the synaptic cleft and consequently depolarizes the postsynaptic membrane. A similar process probably occurs at cholinergic synapses in the central nervous system. In 1950 Fatt and Katz discovered a spontaneous sub-threshold activity (MEPP) of motor nerve endings and were thereby led to the concept that acetylcholine is released in definite units (quanta) of 10^4 to 10^5 molecules.[1] Electron microscopy subsequently revealed characteristic vesicles about 40 nm in diameter, clustered near presynaptic membranes. Subcellular fractionation procedures were devised by Whittaker[2] and de Robertis[3] for the isolation of these vesicles from brain homogenates in sucrose density gradients, and it was soon demonstrated that they were indeed concentrated reservoirs of acetylcholine. The hypothesis that the vesicles discharge the quanta of transmitter became irresistible.

The time course of radioactive labels in proteins from brain fractions provided evidence that synaptic vesicles are entities of long average lifetime in neurons and, therefore, can seldom be destroyed during the processes that release transmitter at the synapse.[4] One is thus led to a picture in which a vesicle strikes an inner presynaptic membrane, discharges its cargo (of acetylcholine, at a cholinergic synapse) through the membrane into the synaptic cleft, and then withdraws from contact with the membrane to refill with transmitter collected from an ambient store in the cytoplasm. Hubbard and Kwanbunbumpen[5] found, however, that treatments which accelerated discharge of quanta of acetylcholine decreased the concentration of vesicles immediately adjacent to the myoneural junction. This observation would be consistent with actual destruction of vesicles as they fuse with the presynaptic membrane. The long lifetime of radioactive protein in vesicles could then be explained by reutilization of labeled proteins to reform vesicles; however, there is the difficulty that the gel electrophoresis pattern of vesicle membrane proteins is idiosyncratic.[9]

ELECTROKINETIC MECHANISM OF MEPP

In 1966 Bass and Moore[7] applied the theory of rate processes as formulated by Henry Eyring to the kinetics of generation of the miniature excitatory postsynaptic potentials (MEPP). The synaptic vesicles were considered to have the electrophoretic properties of charged spherical colloidal particles.[8] The sign of the charge was required to be positive in order to explain the increase in frequency v of the MEPP with depolarization of the presynaptic membrane.[9]

When this electrokinetic theory was proposed, it was generally believed that the vesicles must bear negative charges. Aside from a few observations on isolated vesicles, which would be of limited significance owing to the different environments of vesicles *in vivo*, this conclusion was based on the experiments of Hubbard and Willis[10] on the effect of polarization of the nerve terminals on the amplitude of the endplate potential. In studies on a rat diaphragm preparation they found that if a current was applied through a microelectrode placed very close to the nerve terminal, there was an enormous increase in the EPP if prolonged current was applied in a direction so as to hyperpolarize the nerve terminal (make the inside more negative relative to the outside). This enhancement of the EPP built up slowly over many seconds and slowly declined with a cessation of current. There was no associated increase in the size of MEPP so that the increase in EPP must have been due to an increase in the number of quanta released per impulse. In his analysis of these results Eccles[11] concluded that

The applied current must slowly effect a large building up of the available transmitter. Presumably it does this by causing an electrophoretic migration of the Ach packets up to the presynaptic membrane fronting the presynaptic cleft. The increase in EPP may be as much as 20-fold, so there should be a remarkable concentration of synaptic vesicles close to the synaptic cleft, if indeed synaptic vesicles are the preformed packets of Ach. It would be of great interest to test whether a corresponding concentration of synaptic vesicles can be observed by electron-microscopic examination of such polarized nerve terminals.

The suggested migration of vesicles toward the hyperpolarized membrane would be most consistent with a negative charge on the vesicles, directly contrary to the suggestion of our theory.

In 1969 Landau and Kwanbunbumpen[12] carried out the electron-microscopic study as indicated by Eccles. Polarizing currents were passed through the rat phrenic nerve-diaphragm preparation for 3 min after which a 3%

glutaraldehyde fixative* was added to the preparation without stopping the current flow. The results showed that depolarizing current caused accumulation of vesicles, whereas hyperpolarizing current caused dissipation of vesicles adjacent to the presynaptic membrane. The authors concluded:

These results thus indicate that the effects of conditioning presynaptic current on the transmitter release cannot be attributed to the changes in nerve-terminal morphology proposed by Hubbard and Willis. Our findings support the electrokinetic theory of release outlined by Bass and Moore.

We should emphasize, however, that Eccles was careful to qualify his prediction with the condition, "if indeed synaptic vesicles are the preformed packets of acetylcholine." Some recent experiments and theoretical results have renewed interest in a search for mechanisms for quantal release of transmitter which do not involve the vesicles directly; for example, Whittaker[13] has found that presynaptic membranes can store acetylcholine in bound form. Using the standard order of magnitude figure of 10^{15} adsorption sites/cm^2 for any surface, 10^5 molecules of Ach would require at least 10^{-10} cm^2 of membrane area. The area of an endplate gutter is about 6×10^{-5} cm^2, sufficient for the storage of at most 6×10^5 quanta. With an average of 10^2 quanta per EPP, the membrane could readily store enough Ach for about 10^3 EPP's. The function of the vesicles would then be to act as a sort of buffer to maintain an effectively constant Ach concentration in the terminals. It must be admitted that such a mechanism seems to lack the elegance of the vesicle hypothesis and as yet we do not have a quantitative theory for the hypothetical quantal release from the membrane.

EFFECTIVE CHARGE ON A VESICLE

The effective charge on a vesicle is only partly determined by the fixed charges, if any, at its membrane surface. Since a vesicle is not a solid colloidal particle we must consider also the effective charge due to the potential that arises as a consequence of the semipermeability of the boundary membrane and the difference in ionic activities between the interior of the vesicle and the external cytoplasmic medium. The charge can then be estimated from the actual structure of the electric field set up across the membrane and extending into the diffuse boundary layer outside the membrane wall.

The first problem is to decide which ions are potential-determining in the steady state of the vesicle filled with transmitter. There is good evidence that

---

* Reaction with the aldehyde fixative might shift the charge on the vesicles in a negative direction, by removing uncharged NH_2 groups and thus causing NH_3^+ groups to lose protons by a mass-action effect.

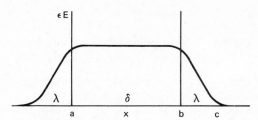

Fig. 1.   Electric displacement vesicular membrane.

the wall of a filled vesicle is virtually impermeable to Ach[14]; hence Ach$^+$ cannot be a potential-determining ion.*

It seems most likely, therefore, that the vesicle membrane behaves similarly to an axonal membrane, in that it is selectively permeable to K$^+$ ions. Since the vesicle is swimming in the cytoplasm, it would be expected to have a high concentration of K$^+$ at the exterior of its membrane and a low concentration of K$^+$ in its internal medium. Thus a vesicle membrane can be considered to be "inside out" as compared to a normal plasma membrane forming a neuronal wall. We should hardly be surprised, therefore, to find that vesicle walls have an effective positive charge, whereas cell membranes appear to have external negative charges.†

Figure 1 depicts the electric displacement ϵE [where E is the field and ϵ the dielectric constant] across the vesicle membrane of thickness δ on the assumption that the media of both sides have equal ionic strengths. The resting potential difference across the membrane, V_N, is of the liquid-junction type but is usually close to a Nernst potential. The Debye lengths in both the ionic media are λ. Inside the membrane $E = E_P$, the constant Planck field. Outside the membrane, the field is assumed to decline in accord with a linear Goüy-Chapman model, $E = E_P e^{-x/\lambda}$, so that $V_N = -E_P(\delta + 2\lambda)$. If we define the surface of shear as $x = x^*$, the effective charge per unit area of the membrane can be obtained as

$$\sigma = \frac{\epsilon}{4\pi} E(x^*) = \frac{\epsilon}{4\pi} E_P e^{-x^*/\lambda} = -\frac{\epsilon}{4\pi}\left(\frac{V_N}{\delta + 2\lambda}\right)e^{-x^*/\lambda}. \tag{1}$$

---

* A high concentration of Ach can be accumulated inside a vesicle against a concentration gradient if (a) it is bound to acceptor sites on the inner membrane wall or (b) a special "acetylcholine pump" exists in the membrane. The former hypothesis is more economical. Binding of Ach$^+$ to the wall will reduce the activity of Ach$^+$ to a low value. If the surface tension γ of the vesicle membrane was sufficiently high, it would be possible to explain the concentration of acetylcholine within the vesicle (radius r) by means of the Laplace equation $\Delta P = 4\gamma/r$, which yields $m = 4\gamma/2\varphi RTr$ for the molality of Ach$^+$Cl$^-$ inside the vesicle, where φ is the osmotic coefficient.

† Segal[15] has found, however, that the apparent surface charge of a perfused lobster axon is not sensitive to the gradient of resting potential.

Equation 1 is a general expression for the estimation of the effective electrophoretic charge when it is due to a Planck diffusion field across a membrane. In the case of the vesicle, since $V_N < 0$, $\sigma > 0$, and the effective charge is positive. Since the displacement ϵE is continuous across jumps in the dielectric constant ϵ, and since most of the potential difference is across the membrane (Fig. 1) we need to use in (1) only the effective dielectric constant of the vesicular membrane. Taking that ϵ by analogy with typical biological membranes, the charge per vesicle would range from $5.6|e|$ for $\epsilon = 20$ to $1.4|e|$ for $\epsilon = 5$, with $x^*/\lambda = 0.5$. This range includes the value $4|e|$ previously estimated from the effect of depolarization on the frequency of MEPP.[7]

ELECTROSTATIC INTERACTION BETWEEN VESICLE AND PRESYNAPTIC MEMBRANE

In the earlier paper,[7] the free energy of activation for the rate process in which a vesicle approached the synaptic membrane was treated as an electrostatic interaction between a positive point charge and the field of the membrane. It would be more appropriate to consider the interaction between two diffuse double layers. Such a problem was considered in detail by Verwey and Overbeek,[16] who obtained the following expression for the potential energy per unit area between two planar double layers having identical surface potentials ψ_0 and separated by a distance d:

$$U = \frac{64NkT}{\kappa} \gamma^2 e^{-2\kappa d},$$

where

$$\gamma = \frac{e^{y/2} - 1}{e^{y/2} + 1}, \quad \text{with} \quad y = \frac{ze\psi_0}{kT}.$$

The reciprocal Debye length is $\kappa = \lambda^{-1}$, the number of ions per unit volume with charge z far from the interface is N.

Since ψ_0 for the vesicle is fixed whereas for the presynaptic membrane it is varied by increments ΔV, in (2) we introduce a rough symmetrization of the Verwey-Overbeek equation:

$$U = \frac{64NkT}{\kappa} \gamma_m \gamma_v e^{-2\kappa d}, \tag{2}$$

where $\gamma_v(\psi_v)$ and $\gamma_m(\psi_m)$ have the significance of $\gamma(\psi_0)$ at the vesicle and terminal membranes, respectively.

Equation 2 is only a rough approximation to the detailed theory of the interaction of two dissimilar surface layers, which has been considered by

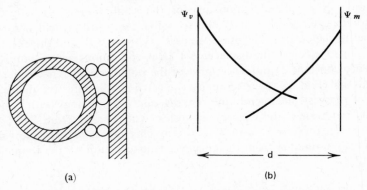

Fig. 2.   Approach of synaptic vesicle to presynaptic membrane.

Derjaguin,[17] by Hogg et al.,[18] and by Devereux and de Bruyn.[19] These treatments show that an attractive interaction can occur even in some cases in which one of the surface potentials is zero, or even when the two potentials have the same sign provided their magnitudes are different and the separation of the double layers is small. For the parameters used in the present application, however, our (2) does not depart from the tabulated values by more than about 20%.

The approach of a vesicle to the membrane is shown schematically in Fig. 2. The effective area of interaction is taken to be the area subtended by the surface of the vesicle between a closest approach to one water-molecule diameter and two such diameters. This area[7] will be $5 \times 10^{-5} \, \mu m^2$ for a vesicle diameter of $4 \times 10^{-2} \, \mu m$.

Let us consider that γ_m applies to the presynaptic membrane of thickness δ. In this case $\psi_m = \alpha V_m$, where V_m is the Planck membrane potential and $\alpha = \epsilon'\lambda/\epsilon\delta$ has been introduced and estimated by Bass and Moore[7] to allow for the effect on the field of the jump in the dielectric constant from ϵ to ϵ' at the membrane boundary; $y \ll 2$ follows and γ_m becomes, with $z = 1$, simply

$$\gamma_m = \frac{y}{4} = \frac{\alpha e V_m}{4kT}. \tag{3}$$

We can now write the electrical contribution to the activation free energy as

$$\Delta G_e^{\ddagger} = 16NA\alpha\kappa^{-1}e\gamma_v e^{-2\kappa d}(V_m - \Delta V),$$

where A is the effective vesicle area, γ_v pertains to the vesicle, and ΔV is the depolarization of presynaptic membrane. Inserting an approximate value for $\gamma_v \simeq e\psi_v/4kT$,

$$\Delta G_e^{\ddagger} = \frac{4NA\alpha\kappa^{-1}e^2\psi_v}{kT} e^{-2\kappa d}(V_m - \Delta V). \tag{4}$$

The frequency of MEPP is then

$$\nu = \nu_0 \exp \left(\frac{-\Delta G_e^{\ddagger} - \Delta G_i^{\ddagger}}{kT} \right) \tag{5}$$

where $\Delta G_i^{\ddagger}$ is the thermal contribution to $\Delta G^{\ddagger}$. It is of interest to compare the $\Delta G_e^{\ddagger}$ calculated from this expression with that previously estimated. The following parameters were used for rat muscle preparations at 305°K:

$$A = 5 \times 10^{-13} \, cm^2, \qquad e^{-2\kappa d} \simeq 0.3$$

$$N = 1.05 \times 10^{20} \, cm^{-3}, \qquad \psi_v = \frac{\lambda}{\delta} V_m = 7 \, mV,$$

$$\lambda = (7.5 \times 10^{-8}) \, cm, \qquad \frac{eV_v}{kT} = \frac{eV_m}{kT} = 2.65.$$

With these we calculate

$$\frac{\Delta G_e^{\ddagger}}{kT} = 31.8\alpha, \qquad \Delta G_e^{\ddagger} = 0.83\alpha \, eV \quad at \quad 305°K.$$

This figure compares with an electrical barrier of 0.28 eV estimated from the point charge model.[7] These values would coincide for $\alpha = 0.34$. Hence the theory based on the Verwey-Overbeek potential is consistent with experiment. From (4) and (5) the experimental plot of ln ν versus ΔV should yield a straight line with the so-called Liley slope. Inserting the numerical values, we find that this slope, with $\alpha = 0.34$, corresponds to a tenfold increase in frequency of MEPP for each 15.0 mV depolarization. The most recent experimental value on the rat muscle preparation was 12.5 mV,[20] and Liley[9] reported about 16 mV.* We can thus conclude that a discussion of the approach of a synaptic vesicle to a presynaptic membrane, which is based on the Verwey-Overbeek theory of the interaction of two double layers, gives reasonable quantitative agreement with the available data on the dependence of ν of MEPP on depolarization of the presynaptic membrane.

The actual discharge of acetylcholine by a vesicle has not yet been described. We suggest that the contact between the "inside-out membrane" of the vesicle and the normal presynaptic membrane produces a momentary fusion of the surfaces. As soon as this happens the electric field across the area of fusion collapses, and acetylcholine is ejected. The mechanism by which collapse of the double layer fields can cause emission of the transmitter is not

---

* Liley measured MEPP 20 min after changing the [K+]. Slopes measured after 5 min would be considerably lower.[21] Since the slope is sensitive to α, these lower values can readily be obtained from the present theory with reasonable values for α.

understood, but it may be analogous to the mechanism of the fast ionic permeability change in an axonal membrane following a critical depolarization. A model based on the Wien dissociation effect may therefore be feasible in this case also.[22]

EFFECTS OF CALCIUM ON DISCHARGE OF SYNAPTIC VESICLES

The principal factors that influence the frequency of discharge of synaptic vesicles are the temperature, the state of polarization of the presynaptic membrane, and the osmotic pressure. The strong dependence of the frequency v of MEPP on these factors is explained by the hypothesis that they all contribute to the free energy of activation in the expression $v = v_0 \exp(-\Delta G^{\ddagger}/kT)$. Another factor that is important in the discharge of synaptic vesicles is the concentration of calcium ions, $[Ca^{2+}]$, in the synaptic cleft. This factor has a definite influence on the v of MEPP, but a more important effect on the amplitude of the end plate potential.* Although MEPP will persist even in the absence of Ca^{2+} ions in the cleft, the EPP cannot be excited without Ca^{2+} ions.

Alternative models for the calcium effects have been advanced by Hubbard, Jones, and Landau[21,23] and by Katz and Miledi.[24] These workers believe that the reaction of calcium with sites on the vesicle or on the presynaptic membrane controls the rate of discharge of synaptic vesicles. The effect of depolarization is then either (a) to facilitate the transport of Ca^{2+} to these specific absorption sites by opening channels for ionic transport, or (b) to facilitate the transport of the calcium complex through the membrane. To what extent can the calcium effect on MEPP be understood within the framework of the basic electrokinetic theory without such new ad hoc assumptions, namely, by means of an effect of $[Ca^{2+}]$ on the free energy of activation $\Delta G^{\ddagger}$?

The ideas upon which this analysis is based are depicted in Fig. 3. In the ordinary treatment of membrane potentials as given by Planck[25] and Schlögl,[26] the field in the membrane will be constant provided the ionic strength of the electrolyte is the same at both edges. This is the situation of the natural axonal membrane, for example, and it is the special case of the Planck-Schlögl theory coinciding with the Goldman[27] or constant-field model. In case the ionic strengths at the two edges are not the same, however, the Planck-Schlögl theory shows that the field will become lower in regions of

* If the postsynaptic potential (EPP) is a summation of MEPP, τ is the width of an MEPP and a its amplitude, and $v\tau \gg 1$ while $(dE/dt)\tau \ll E$, where E is the depolarization, the postsynaptic amplitude will be $A \simeq a\tau v(E)$, so that the amplitude represents a summation of superimposed MEPP.

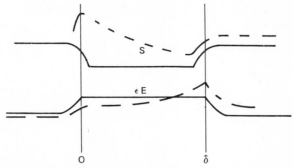

Fig. 3. Ionic strength s and electric displacement ϵE in presynaptic membrane—dashed curves show effect of Ca^{2+}.

higher ionic strength, since by such an adjustment the constancy of ionic flux throughout the membrane can be maintained.[28] If, therefore, a high concentration of Ca^{2+} ions occurs in any region of the membrane, the field is reduced in that region. In particular, Fig. 3 illustrates a situation in which Ca^{2+} ions and negative counterions are concentrated in the region adjacent to the inner edge of the membrane. Such a concentration would be a natural consequence of the distribution of Ca^{2+} ions in the field of the polarized membrane, even in the absence of specific acceptor sites for Ca^{2+} ions. As the field at the inner edge of the presynaptic membrane is reduced, the number of charges in the diffuse layer is also reduced; hence an approaching synaptic vesicle will experience a lowered electrical repulsion. These effects provide a mechanism by which the free energy of activation for the rate of discharge of vesicles may depend on the concentration of Ca^{2+} ions. With this mechanism the effects of $[Ca^{2+}]$ are coupled with the effects of depolarization, since depolarization decreases $[Ca^{2+}]$ at the inner edge of the presynaptic membrane.*

The problem has been treated by solving the electrodiffusion equation for the univalent ions subject to a perturbation caused by Ca^{2+} ions which can enter the membrane but which cannot pass through it. Thus in the system sketched in Fig. 3 the surface $x = 0$ is impermeable to Ca^{2+} ions. The essential evidence for impermeability to Ca^{2+} at $x = 0$ is that when EPP's are lost by Ca deficiency in the cleft, they are *not* restored by Ca injection into the presynaptic region.[29] All the boundary values are taken just inside the membrane, hence include distribution coefficients.

* If we suppose that Mg^{2+} from the cleft penetrates a shorter distance than Ca^{2+} into the presynaptic membrane, we can account for the antagonistic effects of Mg^{2+} by analogous considerations of field redistribution.

The result of the somewhat lengthy perturbation treatment is that the electric field at the inner edge of the synaptic membrane is

$$E = E°(1 - pR),$$ (6)

where

$$R \equiv \frac{1}{\cosh^2 y}, \quad \text{and} \quad p = \frac{n_s}{C(0)} \ll 1;$$

the concentration of sites available to Ca^{2+} in the membrane is n_s, $C(0)$ is the concentration of the mobile ions just inside the membrane at $x = 0$, and

$$y = \frac{e\psi_0}{kT} - \tfrac{1}{2} \ln \frac{n_\delta}{n_s}.$$

Here n_δ/n_s gives the fraction of occupied Ca^{2+} sites in the membrane just adjacent to the cleft. We can now compute the correction to the interaction potential caused by entry of Ca^{2+} ions into the presynaptic membrane, and hence the lowering in the free energy of activation for the MEPP frequencies. (At 305°K a decrease in $\Delta G^{\ddagger}$ of only 0.02 eV will double the frequency ν.) We might say that our theory is based on a model of nonspecific calcium-ion sites. The theory of Hubbard, Jones, and Landau[22] is based on a model of

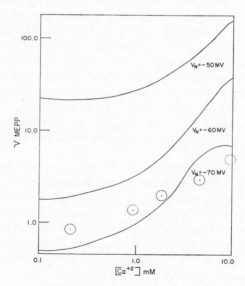

Fig. 4. Theoretical dependence of frequency of MEPP on calcium ion concentration in synaptic cleft—305°K. Experimental points are shown

specific calcium-ion complexes (the most effective one having a stoichiometric ratio of 3 Ca^{2+} per complex) which react in some way with the impinging vesicles.

To test the applicability of (6) would require data on the frequency v of MEPP as a function of $[Ca^{2+}]$ at a series of different fixed depolarizations ΔV, or data on the variation of v with ΔV at different $[Ca^{2+}]$ (Liley plots). Measurements of the effect of $[Ca^{2+}]$ on v at the ordinary resting potential $V_m = -70$ mV have been made on rat neuromuscular junctions[21,23] and Liley plots on the same system at different $[Ca^{2+}]$ have recently become available.[20] Data on MEPP at squid giant synapses[30] also provide information on effects of Ca^{2+}, but transmitter chemistry and vesicular storage in this system are less well understood. We therefore restricted our comparison of (6) with experimental results to the data on rat neuromuscular junctions.

Figure 4 shows the theoretical curves for the effect of $[Ca^{2+}]$ at three different polarizations, -70, -60, and -50 mV. Available experimental points are indicated for the curve at normal polarization. It should be noted that our theory predicts that at sufficiently high $[Ca^{2+}]$ v should decline from a maximum value. From the lowest curve in Fig. 4 we can predict that this decline should become apparent in this preparation between 10 and 20 mM Ca^{2+}.

REFERENCES

1. P. Fatt and B. Katz, *Nature*, **166**, 597 (1950).
2. V. P. Whittaker, *Biochem. J.*, **72**, 694 (1959).
3. E. de Robertis, A. P. de Iraldi, G. Rodriguez, and J. Gomez, *J. Biophys. Biochem. Cytol.*, **9**, 229.
4. K. Von Hungen, H. R. Mahler, and W. J. Moore, *J. Biol. Chem.*, **243**, 1415 (1968).
5. J. I. Hubbard and S. Kwanbunbumpen, *J. Physiol.*, **194**, 407 (1968).
6. C. W. Cotman, H. R. Mahler, and W. J. Moore, *Intern. Soc. Neurochem.*, *First Intern. Mtg.* (*Strasbourg*), *Abstr.*, 1967, p. 45.
7. L. Bass and W. J. Moore, *Proc. Natl. Acad. Sci. U.S.* **55**, 1214 (1966).
8. P. H. Wiersma, A. L. Loeb, and J. T. G. Overbeek, *J. Colloid Interface Sci.*, **22**, 78 (1966).
9. A. W. Liley, *J. Physiol.*, **134**, 427 (1956).
10. J. I. Hubbard and W. D. Willis, *J. Physiol.*, **163**, 115 (1962).
11. J. C. Eccles, *The Physiology of Synapses*, Academic Press, New York, 1964, p. 79.
12. E. M. Landau and S. Kwanbunbumpen, *Nature*, **221**, 271 (1969).
13. V. P. Whittaker, Personal communication, 1969. [See also *Nature*, **224**, 13 (1969).]
14. R. M. Marchbanks, *Biochem. J.*, **106**, 87 (1968).
15. J. R. Segal, *Biophys. J.*, **8**, 470 (1968).
16. E. J. W. Verwey and J. T. G. Overbeek, *Theory of the Stability of Lyophobic Colloids*, Elsevier, Amsterdam, 1948.
17. B. V. Derjaguin, *Disc. Faraday Soc.*, **18**, 85 (1954).
18. R. Hogg, T. W. Healy, and W. D. Furstenau, *Trans. Faraday Soc.*, 62, 1638 (1966).

19. O. F. Devereux and P. L. de Bruyn, *Interaction of Plane Parallel Double Layers*, M.I.T. Press, Cambridge, Mass., 1963.
20. E. M. Landau, Personal communication, 1968.
21. J. I. Hubbard, S. F. Jones, and E. M. Landau, *Ann. N.Y. Acad. Sci.*, **144,** 459 (1967).
22. L. Bass and W. J. Moore, in *Structural Chemistry and Molecular Biology*, A. Rich and N. Davidson, Eds., Freeman, San Francisco, 1968, p. 356.
23. J. I. Hubbard, S. F. Jones, and E. M. Landau, *J. Physiol.*, **194,** 355 (1968).
24. B. Katz and R. Miledi, *J. Physiol.*, **195,** 481 (1968).
25. M. Planck, *Ann. Phys. Chem.*, **39,** 161 (1890).
26. R. Schlögl, *Z. Physik. Chem.*, *N.F.*, **1,** 305 (1954).
27. D. E. Goldman, *J. Gen. Physiol.*, **27,** 37 (1943).
28. L. Bass and W. J. Moore, *Nature*, **214,** 393 (1967).
29. R. Miledi and C. R. Slater, *J. Physiol.*, **184,** 473 (1966).
30. B. Katz and R. Miledi, *J. Physiol.*, **192,** 407 (1967).

Physiologic Actions of Heparin not Related to Blood Clotting[*][†]

THOMAS F. DOUGHERTY, Ph.D. AND
DAVID A. DOLOWITZ, M.D.

*Departments of Anatomy and Surgery, University of Utah College of Medicine,
Salt Lake City*

[*] This work was supported by grants from the American Cancer Society (T33E) and The
U.S. Public Health Service (Leukemogenesis, No. 255-512709).
[†] Reprinted by permission from *American Journal of Cardiology*, Vol. 14, 1964, pp. 18–24.

INTRODUCTION

Some years ago Eyring and Dougherty[12] introduced the idea that sulfated mucopolysaccharides could complex with histamine and thereby inactivate or prevent the cytotoxic effects of this substance. The theory introduced by these authors was that histamine could act as a means of transporting sodium across the membranes of fibroblasts, upsetting the differential balance of sodium inside and outside the cell. This could bring about a fibroblastic cellular damage that could trigger a chain reaction of cyto destruction. The binding of histamine by heparin then could prevent the breakage of cells by preventing the sodium-carrying capacity of histamine.

For this special tribute to Henry Eyring, we have elected to present a paper which has been published before in the *American Journal of Cardiology*. Here we review most of the experimental results on detoxification of histamine through the action of heparin and their clinical applications.

Heparin is usually considered only as an anticoagulant agent. Its more basic antiinflammatory functions, which take place in the connective tissue and result from its local interaction with toxic substances, are often overlooked. The present article deals with the antiinflammatory action of heparin and with its interaction with the steroids and amines.

LOOSE CONNECTIVE TISSUE—THE SITE OF INFLAMMATION

In order to understand the ways heparin inhibits the response of the organism to injurious agents and limits the extent of the inflammatory response, it is necessary first to be familiar with the constituents of loose connective tissue. This tissue acts as a packing material for the blood vessels and cells. The small blood vessels, arterioles, and capillaries are imbedded in an amorphous gelantinous material. There is actually no place in the organism where a blood vessel comes in immediate contact with the parenchymal cell.

635

In the most primitive situation, such as that found in the sinusoidal beds of the liver, adrenal cortex, and pituitary, a reticular fiber and a reticular cell are interposed between the parenchymal cell and the blood. In most areas, however, the loose connective tissue is the material through which diffusion between the vascular and parenchymal compartments must take place. Thus ions, nutritive substances, oxygen, water, and the like must pass through this layer of tissue to the cells; and in reverse the metabolites, secretions of the parenchymal cells, are delivered through this material into the blood. It is precisely in this compartment, this interstitial layer, that inflammation takes place.

The loose connective tissue differs from the denser connective tissues such as tendons and fascia by having a very high proportion of ground substance and a small amount of fibrous material. The cellular population of all connective tissue is very similar qualitatively, although quantitatively it differs considerably from site to site. The cells of loose connective tissue may be categorized as follows: the fibroblasts make up 90% of the cells, whereas lymphocytes, mast cells, eosinophils, and histiocytes make up the remainder of the cells of normal, noninflamed loose connective tissue. Fine collagen fibers and elastic fibers may be found throughout this tissue; however, they constitute so small a portion that the gelatinous matrix and the fluid appearance of loose connective tissue result from the predominance of the mucopolysaccharides comprising the ground substance.

THE GROUND SUBSTANCE

The general character of ground substance is that of a mucopolysaccharide matrix in which the cells and fibers of the loose connective tissue are embedded, somewhat as the constituents of a gelatin salad are encased in the gelatin. The polysaccharides of ground substance vary in their chemical structure from place to place; for example, chondroitin sulfate A may predominate in one site at one time of life and not be present at the same place at another time. Hyaluronic acid may be a major constituent of ground substance in basement membranes and not be present elsewhere. Therefore, in general, we may say that although we know something about the classes of polysaccharides which comprise ground substance, there is considerable quantitative variation in the distribution of these particular macromolecules. There are many aspects of the ground substance which are poorly understood. For example, it is not known exactly how the macromolecules of the polysaccharide are bound to the protein moiety, and the roles of the protein bound to the polysaccharides are not known.

Heparin is found to be one of the principal sulfated polysaccharides in ground substance.[1] Its distribution may not necessarily be limited to one particular area, but it makes up a large portion of the polysaccharides in the ground substance of the lung, intestine, and the wall of the aorta. Since the present subject deals primarily with the role of heparin and possibly synthetic sulfated polysaccharides, less attention will be paid to the distribution, synthesis, and release of the other sulfated and nonsulfated polysaccharides of ground substance, even though they may act similarly.

PINOCYTOSIS

Pinocytosis is a process first described many years ago by which the cell is able to take up, by a sort of active transport, various materials in solution which surround it.[2] Mechanically, the way that the cell does this is by extending cytoplasm which surrounds materials in the ground substance. The ends of the extended cytoplasmic processes then seal together, and the surrounded material moves intracellularly. The process is very similar to phagocytosis, and probably the only workable descriptive difference between the two is that in pinocytosis much smaller particles and substances in solution may be taken up by the cell. Phagocytosis, then, would be a term retained for those cells which are able to take in large particulate matter such as bacteria and foreign bodies. Whereas only some cells are able to phago-cytize, a wide variety of cells including muscle, epithelial cells of the gut tract, and capillary endothelium appear to be capable of pinocytosis. The size of the processes of cytoplasm which are extended and retracted may vary considerably for these different types of cells, so that the observation of the very finest pinocytotic processes may require electron microscopy.

A great deal of research is being done in some European laboratories on the problem of substances (inducers) which bring about pinocytosis on the part of cells.[2] In general, it may be stated that as far as amoebae are concerned many of these inducers are polypeptides and basic amino acids. It is also known that a polysaccharide is involved as a requirement for the inducer to act. Several years ago it was discovered in this laboratory,[3] by the use of phase cinemicrography, that heparin is a very potent stimulator of pinocytosis for mouse fibroblasts of the loose connective tissue. In addition, it was observed that mast cell granules, which are released by dispersion, appear to stimulate not only pinocytosis but also phagocytosis by the surrounding cells.[4] It is of considerable interest that it has been observed recently that heparin plays a similar role as an inducer and stimulator of pseudopod formation in amoebae.[5] At the present time the meaning of these observations is being explored further. A preliminary observation, which is mentioned here because of its possible extreme importance, is that heparin may induce pinocytosis by fibroblasts and produce a destructive effect when cytotoxins are added to the medium.[6] Heparin release, provided cytotoxic substances are available, may enhance the uptake of these substances which then would produce cellular death and stimulate the inflammatory response. Thus heparin, through its biologic exchange capability may, under the appropriate circumstances, be proinflammatory as well as antiinflammatory, depending on the substances present in the medium in which the mast cell granules may be released. The actual mechanism by which heparin stimulates active transport is not as yet understood but is the subject of further investigation. It is likely that it may allow the basic amino acids and polypeptides (inducers) to bring about induction of pinocytosis.

HEPARIN IN CELLS AND IN GROUND SUBSTANCE

It is now many years since it was discovered that heparin is a constituent of the mast cell granule.[7] Because it has been found in mast cells, many assume that this cell is the principal site of synthesis of this polysaccharide. However, there is evidence also that fibroblasts not only make hyaluronic acid, but possibly sulfated chondroitin polysaccharides as well.[8,9] There is not always a distinct correlation between the numbers of mast cells and the content of heparin of a particular tissue. The mast cell is found only in connective and reticular tissues. There is some evidence that mast cells in culture may synthesize heparin.[10] However, another point of view has been put forward: that the fibroblasts may make the polysaccharide and store it for a time and that, when they are full of these storage granules, they are called mast cells.[1]

Mast cells may disrupt (clasmatosis)[4,11] and release their granules into the ground substance, liberating heparin and other chondroitin sulfates. Thus, the mast cell may be a storage form of the fibroblast. This release of granules is similar to secretion of many substances by other cells and would be a merocrine type of secretory activity.

BIOLOGIC EXCHANGE CAPACITY OF SULFATED POLYSACCHARIDES

Some time ago it was proposed that histamine, serotonin, and other basic amines and also basic polypeptides could be bound by ground substance.[12,13] In this way the sulfated polysaccharides of ground substance would act as biologic exchangers; for example, histamine, by complexing with the sulfuric acid groups of the sulfated polysaccharides, would thus not be available to act at the cellular level to bring about destruction of the cell. If, however, the cell should be destroyed, more amines would be released. These, in turn, would act on other cells, and thus a chain reaction of cellular destruction would occur. Such tissue destruction necessitates a removal of the destroyed cellular and fibrous material. Therefore it has been postulated that release of these toxic bases from binding sites brings about cellular destruction followed by inflammation. Exactly how these agents affect cell destruction is not too well understood, but it has been proposed that they do so by influencing transport of sodium ions across the cell membrane, increasing intracellular water and thus breaking the cell.[12]

Of course, these inflaming substances would not only be bound in the ground substance but in mast cell granules as well. One of the earliest changes that occurs in the inflammatory response is the release of mast cell granules. If the mast cell degranulation response is enormous, the surrounding fibroblasts pick up many granules[4]; but also there is a scatter of free granules in the ground substance. Frequently at this time one observes that the granules undergo a change from a deep metachromatic color to a red hue. This is taken as an indication that the free sulfuric acid, which produces the metachromasia, is being masked by the complexing with basic substances. The tinctorial properties of ground substance itself change similarly in the very early stages of nonspecifically induced inflammation.

Mucopolysaccharides bind not only the naturally occurring amines but also substances exogeneously introduced which have a greater affinity to complex with the endogeneous polysaccharides. Such exogeneous substances can "kick off" from their bound sites the naturally occurring substances, such as histamine or serotonin, with ensuing cytotoxic activity of these agents. Thus poisonous substances may bring about inflammation and stress in two ways: (a) by directly producing the local or systemic tissue reaction and (b) by bringing about the release of already bound phlogogenic substances like histamine and serotonin. It seems readily evident that, once a sufficient amount of the exchange media is filled up, relatively small amounts of complexing substances could bring about release of the noxious amines simply because there are inadequate binding sites available. This could be the basis for the chronic inflammatory responses.

In considering the ion exchange function of ground substance and mast cell granules with respect to the transport and the binding of tissue constituents which pass from blood to parenchyma and the reverse, the acid mucopolysaccharides must play a fundamental part from a chemical standpoint. These substances are protein-bound and have anionic properties due to the constituents of uronic acid that are esterified with sulfuric acid. Although the whole molecule of a polymer may not enter into a chemical reaction, it is important that an ion exchanger have groups on the polymers that react. In other words, the polymer provides a support for the reactive group. The polymers contain ionized groups such as phosphate, sulfate, and amino and carboxyl groups. All of these reactive groups provide potential exchangers for cationic substances present in the fluids in which the polymers are contained. With this theoretic explanation of the role played by mucopolysaccharides as cation exchangers, we may consider some of the experimental evidence that has accumulated relative to the complexing of basic amines and polypeptides.

EXPERIMENTAL EVIDENCE FOR BIOLOGIC EXCHANGE CAPACITY

As far as we can determine, Dragstedt and his coworkers[14] in 1942 noted first that heparin inhibited the release of histamine in antigen–antibody reactions. Parrot and Laborde[15] demonstrated in 1951 that the addition of heparin to histamine diminished the biologic activity of histamine. He also found—and this demonstrates the early concept of the biological ion exchange idea—that when putrescine was added to the heparin–histamine solution, the putrescine limited the binding of histamine by heparin. Parrot also showed in his experiments with dialysis that there was a quantitative reaction in the sense that histamine was picked up by the heparin in a quantitative way. Others besides those of our own group, at approximately the same time, also suggested the binding of histamine by heparin and the possibility that other basic substances could either limit the binding of histamine or actually displace histamine from the sulfate complex. MacIntosh,[16] for example, in 1956 correlated the ability of certain basic substances to release histamine with their affinity for heparin, and he and Paton[17,18] suggested an acid–base attraction, so that histamine can be displaced by substances with strong basic properties. The idea that histamine is bound by ground substance was proposed by Eyring and Dougherty in 1955,[12] and in the following years, in a series of papers in collaboration with Higginbotham,[19–21] it was demonstrated in animal experiments that heparin could protect against the lethal effects of the histamine liberator, 48/80. We also demonstrated that fibroblasts participate in detoxification of the amine 48/80 by heparin. The biologic ion-exchange idea was further supported when it was shown that, if one gives a sublethal dose of one base, a sublethal dose of another basic protein such as ACTH can prove lethal to the animal. The reason for this is that the combined effect of "kicking off" the histamine from the ground substance can be an accumulative effect of the two bases even though they may differ chemically.

About the same time Amann and Werle[22] showed the complex of heparin with histamine and other di- and polyamines, and they also suggested a displacement mechanism in the release of histamine from mast cells. Schayer[23] presented experimental evidence that mast cells of rat peritoneal fluid can decarboxylate C^{14}-L-histidine and bind the resulting histamine in a stable form. He found that the optimal temperature for this reaction was 37°C; the optimal time was 3 hr and the optimal pH was 7.4. Amann[24] found that there are two fractions of heparin which move at different rates in paper chromatography and also that the faster one bound more histamine than the slower.

In a series of studies performed in our laboratory Truls Brinck-Johnson demonstrated by *in vitro* studies in which toluidine blue was used as an indicator of degree of metachromasia, that histamine can be released from heparin by various bases. These studies were not published in final form but were reported by Higginbotham and Dougherty[25] in 1957 and extended and confirmed by Higginbotham[13] in 1958. The method used depended upon the optimal absorption of the cationic dye toluidine blue. The absorption is 630 lambda in the free state and shifts to 540 when bound to heparin. The extent, then, to which this cation is displaced from ionizing groups on heparin by other cationic substances can be determined colorimetrically; for example, the nondiffusible heparin and bound dye are placed in a dialysis bag immersed in deionized water. The appropriate exchanger cation can then be added to the outer phase; and when it diffuses across the membrane it replaces the bound dye, which then becomes free and diffuses to the outer phase and can be measured colorimetrically. Higginbotham[26] gives the details concerning the extent to which a whole series of various polypeptides and amino acids will bring about this displacement.

Many of the same substances were then used in *in vivo* animal experiments in order to determine whether the *in vitro* system would predict results obtained in the protection of the animal against the amine release. In general, the *in vitro* method indicated the histamine-releasing capacity of a wide variety of agents. Keller,[27] using a system of dialysis and studying vessel permeability of skin, found certain conditions under which heparin binds histamine and serotonin. He found that the binding is loose and is best in a water solution, pH 5–5.5, and he found that moving the pH toward neutrality or to a higher concentration of salt hindered the binding. The affinity he found is 1 mg of histamine to 4 mg of heparin and 1 mg of serotonin to 15 mg of heparin. It may be seen from this work, for example, that a change in pH itself as it moved upward could set about a release of histamine and thus set off an inflammatory reaction without the intermediation of other noxious histamine-releasing agents. Parrot and Laborde[28] also demonstrated in a very thorough study the displacement of histamine with change in pH; for example, they found that as the pH increased toward neutrality, more and more histamine was displaced. Actually, when the extraction was made at 4-hr intervals, the greatest amount appeared to be displaced at approximately pH 6.

CLINICAL USE OF HEPARIN

Since heparin protected the animals from histamine toxicity in a dose-response, it was thought to be safe to use in human patients for clinical trial. Heparin had of course been used in cardiovascular disease in large doses for many years. It was believed that there would be safety as long as the dosage was kept within the half gram used in cardiovascular disease. Further, since heparin was bound by histamine in the animal studies, it was believed that it would not be free in the blood to cause bleeding difficulties. Lee-White tests were performed on many of the patients and confirmed this finding.

Before using heparin with antibiotics, blood agar plates with various organisms were cultured. The plates were prepared in the ordinary fashion. In a second set of cultures, disks with antibiotics were soaked in heparin, and finally heparin was incorporated into the blood agar alone. Examination of the plates made it obvious that heparin did not inhibit the action of the antibiotics and often seemed to enhance it. It is not known whether heparin is truly an antibiotic.

Weeping Eczema

This was chosen as the first clinical problem to be studied since it presents visible lesions produced by mixtures of inflammation, allergy, and infection, all of which should release histamine. The alkaline antibiotics neomycin, streptomycin, polymyxin, and viomycin were combined with heparin because they could then produce the double-barreled action of an antibiotic complex with the antiinflammatory action of binding the endogeneously released histamine. It is further believed that by releasing the antibiotic slowly, toxic levels might be avoided in the plateau of release.

Of the two clinical types of weeping eczema, one responds to most treatments; the second is resistant to almost all efforts. Patients in the last category were selected after they had failed to respond after two years of therapy. The appropriate antibiotic was determined *in vitro* by culturing the scrapings of the lesions of these patients. If an antibiotic–steroid complex did not help *in vivo*, patients were given local applications of 3% of the appropriate antibiotic, 2.5% of cortisone, and 1% heparin. Of 63 patients so treated, the eczema cleared in all but three. One of these three responded when repeated cultures revealed that another antibiotic was more effective *in vitro*. The eczema recurred as soon as the medication was stopped in two patients who responded to the antibiotic–steroid complex.

Hay Fever

As the results with weeping eczema demonstrated that heparin could be used clinically to bind the histamine, hay fever was selected for the next test. Heparin, 100–200 mg, was injected intravenously into 27 patients during attacks of acute pollinosis. The allergic symptoms of nasal obstruction, discharge, conjunctivitis, redness, and itching of the skin were helped in most cases. In chronic cases with nasal polyps similar treatment was continued for several days. Most polyps were unaffected, but the polynosis symptoms cleared. An attempt to work on catarrhal otitis media showed only transient improvement with the fluid soon returning. On repetition of treatment even the transient benefit was lost.

Asthma

Next, the effects of heparin on acute attacks of asthma were studied. Ten male and 14 female patients were treated with 100–200 mg of intravenous heparin within the first 24 hr of an attack. The wheezing was markedly relieved in 17 cases, moderately in 6, and poorly in one. Coughing was helped in about the same manner, 15 being helped markedly, 8 moderately, and only one poorly. In 23 of the 24 cases the breathing was eased both subjectively and objectively, and all of the patients mentioned a feeling of warmth and relaxation.[29]

Laryngotracheal Bronchitis

To see whether heparin would be of help in infectious liberation of histamine, 10 cases of laryngotracheal bronchitis severe enough to require tracheotomy were also treated with 100–200 mg of heparin administered intravenously. Six of the children were boys and 4 were girls. They ranged in age from 1 to 9 years. All of them responded to heparin early. Tracheotomy was avoided in nine of the cases. The tenth patient developed a temper tantrum 18 hours after relief of symptoms and severe cyanosis appeared to develop. It was believed that if heparin did not produce relief immediately, there would not be time for surgery; hence, an emergency tracheotomy was performed.

We believe that heparin has demonstrated clinically in these pilot efforts its ability to produce immediate relief in acute allergic, infectious, and inflammatory states. We believed too that this is accomplished through the neutralization of the endogenously released histamine and not because of the appearance of thromboembolic phenomena as postulated by others.[30]

OTHER TRIALS OF THE THERAPEUTIC APPLICATION OF HEPARIN

The therapeutic effect of heparin has also been studied by others in allergies, infections, inflammations, and conditions with unknown etiologies.

As the effect of heparin depends not only on its anticoagulant and lipolytic properties, its other actions have to be taken into account also: (a) its antiinflammatory effect (regulation of the properties of ground substance), (b) its inhibitory effect on the spreading of noxious and infectious agents (inhibition of hyaluronidase), and (c) its interaction with steroid hormones, with histamine releasors and with noxious amines.

Johansson[31] suggested that heparin has an antiallergic effect in many. Dahl[32] tried this suggestion experimentally, and he was able to prevent experimentally induced allergic asthma by injecting his patients with heparin just before exposing them to the specific antigen. Videbäk[33] showed heparin effective in a few patients with autoimmune hemolytic anemia.

Donzelot and Kaufmann[34] found that heparin ameliorated bad cases of rheumatism, supposedly by an antiexudative action, but Burkl[35] suggested that it acts by the inhibition of hyaluronidase activity. The action of heparin in rheumatism is an example of its antiinflammatory influence on connective tissue. Other such examples are its ameliorating effect in thrombophlebitis[36] and its action in interstitial cystitis.[37]

Heparin has been used with some success in multiple sclerosis,[38-40] and the authors suggest that this may be due to its action on lipid metabolism.

Heparin was also effective in tinnitus, sudden deafness, Meniere's disease, and Bell's palsy.[41] Here it was used in combination with nicotinic acid and procaine. The authors suggested an antilipemic or unknown action. In cases of viral infections, however, the action of heparin might be due to its antiinflammatory or its antihyaluronidase effects.

The action of a salve which combines a heparinoid with hyaluronidase (which supposedly enhances the heparin effect as it makes possible the fast entrance of heparin into the skin) has been reviewed by Sikorski.[42] This salve has been used with good results in superficial phlebitis, hematomas, varicose veins, inflamed hemorrhoids, frost injuries, early stages of diabetic necrosis, torticollis, and abscesses of sweat glands. It is especially effective in sports injuries and promotes faster healing of distortions, swellings, and mainly hematomas.

SUMMARY

The relation of heparin to inflammation is discussed, as are its relation to ground substance, cells, pinocytosis, and biologic ion exchange.

Experimental evidence is given for biologic exchange capacity of heparin in *in vitro* systems and in the protective effect of heparin in animals.

The clinical use of heparin in inflammatory, allergic, and infectious diseases is reviewed. Intravenous heparin proved effective in eczemas, hay fever, asthma, laryngotracheobronchitis, interstitial cystitis, and many other disease states related to inflammation.

ACKNOWLEDGMENT

We wish to thank Dr. Beata Jencks for her valued and able assistance. The heparin used in these studies was contributed by Abbot Laboratories, North Chicago, Ill., and Riker Laboratories, Northridge, Calif.

REFERENCES

1. T. F. Dougherty and R. D. Higginbotham, "Hormonal influences on Inflammation and Detoxification," in *Fifth Annual Report on Stress*, Acta, Inc. Montreal, 1955/56, p. 117.
2. C. Chapman-Andersen, How amoebae drink, *New Scientist*, **16**, 618 (1962).
3. T. F. Dougherty and G. L. Schneebeli, Film: "Influence of Heparin on Loose Connective Tissue," Department of Anatomy, University of Utah, 1959.
4. T. F. Dougherty and G. L. Schneebeli, Film: "Mast Cells of Loose Connective Tissue," Department of Anatomy, University of Utah, 1959.
5. L. G. Bell and K. W. Jeon, "Stimulation of Cell Locomotion and Pseudopod Formation by Heparin," *Nature*, **195**, 400 (1962).
6. G. L. Schneebeli, personal communication, 1962.
7. J. E. Jorpes, J. Holmgren, and O. Wilander, "Uber das Vorkommen von Heparin in den Gefaesswaenden und in den Augen," *Z. Mikr.-Anat. Forsch.*, **42**, 279 (1937).
8. H. Grossfeld, K. Meyer, and G. Godman, "Differentiation of Fibroblasts in Tissue Culture, as Determined by Mucopolysaccharide Production," *Proc. Soc. Exptl. Biol. Med.*, **88**, 31 (1955).
9. H. Grossfeld, K. Meyer, G. Godman, and A. Linker, "Mucopolysaccharides Produced in Tissue Culture," *J. Biophys. Biochem. Cytol.*, **3**, 391 (1957).
10. E. D. Korn, "The Synthesis of Heparin in Mouse Mast Cell Tumor Slices," *J. Am. Chem. Soc.*, **80**, 1520 (1958).
11. P. B. Carter, R. D. Higginbotham, and T. F. Dougherty, "The Local Response of Tissue Mast Cells to Antigen in Sensitized Mice," *J. Immunol.*, **79**, 259, (1957).
12. H. Eyring and T. F. Dougherty, "Molecular Mechanisms in Inflammation and Stress," I, *Am. Scientist*, **43**, 457 (1955).
13. R. D. Higginbotham, "Studies on the Functional Interrelationship of Fibroblasts and Ground Substance Mucopolysaccharides," *Ann. New York Acad. Sci.*, **73**, 186 (1958).
14. C. A. Dragstedt, J. A. Wells, and M. Rochae Silva, "Inhibitory Effect of Heparin upon Histamine Release by Trypsin, Antigen, and Proteose," *Proc. Soc. Exptl. Biol. Med.*, **51**, 191 (1942).
15. J. L. Parrot and C. Laborde, "Captation de l'histamine par l'heparine," *Compt. Rend Soc. Biol.*, **145**, 1047 (1951).
16. F. C. MacIntosh, "Histamine and Intracellular Particles," in *Ciba Foundation Symp. Histamine*, Churchill, London, 1956, p. 20.
17. F. C. MacIntosh and W. D. M. Paton, "The Liberation of Histamine by Amidines and Other Compounds," *Proc. 17th Intern. Physiol. Congr. 1947*, p. 240.
18. W. D. M. Paton, "The Mechanism of Histamine Release," in ref. 16, p. 59.

19. R. D. Higginbotham and T. F. Dougherty, "Mechanism of Heparin Protection against a Histamine Releasor (48/80)," *Proc. Soc. Exptl. Biol. Med.*, **92**, 493 (1956).
20. R. D. Higginbotham and T. F. Dougherty, "Miscellophagosis," *Reticuloendothelial Bull.*, **3**, 19 (1957).
21. R. D. Higginbotham and T. F. Dougherty, "Potentiation of Polymyxin B Toxicity by ACTH," *Proc. Soc. Exptl. Biol. Med.*, **96**, 466 (1957).
22. R. Amann and E. Werle, "Ueber Komplexe von Heparin mit Histamin und anderen Di- und Polyaminen," *Klin. Wochnschr.*, **34**, 207 (1956).
23. R. W. Schayer, "Formation and Binding of Histamine by Free Mast Cells of Rat Peritoneal Fluid," *Am. J. Physiol.*, **186**, 199 (1956).
24. R. Amann, "Ueber das Verhalten von Heparinen und anderen sulfathaltigen Muco-polysacchariden bei der Papierchromatographie," *Tr. 6th Congr. European Soc. Haematol.*, *Copenhagen*, S547, 1957.
25. R. D. Higginbotham and T. F. Dougherty, "Heparin, a Biologic Ion Exchanger in Protection vs. Noxious Agents" (Abstr. 252), *Federation Proc.*, **16**, 58 (1957).
26. R. D. Higginbotham, "On the Participation of Micellophagosis in the Resistance of Tissue to Injurious Agents," *Intern. Arch. Allergy Applied Immunol.*, **15**, 195 (1959).
27. R. Keller, "Zur Bindung von Histamin and Serotonin in den Mastzellen," *Arznei-mittel-Forsch.*, **8**, 390 (1958).
28. J. L. Parrot, G. Nigot, and C. Laborde, "Captation de l'histamine par l'heparine. Influence du pH," *Compt. Rend Soc. Biol.*, **154**, 1426 (1960).
29. D. A. Dolowitz and T. F. Dougherty, "The Use of Heparin as an Antiinflammatory Agent," *Laryngoscope*, **70**, 873 (1960).
30. M. M. Hartman, "Thrombo-embolic Phenomena in Severe Asthma. Use of Heparin for Prevention and Treatment in Patients Receiving ACTH or Glucosteroids," *California Med.*, **98**, 27 (1963).
31. S.-A. Johansson, "Inhibition of Thrombocytopenia and 5-Hydroxytryptamine Release in Anaphylactic Shock by Heparin," *Acta Physiol. Scand.*, **50**, 95 (1960).
32. Sv. Dahl, "Heparin und Asthma," *Z. Tuberk.*, **118**, 255 (1962).
33. A. Videbäk, *Hamatologie*, Munksgaard, Copenhagen, 1961. (Cited by Sv. Dahl, ref. 32.)
34. E. Donzelot and H. Kaufmann, "Heparine et rhumatisme. L'action antiexsudative de l'heparine," *Presse Med.*, **57**, 989 (1949).
35. W. Burkl, "Uber Mastzellen und ihre Aufgaben," *Wiener Klin. Wochschr.*, **64**, 411 (1952).
36. G. Bauer, "Heparin Therapy in Acute Deep Venous Thrombosis," *J. Am. Med. Assoc.*, **131**, 196 (1946).
37. R. G. Weaver, T. F. Dougherty, and C. A. Natoli, "Recent Concepts of Interstitial Cystitis," *J. Urol.*, **89**, 377 (1963).
38. C. B. Courville, "Multiple Sclerosis as an Incidental Complication of a Disorder of Lipid Metabolism. III. Treatment by Heparin in Acute Exacerbations of the Disease," *Bull. Los Angeles Neurol. Soc.*, **24**, 39 (1959).
39. C. B. Courville, "The Effects of Heparin in Acute Exacerbations of Multiple Sclerosis. Observations and Deductions," *Bull. Los Angeles Neurol. Soc.*, **24**, 187 (1959).
40. J. Maschmeyer, R. Shearer, E. Lonser, and D. K. Spindle, "Heparin Potassium in the Treatment of Chronic Multiple Sclerosis," *Bull. Los Angeles Neurol. Soc.*, **26**, 165 (1961).
41. D. O. Zenker and E. P. Fowler, "Medical Treatment of Sudden Deafness, Meniere's Disease, and Bell's Palsy," *New York State J. Med.*, **63**, 1137 (1963).
42. H. Sikorski, "Hyaluronidase und Heparinoide bei der Behandlung entzuendlicher Stauungen," *Z. Haut. Geschlechtskr.*, **31**, 225 (1961).

Hydrostatic Pressure Reversal of Alcohol Narcosis in Aquatic Animals

FRANK H. JOHNSON

Biology Department, Princeton University, Princeton, New Jersey

A Personal Note

Anyone who has had the privilege of collaborating extensively with Henry Eyring will surely have come to appreciate and respect, among his other attributes, two that are indeed outstanding; namely, his faith in religion and his understanding of reaction rates. The latter becomes obvious in the vast knowledge and intellectual prowess with which he drives relentlessly toward a rigorous solution of fundamental scientific problems; the former is manifested in a variety of ways, including his unswerving abstinence from alcohol and tobacco and his exceptionally relaxed attitude about his prospects for life in the next world. Although it is difficult to say which is the more profound—the depth of his religious convictions or the breadth of his command over reaction rate theory—it seems fair to say that it is almost impossible to win an argument with him in either area. Since the subject of this paper touches on both, it appears appropriate to mention, as an illustration, that at an informal meeting with the medical school faculty, during a discussion of longevity, one of the doctors pointed to statistics which indicated that moderate drinkers live longer than "tee-totalers" and remarked, "So you see, Henry, we are going to live longer than you," to which Henry replied, without a moment's hesitation, "But I am going to a better place!"

648

Application of the theory of absolute reaction rates[1,2] to processes in living cells was first published in 1942 in reference to the enzyme-catalyzed process of light emission in luminous bacteria as influenced by temperature, hydrostatic pressure, and inhibitors, including typical narcotics and certain others.[3-7] The Eyring theory was then already well established in various fields of chemistry but it remained practically unknown among biologists. The purpose of the present paper is to review very briefly a few discoveries, now somewhat of historical interest, which with the aid of the Eyring theory served to bring together several diverse areas of biological research and at the same time to anticipate certain basic results which are illustrated not only by the pressure-sensitive narcosis of animals indicated in the title of this paper, but also by some aspects of present views concerning the mechanism of enzyme activity and inhibition in general.

First, before application of absolute reaction rate theory to bioluminescence, research on temperature relations of biological processes had almost reached an impasse. Some empirically well-known general biological phenomena were at best only partially understood, for example, the "temperature–activity curve" of essentially all biological processes, which universally goes through a maximum or "optimum" at some temperature, familiarly between 25 and 50°C, depending upon various factors, including the particular process and biological species involved. It was frequently found that the rate, even of very complex processes, increased with temperature, through a considerable range of temperatures below the optimum, in accordance with the Arrhenius equation,[8] with a more or less definite, numerical value of μ, the Arrhenius activation energy, typical of single enzyme reactions. Beyond the optimum temperature, a key reactant was apparently destroyed at a rate which might also conform to the Arrhenius equation, but with a much higher value of μ typical of thermal denaturation of proteins. Variations in the observed values for μ of the overall process between the two extremes of temperature were subjects of a large amount of speculation and controversy.

A second fundamentally important area of research, in which the influence of temperature was poorly understood, and in which an influence of hydrostatic pressure was unknown, was the physiological and biochemical action of narcotics and other drugs. Quantitative interpretations were attempted through various hypotheses, based for example on correlations between potency of action and solubility of the drug in olive oil, and distribution of the drug between water and olive oil at different temperatures.[9-11] Other theories were based on partial agreement of quantitative effects with adsorption isotherms[12] or on specific properties of the drug.[13-15]

Finally, a third major field of endeavor, seemingly distinct and unrelated to the two just mentioned, involved the biological action of increased hydrostatic pressure, especially in the range of ca. 100–600 atm, the physiological effects of which were often quantitatively great and, unless maintained too long, easily reversible simply by decompression. No explicit, quantitative theory had been available to apply to the results of experiments, nor was any generally applicable, empirical rule evident; some processes were retarded under pressure, others accelerated, others scarcely affected, and still others showed an optimum, the rate accelerating with rise in pressure but going through a maximum and then decreasing with further rise in pressure.

At the time when the theory of absolute reaction rates appeared in 1935, a vast literature had already accumulated in the above three, seemingly independent areas of research. The seeming independence was emphasized by the contemporary reviews or monographs which were published in each field, with little if any mention of the other two fields (Bělehrádek, Ref. 16, on temperature; Cattell, Ref. 17 on pressure; Henderson, Ref. 14, and Clark, Ref. 15, on narcosis). Soon thereafter, however, the erstwhile separate fields were brought together in the application of the Eyring theory to bacterial luminescence, as alluded to above.

It is of human interest, at least, to record here for once the chance events which triggered this research with luminescence. In December 1940, during a short spell of unseasonably mild weather, at meetings of the American Society of Zoologists in Philadelphia, a small group of participants, including Dugald Brown, Douglas Marsland, the late Professor E. Newton Harvey, and the present author, were gathered at a table of a sidewalk cafe, momentarily escaping the strain of formal scientific sessions, and informally discussing various subjects of biological interest, over liquid refreshments in small glasses containing an olive. Brown mentioned his work[18,19] which had shown that muscle tension is affected by increased pressure in a manner depending not only on the temperature of the experiment but apparently also on the temperature range to which the animal, from which the muscle came, was adapted or acclimatized. Brown remarked that it might prove very useful, for further research on this problem, if he could find one biological process which was basically the same among different species having different temperature–activity relations. On the basis of an intimate acquaintance with luminous bacteria, it was no strain on the author's imagination to suggest that bacterial luminescence might prove to be the ideal process, from the above point of view as well as certain others, including the fact that the brightness of the light provided an unusually convenient, natural indicator of the activity of the limiting system. Moreover, in luminous bacteria under favorable conditions, the system remained in a steady state during convenient intervals of time, characterized by a constant light intensity which, however,

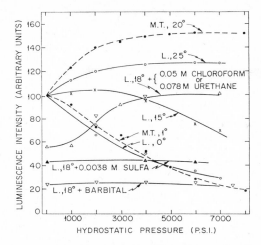

Fig. 1. Changes in intensity of luminescence (*L*) in living cells of a psychrophilic species of luminous bacteria, *Photobacterium phosphoreum*, suspended in an isotonic buffered salt solution at neutral pH as a function of changes in hydrostatic pressure at various temperatures, with and without added drugs in the final concentrations indicated in the figure. Data on muscle tension (M.T.) in excised turtle auricle are also included (dashed lines) for two temperatures (1 and 20°C) without added drugs. For each curve the intensity of a control at normal pressure was arbitrarily taken equal to 100 to show clearly the relative changes induced by increased hydrostatic pressure at different temperatures and with added drugs at one temperature: the actual intensity of luminescence or the actual tension in the muscle varied in the usual manner with changes in temperature at normal pressure. The points shown in this figure are averaged data from experiments by Brown, Marsland, and Johnson, partly in collaborative research, and are replotted from some of the same data illustrated in Figs. 9.8 (p. 310), 9.9 (p. 311), and 10.21 (p. 408) of Ref. 23.

would immediately change, often reversibly, under the influence of factors such as changes in temperature or chemical environment.

In due course Brown, Marsland, and the author collaborated in experiments with luminous bacteria and discovered at once that the effects of pressure on bacterial luminescence were fundamentally similar to the effects which had been found on muscle tension (Fig. 1). With a given species of bacteria, increased hydrostatic pressure had relatively little effect at the normal optimum temperature in that species, whereas at higher temperatures the same pressures increased the rate and at lower temperatures decreased the rate of the overall process. At the optimum temperature in one species, the rate was decreased by pressure in a species having a higher optimum temperature, and increased in a species having a lower optimum temperature. Obviously, these findings brought together as part of a general phenomenon earlier observations concerning the influence of pressure on a variety of processes, and emphasized the basic role of temperature in determining the net result.

Although the investigators were still unaware of the theoretical basis of pressure effects on reaction rates provided by the Eyring theory, it was noted that the increase in luminescence under pressure at temperatures above the normal optimum was never sufficient to exceed appreciably the intensity at the normal optimum; it was as if pressure somehow counterbalanced the thermal inactivation of the system, a view which seemed all the more convincing when the data were plotted as log intensity against reciprocal of the absolute temperature, yielding a family of curves, intersecting near the same maximum, for different constant pressures. If correct, this view practically required that the thermal inactivation of the limiting enzyme was brought about through an equilibrium change, because of the immediate nature of the effects, rather than through an irreversible destruction. Thus it was reasonable to expect that, on very short exposures of the bacterial suspension in a water bath at temperatures high enough drastically to reduce the brightness of luminescence, this reduction would be largely reversible on quickly cooling in an ice bath. Experiments confirmed that such was indeed the case. Of course, if the cells were kept more than momentarily at the high temperature, a process of irreversible destruction became evident, occurring at rates which increased rapidly with rise in temperature, as in the familiar thermal denaturation of proteins.

Though contrary to the view which was widely held at the time, to the effect that, in general, protein denaturation was irreversible, and to the effect also that high hydrostatic pressure brought about, rather than reversed, denaturation of proteins, the conclusion appeared inescapable that, in bacterial luminescence, a shift in an equilibrium between active and reversibly denatured forms of a limiting enzyme, toward the final state with rise in temperature, and back toward the initial state with rise in pressure, was involved in the pressure–temperature parameters. At normal pressure, the chief aspects of the temperature–activity curve could now be accounted for, on the basis of the increase with temperature in rate of the overall process, due to the increase in specific rate of an enzyme reaction, according to the Arrhenius theory, together with the reversible denaturation reaction, which itself was much more sensitive to temperature than was the enzyme reaction. It was comforting to find later that the possibility that such an equilibrium reaction might exercise a controlling influence in a physiological process had in fact been suggested by Anson and Mirsky as early as 1931,[20] and shortly thereafter had been demonstrated in a purified enzyme system by the same authors.[21]

Finally, an obvious question arose as to the possibility that the reversible inhibition of luminescence caused by drugs or other chemical agents might be influenced by hydrostatic pressure. It was convenient, in reference to this question, to investigate first the effects of various drugs at the normal optimum temperature, where, in absence of added drugs, pressure had relatively very little effect. A day's work with portions of a suspension of luminous bacteria in a quart of buffered salt solution was sufficient to show that two types of inhibition could occur under the conditions involved: (I) exemplified by the action of sulfanilamide or barbital, moderate inhibitions by which were practically insensitive to hydrostatic pressure; (II) exemplified by ethyl alcohol, ethyl carbamate (urethane), ether, chloroform, and certain other "lipid soluble" drugs, moderate inhibitions by which could be largely, if not fully, reversed by increased hydrostatic pressure. The temperature relations at normal pressure differed in ways which further research made reasonably clear. For present purposes it suffices to note that pronounced (depending on concentration of the drug) and largely reversible (by dilution or otherwise reducing the effective concentration) increases in overall activation energy and temperature of maximum luminescence intensity occurred under the influence of sulfanilamide, whereas just the opposite effects occurred under the influence of alcohol or urethane. Moreover, the amount of inhibition by sulfanilamide tended to remain constant with time, whereas the amount of inhibition by alcohol or urethane tended to increase with time, at a rate which increased with concentration or temperature but decreased with rise in pressure.

It was again partly through chance circumstances that the investigators were introduced in 1941 by the late Professor Harvey, to Professor Eyring, whose headquarters were only a few hundred yards away in the Chemistry Department. The theoretical basis of pressure effects on chemical reaction rates became at once available to the biological investigators. Quantitative formulations, after many intensive discussions of the data at hand, were derived in accordance with absolute rate theory together with classical thermodynamic equilibrium theory. Although admittedly oversimplified, these formulations succeeded remarkably well in accounting for the various effects of temperature, pressure, and narcotics, considered either separately or in relation to one another, all involving primarily a single species of molecules, namely, the limiting enzyme in bacterial luminescence, in the various states of which it seemed capable; for example, normal and activated, catalytically active, native or catalytically inactive, denatured states, and inhibited by combining with one or another type of inhibitor.

The picture arrived at after exhaustive research with luminous bacteria,[22] and eventually discussed at length with reference to many other types of biological processes,[23] involved molecular volume changes of activation in the enzyme reaction and volume changes of reaction in the reversible denaturation of the protein moiety. Continued research concerning the influence of temperature, pressure, and other factors on muscle contraction, division of fertilized egg cells, sol-gel changes *in vitro*, etc., by Brown, Marsland, and others has found the same general theory useful in accounting for the results obtained.[24-29]

A more specific picture, in terms of molecular structure, was suggested to account for the influence of pressure on the kinetics of bacterial luminescence. Thus changes in "configuration" of the protein, perhaps through a partial unfolding whereby more than a normal amount of hydrophobic groups became exposed to the aqueous solvent, were assumed to accompany the catalytic activity and also the reversible thermal denaturation. In effect, the molecules then occupied a larger volume, by some 50–100 cc/mole, and as a result, increased pressure reduced the rate of the enzyme reaction, or the proportion of reversibly denatured enzyme. Narcotics of the alcohol or urethane type (II) were pictured as combining, in effect, with the hydrophobic groups exposed in the thermal denaturation, thereby furthering the equilibrium toward the denatured side with a volume increase, without necessarily influencing the normal volume change in this reaction. Because the same drugs which furthered a reversible denaturation of the enzyme would seem likely to catalyze a rate of irreversible destruction, the increasing amount of inhibition with time, which was found in the effects of alcohol or urethane, was readily understandable. The pressure-insensitive inhibitions, such as those of the sulfanilamide type (I), on bacterial luminescence were assumed to involve combinations of the inhibitor with the enzyme in a manner that did not influence the thermal denaturation, as if the combination took place with the native and reversibly denatured forms alike, perhaps with, or in place of, a prosthetic group.

For a period of several years neither Eyring nor the author had any special hesitation about extrapolating the effects of pressure on the action of narcotics on bacterial luminescence, to effects of pressure on the action of the same drugs on the behavior of aquatic animals, with which this paper is intended primarily to deal. Technical lectures during this period were made perhaps more interesting, if not clearer and more entertaining, by analogies which, for the time being, were without any direct, experimental support. Thus with disciplined but unbridled imaginations it was postulated, for example, that a fish accustomed to cold temperatures and moderately high pressures in depths of the sea should be able to swim to the surface, get soused with alcohol, then stagger back to the depths, and, because of the decrease in temperature and increase in pressure arrive home in a state of normal lethargy but stone sober. This hypothetical story was repeated to the point that the author felt a slight tinge of embarrassment at the lack of experimental evidence. The time eventually came when it seemed advisable to "put up or shut up," and the following experiments were therefore undertaken forthwith.[30]

The influence of various amounts of increased pressure, from 2000 to 5000 psi, on the activity of unnarcotized tadpoles of *Rana silvatica* and the larvae of *Amblystoma maculatum* were studied at a room temperature of 22–26°C. Such specimens evidenced a minor increase in activity during brief compression at about 2500 psi, and cessation of activity, with subsequent evidence of bodily injury, after brief subjection to 5000 psi. When placed in water containing 2.5–6.5% ethyl alcohol, or 0.08–0.1 M urethane, spontaneous activity as well as response to gentle mechanical stimuli disappeared in specimens of both species. While in the same narcotizing solution in the pressure chamber, seemingly normal activity was restored virtually at once on raising the pressure to between 2000 and 3000 psi (Fig. 2). Recovery occurred somewhat gradually, more readily at low concentrations of the drug, when the specimens were removed to fresh water at normal pressure. Serious injury, however, became evident in the frog (*Rana*) tadpoles which had been exposed to high pressures, in that they died within a day or two, although the salamander larvae usually survived. It is especially to be noted that the swimming motions of the tadpoles under pressure in a narcotizing solution were as beautifully coordinated and normal in appearance as at normal pressure in plain water, strongly suggesting that the observed reversal of narcosis took place within the central nervous system. Moreover, it seems significant that when narcotized in 0.001 M n-amyl carbamate, increased pressure failed to cause appreciable recovery. Similarly, the effect of this drug on bacterial luminescence was found to be practically insensitive to pressure at the optimum temperature.[31] Thus, the parallelism between the effects of pressure on inhibitory action of these drugs on the brightness of bacterial luminescence and on swimming activity (consciousness?) of amphibians is impressive. Moreover, several years later, a critical study concerning the influence of temperature and of hydrostatic pressure up to 10,000 psi on the activity of single neurones, showed that a 30–50% reduction in spike amplitude in the squid giant axon, under the influence of 3% ethanol at 22°C, was reversed to within 5–10% of the normal value simply by cooling to between 6 and 10°C.[32] In the same study it was revealed that increased pressure also partially restored the spike amplitude reduced by ethanol. These results add to the credibility of the interpretation that the counteracting effects of pressure on narcosis in tadpoles involve primarily effects on the nervous system.

The interrelated parameters of temperature, hydrostatic pressure, and drug concentration assumed an added interest through a detailed study[33] of various combinations of drugs which had been found to cause the two types of inhibition of bacterial luminescence. Among the fascinating results encountered was the fact that, at low temperatures, which accentuated the inhibitory action of sulfanilamide (I), the addition of alcohol or urethane (II) whose inhibitory action was minimized at these temperatures, caused the brightness of luminescence to increase severalfold, almost to equal that of a

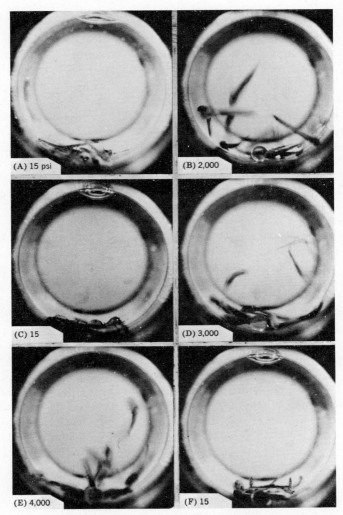

Fig. 2. Activity of salamander (*Amblystoma maculatum*) larvae in 2.5% ethyl alcohol under increased hydrostatic pressure, photographed through Herculite plate glass windows of a stainless steel pressure chamber. Successive changes (*A–F*) in pressure are indicated at the lower left corner of the photographs. These changes required a total period of about 4 min at room temperature (Ref. 34, Fig. 2, p. 20).

drug-free control, as if the sulfanilamide had dissociated from the enzyme and combined with the alcohol or urethane. The concentrations of both types of drugs involved in this experiment were such that each type alone would cause about a 50% reduction in luminescence intensity at the normal optimum temperature. Combinations of drugs causing the same type of inhibition, that is, I with I or II with II, resulted in additive or "synergistic" effects, rather than the "antagonistic" effects just mentioned. Formulations were derived applicable to data from experiments, intended to yield a numerical value for an equilibrium constant in the hypothetical combination of two drugs, such as sulfanilamide and urethane, with each other. For some conditions, the algebra turned out to be complicated and unwieldy, and in the simplest case the numerical answers were so extremely sensitive to experimental error that it was virtually impossible to obtain an accurate value for the equilibrium constant from the kinetic data. By averaging a reasonably large number of values, some in fair agreement but others widely different from each other, a rough value of 50 was obtained, and this value turned out to be sufficiently good that, on the basis of the formulation involved, calculated curves were in close agreement with measured inhibitions of luminescence caused by a wide range in concentrations of urethane mixed with a wide range in concentrations of sulfanilamide (Ref. 33, Fig. 2, p. 13). As pointed out later (Ref. 23, p. 475), the kinetic treatment is the same even if the available evidence is not sufficient to distinguish between antagonisms resulting from combination of two drugs with each other in solution, on the one hand, and combinations of the two drugs on an enzyme to give an intermediate state of the catalyst that is more active than that of the enzyme combined with either drug alone, on the other hand.

Although explicit expressions were found useful and convenient for analyzing the various equilibria involving limiting systems under the influence of temperature, pressure, inhibitors, and combinations of inhibitors referred to in the foregoing discussions, and although a plausible, graphic picture is provided by the idea of "unfolding," resulting in "changes in configuration" of the protein, with a volume increase of activation in the catalytic process, or volume increase of reaction in the "reversible denaturation," a more fundamental, general picture can be drawn with the aid of the theory of absolute reaction rates. Thus, with specific reference to the limiting enzyme in the luminescence system in bacteria, but also in a much more general sense, it has been pointed out that

There is evidence that hydrogen ions, low temperatures, hydrostatic pressure, and sulfanilamide each stabilize the molecule in states characterized by relatively low heat and small volume, whereas hydroxyl ions, high temperature, low presure, and urethane each stabilize the molecule in states characterized by high heat content and large volume, the catalytically active system being intermediate. Apart from quantitative differences among different enzymes, the same concept of an intermediate configuration necessary for activity applies in general and is the basis for the optima that are found under the influence of various factors. (Ref. 23, p. 29).

It is only one of numerous tributes to the prophetic, enduring genius of Henry Eyring that current theories detailing the mechanisms of enzyme action and inhibition, based on X-ray crystallography, and other physical as well as chemical lines of evidence, and using newer phraseology such as "conformational change" and "allosteric effect" (e.g., see review by Koshland and Neet[35]), embody concepts fundamentally similar to those arrived at through such a different avenue of approach as the kinetics of bacterial luminescence.

REFERENCES

1. H. Eyring, *J. Chem. Phys.*, **3**, 107 (1935).
2. S. Glasstone, K. J. Laidler, and H. Eyring, *The Theory of Rate Processes*, McGraw-Hill, New York, 1941.
3. F. H. Johnson, D. E. Brown, and D. A. Marsland, *Science*, **95**, 200 (1942).
4. F. H. Johnson, D. E. Brown, and D. A. Marsland, *J. Cell. Comp. Physiol.*, **20**, 269 (1942).
5. F. H. Johnson, H. Eyring, and R. W. Williams, *J. Cell. Comp. Physiol.*, **20**, 247 (1942).
6. D. E. Brown, F. H. Johnson, and D. A. Marsland, *J. Cell. Comp. Physiol.*, **20**, 151 (1942).
7. H. Eyring and J. L. Magee, *J. Cell. Comp. Physiol.*, **20**, 169 (1942).
8. S. Arrhenius, *Z. Physik. Chem.*, **4**, 226 (1889).
9. H. H. Meyer, *Arch. Exptl. Pathol. Pharm.*, **42**, 109 (1899).
10. H. H. Meyer, *Arch. Exptl. Pathol. Pharm.*, **46**, 338 (1901).
11. E. Overton, *Studien über Narkose, zugleich ein Beitrag zur allgemeinen Pharmakologie*, Fischer, Jena, 1901.
12. O. Warburg and R. Wiesel, *Pflg. Arch. ges. Physiol.*, **144**, 465 (1912).
13. H. Winterstein, *Die Narkose*, Springer, Berlin, 1926.
14. V. E. Henderson, *Physiol. Rev.*, **10**, 171 (1930).
15. A. J. Clark, "General Pharmacology," in *Heffter's Handbuch der Experimentellen Pharmakologie*, Vol. 4, Springer, Berlin, 1937.
16. J. Bělehrádek, *Temperature and Living Matter*, Borntraeger, Berlin, 1935.
17. McK. Cattell, *Biol. Rev. Proc. Cambridge Phil. Soc.*, **11**, 441 (1936).
18. D. E. Brown, *Amer. Jour. Physiol.*, **109**, 16 (1934).
19. D. E. Brown, *Ann. Rep. Tortugas Lab., Car. Inst. Wash.*, 76, 77 (1934–1935).
20. M. L. Anson and A. E. Mirsky, *J. Phys. Chem.*, **35**, 185 (1931).
21. M. L. Anson and A. E. Mirsky, *J. Gen. Physiol.*, **17**, 393 (1933–1934).
22. F. H. Johnson, H. Eyring, R. Steblay, H. Chaplin, C. Huber, and G. Gherardi, *J. Gen. Physiol.*, **28**, 463 (1945).
23. F. H. Johnson, H. Eyring, and M. J. Polissar, *The Kinetic Basis of Molecular Biology*, Wiley, New York, 1954.
24. D. E. Brown, in *Influence of Temperature on Biological Systems*, F. H. Johnson, Ed. Amer. Physiol. Soc., Washington, D.C., pp. 83–110.

25. D. E. Brown, K. F. Guthe, H. C. Lawler, and M. P. Carpenter, *J. Cell. Comp. Physiol.*, **52**, 59 (1958).
26. K. F. Guthe, in *Influence of Temperature on Biological Systems*, F. H. Johnson, Ed., Amer. Physiol. Soc., Washington, D.C., pp. 71–81.
27. K. F. Guthe, and D. E. Brown, *J. Cell. Comp. Physiol.*, **52**, 79 (1958).
28. D. A. Marsland, in *High Pressure Effects on Cellular Processes*, A. M. Zimmerman, Ed., Academic Press, New York, 1970.
29. A. M. Zimmerman, *High Pressure Effects on Cellular Processes*, Academic Press, New York, 1970.
30. F. H. Johnson and E. A. Flagler, *Science*, **112**, 91 (1950).
31. F. H. Johnson, E. A. Flagler, R. Simpson, and K. McGeer, *J. Cell. Comp. Physiol.*, **37**, 1 (1951).
32. I. Tasaki and C. S. Spyropoulos, in *Influence of Temperature on Biological Systems*, F. H. Johnson, Ed., Amer. Physiol. Soc., Washington, D.C., pp. 201–220.
33. F. H. Johnson, H. Eyring, and W. Kearns, *Arch. Biochem.*, **3**, 1 (1943).
34. F. H. Johnson and E. A. Flagler, *J. Cell. Comp. Physiol.*, **37**, 15 (1951).
35. D. E. Koshland and K. E. Neet, *Ann. Rev. Biochem.*, **37**, 359 (1968).

The Dynamics of Life.
IV. Lenticular Cataracts
Induced by Aging and
Irradiation*

BETSY J. STOVER

Departments of Anatomy and Chemistry, University of Utah, Salt Lake City

Abstract

Our recently developed Steady State Theory of Mutation Rates has been used to compare the effects of aging and irradiation on the lens of the eye. A set of data for rats, which included animals irradiated by X-rays or neutrons and nonirradiated animals, was used. It was found that functional survival of the organ could be analyzed separately from survival of the animal. Results indicate that at high doses of radiation two cataractogenic mechanisms are operative: (1) an early-acting independent mechanism, and (2) an acceleration of the aging mechanism. At lower doses the latter mechanism is dominant.

* This work supported by the U.S. Atomic Energy Commission, Contract No. AT (11-1)-119.

661

A Personal Note

"In principle everything in chemistry and physics is calculable, even the blue of the sky." I am certain that the enticement of Henry Eyring's words caught the imagination of many students, as they did mine, and not only swept them up at the time but also served as a remembered goal over the ensuing years.

After I learned to "calculate the blue of the sky," I left to attend graduate school at Berkeley and then returned to Salt Lake City to do research in physical chemistry applied to the biological effects of irradiation and aging. My interest in Dean Eyring's approach to theoretical chemistry continued, and it was my privilege to work with him, Douglas Henderson, and Edward M. Eyring in writing *Statistical Mechanics and Dynamics*. It was during this period that I witnessed the continuous encouragement, praise, and warm understanding that he had for graduate students and all others who were trying to learn.

Over a period of time I had become increasingly aware of the remarkable similarities between statistical biological survival curves and Fermi-Dirac or saturation statistics. On October 21, 1969, I sought Professor Eyring's reaction to this and related observations. Excitingly, my ideas meshed with current ideas of his. During the following weeks we developed a steady state theory of mutation rates and applied it to the survival of irradiated and nonirradiated beagles. In the present paper this theory is applied to the functional survival of the lens of the eye.

INTRODUCTION

From the observation that the statistical nature of survival curves is the same for death from aging, irradiation, and other diseases, and the fact that biological processes proceed at near equilibrium, we have used absolute rate theory to formulate a steady state theory of mutation rates. This theory has been successfully applied to the survival of beagles.[1-4] The groups of dogs considered included protected control animals, groups experiencing intrinsic diseases, and groups subjected to internal irradiation from ^{239}Pu or ^{226}Ra.

Since the steady state theory of mutation rates is basic and thus not limited to a specific application, it should be applicable to many quantal phenomena in biology. In this paper the functional survival of a single organ, the lens of the eye, is considered.

There are several factors which make it interesting to examine the survival of the lens separately from survival of the animal. The lens becomes isolated from the vascular system early in fetal life, and thereafter depends on diffusion for its nutrients. It is an organ that grows throughout life and thus differs from other organs that mature at certain periods of development. Opacification of the lens, as well as embrittlement, is a part of the aging syndrome, and, since ionizing electromagnetic radiation and neutrons can also induce lenticular opacities, the lens is a system in which the effects of aging and irradiation can be compared. Further, there is the experimental advantage that the lens can be observed repeatedly by nonperturbing methods.

EFFECT OF AGING AND IRRADIATION ON THE LENS OF THE RAT

The relative biological effectiveness of neutrons, X-rays, and gamma-rays in the induction of opacities of the lens has been reported by Upton et al.[5] Several species and strains were compared. Their results on the effects of 250-kVp X-rays or cyclotron neutrons (contaminated with gamma-rays to the extent of 5–15% of the dose) have been used. This set was chosen since the lens of the rat was more sensitive to irradiation than were those of mice and guinea pigs, and since the controls showed a higher degree of opacifaction.

In the steady state theory of mutation rates it is assumed that there are r critical sites, one or more per cell, which, if altered, lead to a mutation. If v_i is the rate at which sites are being changed and v_j is the rate at which a changed site is disappearing, and n sites have already been changed, then the steady state equation is

$$v_i(r - n) = v_j n, \tag{1}$$

the fraction of changed sites q is

$$q = \frac{n}{r} = \frac{1}{1 + v_j/v_i}, \tag{2}$$

and the fraction that survives p is

$$p = 1 - q = \frac{1}{1 + v_i/v_j}. \tag{3}$$

From absolute rate theory we obtain expressions for v_i and v_j. Then, since the rate of nonsurvival from a lethal mutation is proportional to the chance that the change has occurred and the rate of cell division of the cells involved, we can calculate survival S as a function of age t:

$$S = (1 + e^{-(a-bt)})^{-1}, \tag{4}$$

where

$$a - bt \equiv \left[\frac{\Delta G_{0i}^{\ddagger} - \Delta G_{0j}^{\ddagger}}{RT} - \sum_i \ln c_{0i} + \sum_j \ln c_{0j}\right] + \left(\sum_i k_i - \sum_j k_j\right)t \tag{5}$$

The terms on the right side of (5) have their usual thermodynamic and kinetic meaning. The subscript i refers to processes in which a critical site is being changed and j refers to those processes acting to eliminate the altered site in order to maintain the biological steady state. If there are multiple independent mechanisms that lead to nonsurvival, then S is given by

$$S = \prod_i S_i, \tag{6}$$

and, if one of these, S_i, acts early in time when all other $S_{i \neq j}$ are close to unity, then (6) reduces to (4). If multiple causes act through the same mechanism, then b in (4) is given by

$$b = \sum_i b_i. \tag{7}$$

The functional survival, that is, transparency, of the lens of the eye with increasing age is shown for control animals and for rats irradiated with

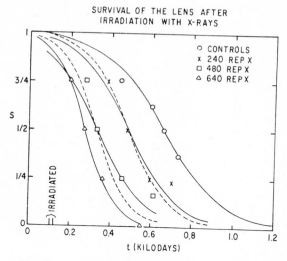

Fig. 1.   Survival of the lens after irradiation with X-rays.

X-rays (Fig. 1) and neutrons (Fig. 2). The points shown are those reported by Upton et al.[5] (with the exception that three were interpolated). The age at irradiation is also shown. The degree of opacification was reported as grade + through grade + + + + (complete opacification). This method places a limitation on the data, but, since there were 10–25 animals per group and there was little variation within a group, the data are probably as valid as is

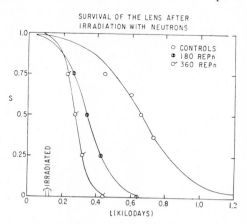

Fig. 2.   Survival of the lens after irradiation with neutrons.

possible in this kind of observation. Another limitation is that neither the controls nor the animals at the lower dose levels of X-rays reached complete opacification in the period reported, and it is presumed that the animals simply did not live long enough for complete opacification to occur. Further it was the survivors of the irradiated groups that were observed, and thus there was a selective factor in these groups that was not operative in the control group. The smooth curves shown are (4) fit to the experimental data. (In Fig. 1 the two broken curves were fit ignoring the last point for each of the two lower dose levels.)

Equation (4) is symmetrical about the point $S = \frac{1}{2}$ and thus describes data that are symmetrically distributed in time about $t_{s=\frac{1}{2}}$. The two sets of data from the neutron irradiations and that of the 640-rep X-ray group are highly symmetrical, and pass through $S = \frac{1}{2}$ at an age when the calculated value of S for controls is ≥ 0.9. The data for the 480-rep X-ray group are less symmetrical but those for the controls and the 240-rep X-ray group fit fairly well, especially considering that complete opacification did not occur during the period of observation.

The parameter a of (4) is a measure of cellular reserves and b is a measure of the rates of the reactions that maintain the steady state. An injurious factor can act to alter the reserves or the rates, or both. Thus a decrease in $(a - bt)$ represents the case in which the rate at which critical changes occur is increasing over the rate at which they disappear. The values of a and b for the curves of Figs. 1 and 2 are given in Table I. From this limited analysis two interesting findings emerge. First, the value of b increases with increasing dose of X-rays or neutrons, and the rate of increase is greater for the more damaging neutrons. Second, there is a change in the value of a at the highest dose of each radiation. This suggests that the cataractogenic effect of both X-rays and neutrons is one of accelerating the aging mechanism (7). But at high doses there is also an early-acting independent mechanism (6). It is interesting that both 640 rep of X-rays and 180 rep of neutrons result in

Table I. Reserves, Rates, and Times to Half-Survival

Group	a	$b \times 10^3$ (d^{-1})	$t_{s=\frac{1}{2}}$ (d)
Control	4.6	7.0	660
X-ray, 240 rep	4.8	9.9	485
X-ray, 480 rep	4.2	12.7	329
X-ray, 640 rep	3.8	13.7	276
Neutron, 180 rep	4.7	13.7	344
Neutron, 360 rep	5.9	25.0	276

approximately the same acceleration of the aging mechanism, but at 640 rep of X-rays the independent mechanism is also effective so that the reduction in time to half-survival is greater. Thus the action of irradiation through multiple mechanisms to produce the same effect provides a possible explanation to the variation with dose of the relative effectiveness of different radiations. Further, these results on cataractogenesis are consistent with the concept that there is a threshold radiation dose for the independent mechanism, but that there is no threshold dose for the acceleration of the aging mechanism. The implication is that some environmental factors may be either threshold or nonthreshold or both.

In summary, this brief analysis suggests that the steady state theory of mutation rates is applicable to the survival of a single organ of an animal, and, in this case, served to relate the effects of aging and irradiation.

REFERENCES

1. B. J. Stover and H. Eyring, "Death from Aging, Cancer, Irradiation and Other Stresses," *Abstr. Papers Eighteenth Ann. Meeting Radiation Res. Soc.*, *Dallas, Texas*, p. 60, 1970.
2. B. J. Stover and H. Eyring, "The Dynamics of Life. I. Death from Internal Irradiation by ^{239}Pu and ^{226}Ra, Aging, Cancer and Other Diseases," *Proc. Natl. Acad. Sci., U.S.*, **66**, 132 (1970),
3. H. Eyring and B. J. Stover, "The Dynamics of Life. II. The Steady State Theory of Mutation Rates," *Proc. Natl. Acad. Sci., U.S.*, **66**, 441 (1970).
4. B. J. Stover and H. Eyring, "The Dynamics of Life. III. Mechanisms of Non-Survival and the Relation of Dose Size," *Proc. Natl. Acad. Sci. U.S.*, **66**, 672 (1970).
5. A. C. Upton, K. W. Christenberry, G. S. Melville, J. Furth, and G. S. Hurst, "The Relative Biological Effectiveness of Neutrons, X-rays, and Gamma Rays for the Production of Lens Opacities: Observations on Mice, Rats, Guinea-Pigs, and Rabbits," *Radiol.*, **67**, 686 (1956).

The Kinetics of Thermal Grooving of Solids

P. V. McALLISTER AND I. B. CUTLER
University of Utah, Salt Lake City

Abstract

Many investigators of the kinetics of thermal grooving of metals and oxides have interpreted their results in terms of surface diffusion. Their data have been reevaluated and the log-log plots that they have used are shown to be erroneous.

Reevaluation leads to the conclusion that groove growth is predominantly by volume diffusion with surface diffusion possibly dominating the initial groove formation. Calculated diffusion coefficients and activation energies are in excellent agreement with tracer diffusion data for volume diffusion. These results provide renewed confidence in the use of thermal grooving as an effective technique for studying the kinetics of atomic transport in material.

A Personal Note

Although the concepts concerning chemical kinetics and their relationship to thermodynamics taught by Dr. Henry Eyring have greatly influenced the scientific contributions of his students, their esteem is even higher for the personal example this great man set for his students to follow. First of all, he has always been very gracious to include anyone as a coauthor in his publications who would lend him even modest help or support. His willingness to give credit to others through references and coauthorship has been a good example for all.

His philosophy of diligent work has been a good influence on his students. First of all, his classes were always taught early in the morning. He did this so that both he and his students would have the rest of the day to devote to research. Then there were Saturday morning seminars. These had a real purpose. To Henry Eyring science is more than a 40 hour per week adventure. Saturdays were great days in which to accomplish significant results and Saturday morning seminars ensured that his students would be in attendance and working on Saturday. His example has always set a great pattern for his students to follow.

On one of these Saturday morning seminars one of the students was discussing a mechanism, the detail of which is now forgotten. The mechanism required the breaking and forming of chemical bonds and did not completely satisfy valence considerations. Henry Eyring, as he was always prone to do, was quick to jump to his feet and challenge the thinking of others. On this occasion he emphasized the dangling bond left unreacted by the proposed mechanism and chided the student by asking, "Now just what are you going to do about this bloody stump?" A roar of laughter from the seminar group made the student acutely aware of the deficiency of his mechanism. Seminars with Henry Eyring were always spiced with humor. He continually amazed his students with agility of thought while on his feet, quickly reaching the heart of the problem and proposing solutions that involved atoms, molecules, chemical bonds, and reaction rates processes.

For his scientific insight and his ability to transfer that insight to his students through the teaching process his students will always be grateful.

INTRODUCTION

Formation of grooves at the grain boundaries on the surface of poly-crystalline solids during annealing has long been observed. This process has often been used to delineate the grain boundaries by "thermal etching." However, it was not until Mullins[1] analytically described thermal grooving, and thus defined the groove shape, that this technique became a useful tool for investigating the kinetics of atomic transport in solid materials. Thermal grooving soon gained the favor of investigators who applied it to the study of the kinetics of diffusion of many materials including copper,[2,3] silver,[4] Al_2O_3,[5,6] MgO,[5-7] NiO_2,[8] and UO_2.[7]

One of the appealing features of the thermal grooving technique is that the curvature of the groove is mathematically exact and therefore not subject to the geometric approximations which are common in sintering models. Theoretically, thermal grooving should be potentially superior to sintering as a technique for determining diffusion mechanisms and activation energies. This potential has not generally been achieved in practice. With few exceptions, investigators have concluded that surface diffusion is the predominant mechanism of groove formation. Their activation energies have been much higher than expected for surface diffusion based on radioactive tracer surface diffusion studies, and they have had to resort to hypothesizing complicated surface models with impurity pinned ledges and extremely long atomic jumps to justify the high activation energies. In addition, the surface diffusion coefficients obtained from thermal grooving experiments are large enough to completely fill in the necks of a powder compact during sintering so that the typical shrinkage observed in everyday experience could never occur. These problems have served to discourage investigators and thermal grooving has fallen into disfavor as a technique for investigation of atomic diffusion in materials.

The authors have completed a new approach to the kinetics of thermal grooving and have concluded that Mullins' theory is correct, but errors have been made in data interpretation that have resulted in incorrect conclusions and these errors have been self-perpetuating. The conventional practice of plotting log w versus log t to get the power dependence of grain boundary width with time is unsatisfactory. A plot of w versus t^n for various values of n is recommended. Incorporation of this correction justifies modification of the reported conclusions of previous thermal grooving research and yields diffusion coefficients in agreement with tracer studies.

THERMAL GROOVING

Theory

The kinetics of thermal grooving have been observed by measuring the width w of a groove as a function of time t at constant temperature T (see Fig. 1). Although not easily accomplished, the depth of the groove might also be followed during isothermal annealing. The contour of the groove produced by either surface or volume diffusion is similar, but surface diffusion is expected to contribute to the early stages of groove formation and volume diffusion to the latter stages[1,2]; that is, the surface diffusion coefficient is normally larger than volume diffusion coefficient but the number of atoms contributing to volume diffusion increases rapidly as w increases until volume diffusion contributes far more than surface diffusion.

Mullins found that the solutions to the differential equations describing mass transport during thermal grooving have the following forms.

For a surface diffusion mechanism

$$w = 4.6(Bt)^{1/4}, \tag{1}$$

$$B = \frac{D_s n \gamma \Omega^2}{kT}. \tag{2}$$

For a volume diffusion mechanism

$$w = 5.0(Ct)^{1/3}, \tag{3}$$

$$C = \frac{D_v \gamma \Omega}{kT} ; \tag{4}$$

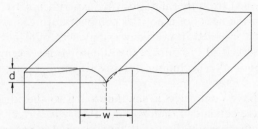

Fig. 1. Schematic of a thermal groove. Dimensions w and d increase with time t during annealing.

D_s = surface diffusion coefficient,

D_v = volume diffusion coefficient,

γ = surface free energy of the gas–solid interface,

n = number of atoms per square centimeter of surface which take part in the diffusion process,

k = Boltzmann's constant,

T = Absolute temperature, °K,

Ω = Atomic volume of the diffusing species.

Since all other terms are assumed constant, a plot of log w versus log t should produce a slope of $\frac{1}{4}$ for surface diffusion and a slope of $\frac{1}{3}$ for volume diffusion if a single mechanism predominates.

Initial Groove Formation

Among the simplifying assumptions accompanying the derivation of (1) and (3) is that atomistic characteristics of the crystal structure can be ignored. No serious objections have been raised to this assumption since it is indeed necessary if we are to have the surface act as a continuum so that macroscopic concepts such as surface curvature, surface tension, and surface free energy may have consistent meanings. This macro concept also permits the small-angle assumption for the slope at the groove root $\gamma_b/2\gamma_s = \sin \theta \simeq \tan \theta$. Since relatively few atoms must be transported before a grain boundary groove approximates the continuum contours, this mass can be safely ignored in determining the diffusion in a developed groove. The shape (width) of the groove in these initial formation stages, however, cannot be neglected without affecting later conclusions.

A grain boundary can be considered to be a plane of several atomic depths which is less ordered and therefore less dense than the bulk crystal. The "corner" atoms at the grain boundary–surface interface are in a very high energy situation and until the corner at the interface can be sufficiently "rounded" to reduce this energy to that of general surface tension, there is considerable driving force to remove these corner atoms by evaporation, diffusion, or even the original polishing process. The contour resulting when the first corner atom is removed and placed on the surface cannot be considered to match the formula continuum contour nor will it match until many atoms are moved and the "steps" are smoothed out. At a width to depth ratio of ~40 (for 19° tilt angle copper), a significant width for the groove at the time the formulas become applicable not only appears logical but is unavoidably mandatory.

In many oxides and halides Debye layers and impurity segregation near free surfaces are known to exist. A polished surface is not likely to be in equilibrium upon insertion into an isothermal atmosphere. During equilibration thermal grooving will decelerate to the rate governed by the steady state approximation.

Data Corrections

Many processes could contribute significantly during the initial groove formation and the groove "profile" would be expected to be controlled by the predominating process at each stage of its development. In a system in which evaporation–condensation or surface diffusion predominated initially with subsequent domination by volume diffusion, a significant width would be obtained by the time steady state growth of the groove is reached and Mullins'

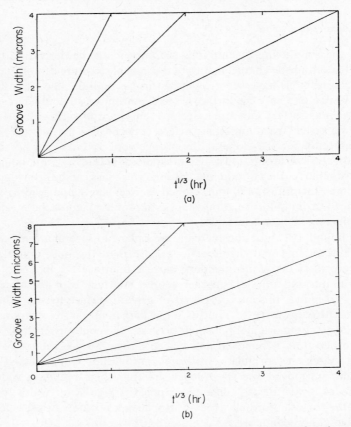

Fig. 2. (a) Synthetic plot of grain boundary width as a function of time for growth by volume diffusion, assuming no initial width; (b) synthetic plot of grain boundary width as a function of time for growth by volume diffusion, assuming an initial rapid growth.

formula becomes applicable. Therefore there should be a finite groove width different from zero which should be added to the theoretical formulas to make the equilibrium configuration grain boundary width effectively nonzero at the beginning of the diffusion growth process (i.e., $t = 0$).

The effect of nonzero initial conditions and different thermal grooving rates on the slope of a log-log plot can be graphically demonstrated. In Fig. 2a a synthetic plot of w as a function of $t^{\frac{1}{3}}$ with initial conditions

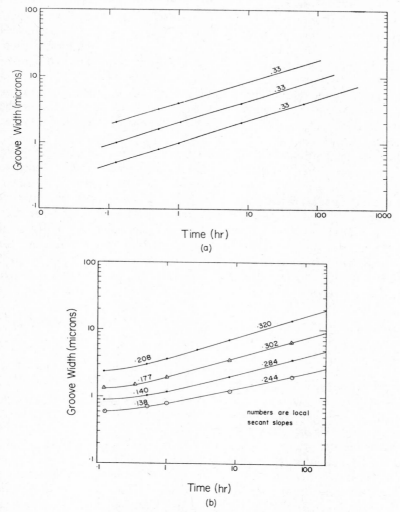

Fig. 3. (a) Replot of Fig. 2a on a log-log scale; (b) replot of Fig. 2b on a log-log scale, illustrating the effect of an initial grain boundary width.

$w = 0$ at $t = 0$ is shown. In Fig. 2b initial conditions are $w = 0.4\,\mu$m at $t = 0$. The data from these curves were replotted on log-log paper as Fig. 3a and 3b. All slopes from Fig. 2a reproduced in Fig. 3a the specific slope of 0.333. Figure 3b shows a plot of data from Fig. 2b. Note that the log-log plots are no longer linear in the short-term region and the slopes of the nearly linear portions at longer times show considerable variation and are less than the specified $\frac{1}{3}$ slope. This disparity is even greater when secant slopes in the short-time region are calculated. The wide variation in this plot makes it meaningless to attempt any conclusions about the transport mechanism from slopes obtained from a log-log type of plot. A nonzero initial condition is the basis for the cautions on the use of log-log plots for analyzing grain growth and sintering data given by Mistler and Coble,[9] Burke and Turnbull,[10] and Johnson and Cutler.[11]

The suggested method of presenting the data from thermal grooving experiments is to plot w versus t^n for various values of n and select, as the actual value of n, that n which results in the most consistent and logical value of w when extrapolated back to $t = 0$.

If the w versus $t^{1/4}$ plot produces a negative value of w when extrapolated to $t = 0$, the use of the $\frac{1}{4}$ exponent should be suspect immediately. A positive intercept can be interpreted as the groove width at which a single mechanism begins to dominate. Most thermal grooving data can be plotted on identically shaped curves on both $n = \frac{1}{4}$ and $n = \frac{1}{3}$, that is, both give straight lines or slightly concave lines, etc. This is additional evidence for the validity of a single mechanism interpretation but it may not be possible to distinguish between mechanisms if both w versus $t^{1/4}$ and w versus $t^{1/3}$ have positive intercepts without evaluating diffusion coefficients and tracer data from other diffusion experiments.

REEVALUATION OF DATA DESCRIBING THERMAL GROOVING

Copper

Thermal grooving of copper has been studied by N. A. Gjostein.[3] In accordance with the usual practice of presenting thermal grooving data, his data were presented in a log-log plot (Fig. 4). Note the striking similarity of these curves to the center portion of the synthetic data in Fig. 3b. Mullins and Shewmon[2] also studied copper and in both these investigations surface diffusion was thought to be the predominant mechanism after an arbitrary volume diffusion correction was subtracted. In a previous publication[12] it was shown that replotting the data from Mullins and Shewmon and from Gjostein on a w versus $t^{1/4}$ curve produces negative initial values of w (see Fig. 5). Replotting as w versus $t^{1/3}$ (Fig. 6) produces the necessary positive

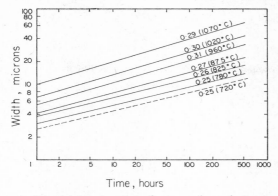

Fig. 4. Log-log plot of copper thermal grooving data.[3]

intercepts and reasonably good convergence to an equilibrium initial width of ~1.2 μ. D_v values were calculated from thermal grooving data and plotted with the tracer diffusion results of Kuper et al.[13] and are shown in Fig. 7. Note that agreement is excellent and that the slightly higher average values of D_v from thermal grooving are in the expected relationship to tracer values since the short half-life of the isotope ^{64}Cu would be expected to bias the D_v values to the low side even though a half-life correction was incorporated in the tracer data.

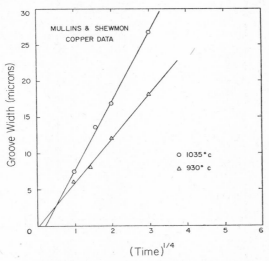

Fig. 5. Plot of copper grain grooving data[2] as surface diffusion.

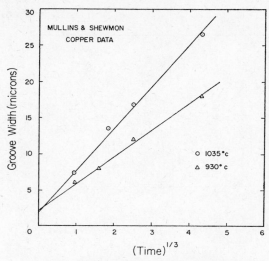

Fig. 6. Plot of copper grain grooving data[2] as volume diffusion.

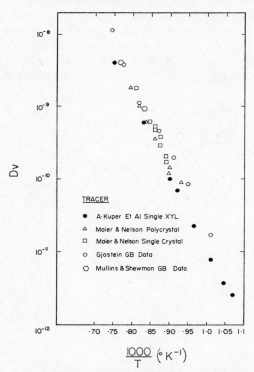

Fig. 7. Superposition of volume diffusion coefficients calculated from Data in Refs. 2 and 3 on tracer diffusion results from Ref. 13.

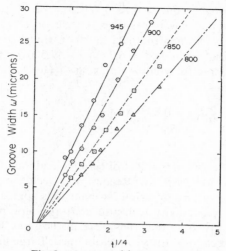

Fig. 8. w versus $t^{1/4}$ plot of silver thermal grooving data from Rhead.[4]

Silver

A similar reanalysis of the work of Rhead[4] on silver gave results that were just as striking. Interpretation of thermal grooving data as the result of surface diffusion yielded a Q_s of 55,500 compared to reference values of $\sim$12,000 for other methods of measuring surface diffusion; D_0 was 10^6

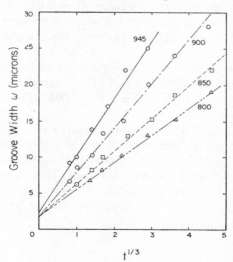

Fig. 9. w versus $t^{1/3}$ plot of silver thermal grooving data from Rhead.[4]

compared with a reference value of 0.13.[14] When w was plotted as a function of $t^{1/4}$ (Fig. 8), the intercepts were negative. A w versus $t^{1/3}$ plot showed good convergence (Fig. 9) and when the data at 850°C and higher were used to calculate D_v, the result was $D_v = 6.7 \exp(-49,500/RT)$, which nearly duplicates the tracer data for volume diffusion of Hoffman and Turnbull,[15] who published the result

$$D_v = 0.895 \exp\left(\frac{-49,500}{RT}\right).$$

Ceramic Materials

A similar situation exists in the field of ceramic materials. With the exception of the nickel oxide work by Readey and Jech,[8] investigators have generally concluded a surface-diffusion mechanism is applicable to the oxide materials. Although Readey and Jech also used a log-log plot, the grooves studied were sufficiently large that they were in the more linear portion of the plot and the correct conclusions were obtained. Large grooves were also measured by Colombo and Howard[16] on iron-silicon alloy so that the correction for the initial groove formation did not unduly influence the slope of the log-log plot.

Magnesium oxide has been investigated by Robertson[5b] and by Henney and Jones.[7] Surface diffusion was identified as the dominant mechanism of transport in both of these studies. The activation energies appeared excessive for this interpretation. In the reevaluation of these data[17] it was shown that interpretation as volume diffusion produced $D_v = 0.02 \exp(-70,000/RT)$, which is in agreement with the tracer and mass spectrometer results compiled by Harrop[18] in which the mean activation energy for Mg and O migration is 70,500 cal/mole.

Similarly, the Henney and Jones[7] activation energy for surface diffusion in $UO_{2.005}$ of 110,000 cal is definitely excessive, whereas reinterpretation as volume diffusion produced $D_v = 0.22 \exp(-71,000/RT)$, which is in good agreement with tracer data.[18]

Aluminum oxide in several varieties and impurity levels has been investigated by Robertson[5] and by Shackelford.[6] The data scatter widely and have internal inconsistency sufficient to prevent the straightforward conclusions which are available with data on other materials. There is ample evidence, however, that the predominant mechanism of thermal grooving in Al_2O_3 is also volume diffusion and that the activation energy for the process is between 100,000 and 200,000 cal/mole.

GROOVE PROFILES

Comparison of experimental groove profiles with the theoretical curves has been used both to verify Mullins' theory and to show the grooving

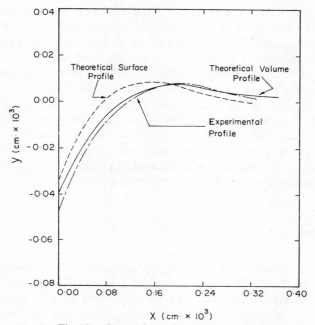

Fig. 10. Comparison of theoretical and experimental profiles of thermal grooves in aluminum oxide.

mechanism. Normalization of the depth of the groove, that is, forcing the theoretical profile through the experimental depth points as done by Mullins and Shewmon[2] reproduces the experimental profile within construction accuracy. However, the same profile is produced by force fitting both the surface and volume diffusion curves. Shackelford and Scott[6] used the measured groove root angle and the time and temperature of thermal etch to construct their theoretical profile. This approach will show a significant difference between the surface diffusion and volume diffusion theoretical curves. However, they were sufficiently converted by the preponderence of log-log plots and surface-diffusion conclusions in the literature that they only plotted the theoretical profile for surface diffusion and then concluded that the experimental profile was within experimental error. Superposition of a theoretical volume curve on Shackelford and Scott's profile is shown in Fig. 10. The superior fit between the experimental and volume diffusion profiles is obvious.

DOPING OF ALUMINA

The addition of minor amounts of manganese oxide to Al_2O_3 has been shown by Keski and Cutler[19] to significantly accelerate the sintering rate and the shrinkage of a powder compact. Volume diffusion was also shown to predominate in sintering. A series of polished samples with various levels of MnO doping was packed in powder which had been doped at a higher level than any of the samples. The samples and powder were placed in a covered crucible, and grooving studies were conducted in air. Under these conditions equilibrium should be established between the volatile manganese oxide in the powder and the sample surfaces at least to the depth considered influential in the models of surface diffusion. Therefore all the samples should have grooved at the same rate regardless of their internal Mn content if surface diffusion were the dominant transport mechanism. At 1500°C, however, the observed grooving rate increased with internal Mn content in agreement with the sintering studies and the volume diffusion hypothesis.

CONCLUSIONS

There is great temptation to use log-log analysis in the investigation of kinetic processes. This is only justified in those cases (if any actually exist) in which the descriptive formulas can be guaranteed to pass through the origin. If there is an initial nonzero condition present, log-log plots of the data can produce sufficiently serious errors to result in grossly incorrect conclusions. This has been shown to be the case with thermal grooving studies in which surface diffusion has been consistently identified as the transport mechanism although activation energies do not correlate with other studies.

Reevaluation of existing thermal grooving studies has shown that volume diffusion predominates after an initial formation period during which many mechanisms can contribute. This conclusion produces diffusion coefficients and activation energies which are in excellent agreement with tracer studies of self-diffusion. This agreement justifies renewed confidence in the use of thermal grooving as a tool for basic investigations of self-diffusion in materials.

REFERENCES

1. (a) W. W. Mullins, *J. Appl. Phys.*, **30**, 77 (1959). (b) W. W. Mullins, *ibid.*, **28**, 333 (1957). (c) W. W. Mullins, *Trans. AIME*, **218**, 354 (1960).
2. W. W. Mullins and P. G. Shewmon, *Acta Met.*, **7**, 163 (1959).
3. N. A. Gjostein, *Trans. AIME*, **221**, 1039 (1961).
4. G. E. Rhead, *Acta Met.*, **11**, 1037 (1963).

5. (a) W. M. Robertson and R. Chang, in *Materials Science Research*, Vol. 3, W. W. Kriefel and H. Palmour, Eds., Plenum, New York, 1966, p. 49; (b) W. M. Robertson, in *Sintering and Related Phenomena*, G. C. Kuezynski et al., Eds., Gordon and Breach, New York, 1967, pp. 215–232; (c) W. M. Robertson and F. E. Elkstrom, *International Symposium on Special Topics in Ceramics: Kinetics of Reactions in Ionic Systems, Alfred University, June 1967*, Plenum, New York, 1969.

6. J. F. Shackelford and W. D. Scott, *J. Amer. Ceram. Soc.*, **51**, 688 (1968).

7. J. Henney and W. S. Jones, *J. Mat. Sci.*, **3**, 158 (1968).

8. D. W. Readey and R. E. Jech, *J. Amer. Ceram. Soc.*, **51**, 8, 472 (1968).

9. R. E. Mistler and R. L. Coble, *J. Amer. Ceram. Soc.*, **51**, 8, 472 (1968).

10. J. E. Burke and D. Turnbull, in *Progress in Metal Physics*, Vol. III, B. Chalmers, Ed., Pergamon, London, 1952, pp. 220–92.

11. D. L. Johnson and I. B. Cutler, *J. Amer. Ceram. Soc.*, **46**(11), 545 (1963).

12. P. V. McAllister and I. B. Cutler, *Met. Trans.* (AIME) **1**, 313 (1970).

13. A. Kuper, H. Letaw, Jr., L. Slifkin, E. Sonder, and C. T. Tomizuka, *Phys. Rev.*, **96**, 1224 (1954).

14. T. Suzuoka, *J. Phys. Soc. Japan*, **20**, 7 (1965).

15. R. E. Hoffman and D. Turnbull, *J. Appl. Phys.*, **22**, 634 (1951).

16. R. L. Colombo and J. Howard, *J. Mat. Sci.*, **4**, 753 (1969).

17. P. V. McAllister and I. B. Cutler, *J. Amer. Ceram. Soc.*, **52**, 348 (1969).

18. P. J. Harrop, *J. Mat. Sci.*, **3**, 206 (1968).

19. J. R. Keski and I. B. Cutler, *J. Amer. Ceram. Soc.*, **51**, 440 (1968).

The Interaction of Cellulose with Simple Salt Solutions and with Dyebaths[*]

MARY MONCRIEFF-YEATES[†] AND HOWARD J. WHITE, Jr.[‡]

Textile Research Institute, Princeton, New Jersey

Abstract

Radioactive tracer techniques have been used to study the absorption of various ions by cellulose from salt solutions and from dyebaths. Measurements have been made on the absorption of Na and SO_4 ions from Na_2SO_4 solutions at 25°C and at 90°C and from Na_2SO_4 solutions containing a direct dye at 90°C. Measurements of the amount of dye absorbed were made using conventional techniques. The absorption of Br from NaBr solutions was also studied.

The results on the absorption of ions from simple salt solutions can be interpreted qualitatively in terms of a simple Donnan equilibrium. The results on the absorption of ions in the presence of dye cannot be interpreted in terms of a Donnan equilibrium as modified by a Langmuir-type dye absorption. Electrostatic interaction between absorbed dye anions is advanced as a possible reason for the discrepancy.

In the course of the work some results have been obtained on the stability of the viscose rayon sample used and the cation-exchange capacity of the cellulose was measured.

* Reprinted by permission from *American Dyestuff Reporter*, Volume 46, 1957, p. 87. Presented in part on September 22, 1955 at the Viking Room, Haddon Hall Hotel, Atlantic City, N.J., during the 1955 National Convention.
† Present address: Vitro Corp. Silver Spring, Md.
‡ Present address: National Bureau of Standards, Washington, D.C. 20237

A Personal Note

H. J. WHITE, JR.

It is only fitting to have an article on a textile topic in this collection, since Henry Eyring was actively associated with the Textile Research Institute from the time of its movement from Washington, D.C. to Princeton, N.J. in 1944 until his move to the University of Utah.

The year 1944–1945 was an active and fertile one. Henry, along with George Halsey, the author, and a host of other graduate students, was busy applying the theory of absolute reaction rates to flow processes in polymers and the mechanical properties of textiles in particular. This work produced added insight into the elementary processes involved in the deformation of textiles and aroused wide interest in textile circles around the world.

The winter and spring were active in other ways also. Textile Research Institute was and still is situated in the old Baker mansion on the shores of Lake Carnegie. At that time it was reached by a winding unpaved drive of perhaps $\frac{1}{2}$ mile length. This drive contained a recently dug utility trench which wandered now on one side, now in the center, now on the other, most of the length of the drive. Although Textile Research Institute took possession at about Christmas time, the utilities company stoutly maintained that they would be unable to compact the trench into a hard surface until after the final thaw and the spring rains were over. The result was a driving challenge which in its own way matched anything that Lime Rock or Indianapolis can offer. The trick was to maintain enough speed not to bog down in casual mud or snow and at the same time hit the boards laid for strategic crossings of the trench, avoid the trench proper, which was a potential well if there ever was one, and the assorted trees, bushes, pot holes, and minor quagmires. The faint-hearted parked on the highway and picked their way in on foot. The more venturesome and the unwary tried the driveway, with inevitable results. Two or three times a week the call would go out that another motorist was in need of aid.

Although Henry in his position as Director of Research could have ignored this problem, he was in fact, one of the more enthusiastic members of the rescue brigade. More than one distinguished foreign scientist stood on the sidelines and thought his own thoughts about the idiosyncrasies of American chemists while Henry, along with Rocky the caretaker and a gaggle of graduate students, shoveled mud, emplaced boards strategically, shouted instructions to the driver, and heaved in unison on bumper, fender, and door post. On one such occasion George Halsey, ever the analyst, voiced the conclusion that the engineering technique being used to free the stuck vehicle was doomed to failure for fundamental reasons. This remark exercised Henry greatly. Although Henry the theorist undoubtedly would have liked to argue the point in engineering mechanics with George, Henry the man of action had to restrain himself from putting a shovel of mud where it would do the most good.

INTRODUCTION

The purpose of this paper is the presentation of work on the interaction of cellulosic fibers with simple salt solutions and with dyebaths.

The interaction of cellulosic fibers with dyebaths has often been studied because of its technological importance. The interaction of such fibers with salt solutions has been less often studied, but is, nonetheless, of interest and importance. In the first place, dyes for cellulose are often salts, and simple salts are often components of dyebaths for cellulose. In the second place, since the behavior of simple salt solutions is better understood than the behavior of dyebaths, it is reasonable to expect that the interaction of such solutions with cellulosic fibers will be simpler and more susceptible to ready interpretation than the interaction of the same fibers with dyebaths. Cellulosic fibers were chosen because they are representative of an important technological class of fibers, and because large, moderately round fibers, which offer advantages for the experimental technique used, were readily available.

The rather extensive literature on the absorption of dyes, especially direct dyes, by cellulose has been reviewed by Vickerstaff.[1] Vickerstaff has also reviewed the literature on salt absorption; however, three papers might be mentioned as pertinent to the work presented here, namely, those of Usher and Wahbi,[2] Neale and Standring,[3] and Farrar and Neale,[4] and particularly the last.

The experiments reported here were formulated with the intention of uncovering, in as much detail as possible, the environment existing within a cellulosic fiber in contact with an ionic aqueous phase. Measurements were made at room temperature ($\sim$25°C) and at 90°C. The salt used most extensively was Na_2SO_4 and measurements were made on the uptake of both cation and anion as a function of the concentration of the bath at 25°C and 90°C and as a function of pH at 25°C. In addition, the uptake of the direct dye, Chrysophenine G, and the effect of the dye on the salt uptake were studied at 90°C. It was also necessary to perform some measurements relating to the stability of the fibers. Finally, some measurements were made on the uptake of the Br ion from NaBr solutions.

689

All of the salt absorption measurements were made using radioactive tracer techniques; dye uptakes were measured by a change-of-concentration method. Details of each method are given below.

Essentially three different measurements were made: room-temperature ion absorption measurements, high-temperature ion absorption measurements, and dye absorption measurements. Both high- and room-temperature salt absorption measurements were made using single fiber samples and radioactive salts; the dye absorption measurements were made using bulk samples.

The use of radioactive tracers offers two advantages: the distribution of individual ions can be determined, and sufficiently small quantities can be measured that single fiber samples can be used. The use of single fibers minimizes the problem of the separation of the fiber sample from entrained solution, which is formidable when bulk samples are used.

Preparation of Materials

Fiber Samples. The fiber samples used were viscose rayon monofils. They had smooth, approximately round cross sections, and each filament had a mass–length ratio of roughly 15 denier. The fibers are made by the American Viscose Corporation and marketed under the name Tufton.

Fibers 10 cm long were extracted in a Soxhlet extractor for 1 hr with thiophene-free benzene and then air dried. They were then soaked in distilled water adjusted to pH 5 with HCl; the water was changed frequently and the soaking was continued until the pH of the water remained constant. This treatment was carried out in an effort to assure that all exchangeable cations within the fiber were hydrogen ions at the start of the experiment.

The fibers can be separated into two groups with respect to preparatory treatment at this point. Some of them were soaked in distilled water for 8 hr. Others were washed three times with distilled water, soaked in distilled water at 90°C for 1 hr, and finally soaked in distilled water for 8 hr at room temperature. This last set of fibers will be referred to, hereafter, as stabilized; the changes brought about by the stabilization process and the reasons for its use will be discussed subsequently.

All of the fibers were then blotted, dried at 85°C for 2 hr (or until all apparent water was removed), and finally dried *in vacuo* over $Mg(ClO_4)_2$ overnight.

The samples were then conditioned for at least 24 hr at 65% RH and 70°C and weighed using a vibroscope.[5] The dry weights were determined from the weights at 65% RH by calculation; the amount of water taken up, 13.4 g/100 g dry fiber, was determined by direct measurement with a bulk sample. It is probable that there are fiber-to-fiber variations in the amount of water absorbed, and that the stabilized fibers have a slightly lower absorptive capacity than the unstabilized fibers. Error from such sources should be about 1 %, which is well within the limits of precision of the experiments. It should also be mentioned that the samples were small (a few hundred fibers) and, hence, did not need the long conditioning times necessary with larger bulk samples.

Salt Solutions.   The radioactive isotopes used were ^{22}Na, ^{82}Br and ^{35}S, which were obtained from Oak Ridge National Laboratory with permission of the Atomic Energy Commission. They were obtained in the form of simple acids or salts and were converted to the desired form by standard methods, such as the use of ion-exchange columns and titration. The salts Na^{82}Br, ^{22}Na$_2$SO$_4$ and Na$_2$^{35}SO$_4$ were prepared.

Dye Solutions.   The direct dye, Chrysophenine G (*Colour Index* Number 365), was used and was purified by repeated salting out with CH$_3$COONa.

Salt Absorption Measurements at Room Temperature

The room temperature measurements were made using a modification of the method of Stam and White.[6] It can be conveniently broken down into four operations: treatment, washing, counting, and calibration.

Treatment.   The fibers were attached to stainless steel spring clips and treated in test tubes or centrifuge tubes. Three to five fibers were treated at a time in about 20 ml of solution, so that there was a negligible change in concentration of the bath during any one treatment. The fibers were separated, one from the other, by a device consisting of glass tubes spaced out from and attached to a central glass rod; this device minimized entanglement of the fibers. The solutions were agitated sporadically by hand.

Washing.   The washing step is designed to separate surface droplets from material that has penetrated within the fibers. One of the reasons for using single fiber samples is the simplification of this separation, which is, in general, the most difficult part of any experiment involving absorption by fibers.

Washing consists of two processes: removal of materal from the surface of the fiber and removal of material from the interior of the fiber. In a successful washing, conditions are so adjusted that the first process is effectively complete before the second starts, and the washing is broken off in the interval between.

Curves for the rate of removal of the various radioactive ions by satisfactory wash liquids are shown in Figs. 1–3. The units of the ordinates in Figs. 1–3 are units of radioactivity and are related to the amount absorbed

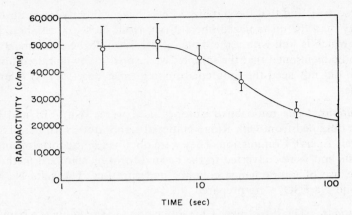

Fig. 1. The rate of desorption of $^{35}SO_4$ from Na_2SO_4-treated monofils into a 75% ethanol–25% water bath at room temperature. The amount of $^{35}SO_4$ in the monofils is specified in terms of its radioactivity.

in a way that will be discussed later. The limits in Figs. 1–3 represent standard deviations.

It is clear that, for a successful washing (as defined above), a complete plot of the desorption in the manner of Fig. 1 would consist of two steps; the first representing removal of surface material, the second removal of

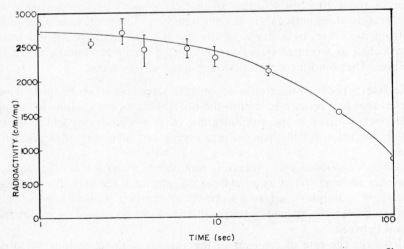

Fig. 2. The rate of desorption of ^{22}Na from Na_2SO_4-treated viscose monofils into a 75% ethanol–25% water bath at room temperature. The amount of ^{22}Na in the monofils is specified in terms of its radioactivity.

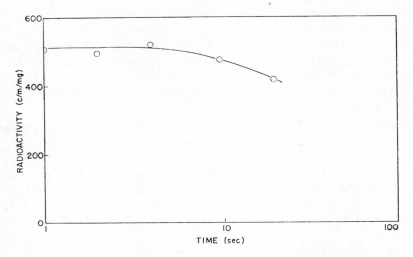

Fig. 3. The rate of desorption of ^{82}Br from NaBr-treated viscose monofils into a 95% ethanol bath. The amount of ^{82}Br in the monofils is specified in terms of its radioactivity.

material from the interior. The latter process is obviously the slower. The existence of only one step in Figs. 1–3 means either that the two processes have been separated, the first occurring in the experimentally unattainable region to the left of the ordinate, or that both occur simultaneously at a moderately slow rate. However, for completely unwashed fibers there is more radioactive material present, the experiments show poor reproducibility, and crystals of salt may be seen on the surface of the dried fiber with a microscope. Washed fibers have no salt crystals on their surfaces and give much more reproducible results. Thus it may be inferred that the two processes have been separated and the washing has been successful. On the basis of the above reasoning the liquids used in Figs. 1–3 have been termed satisfactory wash liquids.

As far as procedure is concerned, the fibers were removed from the treating bath, shaken to remove excess droplets, washed for a few seconds, and mounted. A solution of 75% ethanol–25% water was used with ^{22}Na$_2$SO$_4$ and Na$_2$^{35}SO$_4$; 95% ethanol was used with Na^{82}Br.

Counting.   The fibers were mounted in a loose spiral on tantalum disks using albumen fixative to hold them to the disks. The disks were then heated to harden the albumen and remove water. The samples were counted with an internal flow counter.

Calibration. It is necessary to equate units of radioactivity with units expressing amount absorbed. In general, the chemical composition of a radioactive solution was determined by standard chemical methods; an aliquot of the solution was prepared for counting and its radioactivity was compared with the radioactivity of the treated fibers. Since great attention must be paid to details in this calibration process it is convenient to consider the calibration for each isotope separately.

$Na_2^{35}SO_4$. The calibration factor has been determined as the composite of two factors. For the first factor 1 ml of a standard solution was dried on an albumen-coated tantalum disk, and the disk was counted. The average coating of salt on the disk was much less than 1 mg/cm². A fiber was then placed on the disk and dissolved with concentrated HNO_3 solution, and the disk was dried and again counted. The first factor is the ratio of the first count to the second. A value of 1.12 with a coefficient of variation of the mean of 1.2% was obtained from 11 determinations.

The second factor was obtained by mounting radioactive fibers, counting them, dissolving them with a concentrated HNO_3 solution, and drying and counting the disk on which the fiber was dissolved. The second factor is the ratio of the second count to the first. An average of 11 determinations was 1.45 with a coefficient of variation of the mean of 1.4%.

Combination of the two factors gives an overall calibration factor of 1.63. Thus the radioactivity of a coiled fiber must be multiplied by 1.63 to convert it to the equivalent radioactivity for an aliquot of a standard solution.

$^{22}Na_2SO_4$. The same method was used to determine the calibration factor for $^{22}Na_2SO_4$. The two factors determined were 1.043 and 0.968, respectively. The second factor is less than unity, implying that more of the radioactivity reaches the counter when the fiber is intact than after it has been dissolved. This result is not impossible because of the probability of different geometrical factors in the two cases, but it was unexpected. Since the coefficient of variation of the mean (for 14 fibers) was 0.8%, and a careful check disclosed no possibility for a loss of radioactivity through volatilization or spattering, the factor was accepted.

Combination of the two factors gives an overall calibration factor of 1.01.

$Na^{82}Br$. It is evident that dissolution of the fiber with HNO_3 cannot be used in this case because of the volatility of HBr. Dissolution of the fibers with tetramethylammonium hydroxide and heat also failed because of loss of volatile radioactive material. As a result a leaching technique was used.

Each fiber was washed and then leached for 2 hr or more in 2 ml of 0.001M NaBr solution; 1 ml of this solution was then transferred to a stainless steel disk, dried and counted. The concentration of the leaching solution was chosen so that the average thickness of the salt deposit on the stainless steel disk was less than 1 mg/cm². The radioactivities so determined were compared with the radioactivities of mounted samples.

The calibration factor was determined to be 1.085 with a coefficient of variation of the mean of 1.5%.

Salt Absorption Measurements at 90°C

The measurements at 90°C differed from those at room temperature only with respect to treatment and washing.

Treatment. The temperature was maintained to within +0.1°C with standard equipment. Many of the measurements at 90°C involved dyebaths and in every case the fiber-to-bath ratio was such that exhaustion of the dyebath was negligible.

The method of treatment was the same for 90°C and room temperature except that the precision of the measurements at 90°C was considerably improved if the fibers were preconditioned for 1 hr or more in air which had been bubbled through water. Why such a preconditioning treatment should be effective is obscure; certainly the treating bath contains an excess of water. One possible explanation is the following. Minute occluded air bubbles may be present at the fiber–liquid interface; these can serve as nuclei for drop entrainment when the fiber is removed from the liquid, and thus increase the amount of material which must be washed off, and in this way, the error in washing. Presumably, then, exposure of the fiber to a high relative humidity brings about the adsorption of a surface layer of water on the fiber, and this layer minimizes the occlusion of air bubbles when the fiber is put into the bath.

A time of treatment of 30 min was used. Preliminary experiments showed that no rapid changes took place after 30 min, although slow changes, thought to result from changes in properties of the fiber, occurred more or less continuously. The time chosen was selected as the best compromise between essentially complete absorption and minimum chemical or physical change.

Washing. The fibers were removed from the 90°C bath and immediately immersed in ice-cold 95% ethanol. They were then washed immediately in 75% ethanol at room temperature. The purpose of the first bath was to reduce the temperature and slow down diffusion as quickly as possible; Na_2SO_4 is practically insoluble in 95% ethanol. Using this technique the uptake results for 90°C were found to have an average coefficient of variation of less than 10%, which is only slightly worse than was obtained at room temperature (~7%). Immersion of the fibers directly in a cold 75% ethanol bath led to a greater scatter of results.

From consideration of the rate of diffusion of salt out of the fiber, as shown in Figs. 1–3, and the increase in rate accompanying an increase in temperature, it is evident that washing at 90°C is out of the question for experimental reasons. The fiber must be washed at room temperature, and the cooling must be done as quickly as possible to avoid changes in the equilibrium between the interior and the surface of the fiber resulting from such effects as evaporation of water from the surface droplets. The reproducibility of the washing process outlined above was considerably better than the reproducibilities of other processes involving variations in washing method or no washing at all. This fact suggests that the method used is either a satisfactory method or incorporates a uniform systematic error. However, it should be emphasized that no direct justification of the washing process for high temperature absorption is available.

Dye Absorption Measurements

The absorption of dye was determined from the change in concentration of the dyebath during the dyeing process, which was measured spectrophotometrically. Because of the phototropic character of the dye, each solution was illuminated for 10 min with a 100-W bulb at a distance of 6 in, through 2 in. of water. The results obtained in this way were found to be more precise than those obtained using a 10-min illumination at a distance of 1 in., which was used by previous workers.[7] These latter conditions were shown to cause heating of the solution, which was thought to be a source of error.

Most of the measurements were made using sealed vessels with occasional swirling but no systematic stirring. In some cases, measurements were made in open vessels equipped with stirrers and condensers. The differences between the methods fell within the experimental error. In every case preliminary checks were made to assure that the times allowed were adequate for equilibrium.

RESULTS

Results with Dye Absent

Results on the equilibrium absorption of SO_4 and Na from Na_2SO_4 solutions are given in Tables I and II, respectively. Uptake values are given as a function of bath concentration for both temperatures and for both stabilized and unstabilized fibers. The uptakes are reported in equivalent units; in other words, the uptake of Na is reported in units of Na_2. The average relative coefficient of variation of the uptake values in Tables I and II is 6.6%.

Table I. Uptake of SO₄ by Viscose Rayon from Na₂SO₄ Solutions

Condition of Sample	Temp. (°C)	Concentration Equilibrium of Bath (M/1000 g water)	(M/L)	Uptake (mM/g)
Unstabilized	~25	0.0051	0.0051	0.00203
		0.0102	0.0102	0.00489
		0.0205	0.0205	0.0121
		0.0349	0.0349	0.0233
		0.0772	0.0772	0.0508
		0.123	0.123	0.105
		0.432	0.422	0.381
		0.766	0.754	0.668
		0.991	0.972	0.803
		1.21	1.18	0.929
		1.56	1.51	1.38
Stabilized	~25	0.0120	0.0120	0.00530
		0.0126	0.0126	0.00588
		0.0170	0.0170	0.00756
		0.0193	0.0193	0.0103
		0.0246	0.0246	0.0127
		0.0302	0.0302	0.0148
		0.0310	0.0310	0.0157
		0.0369	0.0369	0.0203
Stabilized	90	0.0126		0.00407
		0.0159		0.00517
		0.0170		0.00603
		0.0193		0.00742
		0.0208		0.00808
		0.0246		0.00941
		0.0279		0.0120
		0.0311		0.0128
		0.0369		0.0163

Various aspects of the results in Tables I and II are shown in Figs. 4–6. Figure 4 shows the effect of the stabilizing treatment and Fig. 5 the effect of temperature on the uptake of Na and SO₄. In Fig. 6 the absorption results for the unstabilized fiber at 25°C are plotted on a log-log basis to cover the entire range of concentration. Results on the uptake of Br from NaBr solutions are also included. These results are also given in Table III.

Results are given in Tables IV and V and in Fig. 7 on the effect of pH on the uptake of Na and SO₄ from Na₂SO₄ solutions by stabilized fibers.

Strictly speaking, the two sets of results were obtained for solutions differing slightly in concentration; however, in view of the coefficients of variation of the results the difference is unimportant.

Table II. Uptake of Na$_2$ by Viscose Rayon from Na$_2$SO$_4$ Solutions

Condition of Sample	Temp. (°C)	Concentration Equilibrium of Bath (M/1000 g water)	(M/L)	Uptake (mM/g)
Unstabilized	~25	0.0025	0.0025	0.0113
		0.0050	0.0050	0.0147
		0.0101	0.0101	0.0194
		0.0151	0.0151	0.0230
		0.0169	0.0169	0.0236
		0.0424	0.0424	0.0441
		0.0753	0.0753	0.0789
		0.118	0.118	0.111
		0.160	0.159	0.159
		0.411	0.407	0.370
		0.842	0.830	0.693
		0.970	0.952	0.768
		1.31	1.27	0.996
		1.93	1.83	1.32
Stabilized	~25	0.0103	0.0103	0.0192
		0.0168	0.0168	0.0242
		0.0240	0.0240	0.0310
		0.0307	0.0307	0.0354
Stabilized	90	0.0105		0.0204
		0.0173		0.0244
		0.0249		0.0310
		0.0327		0.0339

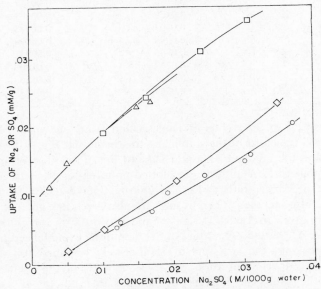

Fig. 4. The uptake of Na$_2$ and SO$_4$ solutions by stabilized and unstabilized viscose monofils at room temperature. △ Na$_2$ unstabilized; □ Na$_2$ stabilized; ◇ SO$_4$ unstabilized; ○ SO$_4$ stabilized.

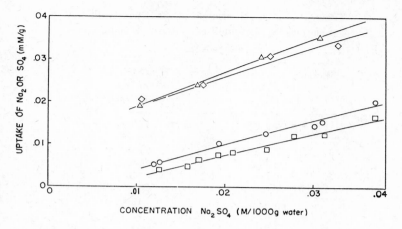

Fig. 5. The uptake of Na_2 and SO_4 from Na_2SO_4 solutions by stabilized viscose monofils as a function of temperature. △ Na_2 25°C; ◇ Na_2 90°C; ○ SO_4 25°C; □ SO_4 90°C.

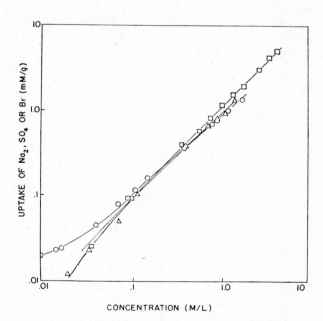

Fig. 6. The uptake of Na_2 and SO_4 from Na_2SO_4 solutions and of Br from NaBr solutions by unstabilized viscose monofils. ○ Na_2; △ SO_4; □ Br.

Table III.   The Absorption of Br from NaBr Solutions by Unstabilized Viscose Monofil at ~25°C

Equilibrium Concentration of Bath (M/L)	Uptake (mM/g fiber)
0.0369	0.0256
0.0965	0.0943
0.108	0.0922
0.386	0.383
0.611	0.570
0.814	0.838
1.09	1.18
1.45	1.53
1.93	1.99
2.83	3.05
3.66	4.13
4.52	4.95

Rate-of-absorption curves for SO_4 and Na from Na_2SO_4 solutions and Br from NaBr solutions are given in Figs. 8–10. These curves are shown to indicate orders of magnitude rather than as quantitative results. In particular, the difference between the curves in Fig. 8 and Fig. 9 is too great to be

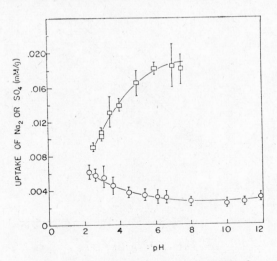

Fig. 7.   The uptake of Na_2 and SO_4 from Na_2SO_4 solutions by stabilized viscose monofils as a function of pH. □ Na_2.

Table IV. Uptake of SO₄ from Na₂SO₄ Solutions by Stabilized Viscose Monofil at ~25°C as a Function of pH (Na₂SO₄ Concentration in Solution 0.0096M)

pH of Solution	Uptake (mM/g)	Coefficient of Variation (%)
2.25	6.23 × 10⁻³	5.8
2.62	5.90	6.7
3.1	5.52	12.1
3.6	4.62	11.6
4.5	3.80	7.9
5.4	3.49	10.3
6.1	3.23	12.7
6.6	3.20	11.3
8.0	2.69	9.9
10.0	2.47	7.5
11.0	2.62	10.5
11.9	3.27	13.1

accounted for by a rapid cation exchange, and probably falls within the error of the measurements.

Results with Dye Present

A set of experiments was run in which the absorption of SO₄ from solutions containing varying amounts of dye and of Na₂SO₄ by stabilized fibers at 90°C was measured. The results are given in Table VI and Fig. 11.

Table V. Uptake of Na₂ from Na₂SO₄ Solutions by Stabilized Viscose Monofil at ~25°C as a Function of pH (Na₂SO₄ Concentration in Solution 0.0098M)

pH of Solution	Uptake (mM/g)	Coefficient of Variation (%)
2.5	9.1 × 10⁻³	6.4
3.0	10.4	6.3
3.0	10.7	8.7
3.5	13.2	13.8
4.0	14.0	4.9
5.0	16.6	9.1
6.0	18.3	3.0
7.0	18.5	14.3
7.5	18.1	9.4

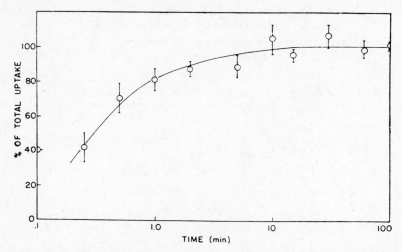

Fig. 8. Rate of uptake of SO_4 from $0.97M$ Na_2SO_4 solution by unstabilized viscose monofils.

In Fig. 11 the solid line represents the results obtained when no dye is present. The dotted lines are theoretical predictions which will be discussed later.

In Table VII and Fig. 12 results on the uptake of Chrysophenine G by the viscose monofil used are given.

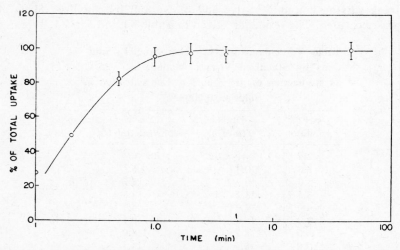

Fig. 9. Rate of uptake of Na from $0.95M$ Na_2SO_4 solution by unstabilized viscose monofils.

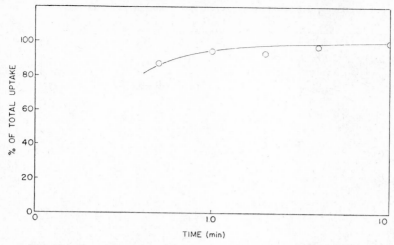

Fig. 10. Rate of uptake of Br from 4.5M NaBr solution by unstabilized viscose monofils.

Table VI. Uptake of SO$_4$ at 90°C from Solutions Containing Na$_2$SO$_4$ and Chrysophenine G

Concentration of Dye mM/1000 g Water (mol wt 680)	Concentration of Salt M/1000 g Water (mol wt 142)	Dye/Salt Ratio	Uptake of SO^4 in mM/g Fiber	Coefficient of Variation (%)
0.290	0.0449	0.00645	0.0186	8.0
0.233	0.0362		0.0126	7.4
0.170	0.0262		0.00875	5.1
0.122	0.0189		0.00623	5.7
0.347	0.0321	0.0108	0.0118	5.6
0.255	0.236		0.00804	6.6
0.214	0.0198		0.00663	1.8
0.171	0.0158		0.00482	3.4
0.121	0.0112		0.00193	4.9
0.062	0.0057		0.00101	13.
0.606	0.0326	0.0186	0.0113	5.1
0.506	0.0272		0.00986	12.
0.430	0.0231		0.00680	5.5
0.392	0.0211		0.00636	18.
0.324	0.0174		0.00491	7.3
1.04	0.0446	0.0234	0.0175	12.
0.798	0.0341		0.0124	15.
0.559	0.0239		0.00799	11.
0.400	0.0171		0.00503	9.1
0.288	0.0123		0.00318	12.
1.08	0.0295	0.0367	0.00877	16.
0.954	0.0260		0.00811	14.
0.837	0.0228		0.00741	8.6
0.653	0.0178		0.00522	10.
0.481	0.0131		0.00353	8.0
0.282	0.00769		0.00147	11.

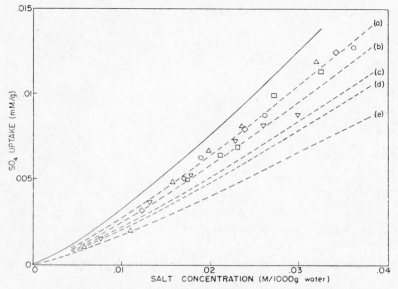

Fig. 11. Effect of dye on the uptake of SO_4 from Na_2SO_4-dye solutions at 90°C.

Dye/salt ratio	Experimental	Theoretical curve
0.0065	○	a
0.0108	△	b
0.0186	□	c
0.0234		d
0.0367		e

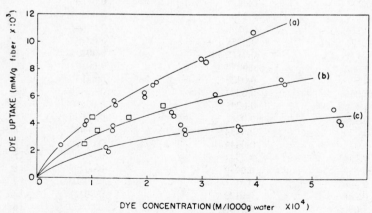

Fig. 12. Uptake of Chrysophenine G by viscose monofils at 90°C for different salt concentrations. Na_2SO_4 concentration; (a) 0.031M; (b) 0.019M; (c) 0.0098M; ○ with stirring; □ without stirring.

Table VII.   The Uptake of Chrysophenine G by Viscose at 90°C as a Function of Dye and Salt Concentration

Condition	Time of Dyeing (hr)	Salt Concentration (M/1000 g Water)	Equilibrium Dye Concentration (M/1000 g Water) $\times 10^4$)	Dye Uptake in mM/g Fiber $\times 10^3$
No stirring	3	0.031	0.424	2.40
			0.424	2.38
			0.893	4.06
			0.888	4.08
			1.42	5.44
			1.40	5.52
			1.94	6.12
			1.95	6.07
			2.11	6.98
			2.15	7.01
			3.02	8.69
			2.97	8.78
			3.91	10.75
			3.91	10.75
Stirring	2¼		0.986	4.51
No stirring	3	0.019	1.38	3.56
			1.36	3.73
			2.47	4.69
			2.44	4.84
	6		3.32	5.66
			3.21	6.17
			4.44	7.23
			4.69	6.92
Stirring	2¼		2.28	5.45
			1.66	4.54
			1.08	3.51
			0.811	2.62
No stirring	3	0.00985	1.27	2.17
			1.29	2.08
	6		2.61	3.96
			2.61	3.97
			2.67	3.58
			2.68	3.38
			3.68	3.72
			3.66	3.86
			5.34	5.11
			5.49	4.20
			5.51	4.09

Table VIII.   Uptake of Dye and SO$_4$ by Stabilized Monofil at 90°C

Salt Concentration (M/1000 g Water)	Dye Concentration (M/1000 g Water × 10^4)	Dye/Salt Ratio × 10^4	SO$_4$ Uptake (mM/g)	Dye Uptake (mM/g)	k
0.031	2.02	65	0.011	0.0069	96.5
0.019	1.24	65	0.0060	0.0035	89.4
0.0099	0.64	65	0.0026	0.0014	83.8
0.031	3.35	108	0.011	0.0094	77.6
0.019	2.05	108	0.0062	0.0047	70.5
0.0099	1.07	108	0.0026	0.0020	70.3
0.019	3.53	186	0.0055	0.0062	60.6
0.0099	1.83	186	0.0022	0.0037	66.2
0.019	4.45	234	0.0058	0.0070	52.0
0.0099	2.31	234	0.0024	0.0031	55.1
0.0099	3.62	365	0.0026	0.0039	40.4

Finally results on the dye and SO$_4$ uptake over a range of concentrations of dye and salt in the solution are given in Table VIII. These results are interpolated from the results previously given. The last column of the table is discussed later in this paper.

DISCUSSION

Stability

Before discussing the mechanism of salt absorption by viscose fibers, it is perhaps desirable to consider the stability of the viscose monofils used.

It was originally found that the absorptive capacity for SO$_4$ ions at room temperature decreased spontaneously with time. It was learned that differences in dyeing resulting from differences in aging had been encountered, and the manufacturer stated that a treatment with steam or hot water successfully minimized such differences. This information was the source of what has been called the stabilizing treatment. Presumably treatment with hot water accelerates the aging process.

An examination of Fig. 4 shows that the stabilization results in a decrease in anion-absorptive capacity and no change in the cation-absorptive capacity. This means a net increase in the cation-exchange capacity.

Such results could ensue from either of two causes. First, there might exist anion-exchange sites within the fiber which would be destroyed by treatment with hot water. Second, crystallization or ordering of the cellulosic chains might occur under the influence of the hot water. If this ordering should be accompanied by the formation of new cation-exchange sites in such a way that the net absorptive capacity for cations remained roughly constant, the observed results would be obtained.

Of the two possibilities mentioned, the second seems the more probable although it is the more complicated. Crystallization in hot water is known to occur for some cellulosic fibers, and new cation-exchange sites could result either from oxidation or from removal of cations such as Zn or Ca which are present in spinning baths and are very hard to remove completely from the fiber during purification. On the other hand, no suitable anion-exchange sites have been found in cellulose.

It might be mentioned that the slight increase in cation-exchange capacity shown in Fig. 5 suggests that the stabilization may not be complete. It is well known that extended treatment with hot water, particularly in the presence of air, produces profound changes in the properties of cellulosic materials.

Absorption of Ions in the Absence of Dye

Cation-Exchange Capacity.   The cation-exchange capacity of the cellulose has already been mentioned in discussing the stability of the sample. It is evident that the curves for each ion in Fig. 4 must start from the origin, separate gradually until the Na concentration becomes so large that the ionizable hydrogens have all been removed and then follow parallel courses. The separation between the parallel curves is a measure of the exchangeable cations, probably primarily the H ion from carboxylic acid groups. The experimental results appear to run together at high salt concentrations, as can be seen from Tables I and II. However, at the high concentrations the amount of each ion absorbed is such that the cation-exchange capacity falls within the experimental error.

The content of exchangeable cations is 0.031 mE/g for the unstabilized material and 0.039 mE/g for the stabilized fibers. These results are comparable in magnitude with values obtained in other ways; for example, Ant-Wuorinen[8] found a value of 0.038 mE/g for a viscose rayon (DP 730) by a titration method, and Farrar and Neale[4] found a value of 0.048 mE/g for a cellophane also by a titration method. It might be mentioned in passing that the tracer method for measuring cation-exchange capacity is potentially superior to titration measurements because it is a direct measurement and should be capable of giving good sensitivity with refinement.

Mechanism of Absorption.   Some consideration can also be given to the mechanism of absorption. Perhaps the simplest reasonable picture of the absorption is one in which the salt is considered to be dissolved in water in the pores of the fiber. In such a picture the amount of salt present within the fiber would depend on the concentration of the external solution, on the volume of the pores, and on the cation-exchange capacity. The actual case considered here is complicated by the fact that two cations, Na and H, are present and that the carboxylic acid groups in the H form may be incompletely ionized. As a first approximation, it may be assumed that only one cation, Na, is present and that ionization is complete. In this case two equations can be written

$$\frac{(Na)^2(SO_4)}{V^3} = [Na]^2[SO_4] \tag{1}$$

and

$$(Na) = 2(SO_4) + E, \tag{2}$$

where parentheses refer to the amount of a given ion in the fiber in mM/g, V refers to the free internal volume of the fiber, square brackets refer to the concentration of the external solution, and E is the cation-exchange capacity of the fiber. Equation 1 is an expression of the thermodynamic equilibrium in the system subject to the assumption that the salt is in an aqueous solution inside the fiber and out. Equation 2 expresses the balance of charge within the fiber which is necessary to prevent the spontaneous generation of large electric potentials on the fiber. The two equations can be combined to give

$$(SO_4)\left[1 + \frac{E}{(SO_4)} + \frac{E^2}{4(SO_4)^2}\right]^{1/3} = V[SO_4]. \tag{3}$$

Consider the region of concentration for which $(SO_4) \gg E$ so that terms above the first power in $E/(SO_4)$ can be neglected and the cube root can be approximated by the first two terms in its expansion. The equation obtained is

$$(SO_4) = V[SO_4] - E/3. \tag{4}$$

Equation 4 is the equation of a straight line with a negative intercept on the uptake axis. The curve for (3) goes through the origin so that for smaller salt concentrations the dependence of (SO_4) on $[SO_4]$ will be convex to the $[SO_4]$ axis. It is obvious that the cation content of the fiber will parallel the anion content and remain a fixed distance (the cation-exchange capacity) from it.

If small amounts of H ion are assumed to be present in all solutions, the basic nature of the anion absorption curve will remain unchanged. The cation absorption curve will only parallel the anion absorption curve for large salt concentrations and will go through the origin. The simple picture just outlined is thus capable of accounting qualitatively for all features of Fig. 4 that do not arise from the instability of the fiber.

It is evident from Fig. 6 that the uptake of Br is similar in character to the uptake of SO_4. Values of V can be found by using (4) and the appropriate portion of the anion uptake curve. Best straight lines were put through the high concentration regions of the linear curves for Br and SO_4 uptake by visual estimate and the slopes of the lines determined. The values of V obtained were 0.92 ml/g in the case of SO_4 and 1.05 ml/g in the case of Br. The values are larger than would be expected from estimates of the water absorptive capacity of viscose rayon.

Effect of pH.   Other consequences of the competition between H and Na can be seen in Fig. 7. As the pH decreases at constant Na_2SO_4 concentration the Na content of the fiber decreases and the SO_4 content increases. The decrease in Na is a natural result of the competition between H and Na. The increase in SO_4 could result from several effects. In the first place as un-ionized COOH groups are formed the osmotic forces restricting absorption of ions are diminished. Also, at the lowest pH values the concentration of SO_4 in the solution is appreciably increased, since H_2SO_4 was used to obtain the desired pH. Finally, a selective absorption of H_2SO_4 at the expense of Na_2SO_4 might be operating. However, the shape of the curve suggests that the first factor is the most important. Thus, between pH 6 and pH 4 there is a 33% increase in SO_4 uptake accompanying a 0.5% increase in SO_4 concentration in the solution.

It is interesting that the SO_4 content appears to decrease as the pH increases beyond pH 6. At pH 6 it is unlikely that there are any COOH groups left in the fiber, so that the decrease must be caused by the removal of some cation with a higher affinity for the COO group than the proton has (or at least a markedly slower rate of desorption) or the removal of anion-exchange sites. In this connection the previous remarks on stability might be noted.

Effect of Temperature.   The effect of temperature on Na and SO_4 absorption is shown in Fig. 5. Temperature coefficients for equilibrium absorption cannot be determined until possible complications resulting from instability are removed. However, the temperature coefficient for cation absorption is certainly small and that for anion absorption would be anticipated to be small also. All in all there is no evidence for any appreciable heat change during the absorption process.

Rates.   Accurate estimates of diffusion coefficients in a fiber cannot be made without an accurate measurement of the fiber's diameter; in the present case the water-swollen diameter of the fiber is involved. No such measurement was made. However, the rate of absorption curve for SO_4 is roughly three orders of magnitude faster than a similar curve for hair[9] from which a diffusion coefficient $D = 4 \times 10^{-11}$ cm²/sec was calculated. The swollen diameter of the viscose monofil was estimated to be about 15% smaller than that of the hair. All in all a diffusion coefficient of order of magnitude 10^{-8} cm²/sec is indicated. A diffusion coefficient of this order of magnitude is compatible with the idea that the salt is present in aqueous solution within the pores of the fiber.

Summary. In summary, when dye is not present, the results all suggest that the mechanism of absorption can be considered to be uptake of aqueous solution in the pores of the fiber with the relative amounts of cation and anion controlled by the cation-exchange capacity of the fiber. Farrar and Neale[4] reached a similar conclusion in a study of the uptake of several salts by cellophane, although certain salts at high concentrations seemed to have a specific affinity for the cellulosic material.

Absorption of Ions in the Presence of Dye

A discussion of the uptake of salt in the presence of dye must logically start from the present theory of dyeing of cellulose which can account for much of what is currently known about dye absorption equilibria to a first approximation.[1] In this theory, dye and salt are considered to exist in an aqueous solution in pores within the fiber. In addition, dye is assumed to be absorbed on absorption sites on the surface of the pores. According to this picture a form of (1) can be written for salt and for "unbound" dye, and in addition the equilibrium between the "bound" and "unbound" dye anions can be expressed by the equation

$$\frac{(D)'}{B - (D)'} = k' \frac{(D)''}{V}, \tag{5}$$

where $(D)'$ refers to the "bound" dye, $(D)''$ to the "unbound," B is the number of absorption sites and k' contains a standard free energy of absorption term and unit-conversion factors. The fiber is assumed to be an equipotential region. It is usually assumed that $B \gg (D)' \gg (D)''$ so that $D = (D)' + (D)'' \cong (D)'$ and (5) can be written

$$\frac{(D)''}{V} = \frac{(D)}{k'B}. \tag{6}$$

Using (6), assuming the salt to be Na_2SO_4 and assuming the dye to be divalent (as Chrysophenine G is), it is possible to write

$$\frac{(Na)^2(SO_4)}{V^2} = [Na]^2[SO_4], \tag{1}$$

$$\frac{(Na)^2 \, D)}{V^3} = k[Na]^2[D], \tag{1'}$$

$$(Na) = 2(SO_4) + 2(D) + E \tag{2'}$$

and

$$[Na] = 2[SO_4] + 2[D], \tag{7}$$

where $k = k'B/V$ and complications introduced by the presence of H ion have been ignored. The same equations can be derived for a model in which the dye anions are considered to be under the influence of a nonlocalized attractive potential within the fiber. In this case the meaning of k is different.

If V is assumed to be a constant and E is neglected, (1), (1'), (2') and (7) can be combined to give two forms capable of being tested experimentally. They are

$$\left(\frac{(SO_4)}{(SO_4)_0}\right)^{3/2} = \frac{1 + [D]/[SO_4]}{1 + k[D]/[SO_4]}, \tag{8}$$

where $(SO_4)_0$ is the amount of SO_4 absorbed when no dye is present, and

$$\frac{(D)}{(SO_4)} \frac{[SO_4]}{[D]} = k. \tag{9}$$

In principle, this second equation provides a powerful method for determining k, because the internal volume, which is difficult to measure, does not occur on the left-hand side of the equation.

The results presented in this paper have been used in checking both (8) and (9). The last colume in Table VIII gives values of k determined by using (9). The values are not constant and there seems to be a trend to lower values of k with increasing dye/salt ratio.

The dotted lines in Fig. 11 have been obtained from a consideration of the equations leading to (8) using a value of $k = 45.7$, which corresponds to an affinity (as defined by Vickerstaff) $\mu_0 = -2800$ cal mol. This value of μ_0 lies in the range of values found for Chrysophenine G on cellulosic fibers as listed by Vickerstaff.[1]

It is evident that the theoretical curves do not give a good representation of the experimental results. However, the trend of the experimental results is only qualitatively apparent and their scatter is such that the validity of distinguishing any one set of results from another or from a theoretical curve might be questioned. In order to obtain a more quantitative interpretation of the results, we made use of statistical methods.

It so happens that, when the results for any given dye/salt ratio are plotted as log uptake against log bath concentration, the points fall along a straight line to a good approximation. This fact has been used in the statistical analysis of the results. Straight lines have been determined for each dye/salt ratio (including zero) by the method of least mean squares, and the various lines have been compared by the methods of analysis of covariance.

The following results emerge from the statistical analysis:

1. It is not possible to distinguish between the slopes of the various lines. The average slope is 1.284.

2. There is less than one chance in one hundred that the results without dye are the same as the results with dye.

3. There is at most one chance in five that the results for the lowest dye/salt ratio are indistinguishable from those for the higher ratios.

4. There is at most one chance in five that the results for the two lowest dye/salt ratios are indistinguishable from the results for the three highest dye/salt ratios.

5. The experimental line for dye/salt ratio 0.0065 and the theoretical line for that ratio are indistinguishable. In other words the theory predicts the results within the limits of precision of the experiment.

6. There are at most five chances in one hundred that the results for dye/salt ratio 0.0108 represent results whose true mean values fall along the theoretical curve for that ratio. In other words, there is only a five percent chance that the theory represents the true results.

7. There is at most one chance in one hundred that the results for dye/salt ratio 0.0234 represent results whose true mean values fall along the theoretical curve.

8. There is even less chance that the true curves for the two largest dye/salt ratios are accurately represented by the theoretical curves.

Thus it is quite certain that, over the range of concentrations and dye/salt ratios used, the presence of dye lowers the salt uptake as compared with the value when no dye is present. It is probable that the amount of salt taken up at a given concentration of salt in the bath decreases as the dye/salt ratio increases. However, for the most part, the differences between the curves for the various dye/salt ratios fall within the precision of the experimental results.

As far as the theoretical curves are concerned, the theory provides an adequate representation of the results only at the lowest dye/salt ratio. As the dye/salt ratio increases, the theory becomes completely inadequate.

The objection might be raised that the value of k used is far from the mean value of the results in Table VIII, which is $k = 69$. However, as k increases the agreement between theory and experiment becomes worse.

Thus both (8) and (9) are inadequate for predicting the observed behavior. The inadequacy is in some way related to the dye/salt ratio in the dyebath. In discussing this inadequacy some attention must be paid to the assumptions made in deriving the equations. The basic assumption that an aqueous solution exists within the fiber has received some experimental justification. The most serious assumptions left would seem to be ignoring the cation-exchange capacity of the fiber and ignoring activity coefficients.

It is obvious that the presence of cation-exchange sites will depress the absorption of either anion but less obvious what will happen to a ratio such as that expressed in (8). Since (9) can be derived without reference to any cation-exchange capacity, it should hold true regardless of the capacity. Thus consideration of the cation-exchange capacity cannot be expected to alter the discrepancy between (9) and the experimental results.

By ignoring the cation-exchange capacity and introducing single-ion activity coefficients f_i for the dyebath and γ_i for the fiber (8) and (9) can be written

$$\left(\frac{(SO_4)}{(SO_4)_0}\right)^{3/2} = \left(\frac{f_{(SO_4)}\gamma_{(SO_4)0}}{f_{(SO_4)0}\gamma_{(SO_4)}}\right)^{3/2}\left(\frac{1 + [D]/[SO_4]}{1 + \dfrac{f^3_{(D)}\gamma^3_{(SO_4)}\ k[D]}{f^3_{(SO_4)}\gamma^3_{(D)}\ [SO_4]}}\right), \qquad (8')$$

and

$$\frac{(D)[SO_4]\gamma_{(D)}f_{(SO_4)}}{(SO_4)[D]\gamma_{(SO_4)}f_{(D)}} = k. \qquad (9')$$

The first term on the right-hand side of (8') can probably be safely considered to be unity. It is interesting that, if the term $\gamma_{(D)}f_{(SO_4)}/\gamma_{(SO_4)}f_{(D)}$ increases with the dye/salt ratio (and its inverse thus decreases), the deviations of the experimental results from (8) and (9) can be qualitatively explained.

Electrostatic repulsion between strongly absorbed dye anions would act to increase the apparent activity coefficient of the dye anions within the fiber and hence would be acting in the right direction. Whether such a repulsion can account quantitatively for the observed results cannot be answered until a satisfactory theory has been evolved, perhaps as an outgrowth of the work of Verwey and Overbeek.[10]

ACKNOWLEDGMENT

This work was undertaken as part of the Dyeing Research Project of Textile Research Institute. The authors appreciate the guidance of the Project's Advisory Committee representing the textile and chemical manufacturing firms who sponsored this work.

We wish to thank Professor J. C. Whitewell for supervising the statistical analysis, and Dr. Lütfü Karasoy and Miss Mildred Stevens for the dye absorption measurements.

Finally, we appreciate the generosity of the American Viscose Corporation, which supplied the fiber sample, and of the Organic Chemicals Department, E. I. du Pont de Nemours & Co. Inc. which supplied the dye.

REFERENCES

1. T. Vickerstaff, *The Physical Chemistry of Dyeing*, 2nd ed., Interscience, New York 1954.
2. F. L. Usher and A. K. Wahbi, *J. Soc. Dyers Col.* **58**, 221 (1942).

3. S. M. Neale and P. T. Standring, *Proc. Roy. Soc. (London)* **A213,** 530 (1952).
4. J. Farrar and S. M. Neale, *J. Colloid Sci.* **7,** 186 (1952).
5. D. J. Montgomery and W. T. Milloway, *Textile Research J.* **22,** 729 (1952).
6. P. B. Stam and H. J. White, Jr., *Textile Research J.* **24,** 785 (1954).
7. A. F. Willis, J. O. Warwicker, H. A. Standing, and A. R. Urquhart, *Shirley Institute Memoirs* **19,** 223 (1944–45).
8. O. Ant-Wuorinen, "Determination of Carboxyl Groups in Cellulose" Report 96, The State Institute for Technical Research, p. 48, Helsinki, Finland (1951).
9. D. L. Underwood and H. J. White, Jr., "The Physical Chemistry of Dyeing and Tanning," Faraday Society Discussion No. 16, p. 66, The Faraday Society, Aberdeen (1954).
10. E. J. W. Verwey and J. Th G. Overbeek, *Theory of the Stability of Lyophobic Colloids* Elsevier, New York, 1948.

Application of Transition State Ideas to Crystal Growth Problems

ANDREW VAN HOOK

Department of Chemistry, College of the Holy Cross,
Worcester, Massachusetts

When I was in Salt Lake City in the fall of 1962 and not involved in sweet things of the sugar world, Henry Eyring and I talked about many things. It is two of these topics which I should like to recall at this time.

NUCLEATION AND CRYSTAL GROWTH

The pertinence of transition state theory to nucleation and growth is obvious. This was first pointed out as applicable to the graining behavior of sugar syrups at the Bristol conference in 1949[1] and considerably elaborated since.[2,3] The continued growth of the crystal, already formed as the critical nucleus or added as seed, is also amenable to the same sort of theoretical treatment; the reaction complex being the two-dimensional nucleus, screw

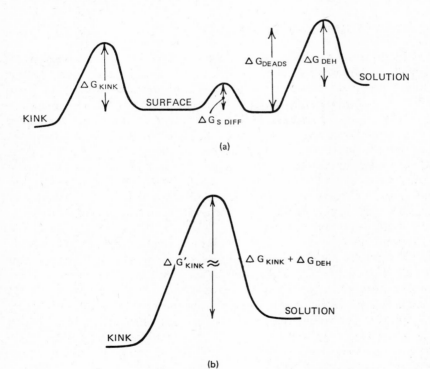

Fig. 1. (a) Growth units enter the adsorption layer first and after surface diffusion enter the kinks; (b) growth units enter directly from the solution into the kinks.

dislocation, or other form of crystallization center. The kinetic problem is similar to that met in adsorption considerations[4,5] and, in the case of sucrose, Bennema[6] considers surface diffusion as the dominating step. His reaction coordinate diagram (Fig. 1) involves three energy barriers, according to which the usual linear growth rate disintegrates to second-order kinetics at small super-saturations. This, of course, is a common feature of any reaction involving an intermediate complex. However, I do not find this falloff at 0°C[7] and visualize the reaction coordinate as one barrier of height equivalent to incorporating the molecule into the crystal lattice. The nonlinearity of the growth curves at higher temperatures results from the transport part of the reaction, the specific rate constant of which is concentration and hydro-dynamically dependent.

This is only one specific instance in which Eyring's ideas have been used; many other[8-12] examples can be found in the literature on crystal growth and related processes.

BRIDGMAN'S ZONE OF INDIFFERENCE

One phenomenon we discussed at great length in connection with organizing high pressure work at Utah was Bridgman's zone of indifference. Here it had been reported that while most polymorphic transitions are prompt, there are a few which apparently stopped at pressures spanning the proper equilibrium values. Bridgman[13] ascribed this to the onset of a different mechanism of reaction, whereas Eyring et al.,[14] accounted for the arrest by a new type of bridge which introduced additional terms in the rate equation. However, since the effect appeared to be the most marked with camphor and phosphorus, which could likely have been the least pure of Bridgman's presumed shelf materials, we looked, with Andy Gabrysh,[15] into the influence of purification of these materials and concluded that the region of indifference is the result of poisoning by impurities.

This is not unexpected; the sluggishness or even unattainability of equilibria being a very old matter in the kinetics of phase transitions. A most recent example is that of $SrSO_4$,[16] whose approach to saturation from either over- or undersaturation is stopped by the presence of minute amounts ($4 \times 10^{-7}M$) of sodium pyrophosphate. The poisoning of the growth of LiF crystals by Fe is another outstanding example,[17] as is the drastic reduction in growth rates of sucrose by the addition of small concentrations of raffinose.[18] In this particular instance even the purest sugar suffers a decline in growth rate at ordinary temperatures as saturation is approached.[19,20]

One experimental factor requiring special consideration in such cases is the steadily changing area of the crystals themselves as growth proceeds. If the crystals remain intact, a correction determined by the $\frac{2}{3}$ power of the amount crystallized is all that is needed, but this is somewhat uncertain on account of a frequently enhanced area due to crystal multiplication or a depleted area due to crystal impedance or agglomeration. To settle either possibility positively, Emmerich's experiment[19] was repeated but with only one large single crystal. Simultaneous measurements of the dimensions of the crystal and refractive index of syrup were taken over the extremely long times required to exhaust the syrup. The results are expressed in Fig. 2. The crystal and syrup curves reproduce each other quite faithfully, so apparently there is no appreciable change in the nature of the surface film as growth proceeds. The decrease in kinetic order as supersaturation falls is also confirmed and, as already intimated, is likely the result of series transport and growth. The details of the mechanism of this reaction will continue to be investigated in this laboratory.

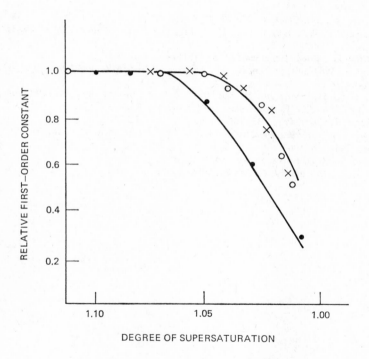

Fig. 2. Progress of adjustment of supersaturated sugar syrups. (●) Emmerich[19]; 20°C, multicrystals. (×) Van Hook; 30°C, single crystal, dimensions. (○) Van Hook, refractometer.

REFERENCES

1. A. VanHook and A. J. Bruno, *Disc. Faraday Soc.*, **5**, 113 (1949).
2. W. J. Dunning, in *Solid State Chemistry*, W. E. Garner, Ed., Butterworths', London, 1955.
3. G. I. Mikhnevich and R. A. Yanchuk, *Growth of Crystals*, Vol. 3, Consultant's Bureau, New York, 1962, p. 90.
4. H. Eyring and J. C. Giddings, *J. Phys. Chem.*, **62**, 365 (1956).
5. H. Eyring and E. M. Eyring, *Modern Chemical Kinetics*, Reinhold, New York, 1963.
6. P. Bennema, *J. Crystal Growth*, **3–4**, 331 (1968); **5**, 29 (1969).
7. A. VanHook, *J. Crystal Growth*, **5**, 305 (1969).
8. R. F. Strickland-Constable, *Kinetics and Mechanism of Crystallization*, Academic, New York, 1968.
9. J. P. Hirth and G. M. Pound, *Condensation and Evaporation*, Macmillan, New York, 1963.
10. D. Turnbull, *J. Chem. Phys.*, **17**, 71 (1949); *Solid State Physics*, Vol. 1, Academic, New York, 1956.
11. R. D. Gretz, *J. Phys. Chem. Solids*, **27**, 1849 (1966).
12. J. E. Manson, F. W. Cagle, Jr., and H. Eyring; *Proc. Natl. Acad. Sci. U.S.*, **44**, 156 (1958).
13. P. W. Bridgman, *Proc. Am. Acad.*, **52**, 57 (1916).
14. H. Eyring, F. W. Cagle, Jr., and C. J. Christensen, *Proc. Natl. Acad. Sci. U.S.*, **44**, 120 (1958).
15. A. F. Gabrysh, H. Eyring, and A. VanHook, *J. Phys. Chem. Solids*, **25**, 129 (1964).
16. J. R. Campbell and G. H. Nancollas, *J. Phys. Chem.*, **73**, 1735 (1969).
17. G. W. Sears, *J. Chem. Phys.*, **33**, 1068 (1960).
18. W. J. Dunning, *Comptes Rend. XIII Assemblée C.I.T.S.*, Falsterbo, Sweden 1967, Tirlemont, Belgium, 1967.
19. A. Emmerich and H. Forth, *Zucker*, **15**, 626 (1962).
20. B. M. Smythe, *Australian J. Chem.*, **20**, 1087 (1967).

The Synthesis of Diamond*

H. TRACY HALL

Brigham Young University, Provo, Utah

* Reprinted from *Journal of Chemical Education*, Vol. 38, page 484, October 1961. Copyright 1961 by Division of Chemical Education, American Chemical Society and reprinted by permission of the copyright owner.

Ever since Antoine Lavoisier in 1792 and Smithson Tenet in 1797 demonstrated that diamond and graphite are allotropic forms of carbon, man has been interested in converting the relatively abundant graphite into the much rarer diamond. Success in this endeavor, however, has been achieved only in recent years, and tiny crystals about $\frac{1}{10}$ mm average diameter, valued at about $6000 per pound, are being quantity produced in direct competition with natural diamonds. These diamonds are used primarily in diamond grinding wheels.

CRYSTAL STRUCTURE INFORMATION

The differences between graphite and diamond at the atomic level were not known until after X-ray diffraction techniques for the elucidation of crystal structures were developed during the 1910–20 decade. Diamond, one of the earliest crystals studied, was shown to consist of carbon atoms arranged in puckered, hexagonal rings lying approximately in the 111 crystallographic plane. (This is the natural cleavage plane of diamond.) These sheets of hexagonal, puckered rings are stacked one above the other in sequences such that atoms in every fourth plane duplicate the position of atoms in the first. By analogy to closest packings, this sequence is designated . . . abcabc . . . . In this structure every carbon atom is surrounded by four other carbon atoms equidistant from the central carbon atom at angles of 109°28′ (the tetrahedral angle) from each other (see Fig. 1).

The arrangement of atoms in crystals of graphite is similar to that of diamond in that it consists of a parallel stacking of layers comprised of carbon atoms forming hexagonal rings. The hexagonal rings in graphite, however, are probably only very slightly puckered. The classical graphite structure indicates an . . . abab . . . stacking sequence for the layers (atoms in alternate layers occupy equivalent positions). According to some inter-pretations of the electron diffraction patterns given by graphite, the . . . abcabc . . . configuration (Fig. 2) is present to an appreciable extent in graphite.[1,2] As with diamond, the cleavage plane is parallel to the stacked layers. However graphite is cleaved with extreme ease compared to diamond.

In diamond all distances between atoms are 1.54 Å. In graphite, the interatomic distances within a layer are all equal (1.42 Å). Individual planes in the graphite lattice, however, are spaced far apart (3.37 Å). This immedi-ately suggests that the bonding within the planes is different from the bonding between atoms of neighboring planes. It is thought that the bonding between planes in the graphite crystal is of the van der Waals type.

723

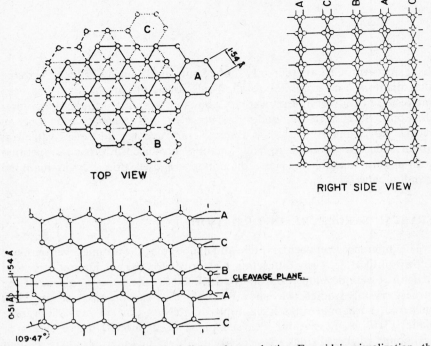

TOP VIEW

RIGHT SIDE VIEW

Fig. 1. Orthographic projections of diamond space lattice. For aid in visualization, the hexagonal rings of layer A are outlined with solid lines in the top view. Layer B hexagons are outlined by dashes, and layer C hexagons with dots.

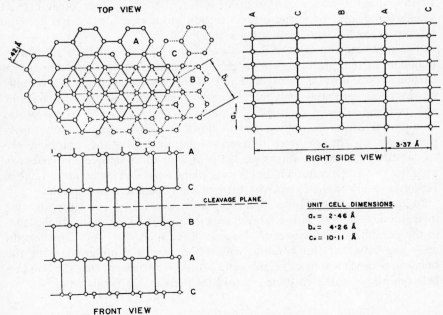

TOP VIEW

RIGHT SIDE VIEW

CLEAVAGE PLANE

FRONT VIEW

UNIT CELL DIMENSIONS.

$a_o = 2.46$ Å
$b_o = 4.26$ Å
$c_o = 10.11$ Å

Fig. 2. Orthographic projections of rhombohedral graphite space lattice. Scale is the same as that used in Fig. 1.

Although the accepted ideas of the crystal structure for diamond have not changed since the earliest determination, there have been periodic questions concerning details of the crystal structure of graphite, and these questions continue to the present day. There is no doubt, however, that the essential features of graphite are those of approximate hexagonal rings of carbon atoms arranged in layers separated a considerable distance apart relative to the distance between adjacent carbon atoms in the hexagonal rings.

To further characterize the differences on the atomic scale between graphite and diamond, it is necessary to look into the nature of the bonding between atoms. In diamond, the bonding is predominantly covalent in nature and is due to the formation of sp^3 hybrid bonds. All bonds within diamond are equivalent and are "aliphatic" in character. The bonds between carbon atoms in the hexagonal rings in graphite apparently have some double bond character. Therefore, the entire layer or sheet consists of a giant, two-dimensional resonating molecule—a molecule that is somewhat "aromatic" in character.[3,4]

Because of the similarities in crystal structures, a cursory examination might lead to the conclusion that graphite could be converted to diamond by brute force; i.e., by applying sufficient pressure, the bonds between graphite layers would be shortened the proper amount and the hexagonal carbon rings would be forced to "pucker," the whole process causing the atoms to conform to the diamond crystal lattice. The years have shown, however, that there are some complicating factors in this direct conversion of graphite to diamond.

THERMODYNAMICS AND KINETICS OF CONVERSION

A modern-day chemist knows that he must be concerned with two problems whenever he wants to bring about a chemical change. (A change from one polymorphic form to another may be regarded as a type of chemical change.) He must consider first the *thermodynamic* problem and second, the *chemical kinetic* or *reaction rate* problem. Chemical thermodynamics is concerned with the relative energies of the reactants and the products of a chemical reaction. Under conditions where the free energy of the reactant(s) F_r is greater than the free energy of the product(s) F_p, the relative energy $F_p - F_r = \Delta F$ is negative and the reaction has thermodynamic permission to proceed toward formation of the products. In the case at hand, the reactant is simply graphite and the product is diamond: $C_{graphite} = C_{diamond}$. On the other hand, if ΔF is positive, the reaction has thermodynamic permission to proceed in the opposite direction; i.e., the products (as the reaction is written) will proceed to transform into the reactants. Under conditions where

ΔF is zero, a stalemate (equilibrium) exists and there is no net tendency for the reaction to proceed in either direction.

Thermodynamics does not give any information about the time required for a reaction to take place. The ΔF may, in many instances, have a large negative value, yet the reaction is found to proceed at an imperceptible rate. An oft-quoted example of this situation is the bulb containing a mixture of hydrogen and oxygen gases. For the reaction, $H_2 + \frac{1}{2}O_2 = H_2O(g)$, the standard change in free energy is $-55,600$ cal. Nevertheless, the bulb may be kept for many years without detecting the formation of any water vapor.

It is the business of chemical kinetics to deal with the rates of chemical reactions. For this purpose theories and techniques for understanding the atomic and molecular processes (reaction mechanisms) taking place in chemical reactions have been developed. Application of the theory, coupled with appropriate experiments, often points the way toward finding means of increasing or decreasing the rate of a given reaction.

FAVORABLE NEGATIVE ΔF AS PRESSURE INCREASES

For the reaction $C_{graphite} \rightarrow C_{diamond}$, $\Delta F^{\circ} = +692$ cal/g-atom at 25°C and 1 atm pressure.[5] This ΔF° is not obtained by direct measurement but, as is the case with most thermochemical data, is calculated from measurements of heats of combustion, specific heats, compressibilities, thermal expansions, etc. The positive ΔF° indicates that diamond is thermodynamically unstable with respect to graphite. However, diamonds have not been known to transform into graphite by any observable amount over periods of hundreds of years under ordinary conditions. The rate of reaction (conversion) must, therefore, be extremely slow. Increased temperature will accelerate most reactions and this is also true for the conversion of diamond to graphite. This transformation begins to proceed at an observable rate at a temperature in the neighborhood of 1200°C at 1 atm. At this pressure and temperature ΔF° has increased to about $+2400$ cal, indicating that higher temperatures decrease the thermodynamic stability of diamond. The ΔF° for the graphite–diamond transition assumes its lowest value at the absolute zero of temperature

$$\Delta F_0^{\circ} = +580 \text{ cal/g-atom at 1 atm.}$$

In order to bring the graphite–diamond reaction into a region where ΔF° is negative, it is necessary to apply pressure. The pressure required depends on the temperature—the higher the temperature, the greater must be the pressure.

Equilibrium is established when the free energy difference between the two allotropes is zero, that is,

$$\Delta F = \Delta H - T\,\Delta S = 0.$$

The manner in which ΔF varies with pressure at a given temperature is given by

$$\left(\frac{\delta \Delta F}{\delta P}\right)_T = \Delta V$$

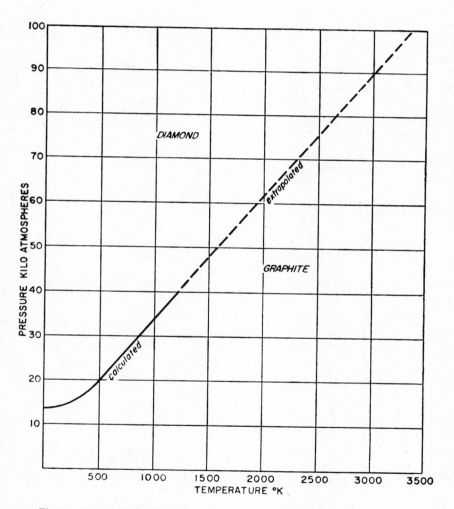

Fig. 3. The graphite–diamond equilibrium curve. [After Berman and Simon (5).]

or by

$$\Delta F_T{}^P - \Delta F^{\circ}_T = \int_0^P \Delta V \, dP,$$

where ΔV is a function of both T and P. From these expressions the free energy difference at any pressure and temperature may be expressed as

$$\Delta F_T{}^P = \Delta H^{\circ}_T - T \Delta S^{\circ}_T + \int_0^P \Delta V \, dP.$$

Data is available for evaluating ΔH°_T and ΔS°_T to 1200°K. Evaluation of the integral, however, involves some reasonable approximations. For additional details, the reader is referred to Berman and Simon.[5]

A graph showing the approximate pressure–temperature relationship is shown in Fig. 3. The line represents the pressure and corresponding temperature for which $\Delta F^{\circ} = 0$. Above the line ΔF° is negative and diamond is the stable carbon allotrope. Conversely, graphite is stable in the positive ΔF° region below the line.

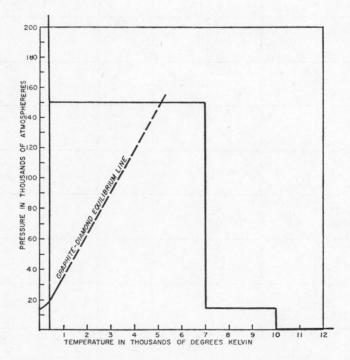

Fig. 4. The approximate pressure-temperature conditions to which graphite has been subjected (area below the solid line).

Graphite, subjected to P–T conditions above the line, should transform to diamond. Graphite has been subjected to conditions lying approximately below the solid line of Fig. 4. Although part of this area overlaps the diamond stable region, diamond has not yet been observed to form directly from graphite under these conditions. Since the thermodynamic criteria for formation of diamond have been met, it must be concluded that kinetic considerations are controlling and are preventing the transformation from occurring in any practical length of time.

LESS FAVORABLE RATE AS PRESSURE INCREASES

Increased temperature might ultimately increase the rate of transformation to a practical value, but increased temperature (to stay within the diamond stable region) calls for an increase in pressure beyond the range of the high pressure devices currently available.* At this impasse an investigation of the rate process is enlightening. Although diamonds cannot yet be made by direct conversion from graphite, graphite can easily be made from diamond. Studies of this reverse transformation under high-pressure, high-temperature conditions can give information concerning the reaction mechanism.

Experiment shows that high pressure retards the rate of transformation of diamond to graphite. The theory of absolute reaction rates indicates for such a situation[6] that

$$\log \text{rate} = \text{constant} - \frac{\Delta V^{\pm} P}{RT}$$

where $\Delta V^{\pm}$ is the molar difference in volume between diamond and the activated complex (the intermediate state or transition state), P is the applied pressure, T is the temperature, and R is the molar gas constant. Although available experimental data are somewhat erratic, a plot of log rate versus P leads to the inescapable conclusion that $\Delta V^{\pm}$ is at least 10 cc. Note that the molar volume of diamond is 3.42 cc, the molar volume of graphite 5.34 cc, and their difference in molar volume ΔV is only 1.92 cc. Obviously the transition state, with a molar volume $V^{\pm}$ of at least $(10 + 3.42)$ cc is a very open or expanded structure. It is reasonable to assume that the transition state for graphite–diamond conversion is the same as for the formation of graphite from diamond. This being the case, there arises the very frustrating but extremely interesting situation that higher pressure leads to more favorable

* P. S. DeCarli and J. C. Jamieson, *Science*, **133**, 1821 (1961), have reported the synthesis of eight-micron-diameter diamonds directly from rhombohedral graphite by means of an explosive shock. Transient pressure was estimated to reach 300,000 atm. An estimate of the temperature achieved was not given.

thermodynamics (a more negative ΔF) for the conversion of graphite to diamond, but, at the same time, leads to less favorable reaction kinetics since it suppresses the formation of the expanded activated complex. At the pressures and temperatures to which graphite has been subjected, to the present time, reaction kinetics apparently have the upper hand and prevent the synthesis of diamond directly from graphite in any practical period of time. Additional information concerning thermodynamics and reaction rates, as applied to high pressure reactions, is set forth in Ref. 7.

SUCCESS: GRAPHITE → DIAMOND

From the above analysis it would seem desirable to take the graphite lattice apart atom by atom and build the atoms one at a time into the diamond lattice. This might be accomplished by means of a solvent. Conceivably the proper solvent would take carbon atoms from graphite into solution as individual entities which, under the influence of proper thermal and concentration gradients would migrate through the solvent and precipitate as diamond. Nature gives some clues as to possible solvents. Diamonds are found imbedded in ferro-magnesium silicates, from which they apparently crystallized, in the famous pipe mines of South Africa. A few diamonds have been found embedded in iron-nickel and in troilite (FeS) constituents of meteorites.[8]

My first synthesis of diamond was based on the idea that diamonds might

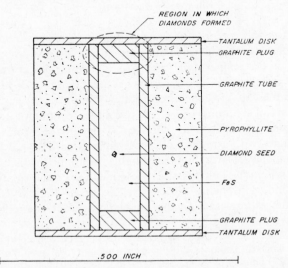

Fig. 5.   Cross section of cell in which first diamonds were grown.

be crystallized from troilite. On December 16, 1954, at the General Electric Research Laboratories in Schenectady, New York, I performed an experiment in the "Belt" high-pressure, high-temperature apparatus with a cell arrangement as shown in Fig. 5.

A brief diversion is necessary at this point to make a few comments concerning the apparatus. The "Belt" apparatus makes use of two opposing, conical pistons thrust into opposite ends of a symmetrical, tapered chamber (Figs. 6, 7). Relative motion of the pistons with respect to the chamber is afforded by a compressible gasket. The solid pressure transmitting material

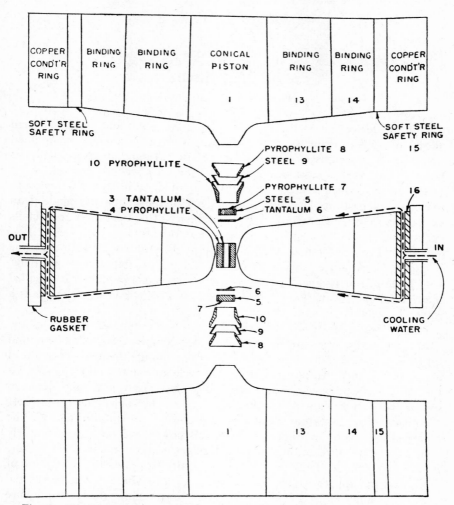

Fig. 6. The "Belt" high-pressure, high-temperature apparatus; "exploded" view.

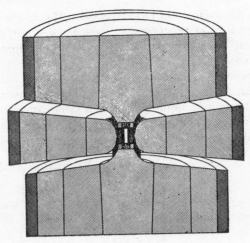

Fig. 7. "Belt" apparatus.

used in the device serves also as electrical and thermal insulation. The design allows pressures greater than 100,000 atm to be continuously maintained simultaneously with temperatures above 2500°C. (Temperatures estimated to be as high as 7000°K have been maintained for a few seconds.) Development of this device opened a large, hitherto inaccessible thermodynamic region to scientific exploration. The relative simplicity of the design has made it an extremely useful and practical tool. (The theory of its operation and details concerning use, design, and development are to be found in Refs. 9–12.)

A second device (actually the first to be publicly disclosed) with the same pressure, temperature, and diamond-synthesizing capabilities as the "Belt" was developed at Brigham Young University during 1956–1957. It is called the "Tetrahedral Anvil Press" and is currently finding considerable use as a research tool.[13]

In the December 16 diamond synthesis run, the tapered pistons of the "Belt" apparatus were forced into the loaded chamber by a 223,000-pound thrust delivered by a hydraulic press. After one minute the load was reduced to 186,000 pounds. An electric current (60-cycle, alternating) was then passed through the assembly via the tapered pistons. Current was gradually increased over a period of 6 min to a maximum value of 346 A. The voltage drop across the pistons was 1.84 V corresponding to a power dissipation of 636 W. The temperature inside the graphite tube at this power level was approximately 1650°C. The pressure, at the time, was thought to have been in the neighborhood of 95,000 atm. There has been some controversy over the calibration of high-pressure apparatus recently, however, that would place the actual pressure as much as 20% below this figure. Even assuming the lower pressure value, however, the system was in a region of thermodynamic stability for

Fig. 8. Photomicrograph of diamonds from the first successful experiment, December 16, 1954.

diamond. Maximum temperature was maintained for about 3 min, after which it was lowered to room temperature in approximately 5 min. After the temperature had been lowered, 18 min was taken to reduce the pressure to 1 atm.

When the cell was broken open, dozens of tiny transparent crystals were found near the tantalum disc at the top end of the cell. Subsequent chemical, physical, and X-ray examinations conducted during the course of the next few days conclusively proved that the material was diamond. Pictures of these diamonds are shown in Fig. 8. A powder X-ray diffraction pattern is shown in Fig. 9, where arrows point to the diamond lines. The extraneous

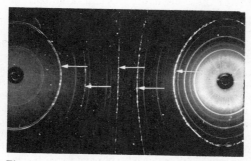

Fig. 9. X-ray diffraction. Powder photographs of some of the diamonds from the first successful synthesis experiment (CuK$_\alpha$ radiation). Arrows point to the diamond lines.

diffraction lines were caused by the presence of graphite and tantalum carbide. The "seed" diamond employed in this run was unchanged at the completion of the experiment. No new diamonds were found in its vicinity, and subsequent runs without a diamond seed were successful.

During the period December 16 through December 31, 1954, I performed 27 high-pressure experiments similar to the successful December 16 experiment with some changes in the various experimental parameters. Diamonds were made in 12 of these runs. Since FeS was known to be a nonstoichiometric type compound, usually with excess iron (Fe_xS, where $x > 1.00$), pure iron was substituted for the FeS and diamonds were successfully grown in that solvent. Substitution of pure sulfur for FeS did not produce diamonds. Microscopic examination of the cell contents following a run with FeS or Fe disclosed that a black coating (removable by acid) covered the diamonds. Carbon atoms, derived from graphite or metallic carbides formed in the reaction mixture, apparently migrate through this film and precipitate as diamond. The film is probably iron. The tantalum (or more particularly tantalum carbide) which formed in the presence of carbon, seemed to aid diamond growth.

Of practical importance in locating the correct temperature for diamond formation was my observation that onset of diamond growth was signalled by a drop in heating current through the sample and a corresponding rise in voltage across the sample. As heating power was controlled by a variable autotransformer, it was only necessary to increase the power output of the transformer until this phenomenon took place. The power setting was then left unchanged for 2 or 3 minutes until diamond growth was about complete.

On December 31, 1954, Dr. H. Hugh Woodbury of the General Electric Laboratory duplicated my December 16 run and thus, to our knowledge, became the first man to duplicate the diamond synthesis claim of another.*

Simple substitution of other transition metals for iron subsequently showed that many of these elements were suitable solvents for the synthesis of diamond.

---

* At this point it was evident that a landmark in science had been reached. In reaching that landmark, I could take particular satisfaction in the following key contributions: the design of the "Belt" apparatus that made experiments in the 100,000-atm, 2500°C temperature range possible. The discovery of a suitable combination of pressure, temperature, chemical ingredients, timing sequence, and cell arrangement for the synthesis of diamonds. It was extremely important that (a) the synthesis could be reproduced by myself and others; (b) diamond growth was very rapid; (c) diamond yields were significant—outstripping a millionfold the concentration, relative to the growing matrix, of diamonds found in nature; (d) the diamonds, though tiny (crystals with triangular faces up to 300 μ on edge), could be observed with the unaided eye and were obtained in sufficient quantity to be "felt" and held in the hand. Items (b) and (c) above obviously foreshadowed the commercial production of diamond.

By February 15, 1955, diamonds had been produced in about 100 separate runs in the General Electric Research Laboratory. On this date a press conference was held and the fact that diamonds had finally been made by man was announced to the world. However, details concerning the synthesis were not released. On October 22, 1957, the company announced that diamond production had successfully been carried through the pilot plant stage—more than 100,000 carats (carat = 0.200 g) having been produced up to that time. Thus, less than three years following the successful conclusion of a 100-plus years' search for a method to convert ordinary black carbon into diamond, man-made diamonds became a commercial product. Details of apparatus and methods of synthesis were finally released in the fall of 1959.

Following the February 15, 1955, synthesis announcement by the General Electric Co., several claims to prior synthesis were voiced. Only one of these claims has appeared in a technical journal. It appeared as an article entitled "Artificial Diamonds" by H. Liander in the *ASEA* (*Allmänna Svenska Elektriska Aktiebolaget*, Västeras, Sweden) *Journal* for May-June of 1955. The method of synthesis was not given but the statement was made that "ASEA produced its first diamonds on 15th February, 1953." The experimental procedures used were disclosed in 1960 and were similar to those used at General Electric.[14]

REFERENCES

1. H. Lipson and A. R. Stokes, *Proc. Roy. Soc.*, **A181**, 101 (1942).
2. J. P. Howe, *J. Am. Ceramic Soc.*, **35**, 275 (1952).
3. W. Huckel, *Structural Chemistry of Inorganic Compounds*, Vol. 2, Elsevier, 1951, p. 537.
4. G. L. Clark, *Applied X-Rays*, 4th ed., McGraw-Hill, New York, 1955, Chapter 20, particularly pp. 584–586.
5. R. Berman and F. Simon, *Z. Elektrochem.*, **59**, 333 (1955).
6. H. Eyring and F. W. Cagle, Jr., *Z. Elektrochem.*, **56**, 480-483 (1952).
7. Proceedings of the Symposium on "High Temperature—A Tool for the Future," Stanford Research Institute, Menlo Park, California, 1956, pp. 161–6 by H. T. Hall.
8. A. E. Foote, *Am. J. Sci.*, **42**, 413–17 (1891).
9. H. T. Hall, *Rev. Sci. Instr.*, **31**, 125–131 (1960).
10. H. T. Hall, U.S. Patent No. 2,941,248.
11. *Proceedings of the Third Conference of Carbon* (held at University of Buffalo, Buffalo, N.Y., June, 1957), Pergamon Press, London, pp. 75–84 by H. T. Hall.
12. F. P. Bundy, W. R. Hibbard, Jr., and H. M. Strong, Eds., *Progress in Very High Pressure Research*, Wiley, New York, 1961, pp. 1–9 by H. T. Hall.
13. H. T. Hall, *Rev. Sci. Instr.*, **29**, 267 (1958).
14. H. Liander and E. Lundblad, *Arkiv för Kemi*, **16**, 139 (1960).

On the Pressure Gradient in a Cylindrical Sample Container under High Pressure*

MASAAKI TAMAYAMA**

Chemistry Division, National Research Council of Canada, Ottawa

* This paper is based on M. Tamayama and H. Eyring, "Discussion on the Determination of the True Pressure on the Sample by Averaging the Gauge Pressures of Up- and Down-stroke in a Piston-Cylinder High Pressure Press," presented at the American Chemical Society Intermountain Division Meeting, May, 1967 at Brigham Young University, Provo, Utah, and ref. 9.
** Present address: Cyanamid (Japan), Ltd. 1-1 Yuraku-cho, Chiyoda-ku, Tokyo.

A Personal Note

Henry Eyring's laboratory, the Institute for the Study of Rate Processes, is a place of freedom. Though not in a separate building, its activities are independent of any university department. He gives his students and colleagues complete freedom to think, to try, and to write. His outstanding contributions to science derive from his insights combined with his generosity toward his students.

Henry's physical command as a lecturer is well known. He is also an elder of the Mormon church, and in that capacity he performed his first marriage at my wedding. On that occasion his command failed him, and his trembling voice and shaking hands showed that he was a good deal more nervous than the bride or groom.

It is not so well known that Henry earned his bachelor's degree in mining engineering. If you worked in his laboratory, you would soon discover that he still has a good sense of engineering. During my $5\frac{1}{2}$ years with him I never saw him work with his hands, except to polish his shoes, but I still felt that his were the sharp but gentle eyes of a veteran foreman whenever he stood beside me. I always enjoyed discussing engineering problems and experimental difficulties with him, and he always offered brief but perceptive suggestions.

INTRODUCTION

For an opposed-anvil type of high pressure press,[1] calibration of the up-stroke gauge pressure has been recommended to estimate the true pressure on a sample. This has been demonstrated by many investigators and extended to belt[2] and cubic[3] high pressure apparatus, whereas a unique pressure calibration has been proposed for a tetrahedral press.[4] The two major reasons for using the up-stroke gauge pressure for calibration are, first, that the up-stroke gauge pressure is more reproducible than that of the down-stroke, and, second, that the average value of the up- and down-stroke gauge pressures is generally far from the transition pressure observed by a piston-cylinder high pressure press enclosing only a sample under investigation. Pressure gradients, asymmetric friction, and plastic deformation of gasket materials (including containers) are the usual explanations of the discrepancy between up- and down-stroke pressure.

A piston-cylinder high pressure press is not excepted from the above problems, unless only the sample is set in its cylinder. There are experimental difficulties even in the simplest high pressure piston-cylinder press concerning the relation among the true pressure on the sample, the up-stroke gauge pressure, and the down-stroke gauge pressure.

In the present paper, the pressure loss—that is, the difference between the gauge pressure and true pressure—in a piston-cylinder high pressure press is investigated by comparing experimental bismuth phase diagrams through a complete up- and down-stroke cycle with the accepted bismuth phase diagram. The effect on the phase diagram of wrapping the sample container with lead foil is also revealed.

EXPERIMENTAL AND RESULTS

The piston-cylinder high pressure press used here is similar to those of previous investigators[5,6] except for one major difference in the cylinder diameter (1 in. in this work; $\frac{1}{2}$ in. in others). The gauge pressure was read with a 16 in. diameter Heise gauge and multiplied by a factor of 100 to obtain

739

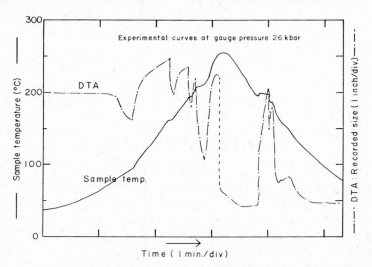

Fig. 1. DTA and sample temperature curves. The DTA baseline was shifted down as cooling began. The transition temperatures during heating were plotted at 26 kbar in Fig. 3a.

the pressure on the piston advancing into the cylinder. The high pressure sample container and the installation of thermocouples were described in the previous note.[7] The volume change of the sample container through the up- and down-strokes was recorded on an X-Y recorder for several runs by plotting the output of a pressure cell in a primary hydraulic oil line against the output of a carbon-film linear potentiometer between an advancing ram and a base plate. Temperatures at phase transitions were measured by a differential thermal analysis method (DTA). Heating was internal. Error in the temperature measurement was a maximum of $\pm 1°$C. Heating rate around 200°C was about 1°C/min.

The sample size and the components of the high pressure furnaces were identical; machining error in the components was less than 0.001 in.

There was only one major difference between the runs; one group of the sample containers had no lead foil around them, hereafter "without lead foil"; the other had a 0.005-in. thick lead foil on the side of the sample container except for 0.05 in. from its bottom edge and 0.25 in. from its top, hereafter "with lead foil." The cylinder and outer surfaces of all the furnaces were wiped with Molykote powder. The bismuth sample (1.973 $\pm$ 0.009 g) was embedded in a silver chloride cup in all cases.

Figure 1 shows an example of the experimental heating and cooling curves. The gauge pressure (calculated) was 26.0 kbar. Four deep valleys characterize the DTA curve of heating. From the left, they correspond to the phase transitions of bismuth (I–II), (II–III), (III–IV), and (IV–liquidus), respectively.

Three peaks are seen on the DTA curve of cooling. They indicate bismuth (liquidus–IV), (IV–III), and (III–II). The bismuth (II–I) phase transition is not seen on the cooling curve because of the slow cooling rate. To plot the phase diagram of bismuth, we usually take the temperature at the beginning of the phase transition or the melting on heating. Our extensive study of the DTA under high pressure[8] shows that the phase transition or melting temperatures corresponding to the peaks of the DTA signals depend strongly on the heating rate, but the temperatures at which these begin are fairly independent of the heating rate.

Figures 2a, b, and c are from the first group of this work: sample containers without lead foil. Five sample setups (in the pressure ranges which follow) were used to draw the figures: 8–26, 24–32, 30–32 and then to 30, 32–4, and 23–4.5 kbar. Figure 2a was drawn for pressure increasing and Fig. 2b for decreasing pressure; Fig. 2c was drawn by calibrating the gauge pressure of Fig. 2a. The calibration values were determined by assuming the midpoint at a particular volume on the piston displacement versus pressure cycle

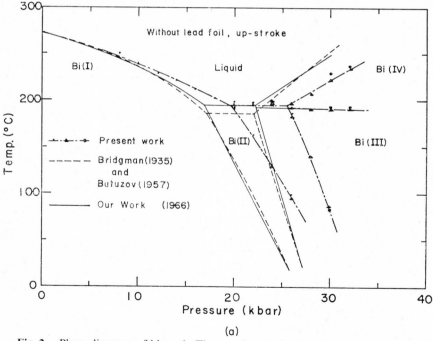

(a)

Fig. 2. Phase diagrams of bismuth. The sample containers were not wrapped with lead foils. (a) Upstroke. Three sample setups were used. (b) Downstroke. Three sample setups were used; however, a run marked 0 was the same sample setup as that in (a). (c) Up-stroke corrected by the volumetric method described in the text. Only a few representative spots were marked on the diagram.

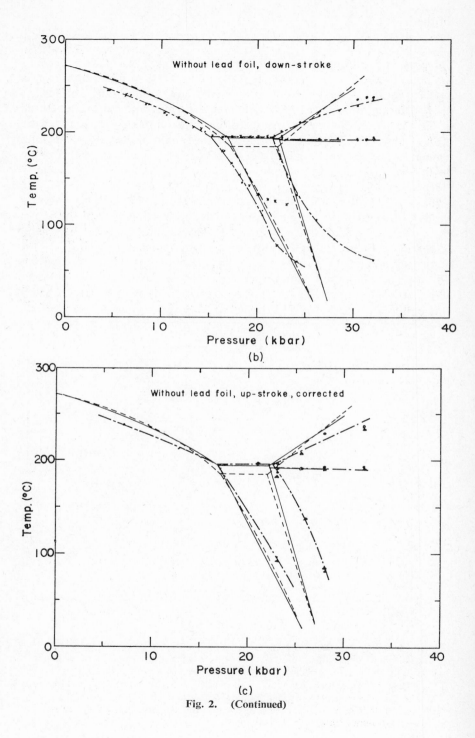

Fig. 2.   (Continued)

Table I. Correction by Volumetric Method

Gauge Pressure (Up-stroke) (kbar)	Correction Values (kbar)	
	Without Lead Foil	With Lead Foil
10.00	2.98	1.75
15.00	3.22	2.00
20.00	3.18	2.00
25.00	2.07	1.80
30.00	2.13	1.50

as the true pressure. The values are listed in Table I. If all pressure losses of the up- and down-strokes are equal, the calibration values are true and the calibrated bismuth phase diagram must fit the standard ones. Figure 2c indicates some disagreement.

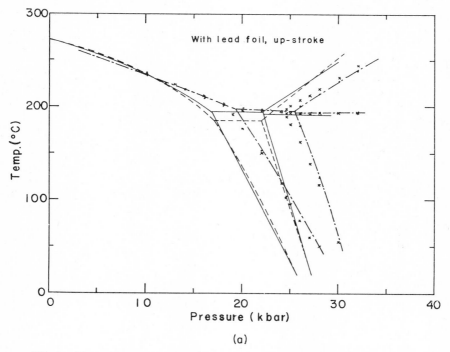

(a)

Fig. 3. Phase diagrams of bismuth. The sample containers were wrapped with lead foils. (*a*) Up-stroke. Two sample setups were used. × is the second run of a sample setup. For discussion the first run, marked · was adopted. (*b*) Down-stroke. One sample setup, which is the same as · (26–32 kbar) and × in (*a*). In this figure · is the first run and × the second. (*c*) Up-stroke corrected by the volumetric method described in the text.

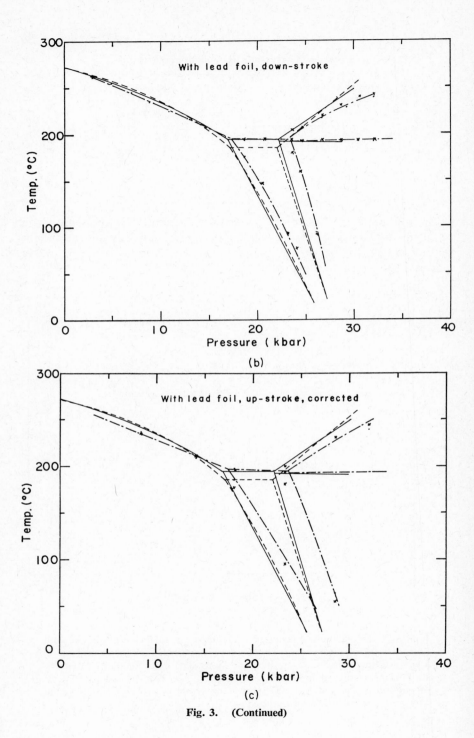

Fig. 3. (Continued)

Figures 3*a*, *b*, and *c* belong to the second group: the cylindrical surface of a sample container was wrapped with lead foil to reduce friction between the cylinder and the container. Two sample setups were used to make the figures: 8–26 and 26–32 then to 2 then 10 to 32 then to 20.5 kbar. Figures 3*a* and *b* were drawn as pressure increased and decreased, respectively, and Fig. 3*c* is the pressure calibrated diagram. The calibration values with lead foil were smaller than those without lead foil, as is seen in Table I.

DISCUSSION

The difference between gauge pressure and true pressure on the sample increases with increasing pressure as seen in Figs. 2*a* and 3*a*. In the lower pressure range, say below 10 kbar, the pressure loss is very small along the bismuth I–liquidus line. The pressure increase by thermal expansion compensates the comparatively small pressure loss around the sample in this pressure-temperature region. Numerical discussion of the thermal expansion cannot be extended to this region because no observable phase-transition lines occur in the lower temperature area. In the middle pressure range, 10–20 kbar, the difference increases sharply and above 20 kbar more gradually. The change in the difference is presumably characteristic of a particular structure including the material of the sample container.

From Figs. 2*a* and 3*a* one can say that the true pressure on the sample is always lower than the gauge pressure and the pressure losses due to friction, nonhydrostaticity, etc. is "positive." The down-stroke phase diagram shows some interesting facts. The up- and down-stroke phase diagrams are not symmetrical to the standard ones. When the pressure was decreased from its highest value (32 kbar), the true pressure of the sample decreased gradually. The pressure loss is still "positive." This contradicts the symmetry of friction or the postulate that pressure losses during up-stroke immediately reverse their direction upon down-stroke. At a certain point the gauge and true pressures become equal. In other words the pressure loss is zero and this happens only once through the entire cycle. Such pressures for samples without lead foil and with lead foil, were about 24.5 and 14.5 kbar, respectively. After the equilibrium point, the pressure loss becomes "negative," which would be expected at much higher pressure than actually observed. What is the role of a lead foil? Comparing the phase diagrams of up-stroke, Figs. 2*a* and 3*a*, one can tell at least one thing, that the lead foil apparently reduces the pressure loss involved in such a sample container. Experimental values are listed in Table II; the last column is the percentage reduction in the pressure loss over that without lead foil. A lead foil more or less reduces the friction between the contacting surfaces of the cylinder and the sample container. It also caused good compression among many small parts which comprised the container; accordingly, a very tight seal around the sample was made. One sample setup with a lead foil was good for 41 heat–cool cycles at different pressures. On the 42nd heat–cool cycle the sample bismuth leaked from the container along the thermocouple wire. The down-stroke phase diagram with a lead foil is closest to the standard ones among the six graphs. The pressure loss is "positive" until about 14.5 kbar, as seen in Fig. 3b. The interpretation for this phenomenon is as follows. The pressure loss can be divided into two major causes: simple frictional loss and complicated loss due to plastic and shearing properties of the components of the sample container. The former is the total loss due to the interfacial frictions between the cylinder and the piston and between the cylinder and the solid pressure-transmitting medium that separates the piston and the specimen (these interfacial frictions were assigned in the previous work[9] as pressure losses I and II, respectively). The latter is the largest among the pressure losses and originates within the sample-containing cross section of the sample container (this is the major part of pressure loss III). The pressure loss of the up-stroke with lead foil is mainly (more than 80%) the latter and the rest is simple frictional loss. When pressure is released, the simple frictional loss will change its working direction and the pressure loss due to plastic properties of the components will be very slowly released. The lead foil promotes the release of the pressure loss. This can be understood by comparing Figs. 2b and 3b lower than 15 kbar and between 20 and 30 kbar below 150°C.

Higher temperature results in lower pressure loss in all cases. The higher temperature not only causes thermal expansion of the components of the sample container, but also promotes plastic flow of the substances under high pressure. However, these effects were not as drastic as those observed by Pistorius et al.[10] The major difference is the diameters of the sample containers: $\frac{1}{2}$ in. in their work and 1 in. in ours. The smaller diameter seems to provide poorer thermal insulation; accordingly, more power is needed for its internal electric heater to obtain a given temperature than in the case of a larger diameter. The observed increase in gauge pressure caused by heating is very small in our experiments. This is seen in Table III, although it should

Table II. Pressure Loss

Gauge Pressure (Up-stroke) (kbar)	Approximate Temperature at Comparison	Pressure Loss (kbar)		Effect of Lead Foil	
		Without Lead Foil	With Lead Foil	Reduced Loss (kbar)	Percentage
15.00	215	1.6	1.3	0.3	19
20.00	185	3.0	2.8	0.2	7
25.00	110	3.8	3.4	0.4	10
30.00	70	4.7	4.0	0.7	15

Table III. Gauge Pressure Increase by Heating

Without Lead Foil			With Lead Foil		
Initial Pressure (kbar)	Final Pressure (kbar)	Temperature (°C)	Initial Pressure (kbar)	Final Pressure (kbar)	Temperature (°C)
Up-stroke					
10.00	10.16	240	10.00	10.15	240
16.00	16.10	210	16.00	16.05	210
20.00	20.10	190	20.00	20.03	200
24.00	24.05	195	24.00	24.03	200
32.00	32.08	235	32.00	32.05	240
Down-stroke					
30.00	30.30	235	30.00	30.45	235
24.00	24.40	190	24.00	24.60	210
20.00	20.40	195	20.00	20.50	195
16.00	16.55	195	16.00	16.60	200
10.00	10.65	220	10.00	10.76	230

be noted that the relationship between the gauge pressure increase and that on the sample was not investigated. The pressures in the first column are those applied to the sample containers before heating. The second column shows the gauge pressures as the temperature of the specimen reached the values listed in the third column. Such pressure increases are much higher in the case of a $\frac{1}{2}$-in. diameter sample container; our best $\frac{1}{2}$-in. sample container showed 1–2 kbar increase over the above pressure and temperature ranges and some as high as 4 kbar. The gauge pressures at down-stroke increase more than those at up-stroke. In other words, higher temperature facilitates the relaxation of the compressed material.

The internal pressure increases caused by heating are seen in Figs. 2a, 2b, 3a, and 3b. The pressure loss decreases with increasing temperature; this diminution is compensated by thermal expansion. Comparisons between experimental data are summarized in Table IV. The transition pressures of bismuth were adopted from Butuzov's phase diagram,[11] which was measured in a liquid pressure-transmitting medium. The present work covers a narrower temperature range than the work of Pistorius et al. Substantial decreases in the percentages in the last column of Table IV are clear as temperature increases. Assuming linearity in the decrease of percentage, very approximate values of the rate of the decrease as a function of temperature were estimated; these values are −0.028 % per degree for samples both without lead foil and with lead foil and −0.052 for Pistorius et al. Calculations based on the rate of the decrease (the same as extrapolating the percentages in the last column to zero percent) give the temperature ranges where the pressure loss becomes zero. The temperature ranges are around 800°C for Pistorius et al., 670–740°C without lead foil, and 580–630°C with lead foil for this work. From

Table IV. Effects of Temperature on Pressure Loss

Temperature (°C)			Present Work (Bi in 1 in. diam. cylinder) Without Lead Foil Up-stroke (kbar)	With Lead Foil Up-stroke (kbar)	Pistorius et al. (KCl in ½ in. diam. cylinder) Up-stroke (kbar)	Transition Pressure (kbar)			$\dfrac{P_{\text{up-stroke}} - P_{\text{transition}}}{P_{\text{transition}}} \times 100$ (%)		
A	B	C	A	B	C	A	B	C	A	B	C
		22			27.1			19.22			41.0
60	51		28.0	28.0		23.8	24.3		17.6	15.2	
80	62		30.0	30.0		25.3	26.0		18.6	15.4	
	123	132		28.0	25.8		24.0	19.19		16.7	34.4
135			28.0			23.5			19.1		
208			28.0			24.8			12.9		
221	214		30.0	28.0		26.2	25.4		14.5	10.2	
	227			30.0			27.0			11.1	
		260			24.0			19.16			25.3
		358			23.5			19.13			22.8
		522			21.9			19.09			14.7

Table IV, presumably the percentage in the last column is a function not only of temperature but also of pressure; it increases as pressure increases at a constant temperature.

The phase diagrams of bismuth corrected by the volumetric method are shown in Figs. 2c and 3c. The corrections were applied to the up-stroke phase diagrams. As seen in the figures the volumetric corrections do not produce a satisfactory result. Moreover, it is difficult to analyze the physical and mathematical relations for the calibration values obtained by the volumetric method to find the true pressures on the sample and/or the pressure losses previously defined. The case treated is confined to that of a sample under pressure in a cylinder. These difficulties cause the author to recommend that for calibration, the up-stroke diagrams should be used.

In concluding this article it is emphasized that the total pressure loss in solid pressure-transmitting media should be formulated as a function of pressure, temperature, the physical properties of the substances comprising the sample container, and its shape.

ACKNOWLEDGMENTS

Sincere gratitude is extended to Dr. Henry Eyring, under whose guidance this work was done at the Institute for the Study of Rate Processes, University of Utah, Salt Lake City, Utah. The author acknowledges support given by the Army Research Office, Durham, N.C., under Grant DA-ARO-D-31-124-G-618. Thanks are due to Dr. Wm. A. Adams, Inland Waters Branch, Department of Energy, Mining and Resources, Ottawa, Canada, for his constructive discussion prior to submission of the manuscript.

REFERENCES

1. A. S. Balchan and H. G. Drickamer, *Rev. Sci. Instr.*, **32**, 308 (1961).
2. F. P. Bundy, in *Modern Very High Pressure Techniques*, R. H. Wentorf, Ed., Butterworth, London, 1962, p. 1.
3. A. Zeitlin and J. Brayman, in *High-Pressure Measurement*, A. A. Giardini and E. C. Lloyd, Eds., Butterworth, Washington, D.C., 1963, p. 301.
4. R. N. Jeffery, J. D. Barnett, H. B. Vanfleet, and H. T. Hall, *J. Appl. Phys.*, **37**, 3172 (1966).
5. F. R. Boyd and J. L. England, *J. Geophys. Res.*, **65**, 741 (1960).
6. G. C. Kennedy and P. N. LaMori, in *Progress in Very High Pressure Research*, F. P. Bundy, W. R. Hibbard, Jr., and H. M. Strong, Eds., Wiley, New York, 1961, p. 304.
7. M. Tamayama and H. Eyring, *Rev. Sci. Instr.*, **38**, 1666 (1967).
8. M. Tamayama and H. Eyring, *Amer. Chem. Soc., Abstr. Papers, 154th Meeting, Chicago, Illinois, 1967*, V-125.
9. M. Tamayama and H. Eyring, *Rev. Sci. Instr.*, **38**, 1009 (1967).
10. C. W. F. T. Pistorius, E. Papoport, and J. B. Clark, *Rev. Sci. Instr.*, **38**, 1741 (1967).
11. V. P. Butuzov, *Soviet Phys. Cryst.*, **2**, 533 (1957).

Analogies between Macromeritic and Molecular Systems and the Mechanical Properties of Sand and Gravel Assemblies

HANS F. WINTERKORN

Professor of Geophysics and Civil Engineering, Princeton University, Princeton, New Jersey

SIMILARITIES BETWEEN MOLECULAR
AND MACROMERITIC SYSTEMS

The similarities existing between the flow of molecular liquids and that of loose assemblies of particles in the visible size range, such as sand, gravel, bird shot, seed grains, and others, have long been known to the practitioners of various arts and have found increasing use in modern industrial technology. They were mentioned by Galilei and Lambert, but their first extensive scientific experimental study was by Delanges in 1788.[1] He called such granular systems "semifluids" because of the visible size of the component particles and "certain and distinct conditions which truly cannot be assigned to the elements of liquids, and not even properly to such fluids as air, fire, etc."*

The "solid" behavior of densely packed, pressure-confined granular systems is usually demonstrated to the novice in soil mechanics by means of an evacuated rubber (hot water) bottle densely packed with sand, under the uniform confining pressure of the external atmosphere. Depending on the degree of packing which can be expressed as the ratio of the volume of voids to that of solid grains (void ratio), the same granular material may, accordingly, exhibit liquid or solid properties or a combination of both within a zone of transition. Searching for a measurable, physically meaningful distinction between solid and liquid behavior, we may use the shear resistance in function of various parameters such as normal pressure σ on the shear plane, rate of shear, or others.

For the viscous or laminar flow of molecular liquids we have, according to Newton,

$$\tau = \eta \frac{dv}{dr},\tag{1}$$

where η = coefficient of viscosity, a material property that is a function of temperature and less of pressure,

$\frac{dv}{dr}$ = change in velocity or rate of flow with increasing distance normal to shear plane,

τ = shear stress.

---

* Quotation from the translation by A. J. D'Atri, C.E., *Proc. Brooklyn Engineers' Club*, **28**, Part I, 13–51, October 1929.

Accordingly, τ increases with increasing rate of shear. Actually at very great rates of shear, brittle fracture may occur even in liquids.[2]

The shear behavior of normal polycrystalline solids may range from brittle fracture as in white cast iron to extreme deformations as in gold, copper, or α-ferrite. The ability to plastically deform increases with degree of symmetry of the atomic arrangement and with decrease of the bond strength between adjoining potential sliding planes. The fact that entire planes are involved in plastic deformation is reflected in the yield strength, which is absent in the case of liquid flow because of its essentially molecular character. For the entire range from brittle to plastic shear, the failure stress can be approximately expressed by

$$\tau = C + \sigma \tan \varphi \tag{2}$$

in which $\tan \varphi$, the coefficient of internal friction, becomes smaller as the crystal structure of the solid becomes more plastic and more highly symmetrical. This equation can also be written

$$\tau = (\sigma_i + \sigma_e) \tan \varphi \tag{3}$$

in which σ_i is the internal pressure or the attraction between sliding planes and σ_e is the externally applied pressure. The product $\sigma_i \tan \varphi$ equals C, which is normally designated as the contribution of the cohesion to the shear resistance. In plastic deformation there may be little or no volume change in the system; in brittle shear failure, however, there is a tendency for the volume to increase during shear.

Following the example of Coulomb,[3] the shear behavior of soils was for a long time expressed by $\tau = C + \sigma \tan \varphi$ for the cohesive varieties and by $\tau \doteq \sigma \tan \varphi$ for the noncohesive granular materials, C and $\tan \varphi$ being considered and treated as material constants. Since these equations are identical with (2) and (3), which are expressive of the behavior of solids, their use implied, or led to, a neglect of the "liquid" aspects of the shear behavior of both granular noncohesive as well as cohesive soils. This continued until the most recent past, even though Reynolds[4] had produced weighty evidence that should have reopened the case of "liquid" behavior occurring in granular systems. He showed that densely packed sands expanded during the shear phenomenon while loosely packed ones collapsed. In between was a state of density, now expressed as critical void ratio (CVR), at which shear occurred without either expansion or collapse.

The CVR was thoroughly investigated by Casagrande[5] and Taylor,[6] but without reference to the work of Delanges. Hence the significance of the CVR with respect to the "liquid" state of granular noncohesive systems was again overlooked. There were a number of good reasons for this. First, the actual CVR value determined on a particular material depended not only on the type of shear test (plane or triaxial) but also on the methodology employed. Thus Taylor[6] listed for the triaxial shear test three methods of determining and thereby practically defining the CVR. These are the Casagrande CVR, the constant σ_3 (minor principal stress) CVR, and the constant volume CVR. These methods give different numerical values for the same material, the differences decreasing with decreasing minor principal stress. Using the same method, the CVR for a particular material was found to decrease linearly with increasing values of the logarithm of the normal stress in the case of plane shear or the minor principal stress in triaxial shear.

The tan φ value for a particular material was found not to be a constant as had been assumed by Coulomb but to increase with decreasing void ratio. The results of the work of Casagrande and Taylor and other facts and relationships known for sands of relatively narrow particle size range were coordinated by Winterkorn[7] into the theory of the "solid" and "liquid" states of large-particle (macromeritic) systems. This theory made use of the following long known facts:

1. The mole volumes of many simple pure chemical substances at the absolute zero point, the melting and boiling points at atmospheric pressure, and the critical point, respectively, are as $1:1.21:1.48:3.78$; hence volume relationships are basic in the physical definition of solid and liquid states.

2. The ratio of the volume of a uniformly sized sand at its CVR to that at its greatest density is approximately 1.2. Accordingly, the CVR may be considered to be a volumetrically defined "melting" point* of a macromeritic system.

3. For simple molecular liquids Batschinski[8] had shown that the viscosity or internal friction at different temperatures could be expressed as a function of the mole volumes at these temperatures in the form

$$\eta = \frac{k}{V - V_0}. \tag{4}$$

This is also one of the bases of the modern hole theory of the viscosity of liquids.[9] A multiplicity of other pertinent analogies became manifest. Of greatest immediate practical importance appeared to be the testing of an equation for tan φ of granular systems formed in analogy to (4), namely,

$$\tan \varphi \, \frac{C}{e - e_{\min}} \tag{5}$$

---

* The inability of physicists to assign a cgs dimension to temperature and the fact that temperature differences are usually measured as volume differences give philosophical support to this concept.

in which C is a material constant, e the void ratio existing in the system at the time of the test, and e_{min} the smallest possible void ratio representing the state of densest packing.

TESTING THE MACROMERITIC SHEAR EQUATION

The validity of (5) was checked against a large body of test data taken from the publications of the best known researchers in this and other countries.[10] The constants C and e_{min} were computed from two test data of the same series. The following could be concluded:

1. Equation 5 yields a good reproduction of the experimental data within the same pressure range, that is, the same range of stored strain energy.

2. The C value remains relatively constant as long as the same method of test (triaxial or direct shear) is used.

3. The e_{min} values decrease in general with increase in strain energy (σ_1 or σ_3) of the granular system in the same sense as the decrease of the CVR. At high pressures the e_{min} values, computed from the results of two tests of the same series, often become even negative without, however, influencing the practical applicability of (5).

Negative e_{min} values possess no physical meaning. The fact that the equation worked even with negative values indicated that the differences actually employed in the equation were meaningful, although the values used in forming the differences had to be larger by a constant value which was sufficient to bring the minimum void ratios to physically meaningful levels. It had been pointed out previously[7] that shear failure in macromeritic systems differs from shear in viscous flow of molecular liquids in that the shear phenomena are concentrated in particular planes or regions whereas in viscous flow shear phenomena occur throughout the entire system. If the density in the shear zone of macromeritic systems exceeds the "melting" point density, then in order that shear may take place holes must migrate from other parts of the system to the shear zone. In other words, the density becomes less in the shear zone and greater in distant parts. The greater the strain energy in the system, the easier is the movement of the holes. This accounts for the decrease of the experimentally determined CVR with increasing σ_n or σ_3, respectively.

The course of CVR as a function of either σ_n or σ_3 can be expressed by the equation

$$\text{CVR} = a - b \log \sigma, \tag{6}$$

in which a and b are constants and σ is either the normal pressure on the shear plane or the minimum principal stress. Unfortunately, the logarithmic character of this relationship does not permit extrapolation to a CVR at zero stress condition in the system. There exist, however, other methods of arriving at theoretically meaningful CVR values for the actual shear zones.

Data presented by Taylor[6] for the three listed types of CVR for Franklin Fall sand (EI-A) as a function of σ_3 were plotted by Winterkorn[7] as a semilogarithmic scale. Although at $\sigma_3 = 60$ psi the respective CVR values were 0.74 (Casagrande), 0.69 (constant σ_3), and 0.65 (constant volume), the line plots converged with decreasing σ_3 and went through the same point at CVR of 1.25 and σ_3 of 0.05 psi. A corresponding plot of data obtained by Dutertre[11] on the #16–#30 U.S. sieve fraction of an Alaska beach sand went through the same point. The value of 1.25 at which the CVR obtained on the Franklin Fall sand by the three different methods coincided must be considered as one at which methodological differences have ceased to have an effect. Another method is to assume a probable density distribution or repartition of voids in the specimen during the shear test. By doing this Dutertre[11] arrived at an extrapolated CVR of 1.3 for the above-named Alaska sand fraction.

Now, the most meaningful (constant volume) CVR for a system of identical spheres has a theoretical value of 0.63 for densest (hexagonal rhombohedric) packing and of 1.31 for the practically loosest (cubic) packing. These values are approached by the experimental data from the Franklin Falls sand at greatest and smallest σ_3 values, respectively. This points to the existence of polymorphism in the case of assemblies of uniformly sized sand grains. As a matter of fact, photographs by Kolbuszewski[12] and others have shown that there exists no random packing in sand systems. Rather such systems have variously oriented well-ordered domains of distinct types of packing (cubic, hexagonal rhombohedric, or other) that are connected by more loosely packed disordered "phases." With increasing void ratios, the size of the well-ordered, more densely packed domains decreases and that of the loose and disordered phases increases. Thus the decrease in the CVR with increasing strain energy and concomitant denser packing finds its logical explanation. But how does this affect the validity of (5)?

Dutertre[11] has shown by means of experimental data and theoretical analysis that the errors caused by the migration of holes during the test into the limited shear zone are of the same magnitude for all three, the CVR, the effective void ratio e, and e_{min}. Hence (5) can be used with confidence in the simple original form even if negative e_{min} values appear in the arithmetic. While this equation was originally proposed intuitively for reasons of analogy, Kezdi[13,14] accomplished an analytical derivation, showing it to be a first approximation of the same theoretical validity as that of Batschinski for simple molecular liquids. He also derived the more rigorous equation:

$$\tan \varphi = \frac{C\sqrt{e - e_{min}}}{\exp\left[(e - e_{min})/a\right]}, \tag{7}$$

which gave excellent agreement with experimental data but requires the additional constant a. He felt, however, that the simpler (5) was appropriate for most purposes.

Evaluation of more recent test data, obtained by means of an improved direct shear tester, yielded the equation[15]

$$\tan \varphi = \tan \varphi_u \frac{e_{crit} - e_{min}}{e - e_{min}}, \tag{8}$$

in which $\tan \varphi$ and $\tan \varphi_u$ are the maximum and ultimate values, respectively, obtained in a particular shear test. The ultimate $\tan \varphi$ corresponds by definition also to that obtained at the CVR for the particular pressure conditions. This equation can be written

$$(\tan \varphi)(e - e_{min}) = (\tan \varphi_{crit})(e_{crit} - e_{min}). \tag{9}$$

BINARY SYSTEMS

Furnas[16] showed that with the same relatively small compactive effort, granular materials composed of particles of two distinct sizes yielded systems of different porosities which were a function of the percentages and the size ratios of the particles. The greater the size ratio, the smaller was the porosity or void ratio obtained and the lower the weight percentage of the smaller particles which yielded the minimum porosity. For a size ratio of $1:0.0x$ the minimum porosity was reached at 32.5% of fines producing a porosity of 24% or a void ratio of 31.6%. Incidentally, the closest possible theoretical packing of single-sized spheres yields a porosity of 25.95%. Similar results were obtained previously by Feret in France and subsequently by other workers. The point of real interest is that the different densities of the different mixtures were obtained by the same relatively small compactive effort, which means that they had the same relative workability or, in other words, the same degree of shear resistance despite their different void ratios. By the same token, if these different mixtures were carefully prepared so that they would have the same porosity, their angle of friction should be the lower the greater the difference between this porosity and the lower one which they easily assumed with small compactive effort. Experimental data by Herbst and Winterkorn showed that this was indeed the case.[15]

TERNARY SYSTEMS

Kezdi and Domjan (see Winterkorn, Ref. 17) investigated ternary systems with respect to the maximum and minimum void ratios obtained by standardized packing methods, and the coefficients of friction at various void ratios.

The materials used were sieved fractions of Danube gravel sand of the following size ranges:

Fraction	$D_{\max}$	$D_{\min}$	$D_{\mean}$
A	1.24	0.58	0.91
B	0.58	0.29	0.435
C	0.29	0.217	0.254

Some of the results of this investigation are shown in Figs. 1 and 2. Figure 1

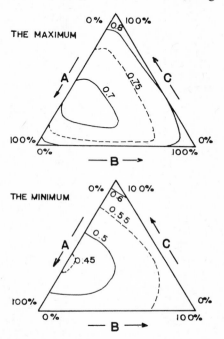

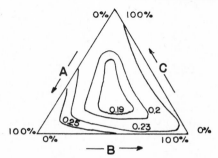

Fig. 1. Maximum and minimum void ratios in ternary systems as a function of composition.

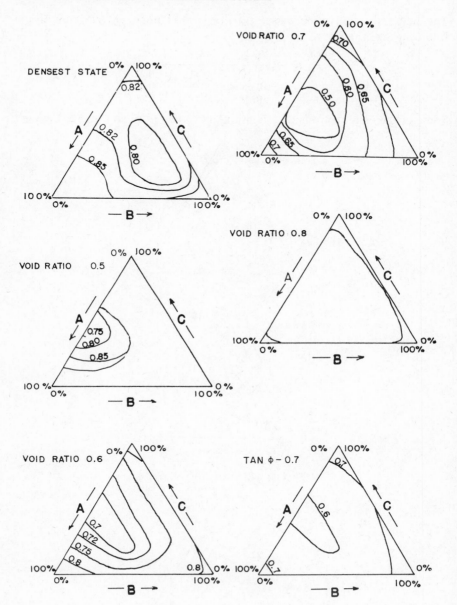

Fig. 2. Tan values of ternary systems at several different void ratios as a function of composition.

contains three graphs depicting, respectively, the maximum and minimum values of density and their differences, obtained for a large number of compositions made with sands A, B, and C. The similarity of these graphs with melting point diagrams of ternary molecular systems is quite impressive and illustrative of the logical extension of the macromeritic solid and liquid state concepts into the area of macromeritic "solutions."

Figure 2 contains five graphs that show tan φ for the ternary compositions for the densest obtainable states and also for the void ratios 0.5, 0.6, 0.7, and 0.8, respectively. It is interesting to see that at specified void ratios, tan φ increases with increasing distance of the composition points from that which easily achieves the maximum absolute density or minimum void ratio. Figure 2 also contains a graph showing the void ratios for which tan φ equals 0.7.

CONTINUOUS GRADINGS

Continuous grading of mineral particles, yielding low porosity mixtures with little compactive effort, are of great importance in concrete technology and soil engineering. Most widely used for this purpose are the gradation curves by Fuller and Thompson[18] and the nearly equivalent formula by Rothfuchs[19]:

$$p = \frac{\sqrt{d} - \sqrt{d_{min}}}{\sqrt{D} - \sqrt{d_{min}}} \times 100,$$

where p = percentage passing sieve of opening d, d_{min} = the smallest particle size, and D = maximum particle size. The greater the range from the maximum to minimum particle size, the less is the porosity of the resulting system. Data published by Goldbeck and Gray[20] for approximate voids in stone and gravel for various nominal size ranges are plotted and extrapolated in Fig. 3.

Herbst and Winterkorn[15] made shear tests on Rothfuchs graded sand employing ratios of D_{max}/d_{min} of 54, 28, and 9.6, respectively. According to Fig. 3, the corresponding void ratios for these gradations were 0.352, 0.395, and 0.45, respectively. The graphs plotted from the experimental data for tan φ as a function of the void ratios for these gradations gave the same tan φ value of about 0.72 for these corresponding void ratios.

The concept of the same tan φ value for corresponding void ratios in the case of various gradation ranges of particles possessing the same degree of sphericity and the same coefficient of material friction appears to be quite well justified by experience with Portland cement concrete but appears to be contrary to experience in soil stabilization. The difference lies in the different

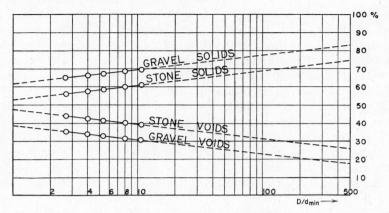

Fig. 3. Relation between normal porosity and size range as expressed by $D/d_{\min}$ in continuous grading.

magnitudes of the ratio of maximum to minimum size. In concrete aggregate, even if we would use as $d_{\min}$ the opening of the $\#200$ sieve (0.074 mm) and as $D_{\max}$ 25 mm we have a size ratio of 356:1 as compared with that of about 10,000:1 in the case of clay–sand–gravel systems. Also, in a concrete, the porosity and therewith the internal friction and workability are controlled by the dosage of water and cement.

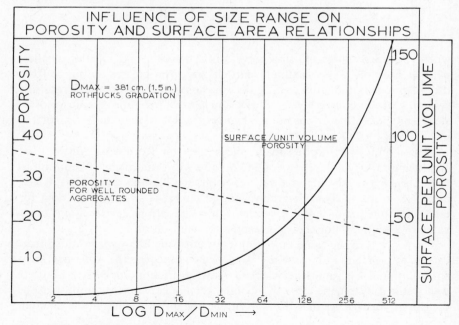

Fig. 4. Influence of size range on porosity and surface area relationships.

The factor that comes more and more into play with widening of the particle size range is the accelerated increase in the ratio of surface area to pore space per unit bulk volume. This is akin to the inverse of a hydraulic radius, and the larger its value the greater the resistance to movement of particles in the system. Figure 4 shows a plot of values calculated for this ratio for systems possessing Rothfuchs gradations in nine different size ranges. They are characterized by ratios of D_{max}/d_{min} of 2, 4, 8, 16, 32, 64, 128, 256, and 512, all having the same maximum size of $1\frac{1}{2}$ in. or 3.81 cm. The figure also gives the curve for the porosity of well-rounded particles for the various size ratios.

Systems possessing maximum density gradations and very large size ranges may be considered as highly viscous macromeritic liquids similar to asphalts which are composed of molecules of a very large size range.

DYNAMIC LIQUID STATES

Dynamic liquid states of macromeritic systems that occur not only in specific shear zones but throughout the system can be produced by adduction of kinetic energy of sufficient intensity to keep the particles in movement

Fig. 5. Perfectural apartment buildings after Niigata earthquake.

Fig. 6.   Uplifted sewage purifier tank after Niigata earthquake.

and to prevent the system from decreasing its porosity below that character-istic for the liquid state. The needed energy may be provided by mechanical devices alone, as by internal vibrators in the placing of harsh concrete mixes, by the pumps employed to load and unload dry powders such as Portland cement, and in wet sand and gravel dredging and transportation. The liquefaction energy may also be provided wholly or in part by molecular liquids or gases flowing through the particulate system in a direction opposite to that of gravity. Examples for the latter are the naturally occurring quick-sands and the fluidization of catalyst beds in the chemical industry. The importance of such "quick" condition occurring in unconsolidated sediments as a result of earthquake shock is well documented in the official report of the Office of the Engineer[21] on the Fukui earthquake that took place on June 28, 1948. Figures 5 and 6 show the results of such liquefaction during the Niigata earthquake of June 1964. Figure 5 shows the Perfectural Apart-ment Buildings at Kawsgishi-Chao in Niigata. Some of these are tilted and partially sunk into the sandy strata that were liquefied by the earthquake shock. Figure 6 shows a sewage purifier tank that was uplifted because its bulk volume weight was smaller than that of the liquefied sand stratum in which it had been buried.

CONCLUSION

I have attempted to show in a rather concentrated manner the pertinency and practical usefulness of various analogies that can be developed between molecular and macromeritic particle systems. I believe that, in addition to the areas in which these concepts have already born theoretical and practical fruit,[14,22,23] they should also prove useful in explaining and predicting, at least semiquantitatively, the energy acceptance and transmission properties of various soil types at different porosities when they are subjected to dynamic loadings deriving from earthquakes, explosions, or other causes. As a matter of fact, the vistas that are opened up by these concepts show so many promising research paths that the greatest difficulty is to make up one's mind what to do next.

ACKNOWLEDGMENTS

The concepts that are presented in this paper together with some of their engineering applications, were born during conversations with Henry Eyring in the later years of his Princeton residence. His continued interest and encouragement contributed greatly to their growth and to their present status in soil and materials engineering.

REFERENCES

1. P. Delanges, "Statics and Mechanics of Semi-fluids," *Memoirs on Mathematics and Physics of the Italian Society, Verona, 1788*, translated by A. T. D'Atri, *Proc. Brooklyn Engineers' Club*, **28**, Part I, pp. 13–51, October 1929.
2. M. Kornfeld, *Elasticity and Strength of Liquids*, Verlag Technik, Berlin, 1952.
3. C. A. Coulomb, "Essay on an Application of the Theory of Maxima and Minima to Some Problems of Statics Pertaining to Architecture," *Memoirs on Mathematics and Physics*, presented at the Royal Academy of Sciences, Paris, France, 1773 (pub. 1776).
4. O. Reynolds, "On the Dilatancy of Media Composed of Rigid Particles in Contact," *Phil. Mag. (Ser. 5)*, **20**, 469 (1885).
5. (a) A. Casagrande, "Characteristics of Cohesionless Soils Affecting the Stability of Slopes and Earth Fills," *J. Boston Soc. Civil Eng.*, **23**(1), 13 (1936); (b) A. Casagrande, "Notes on the Shear Resistance and the Stability of Cohesionless Soils and their Relation to the Design of Earth Dams," *Discussion, Proc. 1st Intern. Conf. Soil Mech. Found. Eng.*, III, 58–60, Cambridge, Mass., 1936.
6. D. W. Taylor, *Fundamentals of Soil Mechanics*, Wiley, New York, 1948.
7. H. F. Winterkorn, "Macromeritic Liquids," *A.S.T.M. Spec. Tech. Publ.* **156**, 77 (1953).
8. A. J. Batschinski, "Untersuchungen Uber die innere Reibung von Flüssigkeiten," *Z. Physik. Chem.*, **84**, 643 (1913).
9. H. Eyring, "Viscosity, Plasticity, and Diffusion as Examples of Absolute Reaction Rates," *J. Chem. Phys.*, **4**, 283 (1936).
10. O. T. Farouki and H. F. Winterkorn, "Mechanical Properties of Granular Systems," *Highway Res. Record*, **52**, 10 (1964).
11. J. C. Dutertre and H. F. Winterkorn, "Shear Phenomena in Natural Granular Materials," *Princeton Soil Eng. Res. Ser. 6*, AFCRL-66-771, Princeton University, Princeton, N.J., 1966.

12. J. Kolbuszewski, "Sand Particles and Their Density," Materials Science Club's Symposium on Densification of Particulate Materials, London, 1965.
13. A. Kezdi, Discussion of paper by Farouki and Winterkorn, *Highway Res. Record*, **52**, 42 (1964).
14. A. Kezdi, "Grundlagen einer allgemeinen Bodenphysik," *VDI Zeit.*, **108**, 5, 161 (Feb. 1966).
15. T. F. Herbst and H. F. Winterkorn, "Shear Phenomena in Granular Random Packings," *Princeton Soil Eng. Res. Ser. 2*, AFCRL-65-370, Princeton University, Princeton, N.J., 1965.
16. C. C. Furnas, "Grading Aggregates. Mathematical Relations for Beds of Broken Solids for Maximum Density," *Ind. Eng. Chem.*, **23**, 1052 (1931).
17. H. F. Winterkorn, "Shear Resistance and Equation of State for Noncohesive, Granular Macromeritic Systems," *Princeton Soil Eng. Res. Ser. 7*, AFCRL-67-0048, Princeton University, Princeton, N.J., 1966.
18. W. B. Fuller and S. E. Thompson, "The Laws of Proportioning Concrete," *Trans. A.S.C.E.*, **59**, 67 (1907).
19. G. Rothfuchs, "How to Obtain Densest Possible Asphaltic and Bituminous Mixtures?" *Bitumen*, **3**, March 1935.
20. A. T. Goldbeck and J. E. Gray, "A Method of Proportioning Concrete for Strength, Workability, and Durability," Bull. No. 11, National Crushed Stone Association, Washington, D.C., 1953.
21. Office of the Engineer, the Fukui earthquake, Hokuriku region, Japan. General Headquarters, Far East Command, February 1949.
22. A. Holl, "Thermodynamics of Granular Systems," *Highway Res. Board, Spec. Rept.* **103**, Washington, D.C., 1969, pp. 91–113.
23. H. F. Winterkorn, "Applications of the Solid and Liquid States Concepts for Macromeritic Systems, *Princeton Soil Eng. Res. Ser. 9*, AFCRL-67-0414, Princeton University, Princeton, N.J., 1967.

A Continuous Fall Method of Growing Crystals from Solution

L. G. TENSMEYER, P. W. LANDIS, AND H. F. COFFEY

The Lilly Research Laboratories, Eli Lilly & Co. Indianapolis, Indiana

Abstract

A method is described for growing crystals from solution as the crystal falls freely through a rotating liquid. The crystal describes a circular orbit about a center displaced from the axis of liquid rotation. A number of small single crystals or one large crystal can be obtained using the equipment and techniques described. Acetoxyphene, propoxyphene, penicillin V, penicillin V sulfoxide, phthalonitrile, glucagon, and testosterone hydrate, among others, have been prepared as single crystals from solution by this method.

A Personal Note

LOWELL TENSMEYER

One of Eyring's characteristics that has had a lasting effect on me is his invitation to all to join in the process of research, including his awareness that people of all ages and experience can make contributions to knowledge. He is always aware that his own work is not perfect; it can and should be improved. One illustration of this attitude occurred several times in his classes:

In the midst of lengthy derivations, as Henry carried them out on the blackboard in front of a class, anyone could feel free to correct an error. Sometimes class members would suggest that one or more assumptions in a theory could be improved. Henry's response after a short discussion was quick and sincere: "Here is some chalk and there is the side board. See what you can do with it!" In this way he avoided bogging the class down into any number of possible by-paths, but left individuals free to "do their own thing" if they chose to do so.

Another: A group had assembled at his car in preparation for a 40-mile trip to Ogden. All were to be speaking at scientific meetings at Weber State College. As he settled into the driver's seat, Henry spoke somewhat as follows: "Gentlemen, I expect each of you to speak out if you see any dangerous situation developing on the highway. I'm driving the car, but I can't see everything. Your lives are in this car just as much as mine is, and you should feel free to help me whether you're in the front seat or the back seat."

768

INTRODUCTION

When excellent single crystals are to be grown from solution, seed crystals are generally supported on a wire at or under the liquid surface, or allowed to grow on the surface of the container.* It is sometimes desirable to grow crystals from solution in such a way that the solution has free access to all surfaces of the crystal. Structure determinations via X-ray crystallography are often hampered because crystals of proper size and shape cannot be grown, as are studies of the optical and electronic properties of compounds in their solid state. When the crystal rests on the bottom of a container the development of one or more faces may be inhibited because of solvent depletion. When the seed is supported by a wire or other thread, at least part of this support material will be included in the as-grown crystal and must be cut away.

Various methods are conceivable for maintaining a growing crystal in contact with solution only. In a dielectric solution, for example, the force of gravity might be overcome by placing electric field plates above and below the solution and crystal and inducing a static charge on the crystal. The position of the crystal in the solution could be controlled by varying the electric field across the plates in a fashion similar to Millikan's oil drop experiments.

In an alternate method the crystal might be falling freely in a solution, while the solution is flowing vertically at the same velocity as the crystal is falling through it. If this "free fall" takes place in a vertical cylinder, however, a gradient of fluid velocity will exist from the axis of flow (where it is greatest) to the surface of the cylinder (where it is zero). A floating crystal will drift toward the cylinder wall and descend quickly unless the fluid velocity field can be adjusted rapidly and reversibly.

DESCRIPTION OF THE CONCEPT

When a right circular cylinder is filled completely with fluid solution and a seed crystal, then rotated at an appropriate rate with its axis horizontal,

* P. H. Egli and L. R. Johnson, in *The Art and Science of Growing Crystals*, J. J. Gilman, Ed., Wiley, New York, 1963, p. 194ff.

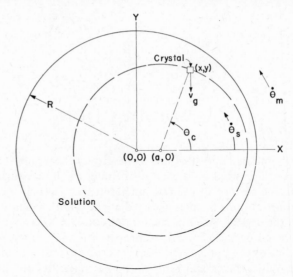

Fig. 1. Crystal falling in a uniformly rotated solution.

the crystal will also fall "freely" and continuously in the solution. As the following analysis will show, this system is a stable one within broad limits, and adjustments need be made only relatively slowly. The mechanics can be very simple.

A crystal falling in a quiescent solution rapidly achieves a terminal velocity v_g, which is a function of its size and shape, the viscosity of the solution η, the difference in density of crystal and solution $(\rho_c - \rho_s)$, and the acceleration due to gravity, g. For a nearly spherical crystal of radius r, for example,

$$v_g \sim \frac{2r^2 g(\rho_c - \rho_s)}{9\eta}. \tag{1}$$

When the cylinder is mechanically accelerated to an angular velocity $\dot{\theta}_m$, shear forces accelerate the liquid also (cf. Fig. 1). In the absence of gas bubbles, the solution eventually achieves an angular velocity of $\dot{\theta}_s$, very near $\dot{\theta}_m$, throughout its entire volume, that is, the liquid and container rotate as a whole. Each point of the solution has a linear velocity v_s, dependent on its distance r from the axis.

$$v_s = r\dot{\theta}_s. \tag{2}$$

At one point, described by $(\theta = 0, r = v_g/\dot{\theta}_s \equiv a)$, the crystal will be falling through the solution at the same velocity as the liquid is rising: $v_s = -v_g$. At every other point the crystal experiences unbalanced forces and moves with respect to any reference point outside the rotating cylinder.

The horizontal and vertical components of solution velocity are, respectively,

$$v_{sh} = v_s \sin \theta_s = r\dot{\theta}_s \sin \theta_s, \tag{3}$$

$$v_{sv} = v_s \cos \theta_s = r\dot{\theta}_s \cos \theta_s. \tag{4}$$

A crystal moving in this liquid, if its density is different from that of the liquid, has an additional vertical component of velocity v_g. There is no horizontal component due to gravity. The components of crystal velocity can be written in polar and in Cartesian coordinates as follows:

$$v_{cv} = r\dot{\theta}_s \cos \theta_s - v_g = \dot{y}_c = x\dot{\theta}_s - v_g, \tag{5}$$

$$v_{ch} = r\dot{\theta}_s \sin \theta_s = \dot{x}_c = y\dot{\theta}_s. \tag{6}$$

The (x, y) coordinates of the crystal at time t, are taken with respect to the axis of fluid rotation. These two simultaneous differential equations can be solved to yield the following expressions for the path of the crystal:

$$x = (x_0 - a) \cos \dot{\theta}_s t - y_0 \sin \dot{\theta}_s t + ra, \tag{7}$$

$$y = y_0 \cos \dot{\theta}_s t + (x_0 - a) \sin \dot{\theta}_s t. \tag{8}$$

"Initial" conditions (x_0, y_0, and $t = 0$) are chosen arbitrarily any time after the liquid achieves constant angular velocity and an essentially steady state exists.

Equations 7 and 8 describe a circle having a center at coordinates $(a, 0)$ and a radius of $[(x_0 - a)^2 + y_0^2]^{1/2}$. The falling crystal follows this circle at the same angular velocity as the fluid rotates; that is, the rotation of the crystal about $(a, 0)$ has the same period as the fluid rotating about $(0, 0)$. Where $x_0 = v_g/\dot{\theta}_s$ and $y_0 = 0$ the crystal carries out a circle of zero radius, a point at $(a, 0)$.

The equations have been verified experimentally in apparatus to be described later. After the cylinder begins rotating, only a few minutes are needed for the falling crystal to establish an orbit. The rotation rate of the crystal matches that of the mechanics, $\dot{\theta}_c = \dot{\theta}_m$, for the 1-mm crystals to within 0.02%. For even the 1-cm crystals, $\dot{\theta}_c$ never lagged $\dot{\theta}_m$ by more than 2%. Because of the fish-bowl optic effects it is difficult to verify that the crystal follows a circular orbit, but the path is repeated on every cycle and is nearly circular if not completely so.

As the crystal grows, v_g changes, usually increasing, and a slightly new circle is carried out. $\dot{\theta}_s$ can be adjusted periodically, or continuously and automatically if desired, to keep the crystal from bumping on the inside surface of the rotating chamber: that is, for an increasing v_g, $\dot{\theta}_s$ is also increased to keep $v_g/\dot{\theta}_s = a$ at nearly the same value. This adjustment need not be made often, because v_g increases approximately as the r^2 of the crystal [see (1)]; for example, a crystal left to grow at $a = 1$ cm and constant $\dot{\theta}_s$ must grow to eight times its volume before its new a value is approximately

4 cm. The latter figure is well within the mechanical radius of the rotating chamber to be described.

We have learned experimentally that the circular path of the crystal decreases in size over a period of days, and approaches a point of suspension on the x axis $(a, 0)$. This situation is not anticipated by the above equations, but it is very convenient. With no circular motion of the crystal to take into account, the value of a can be adjusted anywhere within the range $0 < a < R$ without the crystal bumping the walls, where R is the radius of the rotating chamber.

It should be possible to derive more fundamental equations using the forces acting upon the crystal, rather than the velocities we have assumed. The integrated equations would then include the short-term and long-term transient effects noted above. This approach, however, has not yet been tractable in our hands.

EXPERIMENTAL APPARATUS AND PROCEDURES

Two devices have been constructed for rotating containers about a horizontal axis in constant-temperature baths. In the first device the volume of

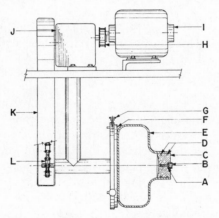

Fig. 2. Schematic of crystallizer. A. O-ring seal for bleeder hole at B. B. Bleeder hole and closure screw. C. Expansion plug. D. O-rings for expansion plug seal. E. Glass crystal-growing chamber. F. Aluminum ring glued to glass chamber. G. Set screws holding ring F to plate E. H. Belt drive. I. Drive motor. J. Variable speed transmission. K. Chain drive guard. L. Chain and sprocket drive.

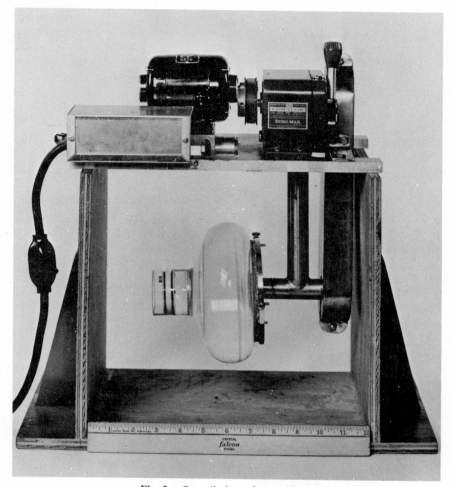

Fig. 3. Overall view of crystallizer.

solution is large, 785 ml. Volumes from 0.5 to 46 ml have been utilized in the second.

The first mechanism, Fig. 2, illustrates the principle rather directly. Its larger volume permits the growth of crystals whose solubility or temperature dependence of solubility is small, as well as of larger crystals, 1 cm³ and larger. The mechanism is shown supported by a wooden frame in Fig. 3, where preparations can be performed before placement in the temperature bath.

The growth chamber of this device is a squat cylindrical glass bowl (*E*, Fig. 2) 16.0 cm in diameter, 5 cm in height. At its axis is a precision bore neck of 5.0-cm i.d. for filling and pressure response. After filling the growth chamber with saturated solution the neck is closed by a Teflon plug, *C*, which has a threaded hole in it for the removal of gas bubbles and the insertion of seed. A small Teflon bolt, *B*, then closes the threaded hole on a small O-ring, *A*. The Teflon plug glides smoothly in or out on two O-ring seals, *D*, in response to volume changes as the temperature is varied to effect crystal growth or dissolution.

Following closure, the bowl is attached to a rotatable stainless steel plate. A ring, *F*, cemented to the glass bowl has three pins which fit into corresponding holes in the plate. Thumbscrews, *G*, clamp the ring and plate together.

The drive shaft and bowl assembly are designed to function under water in the temperature bath. Supported above the water are the drive motor, *I*, timing belt and pulleys, *H*, the speed controller, *J*, and upper section of the chain drive housing, *K*. The use of a glass water bath and glass growth chamber permits the operator to see the crystal in motion. With submillimeter seeds, the initial rotation rate is several rpm and can be increased as needed.

A more frontal view of the bowl in rotation is shown in Fig. 4. This

Fig. 4. Simulated crystal suspended by rotating solution.

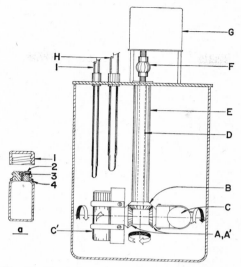

Fig. 5. Second version of crystallizer. A,A'. Miter gears (rotate about horizontal axis). B. Miter gear (fixed to stationary shaft E). C,C. Crystal-growing cells (cross section drawing a). D. Rotating drive shaft. E. Stationary shaft. F. Flexible coupling. G. Drive motor. H. Thermoregulator. I. Thermometer. 1. Screw cap. 2. Teflon plug. 3. O-ring seal (bleed hole to remove air). 4. O-ring seal.

photograph shows the bowl completely filled with water, which sustains a simulated crystal in "suspension" at 30 revolutions per minute.

In the second mechanism, Fig. 5, crystal growing cells C and C' rotate also about horizontal axes. In addition, the entire unit rotates around a vertical axis, stirring a temperature-controlled (water) bath. Vertical rotation is transmitted into horizontal rotation through miter gears B, A, and A'. In this apparatus, both vertical and horizontal rotations have the same angular velocity, $\dot{\theta}_m$.

A first model of this design used a constant-speed 3 rpm motor, fixing $\dot{\theta}_m$ at approximately 0.3 radians sec^{-1}. A variable speed DC motor permits variations in $\dot{\theta}_m$ and the suspension of a greater range of crystal sizes.

CRYSTALS GROWN

The compounds that have been prepared as single crystals by the "free fall" method are listed in Table I. Acetoxyphene and propoxyphene were

Table I. Crystals Grown

Substance	Solvent	Temperature Range (°C)	Dimensions (mm)	v_g (cm sec^{-1})	θ_m (radians sec^{-1})
Phthalonitrile	DMSO	25–21	0.5 × 2.0 needles	0.55	0.3
Acetoxyphene[a]	DMSO	25–15	0.5 × 1.0	0.6	0.3
	55–45% H$_2$O–iso-propanol	24.8–19.6	12 × 8 × 6	17.4	2.7
Propoxyphene[a]	DMSO	30–20	1 × 1 × 2	0.1	0.3
		26–23	9 × 6 × 8	2.1	1.0
	(C$_2$H$_5$)$_2$O	26–24	7 × 4 × 4	13.2	4.2
Penicillin V sulfoxide	DMSO	26.0–25.0	5 × 5 × 3	2.7	1.4
	50–50% DMSO–CH$_3$OH	15.0–11.0	10 × 5 × 4 7 × 7 × 6	6.5	2.9
	20–80% DMSO–CH$_3$OH	25–7.0	2 × 1 × ½	9.2	3.1
Penicillin V	Isopropanol	15.0–10.0	1.7 × 1.0 × 0.6	3.7	1.6
Glucagon	H$_2$O; buffered to pH 10.2 with KH$_2$PO$_4$	27–21	0.2	0.1	0.3
Testosterone·H$_2$O	H$_2$O	33.0–29.5	0.2	0.4	0.3

[a] H. R. Sullivan, J. R. Beck, and A. Pohland, *J. Org. Chem.*, **28,** 2381 (1963). α-*d*-Acetoxy-phene and α-*d*-propoxyphene are the acetyl and propionyl esters, respectively, of (2*S*:3*R*)-4-dimethylamino-1,2-diphenyl-3-methyl-2-hydroxybutane.

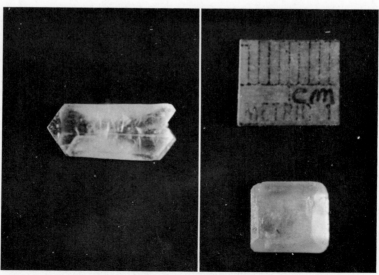

Fig. 6. Acetoxyphene, left, and propoxyphene crystals, ∼2× magnification.

Fig. 7. Phthalonitrile crystal, 50× magnification.

chosen for initial tests of the concept because they readily grow as crystals. Penicillin V sulfoxide was grown to purify the compound and to provide crystals for X-ray structure determinations. The sulfoxide grew large crystals well without any particular difficulties.

The other compounds listed in Table I were prepared primarily for structure determinations and were known to be difficult to obtain as single

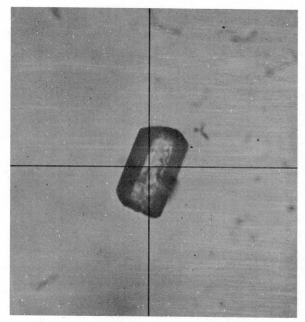

Fig. 8. Testosterone hydrate crystal, 440× magnification.

crystals by other methods. In each case, small individual crystals with fully developed facets were the result. Temperatures were generally lowered at a rate of 0.5°C/hr or less. Occasionally the temperature was maintained constant for 72 hr to allow adequate time for the growth process to occur. Crystals are illustrated in Figs. 6–8.

LIMITATIONS AND EXTENSIONS

The liquids used in crystal growth must be scrupulously clean of dust particles. If the supersaturation is sufficiently great, such particles will act as nuclei additional to the seed crystal that was introduced. Any fine particles will be trapped against the top surface of the crystal by eddy currents. Such currents are produced as the crystal falls relative to the liquid. Ultimately the particles are occluded in the growing crystal itself, deteriorating crystal quality. All the particles in the solution will be caught in this trap; the falling, growing crystal acts as a "getter" for small particles. When the size of two or more crystals grown in the same solution approaches 1 mm the largest of the crystals will also trap the others in its eddy currents. Such crystals will grow together essentially in a random fashion. When large crystals are sought, the method is limited to the growth of one crystal in a chamber at a time.

Imperfections caused by rapid growth in the early stages as by the use of imperfect seeds are greatly reduced in this method. Rapid initial growth is generally prevented by raising the solution above its saturation temperature before or after the seed is introduced, but care must be taken not to dissolve the seed away from its support. In the continuous fall method, the temperature can be maintained appropriately to dissolve the crystal until it is almost invisible. Gradual temperature decrease then allows the tiny crystal to achieve maximum perfection.

In this article the continuous fall approach to growing crystals has been illustrated only with organic compounds near room temperature. The method has application to any precipitating system, however, in any temperature range where the solution can be contained and rotated around a horizontal axis.

ACKNOWLEDGMENTS

We appreciate the excellent technical help of R. F. Miller who provided ideas and skill in glass handling. W. E. Sherman and P. D. Vernon assisted in growing the crystals.

Contributions of Professor Henry Eyring to the Theory of Detonation

MELVIN A. COOK

IRECO Chemicals, Salt Lake City, Utah

779

EYRING SURFACE BURNING MODEL

Perhaps the most significant contribution of Dr. Henry Eyring to detonation theory is his surface burning concept.[1] In a sense this was a rediscovery of "das Piobertsche gesetz" of 1839,[2] except that the Eyring model was characteristically mathematically defined and more carefully articulated. All solid explosives, whether in granular, pressed, or cast form, are composed of grains which in practical systems may vary in size from microns to millimeters. The explosive may comprise grains having a combination of sizes, a few selected sizes, or it may even comprise grains of uniform, single particle size. This model assumes that the surface of each grain is exposed to the high temperature of detonation for such a short time that heat conduction is negligible into the solid grain ahead of the burning surface. On the other hand, conduction in the gas phase is so rapid that the solid surface is considered to be in thermal equilibrium with respect to the gaseous products of detonation surrounding the grain. This thermal profile provides a distinguishing difference between detonation and explosive combustion. In the latter there exists a characteristic—not too steep—thermal gradient outward from the burning surface into the gaseous products. (The steepness of this gradient increases in approximate proportion to the pressure.) Thus the thermal profile may be characterized as infinite in detonation and finite in explosive combustion. Chemical reactions in detonation are confined to the surface of the grain and proceed layer by layer into each individual grain (at a constant radial rate) until it is completely consumed.

HOMOGENEOUS GRANULAR EXPLOSIVES

The theory is most readily illustrated by reference to a pure explosive compound comprising uniform spherical grains of constant size. By studying systems of this character the author and associates[3,4] established beyond reasonable doubt the validity of the Eyring surface burning model.

Consider an explosive comprising uniform spherical grains all of the same size and radius. According to the Eyring model, chemical reaction in the detonation wave occurs via a particular very fast combustion flame proceeding radially inward toward the center of the grain. The rate of reaction thus follows the equation

$$\frac{dn}{dt} = k_r(T)\frac{4\pi r^2}{s},$$ (1)

where r is the radius of the burning grain at time t, s is the effective cross-sectional area of a molecule of the explosive, and $k_r(T)$ is the specific reaction rate constant. The Eyring absolute reaction rate theory defines k_r by the equation

$$k_r = \frac{kT}{\hbar} \exp\left(\frac{-\Delta F^{\ddagger}}{RT}\right),$$ (2)

where k, $\hbar$, R, and $\Delta F^{\ddagger}$ have their usual meanings as Boltzmann's constant, Planck's constant, the gas constant, and the free energy of activation, respectively. The number of molecules reacting per second is the time rate of changes of the number of molecules in the sphere; that is,

$$\frac{dn}{dt} = \frac{d}{dt}\left(\frac{4\pi r^3}{3v}\right) = \frac{-4\pi r^2}{v}\frac{dr}{dt}.$$ (3)

Thus

$$\frac{dr}{dt} = -\frac{k_r v}{s}$$ (4)

v here being the volume of a molecule as distinguished from its usage as specific volume in the equation of state. For isothermal reactions the grain radius r decreases at constant rate. (The phenomenological "detonation head" model outlined below provides some justification for the constant temperature assumption used in the Eyring model.) The diameter, λ, of a molecule of explosive is v/s, and the total reaction time of the homogeneous granular explosive is thus

$$\tau = \frac{r_g}{k_r \lambda},$$ (5)

where r_g is the initial grain radius.

The fraction of actual reaction N completed at a given time is

$$N = 1 - \left(\frac{r}{r_g}\right)^3$$ (6)

so that

$$\frac{dn}{dt} = \frac{3k_r \lambda}{r_g}(1 - N)^{2/3},$$ (7)

which on integration gives

$$N = 1 - \left(1 - \frac{t}{\tau}\right)^3.$$ (8)

Thus the Eyring surface erosion model predicts that chemical reactions in the detonation of spherically grained solid explosives will follow a two-thirds order law.

This theory is singularly useful in the "detonation head" model.[3] This model assumes that the density, pressure, particle velocity, etc., all remain constant within the so-called *detonation head* bounded in front by the detonation wavefront, and in the rear and sides by rear and lateral release waves which move into the products of detonation, the latter starting at the wavefront near the periphery of the charge (one must allow for a small "edge effect") and moving at the velocity $(D + W)/2$ and the former moving forward at the same speed (the Langweiler model[5]). Thus, the detonation head comprises, in steady state, a conical region with base foremost through which material flows while reacting chemically. In unconfined charges propagation must proceed a length of about 3.5 charge diameters before this steady detonation head is attained. Before this steady-state stage the detonation head comprises ever-growing truncated spherical cones of length h proportional to the length of propagation. Incidentally, flash radiographs have clearly revealed outlines of the detonation head. Moreover, the observed plasma character of detonation reaction zones in condensed explosives seems also to justify the ultrahigh thermal conductivity required to account for the thermal, pressure, and particle velocity "flats" in the detonation head.[3,6]

Now, in the detonation head model detonation remains *ideal*, that is, $D = D^*$, where D^* is the maximum or "hydrodynamic" velocity, only as long as the reaction zone length a is equal to or less than the length h of the detonation head at its maximum, or along the cylindrical axis. On the other hand, when the reaction zone exceeds h in length, detonation becomes *nonideal*, that is, $D < D^*$, because the effective reaction, or that part supporting the wave propagation, suddenly cuts off as the burning grain leaves the detonation head. (This does not necessarily mean that chemical reaction does not continue on to completion outside the detonation head.) An equation of state, for example, the $\alpha(v)$ equation of state,

$$pv = nRT + \alpha(v)p \tag{9}$$

may then be used to relate D/D^* to N. The result[3] is

$$\frac{D}{D^*} = N^{1/2}. \tag{10}$$

The time τ that a grain is burning inside the detonation head is

$$\tau = h(D - W) \doteq \frac{4h}{3D}, \tag{11}$$

allowing for the forward motion at particle velocity $W(\sim D/4)$ during the time the particle is inside the detonation head. Therefore from (8), (10), and (11) we obtain

$$\frac{D}{D^*} = \left[1 - \left(\frac{4h}{3D\tau}\right)^3\right]^{1/2}, \tag{12}$$

where, as described above and in the theory of the detonation head,[3] h is geometrically related to the charge diameter by the relations

$$h = h(L, d) = \begin{cases} L/3.5d', & L < L_m, \\ d' & L > L_m, \end{cases} \tag{13}$$

L = charge length, $L_m = 3.5d$, d = charge diameter, $d' = d - 0.6$ cm, where the 0.6 cm is the observed "edge effect"; that is, the effective diameter d' is 0.6 cm less than the real diameter in unconfined solid charges because detonation does not extend all the way to the periphery of the charge. This value was determined from end impulse and radiographic studies.

The velocity–diameter $[D(d)]$ results for granular TNT bear out the Eyring theory remarkably well.[4] (Many other explosives were studied with comparable results.[7]) The results for TNT may be summarized as follows:

1. The velocity–diameter [or $D(d)$] curves computed from (12) and (13) for granular TNT at a given density agreed well with the observed ones for the five different granulations studied, that is, for particle radii r_g ranging from 0.009 to 0.20.

2. For any given D/D^* ratio the corresponding observed diameter d agreed closely with the predicted one of the surface burning model; that is, the effective time a particle will burn within the detonation wave should, in this model, be proportional to the *effective* diameter d' ($= d - 0.6$ cm). Thus

$$d_f = c_f r_g + 0.6 \text{ cm}, \tag{14}$$

where d_f is the charge diameter at which $D/D^* = f$ and c_f is a constant for that particular D. The critical diameter d_c occurs in granular solid explosives at about $D/D^* = 0.75$ because smaller values of D are not generally observed for pure explosives like TNT. As examples, at $D/D^* = 0.75$ we find (for d_f in centimeters) that $c_f = 23.5$; at $D'/D^* = 0.9$ (14) applies with $c_f = 33.5$, etc., and we also find that the minimum diameter for ideal detonation d_m^*, that is, where $D/D^* = 1.0$, follows (14) with $c_f \sim 90$ for granular TNT.

LIQUID EXPLOSIVES AND THE SURFACE BURNING MODEL

In superficial considerations one might question that the surface burning model could play any important role in the chemical reactions supporting wave propagation in liquid explosives. It turns out, however, that the detonation reactions in liquid explosives do indeed follow the Eyring surface burning model; the limitations imposed by high density and lack of free space in liquid explosives give rise to a peculiar but fascinating mode of surface reaction. Framing camera sequences by Mallory and McEwan employing principles of transmission of light in transparent plastics under shock wave compression[8,9] showed that the chemical reactions in the detonation wave of liquid explosives are apparently ignited at points in the body of the liquid created by the detonation wave itself apparently by an orderly mode of cavitation, because these reaction centers occurred in remarkably ordered, lattice-like arrays. Chemical reaction then proceeds by surface burning in spherically expanding propagation from these reaction centers. They are revealed by the fact that the reaction "bubbles" eventually merge into slivers in the final stages of reaction. They are best observed at the surface of the explosive column as remarkably ordered, lattice-like arrays of points and slivering jets. Although the orderly geometrical patterns of initiation centers were not anticipated (they had been recorded but went unobserved in many earlier framing camera sequences), the point initiation mechanism was postulated earlier to explain the peculiar velocity–diameter curves observed in nitromethane and Dithekite (nitric acid–nitrobenzene). (Data on the ratio of the radius of curvature of the wavefront to the charge diameter (R/d) for nitromethane in 6-mil tubing confirmed also the result mentioned below that R/d is about 0.5 at the critical diameter d_c and about 3.5 at $d \gg d_c$ in liquid as well as solid explosives.) In the geometrical model for point initiation in liquid explosives, the condition

$$\frac{D}{D^*} = N^{1/2},$$
(15)

still applies. In the expanding reaction, however, the radius will be r at time t such that

$$\frac{t}{\tau} = \frac{r}{r_g}.$$
(16)

On the other hand, the fraction N of the explosive that has reacted at time t is given in this case by

$$N = \left(\frac{r}{r_g}\right)^3.$$
(17)

Therefore

$$\frac{t}{\tau} = N^{1/3} = \left(\frac{D}{D^*}\right)^{2/3}.$$
(18)

The time t may be eliminated by using again the relation

$$t = \frac{h}{d - W} \simeq \frac{4h}{3D}$$
(11)

of the geometrical model. These equations, of course, apply only when $t < \tau$; if $t > \tau$ detonation is always ideal (or $D/D^* = 1.0$). Therefore, employing the geometrical relationships of the detonation head model ($d = d' - 0.5$ and $h \doteq d'$), we obtain

$$d = \frac{3D^*\tau}{4} \left(\frac{D}{D^*}\right)^{5/3} + 0.5 \text{ cm.} \tag{19}$$

The value of 0.5 cm is the observed edge effect in nitromethane; it may be seen in liquids in high speed framing camera photographs of the actual wavefront.

The importance of reacting surfaces in liquid and gelatin explosives has long been known from experience with liquid nitroglycerine and gelatin dynamites. Aeration is an important aspect of the sensitivity of these liquid and semiliquid explosives.[3] More recently this has been reemphasized by the observed effects of aeration on the sensitivity of slurry explosives sensitized by aluminum or other nonexplosive fuels.[10] The sensitivity, indeed the propagation of the detonation wave, in these slurry explosives is initially associated with small gas bubbles which act as reaction centers for the essential chemical reactions.

These and many other effects of aeration, cavitation, and free surfaces in detonation confirm the apparent complete generality of the Eyring surface erosion concept in the chemical reaction zones of condensed explosives.

The Eyring surface erosion model was applied together with the general absolute reaction rate theory to the rate of mass loss of meteors as they pass through the atmosphere.[11,12] Atmospheric erosion of the "heat shield" of space vehicles upon reentry is a problem of great importance to which this theory is directly applicable.

WAVE SHAPE IN DETONATION

Professor Eyring and associates[1] also introduced novel considerations concerning the relationship between the shape of the detonation wavefront and the rate of chemical reaction in the supporting reaction zone. While the predictions of the "curved front" theory turned out to be incorrect, it may yet point the direction of the correct answer because the discrepancy probably lies only in the basic assumptions, not in the fundamental character of the theory.

Employing the continuity equation, they predicted that the wavefront will be flat when the reaction zone length is negligible and will curve more and more as the length of the reaction zone increases. They reasoned that, at the edge of the charge, a rarefaction wave would be sent into the reaction zone, and since the local velocity of sound in the region should be greater than the detonation velocity, the rarefaction wave should overtake the detonation wave and slow down the velocity of the edge of the wavefront, thus giving rise to a curved front. Accordingly, the detonation velocity would be less than the ideal value because of the curvature of the wave. This process was considered to continue until the angle of intersection of the wavefront with the edge of the charge becomes small enough so that the rarefaction wave is no longer reflected.

It was assumed that any small portion of the wavefront may be considered spherical with a local radius of curvature r_0 and that the hydrodynamic equations appropriate to these conditions, with the exception of the equation of continuity, are, to a good approximation, the same as for a plane wave. The equation of continuity

$$(D - W)\frac{dv}{dr} = \frac{2vW}{r} - v\frac{dW}{dr} \tag{20}$$

is thus the equation for a spherical detonation wave, where D is here the velocity normal to the wave. Since a steady-state solution is desired, that is, the detonation wave is to proceed without change in shape, the further condition

$$D = D_0 \cos \alpha \tag{21}$$

must be imposed, where D_0 is the propagation velocity of the wave itself in the direction parallel to the axis, and α is the angle between the axis of the charge and the normal to the wave at any point.

The normal detonation velocity D_0 is then computed from the perturbed hydrodynamic equations. With an expression for the ideal detonation velocity D^*, it is then possible to relate the ratio D^*/D_0 to the reaction-zone length and the charge radius. The final formulation of this theory by Parlin and Robinson[12] employed the $\alpha(v)$ equation of state. The final results of this formulation were as follows:

$$\frac{y}{a_0} = (M + N)K_c - N, \tag{22}$$

where

$$K_c = \left[\frac{2}{\delta(1 - \delta^2)^{1/2}}\right] \tan^{-1}\left[\frac{(1 + \delta)}{(1 - \delta)}\right]^{1/2} - \frac{\pi}{2\delta},$$

$$\delta = \frac{D}{D^*}; \qquad M' = (1 + L)^{-1}; \qquad N' = 2[(L' + 1) - L]^{-1},$$

$$L = \frac{\acute{C}_v}{nRe^x} + 1; \qquad L = \frac{v_1}{v_2} - 1; \qquad M = 0.7M' \qquad N = 0.7N'.$$

Integration of (20) gives

$$\frac{D - W}{v} = \frac{D}{v_1^{-2}}\int_{r_0}^{r_0 - a_0} \frac{W}{vr}\,dr$$

Defining $D - W = U$ and

$$\Omega = 1 + \frac{2v_1}{D}\int_{r_0}^{r_0 - a_0} \frac{W}{vr}\, dr$$

the equation of continuity if $U/D = v/v_1\Omega$. In the approximation $v_2/v_1 = 0.75$, therefore, $\Omega = 1 - 0.5a_0/r_0$, and Eyring et al. thus obtained the approximate result

$$\frac{D_0}{D^*} = 1 - \frac{a_0}{d}.\tag{23}$$

The value of a_0 may thus be obtained as the slope of the D against d^{-1} curve, which should thus be a straight line.

The curved-front theory permits the derivation of parametric equations defining the shape of the wavefront through the (21) and (22). Hence, the theory is subject to evaluation not only by observed $D(L, d)$ data but also by observed wave shapes.

Actually stable, plane wavefronts do not exist, at least in condensed explosives, as shown by Cook et al.[3,4] Below are summarized some of the important observed results of these studies.

1. The wavefront emerging from the end of an unconfined cylindrical charge is in general spherical in shape (i.e., it was a spherical segment) both in ideal and nonideal explosives, except at the very edge of the charge where slight edge effects may sometimes be observed.

2. The radius of curvature R of the spherical wavefront for point initiation of a cylindrical charge increases at first geometrically ($R = L$) but quickly settles down to a constant or steady-state value R_m significantly at $L \gg L_m$.

3. The steady-state curvature–diameter ratio R_m/d varies from about 0.5 at the critical diameter d_c to a maximum of 3–4 at $d \gg d_c$. (The unconfined critical diameter for propagation of the detonation wave varies from about the edge effect value 0.6 cm for ideal explosives of very high reaction rate to very large values for nonideal explosives of low reaction rate, e.g., it is about 10–15 cm for pure ammonium nitrate of $-65 + 100$ mesh particle size.) The maximum values of R/d in ideal detonation were observed only at diameters well above d_m^*, the maximum diameter for ideal detonation.

4. The wave shape observed at large L/d was independent of the type of initiator used or the initial wave shape. While one may, by the use of appropriate wave-shaping boosters, initiate a charge to propagate initially with almost any desired wave shape, as L increases the shape of the wavefront quickly reverts to the steady-state spherical one of $R = R_m$ characteristic of the explosive.

In ideal explosives R_m/d generally fell between 2.0 and 3.5. Hence one makes no appreciable error in discussions of ideal explosives to treat the wavefront as plane. However, the assumption of plane wavefronts may entail difficulty in nonideal explosives, particularly in discussions concerning the region of the critical diameter where R_m/d approaches 0.5, or in ideal explosives of small charge length.

The above facts permit us to write the following equations pertaining to wave shape:

$$R_i(y_i) = \text{constant}; \qquad y < y', \tag{24}$$

where R_i = radius of curvature of the wave at a particular charge length and at a point on the wavefront a distance y_i perpendicular to the charge axis; y' is the effective radius of the charge, defining effective radius to exclude the slight edge effect. Equation 24 simply expresses the experimental fact that the wavefront is in general spherical in shape. There is no question regarding the validity of this result, especially in large-diameter charges, for example, where $d \geq 7.5$ cm, in which the resolution is especially good.

Next we have

$$\begin{aligned} R &\cong L; & L &> R_m, \\ R &= R_m = \text{constant}; & L &> R_m. \end{aligned} \tag{25}$$

Equation 25 expresses the facts that the spherical wavefront expands geometrically for a length nearly up to R_m and then settles down surprisingly rapidly at $L \sim R_m$ to the steady-state value R_m. For theoretical purposes the assumption of a sharp, discontinuous change from spherical expansion ($R = L$) to the steady-state wavefront ($R = R_m$) is reliable almost within experimental error.

As Eyring predicted, wave shape is indeed dependent on chemical reaction rates in detonation. However, the fact that the detonation wavefront is generally spherical in shape over the entire front irrespective of the reaction rate, and the further fact that it is transient only over a length of propagation ranging from $0.5d$ for the more slowly reacting explosives to about $3.5d$ for the fastest reacting ones, are still not fully understood. It has been suggested that these peculiar effects are associated with the plasma character of the detonation reaction zone and the high free electron concentrations associated therewith. Concentrations of free electrons as high as $10^{21}/\text{cc}$ have been measured in the detonation reaction zone of explosives of very high brisance, and it has been shown that every molecule of explosive generates at least one free electron upon decomposition in the detonation reaction zone. The most significant influence of the plasma condition is, of course, high heat conduction. This appears to be important in justifying the infinite thermal gradient at the reaction surface in the surface burning model. It may also have the effect of slowing down release waves so that they no longer tend to eat into and weaken the wavefront. This would also explain the success of the phenomenological Langweiler wave underlying the detonation head model.

REFERENCES

1. H. Eyring, R. E. Powell, C. H. Duffey, and R. B. Parlin, *Chem. Rev.*, **45,** 16 (199).
2. C Cranz, *Lehrbuch der Ballistik*, Vol. II, 1926, p. 120.
3. M. A. Cook, *The Science of High Explosives*, ACS Monograph No. 139, Reinhold, New York, 1958, Chap. 6.
4. M. A. Cook, G. S. Horsley, W. S. Partridge, and W. O. Ursenbach, *J. Chem. Phys,.* **24,** 60 (1916).
5. H. Langweiler, *Z Tech. Physik*, **19,** 271 (1938).
6. A. Bauer, M. A. Cook, and R. T. Keyes, *Proc. Roy. Soc.* (*London*), **A259,** 508 (1961).
7. R. T. Keyes and M. A. Cook, *J. Chem. Phys.*, **24,** 191 (1956).
8. M. A. Cook, "Explosion Hazards in Propellants," in *Cryogenic Technology*, Wiley, New York, 1963, Chap. 13, pp. 411–416.
9. H. D. Mallory and W. S. McEwan, *J. Appl. Phys.*, **32,** 2421 (1961).
10. M. A. Cook, *Ind. Eng. Chem.*, July, 1968, pp. 44–55.
11. M. A. Cook, H. Eyring, and R. N. Thomas, *Astron. J.*, **113,** 475 (May 1951).
12. R. B. Parlin and D. W. Robinson, "Effect of Charge Radius on Detonation Velocity," TR No. 7, Contract 45107, Project No. 357 239, University of Utah, October, 3, 1952.

The Pore Structure of Catalysts and Catalytic Reaction Rates

IZUMI HIGUCHI

Department of Synthetic Chemistry, Faculty of Engineering, Shizuoka University, Japan

A Personal Note

The 29 months that I spent at Professor Eyring's laboratory were the most memorable of my life. Even after 20 years of teaching and research at Tohoku University I still had doubts concerning the attitudes and way of life of scientists. This was probably because modern science had only recently been introduced to Japan and the stress had been placed on the technical application of scientific knowledge.

Not only did my discussions with Professor Eyring further my study of catalytic reaction rates but my personal contact with him helped me to acquire a viewpoint that continues to guide me through my research life. This was the most important aspect of my experience at the University of Utah.

Once we were discussing a manuscript which described our research and which contained a criticism of earlier researchers. Professor Eyring asked, "Why should we be so concerned with the mistakes of others? I myself make mistakes from time to time. Science progresses with the cooperation of many scientists." This was the only reprimand I received from this great teacher, who usually chided only with a sense of humor. It was this experience with Dr. Eyring that, like an enlightening word of a Zen priest in Japan, freed me from my long-pending doubt.

INTRODUCTION

Even though most solid catalysts are porous bodies, few research reports have been concerned with the relationships between catalytic activity and pore structure. Chemical engineers are now studying the relationships between mass and heat transfer and the pore structure of catalysts used in reactor devices.[1] The usual models of pore structure, however, are too simple to provide fundamental knowledge of the catalyst. The pore structure of a catalyst is related not only to the dispersion state of the solid but also to its surface structure, and so we believe that a study of the porous structure of a solid catalyst is an important fundamental research.

In our efforts to verify deductively our modified capillary condensation theory we found two commercial porous bodies[2,3] in which spherical elementary particles are arranged in a definite pattern. Therefore, if the radius of the elementary particles and their packing are given, a whole model of pore structure is clearly available for these specimens. By using these specimens as a catalyst or a catalyst carrier a series of investigations was carried out on catalytic activity in relation to the pore structure.

The following is a brief summary of our studies in which the catalytic activity of these special porous bodies is discussed in relation to the pore structure of catalysts.

A NEW CAPILLARY CONDENSATION THEORY AND THE PORE STRUCTURE OF SORBENT

During the period 1937–1940[4] Higuchi proposed a modified capillary condensation theory to explain the isotherms of 18 sorbates on titania gel of the same lot. The new theory proposes that in sorption phenomena vapors may be adsorbed in two ways: (a) adsorption due to the surface force of solid sorbents which is usually accomplished by forming a monomolecular film in the relatively low pressure range and (b) capillary condensation of sorbates into pores whose radii are larger than ca. 10 Å and covered by an adsorption film. The capillary condensation is undoubtedly due to the vapor pressure depression of the sorbate liquid described by the Thompson equation.

793

We have confirmed the theory in several ways:

1. Experimental results[5] show a larger dielectric polarization of capillary liquid, compared with that of the adsorption film which has lost its orientational polarization, especially in the lower temperature range.

2. Anomalous dispersion[6] of the capillary liquid appears in a similar temperature range to that observed in bulk liquid.

3. There is good agreement between the theoretical formula derived[7] and the experimental results of freezing point depression of the sorbates condensed in pores as liquid.

The new capillary condensation theory, if essentially valid, claims that the shape of isotherms measured up to saturation, that is, $x = P/P_0 = 1$, is determined by the pore size distribution of porous bodies, and so any theory to explain sorption isotherms by thermodynamic or kinetic mechanisms becomes meaningless except with respect to the formation of monolayer adsorption. Therefore an important problem in sorption is to investigate the pore structure of sorbent specimens, which are easily varied by varying the conditions of their preparation, and to elucidate the pore structure in relation to the material properties.

In order to verify deductively the capillary condensation theory, we calculated the relation between the radius of curvature r of the spherical meniscus subtending between elementary spherical particles and the space volume V in a contact zone of the particles arranged in a definite pattern. On the basis of these values for $r-V$ relations we calculated theoretical condensation isotherms which depend on the type of packing as well as on the radius of elementary catalyst particles. Isotherms were measured on compressed specimens consisting of carbon black or commercial fine silica powder of known radius, and good experimental agreement was found with the theoretically derived isotherms for simple cubic structures. Thus the validity of the capillary condensation theory has been amply verified. In addition, we could conclude that most porous oxides, in view of the shape of their isotherms, should be coagulators consisting of fine elementary particles, leaving pore space in their contact zones. Accordingly, we describe first a series of theoretical and experimental procedures by which we reached such conclusions.

THEORETICAL CAPILLARY CONDENSATION ISOTHERMS AND IDEAL POROUS BODIES CONSISTING OF ELEMENTARY PARTICLES ARRANGED IN A DEFINITE PACKING TYPE

In 1952,[8] Higuchi and Utsugi verified capillary condensation in contact zones of nonporous fine particles by comparing experimental isotherms with

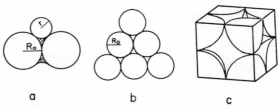

a   b   c

Fig. 1. Capillary condensation in the contact zone of elementary particles of radius R_0 and two particle packing types: (a) shaded zone is a condensing volume enclosed by a hypothetical rolling sphere of radius r in contact with larger spheres of radius R_0; (b) P.P. type; (c) S.C. type.

the theoretical ones and concluded further that fine particles would tend to be packed in a loose state. Many authors[9] worked with this concept to calculate more accurate values for r–V relations. However, Kiselev et al. were the only ones to obtain a theoretical condensation isotherm corresponding to the experimental one. In my laboratory we studied the pore structure of catalysts.

The interstitial volume around contact zones of spheres which are arranged in a definite packing type are calculated as follows: We assume that a sorbate condenses in the contact zones of spheres of radius R_0, as exhibited by the shaded zone in Fig. 1a. The vapor pressure, P, or relative pressure, x, of the condensed liquid of volume V is given by the Thompson equation.

$$x = \frac{P}{P_0} = \exp\left(\frac{-2\gamma V}{rRT}\right). \tag{1}$$

The condensed liquid volume, V, is dependent on R_0 and the radius, r, of a rolling sphere. However, if we take the ratio of the interstitial volume and the volume of particles, that is, V (cc/cc), it is dependent only on $y = r/R_0$ and is independent of R_0. Accordingly, we calculated y–V relationships for four packing types: closest packing (C.P.), simple cubic lattice (S.C.), two-dimensional closest packing (P.P.), and two-dimensional square packing (S.P.).* Calculations for two- or three-dimensional packing are quite complex and are carried out in two or three steps. The calculation formulas and numerical values are given in detail elsewhere.[12]

---

* In the first paper in 1952 we calculated the y–V relationship for the linear packing type. In this case, as criticized by Emmett in a private communication and later by Barrer et al.,[9] an approximation using (1) can lead to serious error. In addition, this type of packing is useless in actual applications, and so it was abandoned. In cases of the denser packing states considered here, errors due to (1) are not so large because a true spherical meniscus soon appears which encloses most of the sorbate liquid.

Theoretical Sorption Isotherms

When R_0 and a packing type are given, we obtain the amount of condensate a_c by (2) from the y–V relation.

$$a_c \,(\mathrm{mg/g}) = 10^3 \, V \, \frac{\rho_L}{\rho_S}, \tag{2}$$

where ρ_L and ρ_S are densities of the sorbate liquid and the sorbent, respectively. The capillary condensation occurs on a monomolecular adsorption layer and so the total sorption amount, a (mg/g), is given by (3):

$$a = a_m + a_c, \tag{3}$$

where a_m is the BET capacity, that is, the amount of sorption needed to form a monomolecular layer. From a value of y corresponding to the chosen value of V, we have $r = yR_0$, and from (1) we obtain an equilibrium relative pressure x corresponding to a. Repeating these calculations, we obtain a theoretical sorption isotherm for any of the packing types.

Theoretical Desorption Isotherms

If there are narrow entrances holding wider holes in a porous body, the Kraemer-McBain mechanism anticipates a hysteresis in the desorption branch. Among the four models given above, P.P. and S.P. porous bodies, which are representative of loose packings, should not show hysteresis; they should give reversible isotherms according to this theory. In a C.P. or S.C. porous body, however, there are narrower entrances and wider inner holes surrounded by 4, 6, or 8 spherical particles. Therefore menisci subtending entrances to spaces between elementary particles in triangular or square arrangements will not be destroyed until equilibrium pressure is reduced sufficiently to break the menisci. The desorption branch, showing practically no desorption at first, suddenly coincides with the sorption branch at the point by abruptly emptying the 4 or 8 R_0 holes. This desorption phenomenon gives a large hysteresis loop in the isotherms.

Theoretical Condensation Isotherms

Calculated theoretical isotherms of methanol at 0°C on porous bodies consisting of spherical carbon blacks of various radii are shown in Fig. 2, where spheres are in C.P. or S.C. configurations and in Fig. 3, where spheres are P.P. or S.P. Values of a_m used in these calculations are reversely given from R_0 by using (4),

$$\Sigma = \frac{3}{\rho_S R_0} = 10^{-3} \, \frac{\sigma N_0 a_m}{M} \tag{4}$$

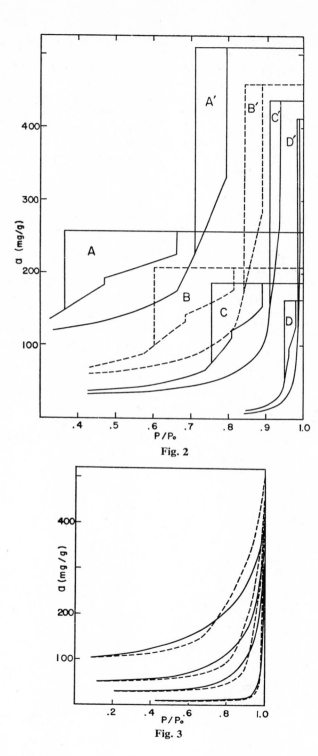

Fig. 2

Fig. 3

where σ, M, and N_0 are the sectional area and molecular weight of the sorbate molecule and Avogadro's number, respectively.

We note that the theoretical isotherms exhibit quite different shapes when the radius of elementary particles and the packing type are changed as shown in Figs. 2 and 3. Therefore in later works we could easily obtain suitable experimental isotherms by choosing proper values of R_0 and a_m and the type of packing by a trial-and-error method, in order to satisfy the experimental observations.

COMPARISON OF EXPERIMENTAL AND THEORETICAL ISOTHERMS

Experimental isotherms of benzene, methanol, and steam at $0°C$ and oxygen at $-190°C$, determined on porous Vycor glass, have convinced us of the validity of the packing model of elementary particles. In the case of benzene [using the experimental BET capacity,[2] $a_m = 31$ mg/g, and the radius of the elementary particles, $R_0 = 180$ Å, calculated from a_m by (2)], the theoretical curve for the C.P. type is shown by the solid line in Fig. 4. Experimental values for sorption and desorption, denoted by open and solid circles, respectively, are in good agreement with the calculated line. It was rather surprising to obtain such satisfactory agreement because there was no arbitrariness in the adopted values of R_0 and a_m. Better agreements are shown in Fig. 5 for the isotherm of methanol, where the theoretical curve was calculated by using the same value of R_0 and experimental a_m obtained by the BET plot. Steam and oxygen isotherms also give satisfactory agreement

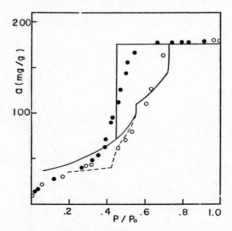

Fig. 4. Theoretical and experimental isotherms on porous Vycor glass.

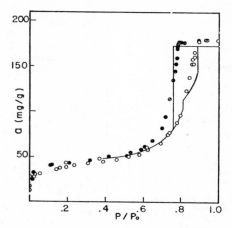

Fig. 5. Theoretical and experimental isotherms on porous Vycor glass.

between theory and experiment. Furthermore, the experimental isotherms of nitrogen, argon, and oxygen obtained by Barrer et al.[10] on a similar specimen agree well with our calculations. Such results lead us naturally to the conclusion that the porous mass Vycor glass spheres might consist of closest-packed elementary particles. It was quite fortunate for us to have used such a system early in our work because conditions of manufacturing the Vycor glass spheres would change the pore volume as well as the pore size distribution.

Comparison of Isotherms on Neobead C-5 Beads

Figure 6 shows isotherms of benzene at 0°C on Neobead C-5, which is a commercial spherical bead of γ-alumina 3.7 mm in diameter, made by Mizusawa Chemical Industries Co. The beads were preheated successively in the range of temperature, 800–1200°C.[11] In Fig. 6 the solid lines are the calculated curves for the S.C. packing type, calculated by using values of R_0 and a'_m chosen so as to obtain the best agreement between experimental and calculated values. Table I gives R_0 and a_m values chosen for the calculations; a_m is obtained from the BET plot and a''_m is calculated from R_0 by (4). When R_0 is larger, all three values of a_m, a'_m, and a''_m are in quite good accord. If R_0 and a'_m are suitably chosen, by giving up the relation of (4), we can achieve satisfactory agreement for the porous bodies consisting of particles of smaller diameter. Abandoning the relation of (4) means either a practical

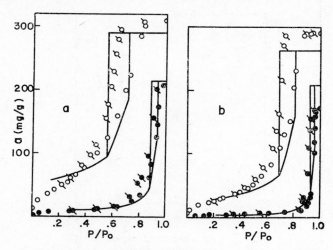

Fig. 6.   Theoretical isotherms for S.C. type and experimental isotherms of benzene at 0°C on Neobead C-5 which were preheated at 800, 1000, 1100, and 1200°C.

extension of application of the theory, or that a_m, obtained by the BET plot, is not necessarily the amount which forms the monomolecular adsorption layer as required by the BET theory.

Isotherms on Loose and Compressed Fine Powders

In the two examples mentioned above, the porous sorbents are already in a compressed state of fine elementary particles. Therefore we extended our work to measuring and calculating isotherms on Cab-o-Sil, Aerosil, Sil-bead, and a number of carbon blacks whose radii were previously determined by electron microscopic or adsorption methods.[12] A few examples are shown in Fig. 7. The theoretical curves for the S.C. type are shown in the figure by the solid lines calculated from the previously known values of R_0 and the

Table I

$T(^{\circ}C)$	a_m (mg/g)	Σ (m²/g)	a'_m (mg/g)	a''_m (mg/g)	R_0 (Å)
300	75.2	174	75.2	54.5	70
800	47.3	102	50	38	100
1000	22.8	55.5	23	23	160
1100	10.0	23.1	10	9.5	400
1200	3.0	6.9	3	4.1	800

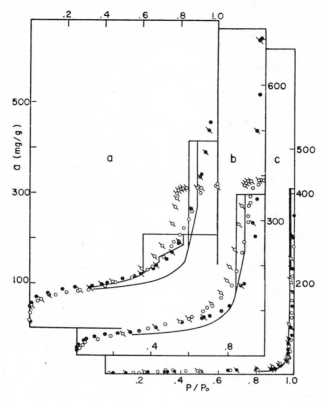

Fig. 7a, b, c

experimental values of a_m. Speaking generally, all isotherms on fine powder specimens denoted by solid circles are reversible or almost reversible; that is, they show little or no hysteresis. On compressing the specimens under ca. 10 tons/cm² , the a_m or isotherms show no change from that in the uncompressed state. However, the saturation amount of sorption, a_s, decreases markedly, and, in the higher pressure range, there appears a hysteresis loop shifting generally to the lower equilibrium pressure. As shown in Fig. 7, some compressed specimens have larger and some have smaller values of a_s than that anticipated for the S.C. type. The compressed Cab-o-Sil gives an experimental isotherm most in accord with the theoretical one. Therefore we may say that compressing these fine powders by 10 tons/cm² , brings them into an ordered state very near that of the S.C. type. These results together with those of the previous sections exhibit clearly that the pores in a porous material are interstices between elementary particles. In addition, we have obtained experimental evidence for the Kraemer-McBain mechanism for hysteresis, because hysteresis has resulted from compression of the powder specimens, leaving the lower part of the isotherm of fine powder unchanged.

On compressing Cab-o-Sil powder by 20 tons/cm², a_s decreases further and the hysteresis loop shifts to a still lower pressure range, approaching the theoretical curve for the C.P. type, even though little or no change is noticeable in the lower pressure range.[13]

Recently, we have obtained another visual evidence of the elementary particle structure of porous Vycor glass and Neobead C-5 by electron microscope photography. These specimens may be dispersed by powerful supersonic methods and give singly dispersed particles of almost the same size as obtained from a_m or aggregates of these particles.

In addition, we can reasonably explain why elementary particles of porous Vycor glass are in the C.P. arrangement and those of Neobead C-5 in the loose S.C. arrangement. In the former case, silica segregates spherically from SiO_2–B_2O_3 glass by annealing, and elementary particles of the segregate can be arranged in the C.P. order if the mixing proportion of component particles is suitable. In the latter case, however, γ-alumina is obtained by hydrolyzing the thermal gel consisting mainly of aluminum sulfate solution. Therefore there exists such a high concentration of sulfate ions and water molecules in the vicinity of the alumina particles that they are unable to arrange in the closest packing, and thus remain more loosely packed.

Further Evidence for the Elementary Particle Structure of Porous Bodies

In the above sections, we have amply shown that capillary condensation occurs in the contact zones of elementary particles and the shape of isotherms measured up to $x = 1$ is determined by existing pore distributions of the porous materials. In addition, it has been shown by sorption methods that porous Vycor glass and Neobead C-5 are ideal porous bodies consisting of spherical elementary particles arranged in characteristic types of packing. Based on these idealized models, we have calculated the effective diffusion coefficients in these porous bodies.

Wheeler[14] gave the effective diffusion coefficient, D_e, of gases in porous bodies by use of the equation

$$D_e = (\epsilon/\tau)D \qquad (5)$$

on a simple model. There was then no other reasonable model available. Later, most researchers commonly used the porosity for the opening fraction, ϵ, and so the tortuosity factor τ lost its real meaning by changing to a mere experimental constant.

With the idealized model mentioned above, the shape and arrangement of the pores are given regularity. Therefore we can define a strict opening fraction, θ, perpendicular to a diffusion direction and a true tortuosity factor L related to the real diffusion path. On introducing a few reasonable assumptions, to simplify calculation of the actual diffusion coefficient, a value of θ/L is easily obtained which corresponds to the experimental value of ϵ/τ in the Wheeler equation, (5). The details of the calculations have been published elsewhere.[15] With porous Vycor glass, a number of investigators[1] unanimously obtained 0.0508 for the value of ϵ/τ and our calculated values for θ/L range from 0.0672 to 0.798. With Neobead, Kawasoe et al.[16] obtained an experimental value of $\epsilon/\tau = 0.117$; our calculated values of θ/L are 0.118–0.131. The source of the differences of ca. 50% at maximum are reasonably attributed to some oversimplification of the nature of the diffusion path adopted. Therefore we may conclude that the validity of the elementary particle structure of porous specimens is confirmed in the diffusion phenomenon.

CATALYTIC ACTIVITY AND THE DISPERSION STATE OF CATALYSTS IN PORES OF CARRIERS

Density of Catalysts Dispersed on a Carrier

In a previous paper[17] concerned with benzene isotherms on active carbons which contained various amounts of deposited silica denoted by f (g/g of carbon), we showed that the pore volume per gram of carbon for each specimen, obtained from the saturation value of sorption, a_s, plus the volume of silica deposited, is equal, as anticipated, to the pore volume of the original active carbon. In addition, we concluded from the change in the shape of isotherms upon increase in f that most silica was deposited in micropores whose radius was smaller than ca. 15 Å. Applying the same experimental method to porous glass or Neobead C-5 containing a number of oxide catalysts deposited in varying amounts, f, we have found that dispersed oxide catalysts have the curious property that the density of some oxides in a dispersed state on a carrier is less than in the bulk state. This might be attributed to the failure of the sorbate molecules to penetrate into the fine crevices of the deposit, or to the solid deposit blocking the entrances to the fine pores of the carrier. However, Itoga[18] obtained similar results using a helium gas densimeter. Therefore we tentatively conclude that some catalysts dispersed on a carrier have densities 20 to 40% less than in the bulk state.

Allocation of Catalysts Deposited on a Carrier

It is generally difficult to know where the catalysts are located when they are deposited on a carrier. After determining the BET capacity and a_m from experimental isotherms on catalysts with different f values, we have proposed at least two typical dispersion states of catalysts on porous Vycor glass.[19] One situation is when the catalyst is deposited homogeneously over the surface of the elementary particles and the other is when the catalyst is accumulated mainly in the contact zones of the elementary particles. For each case we can calculate from the geometry the decrease of specific surface area, or a_m, as a function of f. Comparing the theoretical f–a_m relationships with the experimental a_m obtained from isotherms on specimens different in f, we have shown that cupric oxide, aluminum oxide, and zinc acetate deposited on porous glass are accumulated in the contact zones, whereas nickel oxide and ferric oxide deposited on porous glass are distributed homogeneously over the surfaces of elementary particles. These conclusions are confirmed by calculating the pore size distribution of catalysts in which the fine pores are rapidly decreased in the former case with increase of f.

In addition, catalytic reaction rates measured on specimens in these extreme cases furnish further support for the conclusion mentioned above. Dehydrogenation rates of cyclohexane, r, have been measured by a pulse method on ferric oxide catalysts of various f, supported on porous Vycor glass. With increasing f, r becomes larger and tends rapidly to a constant value where the supposed oxide film, spread homogeneously over the inner surfaces, is thickened to about the lattice constant of the α-ferric oxide crystal, ca. 5.4 Å. Such a result is favorably anticipated by the dispersion state of the oxide. On the other hand, dehydration rates of ethanol on a number of alumina catalysts supported variously on porous glass are almost constant independently of f in the range from 0.004 to 0.08. The carrier itself has catalytic activity, perhaps originating from residual boron oxide. However, deposition of alumina increases catalytic activity about fivefold. On measuring the surface acidity capacity by the Benesi method, the titre of n-butylamine for the specimen of $f = 0.01$ is double that of the original carrier. For specimens with f larger than 0.01 the titre remains constant. From these results we can conclude that a very small amount of alumina at first spreads over the whole surface, and the main portion of alumina is deposited in the contact zones of elementary particles. If alumina surfaces were to have 1.7 times larger activity than the other carrier surfaces covered slightly by alumina, the experimental relation between r and f could be explained.

DIFFUSION IN PORES OF CATALYSTS AND CATALYTIC REACTION RATES

Hydrogenation Rates of Benzene

Hydrogenation rates of benzene have been determined by a flow method over platinum–Neobead C-5 catalysts different in f, and broken fine powder specimens (ca. 100 mesh) of the same catalysts.[20] The theoretical η–ϕ relation given by the Wheeler theory is in good agreement with that obtained between the experimental effectiveness factor, η, and Thiele modulus, ϕ, calculated by using the observed rate constant, k, and the effective diffusion coefficient D_e obtained above (5). Based on these satisfactory results, we conclude that the importance of pore diffusion of a reactant, in this case benzene, is clearly shown when we use a catalyst of large pellet size, in this case 3.8 mm in diameter, and the reaction rates are measured at high temperatures with catalysts of large f.

The reduction rates of nitrobenzene to aniline using a Magen reactor have been determined with platinum–Neobead C-1 catalysts which differ in f from 0.0005 to 0.010.[21] Although the size of this carrier is as small as 100–140 mesh, it has been shown clearly that the reduction rates are decidedly controlled by diffusion of hydrogen in the pores of the catalysts filled with the ethanol solvent provided boundary effects are removed by means of suitable experimental conditions such as the use of a quite small amount of catalyst or catalysts of low f. If we use a catalyst of $f = 0.01$, as usually adopted, our results show that most reduction reactions take place in the surface zones of the catalyst, even though the size of the catalyst particles is as small as 300 mesh.

ACKNOWLEDGMENT

I wish to express my appreciation to Dr. A. T. Ree for his encouragement and kindness in reading this manuscript.

REFERENCES

1. C. N. Satterfield and T. K. Sherwood, *The Role of Diffusion in Catalysis*, Addison-Wesley, London, 1963.
2. I. Higuchi, Y. Ushiki, and R. Suzuki, *Bull. Chem. Soc. Japan*, **34**, 1539 (1961); *J. Chem. Soc. Japan*, **82**, 1620 (1961).
3. S. Kubota, H. Oda, T. Makabe, and I. Higuchi, *J. Chem. Soc. Japan*, **86**, 589 (1965).
4. I. Higuchi, *Bull. Inst. Phys. Chem. Res.*, **16**, 536 (1937); **18**, 657 (1939); **19**, 951 (1940).
5. I. Higuchi, *Bull. Inst. Phys. Chem. Res.*, **20**, 489 (1941); *Sci. Rept. Tohoku Univ.*, *Ser. I*, **33**, 99 (1949).
6. I. Higuchi, *Bull. Inst. Phys. Chem. Res.*, **21**, 1138 (1942).
7. I. Higuchi, *Bull. Inst. Phys. Chem. Res.*, **23**, 382, 565 (1944); *Sci. Rept. Tohoku Univ. Ser. I*, **33**, 174, 231 (1949).
   I. Higuchi and M. Shimizu, *J. Phys. Chem.*, **56**, 198 (1952); I. Higuchi and Y. Iwagami, *ibid.*, **56**, 921 (1952).
8. I. Higuchi and H. Utsugi, *J. Chem. Phys.*, **20**, 1180 (1952); *Sci. Rept. Tohoku Univ. Ser. I*, **36**, 27 (1952).

9. R. M. Barrer, N. McKenzie, and J. S. S. Reay, *J. Colloid Sci.*, **11**, 479 (1956); A. P. Karnaukhov and A. V. Kiselev, *Zh. Fiz. Khim.*, **31**, 3635 (1957); *Chem. Abstr.*, **52**, 5928 (1958); S. Kruyer, *Trans. Faraday Soc.*, **54**, 1758 (1958).
10. R. M. Barrer and J. A. Barrie, *Proc. Roy. Soc. (London)*, **213**, 250 (1952).
11. S. Kubota, H. Oda, and I. Higuchi, *J. Chem. Soc. Japan*, **88**, 43 (1967).
12. I. Higuchi, Y. Ushiki, and R. Suzuki, *J. Chem. Soc. Japan*, **83**, 808 (1962); **84**, 306 (1963).
13. I. Higuchi, *J. Soc. Material Sci. Japan*, **13**, 732 (1966).
14. A. Wheeler, in *Catalysis*, Vol. II, P. Emmett, Ed., Reinhold, New York, 1955, p. 105.
15. I. Higuchi, J. Kobayashi, and H. Katsuzawa, *J. Chem. Soc. Japan*, **90**, 150 (1969).
16. K. Kawazoe and I. Sugiyama, *Kagaku Kogaku (Chem. Eng. Japan)*, **30**, 1007 (1966).
17. I. Higuchi and T. Makabe, *J. Chem. Soc. Japan*, **86**, 35 (1965).
18. K. Itoga, private communication.
19. S. Kubota and I. Higuchi, *J. Chem. Soc. Japan*, **87**, 40 (1966); **88**, 417 (1967).
20. H. Katsuzawa, J. Kobayashi, and I. Higuchi, *J. Chem. Soc. Japan (Ind. Chem. Sect.)*, **72**, 823 (1969).
21. J. Kobayashi, T. Makabe, and I. Higuchi, *J. Chem. Soc. Japan*, **90**, 612 (1969).

AUTHOR INDEX